Grundlehren der mathematischen Wissenschaften 343

A Series of Comprehensive Studies in Mathematics

For further volumes:
http://www.springer.com/series/138

Hajer Bahouri · Jean-Yves Chemin ·
Raphaël Danchin

Fourier Analysis
and Nonlinear Partial
Differential Equations

Springer

Hajer Bahouri
Départment de Mathématiques
Faculté des Sciences de Tunis
Campus Universitaire
Université de Tunis
El Manar
2092 Tunis
Tunisia
hajer.bahouri@fst.rnu.tn

Raphaël Danchin
Centre de Mathématiques
Faculté de Sciences et Technologie
Université Paris XII-Val de Marne
61, avenue du Général de Gaulle
94 010 Créteil Cedex
France
danchin@univ-paris12.fr

Jean-Yves Chemin
Laboratoire Jacques-Louis Lions
Université Pierre et Marie Curie
Boîte courrier 187
75252 Paris Cedex 05
France
chemin@ann.jussieu.fr

ISSN 0072-7830
ISBN 978-3-642-16829-1   e-ISBN 978-3-642-16830-7
DOI 10.1007/978-3-642-16830-7
Springer Heidelberg Dordrecht London New York

Mathematics Subject Classification: 35Q35, 76N10, 76D05, 35Q31, 35Q30

Cover design: VTEX, Vilnius

Printed on acid-free paper

Springer is part of Springer Science+Business Media (www.springer.com)

1006218840

A la mémoire de Noomann Bassou

Preface

Since the 1980s, Fourier analysis methods have become of ever greater interest in the study of linear and nonlinear partial differential equations. In particular, techniques based on *Littlewood–Paley decomposition* have proven to be very efficient in the study of evolution equations. Littlewood–Paley decomposition originates with Littlewood and Paley's works in the early 1930s and provides an elementary device for splitting a (possibly rough) function into a sequence of spectrally well localized smooth functions. In particular, differentiation acts almost as a multiplication on each term of the sequence. However, its systematic use for nonlinear partial differential equations is rather recent. In this context, the main breakthrough was achieved after J.-M. Bony introduced the *paradifferential calculus* in his pioneering 1981 paper (see [39]) and its avatar, the *paraproduct*.

Surprisingly, despite the growing number of authors who now use such techniques, to the best of our knowledge, there is no textbook presenting Fourier analysis tools in such a way that they may be directly used for solving nonlinear partial differential equations.

The aim of this book is threefold. First, we want to give a detailed presentation of harmonic analysis tools that are of constant use for solving nonlinear partial differential equations. Second, we want to convince the reader that the rough frequency splitting supplied by Littlewood–Paley decomposition (which turns out to be much simpler than, e.g., Calderon–Zygmund decomposition or wavelet theory) may still provide elementary and elegant proofs of some classical inequalities (such as Sobolev embedding and Gagliardo–Nirenberg or Hardy inequalities). Third, we give a few examples of how to use these basic Fourier analysis tools to solve linear or nonlinear evolution partial differential equations. We have chosen to present the most popular evolution equations, namely, transport and heat equations, (linear or quasilinear) symmetric hyperbolic systems, (linear, semilinear, or quasilinear) wave equations, and the (linear or semilinear) Schrödinger equation. We place a special emphasis on models coming from fluid mechanics (in particular, on the incompressible Navier–Stokes and Euler equations) for which, historically, the Littlewood-

Paley decomposition was first used. It goes without saying that our methods are also relevant for solving a variety of other equations. In fact, there has been a plethora of recent papers dedicated to more complicated nonlinear partial differential equations in which Littlewood–Paley decomposition proves to be a crucial tool.

This book is almost self-contained, inasmuch as having an undergraduate level understanding of analysis is the only prerequisite. There are rare exceptions where we have had to admit nontrivial mathematical results, in which case references are given. Apart from these, we have postponed references, historical background, and discussion of possible future developments to the end of each chapter. The book does not contain any definitively new results. However, we have tried to provide an exhaustivity that cannot be found in any single paper. Also, we have provided new proofs for some well-known results.

We have also decided not to discuss the theory of wavelets, even though this would be the natural extension of Littlewood–Paley decomposition. Indeed, it turns out that, to the best of our knowledge, there are almost no theoretical results for nonlinear partial differential equations in which wavelets cannot be replaced by a simple Littlewood–Paley decomposition.

When writing this book, we tried as much as possible to make a distinction between what may be proven by means of classical analysis tools and what really does require Littlewood–Paley decomposition (and the paraproduct). In fact, with only a few exceptions, all the material concerning Littlewood–Paley decomposition is contained in Chapter 2 so that the reader who is not accustomed to (or who is afraid of) those techniques may still read a great deal of the book. In fact, the whole of Chapter 1, the first section of Chapter 3, the first half of Chapter 4, Chapter 5 (except for the last section), the first section of Chapter 6, and the first two sections of Chapter 8 may be read completely independently of Chapter 2. In most of the other parts of the book, Chapter 2 may be used freely as a "black box" that does not need to be opened.

Roughly speaking, the book may be divided into two principal parts: Tools are developed in the first two chapters, then applied to a variety of linear and nonlinear partial differential equations (Chapters 3–10). A detailed plan of the book is as follows.

Chapter 1 is devoted to a self-contained elementary presentation of classical Fourier analysis results. Even though none of the results are new, some of the proofs that we present are not the standard ones and are likely to be useful in other contexts. We also pay attention to the construction of explicit examples which illustrate the optimality of some refined estimates.

In Chapter 2 we give a detailed presentation on Littlewood–Paley decomposition and define homogeneous and nonhomogeneous Besov spaces. We should emphasize that we have replaced the usual definition of homogeneous spaces (which are quotient distribution spaces modulo polynomials) by something better adapted to the study of partial differential equations (indeed, dealing with distributions modulo polynomials is not appropriate in this context). We also establish technical results (commutator estimates and functional inequalities, in particular) which will be used in the following chapters.

In Chapter 3 we give a very complete theory of strong solutions for transport and transport-diffusion equations. In particular, we provide a priori estimates which are the key to solving nonlinear systems coming from fluid mechanics. Chapter 4 is devoted to solving linear and quasilinear symmetric systems with data in Sobolev spaces. Blow-up criteria and results concerning the continuity of the flow map are also given. The case of data with critical regularity (in a Besov space) is also investigated.

In Chapter 5 we take advantage of the tools introduced in the previous chapters to establish most of the classical results concerning the well-posedness of the incompressible Navier–Stokes system for data with critical regularity. In order to emphasize the robustness of the tools that have been introduced hitherto in this book, we present in Chapter 6 a nonlinear system of partial differential equations with degenerate parabolicity. In fact, we show that some of the classical results for the Navier–Stokes system may be extended to the case where there is no vertical diffusion. Most of the results of this chapter are based on the use of an *anisotropic* Littlewood–Paley decomposition.

Chapter 7 is the natural continuation of the previous chapter: The diffusion term is removed, leading to the study of the Euler system for inviscid incompressible fluids. Here, we state local (in dimension $d \geq 3$) and global (in dimension two) well-posedness results for data in general Besov spaces. In particular, we study the case where the data belong to Besov spaces for which the embedding in the set of Lipschitz functions is critical. In the two-dimensional case, we also give results concerning the inviscid limit. We stress the case of data with (generalized) vortex patch structure.

Chapter 8 is devoted to *Strichartz estimates* for dispersive equations with a focus on Schrödinger and wave equations. After proving a dispersive inequality (i.e., decay in time of the L^∞ norm in space) for these equations, we present, in a self-contained way, the celebrated $TT^\star$ argument based on a duality method and on bilinear estimates. Some examples of applications to semilinear Schrödinger and wave equations are given at the end of the chapter.

Chapter 9 is devoted to the study of a class of quasilinear wave equations which can be seen as a toy model for the Einstein equations. First, by taking advantage of energy methods in the spirit of those of Chapter 4, we establish local well-posedness for "smooth" initial data (i.e., for data in Sobolev spaces embedded in the set of Lipschitz functions). Next, we weaken our regularity assumptions by taking advantage of the dispersive nature of the wave equation. The key to that improvement is a quasilinear Strichartz estimate and a refinement of the paradifferential calculus. To prove the quasilinear Strichartz estimate, we use a microlocal decomposition of the time interval (i.e., a decomposition in some interval, the length of which depends on the size of the frequency) and geometrical optics.

In Chapter 10 we present a more complicated system of partial differential equations coming from fluid mechanics, the so-called *barotropic compressible*

Navier–Stokes equations. Those equations are of mixed hyperbolic-parabolic type. We show how we may take advantage of the results of Chapter 3 and the techniques introduced in Chapter 2 so as to obtain local (or global) unique solutions with critical regularity. The last part of this chapter is dedicated to the study of the *low Mach number limit* for this system. It is shown that under appropriate assumptions on the data, the limit solution satisfies the incompressible Navier–Stokes system studied in Chapter 5.

In writing this book, we had help from many colleagues. We are particularly indebted to F. Charve, B. Ducomet, C. Fermanian-Kammerer, F. Sueur, B. Texier, and to the anonymous referees for pointing out numerous mistakes and giving suggestions and advice. In addition to J.-M. Bony, our work was inspired by many collaborators and great mathematicians, among them B. Desjardins, I. Gallagher, P. Gérard, E. Grenier, T. Hmidi, D. Iftimie, H. Koch, S. Klainerman, Y. Meyer, M. Paicu, D. Tataru, F. Vigneron, C.J. Xu, and P. Zhang. We would like to express our gratitude to all of them.

Paris

Hajer Bahouri
Jean-Yves Chemin
Raphaël Danchin

Contents

1

Basic Analysis

This chapter is devoted to the presentation of a few basic tools which will be used throughout this book. In the first section we state the Hölder and Minkowski inequalities. Next, we prove convolution inequalities in the general context of locally compact groups equipped with left-invariant Haar measures. The adoption of this rather general framework is motivated by the fact that these inequalities may be used not only in the $\mathbb{R}^d$ and $\mathbb{Z}^d$ cases, but also in other groups such as the Heisenberg group $\mathbb{H}^d$. Both Lebesgue and weak Lebesgue spaces are used. In the latter case, we introduce an atomic decomposition which will help us to establish a bilinear interpolation-type inequality. Finally, we give a few properties of the Hardy–Littlewood maximal operator.

The second section is devoted to a short presentation on the Fourier transform in $\mathbb{R}^d$. The third section is dedicated to homogeneous Sobolev spaces in $\mathbb{R}^d$. There, we state basic topological properties, consider embedding in Lebesgue, bounded mean oscillation, and Hölder spaces, and prove refined Sobolev inequalities. The classical Sobolev inequalities are of course invariant by translation and dilation. The refined versions of the Sobolev inequalities which we prove are, in addition, invariant by translation in the Fourier space. We also present some classes of examples to show that these inequalities are in some sense optimal. In the last section of this chapter, we focus on nonhomogeneous Sobolev spaces, with a special emphasis on trace theorems, compact embedding, and Moser–Trudinger and Hardy inequalities.

1.1 Basic Real Analysis

1.1.1 Hölder and Convolution Inequalities

We begin by recalling the classical Hölder inequality.

Proposition 1.1. *Let (X, μ) be a measure space and (p, q, r) in $[1, \infty]^3$ be such that*

H. Bahouri et al., *Fourier Analysis and Nonlinear Partial Differential Equations*, Grundlehren der mathematischen Wissenschaften 343,
DOI 10.1007/978-3-642-16830-7_1, © Springer-Verlag Berlin Heidelberg 2011

$$\frac{1}{p} + \frac{1}{q} = \frac{1}{r}.$$

If (f, g) belongs to $L^p(X, \mu) \times L^q(X, \mu)$, then fg belongs to $L^r(X, \mu)$ and

$$\|fg\|_{L^r} \le \|f\|_{L^p} \|g\|_{L^q}.$$

Proof. The cases where $p = 1$ or $p = \infty$ being trivial, we assume from now on that p is a real number greater than 1. The concavity of the logarithm function entails that for any positive real numbers a and b and any θ in $[0, 1]$,

$$\theta \log a + (1 - \theta) \log b \le \log(\theta a + (1 - \theta)b),$$

which obviously implies that

$$a^\theta b^{1-\theta} \le \theta a + (1 - \theta)b.$$

Hence, assuming that $\|f\|_{L^p} = \|g\|_{L^q} = 1$, we can write

$$\int_X |fg|^r \, d\mu = \int_X (|f|^p)^{\frac{r}{p}} (|g|^q)^{\frac{r}{q}} \, d\mu$$
$$\le \frac{r}{p} \int_X |f|^p \, d\mu + \frac{r}{q} \int_X |g|^q \, d\mu$$
$$\le \frac{r}{p} + \frac{r}{q} = 1.$$

The proposition is thus proved. □

The following lemma states that Hölder's inequality is in some sense optimal.

Lemma 1.2. *Let (X, μ) be a measure space and $p \in [1, \infty]$. Let f be a measurable function. If*

$$\sup_{\|g\|_{L^{p'}} \le 1} \int_X |f(x)g(x)| \, d\mu(x) < \infty,$$

then f belongs to L^p and[1]

$$\|f\|_{L^p} = \sup_{\|g\|_{L^{p'}} \le 1} \int_X f(x)g(x) \, d\mu(x).$$

Proof. Note that if f is in L^p, then Hölder's inequality ensures that

$$\sup_{\|g\|_{L^{p'}} \le 1} \int_X f(x)g(x) \, d\mu(x) \le \|f\|_{L^p}$$

so that only the reverse inequality has to be proven.

---

[1] Here, and throughout the book, p' denotes the *conjugate exponent* of p, defined by

$$\frac{1}{p} + \frac{1}{p'} = 1, \quad \text{with the rule that} \quad \frac{1}{\infty} = 0.$$

We start with the case $p = \infty$. Let λ be a positive real number such that $\mu(|f| \geq \lambda) > 0$. Writing $E_\lambda \overset{\text{def}}{=} (|f| \geq \lambda)$, we consider a nonnegative function g_0 in L^1, supported in E_λ with integral 1. If we define

$$g(x) = \frac{f(x)}{|f(x)|} g_0,$$

then g is in L^1 so that fg is integrable by assumption, and we have

$$\int_X fg \, d\mu(x) = \int_X |f| g_0 \, d\mu(x) \geq \lambda \int_X g_0 \, d\mu(x) = \lambda.$$

The lemma is proved in this case. We now assume that $p \in \,]1, \infty[$ and consider a nondecreasing sequence $(E_n)_{n \in \mathbb{N}}$ of subsets of finite measure of X, the union of which is X. Let[2]

$$f_n(x) = \mathbf{1}_{E_n \cap (|f| \leq n)} f \quad \text{and} \quad g_n(x) = \frac{f_n(x)|f_n(x)|^{p-1}}{|f_n(x)| \, \|f_n\|_{L^p}^{\frac{p}{p'}}}.$$

It is obvious that f_n belongs to $L^1 \cap L^\infty$ and thus to L^p for any p. Moreover, we have

$$\|g_n\|_{L^{p'}}^{p'} = \frac{1}{\|f_n\|_{L^p}^p} \int_X |f_n(x)|^{(p-1)\frac{p}{p-1}} \, d\mu(x) = 1.$$

The definitions of the functions f_n and g_n ensure that

$$\int_X f(x) \mathbf{1}_{E_n \cap (|f| \leq n)} g_n(x) \, d\mu(x) = \int_X f_n(x) g_n(x) \, d\mu(x)$$
$$= \left(\int_X |f_n(x)|^p \, d\mu(x) \right) \|f_n\|_{L^p}^{-\frac{p}{p'}}$$
$$= \|f_n\|_{L^p}.$$

Thus, we have

$$\|f_n\|_{L^p} \leq \sup_{\|g\|_{L^{p'}} \leq 1} \int_X f(x) g(x) \, d\mu(x).$$

The monotone convergence theorem immediately implies that

$$\|f\|_{L^p} \leq \sup_{\|g\|_{L^{p'}} \leq 1} \int_X f(x) g(x) \, d\mu(x).$$

Finally, in order to treat the case where $p = 1$, we may consider the sequence $(g_n)_{n \in \mathbb{N}}$ defined by

$$g_n(x) = \mathbf{1}_{(f_n \neq 0)}(x) \frac{f_n(x)}{|f_n(x)|}.$$

---

[2] Throughout this book, the notation $\mathbf{1}_A$, where A stands for any subset of X, denotes the *characteristic function* of A.

We obviously have $\|g_n\|_{L^\infty} = 1$ and

$$\int_X f(x)g_n(x)\,d\mu(x) = \int_X |f_n(x)|\,d\mu(x).$$

Using the monotone convergence theorem, we get that

$$\int_X |f(x)|\,d\mu(x) < \infty \quad\text{and}\quad \int_X |f(x)|\,d\mu(x) = \lim_{n\to\infty}\int_X |f_n(x)|\,d\mu(x),$$

which completes the proof of the proposition.   □

We now state *Minkowski's inequality.*

Proposition 1.3. *Let (X_1,μ_1) and (X_2,μ_2) be two measure spaces and f a nonnegative measurable function over $X_1 \times X_2$. For all $1 \leq p \leq q \leq \infty$, we have*

$$\left\| \|f(\cdot,x_2)\|_{L^p(X_1,\mu_1)} \right\|_{L^q(X_2,\mu_2)} \leq \left\| \|f(x_1,\cdot)\|_{L^q(X_2,\mu_2)} \right\|_{L^p(X_1,\mu_1)}.$$

Proof. The result is obvious if $q = \infty$. If q is finite, then, using Fubini's theorem and $r \overset{\text{def}}{=} (q/p)'$, we have

$$\left\| \|f(\cdot,x_2)\|_{L^p(X_1,\mu_1)} \right\|_{L^q(X_2,\mu_2)} = \left(\int_{X_2} \left(\int_{X_1} f^p(x_1,x_2)\,d\mu_1(x_1) \right)^{\frac{q}{p}} d\mu_2(x_2) \right)^{\frac{1}{q}}$$

$$= \left(\sup_{\substack{\|g\|_{L^r(X_2,\mu_2)}=1 \\ g\geq 0}} \int_{X_1\times X_2} f^p(x_1,x_2)g(x_2)\,d\mu_1(x_1)\,d\mu_2(x_2) \right)^{\frac{1}{p}}$$

$$\leq \left(\int_{X_1} \left(\sup_{\substack{\|g\|_{L^r(X_2,\mu_2)}=1 \\ g\geq 0}} \int_{X_2} f^p(x_1,x_2)g(x_2)\,d\mu_2(x_2) \right) d\mu_1(x_1) \right)^{\frac{1}{p}}.$$

Using Hölder's inequality we may then infer that

$$\left\| \|f(\cdot,x_2)\|_{L^p(X_1,\mu_1)} \right\|_{L^q(X_2,\mu_2)} \leq \left(\int_{X_1} \left(\int_{X_2} f^q(x_1,x_2)\,d\mu_2(x_2) \right)^{\frac{p}{q}} d\mu_1(x_1) \right)^{\frac{1}{p}},$$

and the desired inequality follows.   □

The *convolution* between two functions will be used in various contexts in this book. The reader is reminded that convolution makes sense for real- or complex-valued measurable functions defined on some locally compact topological group G equipped with a left-invariant Haar measure[3] μ. The (formal) definition of convolution between two such functions f and g is as follows:

---

[3] This means that μ is a Borel measure on G such that for any Borel set A and element a of G, we have $\mu(a \cdot A) = \mu(A)$.

$$f \star g(x) = \int_G f(y)\, g(y^{-1} \cdot x)\, d\mu(y).$$

We can now state *Young's inequality* for the convolution of two functions.

Lemma 1.4. *Let G be a locally compact topological group endowed with a left-invariant Haar measure μ. If μ satisfies*

$$\mu(A^{-1}) = \mu(A) \quad \text{for any Borel set } A, \tag{1.1}$$

then for all (p,q,r) in $[1,\infty]^3$ such that

$$\frac{1}{p} + \frac{1}{q} = 1 + \frac{1}{r} \tag{1.2}$$

and any (f,g) in $L^p(G,\mu) \times L^q(G,\mu)$, we have

$$f \star g \in L^r(G,\mu) \quad \text{and} \quad \|f \star g\|_{L^r(G,\mu)} \leq \|f\|_{L^p(G,\mu)} \|g\|_{L^q(G,\mu)}.$$

Proof. We first note that, owing to the left invariance and (1.1), for all $x \in G$ and any measurable function h on G, we have

$$\int_G h(y)\, d\mu(y) = \int_G h(y^{-1} \cdot x)\, d\mu(y).$$

Therefore, the case $r = \infty$ reduces to the Hölder inequality which was proven above.

We now consider the case $r < \infty$. Obviously, one can assume without loss of generality that f and g are nonnegative and nonzero. We write

$$(f \star g)(x) = \int_G f^{\frac{r}{r+1}}(y)\, g^{\frac{1}{r+1}}(y^{-1} \cdot x)\, f^{\frac{1}{r+1}}(y)\, g^{\frac{r}{r+1}}(y^{-1} \cdot x)\, d\mu(y).$$

Observing that (1.2) can be written $\dfrac{r}{r+1}\left(\dfrac{1}{p} + \dfrac{1}{q}\right) = 1$, Hölder's inequality implies that

$$(f \star g)(x) \leq \left(\int_G f^p(y) g^{\frac{p}{r}}(y^{-1} \cdot x)\, d\mu(y)\right)^{\frac{r}{(r+1)p}} \left(\int_G f^{\frac{q}{r}}(y) g^q(y^{-1} \cdot x)\, d\mu(y)\right)^{\frac{r}{(r+1)q}}.$$

Applying Hölder's inequality with $\alpha = rq/p$ (resp., $\beta = rp/q$) and the measure $f^p(y)\, d\mu(y)$ [resp., $g^q(y^{-1} \cdot x)\, d\mu(y)$], and using the invariance of the measure μ by the transform $y \mapsto y^{-1} \cdot x$, we get

$$(f \star g)(x) \leq \left(\int_G f^p(y) g^q(y^{-1} \cdot x)\, d\mu(y)\right)^{\frac{1}{r+1}\left(\frac{1}{p}+\frac{1}{q}\right)} \|f\|_{L^p(G,\mu)}^{\frac{r}{r+1}\left(1-\frac{p}{qr}\right)} \|g\|_{L^q(G,\mu)}^{\frac{r}{r+1}\left(1-\frac{q}{pr}\right)}.$$

Hence, raising the above inequality to the power r yields

$$\left| \left(\frac{f}{\|f\|_{L^p}} \star \frac{g}{\|g\|_{L^q}} \right)(x) \right|^r \leq \left(\frac{|f|^p}{\|f\|_{L^p}^p} \star \frac{|g|^q}{\|g\|_{L^q}^q} \right)(x).$$

Since the left invariance of the measure μ combined with Fubini's theorem obviously implies that the convolution maps $L^1(G,\mu) \times L^1(G,\mu)$ into $L^1(G,\mu)$ with norm 1, this yields the desired result in the case $r < \infty$. □

We now state a refined version of Young's inequality.

Theorem 1.5. *Let (G,μ) satisfy the same assumptions as in Lemma 1.4. Let (p,q,r) be in $]1,\infty[^3$ and satisfy (1.2). A constant C exists such that, for any $f \in L^p(G,\mu)$ and any measurable function g on G where*

$$\|g\|_{L_w^q(G,\mu)}^q \overset{def}{=} \sup_{\lambda>0} \lambda^q \mu(|g| > \lambda) < \infty,$$

the function $f \star g$ belongs to $L^r(G,\mu)$, and

$$\|f \star g\|_{L^r(G,\mu)} \leq C \|f\|_{L^p(G,\mu)} \|g\|_{L_w^q(G,\mu)}.$$

Remark 1.6. One can define the *weak L^q space* as the space of measurable functions g on G such that $\|g\|_{L_w^q(G,\mu)}$ is finite. We note that since

$$\lambda^q \mu(|g| > \lambda) \leq \int_{(|g|>\lambda)} |g(x)|^q \, d\mu(x) \leq \|g\|_{L^q(G,\mu)}^q, \tag{1.3}$$

the above theorem leads back to the standard Young inequality (up to a multiplicative constant).

We also that the weak L^q space belongs to the family of *Lorentz spaces* $L^{q,r}(G,\mu)$, which may be defined by means of *real interpolation*:

$$L^{q,r}(G,\mu) = [L^\infty(G,\mu), L^1(G,\mu)]_{1/q,r} \quad \text{for all } 1 < q < \infty \text{ and } 1 \leq r \leq \infty.$$

It turns out that the weak L^q space coincides with $L^{q,\infty}(G,\mu)$. From general real interpolation theory, we can therefore deduce a plethora of Hölder and convolution inequalities for Lorentz spaces (including, of course, the one which was proven above).

We also stress that the above theorem implies the well-known *Hardy–Little-wood–Sobolev inequality* on $\mathbb{R}^d$, given as follows.

Theorem 1.7. *Let α in $]0,d[$ and (p,r) in $]1,\infty[^2$ satisfy*

$$\frac{1}{p} + \frac{\alpha}{d} = 1 + \frac{1}{r}. \tag{1.4}$$

A constant C then exists such that

$$\| |\cdot|^{-\alpha} \star f \|_{L^r(\mathbb{R}^d)} \leq C \|f\|_{L^p(\mathbb{R}^d)}.$$

Our proof of Theorem 1.5 relies on the atomic decomposition that we introduce in the next subsection.

1.1.2 The Atomic Decomposition

The *atomic decomposition* of an L^p function is described by the following proposition, which is valid for any measure space.

Proposition 1.8. *Let (X, μ) be a measure space and p be in $[1, \infty[$. Let f be a nonnegative function in L^p. A sequence of positive real numbers $(c_k)_{k \in \mathbb{Z}}$ and a sequence of nonnegative functions $(f_k)_{k \in \mathbb{Z}}$ (the* atoms*) then exist such that*

$$f = \sum_{k \in \mathbb{Z}} c_k f_k,$$

where the supports of the functions f_k are pairwise disjoint and

$$\mu(\mathrm{Supp}\ f_k) \leq 2^{k+1}, \tag{1.5}$$

$$\|f_k\|_{L^\infty} \leq 2^{-\frac{k}{p}}, \tag{1.6}$$

$$\frac{1}{2}\|f\|_{L^p}^p \leq \sum_{k \in \mathbb{Z}} c_k^p \leq 2\|f\|_{L^p}^p. \tag{1.7}$$

Remark 1.9. As implied by the definition given below, the sequence $(c_k f_k)_{k \in \mathbb{Z}}$ is independent of p and depends only on f.

Proof of Proposition 1.8. Define

$$\lambda_k \stackrel{\mathrm{def}}{=} \inf\left\{\lambda\ /\mu(f > \lambda) < 2^k\right\}, \quad c_k \stackrel{\mathrm{def}}{=} 2^{\frac{k}{p}}\lambda_k, \quad \text{and} \quad f_k \stackrel{\mathrm{def}}{=} c_k^{-1}\mathbf{1}_{(\lambda_{k+1} < f \leq \lambda_k)}f.$$

It is obvious that $\|f_k\|_{L^\infty} \leq 2^{-\frac{k}{p}}$. Moreover, $(\lambda_k)_{k \in \mathbb{Z}}$ is a decreasing sequence which, owing to the fact that f is a nonnegative function in L^p, converges to 0 when k tends to infinity.

By the definition of λ_k, we have $\mu(f > \lambda_k) \leq 2^k$ and thus $\mu(\mathrm{Supp}\ f_k) \leq 2^{k+1}$. This gives

$$\sum_{k \in \mathbb{Z}} c_k^p = \sum_{k \in \mathbb{Z}} 2^k \lambda_k^p$$

$$= p\sum_{k \in \mathbb{Z}} \int_0^\infty 2^k \mathbf{1}_{]0, \lambda_k[}(\lambda)\lambda^{p-1}\, d\lambda.$$

Using Fubini's theorem, we get

$$\sum_{k \in \mathbb{Z}} c_k^p = p\int_0^\infty \lambda^{p-1}\left(\sum_{k\ /\ \lambda_k > \lambda} 2^k\right) d\lambda.$$

By the definition of the sequence $(\lambda_k)_{k \in \mathbb{Z}}$, $\lambda < \lambda_k$ implies that $\mu(f > \lambda) \geq 2^k$. We thus infer that

$$\sum_{k\in\mathbb{Z}} c_k^p \le p \int_0^\infty \lambda^{p-1} \left(\sum_{k\,/\,2^k \le \mu(f>\lambda)} 2^k \right) d\lambda$$

$$\le 2p \int_0^\infty \lambda^{p-1} \mu(f>\lambda)\, d\lambda.$$

The right-hand inequality in (1.7) now follows from the fact that, by Fubini's theorem, we have

$$\|f\|_{L^p}^p = p \int_0^\infty \lambda^{p-1} \mu(|f| > \lambda)\, d\lambda. \tag{1.8}$$

In order to complete the proof of (1.7) it suffices to note that, because the supports of the functions $(f_k)_{k\in\mathbb{Z}}$ are pairwise disjoint, we may write

$$\|f\|_{L^p}^p = \sum_{k\in\mathbb{Z}} c_k^p \|f_k\|_{L^p}^p.$$

Taking advantage of inequalities (1.5) and (1.6), we find that

$$\|f_k\|_{L^p}^p \le 2 \quad \text{for all } k \in \mathbb{Z}.$$

This yields the desired inequality.   □

1.1.3 Proof of Refined Young Inequality

Let f and g be nonnegative measurable functions on (G, μ). Consider a nonnegative function h in $L^{r'}$ and define

$$I(f,g,h) \overset{\text{def}}{=} \int_{G^2} f(y)g(y^{-1}\cdot x)h(x)\, d\mu(x)\, d\mu(y).$$

Arguing by homogeneity, we can assume that $\|f\|_{L^p} = \|g\|_{L_w^q} = \|h\|_{L^{r'}} = 1$. Stating $C_j \overset{\text{def}}{=} \{y \in G,\ 2^j \le g(y) < 2^{j+1}\}$, we can write

$$I(f,g,h) \le 2 \sum_{j\in\mathbb{Z}} 2^j I_j(f,h) \quad \text{with}$$

$$I_j(f,h) \overset{\text{def}}{=} \int_{G^2} f(y)h(x)\mathbf{1}_{C_j}(y^{-1}\cdot x)\, d\mu(x)\, d\mu(y).$$

Because $\|g\|_{L_w^q} = 1$, we have $\|\mathbf{1}_{C_j}\|_{L^s} \le 2^{-j\frac{q}{s}}$ for all $s \in [1,\infty]$. Thus, if we directly apply Young's inequality with p, q, and r, we find that $I_j(f,h) \le 2^{-j}$, so the series $\sum 2^{j+1} I_j(f,h)$ has no reason to converge. In order to bypass this difficulty, we may introduce the atomic decompositions of f and h, as given by Proposition 1.8. We then write

$$I_j(f,h) = \sum_{k,\ell} c_k d_\ell I_j(f_k, h_\ell).$$

Using Young's inequality, for any $(a, b) \in [1, \infty]^2$ such that $b \leq a'$ and for any $(\widetilde{f}, \widetilde{h}) \in L^a \times L^b$, we get

$$I_j(\widetilde{f}, \widetilde{h}) \leq \|\widetilde{f}\|_{L^a} \|\widetilde{h}\|_{L^b} \|\mathbf{1}_{C_j}\|_{L^{c'}} \quad \text{with} \quad \frac{1}{a} + \frac{1}{b} = 1 + \frac{1}{c}.$$

This gives

$$I_j(\widetilde{f}, \widetilde{h}) \leq 2^{-jq\left(2 - \frac{1}{a} - \frac{1}{b}\right)} \|\widetilde{f}\|_{L^a} \|\widetilde{h}\|_{L^b}.$$

Applying this for f_k and h_ℓ and using Proposition 1.8 now yields

$$2^j I_j(f_k, h_\ell) \leq 2^{jq\left(\frac{1}{q} - 2 + \frac{1}{a} + \frac{1}{b}\right)} 2^{k\left(\frac{1}{a} - \frac{1}{p}\right)} 2^{\ell\left(\frac{1}{b} - \frac{1}{r'}\right)}.$$

Using the condition (1.2) on (p, q, r) implies that

$$2^j I_j(f_k, h_\ell) \leq 2^{(jq+k)\left(\frac{1}{a} - \frac{1}{p}\right)} 2^{(jq+\ell)\left(\frac{1}{b} - \frac{1}{r'}\right)}. \tag{1.9}$$

Take a and b such that

$$\frac{1}{a} \stackrel{\text{def}}{=} \frac{1}{p} - 2\varepsilon \operatorname{sg}(jq+k) \quad \text{and} \quad \frac{1}{b} \stackrel{\text{def}}{=} \frac{1}{r'} - 2\varepsilon \operatorname{sg}(jq+\ell) \quad \text{with} \quad \varepsilon \stackrel{\text{def}}{=} \frac{1}{4}\left(\frac{1}{p} - \frac{1}{r}\right),$$

where $\operatorname{sg} n = 1$ if $n \geq 0$, and $\operatorname{sg} n = -1$ if $n < 0$.

As $q > 1$, the condition (1.2) implies that $p < r$. Thus, by the definitions of ε, a, and b, we have $b \leq a'$. With this choice of a and b, (1.9) then becomes, using the triangle inequality,

$$2^j I_j(f_k, h_\ell) \leq 2^{-2\varepsilon|jq+k| - 2\varepsilon|jq+\ell|}$$
$$\leq 2^{-\varepsilon|jq+k| - \varepsilon|jq+\ell| - \varepsilon|k-\ell|}.$$

Using Young's inequality for $\mathbb{Z}$ equipped with the counting measure, we may now deduce that

$$I(f, g, h) \leq C \sum_{j,k,\ell} c_k d_\ell 2^{-\varepsilon|jq+k| - \varepsilon|jq+\ell| - \varepsilon|k-\ell|}$$
$$\leq \frac{C}{\varepsilon} \sum_{k,\ell} c_k d_\ell 2^{-\varepsilon|k-\ell|}$$
$$\leq \frac{C}{\varepsilon^2} \|(c_k)\|_{\ell^p} \|(d_\ell)\|_{\ell^{p'}}.$$

The condition (1.2) implies that $r' \leq p'$ and thus

$$I(f, g, h) \leq \frac{C}{\varepsilon^2} \|(c_k)\|_{\ell^p} \|(d_\ell)\|_{\ell^{r'}}.$$

The theorem is thus proved. $\qquad\square$

1.1.4 A Bilinear Interpolation Theorem

The following interpolation lemma, which will be useful in Chapter 8, provides another example of an application of atomic decomposition.

Proposition 1.10. *Let (X_1, μ_1) and (X_2, μ_2) be two measure spaces. Let $\mathcal{T}$ be a continuous bilinear functional on $L^2(X_1; L^{p_j}(X_2)) \times L^2(X_1; L^{q_j}(X_2))$ for j in $\{0, 1\}$, where (p_j, q_j) is in $[1, 2]^2$ and such that $p_0 \neq p_1$ and $q_0 \neq q_1$. For any $\theta \in [0, 1]$, the bilinear functional $\mathcal{T}$ is then continuous on $L^2(X_1; L^{p_\theta}(X_2)) \times L^2(X_1; L^{q_\theta}(X_2))$ with*

$$\left(\frac{1}{p_\theta}, \frac{1}{q_\theta}\right) = (1 - \theta)\left(\frac{1}{p_0}, \frac{1}{q_0}\right) + \theta\left(\frac{1}{p_1}, \frac{1}{q_1}\right).$$

Proof. Let $f \in L^2(X_1; L^{p_\theta}(X_2))$ and $g \in L^2(X_1; L^{q_\theta}(X_2))$. As in the proof of the refined Young's inequality, we will use the atomic decompositions of f and g. For any $(t, x) \in X_1 \times X_2$, we have

$$f(t, x) = \sum_{k \in \mathbb{Z}} c_k(t) f_k(t, x) \quad \text{and} \quad g(t, x) = \sum_{\ell \in \mathbb{Z}} d_\ell(t) g_\ell(t, x).$$

Let us write that

$$\mathcal{T}(f, g) = \sum_{k, \ell} \mathcal{T}(c_k f_k, d_\ell g_\ell).$$

Using the hypothesis on $\mathcal{T}$ and stating $\alpha \stackrel{\text{def}}{=} \left(\frac{1}{p_0} - \frac{1}{p_1}\right)^{-1}\left(\frac{1}{q_0} - \frac{1}{q_1}\right)$, we get

$$|\mathcal{T}(c_k f_k, d_\ell g_\ell)| \leq C \min_{j \in \{0, 1\}} \|c_k f_k\|_{L^2(X_1; L^{p_j}(X_2))} \|d_\ell g_\ell\|_{L^2(X_1; L^{q_j}(X_2))}$$

$$\leq C \|c_k\|_{L^2(X_1)} \|d_\ell\|_{L^2(X_1)}$$

$$\times \min\left\{2^{-\theta\left(\frac{1}{p_0} - \frac{1}{p_1}\right)(k + \alpha\ell)}, 2^{(1-\theta)\left(\frac{1}{p_0} - \frac{1}{p_1}\right)(k + \alpha\ell)}\right\}.$$

Setting $\varepsilon \stackrel{\text{def}}{=} \left|\frac{1}{p_0} - \frac{1}{p_1}\right| \times \min\{\theta, (1 - \theta)\}$, we deduce that

$$|\mathcal{T}(c_k f_k, d_\ell g_\ell)| \leq C \|c_k\|_{L^2(X_1)} \|d_\ell\|_{L^2(X_1)} 2^{-\varepsilon|k + \alpha\ell|}.$$

Using a weighted Cauchy–Schwarz inequality, we then get

$$|\mathcal{T}(f, g)| \leq C_\varepsilon \left(\sum_k \|c_k\|^2_{L^2(X_1)}\right)^{\frac{1}{2}} \left(\sum_\ell \|d_\ell\|^2_{L^2(X_1)}\right)^{\frac{1}{2}}$$

$$\leq C_\varepsilon \left\|\|(c_k)\|_{\ell^2(\mathbb{Z})}\right\|_{L^2(X_1)} \left\|\|(d_\ell)\|_{\ell^2(\mathbb{Z})}\right\|_{L^2(X_1)}.$$

Using the fact that p_θ and q_θ are less than 2, we infer that

$$|\mathcal{T}(f, g)| \leq C_\varepsilon \left\|\|(c_k)\|_{\ell^{p_\theta}(\mathbb{Z})}\right\|_{L^2(X_1)} \left\|\|(d_\ell)\|_{\ell^{q_\theta}(\mathbb{Z})}\right\|_{L^2(X_1)}.$$

The inequality (1.7) from Proposition 1.8 then implies the proposition.   □

1.1.5 A Linear Interpolation Result

We shall present here a basic result of linear *complex* interpolation theory which will be useful, particularly in Chapter 8.

Lemma 1.11. *Consider three measure spaces $(X_k, \mu_k)_{1 \leq k \leq 3}$ and two elements $(p_j, q_j, r_j)_{j \in \{0,1\}}$ of $[1, \infty]^3$. Further, consider an operator A which continuously maps $L^{p_j}(X_1; L^{q_j}(X_2))$ into $L^{r_j}(X_3)$ for j in $\{0,1\}$. For any θ in $[0,1]$, if*

$$\left(\frac{1}{p_\theta}, \frac{1}{q_\theta}, \frac{1}{r_\theta} \right) \overset{def}{=} (1-\theta) \left(\frac{1}{p_0}, \frac{1}{q_0}, \frac{1}{r_0} \right) + \theta \left(\frac{1}{p_1}, \frac{1}{q_1}, \frac{1}{r_1} \right),$$

then A continuously maps $L^{p_\theta}(X_1; L^{q_\theta}(X_2))$ into $L^{r_\theta}(X_3)$ and

$$\|A\|_{\mathcal{L}(L^{p_\theta}(X_1; L^{q_\theta}(X_2)); L^{r_\theta}(X_3))} \leq \mathcal{A}_\theta \quad with$$

$$\mathcal{A}_\theta \overset{def}{=} \|A\|_{\mathcal{L}(L^{p_0}(X_1; L^{q_0}(X_2)); L^{r_0}(X_3))}^{1-\theta} \|A\|_{\mathcal{L}(L^{p_1}(X_1; L^{q_1}(X_2)); L^{r_1}(X_3))}^{\theta}.$$

Proof. Consider f in $L^{p_\theta}(X_1; L^{q_\theta}(X_2))$ and φ in $L^{r_\theta'}(X_3)$.[4] Using Lemma 1.2, it is enough to prove that

$$\int_{X_3} (Af)(x_3) \varphi(x_3) d\mu_3(x_3) \leq \mathcal{A}_\theta \|f\|_{L^{p_\theta}(L^{q_\theta})} \|\varphi\|_{L^{r_\theta'}}. \tag{1.10}$$

Let z be a complex number in the strip S of complex numbers whose real parts are between 0 and 1. Define

$$f_z(x_1, x_2) \overset{def}{=} \frac{f(x_1, x_2)}{|f(x_1, x_2)|} \left(\frac{|f(x_1, x_2)|}{\|f(x_1, \cdot)\|_{L^{q_\theta}}} \right)^{q_\theta \left(\frac{1-z}{q_0} + \frac{z}{q_1} \right)} \|f(x_1, \cdot)\|_{L^{q_\theta}}^{p_\theta \left(\frac{1-z}{p_0} + \frac{z}{p_1} \right)}$$

and

$$\varphi_z(x_3) = \frac{\varphi(x_3)}{|\varphi(x_3)|} |\varphi(x_3)|^{r_\theta' \left(\frac{1-z}{r_0'} + \frac{z}{r_1'} \right)}.$$

Obviously, we have $f_\theta = f$ and $\varphi_\theta = \varphi$. It can be checked that the function defined by

$$F(z) \overset{def}{=} \int_{X_3} (Af_z)(x_3) \varphi_z(x_3) \, d\mu_3(x_3)$$

is holomorphic and bounded on S and continuous on the closure of S. From the Phragmen–Lindelhöf principle, we infer that

$$F(\theta) \leq M_0^{1-\theta} M_1^\theta \quad with \quad M_j \overset{def}{=} \sup_{t \in \mathbb{R}} |F(j + it)|. \tag{1.11}$$

---

[4] Throughout this proof, we write $L^{p_\theta}(X_1; L^{q_\theta}(X_2))$ simply as $L^{p_\theta}(L^{q_\theta})$ and $L^{r_\theta}(X_3)$ simply as L^{r_θ}.

We have

$$|f_{j+it}(x_1, x_2)| = \left(\frac{|f(x_1, x_2)|}{\|f(x_1, \cdot)\|_{L^{q_\theta}}} \right)^{\frac{q_\theta}{q_j}} \|f(x_1, \cdot)\|_{L^{q_\theta}}^{\frac{p_\theta}{p_j}}.$$

Thus, we have that f_{j+it} belongs to $L^{p_j}(L^{q_j})$ and

$$\|f_{j+it}\|_{L^{p_j}(L^{q_j})} = \|f\|_{L^{p_\theta}(L^{q_\theta})}^{\frac{p_\theta}{p_j}}.$$

In the same way, we get that $|\varphi_{j+it}(x_3)| = |\varphi(x_3)|^{\frac{r'_\theta}{r'_j}}$. Thus, thanks to Hölder's inequality, we get

$$M_j \le \sup_{t \in \mathbb{R}} \left| \int_{X_3} (A f_{j+it})(x_3) \varphi_{j+it}(x_3) \, d\mu_3(x_3) \right|$$

$$\le \|A\|_{\mathcal{L}(L^{p_j}(X_1; L^{q_j}(X_2)); L^{r_j}(X_3))}^{\theta} \|f\|_{L^{p_\theta}(L^q_\theta)}^{\frac{p_\theta}{p_j}} \||\varphi|\|_{L^{r'_\theta}(L^{r'_\theta})}^{\frac{r'_\theta}{r'_j}}.$$

Using (1.11), we then deduce (1.10) and the lemma is proved.   □

From this lemma, taking $X_1 = \{a\}$ and then $X_3 = \{a\}$, we can infer the following two corollaries which will be used in Chapter 8.

Corollary 1.12. *Let $(X_k, \mu_k)_{1 \le k \le 2}$ be two measure spaces and $(p_j, q_j)_{j \in \{0,1\}}$ be two elements of $[1, \infty]^2$. Consider a linear operator A which continuously maps $L^{p_j}(X_1)$ into $L^{q_j}(X_2)$ for $j \in \{0, 1\}$. For any θ in $[0, 1]$, if*

$$\left(\frac{1}{p_\theta}, \frac{1}{q_\theta} \right) \overset{def}{=} (1 - \theta) \left(\frac{1}{p_0}, \frac{1}{q_0} \right) + \theta \left(\frac{1}{p_1}, \frac{1}{q_1} \right),$$

then A continuously maps $L^{p_\theta}(X_1)$ into $L^{q_\theta}(X_2)$ and

$$\|A\|_{\mathcal{L}(L^{p_\theta}(X_1); L^{q_\theta}(X_2))} \le \mathcal{A}_\theta \overset{def}{=} \|A\|_{\mathcal{L}(L^{p_0}(X_1); L^{q_0}(X_2))}^{1-\theta} \|A\|_{\mathcal{L}(L^{p_1}(X_1); L^{q_1}(X_2))}^{\theta}.$$

Corollary 1.13. *Let (X_1, μ_1), (X_2, μ_2) be two measure spaces and (p_0, q_0), (p_1, q_1) be two elements of $[1, \infty]^2$. Let A be a continuous linear functional on $L^{p_j}(X_1; L^{q_j}(X_2))$ for j in $\{0, 1\}$. For any θ in $[0, 1]$, if*

$$\left(\frac{1}{p_\theta}, \frac{1}{q_\theta} \right) \overset{def}{=} (1 - \theta) \left(\frac{1}{p_0}, \frac{1}{q_0} \right) + \theta \left(\frac{1}{p_1}, \frac{1}{q_1} \right),$$

then A is a continuous linear functional on $L^{p_\theta}(X_1; L^{q_\theta}(X_2))$ and

$$\|A\|_{\mathcal{L}(L^{p_\theta}(X_1; L^{q_\theta}(X_2)); \mathbb{C})} \le \mathcal{A}_\theta \quad \text{with}$$

$$\mathcal{A}_\theta \overset{def}{=} \|A\|_{\mathcal{L}(L^{p_0}(X_1; L^{q_0}(X_2)); \mathbb{C})}^{1-\theta} \|A\|_{\mathcal{L}(L^{p_1}(X_1; L^{q_1}(X_2)); \mathbb{C})}^{\theta}.$$

1.1.6 The Hardy–Littlewood Maximal Function

In this subsection, we state a few elementary properties of the maximal function, which will be needed for proving Gagliardo–Nirenberg inequalities on the Euclidean space $\mathbb{R}^d$.

We first recall that the *maximal function* may be defined on any metric space (X, d) endowed with a Borel measure μ. More precisely, if $f : X \mapsto \mathbb{R}$ is in $L^1_{loc}(X, \mu)$, then we define

$$\forall x \in X, \ Mf(x) \overset{\text{def}}{=} \sup_{r>0} \frac{1}{\mu(B(x,r))} \int_{B(x,r)} |f(y)| \, d\mu(y). \tag{1.12}$$

The following well-known continuity result for the maximal function is fundamental in harmonic analysis.

Theorem 1.14. *Assume that the measure metric space (X, d, μ) has the doubling property.[5] There then exists a constant C, depending only on the doubling constant D, such that for all $1 < p \leq \infty$ and $f \in L^p(X, \mu)$, we have $Mf \in L^p(X, \mu)$ and*

$$\|Mf\|_{L^p} \leq \frac{p}{p-1} C^{\frac{1}{p}} \|f\|_{L^p}. \tag{1.13}$$

Proof. First step: M maps L^∞ into L^∞. Indeed, we obviously have

$$\|Mf\|_{L^\infty} \leq \|f\|_{L^\infty} \quad \text{for all} \quad f \in L^\infty(X, \mu). \tag{1.14}$$

Second step: M maps L^1 into L^1_w. We claim that there exists some constant C_1, depending only on D, such that

$$\|Mf\|_{L^1_w} \leq C_1 \|f\|_{L^1} \quad \text{for all} \quad f \in L^1(X, \mu). \tag{1.15}$$

This is a mere consequence of the following *Vitali covering lemma* that we temporarily assume to hold.

Lemma 1.15. *Let (X, d) be a metric space endowed with a Borel measure μ with the doubling property. There then exists a constant c such that for any family $(B_i)_{1 \leq i \leq n}$ of balls, there exists a subfamily $(B_{i_j})_{1 \leq j \leq p}$ of pairwise disjoint balls such that*

$$\mu\left(\bigcup_{j=1}^p B_{i_j}\right) \geq c\,\mu\left(\bigcup_{i=1}^n B_i\right).$$

Fix some $f \in L^1(X, \mu)$ and some $\lambda > 0$. By definition of the function Mf, for any x in the set $E_\lambda \overset{\text{def}}{=} \{Mf > \lambda\}$, we can find some $r_x > 0$ such that

$$\int_{B(x,r_x)} |f| \, d\mu > \lambda \mu(B(x, r_x)). \tag{1.16}$$

---

[5] That is, there exists a positive constant D such that $\mu(B(x, 2r)) \leq D\mu(B(x, r))$ for all $x \in X$ and $r > 0$.

Therefore, if K is a compact subset of E_λ, then we can find a finite covering $(B_i)_{1 \leq i \leq n}$ of K by such balls. Denoting by $(B_{i_j})_{1 \leq j \leq p}$ the subfamily supplied by the Vitali lemma and using (1.16), we can thus write

$$\lambda |K| \leq \frac{\lambda}{c} \mu \left(\bigcup_{j=1}^{p} B_{i_j} \right) \leq \frac{1}{c} \sum_{j=1}^{p} \lambda \mu(B_{i_j}) \leq \frac{1}{c} \sum_{j=1}^{p} \int_{B_{i_j}} |f| \, d\mu \leq \frac{1}{c} \int_X |f| \, d\mu,$$

which obviously leads to (1.15).

Third step: M maps L^p into L^p for all $p \in]1, \infty[$. The proof relies on arguments borrowed from real interpolation. Fix some function f in L^p and $\alpha \in]0, 1[$. Since $M|f| = Mf$, we can assume that $f \geq 0$. Now, for all $\lambda > 0$, we may write

$$f = f_\lambda + f^\lambda \quad \text{with} \quad f^\lambda \overset{\text{def}}{=} (f - \lambda \alpha) \mathbf{1}_{(f \geq \lambda \alpha)}.$$

Note that, thanks to (1.14), we have

$$(Mf > \lambda) \subset (Mf^\lambda > (1 - \alpha)\lambda).$$

Hence the equality (1.8) implies that

$$\|Mf\|_{L^p}^p \leq p \int_0^{+\infty} \lambda^{p-1} \mu \big(Mf^\lambda > (1 - \alpha)\lambda \big) \, d\lambda.$$

According to the inequality (1.15), we have

$$\mu \big(Mf^\lambda > (1 - \alpha)\lambda \big) \leq \frac{C_1}{(1 - \alpha)\lambda} \|f^\lambda\|_{L^1}.$$

So, finally, using the definition of f^λ and Fubini's theorem, we get

$$\|Mf\|_{L^p}^p \leq \frac{C_1 p}{1 - \alpha} \int_0^{+\infty} \lambda^{p-2} \int_{(f \geq \lambda \alpha)} \big(f(x) - \lambda \alpha \big) \, d\mu(x)$$

$$\leq \frac{C_1 p}{1 - \alpha} \left(\int_X f(x) \int_0^{\frac{f(x)}{\alpha}} \lambda^{p-2} \, d\lambda \, d\mu(x) - \alpha \int_X \int_0^{\frac{f(x)}{\alpha}} \lambda^{p-1} \, d\lambda \, d\mu(x) \right)$$

$$\leq \frac{C_1}{(p-1)(1-\alpha)\alpha^{p-1}} \|f\|_{L^p}^p.$$

Choosing $\alpha = (p-1)/p$ completes the proof of the inequality (1.13).   □

Proof of Lemma 1.15. Without loss of generality, we can assume that $B_i = B(x_i, r_i)$ with $r_1 \geq \cdots \geq r_n$. We can now construct the desired subfamily by induction. Indeed, for B_{i_1}, take the largest ball (i.e., B_1). Then, assuming that $B_{i_1}, \ldots, B_{i_k}$ have been chosen, pick up the largest remaining ball which does not intersect the balls which have been taken so far.

Clearly, this process stops within a finite number of steps. In addition, if $i \notin \{i_1, \dots, i_p\}$, then there exists some index i_j such that $i_j < i$ and $B_i \cap B_{i_j}$ is not empty. Therefore, by virtue of the triangle inequality, B_i is included in $B(x_{i_j}, 3r_{i_j})$. This ensures that

$$\bigcup_{i=1}^n B_i \subset \bigcup_{j=1}^p B(x_{i_j}, 3r_{i_j}).$$

As the measure μ has the doubling property, this yields the desired result. $\square$

The following result is of importance for proving Gagliardo–Nirenberg inequalities.

Proposition 1.16. *Let G be a locally compact group with neutral element e, endowed with a distance d such that $d(e, y^{-1} \cdot x) = d(x, y)$ for all $(x, y) \in G^2$ and a left-invariant Haar measure μ satisfying (1.1).*

We assume, in addition, that for all $r > 0$ there exists a positive measure σ_r on the sphere $\Sigma_r \overset{\text{def}}{=} \{x \in G \,/\, d(e, x) = r\}$ such that for any L^1 function g on G, we have

$$\int_G g(z) \, d\mu(z) = \int_0^{+\infty} \left(\int_{\Sigma_r} g(z) \, d\sigma_r(z) \right) dr.$$

For all measurable functions f and any L^1 function K on G such that

$$\forall x \in G, \; K(x) = k(d(e, x))$$

for some nonincreasing function $k : \mathbb{R}^+ \mapsto \mathbb{R}^+$, we then have

$$\forall x \in G, \; |K \star f(x)| \leq \|K\|_{L^1(G, \mu)} \, Mf(x).$$

Proof. Obviously we can restrict the proof to nonnegative functions f. Arguing by density we can also assume that k is C^1 and compactly supported. Owing to our assumptions on d and K, we have

$$\begin{aligned}
K \star f(x) &= \int_G K(y) f(y^{-1} \cdot x) \, d\mu(y) \\
&= \int_0^{+\infty} k(r) \left(\int_{\Sigma_r} f(y^{-1} \cdot x) \, d\sigma_r(y) \right) dr.
\end{aligned}$$

Therefore, integrating by parts with respect to r, we discover that

$$\begin{aligned}
K \star f(x) &= \int_0^{+\infty} (-k'(r)) \left(\int_0^r \int_{\Sigma_s} f(y^{-1} \cdot x) \, d\sigma_s(y) \, ds \right) dr \\
&= \int_0^{+\infty} (-k'(r)) \left(\int_{B(x,r)} f(y) \, d\mu(y) \right) dr \\
&\leq Mf(x) \int_0^{+\infty} (-k'(r)) \mu(B(x, r)) \, dr.
\end{aligned}$$

Finally, since

$$\mu(B(x,r)) = \mu(B(e,r)) = \int_0^r \int_{\Sigma_r} 1 \, d\sigma_r(y) \, dr,$$

performing another integration by parts, we can write that

$$\int_0^{+\infty} (-k'(r))\mu(B(x,r)) \, dr = \int_0^{+\infty} k(r)\left(\int_{\Sigma_r} 1 \, d\sigma_r(y)\right) dr = \|K\|_{L^1(G,\mu)},$$

and the desired inequality follows.   □

Remark 1.17. All the assumptions of the above proposition are satisfied if we take for G the group $(\mathbb{R}^d, +)$ endowed with the usual metric and the Lebesgue measure, or the Heisenberg group $(\mathbb{H}^d, \cdot)$ endowed with the Heisenberg distance and the Lebesgue measure of $\mathbb{R}^{2d+1}$.

We also note the following obvious generalization of the inequality stated in the above proposition:

$$\forall x \in G, \ |K \star f(x)| \leq \left(\int_G \left(\sup_{d(e,y') \geq d(e,y)} |K(y')|\right) dy\right) Mf(x),$$

which holds for any measurable function K on G. In fact, in Chapter 2 we shall use the above inequality rather than the above proposition.

1.2 The Fourier Transform

This section is devoted to a short presentation on the Fourier transform, a key tool in this monograph. In the first subsection we define the Fourier transform of a smooth function with fast decay at infinity. In the second subsection we then extend the definition (by duality) to tempered distributions. We conclude this section with the calculation of the Fourier transforms of some functions which play important roles in the following chapters.

1.2.1 Fourier Transforms of Functions and the Schwartz Space

The *Fourier transform* is defined on $L^1(\mathbb{R}^d)$ by

$$\mathcal{F}f(\xi) = \widehat{f}(\xi) = \int_{\mathbb{R}^d} e^{-i(x|\xi)} f(x) \, dx, \tag{1.17}$$

where $(x|\xi)$ denotes the inner product on $\mathbb{R}^d$. It is a continuous linear map from $L^1(\mathbb{R}^d)$ into $L^\infty(\mathbb{R}^d)$ because, obviously, $|\widehat{f}(\xi)| \leq \|f\|_{L^1}$. It is also clear that for any function $\phi \in L^1$ and automorphism L on $\mathbb{R}^d$, we have

$$\mathcal{F}(\phi \circ L) = \frac{1}{|\det L|} \widehat{\phi} \circ L^{-1}. \tag{1.18}$$

We now introduce the *Schwartz space* $\mathcal{S}(\mathbb{R}^d)$ (also denoted by $\mathcal{S}$ when no confusion is possible), which will be the basic tool for extending the Fourier transform to a very large class of distributions over $\mathbb{R}^d$. Let us first introduce the following notation. If α is a *multi-index* (i.e., an element of $\mathbb{N}^d$), x an element of $\mathbb{R}^d$, and f a smooth function of $\mathbb{R}^d$, then the *length* $|\alpha|$ of α is defined by $|\alpha| \overset{def}{=} \alpha_1 + \cdots + \alpha_d$. We also define $\partial^\alpha f \overset{def}{=} \partial_1^{\alpha_1} \cdots \partial_d^{\alpha_d} f$ and $x^\alpha \overset{def}{=} x^{\alpha_1} \cdots x^{\alpha_d}$.

Definition 1.18. *The Schwartz space $\mathcal{S}(\mathbb{R}^d)$ is the set of smooth functions u on $\mathbb{R}^d$ such that for any $k \in \mathbb{N}$ we have*

$$\|u\|_{k,\mathcal{S}} \overset{def}{=} \sup_{\substack{|\alpha|\leq k \\ x\in\mathbb{R}^d}} (1 + |x|)^k |\partial^\alpha u(x)| < \infty.$$

It is an easy exercise (left to the reader) to prove that, equipped with the family of seminorms $(\|\cdot\|_{k,\mathcal{S}})_{k\in\mathbb{N}}$, the set $\mathcal{S}(\mathbb{R}^d)$ is a Fréchet space and that the space $\mathcal{D}(\mathbb{R}^d)$ of smooth compactly supported functions on $\mathbb{R}^d$ is dense in $\mathcal{S}(\mathbb{R}^d)$.

The way the Fourier transform $\mathcal{F}$ acts on the space $\mathcal{S}$ is described by the following theorem.

Theorem 1.19. *The Fourier transform continuously maps $\mathcal{S}$ into $\mathcal{S}$: For any integer k, there exist a constant C and an integer N such that*

$$\forall \phi \in \mathcal{S}, \ \|\widehat{\phi}\|_{k,\mathcal{S}} \leq C\|\phi\|_{N,\mathcal{S}}.$$

Moreover, the Fourier transform $\mathcal{F}$ is an automorphism of $\mathcal{S}$, the inverse of which is $(2\pi)^{-d}\check{\mathcal{F}}$, where $\check{\mathcal{F}}$ denotes the application $f \longmapsto \{\xi \mapsto (\mathcal{F}f)(-\xi)\}$.

Proof. Let $k \in \mathbb{N}$ and $\alpha \in \mathbb{N}^d$ with length k. Using Lebesgue's theorem and integration by parts, we get that, for any ϕ in $\mathcal{S}$,

$$(i\partial)^\alpha \widehat{f}(\xi) = \mathcal{F}(x^\alpha\phi)(\xi) \quad \text{and} \quad (i\xi)^\alpha \widehat{\phi}(\xi) = \mathcal{F}(\partial^\alpha\phi)(\xi). \qquad (1.19)$$

From this, we deduce that

$$\left| \xi^\beta \partial^\alpha \widehat{\phi}(\xi) \right| \leq \left| \mathcal{F}(\partial^\beta(x^\alpha\phi))(\xi) \right|$$
$$\leq \|\partial^\beta(x^\alpha\phi)\|_{L^1}$$
$$\leq c_d \|(1 + |x|)^{d+1} \partial^\beta(x^\alpha\phi)\|_{L^\infty}.$$

Hence, by the definition of the seminorms, we have $\|\widehat{\phi}\|_{k,\mathcal{S}} \leq C\|\phi\|_{k+d+1,\mathcal{S}}$.

We now prove the inverse formula, namely, $\mathcal{F}^{-1} = (2\pi)^{-d}\check{\mathcal{F}}$. The proof is based on the computation of Fourier transforms of Gaussian functions. If $d = 1$, we have, thanks to (1.19),

$$\frac{d}{d\xi}\left(\mathcal{F}(e^{-x^2})\right)(\xi) = \mathcal{F}(-ixe^{-x^2})(\xi)$$

$$= \mathcal{F}\left(\frac{i}{2}\frac{d}{dx}e^{-x^2}\right)(\xi)$$

$$= -\frac{\xi}{2}\mathcal{F}(e^{-x^2})(\xi).$$

As $\mathcal{F}\left(e^{-x^2}\right)(0) = \int e^{-x^2}\,dx = \pi^{\frac{1}{2}}$, we get that $\mathcal{F}(e^{-x^2})(\xi) = \pi^{\frac{1}{2}}e^{-\frac{\xi^2}{4}}$.

From this and Fubini's theorem, we can now deduce that if d is any positive integer, then $\mathcal{F}\left(e^{-|x|^2}\right)(\xi) = \pi^{\frac{d}{2}}e^{-\frac{|\xi|^2}{4}}$. Using (1.18) we then infer that for any positive real number a,

$$\int_{\mathbb{R}^d} e^{-i(x|\xi)}e^{-a|x|^2}\,dx = \left(\frac{\pi}{a}\right)^{\frac{d}{2}}e^{-\frac{|\xi|^2}{4a}}. \tag{1.20}$$

Let ϕ be a function in $\mathcal{S}(\mathbb{R}^d)$ and ε any positive real number. Fubini's theorem applied to the function $(2\pi)^{-d}e^{i(x-y|\xi)}e^{-\varepsilon|\xi|^2}\phi(y)$, together with (1.20), implies that

$$(2\pi)^{-d}\int_{\mathbb{R}^d} e^{i(x|\xi)}e^{-\varepsilon|\xi|^2}\widehat{\phi}(\xi)\,d\xi = \left(\frac{1}{4\pi\varepsilon}\right)^{\frac{d}{2}}(e^{-\frac{|\cdot|^2}{4\varepsilon}}\star\phi)(x).$$

On the one hand, owing to Lebesgue's dominated convergence theorem, the left-hand side tends to $(2\pi)^{-d}\check{\mathcal{F}}\widehat{\phi}$. On the other hand, the right-hand side is the convolution of ϕ with an approximation of the identity. Letting ε tend to 0 thus completes the proof of the theorem.   □

1.2.2 Tempered Distributions and the Fourier Transform

Definition 1.20. *A* tempered distribution *on $\mathbb{R}^d$ is any continuous linear functional[6] on $\mathcal{S}(\mathbb{R}^d)$. The set of tempered distributions is denoted by $\mathcal{S}'(\mathbb{R}^d)$.*

A sequence $(u_n)_{n\in\mathbb{N}}$ of tempered distributions is said to converge *to u in $\mathcal{S}'(\mathbb{R}^d)$ if*

$$\forall\phi\in\mathcal{S}(\mathbb{R}^d),\ \lim_{n\to\infty}\langle u_n,\phi\rangle = \langle u,\phi\rangle.$$

Remark 1.21. The link with distributions on $\mathbb{R}^d$ is as follows: If T is a distribution on $\mathbb{R}^d$ such that for some integer k and positive real C we have

$$\forall\varphi\in\mathcal{D}(\mathbb{R}^d),\ |\langle T,\varphi\rangle| \le C\|\varphi\|_{k,\mathcal{S}}, \tag{1.21}$$

then, as $\mathcal{D}(\mathbb{R}^d)$ is dense in $\mathcal{S}(\mathbb{R}^d)$, the linear functional T may be uniquely extended to a continuous linear functional. Moreover, if T belongs to $\mathcal{S}'(\mathbb{R}^d)$,

---

[6] That is, u is a tempered distribution if there exist a constant C and an integer k such that $|\langle u,\phi\rangle| \le C\|\phi\|_{k,\mathcal{S}}$ for all $\phi\in\mathcal{S}(\mathbb{R}^d)$.

then the restriction of T to $\mathcal{D}(\mathbb{R}^d)$ defines a distribution on $\mathbb{R}^d$ because, for any positive R and any function φ in $\mathcal{D}(B(0,R))$,

$$|\langle T, \varphi\rangle| \leq C\|\varphi\|_{k,\mathcal{S}} \leq C(1+R)^k \sup_{|\alpha|\leq k} \|\partial^\alpha \varphi\|_{L^\infty}.$$

Thus, the set of distributions T on $\mathbb{R}^d$ which satisfy (1.21) may be identified with $\mathcal{S}'(\mathbb{R}^d)$.

Example 1.22. – Let us denote by $L^1_{\mathcal{M}}$ the space of locally integrable functions f on $\mathbb{R}^d$ such that for some integer N, the function $(1+|x|)^{-N}f(x)$ is integrable. For any $f \in L^1_{\mathcal{M}}$, we can then define the tempered distribution T_f by the formula

$$\langle T_f, \phi\rangle = \int_{\mathbb{R}^d} f(x)\phi(x)\,dx.$$

In other words, we identify the function f with T_f.
 – Any finite Borel measure may be seen as a tempered distribution. Indeed, we may take $k=0$ in (1.21).
 – Any compactly supported distribution may be identified with an element of $\mathcal{S}'$.

Let us use L. Schwartz's idea of duality to define operators on the space of tempered distributions. It is based on the following proposition.

Proposition 1.23. *Let A be a linear continuous map from $\mathcal{S}$ into $\mathcal{S}$.[7] The formula*

$$\langle {}^tA u, \phi\rangle \stackrel{def}{=} \langle u, A\phi\rangle$$

then defines a tempered distribution. Moreover, ${}^tA$ is linear and continuous, in the sense that if $(u_n)_{n\in\mathbb{N}}$ is a sequence of distributions which converges to u in $\mathcal{S}'(\mathbb{R}^d)$, then $({}^tA u_n)_{n\in\mathbb{N}}$ converges to ${}^tA u$.

Proof. By the definition of a tempered distribution, an integer k and a constant C exist such that

$$\forall \theta \in \mathcal{S}, \ |\langle u, \theta\rangle| \leq C\|\theta\|_{k,\mathcal{S}}. \tag{1.22}$$

The linear map A is assumed to be continuous, hence there exist a constant C' and an integer N such that

$$\forall \phi \in \mathcal{S}, \ \|A\phi\|_{k,\mathcal{S}} \leq C'\|\phi\|_{N,\mathcal{S}}.$$

Applying (1.22) with $\theta = A\phi$ and the above inequality, we then get that ${}^tA u$ is a tempered distribution. By the definition of the convergence of a sequence of tempered distributions, we then write

---

[7] That is, for any integer k, there exist a constant C and an integer N such that $\|A\phi\|_{k,\mathcal{S}} \leq C\|\phi\|_{N,\mathcal{S}}$ for all $\phi \in \mathcal{S}(\mathbb{R}^d)$.

$$\langle {}^t A u_n, \phi \rangle = \langle u_n, A\phi \rangle \longrightarrow \langle u, A\phi \rangle = \langle {}^t A u, \phi \rangle.$$

The proposition is thus proved.   □

We now list a few important examples to which Proposition 1.23 applies:

– We may take for A any operator $(-\partial)^\alpha$ or $x^\alpha \mapsto x^\alpha u$ with $\alpha \in \mathbb{N}^d$. Indeed, we have, for all ϕ in $\mathcal{S}$,

$$\|(-\partial)^\alpha \phi\|_{k,\mathcal{S}} \leq \|\phi\|_{k+|\alpha|,\mathcal{S}} \quad \text{and} \quad \|x^\alpha \phi\|_{k,\mathcal{S}} \leq \|\phi\|_{k+|\alpha|,\mathcal{S}}.$$

– Let L be a linear automorphism of $\mathbb{R}^d$ and define

$$A_L \phi \overset{\text{def}}{=} \frac{1}{\det L} \phi \circ L^{-1}.$$

It is clear that A_L satisfies the hypothesis of Proposition 1.23.
– If we denote by $\Theta_\mathcal{M}$ the space of smooth functions on $\mathbb{R}^d$ such that, for any integer k, an integer N exists such that

$$\sup_{x\in\mathbb{R}^d} (1+|x|^k)^{-N} \sup_{|\alpha|\leq k} |\partial^\alpha f(x)| < \infty,$$

then the operator A_f of multiplication by f satisfies the hypothesis of the proposition.
– If θ is a function of $\mathcal{S}$, it is left as an exercise for the reader to check that, for any $\phi \in \mathcal{S}$,

$$\|A_\theta \phi\|_{k,\mathcal{S}} \leq C_k \|\theta\|_{k+d+1,\mathcal{S}} \|\phi\|_{k,\mathcal{S}} \quad \text{with} \quad A_\theta \phi \overset{\text{def}}{=} \check{\theta} \star \phi.$$

– Theorem 1.19 guarantees, in particular, that the Fourier transform $\mathcal{F}$ satisfies the hypothesis of Proposition 1.23.

For all the above operators, we can apply Proposition 1.23. We now check briefly that this is a generalization of classical operations on functions. If u is an $L^1_\mathcal{M}$ function which is also C^1, then we have

$$\forall \phi \in \mathcal{S}, \ \langle {}^t(-\partial_j)u, \phi \rangle = \langle u, -\partial_j \phi \rangle = \int_{\mathbb{R}^d} u(x)(-\partial_j \phi)(x)\, dx.$$

An integration by parts ensures that ${}^t(-\partial_j)u = \partial_j u$, in the classical sense.
   Next, we claim that ${}^t A_L f(y) = f(Ly)$ for all $f \in L^1_\mathcal{M}$. Indeed, a straightforward change of variables ensures that for all $\phi \in \mathcal{S}$ we have

$$\langle {}^t A_L f, \phi \rangle = \frac{1}{|\det L|} \int_{\mathbb{R}^d} f(x)\phi(L^{-1}x)\, dx = \int_{\mathbb{R}^d} f(Ly)\phi(y)\, dy.$$

In the particular case where $Lx = \lambda x$, we denote ${}^t A_L f$ by f_λ, and when $\lambda = -1$, the distribution ${}^t A_L f$ is denoted by $\check{f}$. In passing, let us recall that a tempered distribution f is said to be *homogeneous of degree m* if

$$f_\lambda = \lambda^m f \quad \text{for all} \quad \lambda > 0.$$

It is obvious that the operator A_f generalizes the classical multiplication of functions by f.

Finally, for any L^1 function f, we have, according to Fubini's theorem,

$$\langle {}^t A_\theta f, \phi \rangle = \langle f, \check\theta \star \phi \rangle$$
$$= \int_{\mathbb{R}^d \times \mathbb{R}^d} f(x)\theta(y-x)\phi(y)\, dy\, dx$$
$$= \langle f \star \theta, \phi \rangle.$$

Thus, the notion of convolution between a tempered distribution and a function of $\mathcal{S}$ coincides with the classical definition when the tempered distribution is an L^1 function.

In order to extend the definition of the Fourier transform to tempered distributions, we consider an L^1 function f. By Fubini's theorem and by definition of the Fourier transform on L^1, we have, for all $\phi \in \mathcal{S}$,

$$\langle {}^t \mathcal{F} f, \phi \rangle = \int_{\mathbb{R}^d} f(x)\widehat\phi(x)\, dx$$
$$= \int_{\mathbb{R}^d \times \mathbb{R}^d} f(x) e^{-i(x|\xi)} \phi(\xi)\, dx\, d\xi$$
$$= \langle \widehat f, \phi \rangle.$$

In other words, the operator ${}^t \mathcal{F}$ restricted to L^1 functions coincides with the Fourier transform of functions. Thus, it will also be denoted by $\mathcal{F}$ in all that follows.

Proposition 1.24. *For any (u, θ) in $\mathcal{S}' \times \mathcal{S}$, $\lambda \in \mathbb{R} \setminus \{0\}$ and $(a, \omega) \in \mathbb{R}^d \times \mathbb{R}^d$, we have*[8]

$$(i\partial)^\alpha \widehat u = \mathcal{F}(x^\alpha u), \quad (i\xi)^\alpha \widehat u = \mathcal{F}(\partial^\alpha u), \quad e^{-i(a|\xi)} \widehat u = \mathcal{F}(\tau_a f),$$
$$\tau_\omega \widehat f = \mathcal{F}(e^{i(x|\omega)} f), \quad \lambda^{-d} \widehat f(\lambda^{-1}\xi) = \mathcal{F}(f(\lambda x)), \quad \text{and} \quad \mathcal{F}(u \star \theta) = \widehat\theta\, \widehat u.$$

Proof. The first five equalities readily follow from (1.19) or direct computation once we observe that ${}^t(AB) = {}^t B\, {}^t A$. In order to prove the last identity, it suffices to use the fact that, by definition of the Fourier transform and convolution, we have

$$\langle \mathcal{F}(u \star \theta), \phi \rangle = \langle u \star \theta, \widehat\phi \rangle = \langle u, \check\theta \star \widehat\phi \rangle.$$

Fubini's theorem implies that

---

[8] Below, the notation τ_a stands for the *translation operator* $\tau_a : f \mapsto f(\cdot - a)$.

$$(\check{\theta} \star \widehat{\phi})(\xi) = \int \check{\theta}(\xi - \eta) \left(\int e^{-i(x|\eta)} \phi(x) \, dx \right) d\eta$$

$$= \int e^{-i(x|\xi)} \left(\int e^{-i(x|\eta-\xi)} \theta(\eta - \xi) d\eta \right) \phi(x) \, dx$$

$$= \mathcal{F}(\widehat{\theta}\phi).$$

We infer that $\langle \mathcal{F}(u \star \theta), \phi \rangle = \langle u, \mathcal{F}(\widehat{\theta}\phi) \rangle = \langle \widehat{u}, \widehat{\theta}\phi \rangle = \langle \widehat{\theta}\widehat{u}, \phi \rangle$. The proposition is thus proved. $\qquad \square$

Theorem 1.25 (Fourier–Plancherel formula). *The Fourier transform is an automorphism of $\mathcal{S}'$ with inverse $(2\pi)^{-d}\check{\mathcal{F}}$. Moreover, $\mathcal{F}$ is also an automorphism of $L^2(\mathbb{R}^d)$ which satisfies, for any function f in L^2, $\|\widehat{f}\|_{L^2} = (2\pi)^{\frac{d}{2}}\|f\|_{L^2}$.*

Proof. On the space $\mathcal{S}$, we have $\mathcal{F}\check{\mathcal{F}} = \check{\mathcal{F}}\mathcal{F} = (2\pi)^d \,\mathrm{Id}$. Arguing by transposition, we discover that these two identities remain valid on $\mathcal{S}'$. Next, using the fact that for any function ϕ in $\mathcal{S}$ we have $\overline{\mathcal{F}\phi} = \check{\mathcal{F}}(\overline{\phi})$ and taking advantage of the inverse Fourier formula (see Theorem 1.19), we get, for any function ϕ in $\mathcal{S}$,

$$\|\mathcal{F}\phi\|_{L^2}^2 = \langle \mathcal{F}\phi, \overline{\mathcal{F}\phi} \rangle = \langle \phi, \mathcal{F}\check{\mathcal{F}}\overline{\phi} \rangle = (2\pi)^d \|\phi\|_{L^2}^2.$$

Combining the Riesz representation theorem with the density of $\mathcal{S}$ in L^2 enables us to complete the proof. $\qquad \square$

Finally, let us define a subspace of $\mathcal{S}'(\mathbb{R}^d)$ which will play an important role in the following chapters.

Definition 1.26. *We denote by $\mathcal{S}'_h(\mathbb{R}^d)$ the space of tempered distributions u such that[9]*

$$\lim_{\lambda \to \infty} \|\theta(\lambda D)u\|_{L^\infty} = 0 \quad \text{for any } \theta \text{ in } \mathcal{D}(\mathbb{R}^d).$$

Remark 1.27. It is clear that whether or not a tempered distribution u belongs to $\mathcal{S}'_h$ depends only on low frequencies. As a matter of fact, it is not hard to check that u belongs to $\mathcal{S}'_h(\mathbb{R}^d)$ if and only if one can find some smooth compactly supported function θ satisfying the above equality and such that $\theta(0) \neq 0$.

Examples

- If a tempered distribution u is such that its Fourier transform $\widehat{u}$ is locally integrable near 0, then u belongs to $\mathcal{S}'_h$. In particular, the space $\mathcal{E}'$ of compactly supported distributions is included in $\mathcal{S}'_h$.
- If u is a tempered distribution such that $\theta(D)u \in L^p$ for some $p \in [1, \infty[$ and some function θ in $\mathcal{D}(\mathbb{R}^d)$ with $\theta(0) \neq 0$, then u belongs to $\mathcal{S}'_h$.

---

[9] We agree that if f is a measurable function on $\mathbb{R}^d$ with at most polynomial growth at infinity, then the operator $f(D)$ is defined by $f(D)a \overset{\text{def}}{=} \mathcal{F}^{-1}(f\mathcal{F}a)$.

– A nonzero polynomial P does not belong to $\mathcal{S}'_h$ because for any $\theta \in \mathcal{D}(\mathbb{R}^d)$ with value 1 at 0 and any $\lambda > 0$, we may write $\theta(\lambda D)P = P$. However, if η is in $\mathbb{R}^d \setminus \{0\}$, then $e^{i(\cdot|\eta)}P$ belongs to $\mathcal{S}'_h$ because the support of its Fourier transform is $\{\eta\}$. We note that this example implies that $\mathcal{S}'_h$ is not a closed subspace of $\mathcal{S}'$ for the topology of weak-$\star$ convergence, a fact which must be kept in mind in the applications.

1.2.3 A Few Calculations of Fourier Transforms

This subsection is devoted to the computation of the Fourier transforms of some functions which are definitely not in L^1.

Proposition 1.28. *Let z be a nonzero complex number with nonnegative real part. Then,*

$$\mathcal{F}\left(e^{-z|\cdot|^2}\right)(\xi) = \left(\frac{\pi}{z}\right)^{\frac{d}{2}} e^{-\frac{|\xi|^2}{4z}}$$

with $z^{-\frac{d}{2}} \stackrel{def}{=} |z|^{-\frac{d}{2}} e^{-i\frac{d}{2}\theta}$ if $z = |z|e^{i\theta}$ with $\theta \in [-\pi/2, \pi/2]$.

Proof. Let us remark that for any ξ in $\mathbb{R}^d$, the functions

$$z \longmapsto \int_{\mathbb{R}^d} e^{-i(x|\xi)} e^{-z|x|^2}\, dx \quad \text{and} \quad z \longmapsto \left(\frac{\pi}{z}\right)^{\frac{d}{2}} e^{-\frac{|\xi|^2}{4z}}$$

are holomorphic on the domain D of complex numbers with positive real part. Formula (1.20) states that these two functions coincide on the intersection of the real line with D. Thus, they also coincide on the whole domain D. Now, let $(z_n)_{n \in \mathbb{N}}$ be a sequence of elements of D which converges to it for $t \neq 0$. For any function ϕ in $\mathcal{S}$, we have, by virtue of Lebesgue's dominated convergence theorem,

$$\lim_{n \to \infty} \int_{\mathbb{R}^d} e^{-z_n|x|^2} \phi(x)\, dx = \int_{\mathbb{R}^d} e^{-it|x|^2} \phi(x)\, dx \quad \text{and}$$

$$\lim_{n \to \infty} \int_{\mathbb{R}^d} e^{-\frac{|\xi|^2}{4z_n}} \phi(\xi)\, d\xi = \int_{\mathbb{R}^d} e^{-\frac{|\xi|^2}{4it}} \phi(\xi)\, d\xi.$$

As we have

$$\mathcal{F}\left(e^{-z_n|\cdot|^2}\right) = \left(\frac{\pi}{z_n}\right)^{\frac{d}{2}} e^{-\frac{|\xi|^2}{4z_n}},$$

passing to the limit in $\mathcal{S}'(\mathbb{R}^d)$ when n tends to ∞ gives the result, thanks to Proposition 1.23. $\qquad\square$

Proposition 1.29. *If $\sigma \in\ ]0, d[$, then $\mathcal{F}(|\cdot|^{-\sigma}) = c_{d,\sigma} \cdot |\cdot|^{\sigma-d}$ for some constant $c_{d,\sigma}$ depending only on d and s.*

Proof. We only treat the case $d \geq 2$. The (easier) case $d = 1$ is left to the reader. Defining

$$R \overset{\text{def}}{=} \sum_{j=1}^{d} x_j \partial_j \quad \text{and} \quad Z_{j,k} \overset{\text{def}}{=} x_j \partial_k - x_k \partial_j \,,$$

we have $R(|\cdot|^{-\sigma}) = -\sigma |\cdot|^{-\sigma}$ and $Z_{j,k}(|\cdot|^{-\sigma}) = 0$. Then, using Proposition 1.24, we infer that $Z_{j,k}\mathcal{F}| \cdot |^{-\sigma} = 0$ and

$$R\mathcal{F}| \cdot |^{-\sigma} = \sum_{j=1}^{d} \partial_j \left(\xi_j \mathcal{F}| \cdot |^{-\sigma} \right) - d\mathcal{F}| \cdot |^{-\sigma} = (\sigma - d)\mathcal{F}| \cdot |^{-\sigma}.$$

By restricting to $\mathbb{R}^d \setminus \{0\}$, we then see that

$$R\left(|\cdot|^{d-\sigma} \mathcal{F}| \cdot |^{-\sigma} \right) = Z_{j,k}\left(|\cdot|^{d-\sigma} \mathcal{F}| \cdot |^{-\sigma} \right) = 0 \quad \text{in} \quad \mathcal{D}'(\mathbb{R}^d \setminus \{0\}).$$

We note that for any k,

$$|x|^2 \partial_k = \sum_{j=1}^{d} x_j^2 \partial_k = x_k R + \sum_{j=1}^{d} x_j Z_{j,k}.$$

Therefore, $\nabla\left(|\cdot|^{d-\sigma} \mathcal{F}| \cdot |^{-\sigma} \right)$ is supported in $\mathbb{R}^d \setminus \{0\}$. Because $d \geq 2$, we deduce that there exists some constant $c_{d,\sigma}$ such that $|\cdot|^{d-\sigma} \mathcal{F}| \cdot |^{-\sigma} - c_{d,\sigma}$ is also supported in $\mathbb{R}^d \setminus \{0\}$ and, owing to $\sigma > 0$, so is $\mathcal{F}| \cdot |^{-\sigma} - c_{d,\sigma}|\cdot|^{\sigma-d}$. The conclusion then follows easily from the following lemma.   □

Lemma 1.30. *Let T be a distribution on $\mathbb{R}^d$ supported in $\{0\}$ and such that $RT = sT$ for some real number s.*

- *If s is not an integer less than or equal to $-d$, then $T = 0$.*
- *If s is an integer less than or equal to $-d$, then there exist some real numbers a_α such that*

$$T = \sum_{|\alpha|=-s-d} a_\alpha \partial^\alpha \delta_0.$$

Proof. We first observe that a distribution supported in $\{0\}$ is of the form $T = \sum_{|\alpha| \leq N} a_\alpha \partial^\alpha \delta_0$. We thus have

$$RT = \sum_{j=1}^{d} \sum_{|\alpha| \leq N} a_\alpha x_j \partial_j \partial^\alpha \delta_0$$
$$= - \sum_{|\alpha| \leq N} (d + |\alpha|) a_\alpha \partial^\alpha \delta_0.$$

As $(\partial^\alpha \delta_0)_{\alpha \in \mathbb{N}^d}$ is a family of linearly independent distributions, the fact that $RT = sT$ implies that $(d+|\alpha|)a_\alpha = -sa_\alpha$. The lemma is thus proved.   □

1.3 Homogeneous Sobolev Spaces

This section is concerned with homogeneous Sobolev spaces. We first establish classical properties for these spaces, then we focus on embedding in Lebesgue, BMO and Hölder spaces.

1.3.1 Definition and Basic Properties

Definition 1.31. *Let s be in $\mathbb{R}$. The homogeneous Sobolev space $\dot{H}^s(\mathbb{R}^d)$ (also denoted by $\dot{H}^s$) is the space of tempered distributions u over $\mathbb{R}^d$, the Fourier transform of which belongs to $L^1_{loc}(\mathbb{R}^d)$ and satisfies*

$$\|u\|^2_{\dot{H}^s} \overset{def}{=} \int_{\mathbb{R}^d} |\xi|^{2s} |\widehat{u}(\xi)|^2 \, d\xi < \infty.$$

We note that the spaces $\dot{H}^s$ and $\dot{H}^{s'}$ *cannot* be compared for the inclusion. Nevertheless, we have the following proposition.

Proposition 1.32. *Let $s_0 \leq s \leq s_1$. Then, $\dot{H}^{s_0} \cap \dot{H}^{s_1}$ is included in $\dot{H}^s$, and we have*

$$\|u\|_{\dot{H}^s} \leq \|u\|^{1-\theta}_{\dot{H}^{s_0}} \|u\|^{\theta}_{\dot{H}^{s_1}} \quad with \quad s = (1-\theta)s_0 + \theta s_1.$$

Proof. It suffices to apply Hölder's inequality with $p = 1/(1-\theta)$ and $q = 1/\theta$ to the functions $\xi \mapsto |\xi|^{2(1-\theta)s_0}$, $\xi \mapsto |\xi|^{2\theta s_1}$ and the Borel measure $|\widehat{u}(\xi)|^2 \, d\xi$.
□

Using the Fourier–Plancherel formula, we observe that $L^2 = \dot{H}^0$ and that if s is a positive integer, then $\dot{H}^s$ is the subset of tempered distributions with locally integrable Fourier transforms and such that $\partial^\alpha u$ belongs to L^2 for all α in $\mathbb{N}^d$ of length s.

In the case where s is a negative integer, the Sobolev space $\dot{H}^s$ is described by the following proposition.

Proposition 1.33. *Let k be a positive integer. The space $\dot{H}^{-k}(\mathbb{R}^d)$ consists of distributions which are the sums of derivatives of order k of $L^2(\mathbb{R}^d)$ functions.*

Proof. Let u be in $\dot{H}^{-k}(\mathbb{R}^d)$. Using the fact that for some integer constants A_α, we have

$$|\xi|^{2k} = \sum_{1 \leq j_1,\ldots,j_k \leq d} \xi^2_{j_1} \cdots \xi^2_{j_k} = \sum_{|\alpha|=k} A_\alpha (i\xi)^\alpha (-i\xi)^\alpha, \qquad (1.23)$$

we get that

$$\widehat{u}(\xi) = \sum_{|\alpha|=k} (i\xi)^\alpha v_\alpha(\xi) \quad with \quad v_\alpha(\xi) \overset{def}{=} A_\alpha \frac{(-i\xi)^\alpha}{|\xi|^{2k}} \widehat{u}(\xi).$$

As u is in $\dot{H}^{-k}$, the functions v_α belong to L^2. Defining $u_\alpha \overset{\text{def}}{=} \mathcal{F}^{-1}v_\alpha$, we then obtain

$$u = \sum_{|\alpha|=k} \partial^\alpha u_\alpha \quad \text{with} \quad u_\alpha \in L^2(\mathbb{R}^d).$$

This concludes the proof of the proposition. □

Proposition 1.34. $\dot{H}^s(\mathbb{R}^d)$ *is a Hilbert space if and only if* $s < \dfrac{d}{2}$.

Proof. We first assume that $s < d/2$. Let $(u_n)_{n\in\mathbb{N}}$ be a Cauchy sequence in $\dot{H}^s(\mathbb{R}^d)$. Then, $(\widehat{u}_n)_{n\in\mathbb{N}}$ is a Cauchy sequence in the space $L^2(\mathbb{R}^d; |\xi|^{2s}\, d\xi)$. Because $|\xi|^{2s}\, d\xi$ is a measure on $\mathbb{R}^d$, there exists a function f in $L^2(\mathbb{R}^d; |\xi|^{2s}\, d\xi)$ such that $(\widehat{u}_n)_{n\in\mathbb{N}}$ converges to f in $L^2(\mathbb{R}^d; |\xi|^{2s}\, d\xi)$. Because $s < d/2$, we have

$$\int_{B(0,1)} |f(\xi)|\, d\xi \le \left(\int_{\mathbb{R}^d} |\xi|^{2s}|f(\xi)|^2\, d\xi\right)^{\frac{1}{2}} \left(\int_{B(0,1)} |\xi|^{-2s}\, d\xi\right)^{\frac{1}{2}} < \infty.$$

This ensures that $\mathcal{F}^{-1}(1_{B(0,1)}f)$ is a bounded function. Now, $1_{{}^cB(0,1)}f$ clearly belongs to $L^2(\mathbb{R}^d; (1 + |\xi|^2)^s\, d\xi)$ and thus to $\mathcal{S}'(\mathbb{R}^d)$, so f is a tempered distribution. Define $u \overset{\text{def}}{=} \mathcal{F}^{-1}f$. It is then obvious that u belongs to $\dot{H}^s$ and that $\lim_{n\to\infty} u_n = u$ in the space $\dot{H}^s$.

If $s \ge d/2$, observe that the function

$$N : u \longmapsto \|\widehat{u}\|_{L^1(B(0,1))} + \|u\|_{\dot{H}^s}$$

is a norm over $\dot{H}^s(\mathbb{R}^d)$ and that $(\dot{H}^s(\mathbb{R}^d), N)$ is a Banach space.

Now, if $\dot{H}^s(\mathbb{R}^d)$ endowed with $\|\cdot\|_{\dot{H}^s}$ were also complete, then, according to Banach's theorem, there would exist a constant C such that $N(u) \le C\|u\|_{\dot{H}^s}$. Of course, this would imply that

$$\|\widehat{u}\|_{L^1(B(0,1))} \le C\|u\|_{\dot{H}^s}. \tag{1.24}$$

This inequality is violated by the following example. Let $\mathcal{C}$ be an annulus included in the unit ball $B(0,1)$ and such that $\mathcal{C} \cap 2\mathcal{C} = \emptyset$. Define

$$\Sigma_n \overset{\text{def}}{=} \mathcal{F}^{-1} \sum_{q=1}^{n} \frac{2^{q(s+\frac{d}{2})}}{q} 1_{2^{-q}\mathcal{C}}.$$

We have

$$\|\widehat{\Sigma}_n\|_{L^1(B(0,1))} = C\sum_{q=1}^{n} \frac{2^{q(s-\frac{d}{2})}}{q} \quad \text{and} \quad \|\Sigma_n\|_{\dot{H}^s}^2 \le C\sum_{q=1}^{n} \frac{1}{q^2} \le C_1.$$

As $s \ge d/2$, we deduce that $\|\widehat{\Sigma}_n\|_{L^1(B(0,1))}$ tends to infinity when n goes to infinity. Hence, the inequality (1.24) is false. □

Proposition 1.35. *If $s < d/2$, then the space $\mathcal{S}_0(\mathbb{R}^d)$ of functions of $\mathcal{S}(\mathbb{R}^d)$, the Fourier transform of which vanishes near the origin, is dense in $\dot{H}^s$.*

Proof. Consider u in $\dot{H}^s$ such that

$$\forall \phi \in \mathcal{S}_0(\mathbb{R}^d), \ (u|\phi)_{H^s} = \int_{\mathbb{R}^d} |\xi|^{2s} \widehat{u}(\xi) \overline{\widehat{\phi}(\xi)} \, d\xi = 0.$$

This implies that the L^1_{loc} function $\widehat{u}$ vanishes on $\mathbb{R}^d \setminus \{0\}$. Thus, $\widehat{u} = 0$. Thanks to Theorem 1.25, we infer that $u = 0$. As we are considering the case where $\dot{H}^s$ is a Hilbert space, we deduce that $\mathcal{S}_0(\mathbb{R}^d)$ is dense in $\dot{H}^s$. □

The following proposition explains how the space $\dot{H}^{-s}$ can be considered as the dual space of $\dot{H}^s$.

Proposition 1.36. *If $|s| < d/2$, then the bilinear functional*

$$\mathcal{B} : \begin{cases} \mathcal{S}_0 \times \mathcal{S}_0 \to \mathbb{C} \\ (\phi, \varphi) \mapsto \displaystyle\int_{\mathbb{R}^d} \phi(x) \varphi(x) \, dx \end{cases}$$

can be extended to a continuous bilinear functional on $\dot{H}^{-s} \times \dot{H}^s$. Moreover, if L is a continuous linear functional on $\dot{H}^s$, then a unique tempered distribution u exists in $\dot{H}^{-s}$ such that

$$\forall \phi \in \dot{H}^s, \ \langle L, \phi \rangle = \mathcal{B}(u, \phi) \quad and \quad \|L\|_{(\dot{H}^s)'} = \|u\|_{\dot{H}^{-s}}.$$

Proof. Let ϕ and φ be in $\mathcal{S}_0$. We can write

$$\left| \int_{\mathbb{R}^d} \phi(x) \varphi(x) \, dx \right| = \left| \int_{\mathbb{R}^d} (\mathcal{F}^{-1}\phi)(\xi)(\mathcal{F}\varphi)(\xi) \, d\xi \right|$$

$$= (2\pi)^{-d} \left| \int_{\mathbb{R}^d} |\xi|^{-s} \widehat{\phi}(-\xi) |\xi|^s \widehat{\varphi}(\xi) \, d\xi \right|$$

$$\leq (2\pi)^{-d} \|\phi\|_{\dot{H}^{-s}} \|\varphi\|_{\dot{H}^s}.$$

As $\mathcal{S}_0$ is dense in $\dot{H}^\sigma$ when $|\sigma| < d/2$, we can extend $\mathcal{B}$ to $\dot{H}^{-s} \times \dot{H}^s$. Of course, if $(u, \phi) \in \dot{H}^{-s} \times \mathcal{S}$, then $\mathcal{B}(u, \phi) = \langle u, \phi \rangle$.

Let L be a linear functional on $\dot{H}^s$. Consider the linear functional L_s defined by

$$L_s : \begin{cases} L^2(\mathbb{R}^d) \longrightarrow \mathbb{C} \\ f \longmapsto \langle L, \mathcal{F}^{-1}(|\cdot|^{-s} f) \rangle. \end{cases}$$

It is obvious that

$$\sup_{\|f\|_{L^2}=1} |\langle L_s, f \rangle| = \sup_{\|f\|_{L^2}=1} |\langle L, \mathcal{F}^{-1}(|\cdot|^{-s} f) \rangle|$$

$$= \sup_{\|\phi\|_{\dot{H}^s}=1} |\langle L, \phi \rangle|$$

$$= \|L\|_{(\dot{H}^s)'}.$$

The Riesz representation theorem implies that a function g exists in L^2 such that

$$\forall h \in L^2 \,, \ \langle L_s, h \rangle = \int_{\mathbb{R}^d} g(\xi) h(\xi) \, d\xi.$$

We obviously have $|\cdot|^s g \in L^2(\mathbb{R}^d; |\xi|^{-2s} \, d\xi)$. Now, as $|s| < d/2$, this implies that $|\cdot|^s g$ is in $\mathcal{S}'(\mathbb{R}^d)$ and thus we can define $u \overset{\text{def}}{=} \mathcal{F}(|\cdot|^s g)$. For any ϕ in $\mathcal{S}(\mathbb{R}^d)$, we then have

$$\langle u, \phi \rangle = \int_{\mathbb{R}^d} g(\xi) |\xi|^s \widehat{\phi}(\xi) \, d\xi = \langle L_s, |\cdot|^s \widehat{\phi} \rangle.$$

By the definition of L_s, we have $\langle u, \phi \rangle = \langle L, \phi \rangle$ and the proposition is thus proved. □

For s in the interval $]0, 1[$, the space $\dot{H}^s$ can be described in terms of finite differences.

Proposition 1.37. *Let s be a real number in the interval $]0, 1[$ and u be in $\dot{H}^s(\mathbb{R}^d)$. Then,*

$$u \in L^2_{loc}(\mathbb{R}^d) \quad and \quad \int_{\mathbb{R}^d \times \mathbb{R}^d} \frac{|u(x+y) - u(x)|^2}{|y|^{d+2s}} \, dx \, dy < \infty.$$

Moreover, a constant C_s exists such that for any function u in $\dot{H}^s(\mathbb{R}^d)$, we have

$$\|u\|_{\dot{H}^s}^2 = C_s \int_{\mathbb{R}^d \times \mathbb{R}^d} \frac{|u(x+y) - u(x)|^2}{|y|^{d+2s}} \, dx \, dy.$$

Proof. In order to see that u is in $L^2_{loc}(\mathbb{R}^d)$, it suffices to write

$$u = \mathcal{F}^{-1}\big(1_{B(0,1)} \widehat{u}\big) + \mathcal{F}^{-1}\big(1_{{}^c B(0,1)} \widehat{u}\big).$$

The rest of the proof relies on the Fourier–Plancherel formula (see Theorem 1.25), which implies that

$$\int_{\mathbb{R}^d} \frac{|u(x+y) - u(x)|^2}{|y|^{d+2s}} \, dx = (2\pi)^{-d} \int_{\mathbb{R}^d} \frac{|e^{i(y|\xi)} - 1|^2}{|y|^{d+2s}} |\widehat{u}(\xi)|^2 \, d\xi.$$

Therefore,

$$\int_{\mathbb{R}^d \times \mathbb{R}^d} \frac{|u(x+y) - u(x)|^2}{|y|^{d+2s}} \, dx \, dy = (2\pi)^{-d} \int_{\mathbb{R}^d} F(\xi) |\widehat{u}(\xi)|^2 \, d\xi$$

with

$$F(\xi) \overset{\text{def}}{=} \int_{\mathbb{R}^d} \frac{|e^{i(y|\xi)} - 1|^2}{|y|^{2s}} \frac{dy}{|y|^d}.$$

It may be easily checked that F is a radial and homogeneous function of degree $2s$. This implies that the function $F(\xi)$ is proportional to $|\xi|^{2s}$ and thus completes the proof. □

1.3.2 Sobolev Embedding in Lebesgue Spaces

In this subsection, we investigate the embedding of $\dot{H}^s(\mathbb{R}^d)$ spaces in $L^p(\mathbb{R}^d)$ spaces. We begin with a classical result.

Theorem 1.38. *If s is in $[0, d/2[$, then the space $\dot{H}^s(\mathbb{R}^d)$ is continuously embedded in $L^{\frac{2d}{d-2s}}(\mathbb{R}^d)$.*

Proof. First, let us note that the critical index $p = 2d/(d-2s)$ may be found by using a scaling argument. Indeed, if v is a function on $\mathbb{R}^d$ and v_λ stands for the function $v_\lambda(x) \stackrel{\text{def}}{=} v(\lambda x)$, then we have

$$\|v_\lambda\|_{L^p} = \lambda^{-\frac{d}{p}}\|v\|_{L^p} \quad \text{and} \quad \|v_\lambda\|_{\dot{H}^s} = \lambda^{-\frac{d}{2}+s}\|v\|_{\dot{H}^s}.$$

If an inequality of the type $\|v\|_{L^p} \leq C\|v\|_{\dot{H}^s}$ is true for any smooth function v, then it is also true for v_λ for any λ. Hence, we must have $p = 2d/(d-2s)$.

Consider a function ϕ in $\mathcal{S}_0(\mathbb{R}^d)$. Defining $\widehat{\phi}_s(\xi) \stackrel{\text{def}}{=} |\xi|^s\widehat{\phi}(\xi)$ and using Propositions 1.24 and 1.29, we get that

$$\phi = \frac{(2\pi)^{-d}c_{d,s}}{|\cdot|^{d-s}} \star \phi_s \quad \text{with} \quad \|\phi_s\|_{L^2} = (2\pi)^{-\frac{d}{2}}\|\phi\|_{\dot{H}^s}.$$

Theorem 1.7 thus implies that $\|\phi\|_{L^p} \leq C\|\phi_s\|_{L^2}$. Now, according to Proposition 1.35, the space $\mathcal{S}_0(\mathbb{R}^d)$ is dense in $\dot{H}^s$. The proof is therefore complete. □

Corollary 1.39. *If p belongs to $]1, 2]$, then $L^p(\mathbb{R}^d)$ embeds continuously in $\dot{H}^s(\mathbb{R}^d)$ with $s = \dfrac{d}{2} - \dfrac{d}{p}$.*

Proof. We use the duality between $\dot{H}^s$ and $\dot{H}^{-s}$ described by Proposition 1.36. Write

$$\|a\|_{\dot{H}^s} = \sup_{\|\varphi\|_{\dot{H}^{-s}} \leq 1} \langle a, \varphi \rangle.$$

As $s = d\left(\dfrac{1}{2} - \dfrac{1}{p}\right)$, by Theorem 1.38 we have $\|\varphi\|_{L^{p'}} \leq C\|\varphi\|_{\dot{H}^{-s}}$ and thus

$$\|a\|_{\dot{H}^s} \leq C \sup_{\|\varphi\|_{L^{p'}} \leq 1} \langle a, \varphi \rangle \leq C\|a\|_{L^p}.$$

The corollary is thus proved. □

According to Proposition 1.24, the Fourier transform changes dilation into reciprocal dilation and translation into multiplication by a character $e^{i(x|\omega)}$ (and vice versa). Obviously, the inequality

$$\|u\|_{L^p(\mathbb{R}^d)} \leq C\|u\|_{\dot{H}^s(\mathbb{R}^d)} \quad \text{with} \quad p = 2d/(d-2s)$$

provided by Theorem 1.38 is invariant under translation and dilation.

We claim, however, that it is *not* invariant under multiplication by a character. Indeed, consider a function ϕ in $\mathcal{S}(\mathbb{R}^d)$ such that $\widehat{\phi}$ belongs to $\mathcal{D}(\mathbb{R}^d)$. For all positive ε, define the function

$$\phi_\varepsilon(x) = e^{i\frac{x_1}{\varepsilon}}\phi(x). \tag{1.25}$$

By the definition of $\|\cdot\|_{\dot{H}^s}$, we have

$$\begin{aligned}
\|\phi_\varepsilon\|^2_{\dot{H}^s} &= \int_{\mathbb{R}^d} |\xi|^{2s}\left|\widehat{\phi}\left(\xi - \frac{e_1}{\varepsilon}\right)\right|^2 d\xi \\
&= \int_{\mathbb{R}^d} \left|\xi + \frac{e_1}{\varepsilon}\right|^{2s}|\widehat{\phi}(\xi)|^2 d\xi \quad \text{with} \quad e_1 \overset{\text{def}}{=} (1,0,\dots,0).
\end{aligned}$$

Hence, $\|\phi_\varepsilon\|_{\dot{H}^s}$ is equivalent to ε^{-s} when ε tends to 0, while $\|\phi_\varepsilon\|_{L^p}$ does not depend on ε.

In what follows, we want to improve the estimate of Theorem 1.38 so that it becomes also invariant if u is multiplied by any character $e^{i(x|\omega)}$. In fact, we shall construct a family of Banach spaces E_s, the norm of which is invariant under translation, satisfying

$$\|a(\lambda\cdot)\|_{E_s} \sim \lambda^{s-\frac{d}{2}}\|a\|_{E_s}, \quad f\|a(\lambda\cdot)\|_{E_s} \le C_{s,d}\|a\|_{\dot{H}^s},$$

and, for some real number $\beta \in \ ]0,1[$,

$$\|a\|_{L^p} \le C_{s,d}\|a\|^{1-\beta}_{\dot{H}^s}\|a\|^{\beta}_{E_s}.$$

In order to do this, we introduce the following definition.

Definition 1.40. *Let θ be a function in $\mathcal{S}(\mathbb{R}^d)$ such that $\widehat{\theta}$ is compactly supported, has value 1 near 0, and satisfies $0 \le \widehat{\theta} \le 1$. For u in $\mathcal{S}'(\mathbb{R}^d)$ and $\sigma > 0$, we set*

$$\|u\|_{\dot{B}^{-\sigma}} \overset{\text{def}}{=} \sup_{A>0} A^{d-\sigma}\|\theta(A\cdot) \star u\|_{L^\infty}.$$

It is left to the reader to check that the space $\dot{B}^{-\sigma}$ of tempered distributions u such that $\|u\|_{\dot{B}^{-\sigma}}$ is finite is a Banach space. It is also clear that changing the function θ gives the same space with the equivalent norm. These spaces come up in the next chapter in a more general context. We shall see that $\dot{B}^{-\sigma}$ coincides with the *homogeneous Besov space* $\dot{B}^{-\sigma}_{\infty,\infty}$.

For the time being, we will compare $\dot{B}^{-\sigma}$ with Sobolev spaces.

Proposition 1.41. *For any s less than $d/2$, the space $\dot{H}^s$ is continuously embedded in $\dot{B}^{s-\frac{d}{2}}$ and there exists a constant C, depending only on Supp $\widehat{\theta}$ and d, such that*

$$\|u\|_{\dot{B}^{s-\frac{d}{2}}} \le \frac{C}{\left(\frac{d}{2} - s\right)^{\frac{1}{2}}}\|u\|_{\dot{H}^s} \quad \text{for all} \quad u \in \dot{H}^s.$$

Proof. As $\widehat{u}$ is locally in L^1, the function $\widehat{\theta}(A^{-1}\cdot)\widehat{u}$ is in L^1. The inverse Fourier theorem implies that

$$\|A^d\theta(A\cdot)\star u\|_{L^\infty} \le (2\pi)^{-d}\|\widehat{\theta}(A^{-1}\cdot)\widehat{u}\|_{L^1}$$
$$\le (2\pi)^{-d}\int_{\mathbb{R}^d}\widehat{\theta}(A^{-1}\xi)|\xi|^{-s}|\xi|^s|\widehat{u}(\xi)|\,d\xi.$$

Using the fact that $\widehat{\theta}$ is compactly supported, the Cauchy–Schwarz inequality implies that

$$\|A^d\theta(A\cdot)\star u\|_{L^\infty} \le \frac{C}{\left(\frac{d}{2}-s\right)^{\frac{1}{2}}}A^{\frac{d}{2}-s}\|u\|_{\dot{H}^s}$$

and the proposition is thus proved. $\qquad\square$

The difference between the $\dot{H}^s$ norm the $\dot{B}^{s-\frac{d}{2}}$ norm is emphasized by the following proposition.

Proposition 1.42. *Let* $\sigma \in \ ]0,d]$ *and let* $(\phi_\varepsilon)_{\varepsilon>0}$ *be defined according to (1.25). There then exists a constant* C *such that* $\|\phi_\varepsilon\|_{\dot{B}^{-\sigma}} \le C\varepsilon^\sigma$ *for all* $\varepsilon > 0$.

Proof. By Hölder's inequality, we have

$$A^d\|\theta(A\cdot)\star\phi_\varepsilon\|_{L^\infty} \le \|\theta\|_{L^1}\|\phi\|_{L^\infty}.$$

From this we deduce that if $A\varepsilon \ge 1$, then we have

$$A^{d-\sigma}\|\theta(A\cdot)\star\phi_\varepsilon\|_{L^\infty} \le \varepsilon^\sigma\|\theta\|_{L^1}\|\phi\|_{L^\infty}. \tag{1.26}$$

If $A\varepsilon \le 1$, then we perform integration by parts. More precisely, using the fact that

$$(-i\varepsilon\partial_1)^d e^{i\frac{x_1}{\varepsilon}} = e^{i\frac{x_1}{\varepsilon}}$$

and the Leibniz formula, we get

$$A^d(\theta(A\cdot)\star\phi_\varepsilon)(x) = (iA\varepsilon)^d\int_{\mathbb{R}^d}\partial_{y_1}^d\left(\theta(A(x-y))\phi(y)\right)e^{i\frac{y_1}{\varepsilon}}\,dy$$
$$= (iA\varepsilon)^d\sum_{k\le d}\binom{d}{k}A^k((-\partial_1)^k\theta)(A\cdot)\star(e^{i\frac{y_1}{\varepsilon}}\partial_1^{d-k}\phi)(x).$$

Using Hölder's inequality, we get that

$$A^k\left\|((-\partial_1)^k\theta)(A\cdot)\star(e^{i\frac{y_1}{\varepsilon}}\partial_1^{d-k}\phi)\right\|_{L^\infty} \le \|\partial_1^k\theta\|_{L^{\frac{d}{k}}}\|\partial_1^{d-k}\phi\|_{L^{(\frac{d}{k})'}}.$$

Thus, we get $A^d\|\theta(A\cdot)\star\phi_\varepsilon\|_{L^\infty} \le C(A\varepsilon)^d$. As we are considering the case where $A\varepsilon \le 1$, we get, for any $\sigma \le d$,

$$A^d\|\theta(A\cdot)\star\phi_\varepsilon\|_{L^\infty} \le C(A\varepsilon)^\sigma.$$

Together with (1.26), this concludes the proof of the proposition. $\qquad\square$

We can now state the so-called *refined Sobolev inequalities*.

Theorem 1.43. *Let s be in $]0, d/2[$. There exists a constant C, depending only on d and $\widehat{\theta}$, such that*

$$\|u\|_{L^p} \leq \frac{C}{(p-2)^{\frac{1}{p}}} \|u\|_{\dot{B}^{s-\frac{d}{2}}}^{1-\frac{2}{p}} \|u\|_{\dot{H}^s}^{\frac{2}{p}} \quad \text{with} \quad p = \frac{2d}{d-2s}.$$

Proof. Without loss of generality, we can assume that $\|u\|_{\dot{B}^{s-\frac{d}{2}}} = 1$. As will be done quite often in this book, we shall decompose the function into low and high frequencies. More precisely, we write

$$u = u_{\ell,A} + u_{h,A} \quad \text{with} \quad u_{\ell,A} = \mathcal{F}^{-1}(\widehat{\theta}(A^{-1}\cdot)\widehat{u}), \tag{1.27}$$

where θ is the function from Definition 1.40. The triangle inequality implies that

$$\big(|u| > \lambda\big) \subset \big(|u_{\ell,A}| > \lambda/2\big) \cup \big(|u_{h,A}| > \lambda/2\big).$$

By the definition of $\|\cdot\|_{\dot{B}^{s-\frac{d}{2}}}$ we have $\|u_{\ell,A}\|_{L^\infty} \leq A^{\frac{d}{2}-s}$. From this we deduce that

$$A = A_\lambda \overset{\text{def}}{=} \left(\frac{\lambda}{2}\right)^{\frac{p}{d}} \implies \mu\big(|u_{\ell,A}| > \lambda/2\big) = 0.$$

From the identity (1.8) we deduce that

$$\|u\|_{L^p}^p \leq p \int_0^\infty \lambda^{p-1} \mu\big(|u_{h,A_\lambda}| > \lambda/2\big) \, d\lambda.$$

Using the fact that

$$\mu\big(|u_{h,A_\lambda}| > \lambda/2\big) \leq 4\frac{\|u_{h,A_\lambda}\|_{L^2}^2}{\lambda^2},$$

we get

$$\|u\|_{L^p}^p \leq 4p \int_0^\infty \lambda^{p-3} \|u_{h,A_\lambda}\|_{L^2}^2 \, d\lambda.$$

Because the Fourier transform is (up to a constant) an isometry on $L^2(\mathbb{R}^d)$ and the function $\widehat{\theta}$ has value 1 near 0, we thus get, for some $c > 0$ depending only on $\widehat{\theta}$,

$$\|u\|_{L^p}^p \leq 4p \, (2\pi)^{-d} \int_0^\infty \lambda^{p-3} \int_{(|\xi| \geq cA_\lambda)} |\widehat{u}(\xi)|^2 \, d\xi \, d\lambda. \tag{1.28}$$

Now, by definition of A_λ, we have

$$|\xi| \geq cA_\lambda \iff \lambda \leq C_\xi \overset{\text{def}}{=} 2\left(\frac{|\xi|}{c}\right)^{\frac{d}{p}}.$$

Fubini's theorem thus implies that

$$\|u\|_{L^p}^p \le 4p \, (2\pi)^{-d} \int_{\mathbb{R}^d} \left(\int_0^{C_\xi} \lambda^{p-3} d\lambda \right) |\widehat{u}(\xi)|^2 \, d\xi$$

$$\le (2\pi)^{-d} \frac{p2^p}{p-2} \int_{\mathbb{R}^d} \left(\frac{|\xi|}{c} \right)^{\frac{d(p-2)}{p}} |\widehat{u}(\xi)|^2 \, d\xi.$$

As $s = d\left(\dfrac{1}{2} - \dfrac{1}{p} \right)$, the theorem is proved. □

Remark 1.44. Combining Proposition 1.41 and Theorem 1.43, we see that if $0 < s < d/2$, then we have, for all $u \in \dot{H}^s$,

$$\|u\|_{L^p} \le C_d \frac{p}{\sqrt{p-2}} \|u\|_{\dot{H}^s} \quad \text{with} \quad p = \frac{2d}{d-2s}. \tag{1.29}$$

Of course, since we have $\|u\|_{L^2} = (2\pi)^{-\frac{d}{2}} \|u\|_{\dot{H}^0}$, we do not expect the constant to blow up when p goes to 2. In fact, combining this latter inequality with the inequality (1.29) (with, say, $p = 4$) and resorting to a complex interpolation argument, we get

$$\|u\|_{L^p} \le C_d \sqrt{p} \, \|u\|_{\dot{H}^s} \quad \text{with} \quad p = \frac{2d}{d-2s}. \tag{1.30}$$

By taking advantage of Proposition 1.42 and the computations that follow (1.25), it is not difficult to check that the inequality stated in Theorem 1.43 is indeed invariant (up to an irrelevant constant) under multiplication by a character. We now want to consider whether our refined inequalities are sharp. Obviously, according to Proposition 1.42, we have

$$\lim_{\varepsilon \to 0} \frac{\|\phi_\varepsilon\|_{L^p}}{\|\phi_\varepsilon\|_{\dot{B}^{s-\frac{d}{2}}}^\beta \|\phi_\varepsilon\|_{\dot{H}^s}^{1-\beta}} = +\infty \quad \text{for any} \quad \beta > 1 - 2/p.$$

Therefore, the exponent $1 - 2/p$ cannot be improved. We claim that even under a sign assumption, the above refined Sobolev inequalities are sharp. More precisely, we shall exhibit a sequence $(f_n)_{n \in \mathbb{N}}$ of *nonnegative* functions such that

$$\lim_{n \to \infty} \frac{\|f_n\|_{L^{\frac{2d}{d-2s}}}}{\|f_n\|_{\dot{B}^{s-\frac{d}{2}}}^\beta \|f_n\|_{\dot{H}^s}^{1-\beta}} = +\infty \quad \text{for any} \quad \beta > 1 - 2/p. \tag{1.31}$$

Constructing such a family may be done by means of an iterative process. At each step of the process, we use a linear transform T (defined below) which duplicates any function f into 2^d copies of the same function, at the scale $1/4$.

Definition 1.45. *Define* $Q \overset{def}{=} [-1/2, 1/2]^d$ *and let* $x_J = 3/8 \, J$ *for any element* J *of* $\{-1, 1\}^d$. *We then define the transform* T *by*

$$T : \begin{cases} \mathcal{D}(Q) \longrightarrow \mathcal{D}(Q) \\ f \longmapsto Tf \overset{def}{=} 2^d \displaystyle\sum_{J \in \{-1,1\}^d} T_J f \quad \text{with} \quad T_J f(x) \overset{def}{=} f(4(x - x_J)). \end{cases}$$

For $B \subset Q$, we define $T_J(B) \overset{def}{=} x_J + \frac{1}{4}B$, $T(B) \overset{def}{=} \displaystyle\bigcup_{J \in \{-1,1\}^d} T_J(B)$ and denote $T_J(Q)$ by Q_J.

Using the fact that for any $f \in \mathcal{D}(Q)$, the support of $T_J f$ is included in Q_J and the fact that if $J \neq J'$, then $Q_J \cap Q_{J'} = \emptyset$, we immediately get

$$\|Tf\|_{L^p} = 2^{d\left(1 - \frac{1}{p}\right)} \|f\|_{L^p}. \tag{1.32}$$

For the sake of simplicity we restrict our attention here to the case where s is an integer.[10] Then, observing that

$$\partial_j(Tf)(x) = 2^d \sum_{J \in \{-1,1\}^d} 4(\partial_j f)(4(x - x_J)) = 4T(\partial_j f)(x)$$

and using (1.32), we get

$$\|Tf\|_{\dot{H}^s} = 2^{\frac{d}{2} + 2s} \|f\|_{\dot{H}^s}. \tag{1.33}$$

The estimate of Tf in terms of the $\dot{B}^{-\sigma}$ norm is described by the following proposition.

Proposition 1.46. For $\sigma \in \,]0, d]$, a constant C exists such that

$$\|Tf\|_{\dot{B}^{-\sigma}} \leq 2^{d-2\sigma} \|f\|_{\dot{B}^{-\sigma}} + C\|f\|_{L^1}.$$

Proof. Since, thanks to (1.32), we have

$$\lambda^{d-\sigma} \|\theta(\lambda \cdot) \star (Tf)\|_{L^\infty} \leq \lambda^{d-\sigma} \|\theta\|_{L^\infty} \|Tf\|_{L^1} \leq \lambda^{d-\sigma} \|\theta\|_{L^\infty} \|f\|_{L^1},$$

we get

$$\sup_{\lambda \leq 1} \lambda^{-\sigma} \|\lambda^d \theta(\lambda \cdot) \star (Tf)\|_{L^\infty} \leq \|\theta\|_{L^\infty} \|f\|_{L^1}. \tag{1.34}$$

The case where λ is large (which corresponds to high frequencies) is more intricate. We first estimate $\lambda^d(\theta(\lambda \cdot) \star (Tf))(x)$ when x is not too close to $T(Q)$, namely, $x \in \widetilde{Q}^c \overset{def}{=} \{x \in Q \, / \, d(x, T(Q)) \geq 1/8\}$. As the function θ belongs to $\mathcal{S}(\mathbb{R}^d)$, we have, for any positive integer N,

$$\left| \lambda^d(\theta(\lambda \cdot) \star (Tf))(x) \right| \leq \lambda^d \|\theta\|_{N,\mathcal{S}} \int_{\mathbb{R}^d} \frac{1}{\lambda^N |x - y|^N} |Tf(y)| \, dy$$

$$\leq C\|\theta\|_{N,\mathcal{S}} \lambda^{d-N} \|f\|_{L^1}.$$

---

[10] The general case follows by interpolation.

This gives, for sufficiently large N,

$$\sup_{\lambda \geq 1} \lambda^{-\sigma} \|\lambda^d \theta(\lambda \cdot) \star (Tf)\|_{L^\infty(\widetilde{Q}^c)} \leq C \|\theta\|_{N,\mathcal{S}} \|f\|_{L^1}. \tag{1.35}$$

We now investigate the case where $x \in \widetilde{Q}$. By definition, an element J_x of $\{-1,1\}^d$ and a point y of Q_{J_x} exist such that $d(x,y) \leq 1/8$. For any $J' \neq J_x$, we have

$$d(x, Q_{J'}) \geq d(y, Q_{J'}) - d(x,y) \geq \frac{1}{2} - \frac{1}{8} \geq \frac{3}{8}.$$

We now write

$$\left|\lambda^d \theta(\lambda \cdot) \star (Tf)\right|(x) \leq 2^d \left|\lambda^d \theta(\lambda \cdot) \star (T_{J_x} f)\right|(x)$$
$$+ \sum_{J' \in \{-1,1\}^d \setminus \{J_x\}} 2^d \left|\lambda^d \theta(\lambda \cdot) \star (T_{J'} f)\right|(x).$$

Again using the fact that the function θ belongs to $\mathcal{S}(\mathbb{R}^d)$, we have, for any positive integer N and any $J' \neq J_x$,

$$\left|\lambda^d (\theta(\lambda \cdot) \star (T_{J'} f))(x)\right| \leq \|\theta\|_{N,\mathcal{S}} \lambda^d \int_{\mathbb{R}^d} \frac{1}{\lambda^N |x-y|^N} |T_{J'} f(y)| \, dy$$
$$\leq C \|\theta\|_{N,\mathcal{S}} \lambda^{d-N} \|T_{J'} f\|_{L^1}.$$

Using (1.32), we infer that, for $\lambda \geq 1$ and N sufficiently large,

$$\sum_{J' \in \{-1,1\}^d \setminus \{J_x\}} \left|\lambda^d \theta(\lambda \cdot) \star (T_{J'} f)\right|(x) \leq C \|\theta\|_{N,\mathcal{S}} \sum_{J' \in \{-1,1\}^d \setminus \{J_x\}} \|T_{J'} f\|_{L^1}$$
$$\leq C \|\theta\|_{N,\mathcal{S}} \|f\|_{L^1}. \tag{1.36}$$

For any J, we have, by definition of T_J,

$$\sup_{\lambda > 0} \lambda^{-\sigma} \|\lambda^d \theta(\lambda \cdot) \star (T_J f)\|_{L^\infty} \leq \sup_{\lambda > 0} \lambda^{-\sigma} \left\|\left(\frac{\lambda}{4}\right)^d \theta\left(\frac{\lambda}{4} \cdot\right) \star f\right\|_{L^\infty} \leq 2^{-2\sigma} \|f\|_{\dot{B}^{-\sigma}}.$$

Together with (1.34), (1.35), and (1.36), this gives

$$\sup_{\lambda \geq 1} \lambda^{-\sigma} \|\lambda^d \theta(\lambda \cdot) \star (Tf)\|_{L^\infty} \leq 2^{d-2\sigma} \|f\|_{\dot{B}^{-\sigma}} + C \|f\|_{L^1}.$$

This completes the proof.   □

We can now construct a sequence $(f_n)_{n \in \mathbb{N}}$ of functions satisfying (1.31). For that purpose, we consider a smooth nonnegative function f_0, supported in Q, and define $f_n = T^n f_0$. Iterating the inequality from Proposition 1.46 yields

$$\|f_n\|_{\dot{B}^{-\sigma}} \leq 2^{n(d-2\sigma)} \|f_0\|_{\dot{B}^{-\sigma}} + C \left(\sum_{m=0}^{n-1} 2^{m(d-2\sigma)}\right) \|f_0\|_{L^1}.$$

Taking $\sigma = d/2 - s$ with $s \in \,]0, d/2[$, we deduce that

$$\|f_n\|_{\dot{B}^{s-\frac{d}{2}}} \leq C_{f_0} 2^{2ns}.$$

Using (1.32) and (1.33), we can now conclude that (1.31) is satisfied.

1.3.3 The Limit Case $\dot{H}^{\frac{d}{2}}$

The space $\dot{H}^{\frac{d}{2}}(\mathbb{R}^d)$ is not included in $L^\infty(\mathbb{R}^d)$. We give an explicit counterexample in dimension two. Let the function u be defined by

$$u(x) = \varphi(x)\log(-\log|x|)$$

for some smooth function φ supported in $B(0,1)$ with value 1 near 0. On the one hand, u is not bounded. On the other hand, we have, near the origin,

$$|\partial_j u(x)| \leq \frac{C}{|x|\,|\log|x||}$$

so that u belongs to $\dot{H}^1(\mathbb{R}^2)$.

This motivates the following definition.

Definition 1.47. *The space $BMO(\mathbb{R}^d)$ of bounded mean oscillations is the set of locally integrable functions f such that*

$$\|f\|_{BMO} \overset{def}{=} \sup_B \frac{1}{|B|}\int_B |f - f_B|\,dx < \infty \quad with \quad f_B \overset{def}{=} \frac{1}{|B|}\int_B f\,dx.$$

The above supremum is taken over the set of Euclidean balls.

We point out that the seminorm $\|\cdot\|_{BMO}$ vanishes on constant functions. Therefore, this is not a norm. We now state the critical theorem for Sobolev embedding.

Theorem 1.48. *The space $L^1_{loc}(\mathbb{R}^d) \cap \dot{H}^{\frac{d}{2}}(\mathbb{R}^d)$ is included in $BMO(\mathbb{R}^d)$. Moreover, there exists a constant C such that*

$$\|u\|_{BMO} \leq C\|u\|_{\dot{H}^{\frac{d}{2}}}$$

for all functions $u \in L^1_{loc}(\mathbb{R}^d) \cap \dot{H}^{\frac{d}{2}}(\mathbb{R}^d)$.

Proof. We use the decomposition (1.27) into low and high frequencies. For any Euclidean ball B we have

$$\int_B |u - u_B|\frac{dx}{|B|} \leq \|u_{\ell,A} - (u_{\ell,A})_B\|_{L^2(B,\frac{dx}{|B|})} + \frac{2}{|B|^{\frac{1}{2}}}\|u_{h,A}\|_{L^2}.$$

Let R be the radius of the ball B. We have

$$\|u_{\ell,A} - (u_{\ell,A})_B\|_{L^2(B,\frac{dx}{|B|})} \leq R\|\nabla u_{\ell,A}\|_{L^\infty}$$

$$\leq CR\int_{\mathbb{R}^d} |\xi|^{1-\frac{d}{2}}|\xi|^{\frac{d}{2}}|\widehat{u}_{\ell,A}(\xi)|\,d\xi$$

$$\leq CRA\|u\|_{\dot{H}^{\frac{d}{2}}}.$$

We infer that

$$\int_B |u - u_B|\frac{dx}{|B|} \leq CRA\|u\|_{\dot{H}^{\frac{d}{2}}} + C(AR)^{-\frac{d}{2}}\left(\int_{|\xi|\geq A} |\xi|^d|\widehat{u}(\xi)|^2\,d\xi\right)^{\frac{1}{2}}.$$

Choosing $A = R^{-1}$ then completes the proof. $\qquad\qquad\square$

1.3.4 The Embedding Theorem in Hölder Spaces

Definition 1.49. *Let (k, ρ) be in $\mathbb{N} \times]0, 1]$. The Hölder space $C^{k,\rho}(\mathbb{R}^d)$ (or $C^{k,\rho}$, if no confusion is possible) is the space of C^k functions u on $\mathbb{R}^d$ such that*

$$\|u\|_{C^{k,\rho}} = \sup_{|\alpha| \leq k} \left(\|\partial^\alpha u\|_{L^\infty} + \sup_{x \neq y} \frac{|\partial^\alpha u(x) - \partial^\alpha u(y)|}{|x - y|^\rho} \right) < \infty.$$

Proving that the sets $C^{k,\rho}$ are Banach spaces is left as an exercise. We point out that $C^{0,1}$ is the space of bounded Lipschitz functions.

Theorem 1.50. *If $s > \frac{d}{2}$ and $s - \frac{d}{2}$ is not an integer, then the space $\dot{H}^s(\mathbb{R}^d)$ is included in the Hölder space of index*

$$(k, \rho) = \left(\left[s - \frac{d}{2} \right], s - \frac{d}{2} - \left[s - \frac{d}{2} \right] \right),$$

and we have, for all $u \in \dot{H}^s(\mathbb{R}^d)$,

$$\sup_{|\alpha| = k} \sup_{x \neq y} \frac{|\partial^\alpha u(x) - \partial^\alpha u(y)|}{|x - y|^\rho} \leq C_{d,s} \|u\|_{\dot{H}^s}.$$

Proof. We prove the theorem only in the case where the integer part of $s - d/2$ is 0. As s is greater than $d/2$, writing

$$\widehat{u} = \mathbf{1}_{B(0,1)} \widehat{u} + (1 - \mathbf{1}_{B(0,1)}) \widehat{u},$$

we get that $\widehat{u}$ belongs to $L^1(\mathbb{R}^d)$, and thus u is a bounded continuous function. We again use the decomposition (1.27) into low and high frequencies. The low-frequency part of u is of course smooth. By Taylor's inequality, we have

$$|u_{\ell,A}(x) - u_{\ell,A}(y)| \leq \|\nabla u_{\ell,A}\|_{L^\infty} |x - y|.$$

Using the Fourier inversion formula and the Cauchy–Schwarz inequality, we get

$$\|\nabla u_{\ell,A}\|_{L^\infty} \leq C \int_{\mathbb{R}^d} |\xi| \, |\widehat{u}_{\ell,A}(\xi)| \, d\xi$$

$$\leq C \left(\int_{|\xi| \leq CA} |\xi|^{2-2s} \, d\xi \right)^{\frac{1}{2}} \|u\|_{\dot{H}^s}$$

$$\leq \frac{C}{(1 - \rho)^{\frac{1}{2}}} A^{1-\rho} \|u\|_{\dot{H}^s} \quad \text{with} \quad \rho = s - d/2.$$

Reasoning along exactly the same lines, we also have that

$$\|u_{h,A}\|_{L^\infty} \leq \int_{\mathbb{R}^d} |\widehat{u}_{h,A}(\xi)| \, d\xi$$

$$\leq \left(\int_{|\xi| \geq A} |\xi|^{-2s} \, d\xi \right)^{\frac{1}{2}} \|u\|_{\dot{H}^s}$$

$$\leq \frac{C}{\rho^{\frac{1}{2}}} A^{-\rho} \|u\|_{\dot{H}^s}.$$

It is then obvious that

$$|u(x) - u(y)| \leq \|\nabla u_{\ell,A}\|_{L^\infty} |x - y| + 2\|u_{h,A}\|_{L^\infty}$$
$$\leq C_s \left(|x - y| A^{1-\rho} + A^{-\rho} \right) \|u\|_{\dot{H}^s}.$$

Choosing $A = |x - y|^{-1}$ then completes the proof of the theorem. $\qquad\square$

1.4 Nonhomogeneous Sobolev Spaces on $\mathbb{R}^d$

In this section, we focus on nonhomogeneous Sobolev spaces. As in the previous section, the emphasis is on embedding properties in Lebesgue and Hölder spaces. We also establish a trace theorem and provide an elementary proof for a Hardy inequality.

1.4.1 Definition and Basic Properties

Definition 1.51. *Let s be a real number. The Sobolev space $H^s(\mathbb{R}^d)$ consists of tempered distributions u such that $\widehat{u} \in L^2_{loc}(\mathbb{R}^d)$ and*

$$\|u\|_{H^s}^2 \stackrel{def}{=} \int_{\mathbb{R}^d} (1 + |\xi|^2)^s |\widehat{u}(\xi)|^2 \, d\xi < \infty.$$

As the Fourier transform is an isometric linear operator from the space $H^s(\mathbb{R}^d)$ onto the space $L^2(\mathbb{R}^d; (1 + |\xi|^2)^s \, d\xi)$, the space $H^s(\mathbb{R}^d)$ equipped with the scalar product

$$(u \mid v)_{H^s} \stackrel{def}{=} \int_{\mathbb{R}^d} (1 + |\xi|^2)^s \widehat{u}(\xi) \overline{\widehat{v}}(\xi) \, d\xi \qquad (1.37)$$

is a Hilbert space.

It is obvious that the family of H^s spaces is decreasing with respect to s. Moreover, we have the following proposition, the proof of which is strictly analogous to that of Proposition 1.32.

Proposition 1.52. *If $s_0 \leq s \leq s_1$, then we have*

$$\|u\|_{H^s} \leq \|u\|_{H^{s_0}}^{1-\theta} \|u\|_{H^{s_1}}^{\theta} \quad with \quad s = (1 - \theta)s_0 + \theta s_1.$$

When s is a nonnegative integer, the Fourier–Plancherel formula ensures that the space H^s coincides with the set of L^2 functions u such that $\partial^\alpha u$ belongs to L^2 for any α in $\mathbb{N}^d$ with $|\alpha| \leq s$. In the case where s is a negative integer, the space H^s is described by the following proposition, the proof of which is analogous to that of Proposition 1.33.

Proposition 1.53. *Let k be a positive integer. The space $H^{-k}(\mathbb{R}^d)$ consists of distributions which are sums of an $L^2(\mathbb{R}^d)$ function and derivatives of order k of $L^2(\mathbb{R}^d)$ functions.*

Remark 1.54. The Dirac mass δ_0 belongs to $H^{-\frac{d}{2}-\varepsilon}$ for any positive ε but does not belong to $H^{-\frac{d}{2}}$. Moreover, δ_0 is not in $\dot{H}^s$ for any s.

It is obvious that when s is nonnegative, H^s is included in $\dot{H}^s$, and that the opposite happens when s is negative. Further, $\dot{H}^s \neq H^s$ for $s \neq 0$. In the following proposition, we state that the two spaces coincide for compactly supported distributions and nonnegative s.

Proposition 1.55. *Let s be a nonnegative real number and K a compact subset of $\mathbb{R}^d$. Let $H^s_K(\mathbb{R}^d)$ be the space of those distributions of $H^s(\mathbb{R}^d)$ which are supported in K. There then exists a positive constant C such that*

$$\forall u \in H^s_K(\mathbb{R}^d)\,, \quad \frac{1}{C}\|u\|_{H^s} \leq \|u\|_{\dot{H}^s} \leq \|u\|_{H^s}.$$

Proof. We simply have to prove that $\|u\|_{L^2} \leq C_K \|u\|_{\dot{H}^s}$. Using the Fourier–Plancherel formula and the inverse formula, we have[11]

$$|\widehat{u}(\xi)| \leq \|u\|_{L^1} \leq \sqrt{|K|}\,\|u\|_{L^2} \leq (2\pi)^{-\frac{d}{2}}\sqrt{|K|}\,\|\widehat{u}\|_{L^2}.$$

For any positive ε we then get

$$\|\widehat{u}\|^2_{L^2} \leq (2\pi)^{-d}|K|\|\widehat{u}\|^2_{L^2}|B(0,\varepsilon)| + \int_{\mathbb{R}^d \setminus B(0,\varepsilon)} |\xi|^{-2s}|\xi|^{2s}|\widehat{u}(\xi)|^2\,d\xi$$

$$\leq (2\pi)^{-d}c_d\varepsilon^d\,|K|\,\|\widehat{u}\|^2_{L^2} + \frac{1}{\varepsilon^{2s}}\|u\|^2_{\dot{H}^s}.$$

Taking ε such that $(2\pi)^{-d}c_d\varepsilon^d\,|K| = 1/2$, we see that

$$\|\widehat{u}\|_{L^2} \leq \frac{\sqrt{2}}{(2\pi)^s}\left(2c_d|K|\right)^{\frac{s}{d}}\|u\|_{\dot{H}^s}, \qquad (1.38)$$

and the result follows. □

From the above proposition, we can infer the following Poincaré-type inequality, which is relevant for functions supported in small balls.

---

[11] From now on, we agree that $|K|$ denotes the Lebesgue measure of the set K.

Corollary 1.56. *Let $0 \leq t \leq s$. A constant C exists such that for any positive δ and any function $u \in H^s(\mathbb{R}^d)$ supported in a ball of radius δ, we have*

$$\|u\|_{\dot{H}^t} \leq C\delta^{s-t}\|u\|_{\dot{H}^s} \quad and \quad \|u\|_{H^t} \leq C\delta^{s-t}\|u\|_{H^s}.$$

Proof. Using the fact that the $\|\cdot\|_{H^s}$ norm is invariant under translation, we can suppose that the ball is centered at the origin. If we set $v(x) = u(\delta x)$, then v is supported in the unit ball and obviously satisfies $\|v\|_{H^t} \leq C\|v\|_{H^s}$, hence also $\|v\|_{\dot{H}^t} \leq C\|v\|_{\dot{H}^s}$, due to the previous proposition.

Using the fact that $\hat{v}(\xi) = \delta^{-d}\hat{u}\left(\frac{\xi}{\delta}\right)$, we thus get $\|u\|_{\dot{H}^t} \leq C\delta^{s-t}\|u\|_{\dot{H}^s}$. Using (1.38) we then get the inequality pertaining to nonhomogeneous norms. $\square$

We have the following density result, strictly analogous to Proposition 1.35.

Proposition 1.57. *The space $\mathcal{S}$ is dense in H^s.*

The duality between H^s and H^{-s} is described by the following proposition, the proof of which is analogous to that of Proposition 1.36.

Proposition 1.58. *For any real s, the bilinear functional*

$$\mathcal{B} : \begin{cases} \mathcal{S} \times \mathcal{S} \to \mathbb{C} \\ (\phi, \varphi) \mapsto \displaystyle\int_{\mathbb{R}^d} \phi(x)\varphi(x)\,dx \end{cases}$$

can be extended to a continuous bilinear functional on $H^{-s} \times H^s$. Moreover, if L is a continuous linear functional on H^s, a unique tempered distribution u exists in H^{-s} such that

$$\forall \phi \in \mathcal{S}, \quad \langle L, \phi \rangle = \mathcal{B}(u, \phi).$$

In addition, we have $\|L\|_{(H^s)'} = \|u\|_{H^{-s}}$.

The following proposition can be very easily deduced from Proposition 1.37.

Proposition 1.59. *Let $s = m + \sigma$ with $m \in \mathbb{N}$ and $\sigma \in {]}0,1{[}$. We then have*

$$H^s(\mathbb{R}^d) = \left\{ u \in L^2(\mathbb{R}^d) \, / \, \forall \alpha \in \mathbb{N}^d \, / \, |\alpha| \leq m, \ \partial^\alpha u \in L^2(\mathbb{R}^d) \right.$$

$$\left. and, for \ \alpha \, / \, |\alpha| = m, \ \int_{\mathbb{R}^d \times \mathbb{R}^d} \frac{|\partial^\alpha u(x+y) - \partial^\alpha u(x)|^2}{|y|^{d+2\sigma}} \, dx\,dy < +\infty \right\},$$

and there exists a constant C such that

$$C^{-1}\|u\|_{H^s}^2 \leq \sum_{|\alpha|=m} \int_{\mathbb{R}^d \times \mathbb{R}^d} \frac{|\partial^\alpha u(x+y) - \partial^\alpha u(x)|^2}{|y|^{d+2\sigma}} \, dx\,dy$$

$$+ \sum_{|\alpha| \leq m} \|\partial^\alpha u\|_{L^2}^2 \leq C\|u\|_{H^s}^2.$$

The above characterization of Sobolev spaces is suitable for establishing invariance under diffeomorphism. In what follows, it is understood that a *global k-diffeomorphism* on $\mathbb{R}^d$ is any C^k diffeomorphism φ from $\mathbb{R}^d$ onto $\mathbb{R}^d$ whose derivatives of order less than or equal to k are bounded and which satisfies, for some constant C,

$$\forall (x, y) \in \mathbb{R}^d \times \mathbb{R}^d, \ |\varphi(x) - \varphi(y)| \geq C|x - y|.$$

Corollary 1.60. *Let φ be a global k-diffeomorphism on $\mathbb{R}^d$, $0 \leq s < k$, and $u \in H^s(\mathbb{R}^d)$. Then, $u \circ \varphi \in H^s(\mathbb{R}^d)$.*

Proof. By virtue of the chain rule, it is enough to consider the case where s is in $[0, 1[$. The result follows easily from the identity

$$J(u) \overset{\text{def}}{=} \int_{\mathbb{R}^d \times \mathbb{R}^d} \frac{|u(\varphi(x)) - u(\varphi(y))|^2}{|x - y|^{d+2s}} \, dx \, dy$$

$$= \int_{\mathbb{R}^d \times \mathbb{R}^d} \frac{|u(x) - u(y)|^2}{|\psi(x) - \psi(y)|^{d+2s}} |\det(D\psi(x))|^{-1} |\det(D\psi(y))|^{-1} \, dx \, dy$$

$$\leq C \int_{\mathbb{R}^d \times \mathbb{R}^d} \frac{|u(x) - u(y)|^2}{|x - y|^{d+2s}} \, dx \, dy,$$

where it is understood that $\psi = \varphi^{-1}$. This proves the corollary. $\qquad \square$

The following density theorem will be useful.

Theorem 1.61. *For any real s, the space $\mathcal{D}(\mathbb{R}^d)$ is dense in $H^s(\mathbb{R}^d)$.*

Proof. In order to prove this theorem, we consider a distribution u in $H^s(\mathbb{R}^d)$ such that for any test function φ in $\mathcal{D}(\mathbb{R}^d)$, we have

$$\int_{\mathbb{R}^d} \widehat{\varphi}(\xi)(1 + |\xi|^2)^s \overline{\widehat{u}(\xi)} \, d\xi = 0.$$

Knowing that $\mathcal{D}(\mathbb{R}^d)$ is dense in $\mathcal{S}(\mathbb{R}^d)$ and that the Fourier transform is an automorphism of $\mathcal{S}(\mathbb{R}^d)$, we have, for any function f in $\mathcal{S}(\mathbb{R}^d)$,

$$\int_{\mathbb{R}^d} f(\xi)(1 + |\xi|^2)^s \overline{\widehat{u}(\xi)} \, d\xi = 0.$$

This implies that $(1 + |\cdot|^2)^s \widehat{u} = 0$ as a tempered distribution. Thus, $\widehat{u} = 0$, and then $u = 0$. $\qquad \square$

The Sobolev spaces are not stable under multiplication by C^∞ functions; nevertheless, they are *local*. This is a consequence of the following result.

Theorem 1.62. *Multiplication by a function of $\mathcal{S}(\mathbb{R}^d)$ is a continuous map from $H^s(\mathbb{R}^d)$ into itself.*

Proof. As we know that $\widehat{\varphi u} = (2\pi)^{-d}\widehat{\varphi} \star \widehat{u}$, the proof of Theorem 1.62 is reduced to the estimate of the $L^2(\mathbb{R}^d)$ norm of the function U_s defined by

$$U_s(\xi) \overset{\text{def}}{=} (1+|\xi^2|)^{\frac{s}{2}} \int_{\mathbb{R}^d} |\widehat{\varphi}(\xi - \eta)| \times |\widehat{u}(\eta)|\, d\eta.$$

We will temporarily assume that

$$(1+|\xi|^2)^{\frac{s}{2}} \le 2^{\frac{|s|}{2}}(1+|\xi-\eta|^2)^{\frac{|s|}{2}}(1+|\eta|^2)^{\frac{s}{2}}. \tag{1.39}$$

We then infer that

$$|U_s(\xi)| \le 2^{\frac{|s|}{2}} \int_{\mathbb{R}^d} (1+|\xi-\eta|^2)^{\frac{|s|}{2}} |\widehat{\varphi}(\xi-\eta)|(1+|\eta|^2)^{\frac{s}{2}} |\widehat{u}(\eta)|\, d\eta.$$

Using Young's inequality, we get

$$\|\varphi u\|_{H^s} \le 2^{\frac{|s|}{2}} \|(1+|\cdot|^2)^{\frac{|s|}{2}}\widehat{\varphi}\|_{L^1} \|u\|_{H^s},$$

and the desired result follows.

For the sake of completeness, we now prove the inequality (1.39). Interchanging ξ and η, we see that it suffices to consider the case $s \ge 0$. We have

$$\begin{aligned}(1+|\xi|^2)^{\frac{s}{2}} &\le (1 + 2(|\xi-\eta|^2 + |\eta|^2))^{\frac{s}{2}} \\ &\le 2^{\frac{s}{2}}(1+|\xi-\eta|^2)^{\frac{s}{2}}(1+|\eta|^2)^{\frac{s}{2}}.\end{aligned}$$

This completes the proof of the theorem. □

We will now consider the problem of *trace* and *trace lifting* operators for the Sobolev spaces. Consider the hyperplane $x_1 = 0$ in $\mathbb{R}^d$. Because this has measure zero, we cannot give any reasonable sense to the trace operator γ formally defined by $\gamma u(x') = u(0, x')$ if u belongs to a Lebesgue space. For instance, there exist elements of $L^2(\mathbb{R}^d)$ which are continuous for $x_1 \ne 0$ and tend to infinity when x_1 goes to 0. This obviously precludes us from defining the trace of a general L^2 function.

The following theorem shows that defining γu makes sense for $u \in H^s(\mathbb{R}^d)$ with s greater than $1/2$. Extending the usual trace operator by continuity provides us with the relevant definition.

Theorem 1.63. *Let s be a real number strictly larger than $1/2$. The restriction map γ defined by*

$$\gamma : \begin{cases} \mathcal{S}(\mathbb{R}^d) \longrightarrow \mathcal{S}(\mathbb{R}^{d-1}) \\ \phi \longmapsto \gamma(\phi) : (x_2, \ldots, x_d) \mapsto \phi(0, x_2, \ldots, x_d) \end{cases}$$

can be continuously extended from $H^s(\mathbb{R}^d)$ onto $H^{s-\frac{1}{2}}(\mathbb{R}^{d-1})$.

Proof. We first prove the existence of γ. Arguing by density, it suffices to find a constant C such that

$$\forall \phi \in \mathcal{S}, \ \|\gamma(\phi)\|_{H^{s-\frac{1}{2}}} \leq C\|\phi\|_{H^s}. \tag{1.40}$$

To achieve the above inequality, we may rewrite the trace operator in terms of a Fourier transform:

$$\phi(0, x') = (2\pi)^{-d} \int_{\mathbb{R}^d} e^{i(x'|\xi')} \widehat{\phi}(\xi_1, \xi') \, d\xi_1 \, d\xi'$$

$$= (2\pi)^{1-d} \int_{\mathbb{R}^{d-1}} e^{i(x'|\xi')} \left((2\pi)^{-1} \int_{\mathbb{R}} \widehat{\phi}(\xi_1, \xi') \, d\xi_1 \right) d\xi'.$$

We thus have

$$\widehat{\gamma(\phi)}(\xi') = (2\pi)^{-1} \int_{\mathbb{R}} \widehat{\phi}(\xi_1, \xi') \, d\xi_1.$$

By multiplication and division by $(1 + |\xi_1|^2 + |\xi'|^2)^{\frac{s}{2}}$ and the Cauchy–Schwarz inequality, we have

$$|\widehat{\gamma(\phi)}(\xi')|^2 \leq \frac{1}{4\pi^2} \left(\int_{\mathbb{R}} (1 + \xi_1^2 + |\xi'|^2)^{-s} \, d\xi_1 \right) \left(\int_{\mathbb{R}} (|\widehat{\phi}(\xi)|^2 (1 + |\xi|^2)^s \, d\xi_1 \right).$$

Having $s > \frac{1}{2}$ ensures that the first integral is finite. In order to compute it, we make the change of variables $\xi_1 = (1 + |\xi'|^2)^{\frac{1}{2}} \lambda$. We obtain

$$\int (1 + \xi_1^2 + |\xi'|^2)^{-s} \, d\xi_1 = C_s (1 + |\xi'|^2)^{-s+\frac{1}{2}} \quad \text{with} \quad C_s = \int (1 + \lambda^2)^{-s} d\lambda.$$

We deduce that $\|\gamma(\phi)\|_{H^{s-\frac{1}{2}}}^2 \leq C_s \|\phi\|_{H^s}^2$, which completes the proof of the first part of the theorem.

We now define the trace lifting operator. Let χ be a function in $\mathcal{D}(\mathbb{R})$ such that $\chi(0) = 1$. We define

$$Rv(x) \overset{\text{def}}{=} (2\pi)^{-d+1} \int_{\mathbb{R}^{d-1}} e^{i(x'|\xi')} \chi(x_1 \langle \xi' \rangle) \widehat{v}(\xi') \, d\xi' \quad \text{with} \quad \langle \xi' \rangle = \sqrt{1 + |\xi'|^2}.$$

It is clear that

$$\mathcal{F}Rv(\xi) = \int_{\mathbb{R}} e^{-it\xi_1} \chi(t \langle \xi' \rangle) \widehat{v}(\xi') \, dt$$

$$= \langle \xi' \rangle^{-1} \widehat{\chi} \left(\frac{\xi_1}{\langle \xi' \rangle} \right) \widehat{v}(\xi').$$

Taking N sufficiently large, we deduce that

$$\|Rv\|_{H^s}^2 = \int_{\mathbb{R}^d} (1 + |\xi_1|^2 + |\xi'|^2)^s \langle \xi' \rangle^{-2} |\widehat{\chi}(\langle \xi' \rangle^{-1} \xi_1)|^2 |\widehat{v}(\xi')|^2 \, d\xi$$

$$\leq C_N \int_{\mathbb{R}^{d-1}} \left(\int_{\mathbb{R}} \left(1 + \frac{|\xi_1|^2}{\langle \xi' \rangle^2} \right)^{s-N} \langle \xi' \rangle^{-1} \, d\xi_1 \right) (1 + |\xi'|^2)^{s-\frac{1}{2}} |\widehat{v}(\xi')|^2 \, d\xi'$$

$$\leq C_N \|v\|_{H^{s-\frac{1}{2}}}^2.$$

Of course, we have $\gamma Rv = v$. This completes the proof of the theorem. $\qquad \square$

We infer the following corollary.

Corollary 1.64. *Let $s > m + \frac{1}{2}$ with $m \in \mathbb{N}$. The map*

$$\Gamma : \begin{cases} H^s(\mathbb{R}^d) \longrightarrow \displaystyle\bigoplus_{j=0}^{m} H^{s-j-\frac{1}{2}}(\mathbb{R}^{d-1}) \\ u \longmapsto (\gamma_j(u))_{0 \leq j \leq m} \end{cases}$$

with $\gamma_j(u) = \gamma(\partial_{x_1}^j u)$ is then continuous and onto.

Remark 1.65. More generally, the trace operator γ_Σ may be defined for any smooth hypersurface Σ of $\mathbb{R}^d$. Indeed, according to Theorem 1.62 and Corollary 1.60, the spaces $H^s(\mathbb{R}^d)$ are local and invariant under the action of diffeomorphism, so localizing and straightening Σ reduces the problem to the study of the trace operator defined in Theorem 1.63.

1.4.2 Embedding

In this subsection, we present a few properties concerning embedding in Lebesgue spaces. First, from Theorems 1.38 and 1.50 we can easily deduce the following result.

Theorem 1.66. *The space $H^s(\mathbb{R}^d)$ embeds continuously in:*

– *the Lebesgue space $L^p(\mathbb{R}^d)$, if $0 \leq s < d/2$ and $2 \leq p \leq 2d/(d-2s)$*
– *the Hölder space $C^{k,\rho}(\mathbb{R}^d)$, if $s \geq d/2+k+\rho$ for some $k \in \mathbb{N}$ and $\rho \in \,]0,1[$.*

As in the homogeneous case, the space $H^{\frac{d}{2}}$ fails to be embedded in L^∞. However, the following *Moser–Trudinger inequality* holds.

Theorem 1.67. *There exist two constants, c and C, depending only on the dimension d, such that for any function $u \in H^{\frac{d}{2}}(\mathbb{R}^d)$, we have*

$$\int_{\mathbb{R}^d} \left(\exp\left(c \left(\frac{|f(x)|}{\|f\|_{H^{\frac{d}{2}}}} \right)^2 \right) - 1 \right) dx \leq C.$$

Proof. As usual, arguing by density and homogeneity, it suffices to consider the case where f is in $\mathcal{S}$ and satisfies $\|f\|_{H^{\frac{d}{2}}} = 1$.

Now, the proof is based on the fact that, according to the inequality (1.30) and the definition of nonhomogeneous Sobolev spaces, there exists some constant C_d (depending only on the dimension d) such that

$$\|f\|_{L^{2p}} \leq C_d \sqrt{p} \quad \text{for all} \ \ p \geq 1. \tag{1.41}$$

For all $x \in \mathbb{R}^d$, we may write

$$\exp\left(c|f(x)|^2 \right) - 1 = \sum_{p \geq 1} \frac{c^p}{p!} |f(x)|^{2p}.$$

Integrating over $\mathbb{R}^d$ and using the inequality (1.41) yields

$$\int_{\mathbb{R}^d} \left(\exp\left(c|f(x)|^2 \right) - 1 \right) dx = \sum_{p \geq 1} c^p C_d^{2p} \frac{p^p}{p!}.$$

The theorem then follows from our choosing the constant c sufficiently small.

$\square$

As stated before, the space $H^s(\mathbb{R}^d)$ is included in $H^t(\mathbb{R}^d)$ whenever $t \leq s$. If the inequality is strict, then the following statement ensures that the embedding is locally compact.

Theorem 1.68. *For $t < s$, multiplication by a function in $\mathcal{S}(\mathbb{R}^d)$ is a compact operator from $H^s(\mathbb{R}^d)$ in $H^t(\mathbb{R}^d)$.*

Proof. Let φ be a function in $\mathcal{S}$. We have to prove that for any sequence (u_n) in $H^s(\mathbb{R}^d)$ satisfying $\sup_n \|u_n\|_{H^s} \leq 1$, we can extract a subsequence (u_{n_k}) such that (φu_{n_k}) converges in $H^t(\mathbb{R}^d)$.

As $H^s(\mathbb{R}^d)$ is a Hilbert space, the weak compactness theorem ensures that the sequence $(u_n)_{n \in \mathbb{N}}$ converges weakly, up to extraction, to an element u of $H^s(\mathbb{R}^d)$ with $\|u\|_{H^s} \leq 1$. We continue to denote this subsequence by $(u_n)_{n \in \mathbb{N}}$ and set $v_n = u_n - u$. Thanks to Theorem 1.62, $\sup_n \|\varphi v_n\|_{H^s} \leq C$. Our task is thus reduced to proving that the sequence $(\varphi v_n)_{n \in \mathbb{N}}$ tends to 0 in $H^t(\mathbb{R}^d)$. We now have, for any positive real number R,

$$\int (1 + |\xi|^2)^t |\mathcal{F}(\varphi v_n)(\xi)|^2 \, d\xi \leq \int_{|\xi| \leq R} (1 + |\xi|^2)^t |\mathcal{F}(\varphi v_n)(\xi)|^2 \, d\xi$$

$$+ \int_{|\xi| \geq R} (1 + |\xi|^2)^{t-s} (1 + |\xi|^2)^s |\mathcal{F}(\varphi v_n)(\xi)|^2 \, d\xi$$

$$\leq \int_{|\xi| \leq R} (1 + |\xi|^2)^t |\mathcal{F}(\varphi v_n)(\xi)|^2 \, d\xi + \frac{\|\varphi v_n\|_{H^s}^2}{(1 + R^2)^{s-t}}.$$

As $(\varphi v_n)_{n \in \mathbb{N}}$ is bounded in $H^s(\mathbb{R}^d)$, for a given positive real number ε, we can choose R such that

$$\frac{1}{(1 + R^2)^{s-t}} \|\varphi v_n\|_{H^s}^2 \leq \frac{\varepsilon}{2}.$$

On the other hand, as the function ψ_ξ defined by

$$\psi_\xi(\eta) \stackrel{\text{def}}{=} (2\pi)^{-d} \mathcal{F}^{-1}\left((1 + |\eta|^2)^{-s} \widehat{\varphi}(\xi - \eta) \right)$$

belongs to $\mathcal{S}(\mathbb{R}^d)$, we can write

$$\mathcal{F}(\varphi v_n)(\xi) = (2\pi)^{-d} \int \widehat{\varphi}(\xi - \eta) \widehat{v_n}(\eta) \, d\eta$$

$$= \int (1 + |\eta|^2)^s \widehat{\psi_\xi}(\eta) \widehat{v_n}(\eta) \, d\eta$$

$$= (\psi_\xi \mid v_n)_{H^s}.$$

As $(v_n)_{n\in\mathbb{N}}$ converges weakly to 0 in $H^s(\mathbb{R}^d)$, we can thus conclude that

$$\forall \xi \in \mathbb{R}^d, \quad \lim_{n\to\infty} \mathcal{F}(\varphi v_n)(\xi) = 0.$$

Let us temporarily assume that

$$\sup_{\substack{|\xi|\leq R \\ n\in\mathbb{N}}} |\mathcal{F}(\varphi v_n)(\xi)| \leq M < \infty. \tag{1.42}$$

Lebesgue's theorem then implies that

$$\lim_{n\to\infty} \int_{|\xi|\leq R} (1 + |\xi|^2)^t |\mathcal{F}(\varphi v_n)(\xi)|^2 \, d\xi = 0,$$

which leads to the convergence of the sequence $(\varphi v_n)_{n\in\mathbb{N}}$ to 0 in $H^t(\mathbb{R}^d)$.

To complete the proof of the theorem, let us prove (1.42). It is clear that

$$|\mathcal{F}(\varphi v_n)(\xi)| \leq (2\pi)^{-d} \int_{\mathbb{R}^d} |\widehat{\varphi}(\xi - \eta)| \, |\widehat{v}_n(\eta)| \, d\eta$$

$$\leq (2\pi)^{-d} \|v_n\|_{H^s} \left(\int (1 + |\eta|^2)^{-s} |\widehat{\varphi}(\xi - \eta)|^2 \, d\eta \right)^{\frac{1}{2}}.$$

Now, as $\widehat{\varphi}$ belongs to $\mathcal{S}(\mathbb{R}^d)$, a constant C exists such that

$$|\widehat{\varphi}(\xi - \eta)| \leq \frac{C_{N_0}}{(1 + |\xi - \eta|^2)^{N_0}} \quad \text{with} \quad N_0 = \frac{d}{2} + |s| + 1.$$

We thus obtain

$$\int (1 + |\eta|^2)^{-s} |\widehat{\varphi}(\xi - \eta)|^2 \, d\eta \leq \int_{|\eta|\leq 2R} (1 + |\eta|^2)^{-s} |\widehat{\varphi}(\xi - \eta)|^2 \, d\eta$$

$$+ \int_{|\eta|\geq 2R} (1 + |\eta|^2)^{-s} |\widehat{\varphi}(\xi - \eta)|^2 \, d\eta$$

$$\leq C \int_{|\eta|\leq 2R} (1 + |\eta|^2)^{|s|} \, d\eta$$

$$+ C_{N_0} \int_{|\eta|\geq 2R} (1 + |\eta|^2)^{|s|} (1 + |\xi - \eta|^2)^{-N_0} \, d\eta.$$

Finally, since $|\xi| \leq R$, we always have $|\xi - \eta| \geq \dfrac{|\eta|}{2}$ in the last integral, so we eventually get

$$\int (1 + |\eta|^2)^{-s} |\widehat{\varphi}(\xi - \eta)|^2 \, d\eta \leq C(1 + R^2)^{|s|+\frac{d}{2}} + C \int \frac{d\eta}{(1 + |\eta|^2)^{\frac{d}{2}+1}}.$$

This yields (1.42) and completes the proof of the theorem. $\qquad\square$

From the above theorem, we can deduce the following compactness result.

Theorem 1.69. *For any compact subset K of $\mathbb{R}^d$ and $s' < s$, the embedding of $H_K^s(\mathbb{R}^d)$ into $H_K^{s'}(\mathbb{R}^d)$ is a compact linear operator.*

Proof. It suffices to consider a function φ in $\mathcal{S}(\mathbb{R}^d)$ which is identically equal to 1 in a neighborhood of the compact K and then to apply Theorem 1.68.   □

1.4.3 A Density Theorem

In this subsection we investigate the density of the space $\mathcal{D}(\mathbb{R}^d \setminus \{0\})$ in Sobolev spaces. This result is useful for proving Hardy inequalities and is related to the problem of the pointwise value of a function in $H^s(\mathbb{R}^d)$. Indeed, having $\mathcal{D}(\mathbb{R}^d \setminus \{0\})$ dense in $H^s(\mathbb{R}^d)$ precludes any reasonable definition of the "value at 0" of an element of $H^s(\mathbb{R}^d)$. We now state the result.

Theorem 1.70. *If $s \leq d/2$ (resp., $< d/2$), then the space $\mathcal{D}(\mathbb{R}^d \setminus \{0\})$ is dense in $H^s(\mathbb{R}^d)$ [resp., in $\dot{H}^s(\mathbb{R}^d)$]. If $s > d/2$, then the closure of the space $\mathcal{D}(\mathbb{R}^d \setminus \{0\})$ in $H^s(\mathbb{R}^d)$ is the set of functions u in $H^s(\mathbb{R}^d)$ such that $\partial^\alpha u(0) = 0$ for any $\alpha \in \mathbb{N}^d$ such that $|\alpha| < s - d/2$.*

Proof. As $H^s(\mathbb{R}^d)$ is a Hilbert space it is enough to study the orthogonal complement of $\mathcal{D}(\mathbb{R}^d \setminus \{0\})$ in $H^s(\mathbb{R}^d)$. For u in H^s we define

$$u_s \overset{\text{def}}{=} \mathcal{F}^{-1}((1 + |\xi|^2)^s \hat{u}).$$

If u belongs to the orthogonal complement of $\mathcal{D}(\mathbb{R}^d \setminus \{0\})$, then we have

$$\int_{\mathbb{R}^d} \hat{u}_s(\xi) \overline{\hat{\varphi}}(\xi) \, d\xi = \langle u_s, \varphi \rangle = 0 \quad \text{for any } \varphi \text{ in } \mathcal{D}(\mathbb{R}^d \setminus \{0\}).$$

This implies that the support of u_s is included in $\{0\}$. We infer that a sequence $(a_\alpha)_{|\alpha| \leq N}$ exists such that

$$u_s = \sum_{|\alpha| \leq N} a_\alpha \partial^\alpha \delta_0. \tag{1.43}$$

As u_s belongs to H^{-s}, Remark 1.54 implies that $a_\alpha = 0$ for $|\alpha| \geq s - d/2$. Thus, if $s \leq d/2$, then $u_s = u = 0$ and the density is proved in that case. The proof of the density in the homogeneous case follows the same lines and is left to the reader as an exercise.

When s is greater than $d/2$, the orthogonal complement of the space $\mathcal{D}(\mathbb{R}^d \setminus \{0\})$ is exactly the finite-dimensional vector space $\mathcal{V}_s$ spanned by the functions $(u_\alpha)_{|\alpha| \leq [s-d/2]}$ defined by

$$u_\alpha(x) \overset{\text{def}}{=} (2\pi)^{-d} \int_{\mathbb{R}^d} e^{i(x|\xi)} \frac{(i\xi)^\alpha}{(1 + |\xi|^2)^s} \, d\xi.$$

However, thanks to the relation (1.43), if the partial derivatives of order less than or equal to $s - d/2$ of a function v in H^s vanish at 0, then we have

$$(v|u_\alpha)_{H^s} = \langle v, \partial^\alpha \delta_0 \rangle = 0.$$

Thus, the function v belongs to the orthogonal complement of $\mathcal{V}_s$, which is the closure of $\mathcal{D}(\mathbb{R}^d \setminus \{0\})$. □

Remark 1.71. If $d = 1$, then the above result means that the map $u \mapsto u(0)$ cannot be extended to $H^{\frac{1}{2}}(\mathbb{R})$ functions. More generally, arguing as above, we can prove that the restriction map γ on the hyperplane $x_1 = 0$ cannot be extended to $H^{\frac{1}{2}}(\mathbb{R}^d)$ functions.[12]

1.4.4 Hardy Inequality

This brief subsection is devoted to proving a fundamental inequality with singular weight in Sobolev spaces: the so-called *Hardy inequality*. More general Hardy inequalities will be established in the next chapter (see Theorem 2.57).

Theorem 1.72. *If $d \geq 3$, then*

$$\left(\int_{\mathbb{R}^d} \frac{|f(x)|^2}{|x|^2} \, dx \right)^{\frac{1}{2}} \leq \frac{2}{d-2} \|\nabla f\|_{L^2} \quad \text{for any } f \text{ in } \dot{H}^1(\mathbb{R}^d). \quad (1.44)$$

Proof. Arguing by density, it suffices to prove the inequality for $f \in \mathcal{D}(\mathbb{R}^d \setminus \{0\})$. Let $\mathcal{R}$ be the radial vector field $\mathcal{R} = \sum_{i=1}^{d} x_i \partial_{x_i}$. Because $\mathcal{R}|x|^{-2} = -2|x|^{-2}$, integrating by parts yields

$$\int_{\mathbb{R}^d} \frac{|f(x)|^2}{|x|^2} \, dx = \frac{1}{2} \int_{\mathbb{R}^d} \frac{2f(x)\mathcal{R}f(x)}{|x|^2} \, dx + \frac{d}{2} \int_{\mathbb{R}^d} \frac{|f(x)|^2}{|x|^2} \, dx.$$

Thus, we have, by the Cauchy–Schwarz inequality,

$$\int_{\mathbb{R}^d} \frac{|f(x)|^2}{|x|^2} \, dx = \frac{2}{2-d} \int_{\mathbb{R}^d} \frac{f(x)\mathcal{R}f(x)}{|x|^2} \, dx$$

$$\leq \frac{2}{d-2} \left(\int_{\mathbb{R}^d} \frac{|f(x)|^2}{|x|^2} \, dx \right)^{\frac{1}{2}} \left(\int_{\mathbb{R}^d} \frac{|\mathcal{R}f(x)|^2}{|x|^2} \, dx \right)^{\frac{1}{2}},$$

which implies that

---

[12] In fact, γu makes sense whenever u belongs to the *smaller* space

$$H_{0,0}^{\frac{1}{2}}(\mathbb{R}^d) \stackrel{\text{def}}{=} \left\{ u \in H^{\frac{1}{2}}(\mathbb{R}^d) \,\middle/\, \frac{u}{|x_1|^{\frac{1}{2}}} \in L^2(\mathbb{R}^d) \right\}.$$

$$\left(\int_{\mathbb{R}^d} \frac{|f(x)|^2}{|x|^2}\, dx\right)^{\frac{1}{2}} \leq \frac{2}{d-2}\left(\int_{\mathbb{R}^d} |\nabla f(x)|^2\, dx\right)^{\frac{1}{2}}. \qquad \square$$

Remark 1.73. Let us note that using Lorentz spaces provides an elementary proof of more general Hardy inequalities, namely,

$$\left\|\frac{f}{|x|^s}\right\|_{L^2} \leq C\|f\|_{\dot{H}^s} \quad \text{for } 0 \leq s < \frac{d}{2}.$$

Indeed, using real interpolation we can show that $\dot{H}^s$ not only embeds in the space L^p with $1/p = 1/2 - s/d$, but also in the Lorentz space $L^{p,2}$. Now, it is clear that the function $x \mapsto |\cdot|^{-s}$ belongs to the space $L_w^{d/s}$, so applying generalized Hölder inequalities in Lorentz spaces, we get

$$\left\|\frac{f}{|x|^s}\right\|_{L^2} \leq C\left\|\frac{1}{|\cdot|^s}\right\|_{L_w^{d/s}} \|f\|_{L^{p,2}} \leq C'\|f\|_{\dot{H}^s}.$$

1.5 References and Remarks

The Hölder and Young inequalities belong to mathematical folklore. Refined Young inequalities are special cases of convolution inequalities in Lorentz spaces. An exhaustive list of such inequalities can be found in [171] or the book by P.-G. Lemarié-Rieusset [205]. More about atomic decomposition and bilinear interpolation can be found in the book by L. Grafakos [150].

In the present chapter, we restricted ourselves to the very basic properties of the Fourier transform. For a more complete study of the Fourier transform of harmonic analysis methods for partial differential equations, the reader may refer to the textbooks [40] by J.-M. Bony, [122] by L.C. Evans, [275] by E.M. Stein, [167, vol. 1] by L. Hörmander and [282, 283] by M.E. Taylor.

The Sobolev embedding in Lebesgue spaces was first stated by S. Sobolev himself in [270, 271]. There is now a plethora of generalizations ($W^{s,p}$ spaces, metric spaces, etc.) Basic references for Sobolev spaces may be found in the books [3] by R. Adams and [146] by D. Gilbarg and N. Trudinger. Refined Sobolev inequalities were discovered by P. Gérard, Y. Meyer, and F. Oru in [140]. The proof which has been proposed here is borrowed from [77]. The fractal counterexample comes from [22]. The study of embedding of Sobolev spaces in Hölder spaces goes back to C. Morrey's work in [235]. The BMO space was first introduced by F. John and L. Nirenberg in [174].

Most of the results concerning nonhomogeneous Sobolev spaces are classical. Hardy inequalities go back to the pioneering work by G.H. Hardy in [153, 154]. In the next chapter, we shall state more general Hardy inequalities in Sobolev spaces with *fractional* indices of regularity.

For more details on the Moser–Trudinger inequality, see the pioneering works by J. Moser in [236] and N.S. Trudinger in [290]. For recent developments, see [2].

Note that combining the Sobolev embedding theorem with Theorem 1.68 ensures that the embedding of $\dot{H}^s(\mathbb{R}^d)$ in $L^p(\mathbb{R}^d)$ is locally compact whenever $2 \leq p \leq \infty$ and $s > d/2 - d/p$. In contrast, due to the scaling invariance of the critical Sobolev embedding,[13] the fact that $\dot{H}^s(\mathbb{R}^d) \hookrightarrow L^{p_s}(\mathbb{R}^d)$ when $0 \leq s < d/2$, and that fact that $p_s = 2d/(d-2s)$, no compactness properties may be expected in this case. Indeed, if $u \in \dot{H}^s \setminus \{0\}$, then for any sequence (y_n) of points in $\mathbb{R}^d$ tending to infinity and for any sequence (h_n) of positive real numbers tending to 0 or to infinity, the sequences $(\tau_{y_n} u)$ and $(\delta_{h_n} u)$ converge weakly to 0 in $\dot{H}^s$ but are not relatively compact in L^p since $\|\tau_{y_n} u\|_{L^p} = \|u\|_{L^p}$ and $\|\delta_{h_n} u\|_{L^p} = \|u\|_{L^p}$. The study of this *defect of compactness* was initiated by P.-L. Lions in [212] (see also the paper by P. Gérard [139]). In short, it has been shown that translational and scaling invariance are the only features responsible for the defect of compactness of the embedding of $\dot{H}^s$ into L^p.

---

[13] Throughout this book, we agree that whenever X and Y are Banach spaces, the notation $X \hookrightarrow Y$ means that $X \subset Y$ and that the canonical injection from X to Y is continuous.

2

Littlewood–Paley Theory

In this chapter we introduce most of the Fourier analysis material which will be needed in the next chapters. The main idea is that functions or distributions are easier to deal with if split into countable sums of smooth functions whose Fourier transforms are compactly supported in a ball or an annulus. Littlewood–Paley theory provides such a decomposition.

The first section is dedicated to the study of functions with compactly supported Fourier transforms. We state Bernstein inequalities and study the action of heat flow or of a diffeomorphism over spectrally localized functions. The Littlewood–Paley decomposition is introduced in the second section. Sections 2.3, 2.4, and 2.5 are devoted to the definition of homogeneous Besov spaces and the proofs of some of their properties (basic topological properties, characterizations in terms of heat flow or finite differences, embedding in Lebesgue spaces, and Gagliardo–Nirenberg-type inequalities).

In Section 2.6 we introduce the (homogeneous) paradifferential calculus (after J.-M. Bony in [39]) and state a few results concerning continuity of the paraproduct. We also study the effect of left composition by a smooth function. The next section is devoted to the definition and a few properties of (the more classical) nonhomogeneous Besov spaces. In Section 2.8 we state a paralinearization theorem. Compactness properties of Besov spaces are studied in Section 2.9. In Section 2.10 (which may be skipped at first reading) we give some technical commutator estimates which will be needed in the next chapters. In the last section, we state a few properties for the Zygmund space $B^1_{\infty,\infty}$ and provide some logarithmic-type interpolation inequalities.

2.1 Functions with Compactly Supported Fourier Transforms

Littlewood–Paley theory is a localization procedure in frequency space. The interesting feature of this localization is that the derivatives (or, more generally, Fourier multipliers) act almost as homotheties on distributions whose

H. Bahouri et al., *Fourier Analysis and Nonlinear Partial Differential Equations*, Grundlehren der mathematischen Wissenschaften 343, DOI 10.1007/978-3-642-16830-7_2, © Springer-Verlag Berlin Heidelberg 2011

Fourier transforms are supported in a ball or an annulus. This nice property leads to the so-called Bernstein inequalities and is investigated in the next subsection.

2.1.1 Bernstein-Type Lemmas

Throughout, we shall call a *ball* any set $\{\xi \in \mathbb{R}^d \, / \, |\xi| \leq R\}$ with $R > 0$ and an *annulus* any set $\{\xi \in \mathbb{R}^d \, / \, 0 < r_1 \leq |\xi| \leq r_2\}$ with $0 < r_1 < r_2$.

Lemma 2.1. *Let $\mathcal{C}$ be an annulus and B a ball. A constant C exists such that for any nonnegative integer k, any couple (p, q) in $[1, \infty]^2$ with $q \geq p \geq 1$, and any function u of L^p, we have*

$$\text{Supp } \widehat{u} \subset \lambda B \Longrightarrow \|D^k u\|_{L^q} \overset{def}{=} \sup_{|\alpha|=k} \|\partial^\alpha u\|_{L^q} \leq C^{k+1} \lambda^{k+d(\frac{1}{p}-\frac{1}{q})} \|u\|_{L^p},$$

$$\text{Supp } \widehat{u} \subset \lambda \mathcal{C} \Longrightarrow C^{-k-1} \lambda^k \|u\|_{L^p} \leq \|D^k u\|_{L^p} \leq C^{k+1} \lambda^k \|u\|_{L^p}.$$

Proof. Using a dilation of size λ, we can assume throughout the proof that $\lambda = 1$. Let ϕ be a function of $\mathcal{D}(\mathbb{R}^d)$ with value 1 near B. As $\widehat{u}(\xi) = \phi(\xi)\widehat{u}(\xi)$ we have

$$\partial^\alpha u = \partial^\alpha g \star u \quad \text{with} \quad g = \mathcal{F}^{-1}\phi.$$

Applying Young's inequality we get

$$\|\partial^\alpha g \star u\|_{L^q} \leq \|\partial^\alpha g\|_{L^r} \|u\|_{L^p} \quad \text{with} \quad \frac{1}{r} \overset{def}{=} -\frac{1}{p} + \frac{1}{q} + 1,$$

and the first assertion follows via

$$\begin{aligned}
\|\partial^\alpha g\|_{L^r} &\leq \|\partial^\alpha g\|_{L^\infty} + \|\partial^\alpha g\|_{L^1} \\
&\leq C \|(1 + |\cdot|^2)^d \partial^\alpha g\|_{L^\infty} \\
&\leq C \|(\text{Id} - \Delta)^d ((\cdot)^\alpha \phi)\|_{L^1} \\
&\leq C^{k+1}.
\end{aligned}$$

To prove the second assertion, consider a function $\widetilde{\phi} \in \mathcal{D}(\mathbb{R}^d \setminus \{0\})$ with value 1 on a neighborhood of $\mathcal{C}$. From the algebraic identity (1.23) page 25 and the fact that $\widehat{u} = \widetilde{\phi}\widehat{u}$, we deduce that there exists a family of integers $(A_\alpha)_\alpha \in \mathbb{N}^d$ such that

$$u = \sum_{|\alpha|=k} g_\alpha \star \partial^\alpha u \quad \text{with} \quad g_\alpha \overset{def}{=} A_\alpha \mathcal{F}^{-1}(-i\xi)^\alpha |\xi|^{-2k} \widetilde{\phi}(\xi),$$

and the result follows. □

The following lemma describes the action of *Fourier multipliers* which behave like homogeneous functions of degree m.

Lemma 2.2. *Let $\mathcal{C}$ be an annulus, $m \in \mathbb{R}$, and[1] $k = 2[1 + d/2]$. Let σ be a k-times differentiable function on $\mathbb{R}^d \setminus \{0\}$ such that for any $\alpha \in \mathbb{N}^d$ with $|\alpha| \leq k$, there exists a constant C_α such that*

$$\forall \xi \in \mathbb{R}^d, \ |\partial^\alpha \sigma(\xi)| \leq C_\alpha |\xi|^{m-|\alpha|}.$$

There exists a constant C, depending only on the constants C_α, such that for any $p \in [1, \infty]$ and any $\lambda > 0$, we have, for any function u in L^p with Fourier transform supported in $\lambda \mathcal{C}$,

$$\|\sigma(D)u\|_{L^p} \leq C\lambda^m \|u\|_{L^p} \quad \text{with} \quad \sigma(D)u \overset{def}{=} \mathcal{F}^{-1}(\sigma \widehat{u}).$$

Proof. Consider a smooth function $\widetilde{\varphi}$ supported in an annulus and such that $\widetilde{\varphi} \equiv 1$ on $\mathcal{C}$. It is clear that we have

$$\sigma(D)u = \lambda^d K_\lambda(\lambda \cdot) \star u \quad \text{with} \tag{2.1}$$

$$K_\lambda(x) \overset{def}{=} (2\pi)^{-d} \int_{\mathbb{R}^d} e^{i(x|\xi)} \widetilde{\varphi}(\xi) \sigma(\lambda \xi) \, d\xi.$$

Let $M = [1 + d/2]$. We have

$$(1 + |x|^2)^M K_\lambda(x) = \int (\mathrm{Id} - \Delta_\xi)^M \left(e^{i(x|\xi)} \right) \widetilde{\varphi}(\xi) \sigma(\lambda \xi) \, d\xi$$

$$= \int e^{i(x|\xi)} (\mathrm{Id} - \Delta_\xi)^M \left(\widetilde{\varphi}(\xi) \sigma(\lambda \xi) \right) d\xi$$

$$= \sum_{|\alpha| + |\beta| \leq 2M} c_{\alpha,\beta} \lambda^{|\beta|} \int e^{i(x|\xi)} \partial^\alpha \widetilde{\varphi}(\xi) \, \partial^\beta \sigma(\lambda \xi) \, d\xi$$

for some integers $c_{\alpha,\beta}$ (whose exact values do not matter). The integration may be restricted to Supp $\widetilde{\varphi}$. On this set we have $|\partial^\beta \sigma(\lambda \xi)| \leq C_\beta \lambda^{m-|\beta|}$. Thus, we get

$$(1 + |x|^2)^M |K_\lambda(x)| \leq C_M \lambda^m.$$

As $2M > d$ we may conclude that $\|K_\lambda\|_{L^1} \leq C\lambda^m$. Applying Young's inequality to (2.1) then yields the desired result. □

2.1.2 The Smoothing Effect of Heat Flow

This subsection is devoted to the study of the action of heat flow over spectrally supported functions. Our main result is based on Faà di Bruno's formula, which we recall here for the convenience of the reader.

---

[1] Throughout this book we agree that whenever r is a real number, $[r]$ stands for the *integer part of* r.

Lemma 2.3. *Let $u : \mathbb{R}^d \to \mathbb{R}^m$ and $F : \mathbb{R}^m \to \mathbb{R}$ be smooth functions. For each multi-index α of $\mathbb{N}^d$, we have*

$$\partial^\alpha (F \circ u) = \sum_{\mu,\nu} C_{\mu,\nu} \partial^\mu F \prod_{\substack{1 \le |\beta| \le |\alpha| \\ 1 \le j \le m}} (\partial^\beta u^j)^{\nu_{\beta_j}},$$

where the coefficients $C_{\mu,\nu}$ are nonnegative integers, and the sum is taken over those μ and ν such that $1 \le |\mu| \le |\alpha|$, $\nu_{\beta_j} \in \mathbb{N}^*$,

$$\sum_{1 \le |\beta| \le |\alpha|} \nu_{\beta_j} = \mu_j \quad \text{for} \ \ 1 \le j \le m, \quad \text{and} \quad \sum_{\substack{1 \le |\beta| \le |\alpha| \\ 1 \le j \le m}} \beta \nu_{\beta_j} = \alpha.$$

The following lemma describes the action of the semigroup of the heat equation on distributions with Fourier transforms supported in an annulus.

Lemma 2.4. *Let $\mathcal{C}$ be an annulus. Positive constants c and C exist such that for any p in $[1, \infty]$ and any couple (t, λ) of positive real numbers, we have*

$$\text{Supp} \ \hat{u} \subset \lambda\mathcal{C} \Rightarrow \|e^{t\Delta}u\|_{L^p} \le C e^{-ct\lambda^2} \|u\|_{L^p}.$$

Proof. We again consider a function ϕ in $\mathcal{D}(\mathbb{R}^d \setminus \{0\})$, the value of which is identically 1 near the annulus $\mathcal{C}$. We can also assume without loss of generality that $\lambda = 1$. We then have

$$e^{t\Delta}u = \phi(D)e^{t\Delta}u$$
$$= \mathcal{F}^{-1} \left(\phi(\xi)e^{-t|\xi|^2}\hat{u}(\xi) \right)$$
$$= g(t, \cdot) \star u \quad \text{with} \quad g(t,x) \overset{\text{def}}{=} (2\pi)^{-d} \int e^{i(x|\xi)}\phi(\xi)e^{-t|\xi|^2} d\xi. \quad (2.2)$$

The lemma is proved provided we can find positive real numbers c and C such that

$$\forall t > 0, \ \|g(t, \cdot)\|_{L^1} \le C e^{-ct}. \quad (2.3)$$

To begin, we perform integrations by parts in (2.2). We get

$$g(t,x) = (1 + |x|^2)^{-d} \int (1 + |x|^2)^d e^{i(x|\xi)}\phi(\xi)e^{-t|\xi|^2} d\xi$$
$$= (1 + |x|^2)^{-d} \int \left((\text{Id} - \Delta_\xi)^d e^{i(x|\xi)} \right) \phi(\xi)e^{-t|\xi|^2} d\xi$$
$$= (1 + |x|^2)^{-d} \int_{\mathbb{R}^d} e^{i(x|\xi)}(\text{Id} - \Delta_\xi)^d \left(\phi(\xi)e^{-t|\xi|^2} \right) d\xi.$$

Via Leibniz's formula, we obtain

$$(\text{Id} - \Delta_\xi)^d \left(\phi(\xi) e^{-t|\xi|^2} \right) = \sum_{\substack{\beta \leq \alpha \\ |\alpha| \leq 2d}} C_\beta^\alpha \left(\partial^{(\alpha - \beta)} \phi(\xi) \right) \left(\partial^\beta e^{-t|\xi|^2} \right).$$

From Faà di Bruno's formula (see the above lemma) and the fact that the support of ϕ is included in an annulus, we deduce that there exists a couple (c, C) of positive real numbers such that for any ξ in the support of ϕ,

$$\left| \left(\partial^{(\alpha - \beta)} \phi(\xi) \right) \left(\partial^\beta e^{-t|\xi|^2} \right) \right| \leq C(1 + t)^{|\beta|} e^{-t|\xi|^2}$$
$$\leq C(1 + t)^{|\beta|} e^{-ct}.$$

We have thus proven that $|g(t, x)| \leq C(1 + |x|^2)^{-d} e^{-ct}$, and the inequality (2.3) follows. □

From now on, we agree that if X is a Banach space, I is an interval of $\mathbb{R}$, and p is in $[1, \infty]$, then $L_I^p(X)$ stands for the set of Lebesgue measurable functions u from I to X such that $t \mapsto \|u(t)\|_X$ belongs to $L^p(I)$. If $I = [0, T]$ (resp., $I = \mathbb{R}^+$), then we alternatively use the notation $L_T^p(X)$ [resp., $L^p(X)$]. We shall often use, without justification, the fact that the space $L_I^p(X)$ endowed with the norm

$$\|u\|_{L_I^p(X)} \stackrel{\text{def}}{=} \left(\int_I \|u(t)\|_X^p \, dt \right)^{\frac{1}{p}} \quad \text{if } p < \infty \quad \text{and} \quad \|u\|_{L_I^\infty(X)} \stackrel{\text{def}}{=} \text{ess sup } \|u(t)\|_X$$

is a Banach space.

The following corollary is the key to proving a priori estimates in Besov spaces for the heat equation (see Chapter 3).

Corollary 2.5. *Let $\mathcal{C}$ be an annulus and λ a positive real number. Let u_0 [resp., $f = f(t, x)$] satisfy Supp $\widehat{u}_0 \subset \lambda\mathcal{C}$ (resp., Supp $\widehat{f}(t) \subset \lambda\mathcal{C}$ for all t in $[0, T]$). Consider u, a solution of*

$$\partial_t u - \nu \Delta u = 0 \quad \text{and} \quad u_{|t=0} = u_0,$$

and v, a solution of

$$\partial_t v - \nu \Delta v = f \quad \text{and} \quad v_{|t=0} = 0.$$

There exist positive constants c and C, depending only on $\mathcal{C}$, such that for any $1 \leq a \leq b \leq \infty$ and $1 \leq p \leq q \leq \infty$, we have

$$\|u\|_{L_T^q(L^b)} \leq C(\nu\lambda^2)^{-\frac{1}{q}} \lambda^{d\left(\frac{1}{a} - \frac{1}{b}\right)} \|u_0\|_{L^a},$$
$$\|v\|_{L_T^q(L^b)} \leq C(\nu\lambda^2)^{-1 + \left(\frac{1}{p} - \frac{1}{q}\right)} \lambda^{d\left(\frac{1}{a} - \frac{1}{b}\right)} \|f\|_{L_T^p(L^a)}.$$

Proof. It suffices to use the fact that

$$u(t) = e^{\nu t \Delta} u_0 \quad \text{and} \quad v(t) = \int_0^t e^{\nu(t - \tau)\Delta} f(\tau) \, d\tau.$$

Combining Lemmas 2.1 and 2.4 with Young's inequality now yields the result. The details are left to the reader. □

2.1.3 The Action of a Diffeomorphism

Lemma 2.6. *Let χ be in $\mathcal{S}(\mathbb{R}^d)$. There exists a constant C such that for any $C^{1,1}$ (see Definition 1.26 page 22) global diffeomorphism ψ over $\mathbb{R}^d$ with inverse ϕ, any $u \in \mathcal{S}'(\mathbb{R}^d)$ such that $\widehat{u}$ is supported in $\lambda \mathcal{C}$, any p in $[1, \infty]$, and any (λ, μ) in $]0, \infty[^2$, we have*

$$\left\| \chi(\mu^{-1}D)(u \circ \psi) \right\|_{L^p} \leq C\lambda^{-1} \|J_\phi\|_{L^\infty}^{\frac{1}{p}} \|u\|_{L^p} \left(\|DJ_\phi\|_{L^\infty} \|J_\psi\|_{L^\infty} + \mu \|D\phi\|_{L^\infty} \right),$$

where $J_\phi(z) \overset{def}{=} |\det D\phi(z)|$ and $\chi(\mu^{-1}D)(u \circ \psi) \overset{def}{=} \mathcal{F}^{-1}(\chi(\mu^{-1}\cdot)\mathcal{F}(u \circ \psi))$.

Proof. Using (2.1.1), we get, after rescaling,

$$u = \lambda^{-1} \sum_{k=1}^{d} g_{k,\lambda} \star \partial_k u \quad \text{with} \quad \|\partial^\alpha g_{k,\lambda}\|_{L^1} \leq C\lambda^{|\alpha|}. \tag{2.4}$$

If $\widetilde{h} = \mathcal{F}^{-1}\chi$, we write $\chi(\mu^{-1}D)(u \circ \psi) = \lambda^{-1}U_{\lambda,\mu}$ with

$$U_{\lambda,\mu}(x) \overset{def}{=} \mu^d \sum_{k=1}^{d} \widetilde{h}(\mu\cdot) \star \left((g_{k,\lambda} \star \partial_k u) \circ \psi \right)(x)$$

$$= \mu^d \sum_{k=1}^{d} \int_{\mathbb{R}^d} \widetilde{h}(\mu(x - \phi(z)))\partial_k(g_{k,\lambda} \star u)(z)J_\phi(z)\, dz.$$

Integrating by parts, we get $U_{\lambda,\mu}(x) = U^1_{\lambda,\mu}(x) + U^2_{\lambda,\mu}(x)$ with

$$U^1_{\lambda,\mu}(x) \overset{def}{=} \mu^{d+1} \sum_{k=1}^{d} \int_{\mathbb{R}^d} D\widetilde{h}(\mu(x - \phi(z))) \cdot \partial_k\phi(z)\,(g_{k,\lambda} \star u)(z)J_\phi(z)\, dz,$$

$$U^2_{\lambda,\mu}(x) \overset{def}{=} \mu^d \sum_{k=1}^{d} \int_{\mathbb{R}^d} \widetilde{h}(\mu(x - \phi(z)))(g_{k,\lambda} \star u)(z)\partial_k J_\phi(z)\, dz.$$

We estimate $\|U^1_{\lambda,\mu}\|_{L^p}$. Setting $z = \psi(x - \mu^{-1}y)$, we see that

$$U^1_{\lambda,\mu}(x) = \mu \sum_{k=1}^{d} \int_{\mathbb{R}^d} D\widetilde{h}(y) \cdot \partial_k\phi(\psi(x - \mu^{-1}y))\,(g_{k,\lambda} \star u)(\psi(x - \mu^{-1}y))\, dy.$$

Hence, by Hölder's inequality,

$$|U^1_{\lambda,\mu}(x)| \leq \mu\|D\phi\|_{L^\infty} \left(\int_{\mathbb{R}^d} |D\widetilde{h}(y)|\, dy \right)^{\frac{1}{p'}}$$

$$\times \left(\int_{\mathbb{R}^d} |D\widetilde{h}(y)|\,|(g_{k,\lambda} \star u)(\psi(x - \mu^{-1}y))|^p\, dy \right)^{\frac{1}{p}}.$$

We infer that

$$\|U^1_{\lambda,\mu}\|_{L^p} \le \mu \|D\phi\|_{L^\infty} \|D\widetilde{h}\|_{L^1}^{\frac{1}{p'}}$$
$$\times \left(\int_{\mathbb{R}^d \times \mathbb{R}^d} |D\widetilde{h}(y)| \, |(g_{k,\lambda} \star u)(\psi(x - \mu^{-1}y))|^p \, dx \, dy \right)^{\frac{1}{p}}.$$

Combining the change of variable $x' = \psi(x - \mu^{-1}y)$ with Fubini's theorem, we then get

$$\|U^1_{\lambda,\mu}\|_{L^p} \le \mu \|D\widetilde{h}\|_{L^1} \|D\phi\|_{L^\infty} \|J_\phi\|_{L^\infty}^{\frac{1}{p}} \|g_{k,\lambda} \star u\|_{L^p}$$
$$\le C\mu \|D\phi\|_{L^\infty} \|J_\phi\|_{L^\infty}^{\frac{1}{p}} \|u\|_{L^p}.$$

Following the same lines, we also get

$$\|U^2_{\lambda,\mu}\|_{L^p} \le C \|DJ_\phi\|_{L^\infty} \|J_\phi\|_{L^\infty}^{\frac{1}{p}} \|u\|_{L^p}.$$

The lemma is thus proved. $\qquad\qquad\square$

In the case where the diffeomorphism ϕ preserves the measure, we can get a more accurate result, one which will prove useful for transport and transport-diffusion equations (see Chapter 3).

Lemma 2.7. *Let θ be a smooth function supported in an annulus of $\mathbb{R}^d$. There exists a constant C such that for any $C^{0,1}$ measure-preserving global diffeomorphism ψ over $\mathbb{R}^d$ with inverse ϕ, any tempered distribution u with $\widehat{u}$ supported in $\lambda\mathcal{C}$, any $p \in [1,\infty]$, and any $(\lambda,\mu) \in \,]0,\infty[^2$, we have*

$$\left\| \theta(\mu^{-1}D)(u \circ \psi) \right\|_{L^p} \le C \|u\|_{L^p} \min\left(\frac{\mu}{\lambda} \|D\phi\|_{L^\infty}, \frac{\lambda}{\mu} \|D\psi\|_{L^\infty} \right).$$

Proof. Since $J_\psi = J_\phi \equiv 1$, the fact that

$$\left\| \theta(\mu^{-1}D)(u \circ \psi) \right\|_{L^p} \le C\frac{\mu}{\lambda} \|D\phi\|_{L^\infty} \|u\|_{L^p}$$

is ensured by Lemma 2.6.

In order to prove the other inequality, we use the fact that, owing to the spectral localization of θ, there exists a family of smooth functions $(\theta_1, \ldots, \theta_k)$ with compact support such that

$$\theta(\xi) = i \sum_{k=1}^{d} \xi_k \theta_k(\xi) \quad \text{for all} \quad \xi \in \mathbb{R}^d.$$

Hence,

$$\theta(\mu^{-1}D) = \mu^{-1} \sum_k \partial_k \theta_k(\mu^{-1}D),$$

so we can write

$$\theta(\mu^{-1}D)(u \circ \psi)(x) = \mu^{-1}\mu^d \sum_k \int_{\mathbb{R}^d} \mathcal{F}^{-1}\theta_k(\mu(x-y))\,\partial_k(u \circ \psi)(y)\,dy.$$

From the above equality and the fact that ψ preserves the measure, we easily deduce that

$$\|\theta(\mu^{-1}D)(u \circ \psi)\|_{L^p} \le C\mu^{-1}\|D\psi\|_{L^\infty}\|Du \circ \psi\|_{L^p}$$

$$\le C\mu^{-1}\|D\psi\|_{L^\infty}\|Du\|_{L^p}.$$

Bernstein's lemma yields $\|Du\|_{L^p} \le \lambda\|u\|_{L^p}$. This completes the proof.   □

2.1.4 The Effects of Some Nonlinear Functions

The following lemma describes some properties of powers of functions with Fourier transforms supported in an annulus.

Lemma 2.8. *Let $\mathcal{C}$ be an annulus. A constant C exists such that for any positive real number λ, positive integer p, and function u in L^p whose Fourier transform is supported in $\lambda\mathcal{C}$, we have*

$$\|u^p\|_{L^2} \le C\lambda^{-1}\|\nabla(u^p)\|_{L^2}.$$

Remark 2.9. This lemma is somewhat surprising. Indeed, if $\mathcal{F}u$ is supported in an annulus, then $\mathcal{F}(u^p)$ is *not* supported in an annulus, but rather in a ball. Despite that, the above lemma guarantees that the L^2 norm of u^p may be controlled by the L^2 norm of its gradient.

Proof of Lemma 2.8. As usual, it suffices to consider the case $\lambda = 1$. Owing to the spectral properties of u, we can write

$$u = \sum_{j=1}^d \partial_j u_j \quad \text{with} \quad u_j \stackrel{\text{def}}{=} g_j \star u \quad \text{and} \quad g_j \stackrel{\text{def}}{=} \mathcal{F}^{-1}(-i\xi_j|\xi|^{-2}\widetilde{\phi}(\xi)),$$

where $\widetilde{\phi}$ stands for a smooth function supported in a (suitably large) annulus and with value 1 in a neighborhood of the annulus $\mathcal{C}$.

Using the above decomposition and performing an integration by parts, we thus infer that

$$\int_{\mathbb{R}^d} u^{2p}\,dx = \sum_{j=1}^d \int_{\mathbb{R}^d} \partial_j u_j u^{2p-1}\,dx$$

$$= -(2p-1)\sum_{j=1}^d \int_{\mathbb{R}^d} u_j u^{2p-2}\partial_j u\,dx$$

$$= -\frac{2p-1}{p}\sum_{j=1}^d \int_{\mathbb{R}^d} u_j \partial_j(u^p)u^{p-1}\,dx.$$

Hence, by virtue of the Cauchy–Schwarz inequality,

$$\int_{\mathbb{R}^d} u^{2p}\, dx \le C\|\nabla(u^p)\|_{L^2} \left(\sum_{j=1}^d \int_{\mathbb{R}^d} |u_j|^2 u^{2(p-1)}\, dx \right)^{\frac{1}{2}}.$$

We obviously have $\|u_j\|_{L^{2p}} \le C\|u\|_{L^{2p}}$, so, by Hölder's inequality,

$$\int_{\mathbb{R}^d} u^{2p}\, dx \le C\|\nabla(u^p)\|_{L^2} \|u\|_{L^{2p}}^p,$$

and the result is proved.   □

2.2 Dyadic Partition of Unity

We now define the dyadic partition of unity that we shall use throughout the book.

Proposition 2.10. *Let $\mathcal{C}$ be the annulus $\{\xi \in \mathbb{R}^d \,/\, 3/4 \le |\xi| \le 8/3\}$. There exist radial functions χ and φ, valued in the interval $[0,1]$, belonging respectively to $\mathcal{D}(B(0,4/3))$ and $\mathcal{D}(\mathcal{C})$, and such that*

$$\forall \xi \in \mathbb{R}^d, \ \chi(\xi) + \sum_{j\ge 0} \varphi(2^{-j}\xi) = 1, \tag{2.5}$$

$$\forall \xi \in \mathbb{R}^d \setminus \{0\}, \ \sum_{j\in\mathbb{Z}} \varphi(2^{-j}\xi) = 1, \tag{2.6}$$

$$|j - j'| \ge 2 \Rightarrow \mathrm{Supp}\ \varphi(2^{-j}\cdot) \cap \mathrm{Supp}\ \varphi(2^{-j'}\cdot) = \emptyset, \tag{2.7}$$

$$j \ge 1 \Rightarrow \mathrm{Supp}\ \chi \cap \mathrm{Supp}\ \varphi(2^{-j}\cdot) = \emptyset, \tag{2.8}$$

the set $\widetilde{\mathcal{C}} \overset{def}{=} B(0,2/3) + \mathcal{C}$ is an annulus, and we have

$$|j - j'| \ge 5 \Rightarrow 2^{j'}\widetilde{\mathcal{C}} \cap 2^j \mathcal{C} = \emptyset. \tag{2.9}$$

Further, we have

$$\forall \xi \in \mathbb{R}^d, \ \frac{1}{2} \le \chi^2(\xi) + \sum_{j\ge 0} \varphi^2(2^{-j}\xi) \le 1, \tag{2.10}$$

$$\forall \xi \in \mathbb{R}^d \setminus \{0\}, \ \frac{1}{2} \le \sum_{j\in\mathbb{Z}} \varphi^2(2^{-j}\xi) \le 1. \tag{2.11}$$

Proof. Take α in the interval $]1, 4/3[$ and denote by $\mathcal{C}'$ the annulus with small radius α^{-1} and large radius 2α. Choose a radial smooth function θ with values in $[0, 1]$, supported in $\mathcal{C}$, and with value 1 in the neighborhood of $\mathcal{C}'$. The important point is the following: for any couple of integers (j, j'), we have

$$|j - j'| \geq 2 \Rightarrow 2^{j'} \mathcal{C} \cap 2^j \mathcal{C} = \emptyset. \tag{2.12}$$

Indeed, if $2^{j'} \mathcal{C} \cap 2^j \mathcal{C} \neq \emptyset$ and $j' \geq j$, then $2^{j'} \times 3/4 \leq 4 \times 2^{j+1}/3$, which implies that $j' - j \leq 1$. Now, let

$$S(\xi) = \sum_{j \in \mathbb{Z}} \theta(2^{-j}\xi).$$

Thanks to (2.12), this sum is locally finite on the set $\mathbb{R}^d \setminus \{0\}$. Thus, the function S is smooth on $\mathbb{R}^d \setminus \{0\}$. As α is greater than 1, we have

$$\bigcup_{j \in \mathbb{Z}} 2^j \mathcal{C}' = \mathbb{R}^d \setminus \{0\}.$$

As the function θ is nonnegative and has value 1 near $\mathcal{C}'$, it follows from the above covering property that the function S is positive.

We claim that the function $\varphi \overset{\text{def}}{=} \theta/S$ is suitable. Indeed, it is obvious that φ belongs to $\mathcal{D}(\mathcal{C})$ and that the function $1 - \sum_{j \geq 0} \varphi(2^{-j}\cdot)$ is smooth [use (2.12)]. Further, as Supp $\theta \subset \mathcal{C}$, we have

$$|\xi| \geq \frac{4}{3} \Rightarrow \sum_{j \geq 0} \varphi(2^{-j}\xi) = 1. \tag{2.13}$$

Thus, setting

$$\chi(\xi) = 1 - \sum_{j \geq 0} \varphi(2^{-j}\xi), \tag{2.14}$$

we get the identities (2.5) and (2.7). The identity (2.8) is an obvious consequence of (2.12) and (2.13). We now prove (2.9), which will be useful in Section 2.8. It is clear that the annulus $\widetilde{\mathcal{C}}$ has center 0, small radius $1/12$, and large radius $10/3$. It then turns out that

$$2^k \widetilde{\mathcal{C}} \cap 2^j \mathcal{C} \neq \emptyset \Rightarrow \left(\frac{3}{4} \times 2^j \leq 2^k \times \frac{10}{3} \quad \text{or} \quad \frac{1}{12} \times 2^k \leq 2^j \frac{8}{3}\right),$$

and (2.9) is proved. We now prove (2.10). As χ and φ have their values in $[0, 1]$, it is clear that

$$\chi^2(\xi) + \sum_{j \geq 0} \varphi^2(2^{-j}\xi) \leq 1. \tag{2.15}$$

We bound the sum of squares from below. We have

$$1 = (\Sigma_0(\xi) + \Sigma_1(\xi))^2 \quad \text{with}$$

$$\Sigma_0(\xi) = \sum_{j \text{ even}} \varphi(2^{-j}\xi) \quad \text{and} \quad \Sigma_1(\xi) = \chi(\xi) + \sum_{j \text{ odd}} \varphi(2^{-j}\xi).$$

Obviously, $1 \leq 2(\Sigma_0^2(\xi) + \Sigma_1^2(\xi))$. Now, owing to (2.7) and (2.8), we have

$$\Sigma_0^2(\xi) = \sum_{j \text{ even}} \varphi^2(2^{-j}\xi) \quad \text{and} \quad \Sigma_1^2(\xi) = \chi^2(\xi) + \sum_{j \text{ odd}} \varphi^2(2^{-j}\xi).$$

This yields (2.10). Proving (2.11) proceeds similarly.   □

From now on, we fix two functions χ and φ satisfying the assertions (2.5)–(2.11) and write $h = \mathcal{F}^{-1}\varphi$ and $\widetilde{h} = \mathcal{F}^{-1}\chi$. The *nonhomogeneous dyadic blocks* Δ_j are defined by

$$\Delta_j u = 0 \quad \text{if} \quad j \leq -2, \quad \Delta_{-1} u = \chi(D)u = \int_{\mathbb{R}^d} \widetilde{h}(y) u(x - y)\, dy,$$

$$\text{and} \quad \Delta_j u = \varphi(2^{-j}D)u = 2^{jd} \int_{\mathbb{R}^d} h(2^j y) u(x - y)\, dy \quad \text{if} \quad j \geq 0.$$

The nonhomogeneous low-frequency cut-off operator S_j is defined by

$$S_j u = \sum_{j' \leq j-1} \Delta_{j'} u.$$

The *homogeneous dyadic blocks* $\dot{\Delta}_j$ and the homogeneous low-frequency cut-off operators $\dot{S}_j$ are defined for all $j \in \mathbb{Z}$ by

$$\dot{\Delta}_j u = \varphi(2^{-j}D)u = 2^{jd} \int_{\mathbb{R}^d} h(2^j y) u(x - y)\, dy,$$

$$\dot{S}_j u = \chi(2^{-j}D)u = 2^{jd} \int_{\mathbb{R}^d} \widetilde{h}(2^j y) u(x - y)\, dy\,.$$

Remark 2.11. We also note that the above operators map L^p into L^p with norms *independent* of j and p. This fact will be of constant use in this chapter.

Obviously, we can write the following (formal) *Littlewood–Paley decompositions*:

$$\text{Id} = \sum_j \Delta_j \quad \text{and} \quad \text{Id} = \sum_j \dot{\Delta}_j. \tag{2.16}$$

In the nonhomogeneous case, the above decomposition makes sense in $\mathcal{S}'(\mathbb{R}^d)$.

Proposition 2.12. *Let u be in $\mathcal{S}'(\mathbb{R}^d)$. Then, $u = \lim_{j \to \infty} S_j u$ in $\mathcal{S}'(\mathbb{R}^d)$.*

Proof. Note that $\langle u - S_j u, f \rangle = \langle u, f - S_j f \rangle$ for all f in $\mathcal{S}(\mathbb{R}^d)$ and u in $\mathcal{S}'(\mathbb{R}^d)$, so it suffices to prove that $f = \lim_{j \to \infty} S_j f$ in the space $\mathcal{S}(\mathbb{R}^d)$. Because the Fourier transform is an automorphism of $\mathcal{S}(\mathbb{R}^d)$, we can alternatively prove that $\chi(2^{-j}\cdot)\widehat{f}$ tends to $\widehat{f}$ in $\mathcal{S}(\mathbb{R}^d)$. This is an easy exercise left to the reader.   □

We now state another (somewhat related) result of convergence.

Proposition 2.13. *Let $(u_j)_{j \in \mathbb{N}}$ be a sequence of bounded functions such that the Fourier transform of u_j is supported in $2^j \widetilde{\mathcal{C}}$, where $\widetilde{\mathcal{C}}$ is a given annulus. Assume that, for some integer N, the sequence $(2^{-jN} \|u_j\|_{L^\infty})_{j \in \mathbb{N}}$ is bounded. The series $\sum_j u_j$ then converges in $\mathcal{S}'$.*

Proof. After rescaling, the relation (2.1.1) reads as follows for all integers j and k:

$$u_j = 2^{-jk} \sum_{|\alpha|=k} 2^{jd} g_\alpha(2^j \cdot) \star \partial^\alpha u_j.$$

For any test function ϕ in $\mathcal{S}$, we then write

$$\langle u_j, \phi \rangle = 2^{-jk} \sum_{|\alpha|=k} \langle u_j, 2^{jd} \check{g}_\alpha(2^j \cdot) \star (-\partial)^\alpha \phi \rangle \quad \text{with} \quad \check{g}_\alpha(x) \overset{\text{def}}{=} g_\alpha(-x).$$

We then have

$$|\langle u_j, \phi \rangle| \le C 2^{-jk} \sum_{|\alpha|=k} 2^{jN} \|\partial^\alpha \phi\|_{L^1}.$$

Choose $k > N$. Then, $\sum_j \langle u_j, \phi \rangle$ is a convergent series, the sum of which is less than $C\|\phi\|_{M,\mathcal{S}}$ for some integer M. Thus, the formula

$$\langle u, \phi \rangle \overset{\text{def}}{=} \lim_{j \to \infty} \sum_{j' \le j} \langle u_{j'}, \phi \rangle$$

defines a tempered distribution.   □

Proving the equality (2.16) for the operators $\dot{\Delta}_j$ is not so obvious, even for smooth functions: it clearly fails for nonzero polynomials. However, it holds true for any distribution in the set $\mathcal{S}'_h$ defined on page 22. Indeed, if u belongs to $\mathcal{S}'_h$, then $\dot{S}_j u$ tends uniformly to 0 when j goes to $-\infty$.

The homogeneous version of Proposition 2.13 reads as follows.

Proposition 2.14. *Let $(u_j)_{j \in \mathbb{Z}}$ be a sequence of bounded functions such that the support of $\widehat{u}_j$ is included in $2^j \widetilde{\mathcal{C}}$, where $\widetilde{\mathcal{C}}$ is a given annulus. Assume that, for some integer N, the sequence $(2^{-jN} \|u_j\|_{L^\infty})_{j \in \mathbb{N}}$ is bounded and that the series $\sum_{j<0} u_j$ converges in L^∞. The series $\sum_{j \in \mathbb{Z}} u_j$ then converges to some u in $\mathcal{S}'$, and u belongs to $\mathcal{S}'_h$.*

Proof. Thanks to Proposition 2.13, the series $\sum_{j \in \mathbb{Z}} u_j$ converges to some u in $\mathcal{S}'$. We are therefore left with proving that u belongs to $\mathcal{S}'_h$. We have, for some integer N_0,

$$\|\dot{S}_j u\|_{L^\infty} \le \left\| \dot{S}_j \sum_{j' \le j+N_0} u_{j'} \right\|_{L^\infty} \le C \left\| \sum_{j' \le j+N_0} u_{j'} \right\|_{L^\infty}.$$

As the series $\sum_{j<0} u_j$ converges in L^∞, the proposition is proved.   □

2.3 Homogeneous Besov Spaces

To begin, we define homogeneous Besov spaces.

Definition 2.15. *Let s be a real number and (p,r) be in $[1,\infty]^2$. The homogeneous Besov space $\dot{B}^s_{p,r}$ consists of those distributions u in $\mathcal{S}'_h$ such that*

$$\|u\|_{\dot{B}^s_{p,r}} \overset{def}{=} \left(\sum_{j\in\mathbb{Z}} 2^{rjs} \|\dot{\Delta}_j u\|^r_{L^p} \right)^{\frac{1}{r}} < \infty.$$

Proposition 2.16. *The space $\dot{B}^s_{p,r}$ endowed with $\|\cdot\|_{\dot{B}^s_{p,r}}$ is a normed space.*

Proof. It is obvious that $\|\cdot\|_{\dot{B}^s_{p,r}}$ is a seminorm. Assume that for some u in $\mathcal{S}'_h$, we have $\|u\|_{\dot{B}^s_{p,r}} = 0$. This implies that the support of $\hat{u}$ is included in $\{0\}$ and thus that for any $j \in \mathbb{Z}$, we have $\dot{S}_j u = u$. As u belongs to $\mathcal{S}'_h$, we conclude that $u = 0$. □

Remark 2.17. The definition of the Besov space $\dot{B}^s_{p,r}$ is independent of the function φ used for defining the blocks $\dot{\Delta}_j$, and changing φ yields an equivalent norm. Indeed, if $\tilde{\varphi}$ is another dyadic partition of unity, then an integer N_0 exists such that $|j - j'| \geq N_0$ implies that Supp $\tilde{\varphi}(2^{-j}\cdot) \cap$ Supp $\varphi(2^{-j'}\cdot) = \emptyset$. Thus,

$$2^{js}\|\tilde{\varphi}(2^{-j}D)u\|_{L^p} = 2^{js}\left\| \sum_{|j-j'|\leq N_0} \tilde{\varphi}(2^{-j}D)\dot{\Delta}_{j'} u \right\|_{L^p}$$

$$\leq C 2^{N_0|s|} \sum_{j'} \mathbf{1}_{[-N_0,N_0]}(j-j')2^{j's}\|\dot{\Delta}_{j'}u\|_{L^p}.$$

Young's inequality implies the result.

We also note that a distribution u of $\mathcal{S}'_h$ belongs to $\dot{B}^s_{p,r}$ if and only if there exists some constant C and some nonnegative sequence $(c_j)_{j\in\mathbb{Z}}$ such that

$$\forall j \in \mathbb{Z}, \|\dot{\Delta}_j u\|_{L^p} \leq C c_j 2^{-js} \quad \text{and} \quad \|(c_j)\|_{\ell^r} = 1.$$

This fact will be extensively used throughout the book.

Examples.

– Thanks to (2.11), we can deduce that the (semi)norms $\|\cdot\|_{\dot{H}^s}$ and $\|\cdot\|_{\dot{B}^s_{2,2}}$ are equivalent. Further, it is clear that $\dot{H}^s \subset \dot{B}^s_{2,2}$ and that both spaces coincide if $s < d/2$.
– If $s \in \,]0,1[$, then the Besov space $\dot{B}^s_{\infty,\infty}$ coincides with the space of distributions of $\mathcal{S}'_h$ which are Hölder functions with exponent s (see Theorem 2.36 below).

Homogeneous Besov spaces have nice scaling properties. Indeed, if u is a tempered distribution, then consider the tempered distribution u_N defined by $u_N \overset{\text{def}}{=} u(2^N \cdot)$. We have the following proposition.

Proposition 2.18. *Consider an integer N and a distribution u of $\mathcal{S}_h'$. Then, $\|u\|_{\dot{B}_{p,r}^s}$ is finite if and only if u_N is finite. Moreover, we have*

$$\|u_N\|_{\dot{B}_{p,r}^s} = 2^{N(s-\frac{d}{p})} \|u\|_{\dot{B}_{p,r}^s}.$$

Proof. By definition of $\dot{\Delta}_j$ and by the change of variable $z = 2^N y$, we get

$$\dot{\Delta}_j u_N(x) = 2^{jd} \int h(2^j(x-y)) u(2^N y)\, dy$$

$$= 2^{(j-N)d} \int h(2^{j-N}(2^N x - z)) u(z)\, dz$$

$$= (\dot{\Delta}_{j-N} u)(2^N x).$$

It turns out that $\|\dot{\Delta}_j u_N\|_{L^p} = 2^{-N\frac{d}{p}} \|\dot{\Delta}_{j-N} u\|_{L^p}$. We deduce from this that

$$2^{js} \|\dot{\Delta}_j u_N\|_{L^p} = 2^{N(s-\frac{d}{p})} 2^{(j-N)s} \|\dot{\Delta}_{j-N} u\|_{L^p},$$

and the proposition follows immediately by summation. □

Remark 2.19. More generally, there exists a constant C, depending only on s, such that for all positive λ, we have

$$C^{-1} \lambda^{s-\frac{d}{p}} \|u\|_{\dot{B}_{p,r}^s} \leq \|u(\lambda \cdot)\|_{\dot{B}_{p,r}^s} \leq C \lambda^{s-\frac{d}{p}} \|u\|_{\dot{B}_{p,r}^s}.$$

We emphasize that having u in some homogeneous Besov space $\dot{B}_{p,r}^s$ yields information about both low and high frequencies of u. Thus, if $s_1 \neq s_2$, then we cannot expect any inclusion between the spaces $\dot{B}_{p,r}^{s_1}$ and $\dot{B}_{p,r}^{s_2}$. However, we can state the following theorem, which may be compared with the classical Sobolev embedding theorem (see Theorem 1.38, page 29).

Proposition 2.20. *Let $1 \leq p_1 \leq p_2 \leq \infty$ and $1 \leq r_1 \leq r_2 \leq \infty$. Then, for any real number s, the space $\dot{B}_{p_1,r_1}^s$ is continuously embedded in $\dot{B}_{p_2,r_2}^{s-d\left(\frac{1}{p_1}-\frac{1}{p_2}\right)}$.*

Proof. Lemma 2.1 yields

$$\|\dot{\Delta}_j u\|_{L^{p_2}} \leq C 2^{jd\left(\frac{1}{p_1}-\frac{1}{p_2}\right)} \|\dot{\Delta}_j u\|_{L^{p_1}}.$$

As $\ell^{r_1}(\mathbb{Z})$ is continuously embedded in $\ell^{r_2}(\mathbb{Z})$, the proposition is proved. □

In contrast with the standard function spaces (e.g., Sobolev spaces H^s or L^p spaces with $p < \infty$), homogeneous Besov spaces contain nontrivial homogeneous functions. This is illustrated by the following proposition.

Proposition 2.21. *Let σ be in $]0, d[$. For any p in $[1, \infty]$, the function $|\cdot|^{-\sigma}$ belongs to $\dot{B}_{p,\infty}^{\frac{d}{p}-\sigma}$.*

Proof. Using Proposition 2.20, it is enough to prove that $\rho_\sigma \overset{\text{def}}{=} |\cdot|^{-\sigma}$ belongs to $\dot{B}_{1,\infty}^{d-\sigma}$. In order to do so, we introduce a smooth compactly supported function χ which is identically equal to 1 near the unit ball and we write

$$\rho_\sigma = \rho_0 + \rho_1 \quad \text{with} \quad \rho_0(x) \overset{\text{def}}{=} \chi(x)|x|^{-\sigma} \quad \text{and} \quad \rho_1(x) \overset{\text{def}}{=} (1 - \chi(x))|x|^{-\sigma}.$$

It is obvious that $\rho_0 \in L^1$ and that $\rho_1 \in L^q$ whenever $q > d/\sigma$. This implies that ρ_σ belongs to $\mathcal{S}'_h$. The homogeneity of the function ρ_σ then gives

$$\begin{aligned}
\dot{\Delta}_j \rho_\sigma &= 2^{jd} \rho_\sigma \star h(2^j \cdot) \\
&= 2^{j(d+\sigma)} \rho_\sigma(2^j \cdot) \star h(2^j \cdot) \\
&= 2^{j\sigma} (\dot{\Delta}_0 \rho_\sigma)(2^j \cdot).
\end{aligned}$$

Therefore, $\|\dot{\Delta}_j \rho_\sigma\|_{L^1} = 2^{j(\sigma-d)} \|\dot{\Delta}_0 \rho_\sigma\|_{L^1}$, which reduces the problem to proving that the function $\dot{\Delta}_0 \rho_\sigma$ is in L^1. As ρ_0 is in L^1, $\dot{\Delta}_0 \rho_0$ is also in L^1, thanks to the continuity of the operator $\dot{\Delta}_0$ on Lebesgue spaces. Using Lemma 2.1, we get

$$\|\dot{\Delta}_0 \rho_1\|_{L^1} \leq C_k \|D^k \dot{\Delta}_0 \rho_1\|_{L^1} \leq C_k \|D^k \rho_1\|_{L^1}.$$

By Leibniz's formula, $D^k \rho_1 - (1 - \chi)D^k \rho_\sigma$ is a smooth compactly supported function. We then complete the proof by choosing k such that $k > d - \sigma$. □

Proposition 2.22. *A constant C exists which satisfies the following properties. If s_1 and s_2 are real numbers such that $s_1 < s_2$ and $\theta \in \;]0, 1[$, then we have, for any $(p, r) \in [1, \infty]^2$ and any $u \in \mathcal{S}'_h$,*

$$\|u\|_{\dot{B}_{p,r}^{\theta s_1 + (1-\theta)s_2}} \leq \|u\|_{\dot{B}_{p,r}^{s_1}}^\theta \|u\|_{\dot{B}_{p,r}^{s_2}}^{1-\theta} \quad and$$

$$\|u\|_{\dot{B}_{p,1}^{\theta s_1 + (1-\theta)s_2}} \leq \frac{C}{s_2 - s_1}\left(\frac{1}{\theta} + \frac{1}{1-\theta}\right) \|u\|_{\dot{B}_{p,\infty}^{s_1}}^\theta \|u\|_{\dot{B}_{p,\infty}^{s_2}}^{1-\theta}.$$

Proof. To prove the first inequality, it suffices to write that

$$2^{j(\theta s_1 + (1-\theta)s_2)} \|\dot{\Delta}_j u\|_{L^p} = \left(2^{j s_1} \|\dot{\Delta}_j u\|_{L^p}\right)^\theta \left(2^{j s_2} \|\dot{\Delta}_j u\|_{L^p}\right)^{1-\theta}$$

and to apply Hölder's inequality.

To prove the second one, we shall estimate low and high frequencies of u in a different way. More precisely, we write

$$\|u\|_{\dot{B}_{p,1}^{\theta s_1 + (1-\theta)s_2}} = \sum_{j \leq N} 2^{j(\theta s_1 + (1-\theta)s_2)} \|\dot{\Delta}_j u\|_{L^p} + \sum_{j > N} 2^{j(\theta s_1 + (1-\theta)s_2)} \|\dot{\Delta}_j u\|_{L^p}.$$

By the definition of the Besov norms, we have

$$\begin{cases} 2^{j(\theta s_1 + (1-\theta)s_2)} \|\dot{\Delta}_j u\|_{L^p} \leq 2^{j(1-\theta)(s_2-s_1)} \|u\|_{\dot{B}_{p,\infty}^{s_1}}, \\ 2^{j(\theta s_1 + (1-\theta)s_2)} \|\dot{\Delta}_j u\|_{L^p} \leq 2^{-j\theta(s_2-s_1)} \|u\|_{\dot{B}_{p,\infty}^{s_2}}. \end{cases}$$

We thus infer that

$$\|u\|_{\dot{B}_{p,1}^{\theta s_1 + (1-\theta)s_2}} \leq \|u\|_{\dot{B}_{p,\infty}^{s_1}} \sum_{j \leq N} 2^{j(1-\theta)(s_2-s_1)} + \|u\|_{\dot{B}_{p,\infty}^{s_2}} \sum_{j > N} 2^{-j\theta(s_2-s_1)}$$

$$\leq \|u\|_{\dot{B}_{p,\infty}^{s_1}} \frac{2^{N(1-\theta)(s_2-s_1)}}{2^{(1-\theta)(s_2-s_1)} - 1} + \|u\|_{\dot{B}_{p,\infty}^{s_2}} \frac{2^{-N\theta(s_2-s_1)}}{1 - 2^{-\theta(s_2-s_1)}}.$$

Choosing N such that

$$\frac{\|u\|_{\dot{B}_{p,\infty}^{s_2}}}{\|u\|_{\dot{B}_{p,\infty}^{s_1}}} \leq 2^{N(s_2-s_1)} < 2^{s_2-s_1} \frac{\|u\|_{\dot{B}_{p,\infty}^{s_2}}}{\|u\|_{\dot{B}_{p,\infty}^{s_1}}}$$

completes the proof.   □

The following lemma provides a useful criterion for determining whether the sum of a series belongs to a homogeneous Besov space.

Lemma 2.23. *Let $\mathcal{C}'$ be an annulus and $(u_j)_{j \in \mathbb{Z}}$ be a sequence of functions such that*

$$\text{Supp } \widehat{u}_j \subset 2^j \mathcal{C}' \quad \text{and} \quad \left\| (2^{js} \|u_j\|_{L^p})_{j \in \mathbb{Z}} \right\|_{\ell^r} < \infty.$$

If the series $\sum_{j \in \mathbb{Z}} u_j$ converges in $\mathcal{S}'$ to some u in $\mathcal{S}'_h$, then u is in $\dot{B}_{p,r}^s$ and

$$\|u\|_{\dot{B}_{p,r}^s} \leq C_s \left\| (2^{js} \|u_j\|_{L^p})_{j \in \mathbb{Z}} \right\|_{\ell^r}.$$

Remark 2.24. The above convergence assumption concerns $(u_j)_{j<0}$. We note that if (s, p, r) satisfies the condition

$$s < \frac{d}{p}, \quad \text{or} \quad s = \frac{d}{p} \text{ and } r = 1, \tag{2.17}$$

then, owing to Lemma 2.1, we have

$$\lim_{j \to -\infty} \sum_{j' < j} u_{j'} = 0 \quad \text{in } L^\infty.$$

Hence, $\sum_{j \in \mathbb{Z}} u_j$ converges to some u in $\mathcal{S}'$, and $\dot{S}_j u$ tends to 0 when j goes to $-\infty$. In particular, we have $u \in \mathcal{S}'_h$.

Proof of Lemma 2.23. It is clear that there exists some nonzero integer N_0 such that $\Delta_{j'} u_j = 0$ for $|j' - j| \geq N_0$. Hence,

$$\|\dot{\Delta}_{j'}u\|_{L^p} = \Big\| \sum_{|j-j'|<N_0} \dot{\Delta}_{j'}u_j \Big\|_{L^p}$$

$$\leq C \sum_{|j-j'|<N_0} \|u_j\|_{L^p}.$$

Therefore, we obtain that

$$2^{j's}\|\dot{\Delta}_{j'}u\|_{L^p} \leq C \sum_{|j-j'|\leq N_0} 2^{js}\|u_j\|_{L^p}.$$

We deduce from this that

$$2^{js}\|\dot{\Delta}_j u\|_{L^p} \leq \big((c_k)\star(d_\ell)\big)_j \quad \text{with} \quad c_k = C\mathbf{1}_{[-N_0,N_0]}(k) \quad \text{and} \quad d_\ell = 2^{\ell s}\|u_\ell\|_{L^p}.$$

Applying Young's inequality (namely, Lemma 1.4 page 5 with $G = \mathbb{Z}$) then leads to

$$\|u\|_{\dot{B}^s_{p,r}} \leq C\Big\|(2^{js}\|u_j\|_{L^p})_{j\in\mathbb{Z}}\Big\|_{\ell^r}.$$

As $u \in \mathcal{S}'_h$ by assumption, this proves the lemma. $\qquad\square$

The previous lemma will enable us to establish the following important topological properties of homogeneous Besov spaces.

Theorem 2.25. *Let $(s_1, s_2) \in \mathbb{R}^2$ and $1 \leq p_1, p_2, r_1, r_2 \leq \infty$. Assume that (s_1, p_1, r_1) satisfies the condition (2.17). The space $\dot{B}^{s_1}_{p_1,r_1} \cap \dot{B}^{s_2}_{p_2,r_2}$ endowed with the norm $\|\cdot\|_{\dot{B}^{s_1}_{p_1,r_1}} + \|\cdot\|_{\dot{B}^{s_2}_{p_2,r_2}}$ is then complete and satisfies the Fatou property: If $(u_n)_{n\in\mathbb{N}}$ is a bounded sequence of $\dot{B}^{s_1}_{p_1,r_1} \cap \dot{B}^{s_2}_{p_2,r_2}$, then an element u of $\dot{B}^{s_1}_{p_1,r_1} \cap \dot{B}^{s_2}_{p_2,r_2}$ and a subsequence $u_{\psi(n)}$ exist such that*

$$\lim_{n\to\infty} u_{\psi(n)} = u \quad \text{in} \quad \mathcal{S}' \quad \text{and} \quad \|u\|_{\dot{B}^{s_k}_{p_k,r_k}} \leq C \liminf_{n\to\infty} \|u_{\psi(n)}\|_{\dot{B}^{s_k}_{p_k,r_k}} \quad \text{for} \quad k = 1, 2.$$

Proof. We first prove the Fatou property. According to Lemma 2.1, for any $j \in \mathbb{Z}$, the sequence $(\dot{\Delta}_j u_n)_{n\in\mathbb{N}}$ is bounded in $L^{\min(p_1,p_2)} \cap L^\infty$. Cantor's diagonal process thus supplies a subsequence $(u_{\psi(n)})_{n\in\mathbb{N}}$ and a sequence $(\widetilde{u}_j)_{j\in\mathbb{Z}}$ of C^∞ functions with Fourier transform supported in $2^j\mathcal{C}$ such that, for any $j \in \mathbb{Z}$, $\phi \in \mathcal{S}$, and $k = 1, 2$,

$$\lim_{n\to\infty} \langle \dot{\Delta}_j u_{\psi(n)}, \phi \rangle = \langle \widetilde{u}_j, \phi \rangle \quad \text{and} \quad \|\widetilde{u}_j\|_{L^{p_k}} \leq \liminf_{n\to\infty} \|\dot{\Delta}_j u_n\|_{L^{p_k}}.$$

Now, the sequence $\Big((2^{js_k}\|\dot{\Delta}_j u_{\psi(n)}\|_{L^{p_k}})_j\Big)_{n\in\mathbb{N}}$ is bounded in $\ell^{r_k}(\mathbb{Z})$. Hence, there exists an element $(\widetilde{c}^k_j)_{j\in\mathbb{Z}}$ of ℓ^{r_k} such that (up to an omitted extraction) we have, for any sequence $(d_j)_{j\in\mathbb{Z}}$ of nonnegative real numbers different from 0 for only a finite number of indices j,

$$\lim_{n\to\infty} \sum_{j\in\mathbb{Z}} 2^{js_k}\|\dot{\Delta}_j u_{\psi(n)}\|_{L^{p_k}} d_j = \sum_{j\in\mathbb{Z}} \widetilde{c}^k_j d_j \quad \text{and}$$

$$\|(\widetilde{c}^k_j)_j\|_{\ell^{r_k}} \leq \liminf_{n\to\infty} \|u_{\psi(n)}\|_{\dot{B}^{s_k}_{p_k,r_k}}.$$

Passing to the limit in the sum and using Lemma 1.2 page 2 with $X = \mathbb{Z}$ and μ the counting measure on $\mathbb{Z}$ gives that $(2^{js_k}\|\widetilde{u}_j\|_{L^{p_k}})_j$ belongs to $\ell^{r_k}(\mathbb{Z})$. From the definition of $\widetilde{u}_j$, we easily deduce that $\mathcal{F}\widetilde{u}_j$ is supported in the annulus $2^j\mathcal{C}$ (where $\mathcal{C}$ has been defined in Proposition 2.10). As (s_1, p_1, r_1) satisfies (2.17), Lemma 2.23 thus guarantees that the series $\sum_{j\in\mathbb{Z}}\widetilde{u}_j$ converges to some u in $\mathcal{S}'_h$. Given (2.7), we obviously have, for all $M < N$ and $\phi \in \mathcal{S}$,

$$\left\langle \sum_{j=M}^{N} \dot{\Delta}_j u, \phi \right\rangle = \left\langle \sum_{j=M}^{N} \sum_{|j'-j|\leq 1} \dot{\Delta}_j \widetilde{u}_{j'}, \phi \right\rangle.$$

Hence, by the definition of $\widetilde{u}_j$ and, again, by (2.7), we have

$$\sum_{j=M}^{N} \dot{\Delta}_j u = \lim_{n\to\infty} \sum_{j=M}^{N} \dot{\Delta}_j u_{\psi(n)} \quad \text{in } \mathcal{S}'.$$

Since the condition (2.17) is satisfied by (s_1, p_1, r_1), and $(u_{\psi(n)})_{n\in\mathbb{N}}$ is bounded in $\dot{B}^{s_1}_{p_1,r_1}$, Lemma 2.1 ensures that $\dot{S}_M u_{\psi(n)}$ tends uniformly to 0 when M goes to $-\infty$. Similarly, $(\mathrm{Id} - \dot{S}_N)u_{\psi(n)}$ tend uniformly to 0 in, say, $\dot{B}^{s_2-1}_{p_2,r_2}$. Hence, u is indeed the limit of $(u_{\psi(n)})_{n\in\mathbb{N}}$ in $\mathcal{S}'$, which completes the proof of the Fatou property.

We will now check that $\dot{B}^{s_1}_{p_1,r_1} \cap \dot{B}^{s_2}_{p_2,r_2}$ is complete. Consider a Cauchy sequence $(u_n)_{n\in\mathbb{N}}$. This sequence is of course bounded, so there exists some u in $\dot{B}^{s_1}_{p_1,r_1} \cap \dot{B}^{s_2}_{p_2,r_2}$ and a subsequence $(u_{\psi(n)})_{n\in\mathbb{N}}$ such that $(u_{\psi(n)})_{n\in\mathbb{N}}$ converges to u in $\mathcal{S}'$. Using the fact that for any positive ε, an integer n_ε exists such that

$$n \geq m \geq n_\varepsilon \implies \|u_{\psi(m)} - u_{\psi(n)}\|_{\dot{B}^{s_1}_{p_1,r_1}} + \|u_{\psi(m)} - u_{\psi(n)}\|_{\dot{B}^{s_2}_{p_2,r_2}} < \varepsilon,$$

the Fatou property for $(u_{\psi(m)} - u_{\psi(n)})_{n\in\mathbb{N}}$ ensures that

$$\forall m \geq n_\varepsilon, \quad \|u_{\psi(m)} - u\|_{\dot{B}^{s_1}_{p_1,r_1}} + \|u_{\psi(m)} - u\|_{\dot{B}^{s_2}_{p_2,r_2}} \leq C\varepsilon.$$

Hence, $(u_{\psi(n)})_{n\in\mathbb{N}}$ tends to u in $\dot{B}^{s_1}_{p_1,r_1} \cap \dot{B}^{s_2}_{p_2,r_2}$. This completes the proof. $\square$

Remark 2.26. If $s > d/p$ (or $s = d/p$ and $r > 1$), then $\dot{B}^s_{p,r}$ is no longer a Banach space (Proposition 1.34 may be adapted to the framework of general homogeneous Besov spaces). This is due to a breakdown of convergence for low frequencies, the so-called *infrared divergence*.

There is a way to modify the definition of homogeneous Besov spaces so as to obtain a Banach space, regardless of the regularity index. This is called *realizing* homogeneous Besov spaces. It turns out that realizations coincide with our definition when $s < d/p$, or $s = d/p$ and $r = 1$. In the other cases, however, realizations are defined *up to a polynomial* whose degree depends on $s - d/p$ and r. It goes without saying that solving partial differential equations in such spaces is quite unpleasant.

Proposition 2.27. *If p and r are finite, then the space $\mathcal{S}_0(\mathbb{R}^d)$ of functions in $\mathcal{S}(\mathbb{R}^d)$ whose Fourier transforms are supported away from 0 is dense in $\dot{B}^s_{p,r}(\mathbb{R}^d)$.*

Proof. Let u be in $\dot{B}^s_{p,r}$. Because r is finite, for all $\varepsilon > 0$ we can find some integer N such that

$$\|u - u_N\|_{\dot{B}^s_{p,r}} < \varepsilon/2 \quad \text{with} \quad u_N \overset{\text{def}}{=} \sum_{|j| \leq N} \dot{\Delta}_j u.$$

Fix θ in $\mathcal{C}(B(0,2))$ with value 1 on $B(0,1)$. For $R > 0$ set $\theta_R \overset{\text{def}}{=} \theta(\cdot/R)$. Further, fix an integer M such that $M > N$. We then define

$$u^R_{N,M} \overset{\text{def}}{=} (\text{Id} - \dot{S}_{-M})(\theta_R\, u_N).$$

Because $M > N$, we have $(\text{Id} - \dot{S}_{-M})u_N = u_N$ and hence

$$u^R_{N,M} - u_N = (\text{Id} - \dot{S}_{-M})((\theta_R - 1)u_N).$$

According to Lemma 2.1, we have, for all $j \in \mathbb{N}$ and $k = \max(0, [s] + 2)$,

$$2^{js}\|\dot{\Delta}_j(u^R_{N,M} - u_N)\|_{L^p} \leq 2^{-j} 2^{jk} \|\dot{\Delta}_j((\text{Id} - \dot{S}_{-M})((\theta_R - 1)u_N))\|_{L^p}$$
$$\leq C_s 2^{-j} \|D^k((\theta_R - 1)u_N)\|_{L^p}.$$

If $-M - 1 \leq j \leq -1$, we may write

$$2^{js}\|\dot{\Delta}_j(u^R_{N,M} - u_N)\|_{L^p} \leq C 2^{js}\|(\theta_R - 1)u_N\|_{L^p},$$

and if $j \leq -M - 2$, we have $\dot{\Delta}_j(u^R_{N,M} - u_N) = 0$. So, finally,

$$\|u^R_{N,M} - u_N\|_{\dot{B}^s_{p,r}} \leq C\left(\|D^k((\theta_R - 1)u_N)\|_{L^p} + \sum_{j=-M-1}^{-1} 2^{js}\|(\theta_R - 1)u_N\|_{L^p}\right).$$

Now, by virtue of Leibniz's formula and Lebesgue's dominated convergence theorem (recall that p is finite), the right-hand side of the above inequality tends to 0 when R goes to infinity. Therefore, a positive real number R exists such that

$$\|u^R_{N,M} - u_N\|_{\dot{B}^s_{p,r}} \leq \varepsilon/2.$$

As $u^R_{N,M}$ is a function of $\mathcal{S}_0$, this completes the proof of the proposition. $\quad\square$

Remark 2.28. The same arguments show that when $r = \infty$, the closure of $\mathcal{S}_0$ for the Besov norm $\dot{B}^s_{p,r}$ is the set of distributions in $\mathcal{S}'_h$ such that

$$\lim_{j \to \pm\infty} 2^{js}\|\dot{\Delta}_j u\|_{L^p} = 0.$$

It turns out that Besov spaces have nice duality properties. Observe that in Littlewood–Paley theory, the duality on $\mathcal{S}'_h$ translates, for $\phi \in \mathcal{S}$, into

$$\langle u, \phi \rangle = \sum_{|j-j'| \leq 1} \langle \Delta_j u, \Delta_{j'} \phi \rangle = \sum_{|j-j'| \leq 1} \int_{\mathbb{R}^d} \Delta_j u(x) \Delta_{j'} \phi(x)\, dx.$$

As for the L^p space, we can estimate the norm in $\dot{B}^s_{p,r}$ by duality.

Proposition 2.29. *For all $1 \leq p, r \leq \infty$ and $s \in \mathbb{R}$,*

$$\begin{cases} \dot{B}^s_{p,r} \times \dot{B}^{-s}_{p',r'} \longrightarrow \mathbb{R} \\ (u, \phi) \longmapsto \displaystyle\sum_{|j-j'| \leq 1} \langle \Delta_j u, \Delta_{j'} \phi \rangle \end{cases}$$

defines a continuous bilinear functional on $\dot{B}^s_{p,r} \times \dot{B}^{-s}_{p',r'}$. Denote by $Q^{-s}_{p',r'}$ the set of functions ϕ in $\mathcal{S} \cap \dot{B}^{-s}_{p',r'}$ such that $\|\phi\|_{\dot{B}^{-s}_{p',r'}} \leq 1$. If u is in $\mathcal{S}'_h$, then we have

$$\|u\|_{\dot{B}^s_{p,r}} \leq C \sup_{\phi \in Q^{-s}_{p',r'}} \langle u, \phi \rangle.$$

Proof. For $|j - j'| \leq 1$, we have, thanks to Hölder's inequality,

$$\left| \langle \dot{\Delta}_j u, \dot{\Delta}_{j'} \phi \rangle \right| \leq 2^{|s|} \, 2^{js} \|\dot{\Delta}_j u\|_{L^p} \, 2^{-j's} \|\dot{\Delta}_{j'} \phi\|_{L^{p'}}.$$

Again using Hölder's inequality, we deduce that

$$\left| \langle u, \phi \rangle \right| \leq C \|u\|_{\dot{B}^s_{p,r}} \|\phi\|_{\dot{B}^{-s}_{p',r'}}.$$

In order to prove the second part, for a positive integer N, we denote by $Q^{r'}_N$ the unit ball of the space of sequences of $\ell^{r'}(\mathbb{Z})$ which vanish for indices j such that $|j| > N$. By definition of the Besov norm, we have

$$\|u\|_{\dot{B}^s_{p,r}} = \sup_{N \in \mathbb{N}} \left\| \left(\mathbf{1}_{|j| \leq N} 2^{js} \|\dot{\Delta}_j u\|_{L^p} \right)_j \right\|_{\ell^r}$$

$$= \sup_{N \in \mathbb{N}} \sup_{(\alpha_j) \in Q^{r'}_N} \sum_{|j| \leq N} \|\dot{\Delta}_j u\|_{L^p} 2^{js} \alpha_j.$$

Let ϵ be any positive real number. Lemma 1.2 page 2 ensures that for any j, a function ϕ_j exists in $\mathcal{S}$ such that

$$\|\dot{\Delta}_j u\|_{L^p} \leq \int_{\mathbb{R}^d} \dot{\Delta}_j u(x) \phi_j(x)\, dx + \frac{\epsilon 2^{-js}}{(|\alpha_j| + 1)(1 + |j|^2)}.$$

We define the function Φ_N in $Q^{-s}_{p',r'}$ by

$$\Phi_N \stackrel{\text{def}}{=} \sum_{|j| \leq N} \alpha_j 2^{js} \dot{\Delta}_j \phi_j.$$

Using Lemma 2.23, we infer that $\|\Phi_N\|_{B^{-s}_{p',r'}} \leq C$, independently of N. We then have, for any N,

$$\left\| \left(\mathbb{1}_{|j|\leq N} 2^{js} \|\dot{\Delta}_j u\|_{L^p} \right)_j \right\|_{\ell^r} \leq \langle u, \Phi_N \rangle + \epsilon.$$

The proposition is thus proved. $\square$

Finally, we consider the way that homogeneous Fourier multipliers act on Besov spaces.

Proposition 2.30. *Let σ be a smooth function on $\mathbb{R}^d \setminus \{0\}$ which is homogeneous of degree m. Then, for any $(s_k, p_k, r_k) \in \mathbb{R} \times [1,\infty]^2$ (with $k \in \{1,2\}$) such that $(s_1 - m, p_1, r_1)$ satisfies (2.17), the operator $\sigma(D)$ continuously maps $\dot{B}^{s_1}_{p_1,r_1} \cap \dot{B}^{s_2}_{p_2,r_2}$ into $\dot{B}^{s_1-m}_{p_1,r_1} \cap \dot{B}^{s_2-m}_{p_2,r_2}$.*

Proof. Lemma 2.2 guarantees that $\|\sigma(D)\dot{\Delta}_j u\|_{L^p} \leq C 2^{jm} \|\dot{\Delta}_j u\|_{L^p}$. The fact that $(s_1 - m, p_1, r_1)$ satisfies (2.17) implies that the series $(\sigma(D)\dot{\Delta}_j u)_{j\in\mathbb{Z}}$ converges in $\mathcal{S}'$ to an element of $\mathcal{S}'_h$. Lemma 2.23 then implies the proposition. $\square$

Remark 2.31. We note that this proof is very simple compared with the similar result on L^p spaces when p belongs to $]1,\infty[$. Moreover, as we shall see in the next section, Fourier multipliers do not map L^∞ into L^∞ in general. From this point of view Besov spaces are much easier to handle than classical L^p spaces or Sobolev spaces modeled on L^p.

Corollary 2.32. *Let (s_1, p_1, r_1) and (s_2, p_2, r_2) be in $\mathbb{R} \times [1,\infty]^2$. Assume that $(s_1 + 1, p_1, r_1)$ satisfies the condition (2.17). If v is a vector field with components in $\dot{B}^{s_1-1}_{p_1,r_1} \cap \dot{B}^{s_2-1}_{p_2,r_2}$ which is curl free (i.e., $\partial_j v^k = \partial_k v^j$ for any $1 \leq j, k \leq d$), then a unique function a exists in $\dot{B}^{s_1}_{p_1,r_1} \cap \dot{B}^{s_2}_{p_2,r_2}$ such that $\nabla a = v$ and*

$$C^{-1}\|a\|_{\dot{B}^{s_k}_{p_k,r_k}} \leq \|v\|_{\dot{B}^{s_k-1}_{p_k,r_k}} \leq C\|a\|_{\dot{B}^{s_k}_{p_k,r_k}} \quad \text{for} \quad k = 1, 2$$

with C a positive constant independent of v.

Proof. We define the function[2] $a \stackrel{\text{def}}{=} -(-\Delta)^{-1} \operatorname{div} v$. As the operator $(-\Delta)^{-1} \operatorname{div}$ is homogeneous of degree -1, Proposition 2.30 implies that a belongs to $\dot{B}^{s_1}_{p_1,r_1} \cap \dot{B}^{s_2}_{p_2,r_2}$ and satisfies

$$\|a\|_{\dot{B}^{s_k}_{p_k,r_k}} \leq C\|v\|_{\dot{B}^{s_k-1}_{p_k,r_k}} \quad \text{for} \quad k = 1, 2.$$

As $\operatorname{curl} v = 0$, the classical formula

---

[2] From now on, if $s \in \mathbb{R}$, then $(-\Delta)^s$ denotes the Fourier multiplier with symbol $|\xi|^{2s}$.

$$\Delta w^i = \sum_{j=1}^{d} \partial_j^2 w^i = \sum_{j=1}^{d} \partial_j(\partial_i w^j - \partial_i w^j) + \partial_i \operatorname{div} w$$

ensures that $\Delta v = \nabla \operatorname{div} v$, hence $\nabla a = v$ and $\|v\|_{B_{p_k,r_k}^{s_k-1}} \leq C\|a\|_{\dot{B}_{p_k,r_k}^{s_k}}$ for $k = 1, 2$. The uniqueness of a is obvious because $\mathcal{S}_h'$ does not contain any nonzero constant function.   □

In the case of *negative* indices of regularity, homogeneous Besov spaces may be characterized in terms of operators $\dot{S}_j$, as follows.

Proposition 2.33. *Let $s < 0$ and $1 \leq p, r \leq \infty$. Let u be a distribution in $\mathcal{S}_h'$. Then, u belongs to $\dot{B}_{p,r}^s$ if and only if*

$$(2^{js}\|\dot{S}_j u\|_{L^p})_{j \in \mathbb{Z}} \in \ell^r.$$

Moreover, for some constant C depending only on d, we have

$$C^{-|s|+1}\|u\|_{\dot{B}_{p,r}^s} \leq \left\|(2^{js}\|\dot{S}_j u\|_{L^p})_j\right\|_{\ell^r} \leq C\left(1 + \frac{1}{|s|}\right)\|u\|_{\dot{B}_{p,r}^s}.$$

Proof. We write

$$2^{js}\|\dot{\Delta}_j u\|_{L^p} \leq 2^{js}(\|\dot{S}_{j+1}u\|_{L^p} + \|\dot{S}_j u\|_{L^p})$$
$$\leq 2^{-s}2^{(j+1)s}\|\dot{S}_{j+1}u\|_{L^p} + 2^{js}\|\dot{S}_j u\|_{L^p}.$$

The left inequality is proved. To obtain the right inequality, we write

$$2^{js}\|\dot{S}_j u\|_{L^p} \leq 2^{js} \sum_{j' \leq j-1} \|\dot{\Delta}_{j'}u\|_{L^p}$$
$$\leq \sum_{j' \leq j-1} 2^{(j-j')s} 2^{j's}\|\dot{\Delta}_{j'}u\|_{L^p}.$$

As s is negative, the result follows by convolution.   □

2.4 Characterizations of Homogeneous Besov Spaces

In this section we give characterizations of Besov norms which do not require spectral localization. The first of these concerns negative indices and relies on heat flow.

Theorem 2.34. *Let s be a positive real number and $(p, r) \in [1, \infty]^2$. A constant C exists which satisfies*

$$C^{-1}\|u\|_{\dot{B}_{p,r}^{-2s}} \leq \left\|\|t^s e^{t\Delta}u\|_{L^p}\right\|_{L^r(\mathbb{R}^+, \frac{dt}{t})} \leq C\|u\|_{\dot{B}_{p,r}^{-2s}} \quad \text{for all } u \in \mathcal{S}_h'.$$

Proof. According to Lemma 2.4,

$$\|t^s \dot{\Delta}_j e^{t\Delta} u\|_{L^p} \leq Ct^s 2^{2js} e^{-ct2^{2j}} 2^{-2js} \|\dot{\Delta}_j u\|_{L^p}.$$

Using the fact that u belongs to $\mathcal{S}'_h$ and the definition of the homogeneous Besov seminorm, we have

$$\|t^s e^{t\Delta} u\|_{L^p} \leq \sum_{j \in \mathbb{Z}} \|t^s \dot{\Delta}_j e^{t\Delta} u\|_{L^p}$$

$$\leq C\|u\|_{\dot{B}^{-2s}_{p,r}} \sum_{j \in \mathbb{Z}} t^s 2^{2js} e^{-ct2^{2j}} c_{r,j},$$

where $(c_{r,j})_{j \in \mathbb{Z}}$ denotes (here and throughout this proof) a generic element of the unit sphere of $\ell^r(\mathbb{Z})$. If $r = \infty$, then the inequality readily follows from the next lemma, the proof of which is left to the reader.

Lemma 2.35. *For any positive s, we have*

$$\sup_{t>0} \sum_{j \in \mathbb{Z}} t^s 2^{2js} e^{-ct2^{2j}} < \infty.$$

If $r < \infty$, then using Hölder's inequality with the weight $2^{2js} e^{-ct2^{2j}}$ and the above lemma, we obtain

$$\int_0^\infty t^{rs} \|e^{t\Delta} u\|^r_{L^p} \frac{dt}{t} \leq C\|u\|^r_{\dot{B}^{-2s}_{p,r}} \int_0^\infty \left(\sum_{j \in \mathbb{Z}} t^s 2^{2js} e^{-ct2^{2j}} c_{r,j} \right)^r \frac{dt}{t}$$

$$\leq C\|u\|^r_{\dot{B}^{-2s}_{p,r}} \int_0^\infty \left(\sum_{j \in \mathbb{Z}} t^s 2^{2js} e^{-ct2^{2j}} \right)^{r-1} \left(\sum_{j \in \mathbb{Z}} t^s 2^{2js} e^{-ct2^{2j}} c^r_{r,j} \right) \frac{dt}{t}$$

$$\leq C\|u\|^r_{\dot{B}^{-2s}_{p,r}} \int_0^\infty \sum_{j \in \mathbb{Z}} t^s 2^{2js} e^{-ct2^{2j}} c^r_{r,j} \frac{dt}{t}.$$

Using Fubini's theorem, we infer that

$$\int_0^\infty t^{rs} \|e^{t\Delta} u\|^r_{L^p} \frac{dt}{t} \leq C\|u\|^r_{\dot{B}^{-2s}_{p,r}} \sum_{j \in \mathbb{Z}} c^r_{r,j} \int_0^\infty t^s 2^{2js} e^{-ct2^{2j}} \frac{dt}{t}$$

$$\leq C\Gamma(s)\|u\|^r_{\dot{B}^{-2s}_{p,r}} \quad \text{with} \quad \Gamma(s) \stackrel{\text{def}}{=} \int_0^\infty t^{s-1} e^{-t} \, dt.$$

To prove the other inequality, we use the following identity (which may be easily proven by taking the Fourier transform in x of both sides):

$$\dot{\Delta}_j u = \int_0^\infty t^s (-\Delta)^{s+1} e^{t\Delta} \dot{\Delta}_j u \, dt / \Gamma(s+1). \tag{2.18}$$

As $e^{t\Delta} u = e^{\frac{t}{2}\Delta} e^{\frac{t}{2}\Delta} u$, we can write, using Lemmas 2.1 and 2.4,

$$\|\dot{\Delta}_j u\|_{L^p} \leq C \int_0^\infty t^s 2^{2j(s+1)} e^{-ct2^{2j}} \|\dot{\Delta}_j e^{\frac{t}{2}\Delta} u\|_{L^p} \, dt$$

$$\leq C \int_0^\infty t^s 2^{2j(s+1)} e^{-ct2^{2j}} \|e^{t\Delta} u\|_{L^p} \, dt.$$

If $r = \infty$, then we have

$$\|\dot{\Delta}_j u\|_{L^p} \leq C \Big(\sup_{t>0} t^s \|e^{t\Delta} u\|_{L^p}\Big) \int_0^\infty 2^{2j(s+1)} e^{-ct2^{2j}} \, dt$$

$$\leq C 2^{2js} \Big(\sup_{t>0} t^s \|e^{t\Delta} u\|_{L^p}\Big).$$

If $r < \infty$, we write

$$\sum_{j \in \mathbb{Z}} 2^{-2jsr} \|\dot{\Delta}_j u\|_{L^p}^r \leq C \sum_{j \in \mathbb{Z}} 2^{2jr} \Big(\int_0^\infty t^s e^{-ct2^{2j}} \|e^{t\Delta} u\|_{L^p} \, dt\Big)^r.$$

Hölder's inequality with the weight $e^{-ct2^{2j}}$ implies that

$$\Big(\int_0^\infty t^s e^{-ct2^{2j}} \|e^{t\Delta} u\|_{L^p} \, dt\Big)^r \leq \Big(\int_0^\infty e^{-ct2^{2j}} \, dt\Big)^{r-1} \int_0^\infty t^{rs} e^{-ct2^{2j}} \|e^{t\Delta} u\|_{L^p}^r \, dt$$

$$\leq C 2^{-2j(r-1)} \int_0^\infty t^{rs} e^{-ct2^{2j}} \|e^{t\Delta} u\|_{L^p}^r \, dt.$$

Thanks to Lemma 2.35 and Fubini's theorem, we get

$$\sum_j 2^{-2jsr} \|\dot{\Delta}_j u\|_{L^p}^r \leq C \sum_{j \in \mathbb{Z}} 2^{2j} \int_0^\infty t^{rs} e^{-ct2^{2j}} \|e^{t\Delta} u\|_{L^p}^r \, dt$$

$$\leq C \int_0^\infty \Big(\sum_{j \in \mathbb{Z}} t 2^{2j} e^{-ct2^{2j}}\Big) t^{rs} \|e^{t\Delta} u\|_{L^p}^r \, \frac{dt}{t}$$

$$\leq C \int_0^\infty t^{rs} \|e^{t\Delta} u\|_{L^p}^r \, \frac{dt}{t}.$$

The theorem is thus proved. □

We will now give a characterization of Besov spaces with positive indices in terms of *finite differences*. To simplify the presentation, we only consider the case where the regularity index s is in $]0, 1[$.

Theorem 2.36. *Let s be in $]0, 1[$ and $(p, r) \in [1, \infty]^2$. A constant C exists such that, for any u in $\mathcal{S}_h'$,*

$$C^{-1} \|u\|_{\dot{B}_{p,r}^s} \leq \Big\| \frac{\|\tau_{-y} u - u\|_{L^p}}{|y|^s} \Big\|_{L^r(\mathbb{R}^d; \frac{dy}{|y|^d})} \leq C \|u\|_{\dot{B}_{p,r}^s}.$$

Proof. In order to prove the right-hand inequality, we shall bound the quantity $\|\tau_{-z}\dot{\Delta}_j u - \dot{\Delta}_j u\|_{L^p}$. Note that according to (2.7), we have

$$\dot{\Delta}_j = \sum_{|j'-j|\leq 1} \dot{\Delta}_j \dot{\Delta}_{j'}.$$

Hence, using the definition of $\dot{\Delta}_j$ and Taylor's formula, we get

$$\tau_{-y}\dot{\Delta}_j u - \dot{\Delta}_j u = \sum_{|j'-j|\leq 1} 2^{jd} \int_{\mathbb{R}^d} \left(h(2^j(x+y-z)) - h(2^j(x-z)) \right) \dot{\Delta}_{j'} u(z)\, dz,$$

$$= \sum_{|j'-j|\leq 1} 2^{jd} \sum_{\ell=1}^d 2^j y_\ell \left(\int_0^1 h_{\ell,j}(2^j\cdot, ty)\, dt \right) \star \dot{\Delta}_{j'} u \quad \text{with}$$

$$h_{\ell,j}(X,Y) \overset{\text{def}}{=} \partial_{x_\ell} h(X + 2^j Y).$$

As $\|h_{\ell,j}(\cdot, Y)\|_{L^1} = \|\partial_{x_\ell} h\|_{L^1}$ for any Y, we have

$$\|\tau_{-y}\dot{\Delta}_j u - \dot{\Delta}_j u\|_{L^p} \leq C 2^j |y| \sum_{|j-j'|\leq 1} \|\dot{\Delta}_{j'} u\|_{L^p}$$

$$\leq C c_{r,j} 2^{j(1-s)} |y| \|u\|_{\dot{B}^s_{p,r}},$$

where $(c_{r,j})_{j\in\mathbb{Z}}$ is (as throughout the proof) an element of the unit sphere of $\ell^r(\mathbb{Z})$. We also have

$$\|\tau_{-y}\dot{\Delta}_j u - \dot{\Delta}_j u\|_{L^p} \leq 2\|\dot{\Delta}_j u\|_{L^p}$$

$$\leq C c_{r,j} 2^{-js} \|u\|_{\dot{B}^s_{p,r}}.$$

We infer that for any integer j',

$$\|\tau_{-y} u - u\|_{L^p} \leq C \|u\|_{\dot{B}^s_{p,r}} \left(|y| \sum_{j\leq j'} c_{r,j} 2^{j(1-s)} + \sum_{j>j'} c_{r,j} 2^{-js} \right).$$

We now choose $j' = j_y$ such that $\dfrac{1}{|y|} \leq 2^{j_y} < 2\dfrac{1}{|y|}$. If $r = \infty$, then for any y in $\mathbb{R}^d$, we have

$$\|\tau_{-y} u - u\|_{L^p} \leq C |y|^s \|u\|_{\dot{B}^s_{p,r}}.$$

If $r < \infty$, we write

$$\left\| \frac{\|\tau_{-y} u - u\|_{L^p}}{|y|^s} \right\|^r_{L^r(\mathbb{R}^d;\frac{dy}{|y|^d})} \leq C 2^r \|u\|^r_{\dot{B}^s_{p,r}} (I_1 + I_2) \quad \text{with}$$

$$I_1 \overset{\text{def}}{=} \int_{\mathbb{R}^d} \left(\sum_{j\leq j_y} c_{r,j} 2^{j(1-s)} \right)^r |y|^{-d+r(1-s)}\, dy \quad \text{and}$$

$$I_2 \overset{\text{def}}{=} \int_{\mathbb{R}^d} \left(\sum_{j>j_y} c_{r,j} 2^{-js} \right)^r |y|^{-d-rs}\, dy.$$

Hölder's inequality with the weight $2^{j(1-s)}$ and the definition of j_y together imply that

$$\left(\sum_{j \leq j_y} c_{r,j} 2^{j(1-s)}\right)^r \leq \left(\sum_{j \leq j_y} 2^{j(1-s)}\right)^{r-1} \sum_{j \leq j_y} c_{r,j}^r 2^{j(1-s)}$$

$$\leq C|y|^{-(1-s)(r-1)} \sum_{j \leq j_y} c_{r,j}^r 2^{j(1-s)}.$$

By Fubini's theorem, we deduce that

$$I_1 \leq C \sum_j \left(\int_{B(0,2^{-j+1})} |y|^{-d+1-s}\, dy\right) 2^{j(1-s)} c_{r,j}^r \leq C.$$

Estimating I_2 is strictly analogous.

We will now prove the reverse inequality. As the mean value of the function h is 0, we can write

$$\dot{\Delta}_j u(x) = 2^{jd} \int_{\mathbb{R}^d} h(2^j y) \tau_y u(x)\, dy$$

$$= 2^{jd} \int h(2^j y)(\tau_y u(x) - u(x))\, dy.$$

When $r = \infty$, we have

$$2^{js}\|\dot{\Delta}_j u\|_{L^p} \leq 2^{jd} \int_{\mathbb{R}^d} 2^{js} |h(2^j y)|\, \|\tau_y u - u\|_{L^p}\, dy$$

$$\leq 2^{jd} \int_{\mathbb{R}^d} 2^{js} |y|^s |h(2^j y)|\, dy \sup_{y \in \mathbb{R}^d} \frac{\|\tau_y u - u\|_{L^p}}{|y|^s}$$

$$\leq C \sup_{y \in \mathbb{R}^d} \frac{\|\tau_y u - u\|_{L^p}}{|y|^s}.$$

When $r < \infty$, we write

$$\sum_j 2^{jsr}\|\dot{\Delta}_j u\|_{L^p}^r \leq 2^r(\Sigma_1 + \Sigma_2) \quad \text{with}$$

$$\Sigma_1 \overset{\text{def}}{=} \sum_j 2^{jsr} \left(\int_{2^j|y| \leq 1} 2^{jd} |h(2^j y)|\, \|\tau_y u - u\|_{L^p}\, dy\right)^r \quad \text{and}$$

$$\Sigma_2 \overset{\text{def}}{=} \sum_j 2^{jsr} \left(\int_{2^j|y| \geq 1} 2^{jd} |h(2^j y)|\, \|\tau_y u - u\|_{L^p}\, dy\right)^r.$$

Hölder's inequality implies that

$$\left(\int_{2^j |y| \le 1} 2^{jd} |h(2^j y)| \, \|\tau_y u - u\|_{L^p} \, dy \right)^r \le \left(\int_{2^j |y| \le 1} 2^{jdr'} |h(2^j y)|^{r'} \, dy \right)^{r-1}$$

$$\times \int_{2^j |y| \le 1} \|\tau_y u - u\|_{L^p}^r \, dy$$

$$\le C 2^{jd} \int_{2^j |y| \le 1} \|\tau_y u - u\|_{L^p}^r \, dy.$$

Using Fubini's theorem, we get that

$$\Sigma_1 \le C \int_{\mathbb{R}^d} \left(\sum_{j / 2^j |y| \le 1} 2^{j(rs+d)} \right) \|\tau_y u - u\|_{L^p}^r \, dy$$

$$\le C \int_{\mathbb{R}^d} \frac{\|\tau_y u - u\|_{L^p}^r}{|y|^{rs}} \frac{dy}{|y|^d}.$$

Next, note that applying Hölder's inequality with the measure $|y|^{-d} \, dy$ enables us to bound the general term Σ_2^j of Σ_2 as follows:

$$2^{-jsr} \Sigma_2^j \le 2^{-jr} \left(\int_{2^j |y| \ge 1} |2^j y|^{d+1} |h(2^j y)| \frac{\|\tau_y u - u\|_{L^p}}{|y|} \frac{dy}{|y|^d} \right)^r$$

$$\le 2^{-jr} \int_{2^j |y| \ge 1} \frac{\|\tau_y u - u\|_{L^p}^r}{|y|^r} \frac{dy}{|y|^d}.$$

Using Fubini's theorem and the fact that $s < 1$, we then infer that

$$\Sigma_2 \le C \int_{\mathbb{R}^d} \left(\sum_{j / 2^j |y| \ge 1} 2^{-jr(1-s)} \right) \frac{\|\tau_y u - u\|_{L^p}^r}{|y|^r} \frac{dy}{|y|^d}$$

$$\le C \int_{\mathbb{R}^d} \frac{\|\tau_y u - u\|_{L^p}^r}{|y|^{rs}} \frac{dy}{|y|^d}.$$

The theorem is thus proved.   □

In the limit case $s = 1$, the characterization given in Theorem 2.36 fails. We then have to use finite differences of order two.

Theorem 2.37. Let (p, r) be in $[1, \infty]^2$. A constant C exists such that for any u in $\mathcal{S}'_h$,

$$C^{-1} \|u\|_{\dot{B}^1_{p,r}} \le \left\| \frac{\|\tau_{-y} u + \tau_y u - 2u\|_{L^p}}{|y|} \right\|_{L^r(\mathbb{R}^d; \frac{dy}{|y|^d})} \le C \|u\|_{\dot{B}^1_{p,r}}.$$

Remark 2.38. Applying the above theorem in the case where $p = r = \infty$ shows that the space $\dot{B}^1_{\infty,\infty}$ coincides with the Zygmund class of functions u such that

$$|u(x + y) + u(x - y) - 2u(x)| \le C |y|.$$

Proof of Theorem 2.37. Again using the fact that $\dot\Delta_j = \sum_{|j'-j|\le 1} \dot\Delta_j \dot\Delta_{j'}$, we can write

$$\tau_{-y}\dot\Delta_j u + \tau_y \dot\Delta_j u - 2\dot\Delta_j u = 2^{jd} \sum_{|\alpha|=2} \sum_{|j'-j|\le 1} 2^{2j} y^\alpha \left(\int_0^1 (1-t) h_{\alpha,j}(2^j\cdot, ty)\, dt \right) \star \dot\Delta_{j'} u$$

with $h_{\alpha,j}(X,Y) \overset{\text{def}}{=} \partial^\alpha h(X + 2^j Y)$.

As $\|h_{\alpha,j}(\cdot, Y)\|_{L^1} = \|\partial^\alpha h\|_{L^1}$ for any Y, we have

$$\|\tau_{-y}\dot\Delta_j u + \tau_y \dot\Delta_j u - 2\dot\Delta_j u\|_{L^p} \le C 2^{2j} |y|^2 \sum_{|j-j'|\le 1} \|\dot\Delta_{j'} u\|_{L^p}$$
$$\le C c_{r,j} 2^j |y|^2 \|u\|_{\dot B^1_{p,r}},$$

where $(c_{r,j})_{j\in\mathbb{Z}}$ stands for an element of the unit sphere of $\ell^r(\mathbb{Z})$. We also have

$$\|\tau_{-y}\dot\Delta_j u + \tau_y \dot\Delta_j u - 2\dot\Delta_j u\|_{L^p} \le 4\|\dot\Delta_j u\|_{L^p}$$
$$\le C c_{r,j} 2^{-j} \|u\|_{\dot B^1_{p,r}}.$$

We infer that for any integer j',

$$\|\tau_{-y} u + \tau_y u - 2u\|_{L^p} \le C\|u\|_{\dot B^1_{p,r}} \left(|y|^2 \sum_{j\le j'} c_{r,j} 2^j + \sum_{j>j'} c_{r,j} 2^{-j} \right).$$

The conclusion is strictly analogous to the case where $s \in\ ]0,1[$.

We will now prove the other inequality. Because h is a radial function with mean value 0, we can write

$$\dot\Delta_j u(x) = \frac{1}{2} 2^{jd} \int_{\mathbb{R}^d} h(2^j y)(\tau_y u + \tau_{-y} u)(x)\, dy$$
$$= \frac{1}{2} 2^{jd} \int h(2^j y)(\tau_y u(x) + \tau_{-y} u(x) - u(x))\, dy,$$

and from this point on, we can mimic the proof of Theorem 2.36.

2.5 Besov Spaces, Lebesgue Spaces, and Refined Inequalities

In this section, we compare homogeneous Besov spaces with Lebesgue spaces. We start with an easy (but most useful) result pertaining to Besov spaces with third index 1.

Proposition 2.39. *For any (p,q) in $[1,\infty]^2$ such that $p \le q$, the space $\dot{B}_{p,1}^{\frac{d}{p}-\frac{d}{q}}$ is continuously embedded in L^q. In addition, if p is finite, then $\dot{B}_{p,1}^{\frac{d}{p}}$ is continuously embedded in the space $\mathcal{C}_0$ of continuous functions vanishing at infinity. Finally, for all $q \in [1,\infty]$, the space L^q is continuously embedded in the space $\dot{B}_{q,\infty}^0$, and the space $\mathcal{M}$ of bounded measures on $\mathbb{R}^d$ is continuously embedded in $\dot{B}_{1,\infty}^0$.*

Proof. Let $u \in \dot{B}_{p,1}^{\frac{d}{p}-\frac{d}{q}}$. Because $\dot{B}_{p,1}^{\frac{d}{p}-\frac{d}{q}} \subset \mathcal{S}_h'$, we may write

$$u = \sum_j \dot{\Delta}_j u.$$

Now, according to Bernstein's lemma, we have

$$\|\dot{\Delta}_j u\|_{L^q} \le C 2^{j(\frac{d}{p}-\frac{d}{q})} \|\dot{\Delta}_j u\|_{L^p},$$

so the above series converges in L^q. This yields the first part of the statement. If p is finite, then the space $\mathcal{S}_0$ is dense in $\dot{B}_{p,1}^{\frac{d}{p}}$. This ensures that functions of $\dot{B}_{p,1}^{\frac{d}{p}}$ decay to 0 at infinity. The last part of the statement is easy to prove: It suffices to use the fact that, by definition, $\dot{\Delta}_j u = 2^{jd} h(2^j \cdot) * u$. Hence, Young's inequality (or Fubini's theorem, in the case where u is a bounded measure) gives the result. $\square$

We now compare homogeneous Besov spaces with regularity index 0 and third index 2 to Lebesgue spaces.

Theorem 2.40. *For any p in $[2,\infty[$, $\dot{B}_{p,2}^0$ is continuously included in L^p and $L^{p'}$ is continuously included in $\dot{B}_{p',2}^0$.*

Proof. Arguing by density, we can assume with no loss of generality that u belongs to $\mathcal{S}_0$ (see Proposition 2.27). Therefore, writing $F_p(x) = |x|^p$, we can rewrite $\|u\|_{L^p}^p$ as a telescopic series:

$$\|u\|_{L^p}^p = \sum_{j\in\mathbb{Z}} F_p(\dot{S}_{j+1}u) - F_p(\dot{S}_j u), \quad \text{and hence}$$

$$\|u\|_{L^p}^p = \sum_j \langle \dot{\Delta}_j u, m_j \rangle \quad \text{with} \quad m_j(x) \stackrel{\text{def}}{=} \int_0^1 F_p' \left(\dot{S}_j u(x) + t \dot{\Delta}_j u(x) \right) dt.$$

Using the Fourier–Plancherel formula and denoting by $\widetilde{\Delta}_j$ the convolution operator in terms of the inverse Fourier transform of $\widetilde{\varphi}(2^{-j}\cdot)$, where $\widetilde{\varphi}$ is in $\mathcal{D}(\mathbb{R}^d \setminus \{0\})$ with value 1 near the support of φ, we can write

$$\langle \dot{\Delta}_j u, m_j \rangle = \langle \dot{\Delta}_j u, \widetilde{\Delta}_j m_j \rangle.$$

By Lemma 2.1, we infer that

$$\|\widetilde{\Delta}_j m_j\|_{L^{p'}} \le C2^{-j} \sup_{1 \le \ell \le d} \|\partial_\ell m_j\|_{L^{p'}}. \tag{2.19}$$

The chain rule and Hölder's inequality imply that

$$
\begin{aligned}
\|\partial_\ell m_j\|_{L^{p'}} &\le \int_0^1 \left\| \partial_\ell(\dot{S}_j u + t\dot{\Delta}_j u) F_p''(\dot{S}_j + t\dot{\Delta}_j u) \right\|_{L^{p'}} dt \\
&\le \int_0^1 \|\partial_\ell(\dot{S}_j u + t\dot{\Delta}_j u)\|_{L^p} \|F_p''(\dot{S}_j u + t\dot{\Delta}_j u)\|_{L^{\frac{p}{p-2}}} dt.
\end{aligned}
$$

As $F_p''(x) = p(p-1)|x|^{p-2}$, we immediately get that

$$\forall t \in [0,1], \ \|F_p''(\dot{S}_j u + t\dot{\Delta}_j u)\|_{L^{\frac{p}{p-2}}} \le p(p-1)\|\dot{S}_j u + t\dot{\Delta}_j u\|_{L^p}^{p-2}.$$

Using Lemma 2.1, we infer that for all $t \in [0,1]$,

$$\|F_p''(S_j u + t\Delta_j u)\|_{L^{\frac{p}{p-2}}} \le C^p p(p-1)\|u\|_{L^p}^{p-2}. \tag{2.20}$$

Now, by the definition of $\dot{S}_j$, Lemma 2.1, and Young's inequality for series, we get

$$
\begin{aligned}
\|\partial_\ell(\dot{S}_j u + t\dot{\Delta}_j u)\|_{L^p} &\le \sum_{k \le j} \|\partial_\ell \dot{\Delta}_k u\|_{L^p} \\
&\le 2^j \sum_{k \le j} 2^{k-j} \|\dot{\Delta}_k u\|_{L^p} \\
&\le Cc_j 2^j \|u\|_{\dot{B}_{p,2}^0} \quad \text{with} \quad \sum_j c_j^2 = 1.
\end{aligned}
$$

Combining (2.19) and (2.20), we deduce that

$$\|\widetilde{\Delta}_j m_j\|_{L^{p'}} \le C^p p(p-1)c_j \|u\|_{L^p}^{p-2} \|u\|_{\dot{B}_{p,2}^0} \quad \text{with} \quad \sum_j c_j^2 = 1.$$

As we have $\|u\|_{L^p}^p = \sum_j \langle \dot{\Delta}_j u, \widetilde{\Delta}_j m_j \rangle$, we infer that

$$\|u\|_{L^p}^2 \le C^p p(p-1)\|u\|_{\dot{B}_{p,2}^0} \sum_j c_j \|\dot{\Delta}_j u\|_{L^p} \le C^p p(p-1)\|u\|_{\dot{B}_{p,2}^0}^2. \tag{2.21}$$

This concludes the proof that $\dot{B}_{p,2}^0 \hookrightarrow L^p$. In order to prove the dual result, consider u in $L^{p'}$. For any $\phi \in \mathcal{S}$ such that $\|\phi\|_{\dot{B}_{p,2}^0} \le 1$, we have, thanks to (2.21),

$$|\langle u, \phi \rangle| \le \|u\|_{L^{p'}} \|\phi\|_{L^p} \le C\|u\|_{L^{p'}}.$$

Use of Proposition 2.29 then completes the proof.   □

Theorem 2.41. *For any p in $[1,2]$, the space $\dot{B}^0_{p,p}$ is continuously included in L^p, and $L^{p'}$ is continuously included in $\dot{B}^0_{p',p'}$.*

Proof. We first observe that $\dot{B}^0_{1,1}$ is continuously included in L^1, and $\dot{B}^0_{2,2}$ is equal to L^2. We shall then use a complex interpolation argument to prove that for any $p \in [1,2]$, $\dot{B}^0_{p,p}$ is continuously included in L^p. Consider $f \in \dot{B}^0_{p,p}$ and $\varphi \in L^{p'}$. As in the proof of Lemma 1.11 page 11, we consider a complex number z in the strip S of complex numbers whose real parts are between 0 and 1, and we define, for $\widetilde{\varphi} \in \mathcal{D}(\mathbb{R}^d \setminus \{0\})$ with value 1 near the support of φ,

$$f_z \overset{\text{def}}{=} \sum_{j \in \mathbb{Z}} \widetilde{\varphi}(2^{-j}D)\left(\frac{\Delta_j f}{|\Delta_j f|}|\Delta_j f|^{p\left(1-z+\frac{z}{2}\right)}\right),$$

$$\varphi_z(x) \overset{\text{def}}{=} \frac{\varphi}{|\varphi|}|\varphi|^{\frac{z}{2}p'} \quad \text{and} \quad F(z) \overset{\text{def}}{=} \int_{\mathbb{R}^d} f_z(x)\varphi_z(x)\,dx.$$

Note that $f_\theta = f$ and $\varphi_\theta = \varphi$ if $\theta = 2/p'$. It can be checked that F is holomorphic on S and is continuous and bounded on the closure of S. From the Phragmén–Lindelöf principle, we infer that

$$F(\theta) \leq M_0^{1-\theta} M_1^\theta \quad \text{with} \quad M_j \overset{\text{def}}{=} \sup_{t \in \mathbb{R}} |F(j+it)|. \tag{2.22}$$

We now have, for any $t \in \mathbb{R}$,

$$\begin{aligned}
\|f_{it}\|_{L^1} &\leq \sum_{j \in \mathbb{Z}} \left\| \widetilde{\varphi}(2^{-j}D)\left(\frac{\Delta_j f}{|\Delta_j f|}|\Delta_j f|^{p\left(1-it+\frac{it}{2}\right)}\right) \right\|_{L^1} \\
&\leq C \sum_{j \in \mathbb{Z}} \||\Delta_j f|^p\|_{L^1} \\
&\leq C \sum_{j \in \mathbb{Z}} \|\Delta_j f^p\|_{L^p}^p \\
&\leq C \|f\|_{\dot{B}^0_{p,p}}^p.
\end{aligned} \tag{2.23}$$

In addition, using the "almost orthogonality" of the terms of the series defining f_z, we infer that

$$\begin{aligned}
\|f_{1+it}\|_{L^2}^2 &\leq C \sum_{j \in \mathbb{Z}} \||\Delta_j f|^{\frac{p}{2}}\|_{L^2} \\
&\leq C \sum_{j \in \mathbb{Z}} \|\Delta_j f\|_{L^p}^p \\
&\leq C \|f\|_{\dot{B}^0_{p,p}}^p.
\end{aligned} \tag{2.24}$$

Moreover, $|\varphi_{it}(x)| = 1$ and $|\varphi_{1+it}(x)| = |\varphi(x)|^{\frac{p'}{2}}$. Thus,

$$M_0 \leq C\|f\|_{\dot{B}^0_{p,p}}^p \quad \text{and} \quad M_1 \leq C\|f\|_{\dot{B}^0_{p,p}}^{\frac{p}{2}} \|\varphi\|_{L^{p'}}^{\frac{p'}{2}}.$$

Using (2.22), we infer that

$$\int_{\mathbb{R}^d} f(x)\varphi(x)dx = F(\theta) \leq C\|f\|_{\dot{B}^0_{p,p}} \|\varphi\|_{L^{p'}},$$

and the first result is proved. That $L^{p'}$ embeds continuously in $B^0_{p',p'}$ follows by duality (see Proposition 2.29). □

We now present a generalization of the refined Sobolev embedding stated in Theorem 1.43 page 32.

Theorem 2.42. *Let $1 \leq q < p < \infty$ and α be a positive real number. A constant C exists such that*

$$\|f\|_{L^p} \leq C\|f\|_{\dot{B}^{-\alpha}_{\infty,\infty}}^{1-\theta} \|f\|_{\dot{B}^\beta_{q,q}}^\theta \quad \text{with} \quad \beta = \alpha\left(\frac{p}{q} - 1\right) \quad \text{and} \quad \theta = \frac{q}{p}.$$

Proof. The proof follows along the lines of that of Theorem 1.38, which turns out to be a particular case (take $q = 2$ and $\alpha = d/2 - \beta$). As usual we may assume without loss of generality that $\|f\|_{\dot{B}^{-\alpha}_{\infty,\infty}} = 1$. We write

$$\|f\|_{L^p}^p = p\int_0^\infty \lambda^{p-1}\mu(|f| > \lambda)d\lambda \quad \text{and} \quad f = \dot{S}_j f + (\mathrm{Id} - \dot{S}_j)f.$$

According to Proposition 2.33 we have $\|\dot{S}_j f\|_{L^\infty} \leq C2^{j\alpha}\|f\|_{\dot{B}^{-\alpha}_{\infty,\infty}}$. As

$$\{|f| > \lambda\} \subset \{|\dot{S}_j f| > \lambda/2\} \cup \{|(\mathrm{Id} - \dot{S}_j)f| > \lambda/2\},$$

choosing j_λ in $\mathbb{Z}$ such that

$$\frac{1}{2}\left(\frac{\lambda}{2C}\right)^{\frac{1}{\alpha}} < 2^{j_\lambda} \leq \left(\frac{\lambda}{2C}\right)^{\frac{1}{\alpha}} \tag{2.25}$$

guarantees that $\{|f| > \lambda\} \subset \left\{|(\mathrm{Id} - \dot{S}_{j_\lambda})f| > \lambda/2\right\}$. By the Bienaymé–Chebyshev inequality, we then have

$$\|f\|_{L^p}^p \leq p\int_0^\infty \lambda^{p-1}\mu\left(|(\mathrm{Id} - \dot{S}_{j_\lambda})f| > \lambda/2\right)d\lambda$$

$$\leq p\int_0^\infty \lambda^{p-q-1}\|(\mathrm{Id} - \dot{S}_{j_\lambda})f\|_{L^q}^q \, d\lambda.$$

We now estimate $\|(\mathrm{Id} - \dot{S}_{j_\lambda})f\|_{L^q}$. By the definition of $\|\cdot\|_{\dot{B}^\beta_{q,q}}$, we have

$$\|(\mathrm{Id} - \dot{S}_{j_\lambda})f\|_{L^q} \leq \sum_{j \geq j_\lambda} \|\dot{\Delta}_j f\|_{L^q}$$

$$\leq \sum_{j \geq j_\lambda} 2^{-j\beta}2^{j\beta}\|\dot{\Delta}_j f\|_{L^q}$$

$$\leq C\|f\|_{\dot{B}^\beta_{q,q}} \sum_{j \geq j_\lambda} 2^{-j\beta}c_j \quad \text{with} \quad \|(c_j)\|_{\ell^q} = 1.$$

We thus get

$$\|f\|_{L^p}^p \leq C\|f\|_{\dot{B}_{q,q}^\beta}^q \int_0^\infty \lambda^{p-q-1} \left(\sum_{j\geq j_\lambda} 2^{-j\beta} c_j\right)^q d\lambda.$$

Hölder's inequality with the weight $2^{-j\beta}$ and the definition (2.25) of j_λ together give

$$\left(\sum_{j\geq j_\lambda} 2^{-j\beta} c_j\right)^q \leq \left(\sum_{j\geq j_\lambda} 2^{-j\beta}\right)^{q-1} \sum_{j\geq j_\lambda} 2^{-j\beta} c_j^q$$

$$\leq C 2^{-j_\lambda \beta (q-1)} \sum_{j\geq j_\lambda} 2^{-j\beta} c_j^q$$

$$\leq C\lambda^{-(q-1)\frac{\beta}{\alpha}} \sum_{j\geq j_\lambda} 2^{-j\beta} c_j^q.$$

Hence, it turns out that

$$\|f\|_{L^p}^p \leq C\|f\|_{\dot{B}_{q,q}^\beta}^q \int_0^\infty \left(\sum_j 2^{-j\beta} 1_{j\geq j_\lambda} c_j^q\right) \lambda^{p-q-(q-1)\frac{\beta}{\alpha}-1} d\lambda.$$

Using (2.25) and Fubini's theorem, we end up with

$$\|f\|_{L^p}^p \leq C\|f\|_{\dot{B}_{q,q}^\beta}^q \sum_j 2^{-j\beta} c_j^q \int_0^{2C2^{(j+1)\alpha}} \lambda^{p-q-(q-1)\frac{\beta}{\alpha}-1} d\lambda.$$

Because $p - q - 1 - (q-1)\beta/\alpha = p/q - 1$ is positive, we thus obtain

$$\|f\|_{L^p}^p \leq C\|f\|_{\dot{B}_{q,q}^\beta}^q \sum_j c_j^q 2^{j\left(\alpha\left(\frac{p-q}{q}\right)-\beta\right)}.$$

As $\beta = \alpha\left(\dfrac{p}{q}-1\right)$ and $\|(c_j)\|_{\ell^q} = 1$, we get $\|f\|_{L^p}^p \leq C\|f\|_{\dot{B}_{q,q}^\beta}^q$, and the theorem is proved. □

We now state the analog of the above refined inequalities in the context of Sobolev spaces.

Theorem 2.43. *Let q be in $]1,\infty[$ and s in the interval $]0, d/q[$. A constant C then exists such that*

$$\|u\|_{L^p} \leq C\|u\|_{\dot{B}_{\infty,\infty}^{1-\frac{qs}{d}}}^{1-\frac{qs}{d}} \|u\|_{\dot{W}_q^s}^{\frac{qs}{d}} \quad \text{with} \quad \|u\|_{\dot{W}_q^s} \overset{def}{=} \|(-\Delta)^{\frac{s}{2}} u\|_{L^q}.$$

Proof. We decompose u into low and high frequencies:

$$u = \dot{S}_j u + (\mathrm{Id} - \dot{S}_j)u.$$

Assume that $\|u\|_{\dot{B}^{1-\frac{qs}{d}}_{\infty,\infty}} = 1$. Using the definition of $\|u\|_{\dot{B}_{\infty,\infty}}$, we see that

$$\|\dot{S}_j u\|_{L^\infty} \leq C2^{j(dq-s)}. \tag{2.26}$$

In order to study the high-frequency part, we note that for any smooth homogeneous function a of degree m, we have

$$\dot{\Delta}_j a(D)u = 2^{jm} 2^{jd} h_a(2^j \cdot) \star u \quad \text{with} \quad h_a \overset{\text{def}}{=} \mathcal{F}^{-1}(\varphi a).$$

By Proposition 1.16 page 15 and the remark that follows, we infer that a constant (depending, of course, on a) exists such that for any $j \in \mathbb{Z}$, we have

$$|\dot{\Delta}_j u(x)| \leq C2^{jm}(Mu)(x), \tag{2.27}$$

where Mu denotes the maximal function of u. Thus, we have, for any j in $\mathbb{Z}$,

$$
\begin{aligned}
|(\mathrm{Id} - \dot{S}_j)u(x)| &\leq \sum_{j' \geq j} |\dot{\Delta}_{j'}(-\Delta)^{-\frac{s}{2}}(-\Delta)^{\frac{s}{2}} u(x)| \\
&\leq C\Big(\sum_{j' \geq j} 2^{-j's}\Big)(M(-\Delta)^{\frac{s}{2}} u)(x) \\
&\leq C2^{-js}(M(-\Delta)^{\frac{s}{2}} u)(x).
\end{aligned}
$$

Together with (2.26), this gives, for any $j \in \mathbb{Z}$ and $x \in \mathbb{R}^d$, that

$$|u(x)| \leq C2^{j(dq-s)} + C2^{-js}(M(-\Delta)^{\frac{s}{2}} u)(x).$$

Choosing $2^j \sim (M(-\Delta)^{\frac{s}{2}} u(x))^{\frac{q}{d}}$ then gives

$$|u(x)| \leq C(M(-\Delta)^{\frac{s}{2}} u)(x)^{1-\frac{sq}{d}}.$$

Because the maximal operator maps L^q into L^q continuously (see Theorem 1.14 page 13), the proof is complete. $\qquad\square$

Finally, we establish the so-called *Gagliardo–Nirenberg inequalities*.

Theorem 2.44. *Let (q, r) be in $]1, \infty]^2$ and (σ, s) be in $]0, \infty[^2$ with $\sigma < s$. A constant C exists such that*

$$\|u\|_{\dot{W}^{\sigma}_p} \leq C\|u\|^{\theta}_{L^q}\|u\|^{1-\theta}_{\dot{W}^s_r} \quad \text{with} \quad \frac{1}{p} = \frac{\theta}{q} + \frac{1-\theta}{r} \quad \text{and} \quad \theta = 1 - \frac{\sigma}{s}.$$

Proof. As usual, we decompose u into low and high frequencies:

$$u = \dot{S}_j u + (\mathrm{Id} - \dot{S}_j)u.$$

For the low-frequency part, using (2.27), we may write

$$\begin{aligned} |\dot{S}_j(-\Delta)^{\frac{\sigma}{2}} u(x)| &\leq \sum_{j'<j} |\Delta_{j'}(-\Delta)^{\frac{\sigma}{2}} u(x)| \\ &\leq C\left(\sum_{j'<j} 2^{j'\sigma}\right)(Mu)(x) \\ &\leq C2^{j\sigma}(Mu)(x). \end{aligned} \tag{2.28}$$

For the high-frequency part, again using (2.27), we get

$$\begin{aligned} |(\mathrm{Id}-\dot{S}_j)(-\Delta)^{\frac{\sigma}{2}} u(x)| &\leq \sum_{j'\geq j} |\dot{\Delta}_{j'}(-\Delta)^{\frac{\sigma}{2}-\frac{s}{2}}(-\Delta)^{\frac{s}{2}} u(x)| \\ &\leq C\left(\sum_{j'\geq j} 2^{-j'(s-\sigma)}\right)(M(-\Delta)^{\frac{s}{2}} u)(x) \\ &\leq C2^{-j(s-\sigma)}(M(-\Delta)^{\frac{s}{2}} u)(x). \end{aligned}$$

Together with (2.28), this implies that for any integer $j \in \mathbb{Z}$ and any x in $\mathbb{R}^d$,

$$|(-\Delta)^{\frac{\sigma}{2}} u(x)| \leq C2^{j\sigma}(Mu)(x) + C2^{-j(s-\sigma)}(M(-\Delta)^{\frac{s}{2}} u)(x).$$

Choosing j such that

$$2^j \approx \left(\frac{(M(-\Delta)^{\frac{s}{2}} u)(x)}{(Mu)(x)}\right)^{\frac{1}{s}},$$

we infer that

$$|(-\Delta)^{\frac{\sigma}{2}} u(x)| \leq C(Mu)(x)^{1-\frac{\sigma}{s}} (M(-\Delta)^{\frac{s}{2}} u)(x))^{\frac{\sigma}{s}},$$

from which it follows, by virtue of Hölder's inequality, that

$$\|u\|_{\dot{W}_p^\sigma} \leq C\|Mu\|_{L^q}^{1-\frac{\sigma}{s}} \|M(-\Delta)^{\frac{s}{2}} u\|_{L^r}^{\frac{\sigma}{s}}.$$

As $q > 1$ and $r > 1$, applying Theorem 1.14 page 13 completes the proof. $\square$

2.6 Homogeneous Paradifferential Calculus

In this section, we study the way that the product acts on Besov spaces.

2.6.1 Homogeneous Bony Decomposition

Let u and v be tempered distributions in $\mathcal{S}_h'$. We have

$$u = \sum_{j'} \dot{\Delta}_{j'} u \quad \text{and} \quad v = \sum_j \dot{\Delta}_j v,$$

hence, at least formally,

$$uv = \sum_{j',j} \dot{\Delta}_{j'} u \, \dot{\Delta}_j v.$$

Paradifferential calculus is a mathematical tool for splitting the above sum into three parts:

- The first part concerns the indices (j', j) for which the size of Supp $\mathcal{F}(\dot{\Delta}_{j'} u)$ is small compared to the size of Supp $\mathcal{F}(\dot{\Delta}_j v)$ (i.e., $j' \leq j - N_0$ for some suitable positive integer N_0).
- The second part contains the indices corresponding to those frequencies of u which are large compared with the frequencies of v (i.e., $j' \geq j + N_0$).
- In the last part we keep the indices (j, j') for which Supp $\mathcal{F}(\dot{\Delta}_{j'} u)$ and Supp $\mathcal{F}(\dot{\Delta}_j u)$ have comparable sizes (i.e., $|j - j'| \leq N_0$).

The suitable choice for N_0 depends on the assumptions made on the support of the function φ used in the definition of the dyadic blocks.

In what follows, we shall always assume that φ has been chosen according to Definition 2.10 so that taking $N_0 = 1$ will be appropriate. This leads to the following definition.

Definition 2.45. *The homogeneous paraproduct of v by u is defined as follows:*

$$\dot{T}_u v \stackrel{def}{=} \sum_j \dot{S}_{j-1} u \, \dot{\Delta}_j v.$$

The homogeneous remainder of u and v is defined by

$$\dot{R}(u, v) = \sum_{|k-j| \leq 1} \dot{\Delta}_k u \, \dot{\Delta}_j v.$$

Remark 2.46. It can be checked that $\dot{T}_u v$ makes sense in $\mathcal{S}'$ whenever u and v are in $\mathcal{S}'_h$, and that $\dot{T} : (u, v) \mapsto \dot{T}_u v$ is a bilinear operator. Of course, the remainder operator $\dot{R} : (u, v) \mapsto \dot{R}(u, v)$, when restricted to sufficiently smooth distributions, is also bilinear.

The main motivation for using the operators $\dot{T}$ and $\dot{R}$ is that, at least formally, the following so-called *Bony decomposition* holds true:

$$uv = \dot{T}_u v + \dot{T}_v u + \dot{R}(u, v). \tag{2.29}$$

So, in order to understand how the product operates in Besov spaces, it suffices to investigate the continuity properties of the operators $\dot{T}$ and $\dot{R}$.

To simplify the presentation, it will be understood from now on that whenever the expressions $\dot{T}_u v$ or $\dot{R}(u, v)$ appear in the text, the series with general terms

$$\dot{S}_{j-1} \dot{\Delta}_j v \quad \text{or} \quad \sum_{|\nu| \leq 1} \dot{\Delta}_j u \, \dot{\Delta}_{j-\nu} v$$

converges to some tempered distribution which belongs to $\mathcal{S}'_h$.

We can now state our main result concerning continuity of the homogeneous paraproduct operator $\dot{T}$.

Theorem 2.47. *There exists a constant C such that for any real number s and any (p, r) in $[1, \infty]^2$, we have, for any (u, v) in $L^\infty \times \dot{B}^s_{p,r}$,*

$$\|\dot{T}_u v\|_{\dot{B}^s_{p,r}} \leq C^{1+|s|} \|u\|_{L^\infty} \|v\|_{\dot{B}^s_{p,r}}.$$

Moreover, for any (s, t) in $\mathbb{R} \times \,]-\infty, 0[$ and any (p, r_1, r_2) in $[1, \infty]^3$, we have, for any $(u, v) \in \dot{B}^t_{\infty,r_1} \times \dot{B}^s_{p,r_2}$,

$$\|\dot{T}_u v\|_{\dot{B}^{s+t}_{p,r}} \leq \frac{C^{1+|s+t|}}{-t} \|u\|_{\dot{B}^t_{\infty,r_1}} \|v\|_{\dot{B}^s_{p,r_2}} \quad with \quad \frac{1}{r} \overset{def}{=} \min\left\{1, \frac{1}{r_1} + \frac{1}{r_2}\right\}.$$

Remark 2.48. Thanks to Lemma 2.23 and the remark that follows it, the hypothesis of convergence is satisfied whenever (s, p, r) or $(s + t, p, r)$ satisfies (2.17).

Proof of Theorem 2.47. According to (2.9), $\mathcal{F}(\dot{S}_{j-1}u \dot{\Delta}_j v)$ is supported in $2^j \widetilde{\mathcal{C}}$. Therefore, we are left with proving an appropriate estimate for $\|\dot{S}_{j-1}u \dot{\Delta}_j v\|_{L^p}$. Lemma 2.1 and Proposition 2.33 tell us that for any $j \in \mathbb{Z}$ and $t < 0$,

$$\|\dot{S}_{j-1}u\|_{L^\infty} \leq C\|u\|_{L^\infty} \quad and \quad \|\dot{S}_{j-1}u\|_{L^\infty} \leq \frac{C}{-t} c_{j,r_1} 2^{-jt} \|u\|_{\dot{B}^t_{\infty,r_1}}, \quad (2.30)$$

where $(c_{j,r_1})_{j \in \mathbb{Z}}$ denotes an element of the unit sphere of $\ell^{r_1}(\mathbb{Z})$. Using Lemma 2.23, the estimates concerning the paraproduct are proved. $\square$

We now examine the behavior of the remainder operator $\dot{R}$. Here, we have to consider terms of the type $\dot{\Delta}_j u \, \dot{\Delta}_j v$, the Fourier transforms of which are not supported in annuli, but rather in balls of the type $2^j B$. Thus, to prove that the remainder terms belong to certain Besov spaces, we need the following lemma.

Lemma 2.49. *Let B be a ball in $\mathbb{R}^d$, s a positive real number, and $(p, r) \in [1, \infty]^2$. A constant C exists which satisfies the following. Let $(u_j)_{j \in \mathbb{Z}}$ be a sequence of smooth functions such that*

$$\mathrm{Supp}\, \widehat{u}_j \subset 2^j B \quad and \quad \left\| (2^{js} \|u_j\|_{L^p})_{j \in \mathbb{Z}} \right\|_{\ell^r} < \infty.$$

We assume that the series $\sum_{j \in \mathbb{Z}} u_j$ converges to u in $\mathcal{S}'_h$. We then have

$$u \in \dot{B}^s_{p,r} \quad and \quad \|u\|_{\dot{B}^s_{p,r}} \leq \frac{C}{s} \left\| (2^{js} \|u_j\|_{L^p})_j \right\|_{\ell^r(\mathbb{Z})}.$$

Remark 2.50. Thanks to Lemma 2.49 and the remark that follows it, the hypothesis of convergence is satisfied whenever (s, p, r) satisfies (2.17).

Proof of Lemma 2.49. As $\mathcal{C}$ is an annulus and B is a ball, an integer N_1 exists such that if $j' \geq j + N_1$, then $2^{j'}\mathcal{C} \cap 2^j B = \emptyset$. So, if $j' \geq j + N_1$, then the Fourier transform of $\dot{\Delta}_{j'} u_j$ (and thus $\Delta_{j'} u_j$) is equal to 0. Hence, we may write

$$\|\dot{\Delta}_{j'} u\|_{L^p} \leq \sum_{j > j' - N_1} \|\dot{\Delta}_{j'} u_j\|_{L^p}$$

$$\leq C \sum_{j > j' - N_1} \|u_j\|_{L^p}.$$

We therefore get that

$$2^{j's}\|\dot{\Delta}_{j'} u\|_{L^p} \leq C \sum_{j \geq j' - N_1} 2^{j's}\|u_j\|_{L^p}$$

$$\leq C \sum_{j \geq j' - N_1} 2^{(j'-j)s} 2^{js}\|u_j\|_{L^p}.$$

As s is positive, applying Young's inequality for series completes the proof of the lemma. $\square$

Remark 2.51. The above lemma fails in the limit case $s = 0$. Indeed, fix a nonzero function $f \in L^p$, spectrally supported in some ball B, and a nonnegative real α such that $\alpha r > 1$. Set $u_j = j^{-\alpha} f$ for $j \geq 1$, and $u_j = 0$ otherwise. It is clear that

$$\forall j \in \mathbb{Z}, \ \text{Supp}\, \widehat{u}_j \subset 2^j B \quad \text{and} \quad \left\| (\|u_j\|_{L^p})_{j \in \mathbb{N}} \right\|_{\ell^r} < \infty.$$

If $r > 1$, then we can additionally set $\alpha < 1$ so that the series $\sum_j u_j$ diverges in $\mathcal{S}'$. If $r = 1$, then the series converges to a nonzero multiple of f. As $\dot{B}^0_{p,1}$ is a strict subspace of L^p, the function f need not be in $\dot{B}^0_{p,1}$, so the lemma also fails in this case.

We can now state a result concerning continuity of the remainder operator.

Theorem 2.52. *A constant C exists which satisfies the following inequalities. Let (s_1, s_2) be in $\mathbb{R}^2$ and (p_1, p_2, r_1, r_2) be in $[1, \infty]^4$. Assume that*

$$\frac{1}{p} \stackrel{def}{=} \frac{1}{p_1} + \frac{1}{p_2} \leq 1 \quad and \quad \frac{1}{r} \stackrel{def}{=} \frac{1}{r_1} + \frac{1}{r_2} \leq 1.$$

If $s_1 + s_2$ is positive, then we have, for any (u, v) in $\dot{B}^{s_1}_{p_1, r_1} \times \dot{B}^{s_2}_{p_2, r_2}$,

$$\|\dot{R}(u, v)\|_{\dot{B}^{s_1+s_2}_{p,r}} \leq \frac{C^{|s_1+s_2|+1}}{s_1 + s_2} \|u\|_{\dot{B}^{s_1}_{p_1, r_1}} \|v\|_{\dot{B}^{s_2}_{p_2, r_2}}.$$

When $r = 1$ and $s_1 + s_2 \geq 0$, we have, for any (u, v) in $\dot{B}^{s_1}_{p_1, r_1} \times \dot{B}^{s_2}_{p_2, r_2}$,

$$\|\dot{R}(u, v)\|_{\dot{B}^{s_1+s_2}_{p,\infty}} \leq C^{|s_1+s_2|+1} \|u\|_{\dot{B}^{s_1}_{p_1, r_1}} \|v\|_{\dot{B}^{s_2}_{p_2, r_2}}.$$

Remark 2.53. Thanks to Lemma 2.49 and the remark that follows it, the hypothesis of convergence is satisfied whenever $(s_1 + s_2, p, r)$ or $(s_1 + s_2, p, \infty)$ satisfies (2.17)

Proof of Theorem 2.52. By definition of the homogeneous remainder operator,

$$\dot{R}(u, v) = \sum_j R_j \quad \text{with} \quad R_j = \sum_{|\nu| \leq 1} \dot{\Delta}_{j-\nu} u \dot{\Delta}_j v.$$

Because φ is supported in the annulus $\mathcal{C}$, the Fourier transform of R_j is supported in $2^j B(0, 24)$. So, by construction of the dyadic partition of unity, there exists an integer N_0 such that

$$j' > j + N_0 \Rightarrow \dot{\Delta}_{j'} R_j = 0. \tag{2.31}$$

From this, we deduce that

$$\dot{\Delta}_{j'} \dot{R}(u, v) = \sum_{j \geq j' - N_0} \dot{\Delta}_{j'} R_j.$$

Using Hölder's inequality, we infer that

$$
\begin{aligned}
2^{j'(s_1+s_2)} \| \dot{\Delta}_{j'} \dot{R}(u, v) \|_{L^p} &\leq C 2^{j'(s_1+s_2)} \sum_{\substack{|\nu| \leq 1 \\ j \geq j' - N_0}} \| \dot{\Delta}_{j-\nu} u \dot{\Delta}_j v \|_{L^p} \\
&\leq C 2^{j'(s_1+s_2)} \sum_{\substack{|\nu| \leq 1 \\ j \geq j' - N_0}} \| \dot{\Delta}_{j-\nu} u \|_{L^{p_1}} \| \dot{\Delta}_j v \|_{L^{p_2}} \\
&\leq C \sum_{\substack{|\nu| \leq 1 \\ j \geq j' - N_0}} 2^{-(j-j')(s_1+s_2)} 2^{(j-\nu)s_1} \| \dot{\Delta}_{j-\nu} u \|_{L^{p_1}} 2^{j s_2} \| \dot{\Delta}_j v \|_{L^{p_2}}.
\end{aligned}
$$

Using Hölder's and Young's inequalities for series, we get the theorem in the case where $s_1 + s_2$ is positive.

In the case where $r = 1$ and $s_1 + s_2$ is nonnegative, we use the fact that

$$2^{j'(s_1+s_2)} \| \dot{\Delta}_{j'} \dot{R}(u, v) \|_{L^p} \leq C \sum_{\substack{|\nu| \leq 1 \\ j \geq j' - N_0}} 2^{(j-\nu)s_1} \| \dot{\Delta}_{j-\nu} u \|_{L^{p_1}} 2^{j s_2} \| \dot{\Delta}_j v \|_{L^{p_2}},$$

take the supremum over j', and use Hölder's inequality for series. $\qquad\square$

By taking advantage of Bony's decomposition (2.29), a plethora of results on continuity may be deduced from Theorems 2.47 and 2.52. As an initial example, we derive the following so-called *tame estimates* for the product of two functions in Besov spaces.

Corollary 2.54. *If $(s, p, r) \in \,]0, \infty[\times [1, \infty]^2$ satisfies (2.17), then $L^\infty \cap \dot{B}^s_{p,r}$ is an algebra. Moreover, there exists a constant C, depending only on the dimension d, such that*

$$\|uv\|_{\dot{B}^s_{p,r}} \le \frac{C^{s+1}}{s} \left(\|u\|_{L^\infty} \|v\|_{\dot{B}^s_{p,r}} + \|u\|_{\dot{B}^s_{p,r}} \|v\|_{L^\infty} \right).$$

Proof. Using Bony's decomposition, we have

$$uv = \dot{T}_u v + \dot{T}_v u + \dot{R}(u, v).$$

According to Theorem 2.47, we have

$$\|\dot{T}_u v\|_{\dot{B}^s_{p,r}} \le C^{s+1} \|u\|_{L^\infty} \|v\|_{\dot{B}^s_{p,r}} \quad \text{and} \quad \|\dot{T}_v u\|_{\dot{B}^s_{p,r}} \le C^{s+1} \|u\|_{\dot{B}^s_{p,r}} \|v\|_{L^\infty}.$$

Now, using Theorem 2.52, we get

$$\|\dot{R}(u, v)\|_{\dot{B}^s_{p,r}} \le \frac{C^{s+1}}{s} \|u\|_{\dot{B}^0_{\infty,\infty}} \|v\|_{\dot{B}^s_{p,r}}.$$

Since, obviously, $\|u\|_{\dot{B}^0_{\infty,\infty}} \le C \|u\|_{L^\infty}$, we obtain the desired inequality. $\quad\square$

Our second example deals with the product of two functions in homogeneous Sobolev spaces.

Corollary 2.55. *For any $(s_1, s_2) \in \,]-d/2, d/2[^2$, a constant C exists such that if $s_1 + s_2$ is positive, then we have*

$$\|uv\|_{\dot{B}^{s_1+s_2-\frac{d}{2}}_{2,1}} \le C \|u\|_{\dot{H}^{s_1}} \|v\|_{\dot{H}^{s_2}}.$$

Proof. We again use Bony's decomposition. First, as $\dot{H}^s$ is continuously included in $\dot{B}^{s-\frac{d}{2}}_{\infty,2}$ and $s - d/2 < 0$, Theorem 2.47 implies that

$$\|\dot{T}_u v + \dot{T}_v u\|_{\dot{B}^{s_1+s_2-\frac{d}{2}}_{2,1}} \le C \|u\|_{\dot{H}^{s_1}} \|v\|_{\dot{H}^{s_2}}.$$

Second, as $s_1 + s_2 > 0$, Theorem 2.52 guarantees that

$$\|\dot{R}(u, v)\|_{\dot{B}^{s_1+s_2}_{1,1}} \le C \|u\|_{\dot{H}^{s_1}} \|v\|_{\dot{H}^{s_2}}.$$

As the space $\dot{B}^{s_1+s_2}_{1,1}$ is continuously included in $\dot{B}^{s_1+s_2-\frac{d}{2}}_{2,1}$, the corollary is proved. $\quad\square$

Remark 2.56. The constant in Corollary 2.55 may be bounded by

$$C \min\left\{ \frac{1}{d - 2s_1}, \frac{1}{d - 2s_2}, \frac{1}{s_1 + s_2} \right\}$$

with C depending only on the dimension d.

As an application of Corollary 2.55, we get the following family of Hardy inequalities, which contains the particular case of Theorem 1.72 page 48.

Theorem 2.57. *For any real s in $\left[0, \dfrac{d}{2}\right[$, a constant C exists such that for any f in $\dot{H}^s(\mathbb{R}^d)$,*

$$\int_{\mathbb{R}^d} \frac{|f(x)|^2}{|x|^{2s}}\, dx \le C\|f\|_{\dot{H}^s}^2. \tag{2.32}$$

Proof. The case $s = 0$ being obvious, we assume that $0 < s < d/2$. As $\mathcal{S}_0$ is dense in $\dot{H}^s$, it suffices to prove the above inequality in the case where f belongs to $\mathcal{S}_0$. We define

$$I_s(f) \overset{\text{def}}{=} \int_{\mathbb{R}^d} \frac{|f(x)|^2}{|x|^{2s}}\, dx = \langle |\cdot|^{-2s}, f^2 \rangle.$$

Using Littlewood–Paley decomposition and the fact that f^2 belongs to $\mathcal{S}_h'$, we can write

$$I_s(f) = \sum_{|j-j'|\le 2} \langle \dot{\Delta}_j |\cdot|^{-2s}, \dot{\Delta}_{j'} f^2 \rangle,$$

$$\le C \sum_{|j-j'|\le 2} \left| \langle 2^{j\left(\frac{d}{2}-2s\right)} \dot{\Delta}_j |\cdot|^{-2s}, 2^{-j'\left(\frac{d}{2}-2s\right)} \dot{\Delta}_{j'} f^2 \rangle \right|.$$

By virtue of Proposition 2.21, the function $|\cdot|^{-2s}$ belongs to $\dot{B}_{2,\infty}^{\frac{d}{2}-2s}$. Corollary 2.55 yields $\|f^2\|_{\dot{B}_{2,1}^{2s-\frac{d}{2}}} \le C\|f\|_{\dot{H}^s}^2$. Thus, $I_s(f) \le C\|f\|_{\dot{H}^s}^2$. □

We conclude this section with the statement of some refined Hardy inequalities, in the spirit of the refined Sobolev inequalities (see Theorem 1.43 page 32).

Theorem 2.58. *Let (s, p, q) be a triplet of real numbers such that*

$$0 < s < \frac{d}{2} \quad \text{and} \quad 2 \le q < \frac{2d}{d2s} < p \le \infty.$$

There exists a constant C such that for any function $u \in \dot{B}_{q,2}^{s-d\left(\frac{1}{2}-\frac{1}{q}\right)}$, the following inequality holds:

$$\left(\int \frac{|u(x)|^2}{|x|^{2s}}\, dx\right)^{\frac{1}{2}} \le C\|u\|_{\dot{B}_{p,2}^{s-d\left(\frac{1}{2}-\frac{1}{p}\right)}}^\alpha \|u\|_{\dot{B}_{q,2}^{s-d\left(\frac{1}{2}-\frac{1}{q}\right)}}^{1-\alpha} \quad \text{with} \quad \alpha = \frac{pq}{p-q}\left(\frac{1}{q} - \frac{1}{2} + \frac{s}{d}\right).$$

Proof. Bony's decomposition for u^2 reads

$$u^2 = 2\dot{T}_u u + \dot{R}(u, u),$$

so it suffices to prove the inequality for $\dot{T}_u u$ and $\dot{R}(u,u)$ instead of u^2. Of course, arguing by density, we can assume that u belongs to $\mathcal{S}_0$.

The term $\dot{T}_u u$ is easy to deal with: According to Theorem 2.47, we have

$$\|\dot{T}_u u\|_{\dot{B}^{2s-d}_{\infty,1}} \le C\|u\|^2_{\dot{B}^{s-\frac{d}{2}}_{\infty,2}}.$$

Proposition 2.21 now ensures that the function $|\cdot|^{-2s}$ belongs to $\dot{B}^{d-2s}_{1,\infty}$, so, according to Proposition 2.29, we have

$$\left|\langle|\cdot|^{-2s}, \dot{T}_u u\rangle\right| \le C\|u\|^2_{\dot{B}^{s-\frac{d}{2}}_{\infty,2}}.$$

Since both $\dot{B}^{s-\left(\frac{d}{2}-\frac{d}{q}\right)}_{q,2}$ and $\dot{B}^{s-\left(\frac{d}{2}-\frac{d}{p}\right)}_{p,2}$ are embedded in $\dot{B}^{s-\frac{d}{2}}_{\infty,2}$, we end up with

$$\left|\langle|\cdot|^{-2s}, \dot{T}_u u\rangle\right| \le C\|u\|^{2\alpha}_{\dot{B}^{s-\left(\frac{d}{2}-\frac{d}{q}\right)}_{q,2}}\|u\|^{2-2\alpha}_{\dot{B}^{s-\left(\frac{d}{2}-\frac{d}{p}\right)}_{p,2}}. \tag{2.33}$$

The estimate of $\langle|\cdot|^{-2s}, \dot{R}(u,u)\rangle$ relies on the following interpolation lemma.

Lemma 2.59. *Under the assumptions of Theorem 2.58, there exists a constant C such that for any functions f and g in $L^p \cap L^q$, we have*

$$\langle|\cdot|^{-2s}, fg\rangle \le C\|f\|^\alpha_{L^p}\|g\|^\alpha_{L^p}\|f\|^{1-\alpha}_{L^q}\|g\|^{1-\alpha}_{L^q} \quad \text{with} \quad \alpha = \frac{pq}{p-q}\left(\frac{1}{q} - \frac{1}{2} + \frac{s}{d}\right).$$

Proof. For any positive R, we can write $\langle|\cdot|^{-2s}, fg\rangle = I_1(R) + I_2(R)$ with

$$I_1(R) \stackrel{\text{def}}{=} \int_{|x|\le R} \frac{fg(x)}{|x|^{2s}}\,dx \quad \text{and} \quad I_2(R) \stackrel{\text{def}}{=} \int_{|x|\ge R} \frac{fg(x)}{|x|^{2s}}\,dx.$$

The condition on p and q implies that $|\cdot|^{-2s}$ is locally $L^{\frac{p}{p-2}}$ and is $L^{\frac{q}{q-2}}$ outside any compact neighborhood of 0. By Hölder's inequality, we infer that

$$I_1(R) \le \|\mathbf{1}_{(|\cdot|\le R)}|\cdot|^{-2s}\|_{L^{\frac{p}{p-2}}}\|f\|_{L^p}\|g\|_{L^p},$$

$$I_2(R) \le \|\mathbf{1}_{(|\cdot|\ge R)}|\cdot|^{-2s}\|_{L^{\frac{q}{q-2}}}\|f\|_{L^q}\|g\|_{L^q}.$$

Because the function $|\cdot|^{-2s}$ is homogeneous of order $-2s$, we get

$$\|\mathbf{1}_{(|\cdot|\le R)}|\cdot|^{-2s}\|_{L^{\frac{p}{p-2}}} = R^{d-2s-\frac{2d}{p}}\|\mathbf{1}_{(|\cdot|\le 1)}|\cdot|^{-2s}\|_{L^{\frac{p}{p-2}}},$$

$$\|\mathbf{1}_{(|\cdot|\ge R)}|\cdot|^{-2s}\|_{L^{\frac{q}{q-2}}} = R^{d-2s-\frac{2d}{q}}\|\mathbf{1}_{(|\cdot|\ge 1)}|\cdot|^{-2s}\|_{L^{\frac{q}{q-2}}}.$$

Thus, for any positive R, we have

$$\langle|\cdot|^{-2s}, fg\rangle \le CR^{d-2s}\left(R^{-\frac{2d}{p}}\|f\|_{L^p}\|g\|_{L^p} + R^{-\frac{2d}{q}}\|f\|_{L^q}\|g\|_{L^q}\right).$$

Choosing the best R, namely

$$R = \left(\frac{\|f\|_{L^q}\|g\|_{L^q}}{\|f\|_{L^p}\|g\|_{L^p}}\right)^{\frac{pq}{2d(p-q)}},$$

completes the proof of the lemma. $\qquad\square$

We now resume the proof of Theorem 2.58. By the definition of $\dot{R}(u, u)$, we have

$$\langle |\cdot|^{-2s}, \dot{R}(u, u)\rangle = \sum_{|\ell|\le 1}\sum_{j\in\mathbb{Z}}\langle |\cdot|^{-2s}, \dot{\Delta}_j u \dot{\Delta}_{j-\ell}u\rangle.$$

Lemma 2.59 implies that

$$\left|\langle |\cdot|^{-2s}, \dot{R}(u, u)\rangle\right| \le \sum_{|\ell|\le 1}\sum_{j\in\mathbb{Z}}\left(2^{2j(s-(\frac{d}{2}-\frac{d}{p}))}\|\dot{\Delta}_j u\|_{L^p}\|\dot{\Delta}_{j-\ell}u\|_{L^p}\right)^\alpha$$

$$\times \left(2^{2j(s-(\frac{d}{2}-\frac{d}{q}))}\|\dot{\Delta}_j u\|_{L^q}\|\dot{\Delta}_{j-\ell}u\|_{L^q}\right)^{1-\alpha}.$$

By the definition of the Besov norms, this implies that two sequences, $(c_j)_{j\in\mathbb{Z}}$ and $(c'_j)_{j\in\mathbb{Z}}$, exist in the unit sphere of $\ell^2(\mathbb{Z})$ such that

$$\left|\langle |\cdot|^{-2s}, \dot{R}(u, u)\rangle\right| \le C\|u\|_{\dot{B}^{s-\left(\frac{d}{2}-\frac{d}{q}\right)}_{q,2}}^{2\alpha}\|u\|_{\dot{B}^{s-\left(\frac{d}{2}-\frac{d}{p}\right)}_{p,2}}^{2(1-\alpha)}\sum_{|\ell|\le 1}\sum_{j\in\mathbb{Z}}(c_j c_{j-\ell})^\alpha(c'_j c'_{j-\ell})^{1-\alpha}.$$

From Hölder's inequality, it follows that

$$\left|\langle |\cdot|^{-2s}, \dot{R}(u, u)\rangle\right| \le C\|u\|_{\dot{B}^{s-\left(\frac{d}{2}-\frac{d}{p}\right)}_{p,2}}^{2\alpha}\|u\|_{\dot{B}^{s-\left(\frac{d}{2}-\frac{d}{q}\right)}_{q,2}}^{2(1-\alpha)}.$$

Together with (2.33), this gives Theorem 2.58.   □

Remark 2.60. Theorem 2.58 fails for $p = q_c = \frac{2d}{d-2s}$ since, if it were true, for any function u with Fourier transform supported in $\mathcal{C}$, we would have

$$\int_{\mathbb{R}^d}\frac{|u(x)|^2}{|x|^{2s}}\,dx \le C\|u\|_{\dot{B}^0_{2q_c,2}}^2 \le C\|u\|_{L^{2q_c}(\mathbb{R}^d)}^2. \tag{2.34}$$

In particular, this inequality would be true whenever $u \in \mathcal{S}(\mathbb{R}^d)$ satisfies

$$\text{supp }\widehat{u} \subset \mathcal{B}(\xi_0, \varepsilon) \subset \mathcal{C}.$$

As the inequality (2.34) is invariant under oscillation (i.e., under translation in the Fourier space), we deduce that it is true for any function $u \in \mathcal{S}(\mathbb{R}^d)$ such that supp $\widehat{u} \subset \mathcal{B}(0, \varepsilon)$. The invariance under dilation implies that it is true for any function $u \in \mathcal{S}(\mathbb{R}^d)$ such that supp $\widehat{u} \subset \mathcal{B}(0, R)$ for any $R > 0$. By density, we obtain (2.34) for any function $u \in L^{2q_c}(\mathbb{R}^d)$, but this implies that the singular weight $|x|^{-2s}$ belongs to $L^{\frac{d}{2s}}$, which is false.

2.6.2 Action of Smooth Functions

In this subsection we will consider the action of smooth functions on the space $\dot{B}^s_{p,r}$. More precisely, if f is a smooth function vanishing at 0, and u is a function of $\dot{B}^s_{p,r}$, does $f \circ u$ belong to $\dot{B}^s_{p,r}$? The answer is given by the following theorem.

Theorem 2.61. *Let f be a smooth function on $\mathbb{R}$ which vanishes at 0. Let (s_1, s_2) be a couple of positive real numbers and $(p_1, p_2, r_1, r_2) \in [1, \infty]^2$. Assume that (s_1, p_1, r_1) satisfies the condition (2.17).*

For any real-valued function u in $\dot{B}^{s_1}_{p_1, r_1} \cap \dot{B}^{s_2}_{p_2, r_2} \cap L^\infty$, the function $f \circ u$ belongs to the same space, and we have, for $k = 1$ and $k = 2$,

$$\|f \circ u\|_{\dot{B}^{s_k}_{p_k, r_k}} \leq C(f', \|u\|_{L^\infty}) \|u\|_{\dot{B}^{s_k}_{p_k, r_k}}.$$

Proof. As u is bounded, we can assume without loss of generality that f is compactly supported. The proof then uses the same basic idea as in the proof of Theorem 2.40: We introduce the telescopic series

$$\sum_j f_j \quad \text{with} \quad f_j \overset{\text{def}}{=} f(\dot{S}_{j+1} u) - f(\dot{S}_j u).$$

The convergence of the series is ensured by the following lemma.

Lemma 2.62. *Under the hypotheses of Theorem 2.61, the series $\sum_{j \in \mathbb{Z}} f_j$ converges to $f(u)$ in $\mathcal{S}'$, and we have*

$$f_j = m_j \dot{\Delta}_j u \quad \text{with} \quad m_j \overset{\text{def}}{=} \int_0^1 f'(\dot{S}_j u + t \dot{\Delta}_j u) \, dt. \tag{2.35}$$

Proof. The identity (2.35) readily follows from the mean value theorem, so we will concentrate on the proof of the convergence of the series. We observe that

$$\sum_{j=-N}^{0} f_j = f(\dot{S}_1 u) - f(\dot{S}_{-N} u).$$

As u belongs to $\mathcal{S}'_h$ and $f(0) = 0$, we have that $\|f(\dot{S}_{-N} u)\|_{L^\infty}$ tends to 0 when N tends to infinity. Moreover, for all positive integers M, we have

$$\sum_{j=1}^{M} f_j = f(\dot{S}_M u) - f(\dot{S}_1 u).$$

By virtue of the mean value theorem, we have

$$\|f(u) - f(\dot{S}_M u)\|_{L^{p_2}} \leq \|u - \dot{S}_M u\|_{L^{p_2}} \|f'\|_{L^\infty}.$$

Because $s_2 > 0$, the function $\dot{S}_M u$ tends to u in L^{p_2} when M goes to infinity. Therefore, the series $\sum_{j \in \mathbb{Z}} f_j$ converges to $f(u)$ in $L^\infty + L^{p_2}$.

Next, we prove that $f(u) \in \mathcal{S}'_h$. It suffices to show that $\|\dot{S}_j f(u)\|_{L^\infty} \to 0$ when j goes to $-\infty$. For that, we use the decomposition

$$\dot{S}_j f(u) = \dot{S}_j \sum_{j' < -N} f_{j'} + \dot{S}_j \sum_{j' \geq -N} f_{j'}.$$

Let ε be a positive real number. As the series $\sum_{j<0} f_j$ converges in L^∞, we can choose an integer N_ε such that

$$\left\|\dot{S}_j \sum_{j'<-N_\varepsilon} f_{j'}\right\|_{L^\infty} \le \frac{\varepsilon}{2}.$$

As the f_j's are in L^{p_1} and $\sum_{j\in\mathbb{N}} f_j$ is convergent in L^{p_1}, we then have, using Lemma 2.1,

$$\left\|\dot{S}_j \sum_{j'\ge -N_\varepsilon} f_{j'}\right\|_{L^\infty} \le C_\varepsilon 2^{j\frac{d}{p_1}}.$$

Thus, $\|\dot{S}_j f(u)\|_{L^\infty}$ tends to 0 when j tends to $-\infty$. □

The terms m_j will be handled according to the following lemma.

Lemma 2.63. *Let g be a smooth function from $\mathbb{R}^2$ to $\mathbb{R}$. For $j \in \mathbb{Z}$, define*

$$m_j(g) \overset{def}{=} g(\dot{S}_j u, \dot{\Delta}_j u).$$

For any bounded function u, we then have

$$\forall \alpha \in \mathbb{N}^d,\ \forall j \in \mathbb{Z},\ \|\partial^\alpha m_j(g)\|_{L^\infty} \le C_\alpha(g, \|u\|_{L^\infty}) 2^{j|\alpha|}.$$

Proof. The proof relies on Lemma 2.3, which provides us with the formula

$$\partial^\alpha m_j(g) = \sum_{p_1,p_2,\nu} C^\nu_{p_1,p_2} \left(\prod_{1\le|\beta|\le|\alpha|} (\partial^\beta \dot{S}_j u)^{\nu_{\beta_1}} (\partial^\beta \dot{\Delta}_j u)^{\nu_{\beta_2}} \right) \partial_1^{p_1} \partial_2^{p_2} g(\dot{S}_j u, \dot{\Delta}_j u),$$

where the coefficients $C^\nu_{p_1,p_2}$ are nonnegative integers, and the sum is taken other those p_1, p_2, and ν such that $1 \le p_1 + p_2 \le |\alpha|$,

$$\sum_{1\le|\beta|\le|\alpha|} \nu_{\beta_j} = p_j \ \text{ for } \ j = 1,2, \quad \text{and} \quad \sum_{1\le|\beta|\le|\alpha|} \beta(\nu_{\beta_1} + \nu_{\beta_2}) = \alpha.$$

Note that there exists a constant C such that

$$\max\{\|\dot{\Delta}_j u\|_{L^\infty}, \|\dot{S}_j u\|_{L^\infty}\} \le C\|u\|_{L^\infty} \quad \text{for all } \ j \in \mathbb{Z}.$$

Since g and all its derivatives are bounded on $B(0, C\|u\|_{L^\infty})$, Lemma 2.1 and the above formula thus ensure that

$$\|\partial^\alpha m_j(g)\|_{L^\infty} \le C_\alpha(g, \|u\|_{L^\infty}) 2^{j|\alpha|}.$$

This completes the proof of the lemma. □

In contrast with the situation which was encountered when proving Theorems 2.47 and 2.52, here, the elements f_j of the approximating series $\sum f_j$ are not compactly supported in the Fourier space. This difficulty is overcome by the following lemma.

Lemma 2.64. *Let s be a positive real number and (p, r) be in $[1, \infty]^2$. A constant C_s exists such that if $(u_j)_{j \in \mathbb{Z}}$ is a sequence of smooth functions where $\sum u_j$ converges to some u in $\mathcal{S}'_h$ and*

$$N_s((u_j)_{j \in \mathbb{Z}}) \stackrel{def}{=} \left\| \left(\sup_{|\alpha| \in \{0, [s]+1\}} 2^{j(s-|\alpha|)} \|\partial^\alpha u_j\|_{L^p} \right)_j \right\|_{\ell^r(\mathbb{Z})} < \infty,$$

then u is in $\dot{B}^s_{p,r}$ and $\|u\|_{\dot{B}^s_{p,r}} \le C_s N_s(u)$.

Proof. As the series $\sum u_j$ converges to u in $\mathcal{S}'$, we have

$$\dot{\Delta}_j u = \sum_{j' \le j} \dot{\Delta}_j u_{j'} + \sum_{j' > j} \dot{\Delta}_j u_{j'}.$$

Using the fact that $\|\dot{\Delta}_j u_{j'}\|_{L^p} \le \|u_{j'}\|_{L^p}$, we get

$$2^{js} \left\| \sum_{j' > j} \dot{\Delta}_j u_{j'} \right\|_{L^p} \le 2^{js} \sum_{j' > j} \|u_{j'}\|_{L^p}$$

$$\le \sum_{j' > j} 2^{-(j'-j)s} 2^{j's} \|u_{j'}\|_{L^p}. \qquad (2.36)$$

Using Lemma 2.1, we may then write that

$$\|\dot{\Delta}_j u_{j'}\|_{L^p} \le C 2^{-j([s]+1)} \sup_{|\alpha|=[s]+1} \|\partial^\alpha u_{j'}\|_{L^p},$$

from which it follows that

$$2^{js} \left\| \sum_{j' \le j} \dot{\Delta}_j u_{j'} \right\|_{L^p} \le \sum_{j' \le j} 2^{(j'-j)([s]+1-s)} \sup_{|\alpha|=[s]+1} 2^{j'(s-|\alpha|)} \|\partial^\alpha u_{j'}\|_{L^p}.$$

This inequality, combined with (2.36), implies that

$$2^{js} \|\dot{\Delta}_j u\|_{L^p} \le (a \star b)_j \quad \text{with} \quad \begin{cases} a_j \stackrel{def}{=} \mathbf{1}_{\mathbb{N}}(j) 2^{-js} + \mathbf{1}_{\mathbb{N}}(j) 2^{-j([s]+1-s)}, \\ b_j \stackrel{def}{=} 2^{js} \|u_j\|_{L^p} + \sup_{|\alpha|=[s]+1} 2^{j(s-|\alpha|)} \|\partial^\alpha u_j\|_{L^p}. \end{cases}$$

This proves the lemma. $\qquad \square$

Given the above three lemmas, it is now easy to prove Theorem 2.61. Note that, according to Lemma 2.64, it suffices to establish that

$$N_{s_k}((f_j)_{j \in \mathbb{Z}}) < \infty. \qquad (2.37)$$

Now, using Leibniz's formula, Lemma 2.1, and Lemma 2.63 with the function

$$g(x, y) = \int_0^1 f'(x + ty) \, dt,$$

we get that

$$\|\partial^\alpha f_j\|_{L^p} \leq \sum_{\beta \leq \alpha} C_\beta^\alpha 2^{j|\beta|} C_\beta(f', \|u\|_{L^\infty}) 2^{j(|\alpha|-|\beta|)} \|\dot\Delta_j u\|_{L^p},$$

from which it follows that, for $s = s_1, s_2$,

$$\begin{aligned}
\|\partial^\alpha f_j\|_{L^p} &\leq C_\alpha(f', \|u\|_{L^\infty}) 2^{j|\alpha|} \|\dot\Delta_j u\|_{L^p} \\
&\leq c_j C_\alpha(f', \|u\|_{L^\infty}) 2^{-j(s-|\alpha|)} \|u\|_{\dot B_{p,r}^s} \quad \text{with} \quad \|(c_j)\|_{\ell^r} = 1. \quad (2.38)
\end{aligned}$$

This completes the proof of the theorem.   □

In the case where f belongs to the space $C_b^\infty(\mathbb{R})$ of smooth bounded functions with bounded derivatives of all orders and satisfies $f(0) = 0$, a slightly more accurate estimate may be obtained. Indeed, we have, for $|\alpha_k| \geq 1$ and any j in $\mathbb{Z}$,

$$\max\left(\|\partial^{\alpha_k} \dot S_j u\|_{L^\infty}, \|\partial^{\alpha_k} \dot\Delta_j u\|_{L^\infty}\right) \leq C 2^{j|\alpha_k|} \|\nabla u\|_{\dot B_{\infty,\infty}^{-1}} \leq 2^{j|\alpha_k|} \|u\|_{\dot B_{\infty,\infty}^0}.$$

Arguing as in the proof of Lemma 2.63, we thus get

$$\forall \alpha \in \mathbb{N}^d, \quad \|\partial^\alpha m_j\|_{L^\infty} \leq C_\alpha(f, \|u\|_{\dot B_{\infty,\infty}^0}) 2^{j|\alpha|}. \quad (2.39)$$

We now state the result we have just proven.

Corollary 2.65. *Let f be a function in $C_b^\infty(\mathbb{R})$ such that $f(0) = 0$. Let (s_1, s_2) be in $]0, \infty[^2$ and (p_1, p_2, r_1, r_2) be in $[1, \infty]^4$. Assume that (s_1, p_1, r_1) satisfies the condition (2.17).*

Then, for any real-valued function u in $\dot B_{p_1,r_1}^{s_1} \cap \dot B_{p_2,r_2}^{s_2} \cap \dot B_{\infty,\infty}^0$, the function $f \circ u$ belongs to $\dot B_{p_1,r_1}^{s_1} \cap \dot B_{p_2,r_2}^{s_2}$, and we have

$$\|f \circ u\|_{\dot B_{p_k,r_k}^{s_k}} \leq C(f, \|u\|_{\dot B_{\infty,\infty}^0}) \|u\|_{\dot B_{p_k,r_k}^{s_k}} \quad \text{for} \quad k \in \{1, 2\}.$$

Finally, by combining Corollary 2.54 and Theorem 2.61 with the equality

$$f(v) - f(u) = (v - u) \int_0^1 f'(u + \tau(v - u)) \, d\tau,$$

we readily obtain the following corollary.

Corollary 2.66. *Let f be a smooth function such that $f'(0) = 0$. Let s be a positive real number and (p, r) in $[1, \infty]^2$ be such that (s, p, r) satisfies (2.17). For any couple (u, v) of functions in $\dot B_{p,r}^s \cap L^\infty$, the function $f \circ v - f \circ u$ then belongs to $\dot B_{p,r}^s \cap L^\infty$ and*

$$\begin{aligned}
\|f(v) - f(u)\|_{\dot B_{p,r}^s} \leq C\Big(\|v - u\|_{\dot B_{p,r}^s} \sup_{\tau \in [0,1]} \|u + \tau(v-u)\|_{L^\infty} \\
+ \|v - u\|_{L^\infty} \sup_{\tau \in [0,1]} \|u + \tau(v - u)\|_{\dot B_{p,r}^s}\Big),
\end{aligned}$$

where C depends on f'', $\|u\|_{L^\infty}$, and $\|v\|_{L^\infty}$.

2.6.3 Time-Space Besov Spaces

One of the fundamental ideas in this book is that nonlinear evolution partial differential equations may be treated very efficiently after localization by means of Littlewood–Paley decomposition. Indeed, it is often easier to bound each dyadic block in $L^\rho([0,T]; L^p)$ than to estimate directly the solution of the whole partial differential equation in $L^\rho([0,T]; \dot{B}^s_{p,r})$.

As a final step, we must combine the estimates for each block, then perform a (weighted) ℓ^r summation. In doing so, however, we do not obtain an estimate in a space of type $L^\rho([0,T]; \dot{B}^s_{p,r})$ since the time integration has been performed *before* the summation.

This naturally leads to the following definition.

Definition 2.67. *For $T > 0$, $s \in \mathbb{R}$, and $1 \le r, \rho \le \infty$, we set*

$$\|u\|_{\widetilde{L}^\rho_T(\dot{B}^s_{p,r})} \overset{\text{def}}{=} \left\| 2^{js} \|\dot{\Delta}_j u\|_{L^\rho_T(L^p)} \right\|_{\ell^r(\mathbb{Z})}.$$

We can then define the space $\widetilde{L}^\rho_T(\dot{B}^s_{p,r})$ as the set of tempered distributions u over $(0,T) \times \mathbb{R}^d$ such that $\lim_{j \to -\infty} \dot{S}_j u = 0$ in $L^\rho([0,T]; L^\infty(\mathbb{R}^d))$ and $\|u\|_{\widetilde{L}^\rho_T(\dot{B}^s_{p,r})} < \infty$.

The spaces $\widetilde{L}^\rho_T(\dot{B}^s_{p,r})$ may be linked with the more classical spaces $L^\rho_T(\dot{B}^s_{p,r}) \overset{\text{def}}{=} L^\rho([0,T]; \dot{B}^s_{p,r})$ via the Minkowski inequality: We have

$$\|u\|_{\widetilde{L}^\rho_T(\dot{B}^s_{p,r})} \le \|u\|_{L^\rho_T(\dot{B}^s_{p,r})} \quad \text{if } r \ge \rho, \qquad \|u\|_{\widetilde{L}^\rho_T(\dot{B}^s_{p,r})} \ge \|u\|_{L^\rho_T(\dot{B}^s_{p,r})} \quad \text{if } r \le \rho.$$

The general principle is that all the properties of continuity for the product, composition, remainder, and paraproduct remain true in those spaces. The exponent ρ just has to behave according to Hölder's inequality for the time variable. For instance, we have the time estimate

$$\|uv\|_{\widetilde{L}^\rho_T(\dot{B}^s_{p,r})} \le C \left(\|u\|_{L^{\rho_1}_T(L^\infty)} \|v\|_{\widetilde{L}^{\rho_2}_T(\dot{B}^s_{p,r})} + \|v\|_{L^{\rho_3}_T(L^\infty)} \|u\|_{\widetilde{L}^{\rho_4}_T(\dot{B}^s_{p,r})} \right)$$

whenever $s > 0$, $1 \le p \le \infty$, $1 \le \rho, \rho_1, \rho_2, \rho_3, \rho_4 \le \infty$, and

$$\frac{1}{\rho} = \frac{1}{\rho_1} + \frac{1}{\rho_2} = \frac{1}{\rho_3} + \frac{1}{\rho_4}.$$

It goes without saying that this approach also works in the nonhomogeneous Besov spaces $B^s_{p,r}$ which will be defined in the next section. This leads to function spaces denoted by $\widetilde{L}^\rho_T(B^s_{p,r})$.

2.7 Nonhomogeneous Besov Spaces

This section is devoted to the study of nonhomogeneous Besov spaces. It turns out that most properties which have been proven thus far for homogeneous spaces carry over to the nonhomogeneous framework. The results are basically the same, and the proofs are often simpler since we do not have to worry about the low frequencies. Therefore, we shall omit the proofs whenever a similar statement has been proven in the homogeneous setting.

Definition 2.68. *Let $s \in \mathbb{R}$ and $1 \leq p, r \leq \infty$. The nonhomogeneous Besov space $B^s_{p,r}$ consists of all tempered distributions u such that*

$$\|u\|_{B^s_{p,r}} \overset{def}{=} \left\| \left(2^{js} \|\Delta_j u\|_{L^p} \right)_{j \in \mathbb{Z}} \right\|_{\ell^r(\mathbb{Z})} < \infty.$$

Examples.

- Nonhomogeneous Besov spaces contain Sobolev spaces. Indeed, by (2.10) and the Fourier–Plancherel formula, we find that the Besov space $B^s_{2,2}$ coincides with the Sobolev space H^s defined on page 38.
- In the case where $s \in \mathbb{R}^+ \setminus \mathbb{N}$, we can show that $B^s_{\infty,\infty}$ coincides with the Hölder space $C^{[s], s-[s]}$ of bounded functions u whose derivatives of order $|\alpha| \leq [s]$ are bounded and satisfy

$$|\partial^\alpha u(x) - \partial^\alpha u(y)| \leq C|x - y|^{s-[s]} \quad \text{for} \quad |x - y| \leq 1.$$

We emphasize, however, that in the case $s \in \mathbb{N}$, the space $B^s_{\infty,\infty}$ is *strictly larger* than the space C^s (and than $C^{s-1,1}$, if $s \in \mathbb{N}^*$).

The first point to look at is the invariance with respect to the choice of Littlewood–Paley decomposition. This fundamental property is based on the following lemma, the proof of which is analogous to that of Lemma 2.23.

Lemma 2.69. *Let $\mathcal{C}'$ be an annulus of $\mathbb{R}^d$, s be a real number, and $(p, r) \in [1, \infty]^2$. Let $(u_j)_{j \in \mathbb{N}}$ be a sequence of smooth functions such that*

$$\text{Supp } \widehat{u_j} \subset 2^j \mathcal{C}' \quad \text{and} \quad \left\| \left(2^{js} \|u_j\|_{L^p} \right)_{j \in \mathbb{N}} \right\|_{\ell^r(\mathbb{N})} < \infty.$$

We then have

$$u \overset{def}{=} \sum_{j \in \mathbb{N}} u_j \in B^s_{p,r} \quad \text{and} \quad \|u\|_{B^s_{p,r}} \leq C_s \left\| \left(2^{js} \|u_j\|_{L^p} \right)_{j \in \mathbb{N}} \right\|_{\ell^r(\mathbb{N})}.$$

This immediately implies the following corollary.

Corollary 2.70. *The space $B^s_{p,r}$ does not depend on the choice of the functions χ and φ used in Definition 2.68.*

The following result is the equivalent of the Sobolev embedding (see Theorem 1.38 page 29) for nonhomogeneous Besov spaces.

Proposition 2.71. *Let $1 \leq p_1 \leq p_2 \leq \infty$ and $1 \leq r_1 \leq r_2 \leq \infty$. Then, for any real number s, the space $B^s_{p_1,r_1}$ is continuously embedded in $B^{s-d\left(\frac{1}{p_1}-\frac{1}{p_2}\right)}_{p_2,r_2}$.*

Proof. It suffices to apply Lemma 2.1, which yields

$$\|S_0 u\|_{L^{p_2}} \leq C \|S_0 u\|_{L^{p_1}} \text{ and } \|\Delta_j u\|_{L^{p_2}} \leq C 2^{jd(\frac{1}{p_1} - \frac{1}{p_2})} \|\Delta_j u\|_{L^{p_1}} \text{ for all } j \in \mathbb{N}.$$

As $\ell^{r_1}(\mathbb{Z})$ is continuously embedded in $\ell^{r_2}(\mathbb{Z})$, the result is proved. $\square$

Theorem 2.72. *The set $B_{p,r}^s$ is a Banach space and satisfies the Fatou property, namely, if $(u_n)_{n \in \mathbb{N}}$ is a bounded sequence of $B_{p,r}^s$, then an element u of $B_{p,r}^s$ and a subsequence $u_{\psi(n)}$ exist such that*

$$\lim_{n \to \infty} u_{\psi(n)} = u \quad \text{in} \quad \mathcal{S}' \quad \text{and} \quad \|u\|_{B_{p,r}^s} \leq C \liminf_{n \to \infty} \|u_{\psi(n)}\|_{B_{p,r}^s}.$$

The following result will help us to prove that the set of test functions is densely embedded in Besov spaces $B_{p,r}^s$ with finite r.

Lemma 2.73. *If r is finite, then for any u in $B_{p,r}^s$, we have*

$$\lim_{j \to \infty} \|S_j u - u\|_{B_{p,r}^s} = 0.$$

Proof. Let u be in $B_{p,r}^s$. Because r is finite, we have

$$\lim_{j \to \infty} \sum_{j' \geq j} 2^{j' s r} \|\Delta_{j'} u\|_{L^p}^r = 0.$$

This obviously implies that $\lim_{j \to \infty} S_j u = u$ in $B_{p,r}^s$. $\square$

We can now state a very useful density result.

Proposition 2.74. *If p and r are finite, then $\mathcal{D}(\mathbb{R}^d)$ is dense in $B_{p,r}^s(\mathbb{R}^d)$.*

Proof. Assume that p and r are finite. Let ε be a positive real number. According to Lemma 2.73, there exists an integer N such that

$$\|u - S_N u\|_{B_{p,r}^s} < \varepsilon/2.$$

Fix a smooth positive function θ supported in $B(0, 2)$ and with value 1 on the ball $B(0, 1)$. For $R > 0$, set $\theta_R \overset{\text{def}}{=} \theta(\cdot/R)$. Let $k = \max(0, [s] + 2)$. Arguing as in the proof of Proposition 2.27, we deduce that for all $j \in \mathbb{N}$, we have

$$2^{js} \|\Delta_j (\theta_R S_N u - S_N u)\|_{L^p} \leq C_s 2^{-j} \|D^k (\theta_R S_N u - S_N u)\|_{L^p}.$$

From the above inequality, we get that

$$\|\theta_R S_N u - S_N u\|_{B_{p,r}^s} \leq C_s \big(\|D^k (\theta_R S_N u - S_N u)\|_{L^p} + \|\theta_R S_N u - S_N u\|_{L^p} \big).$$

Because p is finite, combining Leibniz's formula and Lebesgue's dominated convergence theorem ensures that there exists some $R > 0$ such that

$$\|\theta_R S_N u - S_N u\|_{B_{p,r}^s} < \varepsilon/2.$$

As $S_N u$ is a C^∞ function, we have proven that $\mathcal{D}$ is dense in $B_{p,r}^s$. $\square$

Remark 2.75. When $r = \infty$, it is obvious that the closure of $\mathcal{D}$ for the Besov norm $B^s_{p,r}$ is the space of tempered distributions such that

$$\lim_{j \to \infty} 2^{js} \|\Delta_j u\|_{L^p} = 0.$$

Nonhomogeneous Besov spaces have nice properties of duality: The space $B^{-s}_{p',r'}$ may be identified with the dual space of the completion $\mathcal{B}^s_{p,r}$ of $\mathcal{D}$ for the norm $B^s_{p,r}$. In this book, we shall only use the following, much simpler, result, the proof of which is similar to that of Proposition 2.29.

Proposition 2.76. *For all $1 \leq p, r \leq \infty$ and $s \in \mathbb{R}$,*

$$\begin{cases} B^s_{p,r} \times B^{-s}_{p',r'} \longrightarrow \mathbb{R} \\ (u, \phi) \longmapsto \displaystyle\sum_{|j-j'| \leq 1} \langle \Delta_j u, \Delta_{j'} \phi \rangle \end{cases}$$

defines a continuous bilinear functional on $B^s_{p,r} \times B^{-s}_{p',r'}$. Denote by $Q^{-s}_{p',r'}$ the set of functions ϕ in $\mathcal{S}$ such that $\|\phi\|_{B^{-s}_{p',r'}} \leq 1$. If u is in $\mathcal{S}'$, then we have

$$\|u\|_{B^s_{p,r}} \leq C \sup_{\phi \in Q^{-s}_{p',r'}} \langle u, \phi \rangle.$$

We will now examine the way Fourier multipliers act on nonhomogeneous Besov spaces. Before stating our result, we need to define the multipliers we are going to consider.

Definition 2.77. *A smooth function $f : \mathbb{R}^d \to \mathbb{R}$ is said to be an S^m-multiplier if, for each multi-index α, there exists a constant C_α such that*

$$\forall \xi \in \mathbb{R}^d, \ |\partial^\alpha f(\xi)| \leq C_\alpha (1 + |\xi|)^{m-|\alpha|}.$$

Proposition 2.78. *Let $m \in \mathbb{R}$ and f be a S^m-multiplier. Then, for all $s \in \mathbb{R}$ and $1 \leq p, r \leq \infty$, the operator $f(D)$ is continuous from $B^s_{p,r}$ to $B^{s-m}_{p,r}$.*

Proof. According to Lemma 2.69 it suffices to prove that

$$\forall j \geq -1, \ 2^{j(s-m)} \|f(D)\Delta_j u\|_{L^p} \leq C 2^{js} \|\Delta_j u\|_{L^p}. \tag{2.40}$$

Obviously, we can find some smooth function σ satisfying the assumptions of Lemma 2.2 and such that

$$\forall j \geq 0, \ \Delta_j f(D) u = \sigma(D) \Delta_j u.$$

Hence, Lemma 2.2 guarantees that (2.40) is satisfied for $j \geq 0$.

Next, introducing θ in $\mathcal{D}(\mathbb{R}^d)$ such that $\theta \equiv 1$ on Supp χ, we see that

$$\Delta_{-1} f(D) u = (\theta f)(D) \Delta_{-1} u.$$

As $\mathcal{F}^{-1}(\theta f)$ is in L^1, convolution inequalities yield (2.40) for $j = -1$. This completes the proof. $\qquad \square$

Proposition 2.79. *Let $s < 0$, $1 \leq p, r \leq \infty$, and u be a tempered distribution. Then, u belongs to $B^s_{p,r}$ if and only if*

$$(2^{js}\|S_j u\|_{L^p})_{j\in\mathbb{N}} \in \ell^r.$$

Moreover, a constant C exists such that

$$C^{-|s|+1}\|u\|_{B^s_{p,r}} \leq \left\|(2^{js}\|S_j u\|_{L^p})_j\right\|_{\ell^r} \leq C\left(1 + \frac{1}{|s|}\right)\|u\|_{B^s_{p,r}}.$$

The proof is very close to that proof of Proposition 2.33 and is thus omitted.

We conclude this section with the statement of interpolation inequalities.

Theorem 2.80. *A constant C exists which satisfies the following properties. If s_1 and s_2 are real numbers such that $s_1 < s_2$, $\theta \in \,]0,1[$, and (p,r) is in $[1,\infty]$, then we have*

$$\|u\|_{B^{\theta s_1 + (1-\theta)s_2}_{p,r}} \leq \|u\|^{\theta}_{B^{s_1}_{p,r}}\|u\|^{1-\theta}_{B^{s_2}_{p,r}} \quad and$$

$$\|u\|_{B^{\theta s_1 + (1-\theta)s_2}_{p,1}} \leq \frac{C}{s_2 - s_1}\left(\frac{1}{\theta} + \frac{1}{1-\theta}\right)\|u\|^{\theta}_{B^{s_1}_{p,\infty}}\|u\|^{1-\theta}_{B^{s_2}_{p,\infty}}.$$

2.8 Nonhomogeneous Paradifferential Calculus

In this section, we are going to study the way the product acts on *nonhomogeneous* Besov spaces. Our approach will follow the one that we used in the homogeneous framework and most proofs will be omitted. Of course, we shall now use the nonhomogeneous Littlewood–Paley decomposition constructed in Section 2.2.

2.8.1 The Bony Decomposition

The basic idea of nonhomogeneous paradifferential calculus is the same as in Section 2.6: Considering two tempered distributions u and v, we have

$$uv = \sum_{j',j} \Delta_{j'}u \, \Delta_j v.$$

We then split the sum into three parts: The first corresponds to the low frequencies of u multiplied by the high frequencies of v, the second is the symmetric counterpart of the first, and the third part concerns the indices j and j' which are comparable. This leads to the following definition.

Definition 2.81. *The nonhomogeneous paraproduct of v by u is defined by*

$$T_u v \stackrel{def}{=} \sum_j S_{j-1}u \, \Delta_j v.$$

The nonhomogeneous remainder of u and v is defined by

$$R(u,v) = \sum_{|k-j|\leq 1} \Delta_k u \, \Delta_j v.$$

At least formally, the operators T and R are bilinear, and we have the following *Bony decomposition*:

$$uv = T_u v + T_v u + R(u,v). \tag{2.41}$$

We shall sometimes also use the following simplified decomposition:

$$uv = T_u v + T'_v u \quad \text{with} \quad T'_v u \overset{\text{def}}{=} \sum_j S_{j+2} v \, \Delta_j u. \tag{2.42}$$

The main continuity properties of the paraproduct are described below.

Theorem 2.82. *A constant C exists which satisfies the following inequalities for any couple of real numbers (s,t) with t negative and any (p, r_1, r_2) in $[1,\infty]^3$:*

$$\|T\|_{\mathcal{L}(L^\infty \times B^s_{p,r};B^s_{p,r})} \leq C^{|s|+1},$$

$$\|T\|_{\mathcal{L}(B^t_{\infty,r_1} \times B^s_{p,r_2};B^{s+t}_{p,r})} \leq \frac{C^{|s+t|+1}}{-t} \quad \text{with} \quad \frac{1}{r} \overset{\text{def}}{=} \min\left\{1, \frac{1}{r_1} + \frac{1}{r_2}\right\}.$$

The proof of this theorem is analogous to that of Theorem 2.47 and is thus omitted.

Remark 2.83. In fact, due to $S_j u = 0$ for $j < 0$ and the property (2.7), we have

$$T_u v = \sum_{j\geq 1} S_{j-1} u \, \Delta_j \big((\mathrm{Id} - \chi(D)) v \big).$$

Lemma 2.1 thus provides a slightly more accurate estimate: Under the assumptions of the above theorem, we have, for all $k \in \mathbb{N}$,

$$\|T_u v\|_{B^s_{p,r}} \leq C \|u\|_{L^\infty} \|D^k v\|_{B^{s-k}_{p,r}} \quad \text{and} \quad \|T_u v\|_{B^{s+t}_{p,r}} \leq C \|u\|_{B^t_{\infty,r_1}} \|D^k v\|_{B^{s-k}_{p,r_2}}.$$

Next, we want to study the continuity properties of the remainder operator R. As in the homogeneous case, we have to consider terms of the type $\Delta_j u \Delta_j v$ whose Fourier transforms are not supported in annuli but in balls $2^j B$. We thus need the following nonhomogeneous version of Lemma 2.49.

Lemma 2.84. *Let B be a ball in $\mathbb{R}^d$, s be a positive real number, and $(p,r) \in [1,\infty]^2$. Let $(u_j)_{j\in\mathbb{N}}$ be a sequence of smooth functions such that*

$$\mathrm{Supp} \ \widehat{u}_j \subset 2^j B \quad \text{and} \quad \left\| (2^{js} \|u_j\|_{L^p})_{j\in\mathbb{N}} \right\|_{\ell^r} < \infty.$$

We then have

$$u \overset{\text{def}}{=} \sum_{j\in\mathbb{N}} u_j \in B^s_{p,r} \quad \text{and} \quad \|u\|_{B^s_{p,r}} \leq C_s \left\| (2^{js} \|u_j\|_{L^p})_{j\in\mathbb{N}} \right\|_{\ell^r}.$$

Theorem 2.85. *A constant C exists which satisfies the following inequalities. Let (s_1, s_2) be in $\mathbb{R}^2$ and (p_1, p_2, r_1, r_2) be in $[1, \infty]^4$. Assume that*

$$\frac{1}{p} \overset{def}{=} \frac{1}{p_1} + \frac{1}{p_2} \le 1 \quad and \quad \frac{1}{r} \overset{def}{=} \frac{1}{r_1} + \frac{1}{r_2} \le 1.$$

If $s_1 + s_2 > 0$, then we have, for any (u, v) in $B^{s_1}_{p_1, r_1} \times B^{s_2}_{p_2, r_2}$,

$$\|R(u, v)\|_{B^{s_1 + s_2}_{p, r}} \le \frac{C^{|s_1 + s_2| + 1}}{s_1 + s_2} \|u\|_{B^{s_1}_{p_1, r_1}} \|v\|_{B^{s_2}_{p_2, r_2}}.$$

If $r = 1$ and $s_1 + s_2 = 0$, then we have, for any (u, v) in $B^{s_1}_{p_1, r_1} \times B^{s_2}_{p_2, r_2}$,

$$\|R(u, v)\|_{B^0_{p, \infty}} \le C^{|s_1 + s_2| + 1} \|u\|_{B^{s_1}_{p_1, r_1}} \|v\|_{B^{s_2}_{p_2, r_2}}.$$

From this theorem, we infer the following tame estimate.

Corollary 2.86. *For any positive real number s and any (p, r) in $[1, \infty]^2$, the space $L^\infty \cap B^s_{p, r}$ is an algebra, and a constant C exists such that*

$$\|uv\|_{B^s_{p, r}} \le \frac{C^{s+1}}{s} \left(\|u\|_{L^\infty} \|v\|_{B^s_{p, r}} + \|u\|_{B^s_{p, r}} \|v\|_{L^\infty} \right).$$

The proof simply involves the systematic use of Bony's decomposition (2.41) combined with Theorems 2.82 and 2.85.

2.8.2 The Paralinearization Theorem

In this subsection we investigate the effect of left composition by smooth functions on Besov spaces $B^s_{p, r}$. We state an initial result.

Theorem 2.87. *Let f be a smooth function vanishing at 0, s be a positive real number, and $(p, r) \in [1, \infty]^2$. If u belongs to $B^s_{p, r} \cap L^\infty$, then so does $f \circ u$, and we have*

$$\|f \circ u\|_{B^s_{p, r}} \le C(s, f', \|u\|_{L^\infty}) \|u\|_{B^s_{p, r}}.$$

This theorem can be proven along the same lines as the proof of Theorem 2.61. We note that it is based on the following lemma, the proof of which is left to the reader.

Lemma 2.88. *Let s be a positive real number and (p, r) be in $[1, \infty]^2$. A constant C_s exists such that if $(u_j)_{j \in \mathbb{N}}$ is a sequence of smooth functions which satisfies*

$$\left(\sup_{|\alpha| \le [s]+1} 2^{j(s-|\alpha|)} \|\partial^\alpha u_j\|_{L^p} \right)_j \in \ell^r(\mathbb{N}),$$

then we have

$$u \overset{def}{=} \sum_{j \in \mathbb{N}} u_j \in B^s_{p, r} \quad and \quad \|u\|_{B^s_{p, r}} \le C_s \left\| \left(\sup_{|\alpha| \le [s]+1} 2^{j(s-|\alpha|)} \|\partial^\alpha u_j\|_{L^p} \right)_j \right\|_{\ell^r}.$$

In the case where the function f belongs to $C_b^\infty(\mathbb{R})$, Theorem 2.87 may be slightly improved.

Theorem 2.89. *Let f be in $C_b^\infty(\mathbb{R})$ and satisfy $f(0) = 0$. Let s be positive and (p,r) be in $[1,\infty]^2$. If u belongs to $B_{p,r}^s$ and the first derivatives of u belong to $B_{\infty,\infty}^{-1}$, then $f \circ u$ belongs to $B_{p,r}^s$, and we have*

$$\|f \circ u\|_{B_{p,r}^s} \le C(s, f, \|\nabla u\|_{B_{\infty,\infty}^{-1}})\|u\|_{B_{p,r}^s}.$$

Remark 2.90. If u belongs to the space $B_{p,r}^{\frac{d}{p}}$, then the first order derivative of u belongs to $B_{\infty,\infty}^{-1}$. Thus, the space $B_{p,r}^{\frac{d}{p}}$ is stable under left composition by functions of C_b^∞ vanishing at 0. This result applies in particular to the Sobolev space $H^{\frac{d}{2}} = B_{2,2}^{\frac{d}{2}}$.

Finally, we state the nonhomogeneous counterpart of Corollary 2.66.

Corollary 2.91. *Let f be a smooth function such that $f'(0) = 0$. Let $s > 0$ and $(p,r) \in [1,\infty]^2$. For any couple (u,v) of functions in $B_{p,r}^s \cap L^\infty$, the function $f \circ v - f \circ u$ then belongs to $B_{p,r}^s \cap L^\infty$ and*

$$\|f(v) - f(u)\|_{B_{p,r}^s} \le C\Big(\|v - u\|_{B_{p,r}^s} \sup_{\tau \in [0,1]} \|u + \tau(v - u)\|_{L^\infty}$$

$$+ \|v - u\|_{L^\infty} \sup_{\tau \in [0,1]} \|u + \tau(v - u)\|_{B_{p,r}^s}\Big),$$

where C depends on f'', $\|u\|_{L^\infty}$, and $\|v\|_{L^\infty}$.

When the function u has enough regularity, we can obtain more information on $f \circ u$. In the following theorem, we state that, up to an error term which proves to be more regular than u, $f \circ u$ may be written as a paraproduct involving u and $f' \circ u$.

Theorem 2.92. *Let s and ρ be positive real numbers and f be a smooth function. Assume that ρ is not an integer. Let p, r_1, and r_2 be in $[1,\infty]$ and such that $r_2 \ge r_1$. Let $r \in [1,\infty]$ be defined by $1/r = \min(1, 1/r_1 + 1/r_2)$. For any function u in $B_{p,r_1}^s \cap B_{\infty,r_2}^\rho$, we then have*

$$\|f \circ u - T_{f' \circ u} u\|_{B_{p,r}^{s+\rho}} \le C(f'', \|u\|_{L^\infty})\|u\|_{B_{\infty,r_2}^\rho}\|u\|_{B_{p,r_1}^s}.$$

Proof. To prove this theorem, we again write that

$$f(u) = \sum_j f_j \quad \text{with} \quad f_j \stackrel{\text{def}}{=} f(S_{j+1}u) - f(S_j u).$$

According to the second order Taylor formula, we have

$$f_j = f'(S_j u)\Delta_j u + M_j(\Delta_j u)^2 \quad \text{with} \quad M_j \stackrel{\text{def}}{=} \int_0^1 (1-t)f''(S_j u + t\Delta_j u)\, dt.$$

Applying Lemma 2.63 with $g(x, y) = \int_0^1 (1 - t) f''(x + ty)\, dt$ gives

$$\forall \alpha \in \mathbb{N}^d, \ \|\partial^\alpha M_j\|_{L^\infty} \leq C_\alpha(f'', \|u\|_{L^\infty}) 2^{j|\alpha|}. \tag{2.43}$$

Using Leibniz's formula, we can write

$$\partial^\alpha (M_j(\Delta_j u)^2) = \sum_{\gamma \leq \beta \leq \alpha} C_\alpha^\beta C_\beta^\gamma \partial^{\alpha - \beta} M_j\, \partial^{\beta - \gamma} \Delta_j u\, \partial^\gamma \Delta_j u.$$

Using Lemma 2.1 and the inequality (2.43), we get

$$\|\partial^{\alpha - \beta} M_j\, \partial^{\beta - \gamma} \Delta_j u\, \partial^\gamma \Delta_j u\|_{L^p} \leq C_\alpha(f'', \|u\|_{L^\infty}) 2^{j|\alpha|} \|\Delta_j u\|_{L^\infty} \|\Delta_j u\|_{L^p}.$$

Thus, according to the definition of Besov spaces, we have, for some sequence $(c_{j,\alpha})_{j \geq -1}$ satisfying $\|(c_j)\|_{\ell^r} = 1$,

$$2^{j(s + \rho - |\alpha|)} \|\partial^\alpha (M_j(\Delta_j u)^2)\|_{L^p} \leq C_\alpha(f'', \|u\|_{L^\infty}) c_{j,\alpha} \|u\|_{B_{\infty, r_2}^\rho} \|u\|_{B_{p, r_1}^s}. \tag{2.44}$$

We now focus on the term $f'(S_j u) \Delta_j u$. Clearly, it is *not* the desired paraproduct involving u. Therefore, we consider

$$\mu_j \overset{\text{def}}{=} f'(S_j u) - S_{j-1}(f' \circ u).$$

Obviously, we have

$$f_j = S_{j-1}(f' \circ u) \Delta_j u + \mu_j \Delta_j u + M_j(\Delta_j u)^2.$$

We temporarily assume that

$$2^{j(\rho - |\alpha|)} \|\partial^\alpha \mu_j\|_{L^\infty} \leq c_{j,\alpha} C_\alpha(f'', \|u\|_{L^\infty}) \|u\|_{B_{\infty, r_2}^\rho} \text{ with } \|(c_{j,\alpha})\|_{\ell^{r_2}} = 1. \tag{2.45}$$

Using (2.44), we then have, for some sequence $(c_{j,\alpha})_{j \geq -1}$ belonging to the unit ball of ℓ^r,

$$2^{j(s + \rho - |\alpha|)} \|\partial^\alpha (f_j - S_{j-1}(f' \circ u) \Delta_j u)\|_{L^\infty} \leq C_\alpha(f'', \|u\|_{L^\infty}) c_{j,\alpha} \|u\|_{B_{\infty, r_2}^\rho} \|u\|_{B_{p, r_1}^s}.$$

Applying Lemma 2.88 then yields the desired result.

In order to complete the proof of the theorem, we have to justify the inequality (2.45). First, we investigate the case where $|\alpha| < \rho$. We have

$$\mu_j = \mu_j^{(1)} + \mu_j^{(2)} \qquad \text{with} \quad \begin{cases} \mu_j^{(1)} \overset{\text{def}}{=} f'(S_j u) - f'(u), \\ \mu_j^{(2)} \overset{\text{def}}{=} f'(u) - S_{j-1}(f'(u)). \end{cases}$$

Using the fact that $(S_j u)_{j \in \mathbb{N}}$ converges to u in L^∞, we get

$$f'(u) - f'(S_j u) = \sum_{j' \geq j} \widetilde{f}_{j'} \text{ with } \widetilde{f}_{j'} \overset{\text{def}}{=} f'(S_{j'+1} u) - f'(S_{j'} u). \tag{2.46}$$

Applying (2.38) yields, for some sequence $(c_{j,\alpha})_{j\geq-1}$ with $\|(c_{j,\alpha})\|_{\ell^{r_2}}=1$,

$$2^{j'(\rho-|\alpha|)}\|\partial^\alpha\widetilde{f}_{j'}\|_{L^\infty} \leq c_{j',\alpha}C_\alpha(f'',\|u\|_{L^\infty})\|u\|_{B^\rho_{\infty,r_2}}. \qquad (2.47)$$

By summation, we then infer that, when $|\alpha| < \rho$,

$$2^{j(\rho-|\alpha|)}\|\partial^\alpha(\mu_j^{(1)})\|_{L^\infty} \leq c_{j,\alpha}C_\alpha(f'',\|u\|_{L^\infty})\|u\|_{B^\rho_{\infty,r_2}} \quad \text{with} \quad \|(c_{j,\alpha})\|_{\ell^{r_2}}=1.$$

Next, thanks to Theorem 2.87, we have

$$\partial^\alpha f'(u) \in B^{\rho-|\alpha|}_{\infty,r_2} \quad \text{and} \quad \|\partial^\alpha f'(u)\|_{B^{\rho-|\alpha|}_{\infty,r_2}} \leq C_\alpha(f'',\|u\|_{L^\infty})\|u\|_{B^\rho_{\infty,r_2}}.$$

Thus, we can write that

$$2^{j(\rho-|\alpha|)}\|\partial^\alpha\mu_j^{(2)}\|_{L^\infty} \leq 2^{j(\rho-|\alpha|)}\sum_{j'\geq j-1}\|\Delta_{j'}\partial^\alpha f'(u)\|_{L^\infty}$$

$$\leq C_\alpha(f,\|u\|_{L^\infty})\|u\|_{B^\rho_{\infty,r_2}}\sum_{j'\geq j-1}c_{j',\alpha}2^{(j-j'()\rho-|\alpha|)}$$

$$\leq c_{j,\alpha}C_\alpha(f,\|u\|_{L^\infty})\|u\|_{B^\rho_{\infty,r_2}} \quad \text{with} \quad \|(c_{j,\alpha})\|_{\ell^{r_2}}=1.$$

This completes the proof of (2.45) when $|\alpha| < \rho$.

The case when $|\alpha| > \rho$ is treated differently.[3] As $\partial^\alpha f'(u)$ belongs to $B^{\rho-|\alpha|}_{\infty,r_2}$, we have, using Proposition 2.79 and Theorem 2.87,

$$2^{j(\rho-|\alpha|)}\|\partial^\alpha S_{j-1}f'(u)\|_{L^\infty} \leq c_{j,\alpha}C_\alpha(f'',\|u\|_{L^\infty})\|u\|_{B^\rho_{\infty,r_2}} \quad \text{with} \quad \|(c_{j,\alpha})\|_{\ell^{r_2}}=1.$$

We now estimate $\partial^\alpha f'(S_j u)$. Again using the fact that $(S_j u)_j$ converges to u in L^∞, we can write that

$$f'(S_j(u)) = \sum_{j'\leq j-1}\widetilde{f}_{j'} \quad \text{with} \quad \widetilde{f}_{j'} \stackrel{\text{def}}{=} f'(S_{j'+1}u) - f'(S_{j'}u).$$

Using (2.47), we then get

$$2^{j(\rho-|\alpha|)}\|\partial^\alpha f'(S_j u)\|_{L^\infty} \leq 2^{j(\rho-|\alpha|)}\sum_{j'\leq j-1}\|\partial^\alpha\widetilde{f}_{j'}\|_{L^\infty}$$

$$\leq C_\alpha(f'',\|u\|_{L^\infty})\|u\|_{B^\rho_{\infty,r_2}}\sum_{j'\leq j-1}c_{j',\alpha}2^{(j-j')(\rho-|\alpha|)}$$

$$\leq C_\alpha(f'',\|u\|_{L^\infty})\|u\|_{B^\rho_{\infty,r_2}}c_{j,\alpha} \quad \text{with} \quad \|(c_{j,\alpha})\|_{\ell^{r_2}}=1.$$

The inequality (2.45), and thus Theorem 2.92, is proved.   □

---

[3] Recall that ρ is not an integer, so $|\alpha| \neq \rho$.

2.9 Besov Spaces and Compact Embeddings

This section is devoted to the statement of (locally) compact embeddings for Besov spaces, properties which prove to be of importance for solving certain partial differential equations in the following chapters.

The following statement is an extension of Proposition 1.55 to general Besov spaces.

Proposition 2.93. *Let K be a compact subset of $\mathbb{R}^d$. Denote by $B^s_{p,r}(K)$ [resp., $\dot{B}^s_{p,r}(K)$] the set of distributions u in $B^s_{p,r}$ (resp., $\dot{B}^s_{p,r}$), the support of which is included in K. If $s > 0$, then the spaces $B^s_{p,r}(K)$ and $\dot{B}^s_{p,r}(K)$ coincide. Moreover, a constant C exists such that for any u in $\dot{B}^s_{p,r}(K)$,*

$$\|u\|_{B^s_{p,r}} \leq C\big(1 + |K|\big)^{\frac{s}{d}} \|u\|_{\dot{B}^s_{p,r}}.$$

Proof. For any j in $\mathbb{Z}$, we write $u = \dot{S}_j u + (\mathrm{Id} - \dot{S}_j)u$. As u belongs to $\dot{B}^s_{p,r}$, the function $\dot{S}_j u$ belongs to L^∞, and $(\mathrm{Id} - \dot{S}_j)u$ belongs to L^p. This implies that $\dot{B}^s_{p,r}$ is included in L^p_{loc} and thus that $\dot{B}^s_{p,r}(K)$ is included in L^p. In order to prove the inequality, we write, for any u in $\dot{B}^s_{p,r}(K)$ and $j \in \mathbb{Z}$,

$$\|u\|_{L^p(K)} \leq \|\dot{S}_j u\|_{L^p(K)} + \|(\mathrm{Id} - \dot{S}_j u)\|_{L^p}$$
$$\leq |K|^{\frac{1}{p}} \|\dot{S}_j u\|_{L^\infty} + C 2^{-js} \|u\|_{\dot{B}^s_{p,r}}.$$

Using Bernstein's inequalities and, again, the fact that Supp $u \subset K$, we get

$$\|u\|_{L^p} \leq C|K|^{\frac{1}{p}} 2^{jd} \|u\|_{L^1} + C 2^{-js} \|u\|_{\dot{B}^s_{p,r}}$$
$$\leq C|K| 2^{jd} \|u\|_{L^p} + C 2^{-js} \|u\|_{\dot{B}^s_{p,r}}.$$

If j is chosen in $\mathbb{Z}$ such that $1/4 \leq |K| 2^{jd} \leq 1/2$, then the first term of the right-hand side may be absorbed by the left-hand side, and we can infer that

$$\|u\|_{L^p} \leq C|K|^{\frac{s}{d}} \|u\|_{\dot{B}^s_{p,r}}.$$

Because s is positive, we have $B^s_{p,r} = \dot{B}^s_{p,r} \cap L^p$. This completes the proof of the proposition.   □

Theorem 2.94. *If $s' < s$, then for all ϕ in $\mathcal{S}(\mathbb{R}^d)$, multiplication by ϕ is a compact operator from $B^s_{p,\infty}$ to $B^{s'}_{p,1}$.*

Proof. Let $(u_n)_{n\in\mathbb{N}}$ be a bounded sequence of $B^s_{p,\infty}$. Thanks to Theorem 2.72, a subsequence $(u_{\psi(n)})_{n\in\mathbb{N}}$ and a function u exist in $B^s_{p,\infty}$ such that $(u_{\psi(n)})_{n\in\mathbb{N}}$ converges to u in $\mathcal{S}'$. Thus, we are reduced to proving that if $(u_n)_{n\in\mathbb{N}}$ is a bounded sequence of $B^s_{p,\infty}$ which tends to 0 in $\mathcal{S}'$, then $\|\phi u_n\|_{B^{s'}_{p,1}}$ tends to 0.

By virtue of product laws in nonhomogeneous Besov spaces (see Theorems 2.82 and 2.85), the sequence $(\phi u_n)_{n \in \mathbb{N}}$ is bounded in $B_{p,\infty}^s$. We then write

$$\|\phi u_n\|_{B_{p,1}^{s'}} = \sum_j 2^{js'} \|\Delta_j(\phi u_n)\|_{L^p}$$

$$\leq \sum_{j \leq j_0} 2^{js'} \|\Delta_j(\phi u_n)\|_{L^p} + \sum_{j > j_0} 2^{-j(s-s')} 2^{js} \|\Delta_j(\phi u_n)\|_{L^p}$$

$$\leq \sum_{j \leq j_0} 2^{js'} \|\Delta_j(\phi u_n)\|_{L^p} + C_{s,s'} 2^{-j_0(s-s')} \sup_n \|\phi u_n\|_{B_{p,\infty}^s}.$$

A positive ε being given, we choose j_0 such that

$$C_{s,s'} 2^{-j_0(s-s')} \sup_n \|\phi u_n\|_{B_{p,\infty}^s} \leq \varepsilon/2.$$

We then simply have to prove that

$$\lim_{n \to \infty} \|\Delta_j(\phi u_n)\|_{L^p} = 0 \quad \text{for all} \ \ j \geq -1. \tag{2.48}$$

Actually, it suffices to consider the case where $p = 1$. Indeed, first, since ϕ is in (say) $B_{p',\infty}^{|s|+1}$ and $(u_n)_{n \in \mathbb{N}}$ is bounded in $B_{p,\infty}^s$, it is not difficult to check that $(\phi u_n)_{n \in \mathbb{N}}$ is bounded in $B_{1,\infty}^s$ (use Theorems 2.82 and 2.85). Second, Bernstein's lemma guarantees that

$$\|\Delta_j(\phi u_n)\|_{L^p} \leq C 2^{jd/p'} \|\Delta_j(\phi u_n)\|_{L^1}.$$

We therefore assume from now on that $p = 1$. We only treat the case where $j \in \mathbb{N}$, the case $j = -1$ being similar. By the definition of Δ_j, we then have

$$\Delta_j(\phi u_n)(x) = 2^{jd} \int_{\mathbb{R}^d} h(2^j(x-y))\phi(y)u_n(y) \, dy$$

$$= 2^{jd} \langle u_n, \tau_{-x} \check{h}(2^j \cdot)\phi \rangle.$$

As u_n tends to 0 in $\mathcal{S}'$, the above equation ensures that the function $\Delta_j(\phi u_n)$ tends to 0 pointwise. Moreover, according to Proposition 2.76,

$$|\Delta_j(\phi u_n)(x)| \leq C 2^{jd} \left(\sup_n \|u_n\|_{B_{1,\infty}^s} \right) \|\tau_{-x} \check{h}(2^j \cdot)\phi\|_{B_{\infty,1}^{-s}}.$$

Hence, thanks to Lebesgue's dominated convergence theorem, proving (2.48) reduces to the following lemma.

Lemma 2.95. *For any (f, g) in $\mathcal{S}^2$ and any (σ, p, r) in $\mathbb{R} \times [1, \infty]^2$, the map*

$$z \longmapsto \|(\tau_z f)g\|_{B_{p,r}^{\sigma}}$$

belongs to $L^1(\mathbb{R}^d)$.

Proof. Observe that for $j \geq 0$, by using a rescaled version of the relation (2.1.1) and Leibniz's formula, we get, for any positive integer N and some functions h_α in $\mathcal{S}(\mathbb{R}^d)$,

$$\Delta_j(\tau_z f\, g) = 2^{-jN} \sum_{\substack{|\alpha|=N \\ \beta \leq \alpha}} 2^{jd} h_\alpha(2^j \cdot) \star \left(\partial^{\alpha-\beta}\tau_z f\, \partial^\beta g\right).$$

Thus, using Bernstein's inequalities, we infer that

$$\|\Delta_j(\tau_z f\, g)\|_{L^p} \leq C_N 2^{-j(N-\frac{d}{p'})} \sup_{|\alpha+\beta| \leq N} \|\partial^\alpha \tau_z f\, \partial^\beta g\|_{L^1}$$

$$\leq C_N 2^{-j(N-\frac{d}{p'})} (f_N \star g_N)(z)$$

with $f_N(x) \stackrel{\text{def}}{=} \sup_{|\alpha| \leq N} |\partial^\alpha f(x)|$ and $g_N(x) \stackrel{\text{def}}{=} \sup_{|\alpha| \leq N} |\partial^\alpha g(x)|$. Choosing N greater than $d + \sigma + 1$, we infer that

$$\|\tau_z f\, g\|_{B_{p,r}^\sigma} \leq C(f_N \star g_N)(z).$$

Observing that the convolution maps $L^1 \times L^1$ into L^1 completes the proof of the lemma. $\square$

Theorem 2.94 immediately implies the following corollary.

Corollary 2.96. *For any (s', s) in $\mathbb{R}^2$ such that $s' < s$ and any compact set K of $\mathbb{R}^d$, the space $B_{p,\infty}^s(K)$ is compactly embedded in $B_{p,1}^{s'}(K)$.*

2.10 Commutator Estimates

This section is devoted to various commutator estimates which will be used in the next chapters. The following basic lemma will be of constant use in this section.

Lemma 2.97. *Let θ be a C^1 function on $\mathbb{R}^d$ such that $(1 + |\cdot|)\widehat{\theta} \in L^1$. There exists a constant C such that for any Lipschitz function a with gradient in L^p and any function b in L^q, we have, for any positive λ,*

$$\|[\theta(\lambda^{-1}D), a]b\|_{L^r} \leq C\lambda^{-1}\|\nabla a\|_{L^p}\|b\|_{L^q} \quad \text{with} \quad \frac{1}{p} + \frac{1}{q} = \frac{1}{r}.$$

Proof. In order to prove this lemma, it suffices to rewrite $\theta(\lambda^{-1}D)$ as a convolution operator. Indeed,

$$\left([\theta(\lambda^{-1}D), a]b\right)(x) = \theta(\lambda^{-1}D)(ab)(x) - a(x)\theta(\lambda^{-1}D)b(x)$$

$$= \lambda^d \int_{\mathbb{R}^d} k(\lambda(x-y))(a(y)-a(x))b(y)\,dy \quad \text{with} \quad k = \mathcal{F}^{-1}\theta.$$

Let $k_1(z) \overset{\text{def}}{=} |z|\,|k(z)|$. From the first order Taylor formula, we deduce that

$$\left|\left([\theta(\lambda^{-1}D), a]b\right)(x)\right| \le \lambda^{-1} \int_{[0,1]\times\mathbb{R}^d} \lambda^d |k_1(\lambda z)| |\nabla a(x - \tau z)|\,|b(x - z)|\,dz\,d\tau.$$

Now, taking the L^r norm of the above inequality, using the fact that the norm of an integral is less than the integral of the norm, and using Hölder's inequality, we get

$$\left\|[\theta(\lambda^{-1}D), a]b\right\|_{L^r} \le \lambda^{-1} \int_0^1 \int_{\mathbb{R}^d} \lambda^d k_1(\lambda z) \|\nabla a(\cdot - \tau z)\|_{L^p} \|b(\cdot - z)\|_{L^q}\,d\tau\,dz.$$

The translation invariance of the Lebesgue measure then ensures that

$$\left\|[\theta(\lambda^{-1}D), a]b\right\|_{L^r} \le \lambda^{-1} \|k_1\|_{L^1} \|\nabla a\|_{L^p} \|b\|_{L^q},$$

which is the desired result. □

Remark 2.98. If we take $\theta = \varphi$ and $\lambda = 2^j$, then this lemma can be interpreted as a gain of one derivative by commutation between the operator Δ_j and the multiplication by a function with gradient in L^p.

Lemma 2.99. *Let f be a smooth function on $\mathbb{R}^d$. Assume that f is homogeneous of degree m away from a neighborhood of 0. Let ρ be in $]0,1[$, s be in $\mathbb{R}$, and (p,r) be in $[1,\infty]^2$. There exists a constant C, depending only on s, ρ, and d, such that if $(p_1, p_2) \in [1,\infty]^2$ satisfies $1/p = 1/p_1 + 1/p_2$, then the following estimate holds true:*

$$\|[T_a, f(D)]u\|_{B^{s-m+\rho}_{p,r}} \le C\|\nabla a\|_{B^{\rho-1}_{p_1,\infty}} \|u\|_{B^s_{p_2,r}}. \tag{2.49}$$

In the limit case $\rho = 1$, we have

$$\|[T_a, f(D)]u\|_{B^{s-m+1}_{p,r}} \le C\|\nabla a\|_{L^{p_1}} \|u\|_{B^s_{p_2,r}}. \tag{2.50}$$

Proof. We only treat the case $\rho < 1$. The limit case $\rho = 1$ stems from similar arguments. Let $\widetilde{\varphi}$ be a smooth function supported in an annulus and with value 1 on a neighborhood of Supp φ + Supp $\chi(\cdot/4)$. We have

$$[T_a, f(D)]u = \sum_{j\ge 1} S_{j-1}a\,f(D)\Delta_j u - f(D)\big(S_{j-1}a\Delta_j u\big)$$

$$= \sum_{j\ge 1} [S_{j-1}a, f(D)\widetilde{\Delta}_j]\Delta_j u \quad \text{with} \quad \widetilde{\Delta}_j \overset{\text{def}}{=} \widetilde{\varphi}(2^{-j}D).$$

Note that the general term of the above series is spectrally supported in dyadic annuli. Hence, according to Lemma 2.69, it suffices to prove that

$$\left\| 2^{j(s-m+\rho)} \|[S_{j-1}a, f(D)\widetilde{\Delta}_j]\Delta_j u\|_{L^p} \right\|_{\ell^r} \le C\|\nabla a\|_{B^{\rho-1}_{p_1,\infty}} \|u\|_{B^s_{p_2,r}}. \tag{2.51}$$

Owing to the homogeneity of the function f away from 0, there exists an integer N_0 such that

$$\forall j \geq N_0, \ f(D)\widetilde{\Delta}_j = 2^{jm}(f\widetilde{\varphi})(2^{-j}D).$$

Taking advantage of Lemma 2.97, we thus infer that for any $j \geq N_0$,

$$\|[S_{j-1}a, f(D)\widetilde{\Delta}_j]\Delta_j u\|_{L^p} \leq C2^{j(m-1)}\|S_{j-1}a\|_{L^{p_1}}\|\Delta_j u\|_{L^{p_2}}.$$

Of course, if $1 \leq j < N_0$, we can still write, according to Lemma 2.97,

$$\|[S_{j-1}a, f(D)\widetilde{\Delta}_j]\Delta_j u\|_{L^p} \leq C2^{-j}\|\nabla S_{j-1}a\|_{L^{p_1}}\|\Delta_j u\|_{L^{p_2}}$$
$$\leq C2^{N_0|m|}2^{j(m-1)}\|\nabla S_{j-1}a\|_{L^{p_1}}\|\Delta_j u\|_{L^{p_2}}.$$

Because $\|\nabla S_{j-1}a\|_{L^{p_1}} \leq C2^{j(1-\rho)}\|\nabla a\|_{B^{\rho-1}_{p_1,\infty}}$ if $\rho < 1$, we can now conclude that (2.51) is satisfied, completing the proof. $\qquad\square$

The following corollary will be important in the next chapter.

Lemma 2.100. *Let $\sigma \in \mathbb{R}$, $1 \leq r \leq \infty$, and $1 \leq p \leq p_1 \leq \infty$. Let v be a vector field over $\mathbb{R}^d$. Assume that*

$$\sigma > -d\min\left\{\frac{1}{p_1},\frac{1}{p'}\right\} \quad \text{or} \quad \sigma > -1 - d\min\left\{\frac{1}{p_1},\frac{1}{p'}\right\} \quad \text{if} \quad \operatorname{div} v = 0. \quad (2.52)$$

Define $R_j \overset{def}{=} [v \cdot \nabla, \Delta_j]f$ (or $R_j \overset{def}{=} \operatorname{div}([v, \Delta_j]f)$, if $\operatorname{div} v = 0$). There exists a constant C, depending continuously on p, p_1, σ, and d, such that

$$\left\|\left(2^{j\sigma}\|R_j\|_{L^p}\right)_j\right\|_{\ell^r} \leq C\|\nabla v\|_{B^{\frac{d}{p}}_{p_1,\infty}\cap L^\infty}\|f\|_{B^\sigma_{p,r}} \quad \text{if} \quad \sigma < 1 + \frac{d}{p_1}. \quad (2.53)$$

Further, if $\sigma > 0$ (or $\sigma > -1$, if $\operatorname{div} v = 0$) and $\frac{1}{p_2} = \frac{1}{p} - \frac{1}{p_1}$, then

$$\left\|\left(2^{j\sigma}\|R_j\|_{L^p}\right)_j\right\|_{\ell^r} \leq C\left(\|\nabla v\|_{L^\infty}\|f\|_{B^\sigma_{p,r}} + \|\nabla f\|_{L^{p_2}}\|\nabla v\|_{B^{\sigma-1}_{p_1,r}}\right). \quad (2.54)$$

In the limit case $\sigma = -\min\left(\frac{d}{p_1},\frac{d}{p'}\right)$ [or $\sigma = -1 - \min\left(\frac{d}{p_1},\frac{d}{p'}\right)$, if $\operatorname{div} v = 0$], we have

$$\sup_{j\geq-1} 2^{j\sigma}\|R_j\|_{L^p} \leq C\|\nabla v\|_{B^{\frac{d}{p_1}}_{p_1,1}}\|f\|_{B^\sigma_{p,\infty}}. \quad (2.55)$$

Proof. In order to show that only the gradient part of v is involved in the estimates, we shall split v into low and high frequencies: $v = S_0 v + \widetilde{v}$. Obviously, there exists a constant C such that

$$\forall a \in [1,\infty], \ \|S_0\nabla v\|_{L^a} \leq C\|\nabla v\|_{L^a} \quad \text{and} \quad \|\nabla\widetilde{v}\|_{L^a} \leq C\|\nabla v\|_{L^a}. \quad (2.56)$$

Further, as $\widetilde{v}$ is spectrally supported away from the origin, Lemma 2.1 ensures that

$$\forall a \in [1,\infty], \ \forall j \geq -1, \ \|\Delta_j\nabla\widetilde{v}\|_{L^a} \approx 2^j\|\Delta_j\widetilde{v}\|_{L^a}. \quad (2.57)$$

We now have (with the summation convention over repeated indices):

$$R_j = v \cdot \nabla \Delta_j f - \Delta_j (v \cdot \nabla f)$$
$$= [\widetilde{v}^k, \Delta_j] \partial_k f + [S_0 v^k, \Delta_j] \partial_k f.$$

Hence, writing Bony's decomposition for $[\widetilde{v}^k, \Delta_j] \partial_k f$, we end up with $R_j = \sum_{i=1}^{8} R_j^i$, where

$$
\begin{aligned}
R_j^1 &= [T_{\widetilde{v}^k}, \Delta_j] \partial_k f, & R_j^2 &= T_{\partial_k \Delta_j f} \widetilde{v}^k, \\
R_j^3 &= -\Delta_j T_{\partial_k f} \widetilde{v}^k, & R_j^4 &= \partial_k R(\widetilde{v}^k, \Delta_j f), \\
R_j^5 &= -R(\operatorname{div} \widetilde{v}, \Delta_j f), & R_j^6 &= -\partial_k \Delta_j R(\widetilde{v}^k, f), \\
R_j^7 &= \Delta_j R(\operatorname{div} \widetilde{v}, f), & R_j^8 &= [S_0 v^k, \Delta_j] \partial_k f.
\end{aligned}
$$

In the following computations, the constant C depends continuously on σ, p, p_1, and d, and we denote by $(c_j)_{j \geq -1}$ a sequence such that $\|(c_j)\|_{\ell^r} \leq 1$.

Bounds for $2^{j\sigma} \|R_j^1\|_{L^p}$. By virtue of Proposition 2.10, we have

$$R_j^1 = \sum_{|j-j'| \leq 4} [S_{j'-1} \widetilde{v}^k, \Delta_j] \partial_k \Delta_{j'} f.$$

Hence, according to Lemma 2.97 and the inequality (2.56),

$$
\begin{aligned}
2^{j\sigma} \|R_j^1\|_{L^p} &\leq C \|\nabla v\|_{L^\infty} \sum_{|j'-j| \leq 4} 2^{j'\sigma} \|\Delta_{j'} f\|_{L^p} \\
&\leq C c_j \|\nabla v\|_{L^\infty} \|f\|_{B_{p,r}^\sigma}.
\end{aligned}
\tag{2.58}
$$

Bounds for $2^{j\sigma} \|R_j^2\|_{L^p}$. By virtue of Proposition 2.10, we have

$$R_j^2 = \sum_{j' \geq j-3} S_{j'-1} \partial_k \Delta_j f \, \Delta_{j'} \widetilde{v}^k.$$

Hence, using inequalities (2.56) and (2.57) yields

$$2^{j\sigma} \|R_j^2\|_{L^p} \leq C c_j \|\nabla v\|_{L^\infty} \|f\|_{B_{p,r}^\sigma}.
\tag{2.59}$$

Bounds for $2^{j\sigma} \|R_j^3\|_{L^p}$. We proceed as follows:

$$R_j^3 = -\sum_{|j'-j| \leq 4} \Delta_j \left(S_{j'-1} \partial_k f \Delta_{j'} \widetilde{v}^k \right)
\tag{2.60}$$

$$= -\sum_{\substack{|j'-j| \leq 4 \\ j'' \leq j'-2}} \Delta_j \left(\Delta_{j''} \partial_k f \Delta_{j'} \widetilde{v}^k \right).
\tag{2.61}$$

Therefore, writing $1/p_2 = 1/p - 1/p_1$ and using (2.56) and (2.57), we have

$$2^{j\sigma} \left\| R_j^3 \right\|_{L^p} \leq C \sum_{\substack{|j'-j| \leq 4 \\ j'' \leq j'-2}} 2^{j\sigma} \left\| \Delta_{j''} \partial_k f \right\|_{L^{p_2}} \left\| \Delta_{j'} \widetilde{v}^k \right\|_{L^{p_1}}$$

$$\leq C \sum_{\substack{|j'-j| \leq 4 \\ j'' \leq j'-2}} 2^{(j-j'')(\sigma-1-\frac{d}{p_1})} 2^{j''\sigma} \left\| \Delta_{j''} f \right\|_{L^p} 2^{j'\frac{d}{p_1}} \left\| \Delta_{j'} \nabla v \right\|_{L^{p_1}}.$$

Hence, if $\sigma < 1 + d/p_1$, then

$$2^{j\sigma} \left\| R_j^3 \right\|_{L^p} \leq C c_j \|\nabla v\|_{B_{p_1,\infty}^{\frac{d}{p_1}}} \|f\|_{B_{p,r}^{\sigma}}. \tag{2.62}$$

Note that, starting from (2.60), we can alternatively get

$$2^{j\sigma} \left\| R_j^3 \right\|_{L^p} \leq C \sum_{|j'-j| \leq 4} \|\nabla S_{j'-1} f\|_{L^{p_2}} 2^{j'(\sigma-1)} \|\Delta_{j'} \nabla v\|_{L^{p_1}},$$

from which it follows that

$$2^{j\sigma} \left\| R_j^3 \right\|_{L^p} \leq C c_j \|\nabla f\|_{L^{p_2}} \|\nabla v\|_{B_{p_1,r}^{\sigma-1}}. \tag{2.63}$$

Bounds for $2^{j\sigma} \left\| R_j^4 \right\|_{L^p}$ *and* $2^{j\sigma} \left\| R_j^5 \right\|_{L^p}$. Defining $\widetilde{\Delta}_{j'} \overset{\text{def}}{=} \Delta_{j'-1} + \Delta_{j'} + \Delta_{j'+1}$, we have

$$R_j^4 = \sum_{|j'-j| \leq 2} \partial_k (\Delta_{j'} \widetilde{v}^k \, \Delta_j \widetilde{\Delta}_{j'} f).$$

Hence, by virtue of (2.57), we get

$$2^{j\sigma} \left\| R_j^4 \right\|_{L^p} \leq C c_j \|\nabla v\|_{L^\infty} \|f\|_{B_{p,r}^{\sigma}}. \tag{2.64}$$

A similar bound holds for R_j^5.

Bounds for $2^{j\sigma} \left\| R_j^6 \right\|_{L^p}$ *and* $2^{j\sigma} \left\| R_j^7 \right\|_{L^p}$. We first consider the case where $1/p + 1/p_1 \leq 1$. Let p_3 satisfy $1/p_3 \overset{\text{def}}{=} 1/p + 1/p_1$. Then, under the condition $\sigma > -1 - d/p_1$, Proposition 2.85, combined with the embedding $B_{p_3,r}^{\sigma+\frac{d}{p_1}} \hookrightarrow B_{p,r}^{\sigma}$, yields

$$2^{j\sigma} \left\| R_j^6 \right\|_{L^p} \leq C c_j \|\widetilde{v}\|_{B_{p_1,\infty}^{\frac{d}{p_1}+1}} \|f\|_{B_{p,r}^{\sigma}}. \tag{2.65}$$

Now, if $1/p + 1/p_1 > 1$, then the above argument has to be applied with p' instead of p_2, and we still get (2.65), provided that $\sigma > -1 - \frac{d}{p'}$. Appealing to (2.56), we eventually get

$$2^{j\sigma} \left\| R_j^6 \right\|_{L^p} \leq C c_j \|\nabla v\|_{B_{p_1,\infty}^{\frac{d}{p_1}}} \|f\|_{B_{p,r}^{\sigma}}. \tag{2.66}$$

Note that in the limit case $\sigma = -1 - \min(\frac{d}{p_1}, \frac{d}{p'})$, Proposition 2.85 yields

$$\sup_j 2^{j\sigma} \left\| R_j^6 \right\|_{L^p} \leq C \|\nabla v\|_{B_{p_1,\infty}^{\frac{d}{p_1}}} \|f\|_{B_{p,\infty}^{\sigma}}. \tag{2.67}$$

Similar arguments lead to

$$2^{j\sigma} \left\| R_j^7 \right\|_{L^p} \le Cc_j \|\nabla v\|_{B_{p_1,\infty}^{\frac{d}{p_1}}} \|f\|_{B_{p,r}^\sigma}, \quad \text{if } \sigma > -\min(\tfrac{d}{p_1}, \tfrac{d}{p'}), \tag{2.68}$$

$$2^{j\sigma} \left\| R_j^7 \right\|_{L^p} \le C \|\nabla v\|_{B_{p_1,1}^{\frac{d}{p_1}}} \|f\|_{B_{p,\infty}^\sigma}, \quad \text{if } \sigma = -\min(\tfrac{d}{p_1}, \tfrac{d}{p'}) \text{ and } r = \infty. \tag{2.69}$$

Finally, we stress that if $\sigma > -1$, then the standard continuity results for the remainder, combined with the embedding $L^\infty \hookrightarrow B_{\infty,\infty}^0$, yield

$$2^{j\sigma} \left\| R_j^6 \right\|_{L^p} \le Cc_j \|\nabla v\|_{L^\infty} \|f\|_{B_{p,r}^\sigma}. \tag{2.70}$$

Of course, the same inequality holds true for R_j^7 if $\sigma > 0$.

Bounds for $2^{j\sigma} \left\| R_j^8 \right\|_{L^p}$. As $R_j^8 = \sum_{|j'-j|\le 1} [\Delta_j, \Delta_{-1}v] \cdot \nabla \Delta_{j'} f$, Lemma 2.97 yields

$$2^{j\sigma} \left\| R_j^8 \right\|_{L^p} \le C \sum_{|j'-j|\le 1} \|\nabla \Delta_{-1}v\|_{L^\infty} 2^{j'\sigma} \|\Delta_{j'} f\|_{L^p}$$
$$\le Cc_j \|\nabla v\|_{L^\infty} \|f\|_{B_{p,r}^\sigma}. \tag{2.71}$$

Combining inequalities (2.58), (2.59), (2.62) or (2.63), (2.64), (2.66) or (2.67), (2.68), (2.69) or (2.70), and (2.71) yields (2.53), (2.54), and (2.55). □

Remark 2.101. Assume that $\sigma > 1 + \frac{d}{p_1}$, or $\sigma = 1 + \frac{d}{p_1}$ and $r = 1$. We note that $B_{p,r}^{\sigma-1} \hookrightarrow L^{p_2}$, so the inequality (2.54) ensures that

$$\left\| 2^{j\sigma} \|R_j\|_{L^p} \right\|_{\ell^r} \le C \|\nabla v\|_{B_{p_1,r}^{\sigma-1}} \|f\|_{B_{p,r}^\sigma}.$$

Remark 2.102. There are a number of variations on the statement of Lemma 2.100. For instance, the inequalities (2.53), (2.54), and (2.55) are also valid in the homogeneous framework (i.e., with $\dot{\Delta}_j$ instead of Δ_j and with homogeneous Besov norms instead of nonhomogeneous ones), provided (2.17) is satisfied by (p, r, σ). The proof follows along the lines of the proof of Lemma 2.100. It is simply a matter of replacing the nonhomogeneous blocks by homogeneous ones.

Remark 2.103. In Section 3.4 of the next chapter, we shall also make use of the fact that the inequalities (2.53), (2.54), and (2.55) are still true for the commutator

$$\dot{S}_{j+N_0} v \cdot \nabla \dot{\Delta}_j f - \dot{\Delta}_j (v \cdot \nabla f),$$

where N_0 is any fixed integer. Indeed, it suffices to note that for all $j \ge -1$, we have

$$\left\| (\dot{S}_{j+N_0} v - v) \cdot \nabla \dot{\Delta}_j f \right\|_{L^p} \le C 2^j \left\| \dot{S}_{j+N_0} v - v \right\|_{L^\infty} \left\| \dot{\Delta}_j f \right\|_{L^p}$$
$$\le C \sum_{j' \ge j+N_0} 2^{j-j'} \left\| \nabla \dot{\Delta}_{j'} v \right\|_{L^\infty} \left\| \dot{\Delta}_j f \right\|_{L^p}$$
$$\le C \|\nabla v\|_{\dot{B}_{\infty,\infty}^0} \|\dot{\Delta}_j f\|_{L^p}.$$

2.11 Around the Space $B^1_{\infty,\infty}$

The space $B^1_{\infty,\infty}$ will play an important role in Chapter 7 when dealing with the incompressible Euler equations. This section is devoted to proving various logarithmic interpolation inequalities involving that space. We start with the most elementary of these.

Proposition 2.104. *Let ε be in $]0,1[$. A constant C exists such that for any f in $B^\varepsilon_{\infty,\infty}$,*

$$\|f\|_{L^\infty} \leq \frac{C}{\varepsilon}\|f\|_{B^0_{\infty,\infty}}\left(1 + \log\frac{\|f\|_{B^\varepsilon_{\infty,\infty}}}{\|f\|_{B^0_{\infty,\infty}}}\right).$$

Proof. In order to prove this, we write the function f as the sum of the dyadic blocks $\Delta_j f$. For any positive integer N, we have

$$\sum_{j\geq-1}\|\Delta_j f\|_{L^\infty} \leq \sum_{-1\leq j\leq N-1}\|\Delta_j f\|_{L^\infty} + \sum_{j\geq N}\|\Delta_j f\|_{L^\infty}$$

$$\leq (N+1)\|f\|_{B^0_{\infty,\infty}} + \frac{2^{-(N-1)\varepsilon}}{2^\varepsilon - 1}\|f\|_{B^\varepsilon_{\infty,\infty}}.$$

As $\|f\|_{L^\infty} \leq \sum_{j\geq-1}\|\Delta_j f\|_{L^\infty}$, taking

$$N = 1 + \left[\frac{1}{\varepsilon}\log_2\frac{\|f\|_{B^\varepsilon_{\infty,\infty}}}{\|f\|_{B^0_{\infty,\infty}}}\right]$$

yields the result   □

Remark 2.105. In fact, the above proof gives the following, slightly more accurate, estimate:

$$\|f\|_{B^0_{\infty,1}} \leq \frac{C}{\varepsilon}\|f\|_{B^0_{\infty,\infty}}\left(1 + \log\frac{\|f\|_{B^\varepsilon_{\infty,\infty}}}{\|f\|_{B^0_{\infty,\infty}}}\right).$$

We now define the space LL of log-Lipschitz functions.

Definition 2.106. *The space LL consists of those bounded functions f such that*

$$\|f\|_{LL} \overset{def}{=} \sup_{0<|x-x'|\leq 1}\frac{|f(x)-f(x')|}{|x-x'|(1-\log|x-x'|)} < \infty.$$

$B^1_{\infty,\infty}$ is a subspace of the space LL of log-Lipschitz functions. More precisely, we have the following.

Proposition 2.107. *A constant C exists such that for any function u in $B^1_{\infty,\infty}$ and any x,y in $\mathbb{R}^d$ such that $|x-y|\leq 1$, we have*

$$|u(x)-u(y)| \leq C\|\nabla u\|_{B^0_{\infty,\infty}}|x-y|(1-\log|x-y|).$$

Proof. The proof is similar to the one above. We write that, for $|x - y| \leq 1$,

$$u(x) - u(y) = \sum_{j<N} \Delta_j u(x) - \Delta_j u(y) + \sum_{j\geq N} \Delta_j u(x) - \Delta_j u(y).$$

By combining the mean value inequality and Bernstein's lemma, we get

$$|u(x) - u(y)| \leq C|x - y| \sum_{j<N} \|\Delta_j \nabla u\|_{L^\infty} + 2 \sum_{j\geq N} 2^{-j} \|\nabla \Delta_j u\|_{L^\infty},$$

from which it follows, by the definition of the space $B^0_{\infty,\infty}$, that

$$|u(x) - u(y)| \leq C\|\nabla u\|_{B^0_{\infty,\infty}} ((N+1)|x - y| + 2^{-N}).$$

As above, choosing $N = [-\log_2 |x - y|] + 1$ completes the proof.   □

We have just established a relationship between the modulus of continuity and the growth of L^∞ norms of dyadic blocks in the special case of log-Lipschitz functions. A similar connection may be established for a more general class of moduli of continuity given in the following definition.

Definition 2.108. *Let a be in $]0, 1]$. A* modulus of continuity *is any nondecreasing nonzero continuous function $\mu : [0, a] \to \mathbb{R}_+$ such that $\mu(0) = 0$. The modulus of continuity μ is* admissible *if, in addition, the function Γ defined for $y \geq 1/a$ by*

$$\Gamma(y) \stackrel{def}{=} y\mu\left(\frac{1}{y}\right)$$

is nondecreasing and satisfies, for some constant C and all $x \geq 1/a$,

$$(A) \qquad \int_x^\infty \frac{1}{y^2} \Gamma(y) \, dy \leq C \frac{\Gamma(x)}{x}.$$

Examples. If $\alpha \in \ ]0, 1]$, then the functions $\mu(r) = r^\alpha$, $\mu(r) = r(-\log r)^\alpha$, and $\mu(r) = r(-\log r)(\log(-\log r))^\alpha$ are admissible moduli of continuity.

Definition 2.109. *Let μ be a modulus of continuity and (X, d) a metric space. We denote by $C_\mu(X)$ the set of bounded, continuous, real-valued functions u over X such that*

$$\|u\|_{C_\mu} \stackrel{def}{=} \|u\|_{L^\infty(X)} + \sup_{0<d(x,y)\leq a} \frac{|u(x) - u(y)|}{\mu(d(x, y))} < \infty.$$

Examples. When $\mu(r) = r^\alpha$ for some $\alpha \in]0, 1]$, the space $C_\mu(X)$ coincides with the Hölder space $C^\alpha(X)$. If $\mu(r) = r(1 - \log r)$, then $C_\mu(X)$ is the space $LL(X)$ of log-Lipschitz functions on X.

Definition 2.110. *Let Γ be a nondecreasing function on $[1, \infty[$. We denote by $B_\Gamma(\mathbb{R}^d)$ the set of bounded real-valued continuous functions u over $\mathbb{R}^d$ such that*

$$\|u\|_{B_\Gamma} \stackrel{def}{=} \|u\|_{L^\infty} + \sup_{j\geq 0} \frac{\|\nabla S_j u\|_{L^\infty}}{\Gamma(2^j)} < \infty.$$

Example. When $\Gamma(y) = y^{1-\alpha}$ [hence $\mu(r) = r^\alpha$], the space B_Γ is equal to $B^\alpha_{\infty,\infty}$. This is a consequence of Proposition 2.79, which ensures that ∇u is in $B^{\alpha-1}_{\infty,\infty}$ if and only if $\sup_{j\in\mathbb{N}} 2^{j(\alpha-1)} \|S_j \nabla u\|_{L^\infty}$ is finite. Therefore, we see that in this particular case, the spaces C_μ and B_Γ coincide.

The following proposition states that this is still true in a much more general framework.

Proposition 2.111. *Let μ be an admissible modulus of continuity and let Γ be defined as in Definition 2.109. Then, $C_\mu(\mathbb{R}^d) = B_\Gamma(\mathbb{R}^d)$.*

Proof. Assume that u belongs to B_Γ. According to the identity (1.23) page 25, there exists a family of functions $(\varphi_k)_{1\leq k\leq d}$ in $\mathcal{D}(\mathbb{R}^d\setminus\{0\})$ such that

$$\Delta_j = \sum_{k=1}^d 2^{-j}\varphi_k(2^{-j}D)\partial_k \Delta_j.$$

This implies that

$$\|\Delta_j u\|_{L^\infty} \leq C 2^{-j}\Gamma(2^j)\|u\|_{B_\Gamma}. \tag{2.72}$$

We now write, for $|x - x'| \leq a$,

$$|u(x) - u(x')| \leq \|\nabla S_j u\|_{L^\infty}|x - x'| + 2\sum_{j'\geq j} \|\Delta_{j'} u\|_{L^\infty}$$

$$\leq \|\nabla S_j u\|_{L^\infty}|x - x'| + C\|u\|_{B_\Gamma}\sum_{j'\geq j} 2^{-j'}\Gamma(2^{j'}).$$

Using the condition (A), the fact that Γ is nondecreasing, and the definition of $\|\cdot\|_{B_\Gamma}$, we get

$$|u(x) - u(x')| \leq \|u\|_{B_\Gamma}\left(\Gamma(2^j)|x - x'| + C\int_{2^j}^\infty \frac{1}{y^2}\Gamma(y)\,dy\right)$$

$$\leq \|u\|_{B_\Gamma}\left(\Gamma(2^j)|x - x'| + C2^{-j}\Gamma(2^j)\right).$$

Choosing j such that $2^{-j} \approx |x-x'|$ and using the definition of Γ gives $u \in C_\mu$.

Now, assume that u belongs to C_μ and let $\widetilde{h} = \mathcal{F}^{-1}\chi$. By the definition of S_j and the fact that $\int_{\mathbb{R}^d} \partial_k\widetilde{h}(y)\,dy = 0$, we may write

$$\partial_k S_j u(x) = 2^{jd}2^j \int_{\mathbb{R}^d} \partial_k\widetilde{h}(2^j(x - y))(u(y) - u(x))\,dy.$$

Therefore, using the definition of $\|u\|_{C_\mu}$ and splitting the integral into $|y-x| \leq a$ and $|y - x| > a$, we get

$$|\partial_k S_j u(x)| \leq 2^{jd}2^j\left(\|u\|_{C_\mu}\int_{|y-x|\leq a} \partial_k\widetilde{h}(2^j(x - y))|\mu(|y - x|)\,dy\right.$$

$$\left. + 2\|u\|_{L^\infty}\int_{|y-x|>a} |\partial_k\widetilde{h}(2^j(x - y))|\,dy\right).$$

Setting $z = 2^j(x - y)$ and splitting the first integral into two parts yields

$$|\partial_k S_j u(x)| \leq \|u\|_{C_\mu} 2^j \int_{|z| \leq 1} |\partial_k \widetilde{h}(z)| \mu(2^{-j}|z|)\, dz$$

$$+ \|u\|_{C_\mu} \int_{1 \leq |z| \leq 2^j a} |\partial_k \widetilde{h}(z)|\, |z| \Gamma\left(\frac{2^j}{|z|}\right) dz + 2\|u\|_{L^\infty} \int_{|z| \geq 2^j a} |\partial_k \widetilde{h}(z)|\, dz.$$

As μ and Γ are nondecreasing functions, we have, for any z such that $|z| \leq 1$ (resp., $|z| \geq 1$), $\mu(2^{-j}|z|) \leq \mu(2^{-j})$ [resp., $\Gamma\left(\frac{2^j}{|z|}\right) \leq \Gamma(2^j)$]. Thus, we get

$$|\partial_k S_j u(x)| \leq \|u\|_{C_\mu} 2^j \mu(2^{-j}) \int_{|z| \leq 1} |\partial_k \widetilde{h}(z)|\, dz$$

$$+ \|u\|_{C_\mu} \Gamma(2^j) \int_{1 \leq |z| \leq 2^j a} |\partial_k \widetilde{h}(z)|\, |z|\, dz + C\|u\|_{L^\infty}.$$

As the last term may obviously be bounded by $C'\Gamma(2^j)\|u\|_{L^\infty}$ for some constant C' independent of j and u, we end up with

$$\|\nabla S_j u\|_{L^\infty} \leq C\|u\|_{C_\mu} \Gamma(2^j), \tag{2.73}$$

and the proposition is proved. $\qquad\square$

Example. If we take $\mu(r) = r(1 - \log r)$, then we get $\Gamma(y) = 1 + \log y$. Hence, the above proposition shows that LL coincides with the set of bounded functions u such that

$$\sup_{j \in \mathbb{N}} \frac{\|\nabla S_j u\|_{L^\infty}}{j + 1} < \infty.$$

Proposition 2.104 extends to general C_μ spaces in the following way.

Proposition 2.112. *Let μ be an admissible modulus of continuity. There exists a constant C such that for any $\varepsilon \in \,]0, 1]$, u in $C^{1,\varepsilon}$, and positive Λ, we have*

$$\|\nabla u\|_{L^\infty} \leq C\left(\frac{\|u\|_{C_\mu} + \Lambda}{\varepsilon} + \|u\|_{C_\mu} \Gamma\left(\left(\frac{\|\nabla u\|_{C^{0,\varepsilon}}}{\|u\|_{C_\mu} + \Lambda}\right)^{\frac{1}{\varepsilon}}\right)\right)$$

whenever $\|u\|_{C_\mu} + \Lambda \leq \left(\frac{a}{2}\right)^\varepsilon \|\nabla u\|_{C^{0,\varepsilon}}$.

Proof. Write

$$\nabla u = S_j \nabla u + (\mathrm{Id} - S_j)\nabla u.$$

By definition of $C^{0,\varepsilon}$, for any $j \in \mathbb{N}$ such that $2^j \geq 1/a$, we have, using (2.73),

$$\|\nabla u\|_{L^\infty} \leq C\|u\|_{C_\mu} \Gamma(2^j) + C\varepsilon^{-1} 2^{-j\varepsilon} \|\nabla u\|_{C^{0,\varepsilon}}.$$

Choosing j such that

$$\frac{1}{2}\left(\frac{\|\nabla u\|_{C^{0,\varepsilon}}}{\|u\|_{C_\mu} + \Lambda}\right)^{\frac{1}{\varepsilon}} \leq 2^j < \left(\frac{\|\nabla u\|_{C^{0,\varepsilon}}}{\|u\|_{C_\mu} + \Lambda}\right)^{\frac{1}{\varepsilon}}$$

and using the fact that Γ is nondecreasing gives the result. $\qquad\square$

2.12 References and Remarks

Bernstein's inequality and the lemma that follows it belong to the mathematical folklore. The statement of Lemma 2.1 is borrowed from [69]. The subsection devoted to the action of a diffeomorphism over functions with localized Fourier transforms was inspired by the work of M. Vishik in [296] (see also [156] and [103]). The smoothing effect for the heat flow which is pointed out in Lemma 2.4 was first stated in [71, 72]. A proof of Faà di Bruno's formula (Lemma 2.3) may be found in [298]. Lemma 2.8 has been stated in [90] (under slightly more restrictive assumptions over the support), then extended in [251] and [95] for any $p \in]1, \infty[$.

Littlewood–Paley theory first appeared in the context of one-dimensional Fourier series (see the works by J. Littlewood and R. Paley in [218, 219]). The presentation adopted in Section 2.2 follows that of J.-Y. Chemin in [69]. We mention in passing that a number of more sophisticated decompositions of the phase space (x, ξ) have recently been proposed. The most celebrated of these is probably the *wavelet decomposition* introduced by Y. Meyer in [230–232]. In the present book, we restrict ourselves to the cruder Littlewood–Paley decomposition, which proves to be sufficient for tackling most problems related to nonlinear partial differential equations.

Besov spaces were named after O. Besov who introduced them in [37] for the study of the embedding and trace of functions with derivatives in L^p. The general definition is due to J. Peetre in [249]. The characterizations in terms of finite difference or heat flow are standard. More properties for these spaces may be found in [34, 204, 240, 250, 254, 288, 289]. There is no consensus surrounding the definition of *homogeneous* Besov spaces. In the above references, they are defined modulo polynomials of arbitrary degree. In [41], G. Bourdaud showed that homogeneous Besov spaces $\dot{B}^s_{p,r}$ may be realized, that is, embedded in some Banach space. When $s \geq d/p$ (or $s > d/p$, if $r = 1$), however, that Banach space is a function space modulo polynomials of degree less than or equal to $s - d/p$ if $r > 1$ (less than $s - d/p$, if $r = 1$). We believe that the presentation adopted in this chapter is the most suitable one for the study of partial differential equations.

The refined Sobolev inequality was discovered by P. Gérard, Y. Meyer, and F. Oru (see [140]). The approach presented here is taken from [77]. The embedding property $\dot{B}^0_{p,2} \hookrightarrow L^p$ for $2 \leq p < \infty$ is sharp. It may actually be shown that for any $p \in]1, \infty[$, the Lebesgue space L^p coincides with the Triebel–Lizorkin space $F^0_{p,2}$ (see [150, 273]). The proof relies on general results for vector-valued singular integrals which are beyond the scope of this book. Gagliardo–Nirenberg inequalities arise from the works by E. Gagliardo in [131] and L. Nirenberg in [241].

Paradifferential calculus was invented by J.-M. Bony in [39] for proving a priori estimates for quasilinear hyperbolic partial differential equations in nonhomogeneous Sobolev spaces. The *discrete* version of paradifferential calculus that we chose to present here is due to P. Gérard and J. Rauch [141] (in the nonhomogeneous framework). More results on continuity may be found in, for instance, [254] or [285]. The proof of the Hardy and refined Hardy inequalities is borrowed from [22]. More general refined Hardy inequalities have been proved in [23, 24].

There is an extensive literature on the properties of Besov spaces with respect to left composition (see, in particular, [42, 43], and [254]). The proof which is presented in Section 2.6.2 is an adaptation of the so-called paralinearization Meyer method (see [11] and [232]) to the homogeneous functional framework. The paralinearization theorem stated on page 105 was inspired by the work of S. Alinhac in [5] and Y. Meyer in [232].

The compactness properties of Besov spaces presented in Section 2.9 belong to the mathematical folklore; however, we did not find any comprehensive and self-contained proof in the literature. Those properties are fundamental for proving the existence for some of the nonlinear partial differential equations which will be studied in the next chapters.

Section 2.10 provides the reader with various commutator estimates which will be used throughout the book. Lemma 2.100 gathers different estimates which have been proven in [69, 99], and [103] and is likely to be useful for investigating a number of partial differential equations.

The Zygmund space (here denoted by $B^1_{\infty,\infty}$) was introduced by A. Zygmund in [304]. The logarithmic interpolation inequalities were discovered by H. Brézis and T. Gallouët in [50]. They will be used in Chapters 3, 4, and 7 for proving continuation criteria for different types of nonlinear partial differential equations.

3

Transport and Transport-Diffusion Equations

This chapter is devoted to the study of the following class of *transport equations*:

$$(T) \qquad \begin{cases} \partial_t f + v \cdot \nabla f + A \cdot f = g \\ f_{|t=0} = f_0, \end{cases}$$

where the functions $v : \mathbb{R} \times \mathbb{R}^d \to \mathbb{R}^d$, $A : \mathbb{R} \times \mathbb{R}^d \to \mathcal{M}_N(\mathbb{R})$, $f_0 : \mathbb{R}^d \to \mathbb{R}^N$, and $g : \mathbb{R} \times \mathbb{R}^d \to \mathbb{R}^N$ are given.

Transport equations arise in many mathematical problems and, in particular, in most partial differential equations related to fluid mechanics. Although the velocity field v and the source term g may depend (nonlinearly) on f, having a good theory for linear transport equations is an important first step for studying such partial differential equations.

The first section is devoted to the study of ordinary differential equations. The emphasis is on generalizations of the classical Cauchy–Lipschitz theorem. When the vector field v is Lipschitz, there is an obvious correspondence between the ordinary differential equation associated with v and the transport equation (T). Moreover, this study will provide an opportunity to establish some very simple blow-up criteria for ordinary differential equations that will act as guidelines for proving blow-up criteria in evolution partial differential equations (see Chapters 4, 5, 7, and 10).

In the second section we focus on the transport equation (T) in the case where the vector field v is at least Lipschitz with respect to the space variable. As an application of the results established, we solve the Cauchy problem for a shallow water equation. The main focus of the third section is the proof of estimates of propagation of regularity with loss when the vector field is not Lipschitz. The particular case of log-Lipschitz vector fields plays an important role in the study of two-dimensional incompressible fluids (see Chapter 7).

Finally, in the last section of this chapter, we prove a few estimates for the solution of the transport-diffusion equation. This type of equation appears, in particular, in the study of the problem of vortex patches with vanishing viscosity (see Chapter 7).

H. Bahouri et al., *Fourier Analysis and Nonlinear Partial Differential Equations*, Grundlehren der mathematischen Wissenschaften 343, DOI 10.1007/978-3-642-16830-7_3, © Springer-Verlag Berlin Heidelberg 2011

3.1 Ordinary Differential Equations

This section recalls some basic facts about ordinary differential equations.

3.1.1 The Cauchy–Lipschitz Theorem Revisited

To begin, we establish a generalization of the classical Cauchy–Lipschitz theorem. The underlying concept is the *Osgood condition,* defined below.

Definition 3.1. *Let $a > 0$ and μ be a modulus of continuity defined on $[0, a]$ (see Definition 2.108). We say that μ is an* Osgood *modulus of continuity if*

$$\int_0^a \frac{dr}{\mu(r)} = \infty.$$

Examples. The function $r \mapsto r$ is an Osgood modulus of continuity, as are the functions

$$r \longmapsto r \left(\log \frac{1}{r} \right)^\alpha \quad \text{and} \quad r \longmapsto r \log \frac{1}{r} \left(\log \log \frac{1}{r} \right)^\alpha \quad \text{if} \quad \alpha \leq 1.$$

The function $r \mapsto r^\alpha$ with $\alpha < 1$, however, is not an Osgood modulus of continuity. Neither are the functions

$$r \longmapsto r \left(\log \frac{1}{r} \right)^\alpha \quad \text{and} \quad r \longmapsto r \log \frac{1}{r} \left(\log \log \frac{1}{r} \right)^\alpha \quad \text{with} \quad \alpha > 1.$$

The relevance of Definition 3.1 is illustrated by the following theorem.

Theorem 3.2. *Let E be a Banach space, Ω an open subset of E, I an open interval of $\mathbb{R}$, and (t_0, x_0) an element of $I \times \Omega$. Let F be in $L^1_{loc}(I; C_\mu(\Omega; E))$, where μ is an Osgood modulus of continuity and $C_\mu(\Omega; E)$ is the Banach space introduced in Definition 2.109. There then exists an open interval $J \subset I$ such that the equation*

$$(ODE) \qquad\qquad x(t) = x_0 + \int_{t_0}^t F(t', x(t')) \, dt'$$

has a unique continuous solution on J.

Proof. We first establish the uniqueness of the trajectories of the equation. Let x_1 and x_2 be solutions of the equation (ODE) defined on a neighborhood $\widetilde{J}$ of t_0 with the same initial data x_0. Define $\delta(t) \overset{\text{def}}{=} \|x_1(t) - x_2(t)\|$. Because $F \in L^1_{loc}(I; C_\mu(\Omega; E))$, we have

$$0 \leq \delta(t) \leq \int_{t_0}^t \gamma(t') \mu(\delta(t')) \, dt' \quad \text{with} \quad \gamma \in L^1_{loc}(I) \quad \text{and} \quad \gamma \geq 0. \qquad (3.1)$$

The key to uniqueness is the so-called *Osgood lemma,* a generalization of the Gronwall lemma. For the reader's convenience, we first recall the Gronwall lemma.

Lemma 3.3. *Let f and g be two C^0 (resp., C^1) nonnegative functions on $[t_0, T]$. Let $\mathcal{A}$ be a continuous function on $[t_0, T]$. Suppose that, for t in $[t_0, T]$,*

$$\frac{1}{2}\frac{d}{dt}g^2(t) \leq \mathcal{A}(t)\,g^2(t) + f(t)g(t). \tag{3.2}$$

For any time t in $[t_0, T]$ we then have

$$g(t) \leq g(t_0)\exp\int_{t_0}^t \mathcal{A}(t')\,dt' + \int_{t_0}^t f(t')\exp\left(\int_{t'}^t \mathcal{A}(t'')\,dt''\right)dt'.$$

Proof. Define

$$g_{\mathcal{A}}(t) \overset{\text{def}}{=} g(t)\exp\left(-\int_{t_0}^t \mathcal{A}(t')\,dt'\right) \quad \text{and} \quad f_{\mathcal{A}}(t) \overset{\text{def}}{=} f(t)\exp\left(-\int_{t_0}^t \mathcal{A}(t')\,dt'\right).$$

Obviously, we have $\dfrac{1}{2}\dfrac{d}{dt}g_{\mathcal{A}}^2 \leq f_{\mathcal{A}}g_{\mathcal{A}}$, so for any positive ε,

$$\frac{d}{dt}(g_{\mathcal{A}}^2 + \varepsilon)^{\frac{1}{2}} \leq \frac{g_{\mathcal{A}}}{(g_{\mathcal{A}}^2 + \varepsilon^2)^{\frac{1}{2}}}f_{\mathcal{A}} \leq f_{\mathcal{A}}.$$

By integration we get, for all $t \in [t_0, T]$,

$$(g_{\mathcal{A}}^2(t) + \varepsilon^2)^{\frac{1}{2}} \leq (g_{\mathcal{A}}^2(t_0) + \varepsilon^2)^{\frac{1}{2}} + \int_{t_0}^t f_{\mathcal{A}}(t')\,dt'.$$

Letting ε tend to 0 then gives the result. □

We now state the Osgood lemma.

Lemma 3.4. *Let ρ be a measurable function from $[t_0, T]$ to $[0, a]$, γ a locally integrable function from $[t_0, T]$ to $\mathbb{R}^+$, and μ a continuous and nondecreasing function from $[0, a]$ to $\mathbb{R}^+$. Assume that, for some nonnegative real number c, the function ρ satisfies*

$$\rho(t) \leq c + \int_{t_0}^t \gamma(t')\mu(\rho(t'))\,dt' \quad \text{for a.e.}^1\ t \in [t_0, T]. \tag{3.3}$$

– *If c is positive, then we have, for a.e. $t \in [t_0, T]$,*

$$-\mathcal{M}(\rho(t)) + \mathcal{M}(c) \leq \int_{t_0}^t \gamma(t')\,dt' \quad \text{with} \quad \mathcal{M}(x) = \int_x^a \frac{dr}{\mu(r)}. \tag{3.4}$$

– *If $c = 0$ and μ is an Osgood modulus of continuity, then $\rho = 0$ a.e.*

If we assume this lemma to hold, then we get $\delta \equiv 0$ in (3.1), from which uniqueness follows.

---

[1] From now on, the abbreviation "a.e." means "almost every."

In order to prove existence in Theorem 3.2, we use the classical Picard scheme:

$$x_{k+1}(t) = x_0 + \int_{t_0}^t F(t', x_k(t')) \, dt'.$$

We skip the fact that for J sufficiently small, the sequence $(x_k)_{k \in \mathbb{N}}$ is well defined and bounded in the space $\mathcal{C}_b(J, \Omega)$. Let $\rho_{k,n}(t) \overset{\text{def}}{=} \sup_{t_0 \leq t' \leq t} \|x_{k+n}(t') - x_k(t')\|$. It is obvious that

$$0 \leq \rho_{k+1,n}(t) \leq \int_{t_0}^t \gamma(t') \mu(\rho_{k,n}(t')) \, dt'.$$

Because the function μ is nondecreasing, we deduce that $\rho_k \overset{\text{def}}{=} \sup_n \rho_{k,n}$ satisfies

$$0 \leq \rho_{k+1}(t) \leq \int_{t_0}^t \gamma(t') \mu(\rho_k(t')) \, dt,$$

from which it follows that

$$\widetilde{\rho}(t) \overset{\text{def}}{=} \limsup_{k \to +\infty} \rho_k(t) \leq \int_{t_0}^t \gamma(t') \mu(\widetilde{\rho}(t')) \, dt'.$$

Lemma 3.4 implies that $\widetilde{\rho}(t) \equiv 0$ near t_0; in other words, $(x_k)_{k \in \mathbb{N}}$ is a Cauchy sequence in $\mathcal{C}_b(J; \Omega)$. This completes the proof of Theorem 3.2.   □

Proof of Lemma 3.4. Arguing by density, it suffices to consider the case where the functions γ and ρ are continuous. Now, consider the following continuous function:

$$R_c(t) \overset{\text{def}}{=} c + \int_{t_0}^t \gamma(t') \mu(\rho(t')) \, dt'.$$

Because μ is nondecreasing, we have

$$\frac{dR_c}{dt} = \gamma(t) \mu(\rho(t))$$
$$\leq \gamma(t) \mu(R_c(t)). \tag{3.5}$$

First, we assume that c is positive. As the function R_c is also positive, we infer from the inequality (3.5) that

$$-\frac{d}{dt} \mathcal{M}(R_c(t)) = \frac{1}{\mu(R_c(t))} \frac{dR_c}{dt} \leq \gamma(t).$$

Integrating, we thus get (3.4).

Finally, suppose that $c = 0$ and that ρ is not identically 0 near t_0. As the function μ is nondecreasing, it is possible to replace the function ρ by the function $\widetilde{\rho}(t) \overset{\text{def}}{=} \sup_{t' \in [t_0, t]} \rho(t')$. A real number t_1 greater than t_0 exists such that $\rho(t_1)$ is positive. As the function ρ satisfies (3.3) with $c = 0$, it also satisfies this inequality for any positive c' less than $\rho(t_1)$. The inequality (3.4) thus entails that

$$\forall c' \in \ ]0, \rho(t_1)] \, , \ \mathcal{M}(c') \leq \int_{t_0}^{t_1} \gamma(t') \, dt' + \mathcal{M}(\rho(t_1)),$$

which implies that $\displaystyle\int_0^a \frac{dr}{\mu(r)} < \infty.$ $\square$

The following corollary will enable us to compute the modulus of continuity of the flow of a vector field satisfying the Osgood condition.

Corollary 3.5. *Let μ be an Osgood modulus of continuity defined on $[0, a]$ and $\mathcal{M}$ the function defined by (3.4). Let ρ be a measurable function such that*

$$\rho(t) \leq \rho(t_0) + \int_{t_0}^t \gamma(t') \mu(\rho(t')) \, dt'.$$

If t is such that $\displaystyle\int_{t_0}^t \gamma(t) \, dt \leq \mathcal{M}(\rho(t_0))$, then we have

$$\rho(t) \leq \mathcal{M}^{-1}\left(\mathcal{M}(\rho(t_0)) - \int_{t_0}^t \gamma(t') \, dt' \right).$$

Proof. The inequality (3.4) can be written

$$\mathcal{M}(\rho(t)) \geq \mathcal{M}(\rho(t_0)) - \int_{t_0}^t \gamma(t') \, dt'.$$

The fact that μ satisfies the Osgood condition implies that the function $\mathcal{M}$ is one-to-one from $]0, a]$ to $[0, +\infty[$. Thus, as the function $\mathcal{M}$ is nonincreasing, the corollary follows by applying $\mathcal{M}^{-1}$ to both sides of the above inequality. $\square$

Corollary 3.6. *Let v be a vector field satisfying the hypothesis of Theorem 3.2. Assume that*

$$x_j(t) = x_j + \int_{t_0}^t v(t', x_j(t')) \, dt' \quad \text{for } j = 1, 2.$$

If $\displaystyle\int_{t_0}^t \gamma(t') \, dt' \leq \mathcal{M}(\|x_1 - x_2\|)$, then we have

$$\|x_1(t) - x_2(t)\| \leq \mathcal{M}^{-1}\left(\mathcal{M}(\|x_1 - x_2\|) - \int_{t_0}^t \gamma(t') \, dt' \right).$$

Applying this corollary with $\mu(r) = r(1 - \log r)$ [in which case we have $a = 1$ and $\mathcal{M}(x) = \log(1 - \log x)$], we get the following result which will be useful in the study of the incompressible Euler system (see Chapter 7).

Theorem 3.7. *Let v be a time-dependent vector field[2] in $L^1_{loc}(\mathbb{R}^+; LL)$. There exists a unique continuous map ψ from $\mathbb{R}^+ \times \mathbb{R}^d$ to $\mathbb{R}^d$ such that*

$$\psi(t, x) = x + \int_0^t v(t', \psi(t', x)) \, dt'.$$

Moreover, for any positive time t, the function $\psi_t : x \mapsto \psi(t, x)$ is such that

$$\psi_t - \mathrm{Id} \in C^{\exp(-V_{LL}(t))} \quad \text{with} \quad V_{LL}(t) \overset{def}{=} \int_0^t \|v(t')\|_{LL} \, dt'.$$

More precisely, if $|x - y| \le e^{1 - \exp V_{LL}(t)}$, then we have

$$|\psi(t, x) - \psi(t, y)| \le |x - y|^{\exp(-V_{LL}(t))} e^{1 - \exp(-V_{LL}(t))}.$$

Corollary 3.5 provides a control of ρ on a small time interval. In order to control ρ on larger time intervals, we now establish a dual version of the Osgood lemma [involving the function $\Gamma(y) = y\mu\left(\frac{1}{y}\right)$ introduced in Definition 2.108 page 117].

Lemma 3.8. *Let μ in $C([0, a]; \mathbb{R}^+)$ be an Osgood modulus of continuity. Let ρ be a measurable function on $[t_0, T]$ with values in $[a^{-1}, \infty[$ and γ a nonnegative locally integrable function on $[t_0, T]$. Assume that*

$$\rho(t) \le \rho(t_0) + \int_{t_0}^t \gamma(t') \Gamma(\rho(t')) \rho(t') \, dt' \quad \text{for a.e. } t \in [t_0, T].$$

The function $\mathcal{G}(y) \overset{def}{=} \int_{1/a}^y \frac{dy'}{y' \Gamma(y')}$ then maps $[a^{-1}, +\infty[$ onto and one-to-one $[0, +\infty[$, and we have

$$\rho(t) \le \mathcal{G}^{-1}\left(\mathcal{G}(\rho(t_0)) + \int_{t_0}^t \gamma(t') \, dt'\right) \quad \text{for a.e. } t \in [t_0, T].$$

Proof. The proof of this lemma is very similar to that of the previous one. The fact that $\mathcal{G}$ maps $[a^{-1}, +\infty[$ onto and one-to-one $[0, +\infty[$ follows immediately from the fact that μ is an Osgood modulus of continuity. We now introduce the function

$$R(t) \overset{def}{=} \rho(t_0) + \int_{t_0}^t \gamma(t') \Gamma(\rho(t')) \rho(t') \, dt'.$$

Because the function Γ is nondecreasing, we have (assuming that ρ and γ are continuous) that

---

[2] See page 116 for the definition of the set LL of log-Lipschitz functions.

$$\frac{dR}{dt} = \gamma(t)\Gamma(\rho(t))\rho(t)$$
$$\leq \gamma(t)\Gamma(R(t))R(t),$$

and thus $\dfrac{d}{dt}\mathcal{G}(R(t)) \leq \gamma(t)$. Integrating then completes the proof. □

Finally, we consider the way the flow depends on its generating vector field.

Proposition 3.9. *Let μ be an Osgood modulus of continuity. Let $(v_n)_{n\in\mathbb{N}}$ be a bounded sequence of time-dependent vector fields in $L^1([0,T];C_\mu)$ converging to v in $L^1([0,T];L^\infty)$, and let ψ_n (resp., ψ) denote the flow of v_n (resp., v). We then have*

$$\lim_{n\to\infty} \|\psi_n - \psi\|_{L^\infty([0,T];L^\infty)} = 0.$$

Proof. By the definitions of ψ and ψ_n, we have, for all $n \in \mathbb{N}$,

$$\psi_n(t,x) - \psi(t,x) = \int_0^t \left(v_n(t',\psi_n(t',x)) - v(t',\psi(t',x))\right) dt'.$$

Hence, defining $\rho_n(t) \overset{\text{def}}{=} \|\psi_n(t,x) - \psi(t,x)\|_{L^\infty}$, we deduce that there exists some integrable function γ such that for all $t \in [0,T]$, we have

$$\rho_n(t) \leq \varepsilon_n + \int_0^t \gamma(t')\mu(\rho_n(t')) \, dt' \quad \text{with} \quad \varepsilon_n \overset{\text{def}}{=} \int_0^T \|v_n(t) - v(t)\|_{L^\infty} \, dt.$$

According to the Osgood lemma, we thus have, for all $t \in [0,T]$,

$$\rho_n(t) \leq \varepsilon_n \quad \text{or} \quad \int_{\varepsilon_n}^{\rho_n(t)} \frac{dr}{\mu(r)} \leq \int_0^t \gamma(t') \, dt'.$$

Since the Osgood condition is satisfied, we can now conclude that $(\rho_n)_{n\in\mathbb{N}}$ goes to zero uniformly on $[0,T]$ when n tends to infinity. □

3.1.2 Estimates for the Flow

In this section, we recall a few estimates for the flow of a smooth vector field. These estimates will be needed in the study of transport-diffusion equations (see Section 3.4 below).

Proposition 3.10. *Let v be a smooth time-dependent vector field with bounded first order space derivatives. Let ψ_t satisfy*

$$\psi_t(x) = x + \int_0^t v(t',\psi_{t'}(x)) \, dt'.$$

Then, for all $t \in \mathbb{R}^+$, the flow ψ_t is a C^1 diffeomorphism over $\mathbb{R}^d$, and we have

$$\left\|D\psi_t^{\pm 1}\right\|_{L^\infty} \le \exp V(t), \tag{3.6}$$

$$\left\|D\psi_t^{\pm 1} - \mathrm{Id}\right\|_{L^\infty} \le \exp V(t) - 1, \tag{3.7}$$

$$\left\|D^2\psi_t^{\pm 1}\right\|_{L^\infty} \le \exp V(t) \int_0^t \left\|D^2 v(t')\right\|_{L^\infty} \exp V(t') \, dt' \tag{3.8}$$

with, as throughout this chapter, $V(t) \stackrel{def}{=} \int_0^t \|Dv(t')\|_{L^\infty} \, dt'$.

Proof. Let $(t, t', x) \mapsto X(t, t', x)$ be (uniquely) defined by

$$X(t, t', x) = x + \int_{t'}^t v\big(t'', X(t'', t', x)\big) \, dt''. \tag{3.9}$$

Uniqueness for ordinary differential equations entails that

$$X(t, t'', X(t'', t', x)) = X(t, t', x).$$

Hence, $\psi_t = X(t, 0, \cdot)$ and $\psi_t^{-1} = X(0, t, \cdot)$. Differentiating (3.9) with respect to x, we get, by virtue of the chain rule,

$$\partial_j X^k(t, t', x) = \delta_{j,k} + \int_{t'}^t \partial_\ell v^k(t'', X(t'', t', x)) \partial_j X^\ell(t'', t', x) \, dt''. \tag{3.10}$$

Taking the modulus and applying the Gronwall lemma thus leads to

$$\left|DX(t, t', x)\right| \le \exp\left|\int_{t'}^t |Dv(t'', X(t'', t', x))| \, dt''\right|$$

$$\le \exp\left|\int_{t'}^t \|Dv(t'')\|_{L^\infty} \, dt''\right|,$$

which obviously yields (3.6).

The proof of (3.7) is similar. This is just a matter of subtracting the identity matrix from (3.10). To prove (3.8), we differentiate (3.9) twice. This yields (with the summation convention over repeated indices)

$$\partial_j \partial_k X^i(t, t', x) = \int_{t'}^t \partial_\ell v^i(t'', X(t'', t', x)) \partial_j \partial_k X^\ell(t'', t', x) \, dt''$$

$$+ \int_{t'}^t \partial_\ell \partial_m v^i(t'', X(t'', t', x)) \partial_k X^m(t'', t', x) \partial_j X^\ell(t'', t', x) \, dt''.$$

Taking advantage of (3.6) and the Gronwall lemma once again, we easily get, for all nonnegative t and t', and all $x \in \mathbb{R}^d$,

$$\left|\partial_j \partial_k X^i(t, t', x)\right| \le e^{\left|\int_{t'}^t |Dv(t'', X(t'', t', x))| \, dt''\right|}$$

$$\times \left|\int_{t'}^t |D^2 v(t'', X(t'', t', x))| e^{\left|\int_{t'}^t |Dv(t'', X(t'', s, x))| \, ds\right|} \, dt''\right|,$$

which clearly entails (3.8). $\square$

3.1.3 A Blow-up Criterion for Ordinary Differential Equations

We emphasize that Theorem 3.2 is only a *local-in-time* statement. This section is devoted to blow-up statements for ordinary differential equations.

Proposition 3.11. *Let $F : \mathbb{R} \times E \to E$ satisfy the hypothesis of Theorem 3.2. Assume, further, that a locally bounded function $M : \mathbb{R}^+ \to \mathbb{R}^+$ and a locally integrable function $\beta : \mathbb{R} \to \mathbb{R}^+$ exist such that*

$$\|F(t, u)\| \le \beta(t) M(\|u\|). \tag{3.11}$$

Let $]T_\star, T^\star[$ be the maximal interval of existence of an integral curve u of the equation (ODE). If $T_\star$ (resp., $T^\star$) is finite, then we have

$$\limsup_{t \to T_\star} \|u(t)\| = \infty \qquad \left(\text{resp., } \limsup_{t \le T^\star} \|u(t)\| = \infty\right).$$

Proof. We shall only prove the result for the upper bound $T^\star$. Consider two times $T_0 < T$ such that $\|u(t)\|$ is bounded on $[T_0, T[$. We deduce from (3.11) that for any time t in $[T_0, T[$, we have

$$\|F(t, u(t))\| \le C\beta(t).$$

As the function β is integrable on the interval $[T_0, T]$, we deduce that for any positive ε, there exists some $\eta > 0$ such that

$$\|u(t) - u(t')\| < \varepsilon \quad \text{for any } (t, t') \in [T_0, T[^2 \text{ verifying } |t - t'| < \eta.$$

As E is a Banach space, we deduce that there exists some u_T in E such that

$$\lim_{t \to T} u(t) = u_T.$$

Applying Theorem 3.2, we can now construct a solution $\widetilde{u}$ of (ODE) on some interval $[T - \tau, T + \tau]$ such that $\widetilde{u}(T) = u_T$.

By virtue of uniqueness, $\widetilde{u}$ coincides with u on $[T - \tau, T[$ and is hence a continuation of u beyond T. We can thus conclude that $T < T^\star$. $\qquad \square$

Corollary 3.12. *With the notation and hypothesis of Proposition 3.11, let x be a maximal integral curve of (ODE). If F satisfies*

$$\|F(t, u)\| \le M \|u\|^2$$

for some constant M, then for any $t_0 \in \;]T_\star, T^\star[$, we have

$$\int_{T_\star}^{t_0} \|x(t)\| \, dt - T_\star = T^\star + \int_{t_0}^{T^\star} \|x(t)\| \, dt = \infty.$$

Proof. The solution satisfies

$$\|x(t)\| \leq \|x_0\| + M \left| \int_{t_0}^t \|x(t')\|^2 \, dt' \right|.$$

The Gronwall lemma implies that

$$\|x(t)\| \leq \|x_0\| \exp \left(M \left| \int_{t_0}^t \|x(t')\| \, dt' \right| \right),$$

which completes the proof of the corollary.   □

3.2 Transport Equations: The Lipschitz Case

This section is devoted to the study of the transport equation (T) in the case where the time-dependent vector field v is at least Lipschitz with respect to the space variable. To simplify the presentation, we focus on the evolution for *nonnegative* times and assume that there is no 0-order term (i.e., $A \equiv 0$). Similar results may be obtained for negative times and for nonzero A (see Remarks 3.17 and 3.20).

The basic idea is that the Lipschitz assumption should ensure that the initial regularity is preserved by the flow. The importance of the Lipschitz condition becomes obvious if we consider Hölder regularity. Indeed, assume that $f_0 \in C^{0,\varepsilon}$ for some $\varepsilon \in \,]0,1]$ and that $A \equiv 0$ and $g \equiv 0$ (to simplify matters). Since v is Lipschitz, the flow ψ of v is also Lipschitz, and we have, for all $(x,y) \in \mathbb{R}^d \times \mathbb{R}^d$ and $t \in [0,T]$,

$$f(t,y) - f(t,x) = f_0(\psi_t^{-1}(y)) - f_0(\psi_t^{-1}(x)).$$

Therefore, by virtue of the first inequality of Proposition 3.10,

$$\begin{aligned}
\left| f(t,y) - f(t,x) \right| &\leq \|f_0\|_{C^{0,\varepsilon}} \left| \psi_t^{-1}(y) - \psi_t^{-1}(x) \right|^\varepsilon \\
&\leq \|f_0\|_{C^{0,\varepsilon}} \exp(\varepsilon V(t)) \, |y - x|^\varepsilon.
\end{aligned}$$

From this, we deduce that Hölder regularity is preserved during the evolution. In this section, we seek to generalize this basic result to general Besov spaces.

3.2.1 A Priori Estimates in General Besov Spaces

Let us first explain what a solution to (T) is.

Definition 3.13. *Assume that $f_0 \in (\mathcal{S}'(\mathbb{R}^d))^N$ and $g \in L^1\big([0,T]; (\mathcal{S}'(\mathbb{R}^d))^N\big)$. A function f in $\mathcal{C}\big([0,T]; (\mathcal{S}'(\mathbb{R}^d))^N\big)$ is called a solution to (T) if $A \cdot f$, $f \otimes v$, and $f \operatorname{div} v$ are in $L^1\big([0,T]; (\mathcal{S}'(\mathbb{R}^d))^N\big)$, and, for all $\phi \in \mathcal{C}^1\big([0,T]; (\mathcal{S}(\mathbb{R}^d))^N\big)$,*

$$\sum_i \left(\int_0^t \langle f^i, \partial_t \phi_i \rangle \, dt' + \langle f^i \operatorname{div} v, \phi_i \rangle \, dt' \right) + \sum_{i,j} \int_0^t \langle f^i v^j, \partial_j \phi_i \rangle \, dt'$$

$$+ \sum_{i,j} \int_0^t \langle A_j^i f^j, \phi_i \rangle \, dt' = \sum_i \Big(\langle f^i(t), \phi_i(t) \rangle - \langle f_0^i, \phi_i(0) \rangle \Big).$$

This section is devoted to the proof of the following result pertaining to the case where $A \equiv 0$ (a more general statement will be given in Remark 3.17).

Theorem 3.14. *Let $1 \le p \le p_1 \le \infty$, $1 \le r \le \infty$. Assume that*

$$\sigma \ge -d \min \left(\frac{1}{p_1}, \frac{1}{p'} \right) \quad \text{or} \quad \sigma \ge -1 - d \min \left(\frac{1}{p_1}, \frac{1}{p'} \right) \quad \text{if} \quad \operatorname{div} v = 0 \quad (3.12)$$

with strict inequality if $r < \infty$.

There exists a constant C, depending only on d, p, p_1, r, and σ, such that for all solutions $f \in L^\infty([0,T]; B_{p,r}^\sigma)$ of (T) with $A \equiv 0$, initial data f_0 in $B_{p,r}^\sigma$, and g in $L^1([0,T]; B_{p,r}^\sigma)$, we have, for a.e. $t \in [0,T]$,

$$\|f\|_{\widetilde{L}_t^\infty(B_{p,r}^\sigma)} \le \left(\|f_0\|_{B_{p,r}^\sigma} \right.$$

$$\left. + \int_0^t \exp(-CV_{p_1}(t')) \|g(t')\|_{B_{p,r}^\sigma} \, dt' \right) \exp(CV_{p_1}(t)) \quad (3.13)$$

with, if the inequality is strict in (3.12),

$$V_{p_1}'(t) \stackrel{def}{=} \begin{cases} \|\nabla v(t)\|_{B_{p_1,\infty}^{\frac{d}{p_1}} \cap L^\infty}, & \text{if } \sigma < 1 + \frac{d}{p_1}, \\[2mm] \|\nabla v(t)\|_{B_{p_1,r}^{\sigma-1}}, & \text{if } \sigma > 1 + \frac{d}{p_1} \quad \text{or} \quad \left\{ \sigma = 1 + \frac{d}{p_1} \text{ and } r = 1 \right\}, \end{cases}$$

and, if equality holds in (3.12) and $r = \infty$,

$$V_{p_1}'(t) \stackrel{def}{=} \|\nabla v(t)\|_{B_{p_1,1}^{\frac{d}{p_1}}}.$$

If $f = v$, then for all $\sigma > 0$ ($\sigma > -1$, if $\operatorname{div} v = 0$), the estimate (3.13) holds with

$$V_{p_1}'(t) = \|\nabla v(t)\|_{L^\infty}.$$

Proof. To prove this theorem, we (as quite often in this book) perform a spectral localization of the equation under consideration. More precisely, applying Δ_j to (T) yields

$$(T_j) \quad \begin{cases} (\partial_t + v \cdot \nabla)\Delta_j f = \Delta_j g + R_j \\ \Delta_j f_{|t=0} = \Delta_j f_0 \end{cases} \quad \text{with} \quad R_j \stackrel{def}{=} v \cdot \nabla \Delta_j f - \Delta_j(v \cdot \nabla f).$$

Since $\nabla v \in L^1([0,T]; L^\infty)$, we readily obtain

$$\|\Delta_j f(t)\|_{L^p} \leq \|\Delta_j f_0\|_{L^p} + \int_0^t \|\Delta_j g(t')\|_{L^p}\, dt'$$
$$+ \int_0^t \left(\|R_j(t')\|_{L^p} + \frac{1}{p} \|\operatorname{div} v(t')\|_{L^\infty} \|\Delta_j f(t')\|_{L^p} \right) dt'. \quad (3.14)$$

This may be proven by writing an explicit formula for $\Delta_j f$ in terms of the flow of v and of the data, or by multiplying both sides of $(\mathcal{T}_j)$ by $\operatorname{sgn}(\Delta_j f)|\Delta_j f|^{p-1}$ (in the scalar case) and integrating over $\mathbb{R}^d$. We note that $\|\Delta_j f(t)\|_{L^p}$ may be replaced by $\sup_{t' \in [0,t]} \|\Delta_j f(t')\|_{L^p}$ in the left-hand side.

According to Lemma 2.100 page 112, there exists some constant C, independent of v and f, such that

$$\|R_j(t)\|_{L^p} \leq Cc_j(t)2^{-j\sigma}V'_{p_1}(t)\|f(t)\|_{B^\sigma_{p,r}} \quad \text{with} \quad \|c_j(t)\|_{\ell^r} = 1, \quad (3.15)$$

where V'_{p_1} is defined as in Theorem 3.14 [note that if $f = v$, then we can apply the inequality (2.54) page 112 with $p_1 = p$ and $p_2 = \infty$].

Take the ℓ^r norm in (3.14). Using (3.15) and the fact that

$$\|f\|_{L^\infty_t(B^\sigma_{p,r})} \leq \|f\|_{\widetilde{L}^\infty_t(B^\sigma_{p,r})} \quad \text{and} \quad \|g\|_{\widetilde{L}^1_t(B^\sigma_{p,r})} \leq \|g\|_{L^1_t(B^\sigma_{p,r})},$$

we get

$$\|f\|_{\widetilde{L}^\infty_t(B^\sigma_{p,r})} \leq \|f_0\|_{B^\sigma_{p,r}} + \int_0^t \left(\|g(t')\|_{B^\sigma_{p,r}} + CV'_{p_1}(t')\|f\|_{\widetilde{L}^\infty_{t'}(B^\sigma_{p,r})} \right) dt'. \quad (3.16)$$

Applying the Gronwall lemma completes the proof of the theorem. □

Remark 3.15. Actually, the above proof yields

$$\|f\|_{\widetilde{L}^\infty_t(B^\sigma_{p,r})} \leq \|f_0\|_{B^\sigma_{p,r}} + C\int_0^t V'_{p_1}(t')\|f\|_{\widetilde{L}^\infty_{t'}(B^\sigma_{p,r})}\, dt' + \|g\|_{\widetilde{L}^1_t(B^\sigma_{p,r})},$$

and we thus have a slightly more accurate estimate, namely,

$$\|f\|_{\widetilde{L}^\infty_t(B^\sigma_{p,r})} \leq \left(\|f_0\|_{B^\sigma_{p,r}} + \|g\|_{\widetilde{L}^1_t(B^\sigma_{p,r})} \right) \exp(CV_{p_1}(t)). \quad (3.17)$$

Remark 3.16. By taking advantage of Remark 2.102, we can extend Theorem 3.14 to the homogeneous framework under the additional condition that

$$\sigma < 1 + \frac{d}{p_1} \quad \text{or} \quad \sigma \leq 1 + \frac{d}{p_1} \quad \text{if} \quad r = 1. \quad (3.18)$$

Remark 3.17. In the general case where A is nonzero and satisfies

$$\|(A \cdot f)(t)\|_{B^\sigma_{p,r}} \leq \mathcal{A}(t)\|f(t)\|_{B^\sigma_{p,r}} \quad \text{a.e. for some} \quad \mathcal{A} \in L^1([0,T]), \quad (3.19)$$

an easy variation on the proof of (3.14) leads to the inequality (3.13) with V'_{p_1} replaced by $V'_{p_1} + \mathcal{A}$.

Note that the inequality (3.19) is satisfied whenever A belongs to a Besov space with a suitably large index of regularity.

3.2.2 Refined Estimates in Besov Spaces with Index 0

If the vector field v is divergence-free, then the flow of v is measure-preserving so that there is no exponential term involving $\|\nabla v\|_{L^\infty}$ in the estimates of $\|f\|_{L^p}$: We have

$$\|f(t)\|_{L^p} \leq \|f_0\|_{L^p} + \int_0^t \|g(t')\|_{L^p} \, dt' \quad \text{for } t \geq 0. \tag{3.20}$$

As a consequence, we shall see that the exponential term may be replaced by a *linear term* in the inequality (3.13). This improvement will be the cornerstone of the proof of global existence for two-dimensional incompressible Euler equations with data having critical regularity (see Chapter 7).

Below, we give a statement pertaining to nonhomogeneous Besov spaces of index 0. It goes without saying that a similar statement holds true in the homogeneous framework.

Theorem 3.18. *Assume that v is divergence-free and that f satisfies the transport equation (T) with $A \equiv 0$. There exists a constant C, depending only on d, such that for all $1 \leq p, r \leq \infty$ and $t \in [0, T]$, we have*

$$\|f\|_{\widetilde{L}^\infty_t(B^0_{p,r})} \leq C\big(\|f_0\|_{B^0_{p,r}} + \|g\|_{\widetilde{L}^1_t(B^0_{p,r})}\big)\big(1 + V(t)\big) \quad \text{with}$$

$$V(t) \stackrel{def}{=} \int_0^t \|\nabla v(t')\|_{L^\infty} \, dt'.$$

Proof. In order to simplify the presentation, we shall only consider the case where $r = 1$. First, by virtue of the uniqueness of the transport equation (see Theorem 3.18 below), we can write $f = \sum_{k \geq -1} f_k$ with f_k satisfying

$$\begin{cases} \partial_t f_k + v \cdot \nabla f_k = \Delta_k g \\ f_{k|t=0} = \Delta_k f_0. \end{cases} \tag{3.21}$$

We obviously have

$$\|f\|_{B^0_{p,1}} \leq \sum_{j,k \geq -1} \|\Delta_j f_k\|_{L^p} = \sum_{|j-k|<N} \|\Delta_j f_k\|_{L^p} + \sum_{|j-k| \geq N} \|\Delta_j f_k\|_{L^p}, \tag{3.22}$$

where N stands for some positive integer to be fixed hereafter.

Because $\operatorname{div} v = 0$, using (3.20) yields

$$\|f_k(t)\|_{L^p} \leq \|\Delta_k f_0\|_{L^p} + \int_0^t \|\Delta_k g\|_{L^p} \, dt'.$$

Hence,

$$\sum_{|j-k|<N} \|\Delta_j f_k(t)\|_{L^p} \leq C \sum_{|j-k|<N} \|f_k(t)\|_{L^p}$$

$$\leq CN \sum_k \left(\|\Delta_k f_0\|_{L^p} + \|\Delta_k g\|_{L_t^1(L^p)} \right)$$

$$\leq CN \left(\|f_0\|_{B_{p,1}^0} + \|g\|_{L_t^1(B_{p,1}^0)} \right).$$

The last sum in (3.22) may be dealt with by taking advantage of the estimates in the space $B_{p,1}^{\pm\varepsilon}$ (where ε is chosen in $]0,1[$) supplied by Theorem 3.14 for f_k. We thus have

$$\|f_k(t)\|_{B_{p,1}^{\pm\varepsilon}} \leq \left(\|\Delta_k f_0\|_{B_{p,1}^{\pm\varepsilon}} + \|\Delta_k g\|_{L_t^1(B_{p,1}^{\pm\varepsilon})} \right) \exp(CV(t)),$$

from which it follows, for some nonnegative sequence $(a_j)_{j\geq-1}$ such that $\sum_{j\geq-1} a_j = 1$, that

$$\|\Delta_j f_k(t)\|_{L^p} \leq 2^{-\varepsilon|k-j|} a_j \left(\|\Delta_k f_0\|_{L^p} + \|\Delta_k g\|_{L_t^1(L^p)} \right) \exp(CV(t)).$$

From this latter inequality, we deduce that

$$\sum_{|j-k|\geq N} \|\Delta_j f_k(t)\|_{L^p} \leq C 2^{-N\varepsilon} \left(\|f_0\|_{B_{p,1}^0} + \|g\|_{L_t^1(B_{p,1}^0)} \right) \exp(CV(t)).$$

Choosing N such that[3] $N\varepsilon \log 2 \approx 1 + CV(t)$ then completes the proof. □

3.2.3 Solving the Transport Equation in Besov Spaces

We now state an existence result for the transport equation with data in Besov spaces. To simplify the presentation, we assume here that there is no 0-order term in (T) (see Remark 3.20 for the general case).

Theorem 3.19. *Let p, p_1, r, and σ be as in the statement of Theorem 3.14 with strict inequality in 3.12. Let $f_0 \in B_{p,r}^\sigma$, $g \in L^1([0,T]; B_{p,r}^\sigma)$, and v be a time-dependent vector field such that $v \in L^\rho([0,T]; B_{\infty,\infty}^{-M})$ for some $\rho > 1$ and $M > 0$, and*

$$\nabla v \in L^1([0,T]; B_{p_1,\infty}^{\frac{d}{p_1}} \cap L^\infty), \quad \text{if } \sigma < 1 + \frac{d}{p_1},$$

$$\nabla v \in L^1([0,T]; B_{p_1,r}^{\sigma-1}), \quad\quad \text{if } \sigma > 1 + \frac{d}{p_1}, \text{ or } \sigma = 1 + \frac{d}{p_1} \text{ and } r = 1.$$

The equation (T) with $A \equiv 0$ then has a unique solution f in

---

[3] From now on, the notation $A \approx B$ means that $C^{-1}A \leq B \leq CA$ for some irrelevant positive constant C.

- the space $\mathcal{C}([0,T]; B_{p,r}^{\sigma})$, if $r < \infty$,
- the space $\left(\bigcap_{\sigma' < \sigma} \mathcal{C}([0,T]; B_{p,\infty}^{\sigma'})\right) \bigcap \mathcal{C}_w([0,T]; B_{p,\infty}^{\sigma})$, if $r = \infty$.[4]

Moreover, the inequalities of Theorem 3.14 hold true.

Proof. Uniqueness readily follows from Theorem 3.14, so we focus on the proof of the existence. For the sake of conciseness, we treat only the case $\sigma < 1 + \frac{d}{p_1}$.

We first smooth out the data and the velocity field v by setting

$$f_0^n \overset{\text{def}}{=} S_n f_0, \quad g^n \overset{\text{def}}{=} \rho_n *_t S_n g, \quad \text{and} \quad v^n \overset{\text{def}}{=} \rho_n *_t S_n v,$$

where $\rho_n \overset{\text{def}}{=} \rho_n(t)$ stands for a sequence of mollifiers with respect to the time variable.[5]

We clearly have $f_0^n \in B_{p,r}^{\infty}$, $g \in \mathcal{C}([0,T]; B_{p,r}^{\infty})$, $v^n \in \mathcal{C}_b([0,T] \times \mathbb{R}^d)$, and Dv^n in $\mathcal{C}([0,T]; B_{p,r}^{\infty})$ with $B_{p,r}^{\infty} \overset{\text{def}}{=} \bigcap_{s \in \mathbb{R}} B_{p,r}^s$. Moreover, $(f_0^n)_{n \in \mathbb{N}}$ is bounded in $B_{p,r}^{\sigma}$, $(g^n)_{n \in \mathbb{N}}$ is bounded in $L^1([0,T]; B_{p,r}^{\sigma})$, $(v^n)_{n \in \mathbb{N}}$ is bounded in $L^p([0,T]; B_{\infty,\infty}^{-M})$, and $(Dv^n)_{n \in \mathbb{N}}$ is bounded in $L^1([0,T]; B_{p_1,\infty}^{\frac{d}{p_1}} \cap L^{\infty})$.

The standard Cauchy–Lipschitz theorem ensures that v^n has a smooth flow ψ^n defined on $[0,T] \times \mathbb{R}^d$. Hence, the function $f^n : [0,T] \times \mathbb{R}^d \to \mathbb{R}^N$ defined by

$$f^n(t,x) = f_0^n(\psi_t^{-1}(x)) + \int_0^t g^n(\tau, \psi_\tau(\psi_t^{-1}(x)))\, d\tau$$

is a solution to

$$\partial_t f^n + v^n \cdot \nabla f^n = g^n, \qquad f_{|t=0}^n = f_0^n.$$

Further, as all the functions are smooth, we have, according to Theorem 3.14,

$$\|f^n(t)\|_{B_{p,r}^{\sigma}} \le e^{C \int_0^t V^n(t')\, dt'} \left(\|f_0^n\|_{B_{p,r}^{\sigma}} + \int_0^t e^{-C \int_0^{t'} V^n(t'')\, dt''} \|f^n(t')\|_{B_{p,r}^{\sigma}}\, dt' \right)$$

with $V^n(t) \overset{\text{def}}{=} \|\nabla v^n(t)\|_{B_{p_1,\infty}^{\frac{d}{p_1}} \cap L^{\infty}}$.

Thus, the sequence $(f^n)_{n \in \mathbb{N}}$ is uniformly bounded in $\mathcal{C}([0,T]; B_{p,r}^{\sigma})$.

In order to prove the convergence of a subsequence, we appeal to compactness arguments. First, because

$$\partial_t f^n - g^n = -v^n \cdot \nabla f^n, \tag{3.23}$$

---

[4] In what follows, if X is a Banach space with predual X^*, then we denote by $\mathcal{C}_w([0,T]; X)$ the set of measurable functions $f : [0,T] \to X$ such that for any $\phi \in X^*$, the function $t \mapsto \langle f(t), \phi \rangle_{X \times X^*}$ is continuous over $[0,T]$.

[5] With no loss of generality, we can assume that v and g are defined on $\mathbb{R} \times \mathbb{R}^d$.

we can deduce that $(\partial_t f^n - g^n)_{n \in \mathbb{N}}$ is bounded in $L^p([0, T]; B_{p,\infty}^{-m})$ for some sufficiently large $m > 0$: It suffices to use the bounds for v^n and f^n and to take advantage of the results on continuity of the paraproduct and the remainder stated in Section 2.8.

Second, introducing the functions $\overline{f}^n(t) \overset{\text{def}}{=} f^n(t) - \int_0^t g^n(t') \, dt'$, we thus deduce that there exists some $\beta > 0$ such that the sequence $(\overline{f}^n)_{n \in \mathbb{N}}$ is uniformly bounded in $\mathcal{C}^\beta([0, T]; B_{p,\infty}^{-m})$ and hence uniformly equicontinuous with values in $B_{p,\infty}^{-m}$. Now, if m is large enough, then Theorem 2.94 guarantees that for all $\varphi \in \mathcal{C}_c^\infty$, the map $u \mapsto \varphi u$ is compact from $B_{p,r}^\sigma$ to $B_{p,\infty}^{-m}$. Combining Ascoli's theorem and the Cantor diagonal process thus ensures that, up to a subsequence, the sequence $(\overline{f}^n)_{n \in \mathbb{N}}$ converges in $\mathcal{S}'$ to some distribution $\overline{f}$ such that $\varphi \overline{f}$ belongs to $\mathcal{C}([0, T]; B_{p,\infty}^{-m})$ for all $\varphi \in \mathcal{D}$.

Finally, appealing once again to the uniform bounds in $L^\infty([0, T]; B_{p,r}^\sigma)$ and the Fatou property for Besov spaces (see Theorem 2.25), we get $\overline{f} \in L^\infty([0, T]; B_{p,r}^\sigma)$. By interpolating the above results on convergence with the bounds in $L^\infty([0, T]; B_{p,r}^\sigma)0$ for $(\overline{f}^n)_{n \in \mathbb{N}}$, we find that $\varphi \overline{f}^n$ tends to $\varphi \overline{f}$ in $\mathcal{C}([0, T]; B_{p,\infty}^{\sigma - \varepsilon})$ for all $\varepsilon > 0$ and $\varphi \in \mathcal{D}$ so that we may pass to the limit in the equation for f^n, in the sense of Definition 3.13.[6] That the sequences $(f_0^n)_{n \in \mathbb{N}}$, $(g^n)_{n \in \mathbb{N}}$, and $(v^n)_{n \in \mathbb{N}}$ converge respectively to f_0, g, and v may be easily deduced from their definitions. We conclude that the function $f \overset{\text{def}}{=} \overline{f} + \int_0^t g(t') \, dt'$ is a solution of (T).

We still have to prove that $f \in \mathcal{C}([0, T]; B_{p,r}^\sigma)$ in the case where r is finite. From the equation (T), we deduce that $\partial_t f \in L^1([0, T]; B_{p,\infty}^{-M'})$ for some sufficiently large M'. Hence, f belongs to $\mathcal{C}([0, T]; B_{p,\infty}^{-M'})$. Therefore, $\Delta_j f \in \mathcal{C}([0, T]; L^p)$ for any $j \geq -1$, from which it follows that $S_j f \in \mathcal{C}([0, T]; B_{p,r}^\sigma)$ for all $j \in \mathbb{N}$.

We claim that the sequence of continuous $B_{p,r}^\sigma$-valued functions $(S_j f)_{j \in \mathbb{N}}$ converges uniformly on $[0, T]$. Indeed, according to Proposition 2.10, we have

$$\Delta_{j'}(f - S_j f) = \sum_{\substack{|j'' - j'| \geq 1 \\ j'' \geq j}} \Delta_{j'} \Delta_{j''} f,$$

from which it follows that

$$\|f - S_j f\|_{B_{p,r}^\sigma} \leq C \left(\sum_{j' \geq j-1} 2^{j' \sigma r} \|\Delta_{j'} f\|_{L^p}^r \right)^{\frac{1}{r}}. \tag{3.24}$$

Using the inequalities (3.14) and (3.15) to bound the right-hand side of (3.24), we deduce that, for some sequence $(c_{j'})_{j' \geq -1}$ such that $\sum_{j' \geq -1} c_{j'}^r(t) = 1$ for all $t \in [0, T]$, we have

---

[6] In order to pass to the limit in $f^i v^j$ and $f^i \operatorname{div} v$, we use the fact that strict inequality has been assumed in the condition (3.12).

$$\|f - S_j f\|_{L_T^\infty(B_{p,r}^\sigma)} \leq C\left(\left(\sum_{j' \geq j-1} \left(2^{j'\sigma}\|\Delta_{j'}f_0\|_{L^p}\right)^r\right)^{\frac{1}{r}}\right.$$

$$+ \int_0^T \left(\sum_{j' \geq j-1} \left(2^{j'\sigma}\|\Delta_{j'}g(t)\|_{L^p}\right)^r\right)^{\frac{1}{r}} dt$$

$$\left. + \|f\|_{L_T^\infty(B_{p,r}^\sigma)} \int_0^T \left(\sum_{j' \geq j-1} c_{j'}^r(t)\right)^{\frac{1}{r}} V_{p_1}'(t) dt\right).$$

The first term clearly tends to zero when j goes to infinity. The terms in the integrals also tend to zero for almost every t. Hence, by virtue of Lebesgue's dominated convergence theorem, $\|f - S_j f\|_{L_T^\infty(B_{p,r}^\sigma)}$ tends to zero when j goes to infinity. This completes the proof that $f \in \mathcal{C}([0,T]; B_{p,r}^\sigma)$ in the case $r < \infty$.

When $r = \infty$, we can use the embedding $B_{p,\infty}^\sigma \hookrightarrow B_{p,1}^{\sigma'}$ for all $\sigma' < \sigma$ and the previous argument applied to the space $B_{p,1}^{\sigma'}$ in order to show that f belongs to $\mathcal{C}([0,T]; B_{p,1}^{\sigma'})$ for all $\sigma' < \sigma$. As a matter of fact, we can also prove that $f \in \mathcal{C}_w([0,T]; B_{p,\infty}^\sigma)$. Indeed, for fixed $\phi \in \mathcal{S}(\mathbb{R}^d)$ we write

$$\langle f(t), \phi\rangle = \langle S_j f(t), \phi\rangle + \langle(\mathrm{Id} - S_j)f, \phi\rangle$$
$$= \langle S_j f(t), \phi\rangle + \langle f, (\mathrm{Id} - S_j)\phi\rangle.$$

Since $f \in \mathcal{C}([0,T]; B_{p,\infty}^{\sigma-1})$, for all $j \in \mathbb{N}$, the function $t \mapsto \langle S_j f(t), \phi\rangle$ is continuous. Now, by duality (see Proposition 2.76), we have

$$|\langle f, (\mathrm{Id} - S_j)\phi\rangle| \leq \|f\|_{B_{p,\infty}^\sigma}\|\phi - S_j\phi\|_{B_{p',1}^{-\sigma}},$$

hence $\langle f, (\mathrm{Id} - S_j)\phi\rangle$ goes to 0 uniformly on $[0,T]$ when j tends to infinity. We can thus conclude that $t \mapsto \langle f(t), \phi\rangle$ is continuous. This completes the proof of weak continuity in the case $r = \infty$.   □

Remark 3.20. Theorem 3.19 extends to the case of nonzero functions A with sufficient regularity. Indeed, the above proof may be adapted to the case where A may be approximated by a sequence of smooth functions A^n satisfying the inequality (3.19). The obtained solution f is unique and satisfies the regularity properties described in Theorem 3.19 and the inequality of Remark 3.17.

The main point is that if A^n is smooth, then

$$\partial_t f^n + v^n \cdot \nabla f^n + A^n \cdot f^n = g^n, \qquad f^n_{|t=0} = f_0^n$$

has a unique smooth solution given by the formula

$$f^n(t,x) = \exp\left(-\int_0^t A^n(\tau, \psi_\tau(\psi_t^{-1}(x))) d\tau\right) \cdot \left[f_0^n(\psi_t^{-1}(x))\right.$$

$$\left. + \int_0^t \exp\left(\int_0^\tau A^n(\tau', \psi_{\tau'}'(\psi_t^{-1}(x))) d\tau'\right) \cdot g^n(\tau, \psi_\tau(\psi_t^{-1}(x))) d\tau\right].$$

3.2.4 Application to a Shallow Water Equation

The a priori estimates stated in Theorem 3.19 are a good starting point for the study of equations of the type

$$\partial_t u + f(u) \cdot \nabla u = g(u).$$

As an example, we here solve a nonlinear one-dimensional shallow water equation which has recently received a lot of attention: the so-called Camassa–Holm equation,

$$\partial_t u - \partial_{txx}^3 u + 3u\,\partial_x u = 2\partial_x u\,\partial_{xx}^2 u + u\,\partial_{xxx}^3 u. \tag{3.25}$$

Above, the scalar function $u = u(t, x)$ stands for the fluid velocity at time $t \geq 0$ in the x direction. We assume that x belongs to $\mathbb{R}$, but (as usual in this book) similar results may be proven if x belongs to the circle.

We address the question of existence and uniqueness for the initial value problem. For simplicity, we restrict ourselves to the evolution for positive times. (We would get similar results for negative times: just change the initial condition u_0 to $-u_0$.)

At this point, the reader may wonder which regularity assumptions are relevant for u_0 so that the initial value problem is well posed in the sense of Hadamard [i.e., (3.25) has a unique local solution in a suitable functional setting with continuity with respect to the initial data].

Note that applying the pseudodifferential operator $(1 - \partial_x^2)^{-1}$ to (3.25) yields

$$(CH) \quad \begin{cases} \partial_t u + u\partial_x u = P(D)\left(u^2 + \tfrac{1}{2}(\partial_x u)^2\right) \\ u_{|t=0} = u_0 \end{cases} \quad \text{with } P(D) \stackrel{\text{def}}{=} -\partial_x(1 - \partial_x^2)^{-1}.$$

Hence, the Camassa–Holm equation is nothing but a generalized Burgers equation with an additional nonlocal nonlinearity of order 0. In light of Proposition 3.19, we thus expect that having data in some subset E of the space $C^{0,1}$ is a necessary condition for well-posedness. Moreover, as the solution u will be in $\mathcal{C}([0, T]; E)$ (a gain of regularity cannot be expected in a Burgers-like equation), the application

$$G : u \mapsto P(D)\left(u^2 + \tfrac{1}{2}(\partial_x u)^2\right)$$

should map E to E continuously.

If we restrict our attention to nonhomogeneous Besov spaces $B_{p,r}^s$, then the condition $E \subset C^{0,1}$ is equivalent to $s > 1 + 1/p$ (or $s \geq 1 + 1/p$, if $r = 1$), and no further restrictions are needed for the continuity of the map G (up to the endpoint $r = 1$, $s = 1$, $p = \infty$, which has to be avoided). We shall see, however, that for proving uniqueness, our method requires that we additionally have $s > \max(1 + 1/p, 3/2)$.

Before stating our local existence result, we introduce the following function spaces:

$$E_{p,r}^s(T) \overset{\text{def}}{=} \mathcal{C}([0,T]; B_{p,r}^s) \cap \mathcal{C}^1([0,T]; B_{p,r}^{s-1}) \quad \text{if} \quad r < \infty,$$

$$E_{p,\infty}^s(T) \overset{\text{def}}{=} \mathcal{C}_w(0,T; B_{p,\infty}^s) \cap C^{0,1}([0,T]; B_{p,\infty}^{s-1})$$

with $T > 0$, $s \in \mathbb{R}$, and $1 \leq p,r \leq \infty$.

Theorem 3.21. *Let $1 \leq p,r \leq \infty$, $s > \max(3/2, 1 + 1/p)$, and $u_0 \in B_{p,r}^s$. There exists a time $T > 0$ such that (CH) has a unique solution u in $E_{p,r}^s(T)$.*

The proof relies heavily on the following lemma.

Lemma 3.22. *Let $1 \leq p,r \leq \infty$ and $(\sigma_1, \sigma_2) \in \mathbb{R}^2$ be such that*

$$B_{p,r}^{\sigma_2} \hookrightarrow C^{0,1}, \quad \sigma_1 \leq \sigma_2, \quad \text{and} \quad \sigma_1 + \sigma_2 > 2 + \max\left\{0, \frac{2}{p} - 1\right\}.$$

Then, $\mathcal{B} : (f,g) \mapsto P(D)\big(fg + \frac{1}{2}\partial_x f\, \partial_x g\big)$ maps $B_{p,r}^{\sigma_1} \times B_{p,r}^{\sigma_2}$ into $B_{p,r}^{\sigma_1}$.

Proof. We note that $P(D)$ is a multiplier of degree -1, in the sense of Proposition 2.78. It hence suffices to prove that

$$H : (f,g) \mapsto fg + \tfrac{1}{2}\partial_x f\, \partial_x g$$

maps $B_{p,r}^{\sigma_1} \times B_{p,r}^{\sigma_2}$ into $B_{p,r}^{\sigma_1 - 1}$.

The term fg is easy to handle, so we focus on the study of $\partial_x f \partial_x g$. By virtue of Bony's decomposition, we have

$$\partial_x f \partial_x g = T_{\partial_x f} \partial_x g + T_{\partial_x g} \partial_x f + R(\partial_x f, \partial_x g).$$

Proposition 2.82 ensures that the map $(f,g) \mapsto T_{\partial_x f} \partial_x g$ is continuous from $B_{p,r}^{\sigma_1} \times B_{p,r}^{\sigma_2}$ to

- the space $B_{p,r}^{\sigma_1 + \sigma_2 - 2 - \frac{1}{p}}$, if $\sigma_1 < 1 + \frac{1}{p}$,
- the space $B_{p,r}^{\sigma_2 - 1 - \varepsilon}$ for all $\varepsilon > 0$, if $\sigma_1 = 1 + \frac{1}{p}$ and $r > 1$,
- the space $B_{p,r}^{\sigma_2 - 1}$, if $\sigma_1 = 1 + \frac{1}{p}$ and $r = 1$, or $\sigma_1 > 1 + \frac{1}{p}$.

According to our assumptions on σ_1, σ_2, p, and r, we thus can conclude that $(f,g) \mapsto T_{\partial_x f} \partial_x g$ maps $B_{p,r}^{\sigma_1} \times B_{p,r}^{\sigma_2}$ into $B_{p,r}^{\sigma_1 - 1}$. Since $B_{p,r}^{\sigma_2 - 1}$ is continuously included in L^∞, Proposition 2.82 readily yields the continuity of $(f,g) \mapsto T_{\partial_x g} \partial_x f$ from $B_{p,r}^{\sigma_1} \times B_{p,r}^{\sigma_2}$ to $B_{p,r}^{\sigma_1 - 1}$. Finally, according to Proposition 2.85, the remainder term maps $B_{p,r}^{\sigma_1} \times B_{p,r}^{\sigma_2}$ into $B_{p,r}^{\sigma_1 + \sigma_2 - 2 - \frac{1}{p}}$ (and thus to $B_{p,r}^{\sigma_1 - 1}$), provided that

$$\sigma_1 + \sigma_2 > 2 + \max\left\{0, \frac{2}{p} - 1\right\}.$$

$\square$

Uniqueness in Theorem 3.21 is a straightforward corollary of the following proposition.

Proposition 3.23. *Let $1 \leq p, r \leq \infty$ and $s > \max(1 + 1/p, 3/2)$. Suppose that we are given*

$$(u, v) \in \left(L^\infty([0, T]; B_{p,r}^s) \cap \mathcal{C}([0, T]; B_{p,r}^{s-1}) \right)^2,$$

two solutions of (CH) with initial data $u_0, v_0 \in B_{p,r}^s$. We then have, for every $t \in [0, T]$ and some constant C, depending only on s, p, and r,

$$\|u(t) - v(t)\|_{B_{p,r}^{s-1}} \leq \|u_0 - v_0\|_{B_{p,r}^{s-1}} \exp\left(C \int_0^t \left(\|u(t')\|_{B_{p,r}^s} + \|v(t')\|_{B_{p,r}^s} \right) dt' \right).$$

Proof. It is obvious that $w \stackrel{\text{def}}{=} v - u$ solves the transport equation

$$\partial_t w + u \partial_x w = -w \partial_x v + \mathcal{B}(w, u+v).$$

According to Theorem 3.14, the following inequality holds true:

$$\|w(t)\|_{B_{p,r}^{s-1}} \leq \|w_0\|_{B_{p,r}^{s-1}} e^{C \int_0^t \|\partial_x u\|_{B_{p,r}^{s-1}} dt''} + C \int_0^t e^{C \int_{t'}^t \|\partial_x u\|_{B_{p,r}^{s-1}} dt''}$$
$$\times \left(\|w \partial_x v\|_{B_{p,r}^{s-1}} + \|\mathcal{B}(w, u+v)\|_{B_{p,r}^{s-1}} \right) dt'. \quad (3.26)$$

Since $s > \max\{\frac{3}{2}, 1 + \frac{1}{p}\}$, we have, according to Lemma 3.22 and the product laws in Besov spaces,

$$\|\mathcal{B}(w, u+v)\|_{B_{p,r}^{s-1}} \leq C \|w\|_{B_{p,r}^{s-1}} \left(\|u\|_{B_{p,r}^s} + \|v\|_{B_{p,r}^s} \right).$$

Plugging this last inequality into (3.26) and applying the Gronwall lemma completes the proof. □

In order to prove the existence of a solution for (CH), we shall proceed as follows:

- First, we construct approximate solutions of (CH) which are smooth solutions of some *linear* transport equation.
- Second, we find a positive T for which those approximate solutions are uniformly bounded in $E_{p,r}^s(T)$.
- Third, we prove that the sequence of approximate solutions is a Cauchy sequence in some superspace of $E_{p,r}^s(T)$.
- Finally, we check that the limit is indeed a solution and has the desired regularity.

First Step: Constructing Approximate Solutions

Starting from $u^0 \stackrel{\text{def}}{=} 0$ we define by induction a sequence $(u^n)_{n \in \mathbb{N}}$ of smooth functions by solving the following linear transport equation:

$$(T_n) \qquad \begin{cases} (\partial_t + u^n \partial_x) u^{n+1} = P(D)\left((u^n)^2 + \tfrac{1}{2}(\partial_x u^n)^2\right) \\ u^{n+1}{}_{|t=0} = u_0. \end{cases}$$

Assuming that $u^n \in E^s_{p,r}(T)$ for all positive T, Lemma 3.22 guarantees that the right-hand side of the equation (T_n) is in $L^\infty_{loc}(\mathbb{R}^+; B^s_{p,r})$. Hence, applying Theorem 3.19 ensures that (T_n) has a global solution u^{n+1} which belongs to $E^s_{p,r}(T)$ for all positive T.

Second Step: Uniform Bounds

We define $U^n \stackrel{\text{def}}{=} \int_0^t \|u^n(t')\|_{B^s_{p,r}} \, dt'$. According to Theorem 3.19 and Lemma 3.22, we have the following inequality for all $n \in \mathbb{N}$:

$$\|u^{n+1}(t)\|_{B^s_{p,r}} \le e^{CU^n(t)}\left(\|u_0\|_{B^s_{p,r}} + C\int_0^t e^{-CU^n(t')}\|u^n(t')\|^2_{B^s_{p,r}} \, dt'\right). \quad (3.27)$$

We fix a $T > 0$ such that $2C\|u_0\|_{B^s_{p,r}}T < 1$ and suppose that

$$\forall t \in [0,T], \ \|u^n(t)\|_{B^s_{p,r}} \le \frac{\|u_0\|_{B^s_{p,r}}}{1 - 2Ct\|u_0\|_{B^s_{p,r}}}. \quad (3.28)$$

Plugging (3.28) into (3.27) yields

$$\|u^{n+1}(t)\|_{B^s_{p,r}} \le (1 - 2Ct\|u_0\|_{B^s_{p,r}})^{-\frac{1}{2}}\left(\|u_0\|_{B^s_{p,r}} \right.$$
$$\left. + C\|u_0\|^2_{B^s_{p,r}}\int_0^t \frac{dt'}{(1 - 2Ct\|u_0\|_{B^s_{p,r}})^{\frac{3}{2}}}\right)$$
$$\le \frac{\|u_0\|_{B^s_{p,r}}}{1 - 2Ct\|u_0\|_{B^s_{p,r}}}.$$

Therefore, $(u^n)_{n \in \mathbb{N}}$ is bounded in $L^\infty([0,T]; B^s_{p,r})$. This clearly entails that $u^n \partial_x u^n$ is bounded in $L^\infty([0,T]; B^{s-1}_{p,r})$. As the right-hand side of (T_n) is bounded in $L^\infty([0,T]; B^s_{p,r})$, we can conclude that the sequence $(u^n)_{n \in \mathbb{N}}$ is bounded in $E^s_{p,r}(T)$.

Third Step: Convergence

We are going to show that $(u^n)_{n \in \mathbb{N}}$ is a Cauchy sequence in $\mathcal{C}([0,T]; B^{s-1}_{p,r})$. For that purpose, we note that for all $(m,n) \in \mathbb{N}^2$, we have

$$(\partial_t + u^{n+m}\partial_x)(u^{n+m+1} - u^{n+1}) = (u^n - u^{n+m})\partial_x u^{n+1} + \mathcal{B}(u^{n+m} - u^n, u^{n+m} + u^n).$$

Applying Theorem 3.19 and Lemma 3.22, and using the fact that $B^{s-1}_{p,r}$ is an algebra yields, for any t in $[0, T]$,

$$\|(u^{n+m+1} - u^{n+1})(t)\|_{B^{s-1}_{p,r}} \leq Ce^{CU^{n+m}(t)}$$

$$\times \int_0^t e^{-CU^{n+m}(t')}\|u^{n+m} - u^n\|_{B^{s-1}_{p,r}}\left(\|u^n\|_{B^s_{p,r}} + \|u^{n+1}\|_{B^s_{p,r}} + \|u^{n+m}\|_{B^s_{p,r}}\right) dt'.$$

Since $(u^n)_{n \in \mathbb{N}}$ is bounded in $E^s_{p,r}(T)$, we finally get a constant C_T, independent of n and m, and such that for all t in $[0, T]$, we have

$$\|(u^{n+m+1} - u^{n+1})(t)\|_{B^{s-1}_{p,r}} \leq C_T \int_0^t \|(u^{n+m} - u^n)(t')\|_{B^{s-1}_{p,r}} dt'.$$

Hence, arguing by induction, we get

$$\|u^{n+m+1} - u^{n+1}\|_{L^\infty_T(B^{s-1}_{p,r})} \leq \frac{(TC_T)^{n+1}}{(n+1)!}\|u^m\|_{L^\infty_T(B^s_{p,r})}.$$

As $\|u^m\|_{L^\infty_T(B^s_{p,r})}$ may be bounded independently of m, we can now guarantee the existence of some new constant C'_T such that

$$\|u^{n+m} - u^n\|_{L^\infty_T(B^{s-1}_{p,r})} \leq C'_T 2^{-n}.$$

Hence, $(u^n)_{n \in \mathbb{N}}$ is a Cauchy sequence in $\mathcal{C}([0, T]; B^{s-1}_{p,r})$ and converges to some limit function $u \in \mathcal{C}([0, T]; B^{s-1}_{p,r})$.

Final Step: Conclusion

We have to check that u belongs to $E^s_{p,r}(T)$ and satisfies (CH). Since $(u^n)_{n \in \mathbb{N}}$ is bounded in $L^\infty([0, T]; B^s_{p,r})$, the Fatou property for Besov spaces guarantees that u also belongs to $L^\infty([0, T]; B^s_{p,r})$. Now, as $(u^n)_{n \in \mathbb{N}}$ converges to u in $\mathcal{C}([0, T]; B^{s-1}_{p,r})$, an interpolation argument ensures that convergence actually holds true in $\mathcal{C}([0, T]; B^{s'}_{p,r})$ for any $s' < s$. It is then easy to pass to the limit in (T_n) and to conclude that u is indeed a solution of (CH).

Finally, because u belongs to $L^\infty([0, T]; B^s_{p,r})$, the right-hand side of the equation

$$\partial_t u + u\partial_x u = P(D)(u^2 + \tfrac{1}{2}(\partial_x u)^2)$$

also belongs to $L^\infty([0, T]; B^s_{p,r})$. Hence, according to Theorem 3.19, the function u belongs to $\mathcal{C}([0, T]; B^s_{p,r})$ (resp., $\mathcal{C}_w([0, T]; B^s_{p,r})$) if $r < \infty$ (resp., $r = \infty$). Again using the equation, we see that $\partial_t u$ is in $\mathcal{C}([0, T]; B^{s-1}_{p,r})$ if r is finite, and in $L^\infty([0, T]; B^{s-1}_{p,r})$ otherwise, so u belongs to $E^s_{p,r}(T)$.

Remark 3.24. If v_0 belongs to a small neighborhood of u_0 in $B_{p,r}^s$, then the arguments above give the existence of a solution $v \in E_{p,r}^s(T)$ of (CH) with initial data v_0. Proposition 3.23, combined with an obvious interpolation, ensures continuity with respect to the initial data in $\mathcal{C}([0,T]; B_{p,r}^{s'}) \cap \mathcal{C}^1([0,T]; B_{p,r}^{s'-1})$ for any $s' < s$. In fact, continuity holds up to exponent s whenever r is finite. This may be proven by adapting the method presented in Section 4.5.

Finally, we state a blow-up criterion for (CH). In what follows, we define the *lifespan* $T_{u_0}^{\star}$ of the solution u of (CH) with initial data u_0 as the supremum of positive times T such that (CH) has a solution $u \in E_{p,r}^s(T)$ on $[0,T] \times \mathbb{R}$. We have the following result.

Theorem 3.25. *Let u_0 be as in Theorem 3.21 and u the corresponding solution. If $T_{u_0}^{\star}$ is finite, then we have*

$$\int_0^{T_{u_0}^{\star}} \|\partial_x u(t')\|_{L^\infty}\, dt' = \infty \quad \text{and} \quad \int_0^{T_{u_0}^{\star}} \|u(t')\|_{B_{\infty,\infty}^1}\, dt' = \infty.$$

The proof is based on the following lemma.

Lemma 3.26. *Let $1 \leq p, r \leq \infty$ and $s > 1$. Let $u \in L^\infty([0,T]; B_{p,r}^s)$ solve (CH) on $[0,T[\times \mathbb{R}$ with $u_0 \in B_{p,r}^s$ as initial data. There exist a constant C, depending only on s and p, and a universal constant C' such that for all $t \in [0,T[$, we have*

$$\|u(t)\|_{B_{p,r}^s} \leq \|u_0\|_{B_{p,r}^s} e^{C \int_0^t \|u(t')\|_{C^{0,1}}\, dt'}, \tag{3.29}$$

$$\|u(t)\|_{C^{0,1}} \leq \|u_0\|_{C^{0,1}} e^{C' \int_0^t \|\partial_x u(t')\|_{L^\infty}\, dt'}. \tag{3.30}$$

Proof. Applying the last part of Theorem 3.14 to (CH) and using the fact that $P(D)$ is a multiplier of order -1 yields

$$e^{-C \int_0^t \|\partial_x u\|_{L^\infty}\, dt'} \|u(t)\|_{B_{p,r}^s} \leq \|u_0\|_{B_{p,r}^s}$$
$$+ C \int_0^t e^{-C \int_0^{t'} \|\partial_x u\|_{L^\infty}\, dt''} \left(\|u^2\|_{B_{p,r}^{s-1}} + \|(\partial_x u)^2\|_{B_{p,r}^{s-1}} \right) dt'.$$

As $s - 1 > 0$, we have

$$\|u^2\|_{B_{p,r}^{s-1}} + \|(\partial_x u)^2\|_{B_{p,r}^{s-1}} \leq C \|u\|_{C^{0,1}} \|u\|_{B_{p,r}^s}.$$

Therefore,

$$e^{-C \int_0^t \|\partial_x u\|_{L^\infty}\, dt'} \|u(t)\|_{B_{p,r}^s} \leq \|u_0\|_{B_{p,r}^s}$$
$$+ C \int_0^t e^{-C \int_0^{t'} \|\partial_x u\|_{L^\infty}\, dt''} \|u\|_{B_{p,r}^s} \|u\|_{C^{0,1}}\, dt'.$$

Applying the Gronwall lemma completes the proof of (3.29).

By differentiating equation (CH) once with respect to x and applying the L^∞ estimate for transport equations, we get

$$e^{-\int_0^t \|\partial_x u\|_{L^\infty} dt'} \|u(t)\|_{C^{0,1}} \leq \|u_0\|_{C^{0,1}}$$
$$+ \int_0^t e^{-\int_0^{t'} \|\partial_x u\|_{L^\infty} dt'} \|P(D)(u^2 + \tfrac{1}{2}(\partial_x u)^2)(t')\|_{C^{0,1}} dt'.$$

Now, by using the fact that the operator $(1-\partial_{xx}^2)^{-1}$ coincides with convolution by the function $x \mapsto \frac{1}{2}e^{-|x|}$, it may be easily proven that for some universal constant C', we have

$$\|P(D)(u^2 + \tfrac{1}{2}(\partial_x u)^2)\|_{C^{0,1}} \leq C' \|u\|_{C^{0,1}} \|\partial_x u\|_{L^\infty}.$$

Hence, the Gronwall lemma gives the inequality (3.30). $\qquad\square$

Proof of Theorem 3.25. Let $u \in \bigcap_{T<T^\star} E_{p,r}^s(T)$. We want to show that if

$$\int_0^{T^\star} \|\partial_x u(t')\|_{L^\infty} dt' < \infty,$$

then no blow-up occurs at time T.

According to the inequality (3.30), $\int_0^{T^\star} \|u(t')\|_{C^{0,1}} dt'$ is also finite. Therefore, the inequality (3.29) ensures that

$$\forall t \in [0, T^\star[, \ \|u(t)\|_{B_{p,r}^s} \leq M_{T^\star} \stackrel{\text{def}}{=} \|u_0\|_{B_{p,r}^s} e^{C \int_0^{T^\star} \|u(t')\|_{C^{0,1}} dt'} < \infty. \quad (3.31)$$

Let $\varepsilon > 0$ be such that $2C^2 \varepsilon M_{T^\star} < 1$, where C is the constant used in the proof of Theorem 3.19. We then have a solution $\tilde{u} \in E_{p,r}^s(\varepsilon)$ of (CH) with initial data $u(T^\star - \varepsilon/2)$. By uniqueness, we have $\tilde{u}(t) = u(t + T^\star - \varepsilon/2)$ on $[0, \varepsilon/2[$ so that $\tilde{u}$ extends the solution u beyond $T^\star$. We conclude that $T^\star < T_{u_0}^\star$.

We can now conclude that if $T_{u_0}^\star$ is finite, then we must have

$$\int_0^{T_{u_0}^\star} \|u(t')\|_{B_{\infty,\infty}^1} dt' = \infty.$$

This simply follows from the logarithmic interpolation inequality

$$\|u\|_{C^{0,1}} \leq C\left(1 + \|u\|_{B_{\infty,\infty}^1} \log\left(e + \|u\|_{B_{p,r}^s}\right)\right), \quad (3.32)$$

which holds true whenever $s > 1 + 1/p$ and which may be deduced from Proposition 2.104 combined with the embedding $B_{p,1}^{s-1} \hookrightarrow B_{\infty,\infty}^{s-1-\frac{1}{p}}$.

Now, plugging (3.32) into (3.31), we get

$$\|u(t)\|_{B_{p,r}^s} \leq \|u_0\|_{B_{p,r}^s} e^{Ct} \exp\left(C \int_0^t \|u\|_{B_{\infty,\infty}^1} \log(e + \|u\|_{B_{p,r}^s}) dt'\right).$$

Therefore, easy calculations lead to

$$\log\left(e+\|u(t)\|_{B_{p,r}^s}\right) \leq \log\left(e+\|u_0\|_{B_{p,r}^s}\right)+Ct+C\int_0^t \|u\|_{B_{\infty,\infty}^1} \log\left(e+\|u\|_{B_{p,r}^s}\right) dt'.$$

The Gronwall lemma thus yields

$$\log\left(e+\|u(t)\|_{B_{p,r}^s}\right) \leq \left(\log\left(e+\|u_0\|_{B_{p,r}^s}\right)+Ct\right)e^{C\int_0^t \|u\|_{B_{\infty,\infty}^1}\, dt'}.$$

Therefore, $\int_0^T \|u\|_{B_{\infty,\infty}^1}\, dt < \infty$ implies that $u \in L^\infty([0,T]; B_{p,r}^s)$. Arguing as above completes the proof of Theorem 3.25. $\qquad\square$

Remark 3.27. The fact that $\|\partial_x u\|_{L^\infty}$ may be replaced by the weaker norm $\|\partial_x u\|_{B_{\infty,\infty}^0}$ is not particularly sensitive to the structure of the equation. In fact, a similar criterion may be stated for the incompressible Euler equations (see Chapter 7) and for quasilinear symmetric systems (see Chapter 4).

3.3 Losing Estimates for Transport Equations

In this section, we consider transport equations associated with vector fields which *are not* Lipschitz with respect to the space variable. Since we still intend to obtain regularity theorems, those vector fields cannot be too rough. The minimal requirement seems to be that the vector field v is log-Lipschitz, in the sense of Definition 2.106. We shall see that if v is not Lipschitz, then loss of regularity may occur, going from linear loss of regularity to arbitrarily small loss of regularity, depending on how far from Lipschitz v is. In order to precisely measure the regularity of the vector field v, we shall introduce the following notation, used throughout this section:

$$V_{p_1,\alpha}'(t) \overset{\text{def}}{=} \sup_{j\geq 0} \frac{2^{j\frac{d}{p_1}} \|\nabla S_j v(t)\|_{L^{p_1}}}{(j+1)^\alpha} < \infty. \tag{3.33}$$

We note that if $p_1 = \infty$, then $V_{p_1,\alpha}'(t)$ is exactly the norm $\|\cdot\|_{B_\Gamma}$ of Definition 2.110 page 117 in the case where $\Gamma(r) = (\log r)^\alpha$.

Those results have many applications in problems related to fluid mechanics (see Chapter 7 and the last part of this section).

3.3.1 Linear Loss of Regularity in Besov Spaces

This section is devoted to the statement of estimates with *linear* loss of regularity. Recall that, according to Proposition 2.111, v is log-Lipschitz if and only if there exists some constant C such that

$$\|\nabla S_j u\|_{L^\infty} \leq C(j+1) \quad \text{for all} \quad j \geq -1.$$

This motivates the following statement.

Theorem 3.28. *Let* $1 \leq p \leq p_1 \leq \infty$ *and suppose that* $s_1 \in \mathbb{R}$ *satisfies* (3.12). *Let* σ *be in* $]s_1, 1 + \frac{d}{p_1}[$ *and* v *be a vector field. There then exists a constant* C, *depending only on* p, p_1, σ, s_1, *and* d, *such that for any* $\lambda > C$, $T > 0$, *and any nonnegative integrable function* W *over* $[0, T]$ *such that* $\sigma_T \geq s_1$ *with*

$$\sigma_t \stackrel{def}{=} \sigma - \lambda \int_0^t \left(V'_{p_1,1}(t') + W(t') \right) dt',$$

the following property holds true.

Let $f_0 \in B^{\sigma}_{p,\infty}$ *and* $g = g_1 + g_2$ *with, for all* $t \in [0, T]$, $g_1(t) \in B^{\sigma_t}_{p,\infty}$, *and*

$$\forall j \geq -1, \ \|\Delta_j g_2(t)\|_{L^p} \leq 2^{-j\sigma_t}(j + 2)W(t)\|f(t)\|_{B^{\sigma_t}_{p,\infty}}.$$

Let $f \in \mathcal{C}([0, T]; B^{s_1}_{p,\infty})$ *be a solution of* (T) *with* $A \equiv 0$ *such that* $f(t) \in B^{\sigma_t}_{p,\infty}$ *for all* $t \in [0, T]$. *The following estimate then holds:*

$$\sup_{t \in [0,T]} \|f(t)\|_{B^{\sigma_t}_{p,\infty}} \leq \frac{\lambda}{\lambda - C} \left(\|f_0\|_{B^{\sigma}_{p,\infty}} + \int_0^T \|g_1(t)\|_{B^{\sigma_t}_{p,\infty}} \, dt \right).$$

Proof. Applying the operator Δ_j to the equation (T), we see that for all $j \geq -1$, the function $\Delta_j f$ is a solution of

$$(T_j) \qquad \begin{cases} \partial_t f_j + S_{j+1} v \cdot \nabla f_j = \Delta_j g - \widetilde{R}_j \\ f_{j|t=0} = \Delta_j f_0 \end{cases}$$

with $\widetilde{R}_j \stackrel{def}{=} \Delta_j (v \cdot \nabla f) - S_{j+1} v \cdot \nabla \Delta_j f$.

We shall now temporarily assume the following result.

Lemma 3.29. *Let* $\sigma \in \mathbb{R}$, $\alpha \geq 0$, *and* $1 \leq p \leq p_1 \leq \infty$. *Assume that* (3.12) *is satisfied and that* $\sigma < 1 + \frac{d}{p_1}$. *There then exists a constant* C, *depending continuously on* p, p_1, σ, *and* d, *such that*

$$\sup_{j \geq -1} 2^{j\sigma} \|\widetilde{R}_j\|_{L^p} \leq C(j+2)^{\alpha} V'_{p_1,\alpha}(t) \|f(t)\|_{B^{\sigma}_{p,\infty}}.$$

The proof of the theorem is now easy. Indeed, as $\Delta_j f$ is a solution of (T_j), we have

$$\Delta_j f(t, x) = \Delta_j f_0(\psi_j^{-1}(t, x)) + \int_0^t \Delta_j g(t', \psi_j(t', \psi_j^{-1}(t, x))) \, dt'$$

$$- \int_0^t \widetilde{R}_j(t', \psi_j(t', \psi_j^{-1}(t, x))) \, dt',$$

where we have denoted by ψ_j the flow of the vector field $S_{j+1} v$.

From inequality (3.6) and the Bernstein inequality, we get

$$\sup_{x \in \mathbb{R}^d} |\det D_x \psi_j(t', \psi_j^{-1}(t, x))|^{-1} \leq 2^{C(2+j) \int_{t'}^t V'_{p_1,1}(t'') \, dt''}.$$

We deduce that

$$\|\Delta_j f(t)\|_{L^p} \le \|\Delta_j f_0\|_{L^p} 2^{C(2+j)V_{p_1,1}(t)}$$
$$+ \int_0^t \|\Delta_j g_1(t')\|_{L^p} 2^{C(2+j)\int_{t'}^t V'_{p_1,1}(t'')\,dt''}\,dt'$$
$$+ C \int_0^t (2+j)(V'_{p_1,1}+W)(t') 2^{(2+j)(C\int_{t'}^t V'_{p_1,1}(t'')\,dt''-\sigma_{t'})} \|f(t')\|_{B^{\sigma_{t'}}_{p,\infty}}\,dt'.$$

Next, we multiply the above inequality by $2^{(2+j)\sigma_t}$ and take the ℓ^∞ norm of both sides. As

$$\sigma_t = \sigma_{t'} - \lambda \int_{t'}^t (V'_{p_1,1}(t'') + W)(t'')\,dt'',$$

we get

$$\|f(t)\|_{B^{\sigma_t}_{p,\infty}} \le \|f_0\|_{B^\sigma_{p,\infty}} + \int_0^t \|g_1(t')\|_{B^{\sigma_{t'}}_{p,\infty}}\,dt'$$
$$+ C \int_0^t (2+j)(V'_{p_1,1}+W)(t') 2^{(2+j)(C-\lambda)\int_{t'}^t (V'_{p_1,1}+W)(t'')\,dt''} \|f(t')\|_{B^{\sigma_{t'}}_{p,\infty}}\,dt'.$$

Straightforward calculations show that the second integral in the above inequality is bounded by

$$\frac{C}{(\lambda-C)\log 2} \sup_{t\in[0,T]} \|f(t)\|_{B^{\sigma_t}_{p,\infty}}.$$

Therefore, changing C if necessary, we get, for any $\lambda > C$,

$$\|f(t)\|_{B^{\sigma_t}_{p,\infty}} \le \|f_0\|_{B^s_{p,\infty}} + \int_0^t \|g_1(t')\|_{B^{\sigma_{t'}}_{p,\infty}}\,dt' + \frac{C}{\lambda - C} \sup_{t\in[0,T]} \|f(t)\|_{B^{\sigma_t}_{p,\infty}},$$

which leads to the theorem. $\qquad\square$

Proof of Lemma 3.29. It suffices to observe that

$$\widetilde{R}_j = [\Delta_j, S_{j+1}v] \cdot \nabla f + \Delta_j\big((v - S_{j+1}v) \cdot \nabla f\big).$$

Now, on the one hand, we have, according to Lemma 2.100 page 112,

$$\sup_{j\ge -1} 2^{j\sigma} \|[\Delta_j, S_{j+1}v] \cdot \nabla f\|_{L^p} \le C\|\nabla S_{j+1}v\|_{B^{\frac{d}{p_1}}_{p_1,\infty}\cap L^\infty} \|f\|_{B^\sigma_{p,\infty}}$$
$$\le C(j+2)^\alpha V'_{p_1,\alpha} \|f\|_{B^\sigma_{p,\infty}}.$$

On the other hand, we have (with the summation convention)

$$\Delta_j\big((v - S_{j+1}v) \cdot \nabla f\big) = \underbrace{\Delta_j T_{v^i - S_{j+1}v^i}\partial_i f}_{\widetilde{R}^1_j}$$

$$+ \underbrace{\partial_i \Delta_j R(v^i - S_{j+1}v^i, f)}_{\widetilde{R}^2_j} + \underbrace{R(S_{j+1}\,\mathrm{div}\,v - \mathrm{div}\,v, f)}_{\widetilde{R}^3_j} + \underbrace{T_{\partial_i f}(v^i - S_{j+1}v^i)}_{\widetilde{R}^4_j}.$$

Continuity results for the paraproduct (see Proposition 2.82) ensure that

$$2^{j\sigma}\|\widetilde{R}_j^1\|_{L^p} \leq C2^j \|v - S_{j+1}v\|_{L^\infty} \|f\|_{B_{p,\infty}^\sigma} \quad \text{for all} \ \ j \geq -1.$$

Now, observing that

$$
\begin{aligned}
\|v - S_{j+1}v\|_{L^\infty} &\leq C \sum_{j'>j} 2^{-j'} \|\nabla\Delta_{j'}v\|_{L^\infty} \\
&\leq C \sum_{j'\geq j} 2^{-j'} 2^{j'\frac{d}{p_1}} \|\nabla S_{j'+1}v\|_{L^{p_1}} \\
&\leq C V'_{p_1,\alpha} \sum_{j'\geq j} 2^{-j'}(2+j')^\alpha \\
&\leq C V'_{p_1,\alpha}(2+j)^\alpha 2^{-j},
\end{aligned}
$$

we get the desired inequality for $\widetilde{R}_j^1$.

Next, setting $1/p_2 = 1/p + 1/p_1$ and $\widetilde{\Delta}_j = \Delta_{j-1} + \Delta_j + \Delta_{j+1}$, we have, if $1/p + 1/p_1 \leq 1$,

$$
\begin{aligned}
2^{j(\sigma+\frac{d}{p_1})}\|\widetilde{R}_j^2\|_{L^{p_2}} &\leq C2^{j(1+\sigma+\frac{d}{p_1})} \sum_{j'\geq j} \|\Delta_{j'}(\mathrm{Id}-S_{j+1})v\|_{L^{p_1}} \|\widetilde{\Delta}_{j'}f\|_{L^p} \\
&\leq C \sum_{j'\geq j} 2^{(j-j')(\sigma+1+\frac{d}{p_1})} 2^{j'\frac{d}{p_1}} \|\Delta_{j'}\nabla v\|_{L^{p_1}} 2^{j'\sigma}\|\widetilde{\Delta}_{j'}f\|_{L^p} \\
&\leq C V'_{p_1,\alpha} \Bigg((j+2)^\alpha \sum_{j'\geq j} 2^{(j-j')(\sigma+1+\frac{d}{p_1})} 2^{j'\sigma}\|\widetilde{\Delta}_{j'}f\|_{L^p} \\
&\qquad\qquad + \sum_{j'\geq j} (j'-j)^\alpha 2^{(j-j')(\sigma+1+\frac{d}{p_1})} 2^{j'\sigma}\|\widetilde{\Delta}_{j'}f\|_{L^p} \Bigg) \\
&\leq C V'_{p_1,\alpha}(2+j)^\alpha 2^{-j\sigma}\|f\|_{B_{p,\infty}^\sigma}.
\end{aligned}
$$

Hence, taking advantage of the Bernstein inequality, we get

$$\|\widetilde{R}_j^2\|_{L^p} \leq C(2+j)^\alpha 2^{-j\sigma} V'_{p_1,\alpha}\|f\|_{B_{p,\infty}^\sigma} \quad \text{if} \ \ \sigma+1+\frac{d}{p_1} > 0. \tag{3.34}$$

In the case where $1/p + 1/p_1 > 1$, we replace p_1 with p' in the above computations and we still get (3.34), provided $\sigma + 1 + \frac{d}{p'} > 0$.

A similar bound may be proven for $\widetilde{R}_j^3$ if $\sigma > -d\min(\frac{1}{p_1},\frac{1}{p'})$. Finally, we note that

$$\widetilde{R}_j^4 = - \sum_{\substack{|j'-j|\leq 4 \\ j''\leq j'-2}} \Delta_j\Big(\Delta_{j''}\partial_i f \Delta_{j'}(v^i - S_{j+1}v^i)\Big).$$

Therefore, writing $1/p_3 = 1/p - 1/p_1$, we have

$$2^{j\sigma}\|\widetilde{R}_j^4\|_{L^p} \leq C \sum_{\substack{|j'-j|\leq 4 \\ j''\leq j'-2}} 2^{j\sigma} \|\Delta_{j''}\partial_i f\|_{L^{p_3}} \|\Delta_{j'}(v^i - S_{j+1}v^i)\|_{L^{p_1}}.$$

Because, for $j \geq -1$, the function $\mathcal{F}(v - S_{j+1}v)$ is supported away from 0, we can write, thanks to Lemma 2.1,

$$\|v - S_{j+1}v\|_{L^{p_1}} \leq C \sum_{j'>j} 2^{-j'} \|\nabla \Delta_{j'}v\|_{L^{p_1}}$$

$$\leq CV'_{p_1,\alpha} \sum_{j' \geq j} (j'+2)^\alpha 2^{-j'(1+\frac{d}{p_1})}$$

$$\leq CV'_{p_1,\alpha}(j+2)^\alpha 2^{-j(1+\frac{d}{p_1})}.$$

Hence, as $\sigma < 1 + d/p_1$, we conclude that

$$2^{j\sigma}\|\widetilde{R}_j^4\|_{L^p} \leq CV'_{p_1,\alpha}(j+2)^\alpha \|f\|_{B_{p,\infty}^\sigma}.$$

This completes the proof of the lemma.   □

3.3.2 The Exponential Loss

In this section, we give an example of a *global* result with exponential loss of regularity for transport equations. Before stating our main result, we have to introduce some new function spaces.

Definition 3.30. *Let $p \in [1,\infty]$ and $s \in]0,1[$. We denote by F_p^s the space of functions u in $L^p(\mathbb{R}^d)$ such that for any couple $(x, x') \in \mathbb{R}^d \times \mathbb{R}^d$,*

$$\frac{|u(x) - u(x')|}{|x - x'|^s} \leq U(x) + U(x') \tag{3.35}$$

for some function U in $L^p(\mathbb{R}^d)$.

Endowed with the norm

$$\|u\|_{F_p^s} = \|u\|_{L^p} + \inf\{\|U\|_{L^p}, \ U \text{ satisfying } (3.35)\},$$

the space F_p^s is complete. In the case $p > 1$, it may be proven that F_p^s belongs to the family of so-called *Triebel–Lizorkin spaces* (in fact, $F_p^s = F_{p,\infty}^s$) and that $B_{p,1}^s \hookrightarrow F_p^s \hookrightarrow B_{p,\infty}^s$.

In the present section, we shall just use the following, easy, lemma.

Lemma 3.31. *For all $p \in [1,\infty]$ and $s \in]0,1[$, the space F_p^s is continuously embedded in $B_{p,\infty}^s$.*

Proof. It suffices to use the characterization of $B_{p,\infty}^s$ in terms of finite differences. Indeed, since $B_{p,\infty}^s = L^p \cap \dot{B}_{p,\infty}^s$, Theorem 2.36 page 74 guarantees that

$$\|u\|_{B_{p,\infty}^s} \approx \|u\|_{L^p} + \sup_{0<|h|<1} |h|^{-s}\|\tau_h u - u\|_{L^p}.$$

Now, if u belongs to F_p^s, we have, for all $(x, h) \in \mathbb{R}^d$,

$$|u(x-h) - u(x)| \leq (U(x-h) + U(x))|h|^s \quad \text{with} \quad U \in L^p.$$

This obviously ensures that $\sup_{0<|h|<1} |h|^{-s}\|\tau_h u - u\|_{L^p}$ is finite.   □

Theorem 3.32. *Let $s \in \;]0, 1[$ and let v be a divergence-free, time-dependent vector field with coefficients in $L^1_{loc}(\mathbb{R}^+; LL)$. Let*

$$\sigma(s, t) \overset{def}{=} s \exp(-V_{LL}(t)) \quad with \quad V_{LL}(t) \overset{def}{=} \int_0^t \|v(t')\|_{LL} \, dt',$$

with $\| \cdot \|_{LL}$ as defined on page 116. Let f_0 be in F_p^s and g be such that the function $t \mapsto \|g(t)\|_{F_p^{\sigma(s,t)}}$ belongs to $L^1_{loc}(\mathbb{R}^+)$.

There then exists a unique solution of (T) with $A \equiv 0$ such that $t \mapsto \|f(t)\|_{F_p^{\sigma(s,t)}}$ belongs to $L^\infty_{loc}(\mathbb{R}^+)$.

Proof. We consider a sequence $(v_n)_{n \in \mathbb{N}}$ of smooth, bounded, divergence-free vector fields satisfying

$$\|v_n(t)\|_{LL} \leq \|v(t)\|_{LL} \quad and \quad \lim_{n \to \infty} v_n = v \quad in \quad L^1_{loc}(\mathbb{R}^+; L^\infty).$$

Let f_n be the solution of the equation

$$\begin{cases} \partial_t f_n + v_n \cdot \nabla f_n = g \\ f_{n|t=0} = f_0. \end{cases}$$

Denoting by ψ_n the flow of v_n, we may write

$$f_n(t, x) = f_0(\psi_n^{-1}(t, x)) + \int_0^t g(t', \psi_n(t', \psi_n^{-1}(t, x))) \, dt'. \tag{3.36}$$

In the light of Theorem 3.7, the problem reduces to the study of how the right composition by a $C^{0,\alpha}$ measure-preserving homeomorphism operates on F_p^s. Let θ be such a homeomorphism and u be in F_p^s. For $|x - x'| \leq 1$, we have, with the notation of Definition 3.30,

$$\frac{|u(\theta(x)) - u(\theta(x'))|}{|x - x'|^{s\alpha}} \leq \frac{|u(\theta(x)) - u(\theta(x'))|}{|\theta(x) - \theta(x')|^s} \frac{|\theta(x) - \theta(x')|^s}{|x - x'|^{s\alpha}}$$

$$\leq \|\theta\|_{C^{0,\alpha}}^s \left(U(\theta(x)) + U(\theta(x')) \right).$$

As θ preserves the measure, we thus have

$$\|u \circ \theta\|_{F_p^{s\alpha}} \leq (1 + \|\theta\|_{C^{0,\alpha}}^s) \|u\|_{F_p^s}.$$

Applying the above inequality at each time to (3.36), we get, for some non-decreasing locally bounded function A,

$$\|f_n(t)\|_{F_p^{\sigma(s,t)}} \leq A(t) \left(\|f_0\|_{F_p^s} + \int_0^t \|g(\tau)\|_{F_p^{\sigma(s,\tau)}} \, d\tau \right). \tag{3.37}$$

Therefore, the sequence $(\|f_n(t)\|_{F_p^{\sigma(s,t)}})_{n \in \mathbb{N}}$ is bounded in $L^\infty_{loc}(\mathbb{R}^+)$.

We shall now prove that $(f_n)_{n\in\mathbb{N}}$ is a Cauchy sequence in some Besov space with *negative* index. For that, we shall take advantage of Theorem 3.28. For all $(m,n) \in \mathbb{N}^2$, we have

$$\begin{cases} \partial_t(f_n - f_m) + \operatorname{div}((f_n - f_m)v_n) = \operatorname{div}\big(f_m(v_m - v_n)\big) \\ (f_n - f_m)_{|t=0} = 0. \end{cases}$$

Fix $T > 0$ and set

$$s_T = \frac{s}{2}\exp(-V_{LL}(t)).$$

For sufficiently large λ, let

$$\sigma_t \stackrel{\text{def}}{=} s_T - 1 - \lambda V_{LL}(t).$$

Applying Theorem 3.28 with $s_1 = s_T/2$, $g_1 \stackrel{\text{def}}{=} (v_m - v_n)\cdot\nabla f_m$, and $g_2 \stackrel{\text{def}}{=} 0$, we get, for any t such that $\sigma_t \geq s_1 - 1$,

$$\|(f_n - f_m)(t)\|_{B_{p,\infty}^{\sigma_t}} \leq C_T \int_0^t \|\big((v_m - v_n)\cdot\nabla f_m\big)(t')\|_{B_{p,\infty}^{\sigma_{t'}}}\, dt'.$$

However, owing to Theorems 2.82 and 2.85, we have

$$\|T_{v_m^i - v_n^i}\partial_i f_m\|_{B_{p,\infty}^{-1+s_T}} \leq C_T\|v_m - v_n\|_{L^\infty}\|f_m\|_{B_{p,\infty}^{s_T}},$$

$$\|T_{\partial_i f_m}(v_m^i - v_n^i)\|_{B_{p,\infty}^{-1+s_T}} \leq C_T\|v_m - v_n\|_{L^\infty}\|f_m\|_{B_{p,\infty}^{s_T}},$$

$$\|\partial_i R(v_m^i - v_n^i, f_m)\|_{B_{p,\infty}^{-1+s_T}} \leq C_T\|v_m - v_n\|_{L^\infty}\|f_m\|_{B_{p,\infty}^{s_T}}.$$

Because $F_p^{s_T} \hookrightarrow B_{p,\infty}^{s_T}$ and (3.37) is satisfied, we find that the sequence $(f_n)_{n\in\mathbb{N}}$ belongs to the space $L^\infty([0,T]; B_{p,\infty}^{s_T})$. Hence, for any $t \in [0,T]$ such that $\sigma_t \geq s_1 - 1$, we have

$$\|\big((v_m - v_n)\cdot\nabla f_m\big)(t')\|_{B_{p,\infty}^{\sigma_{t'}}} \leq C_T\|(v_n - v_m)(t')\|_{L^\infty}.$$

Therefore, for small enough T_0, the sequence $(f_n)_{n\in\mathbb{N}}$ satisfies the Cauchy criterion in the space $\mathcal{C}([0,T_0]; B_{p,\infty}^{s_T/2-1})$. This proves the theorem on the interval $[0,T_0]$. Note that the argument may be applied again, starting from T_0. After a finite number of steps, we finally prove convergence on $[0,T]$. Since T has been fixed arbitrarily, we end up with a global existence result.   □

3.3.3 Limited Loss of Regularity

In this section, we make the assumption that there exists some $\alpha \in\,]0,1[$ such that the function $V'_{p_1,\alpha}$ defined in (3.33) is locally integrable.

Recall that in the limit case $\alpha = 0$ (treated in Theorem 3.14), there is no loss of regularity and that if $\alpha = 1$, then a *linear* loss of regularity may occur (see Theorem 3.28). In the theorem below, we state that if $\alpha \in\,]0,1[$, then the loss of regularity in the estimates is *arbitrarily small*.

Theorem 3.33. *Let (p, p_1) be in $[1, \infty]^2$ such that $1 \leq p \leq p_1$ and suppose that σ satisfies (3.12). Assume that σ is less than $1 + \frac{d}{p_1}$ and that $V'_{p_1, \alpha}$ is in $L^1([0, T])$ for some $\alpha \in]0, 1[$. Let f_0 be in $B^\sigma_{p, \infty}$ and g be in $\widetilde{L}^1_T(B^\sigma_{p, \infty})$. The equation (T) with $A \equiv 0$ then has a unique solution f in $\mathcal{C}([0, T]; \bigcap_{\sigma' < \sigma} B^{\sigma'}_{p, \infty})$, and the following estimate holds for all small enough ε:*

$$\|f\|_{\widetilde{L}^\infty_T(B^{\sigma-\varepsilon}_{p,\infty})} \leq C \left(\|f_0\|_{B^\sigma_{p,\infty}} + \|g\|_{\widetilde{L}^1_T(B^\sigma_{p,\infty})} \right) \exp\left(\frac{C}{\varepsilon^{\frac{\alpha}{1-\alpha}}} \left(V_{p_1, \alpha}(T) \right)^{\frac{1}{1-\alpha}} \right),$$

where C depends only on α, p, p_1, σ, and d.

Remark 3.34. Theorem 3.33 applies with $\alpha = 1 - 1/r$ whenever ∇v belongs to $L^1_T(B^{\frac{d}{p_1}}_{p_1, r})$.

Proof. We focus on the proof of the a priori estimate. Existence may be obtained by arguing as in Theorem 3.32.

Fix a small enough ε and let $\eta > 0$ satisfy $\varepsilon = \eta V_{p_1, \alpha}(T)$. We define, for t in $[0, T]$,

$$\sigma_t \stackrel{\text{def}}{=} \sigma - \eta V_{p_1, \alpha}(t).$$

Following the lines of the proof of Theorem 3.28, we now get

$$2^{(2+j)\sigma_t} \|\Delta_j f(t)\|_{L^p} \leq 2^{(2+j)\sigma} \|\Delta_j f_0\|_{L^p} 2^{-\eta(2+j) V_{p_1, \alpha}(t)}$$

$$+ \int_0^t 2^{(2+j)\sigma_{t'}} \|\Delta_j g(t')\|_{L^p} 2^{-\eta(2+j) \int_{t'}^t V'_{p_1, \alpha}(t'') \, dt''} \, dt'$$

$$+ C \int_0^t (2 + j)^\alpha V'_{p_1, \alpha}(t') 2^{-\eta(2+j) \int_{t'}^t V'_{p_1, \alpha}(t'') \, dt''} \|f(t')\|_{B^{\sigma_{t'}}_{p,\infty}} \, dt'.$$

On the one hand, if j is so large as to satisfy

$$2 + j \geq \left(\frac{2C}{\eta \log 2} \right)^{\frac{1}{1-\alpha}}, \tag{3.38}$$

then we have

$$C \int_0^t (2 + j)^\alpha V'_{p_1, \alpha}(t') 2^{-\eta(2+j) \int_{t'}^t V'_{p_1, \alpha}(t'') \, dt''} \, dt' \leq \frac{1}{2},$$

from which it follows that

$$2^{(2+j)\sigma_t} \|\Delta_j f(t)\|_{L^p} \leq 2^{(2+j)\sigma} \|\Delta_j f_0\|_{L^p}$$

$$+ \int_0^t 2^{(2+j)\sigma_{t'}} \|\Delta_j g(t')\|_{L^p} \, dt' + \frac{1}{2} \sup_{t' \in [0, t]} \|f(t')\|_{B^{\sigma_{t'}}_{p,\infty}}.$$

On the other hand, if (3.38) is not satisfied, then we can write

$$2^{(2+j)\sigma_t}\|\Delta_j f(t)\|_{L^p} \leq 2^{(2+j)\sigma}\|\Delta_j f_0\|_{L^p}$$

$$+ \int_0^t 2^{(2+j)\sigma_{t'}}\|\Delta_j g(t')\|_{L^p}\, dt' + C\left(\frac{2C}{\eta \log 2}\right)^{\frac{\alpha}{1-\alpha}} \int_0^t V'_{p_1,\alpha}(t')\|f(t')\|_{B^{\sigma_{t'}}_{p,\infty}}\, dt'.$$

Combining these two inequalities and using the fact that $\sigma_{t'} \leq \sigma$ for $t' \in [0,T]$, we deduce that

$$\sup_{t'\in[0,t]} \|f(t')\|_{B^{\sigma_{t'}}_{p,\infty}} \leq 2\|f_0\|_{B^\sigma_{p,\infty}} + 2\|g\|_{\widetilde{L}^1_t(B^\sigma_{p,\infty})} + C\eta^{\frac{\alpha}{\alpha-1}} \int_0^t V'_{p_1,\alpha}(t')\|f(t')\|_{B^{\sigma_{t'}}_{p,\infty}}\, dt',$$

from which it follows, according to the Gronwall lemma, that

$$\sup_{t'\in[0,t]} \|f(t')\|_{B^{\sigma_{t'}}_{p,\infty}} \leq 2e^{C\eta^{\frac{\alpha}{\alpha-1}}\int_0^t V'_{p_1,\alpha}(t')\, dt'}\left(\|f_0\|_{B^\sigma_{p,\infty}} + \|g\|_{\widetilde{L}^1_t(B^\sigma_{p,\infty})}\right).$$

Taking $t = T$ and using the definition of η completes the proof. Indeed, we obviously have $\sigma_t \geq \sigma - \varepsilon$ for all $t \in [0,T]$. □

Remark 3.35. The estimate stated in Theorem 3.33 may be generalized to the case where a small loss of regularity occurs in the source term. More precisely, if $g = g_1 + g_2$ with $g_1 \in \widetilde{L}^1_T(B^\sigma_{p,\infty})$ and if, for some integrable function W, we have, for all $t \in [0,T]$,

$$\forall j \geq -1, \forall \sigma' \in\ ]\sigma - \varepsilon, \sigma[,\ \|\Delta_j g_2(t)\|_{L^p} \leq 2^{-j\sigma'}(j+2)^\alpha W(t)\|f(t)\|_{B^{\sigma'}_{p,r}},$$

then the following estimate holds:

$$\|f\|_{\widetilde{L}^\infty_T(B^{\sigma-\varepsilon}_{p,\infty})} \leq C\left(\|f_0\|_{B^\sigma_{p,\infty}} + \|g_1\|_{\widetilde{L}^1_T(B^\sigma_{p,\infty})}\right)$$

$$\times \exp\left(\frac{C}{\varepsilon^{\frac{\alpha}{1-\alpha}}}\left(\int_0^T (V'_{p_1,\alpha}(t) + W)(t)\, dt\right)^{\frac{1}{1-\alpha}}\right).$$

3.3.4 A Few Applications

Theorem 3.28 will help us to prove uniqueness for the incompressible Euler system with minimal regularity assumptions. The reader is referred to Chapter 7 for more details. It may also be used to establish the global well-posedness of the density-dependent incompressible Navier–Stokes equations in the two-dimensional case (see the last section of this chapter).

In this subsection, we shall use Theorem 3.28 to obtain the following uniqueness result for linear transport equations.

Theorem 3.36. *Let v be a divergence-free vector field in $L^1_{loc}(\mathbb{R}^+; LL(\mathbb{R}^d))$. Consider a distribution g in $L^1_{loc}(\mathbb{R}^+; \mathcal{M})$ and a measure[7] f_0 of $\mathcal{M}$. There exists a unique solution of*

---

[7] Recall that $\mathcal{M}$ denotes the set of bounded measures on $\mathbb{R}^d$.

$$(T) \qquad \begin{cases} \partial_t f + \operatorname{div}(fv) = g \\ \quad\quad f_{|t=0} = f_0 \end{cases}$$

in the space $L^\infty_{loc}(\mathbb{R}^+; \mathcal{M})$.

Proof. To prove existence, it suffices to smooth out the data and the vector field v. More precisely, let $(v_n)_{n \in \mathbb{N}}$ be a sequence of $\mathcal{C}^\infty_b$ vector fields with null divergence, uniformly bounded in the space $L^1_{loc}(\mathbb{R}^+; LL)$ and such that

$$\lim_{n \to \infty} v_n = v \quad \text{in} \quad L^1_{loc}(\mathbb{R}^+, B^\varepsilon_{\infty,\infty}) \quad \text{for all} \quad \varepsilon < 1.$$

Also, consider a bounded sequence of functions $(f_{0,n})_{n \in \mathbb{N}}$ in $\mathcal{S}$ satisfying

$$\lim_{n \to \infty} f_{0,n} = f_0 \quad \text{in} \quad \mathcal{M}.$$

Finally, take a bounded sequence $(g_n)_{n \in \mathbb{N}}$ in $L^1_{loc}(\mathbb{R}^+; L^1)$ such that

$$\lim_{n \to \infty} g_n = g \quad \text{in} \quad L^1_{loc}(\mathbb{R}^+; \mathcal{M}).$$

Let f_n be the solution of

$$\begin{cases} \partial_t f_n + v_n \cdot \nabla f_n = g_n \\ \quad\quad f_{n|t=0} = f_{0,n}. \end{cases}$$

It is clear that $(f_n)_{n \in \mathbb{N}}$ is a bounded sequence of $L^\infty_{loc}(\mathbb{R}^+; L^1)$. We can then extract a weakly convergent subsequence. The limit distribution f belongs to $L^\infty_{loc}(\mathbb{R}^+; \mathcal{M})$. It is then obvious that $f_n v_n$ tends weakly to fv. Hence, the equation (T) is satisfied.

Finally, as ensured by Proposition 2.39 , the space $L^\infty_{loc}(\mathbb{R}^+; \mathcal{M})$ is embedded in $L^\infty_{loc}(\mathbb{R}^+; B^0_{1,\infty})$. Theorem 3.28 ensures uniqueness. $\qquad\square$

3.4 Transport-Diffusion Equations

In a number of physical models, both convective and diffusive phenomena occur. This is particularly the case in most models coming from fluid mechanics. At the mathematical level, it means that the corresponding partial differential equations contain both a transport term of the type $v \cdot \nabla f$ and a diffusion term which, in the simplest case, reduces to $\nu \Delta f$ for some nonnegative constant ν.

There is a profuse mathematical literature on the transport and heat equations. We must note, however, that most methods which suit transport equations fail to treat the heat equation efficiently, and vice versa. In the present section, we consider equations of the type

$$(TD_\nu) \qquad \begin{cases} \partial_t f + v \cdot \nabla f - \nu \Delta f = g \\ f_{|t=0} = f_0, \end{cases}$$

where f_0, g, and v stand for given initial data, external force, and vector field, respectively. We aim to state a priori estimates which apply for all possible values of $\nu \geq 0$ and Lipschitz vector fields v.

3.4.1 A Priori Estimates

We focus on the study of (TD_ν) in the whole space $\mathbb{R}^d$ (although our approach also works in the torus $\mathbb{T}^d$) and, in order to simplify the presentation, we restrict our attention to estimates in *homogeneous* Besov spaces.

On the one hand, if there is no convection (i.e., $v \equiv 0$), then (TD_ν) reduces to the standard heat equation with constant diffusion so that applying $\dot\Delta_j$ to the equation yields

$$\partial_t \dot\Delta_j f - \nu\Delta\dot\Delta_j f = \dot\Delta_j g, \qquad \dot\Delta_j f_{|t=0} = \dot\Delta_j f_0.$$

Since $\dot\Delta_j f_0$ and $\dot\Delta_j g$ are spectrally localized in the annulus $2^j \mathcal{C}$, we have, by virtue of Lemma 2.4 and Corollary 2.5,

$$\|\dot\Delta_j f\|_{L_T^\rho(L^p)} \le C\left(\left(\nu 2^{2j}\right)^{-\frac{1}{\rho}}\|\dot\Delta_j f_0\|_{L^p} + \left(\nu 2^{2j}\right)^{-1+\frac{1}{\rho_1}-\frac{1}{\rho}}\|\dot\Delta_j g\|_{L_T^{\rho_1}(L^p)}\right)$$

for all real numbers ρ, ρ_1, p, and r such that $1 \le p, r \le \infty$ and $1 \le \rho_1 \le \rho \le \infty$. Therefore, multiplying both sides by $\nu^{\frac{1}{\rho}} 2^{\frac{2j}{\rho}} 2^{js}$ and performing an ℓ^r summation, we get

$$\nu^{\frac{1}{\rho}}\|f\|_{\widetilde{L}_T^\rho(\dot B_{p,r}^{s+\frac{2}{\rho}})} \le C\left(\|f_0\|_{\dot B_{p,r}^s} + \nu^{\frac{1}{\rho_1}-1}\|g\|_{\widetilde{L}_T^{\rho_1}(\dot B_{p,r}^{s-2+\frac{2}{\rho_1}})}\right) \tag{3.39}$$

for some universal constant C.

On the other hand, if there is convection but no diffusion in (TD_ν), then, as stated in Theorem 3.14 and Remark 3.16, we have

$$\|f\|_{\widetilde{L}_T^\infty(\dot B_{p,r}^{s+\frac{2}{\rho}})} \le e^{CV_{p_1}(T)}\left(\|f_0\|_{\dot B_{p,r}^s} + \|g\|_{\widetilde{L}_T^1(\dot B_{p,r}^s)}\right)$$

with $V_{p_1}(T) \stackrel{\text{def}}{=} \int_0^T \|\nabla v(t)\|_{\dot B_{p_1,\infty}^{\frac{d}{p_1}}\cap L^\infty}\, dt$, subject to some restrictions on the indices p, p_1, r, and s.

This section aims to unify the above two estimates for (TD_ν). This is achieved in the following theorem.

Theorem 3.37. *Let $1 \le p \le p_1 \le \infty$ and $1 \le \rho_1, r \le \infty$. Let $s \in \mathbb{R}$ satisfy*

$$\begin{cases} s < 1 + \dfrac{d}{p_1} \quad \text{or} \quad s \le 1 + \dfrac{d}{p_1}, \text{ if } r = 1, \\ s > -d\min\left\{\dfrac{1}{p_1}, \dfrac{1}{p'}\right\} \quad \text{or} \quad s > -1 - d\min\left\{\dfrac{1}{p_1}, \dfrac{1}{p'}\right\}, \text{ if } \operatorname{div} v = 0. \end{cases} \tag{3.40}$$

There exists a constant C, depending only on d, r, s, and $s - 1 - \frac{d}{p_1}$, such that for any smooth solution f of (TD_ν) with $\nu \ge 0$, and $\rho \in [\rho_1, \infty]$, we have the following a priori estimate:

$$\nu^{\frac{1}{\rho}}\|f\|_{\widetilde{L}^\rho_T(\dot{B}^{s+\frac{2}{\rho}}_{p,r})} \le e^{CV_{p_1}(T)}\left(\|f_0\|_{\dot{B}^s_{p,r}} + \nu^{\frac{1}{\rho_1}-1}\|g\|_{\widetilde{L}^{\rho_1}_T(\dot{B}^{s-2+\frac{2}{\rho_1}}_{p,r})}\right)$$

with
$$\begin{cases} V_{p_1}(T) \stackrel{def}{=} \int_0^T \|\nabla v(t)\|_{\dot{B}^{\frac{d}{p_1}}_{p_1,\infty}\cap L^\infty}\, dt, & \text{if } s < \frac{d}{p_1}+1, \\ V_{p_1}(T) \stackrel{def}{=} \int_0^T \|\nabla v(t)\|_{\dot{B}^{\frac{d}{p_1}}_{p_1,1}}\, dt, & \text{if } s = \frac{d}{p_1}+1. \end{cases}$$

Note that a standard energy method provides such estimates in the framework of Sobolev spaces H^s (at least in the case $\rho \ge 2$, where no tilde spaces are needed). Also, note that by taking $\rho = \infty$ and $\rho_1 = 1$, we find a family of estimates which are independent of ν and coincide with those of Theorem 3.14. If the vector field v is equal to 0, then $V_{p_1} = 0$, and we recover the inequality (3.39) exactly.

The proof of Theorem 3.37 is based on a Lagrangian approach (after suitable localization in Fourier space) which amounts to canceling out the bad convection term. Of course, in the Lagrangian formulation the good Laplace operator Δ is no longer "flat". It turns out, however, that it remains "almost" flat at small time so that it is still possible to take advantage of the inequality (3.39).

Proof of Theorem 3.37. Let $f_j \stackrel{def}{=} \dot{\Delta}_j f$ and $f_j \stackrel{def}{=} \dot{\Delta}_j g$. Applying $\dot{\Delta}_j$ to (TD_ν) yields

$$\partial_t f_j + \dot{S}_{j-1}v \cdot \nabla f_j - \nu\Delta f_j = g_j + R_j$$

with $R_j \stackrel{def}{=} (\dot{S}_{j-1}v - v) \cdot \nabla f_j - [\dot{\Delta}_j, v \cdot \nabla]f$.

Let ψ_j be the flow of $\dot{S}_{j-1}v$ and $\phi_j \stackrel{def}{=} \psi_j^{-1}$. Define $\widetilde{f}_j \stackrel{def}{=} f_j \circ \psi_j$, $\widetilde{g}_j \stackrel{def}{=} g_j \circ \psi_j$, and $\widetilde{R}_j \stackrel{def}{=} R_j \circ \psi_j$. We have

$$\partial_t \widetilde{f}_j - \nu\Delta\widetilde{f}_j = \widetilde{g}_j + \widetilde{R}_j + \nu T_j \quad \text{with} \quad T_j \stackrel{def}{=} \Delta f_j \circ \psi_j - \Delta\widetilde{f}_j. \tag{3.41}$$

Applying $\dot{\Delta}_{j'}$ to (3.41) and using Lemma 2.4, we get

$$\|\dot{\Delta}_{j'}\widetilde{f}_j(t)\|_{L^p} \le Ce^{-\kappa\nu t2^{2j}}\|\dot{\Delta}_{j'}f_{0,j}\|_{L^p}$$
$$+C\int_0^t e^{-\kappa\nu(t-t')2^{2j}}\left(\|\dot{\Delta}_{j'}\widetilde{g}_j\|_{L^p}+\|\dot{\Delta}_{j'}\widetilde{R}_j\|_{L^p}+\nu\|\dot{\Delta}_{j'}T_j\|_{L^p}\right)dt'. \tag{3.42}$$

We first focus on the term $\dot{\Delta}_{j'}T_j$. We have[8]

$$T_j = \Delta f_j \circ \psi_j - \text{tr}\big(\nabla\psi_j \cdot D^2 f_j \circ \psi_j \cdot D\psi_j\big) - Df_j \circ \psi_j \cdot \Delta\psi_j$$
$$= \text{tr}\big((\text{Id}-\nabla\psi_j) \cdot D^2 f_j \circ \psi_j \cdot D\psi_j\big)$$
$$\qquad - \text{tr}\big(D^2 f_j \circ \psi_j \cdot (\text{Id}-D\psi_j)\big) - Df_j \circ \psi_j \cdot \Delta\psi_j.$$

---

[8] Here, DF denotes the Jacobean matrix of F, and ∇F denotes the transposed matrix of DF. If F has d components, then we define $J_F \stackrel{def}{=} \det DF$.

Therefore,

$$\|\Delta_{j'}T_j\|_{L^p} \le C\big(\|D\psi_j\|_{L^\infty} + 1\big)\|\mathrm{Id} - D\psi_j\|_{L^\infty}\|D^2 f_j \circ \psi_j\|_{L^p}$$
$$+ C\|\Delta\psi_j\|_{L^\infty}\|Df_j \circ \psi_j\|_{L^p}.$$

Combining Bernstein's inequality with an obvious change of variable when computing the L^p norm, we infer that

$$\|Df_j \circ \psi_j\|_{L^p} \le C2^j\|J_{\phi_j}\|_{L^\infty}^{\frac{1}{p}}\|f_j\|_{L^p},$$
$$\|D^2 f_j \circ \psi_j\|_{L^p} \le C2^{2j}\|J_{\phi_j}\|_{L^\infty}^{\frac{1}{p}}\|f_j\|_{L^p}.$$

Hence, appealing to Proposition 3.10, we get, for all t in $[0, T]$,

$$\|\dot\Delta_{j'}T_j(t)\|_{L^p} \le C2^{2j}\left(e^{CV(t)} - 1\right)\|f_j(t)\|_{L^p} \quad \text{with} \tag{3.43}$$
$$V(t) \stackrel{\mathrm{def}}{=} \int_0^t \|\nabla v(t')\|_{L^\infty}\,dt'.$$

Next, we treat $\dot\Delta_{j'}\widetilde{g}_j$. According to Bernstein's lemma, we have

$$\|\dot\Delta_{j'}\widetilde{g}_j\|_{L^p} \approx 2^{-j'}\|\dot\Delta_{j'}D\widetilde{g}_j\|_{L^p}.$$

We also have $D\widetilde{g}_j = Dg_j \circ \psi_j \cdot D\psi_j$. Hence, according to Bernstein's inequality and Proposition 3.10,

$$\|\dot\Delta_{j'}\widetilde{g}_j(t)\|_{L^p} \le Ce^{CV(t)}2^{j-j'}\|g_j(t)\|_{L^p}. \tag{3.44}$$

From similar arguments, we get

$$\|\dot\Delta_{j'}\widetilde{R}_j(t)\|_{L^p} \le Ce^{CV(t)}2^{j-j'}\|R_j(t)\|_{L^p}.$$

The term $\|R_j(t)\|_{L^p}$ may be bounded according to Remark 2.103, and we eventually get

$$\|\dot\Delta_{j'}\widetilde{R}_j(t)\|_{L^p} \le C2^{j-j'}c_j(t)2^{-js}V'_{p_1}(t)e^{CV(t)}\|f(t)\|_{\dot B^s_{p,r}} \tag{3.45}$$

with $\|c_j(t)\|_{\ell^r} = 1$ and V'_{p_1} as defined in the statement of Theorem 3.37.

Plugging (3.43), (3.44), and (3.45) into (3.42), taking the L^ρ norm over $[0, t]$, and multiplying by $\nu^{\frac{1}{\rho}}2^{\frac{2j'}{\rho}}$, we thus get

$$\nu^{\frac{1}{\rho}}2^{\frac{2j'}{\rho}}\|\dot\Delta_{j'}\widetilde{f}_j\|_{L^\rho_t(L^p)} \le C\bigg(\|\dot\Delta_{j'}f_{0,j}\|_{L^p} + 2^{j-j'}\nu^{-\frac{1}{\rho_1}}2^{-\frac{2j'}{\rho_1}}e^{CV(t)}\|g_j\|_{L^{\rho_1}_t(L^p)}$$
$$+ 2^{2(j-j')}\nu^{\frac{1}{\rho}}2^{\frac{2j'}{\rho}}\left(e^{CV(t)} - 1\right)\|f_j\|_{L^\rho_t(L^p)} + 2^{j-j'}\int_0^t c_j 2^{-js}V'_{p_1}e^{CV}\|f\|_{\dot B^s_{p,r}}\,dt'\bigg),$$

where ρ'_1 stands for the conjugate exponent of ρ_1.

After multiplying both sides by $2^{js}2^{\frac{2}{p}(j-j')}$, we obtain

$$\nu^{\frac{1}{p}}2^{j\left(s+\frac{2}{p}\right)}\|\dot{\Delta}_{j'}\widetilde{f}_j\|_{L_t^\rho(L^p)} \le C\Bigg(2^{\frac{2}{p}(j-j')}2^{js}\|\dot{\Delta}_{j'}f_{0,j}\|_{L^p}$$

$$+\nu^{-\frac{1}{\rho_1}}2^{\left(1+\frac{2}{p}+\frac{2}{\rho_1}\right)(j-j')}e^{CV(t)}2^{j\left(s-\frac{2}{\rho_1}\right)}\|g_j\|_{L_t^{\rho_1}(L^p)}$$

$$+2^{2(j-j')}\nu^{\frac{1}{p}}2^{j\left(s+\frac{2}{p}\right)}\left(e^{CV(t)}-1\right)\|f_j\|_{L_t^\rho(L^p)}$$

$$+2^{\left(1+\frac{2}{p}\right)(j-j')}\int_0^t c_j V_{\rho_1}' e^{CV}\|f\|_{\dot{B}_{p,r}^s}\,dt'\Bigg). \quad (3.46)$$

Let $N_0 \in \mathbb{N}$, to be fixed hereafter. Because

$$f_j = \dot{S}_{j-N_0}\widetilde{f}_j \circ \phi_j + \sum_{j' \ge j-N_0} \dot{\Delta}_{j'}\widetilde{f}_j \circ \phi_j,$$

we have, for all $t \in [0,T]$,

$$\|f_j\|_{L_t^\rho(L^p)} \le e^{CV(t)}\left(\|\dot{S}_{j-N_0}\widetilde{f}_j\|_{L_t^\rho(L^p)} + \sum_{j' \ge j-N_0}\|\dot{\Delta}_{j'}\widetilde{f}_j\|_{L_t^\rho(L^p)}\right). \quad (3.47)$$

In order to bound the term $\|\dot{S}_{j-N_0}\widetilde{f}_j\|_{L_t^\rho(L^p)}$, we use Lemma 2.6 page 56 with $\lambda = 2^j$ and $\mu = 2^{j-N_0}$. This implies that for any t in $[0,T]$,

$$\|\dot{S}_{j-N_0}\widetilde{f}_j(t)\|_{L^p} \le C2^{-j}\|J_{\phi_j}\|_{L^\infty}\|f_j(t)\|_{L^p}$$
$$\times \left(\|DJ_{\phi_j}\|_{L^\infty}\|J_{\psi_j}\|_{L^\infty} + 2^{j-N_0}\|D\phi_j\|_{L^\infty}\right).$$

Thanks to Proposition 3.10, we have that $\|J_{\phi_j}\|_{L^\infty}$ and $\|J_{\psi_j}\|_{L^\infty}$ are bounded by $e^{CV(t)}$. Moreover,

$$DJ_{\phi_j} \cdot h = D(\det D\phi_j) \cdot h = \sum_{\ell=1}^d \det(D\phi_j^1, \ldots, D^2\phi_j^\ell \cdot h, \ldots, D\phi_j^d).$$

Therefore, again using Proposition 3.10, we infer that

$$\|DJ_{\phi_j}\|_{L^\infty} \le e^{CV(t)}\int_0^t \|D^2\dot{S}_{j-1}v(t')\|_{L^\infty}e^{CV(t')}\,dt'$$
$$\le Ce^{CV(t)}2^j\int_0^t \|\nabla v(t')\|_{L^\infty}e^{CV(t')}\,dt'$$
$$\le Ce^{CV(t)}2^j(e^{CV(t)}-1).$$

Thus, we get

$$\|\dot{S}_{j-N_0}\widetilde{f}_j\|_{L_t^\rho(L^p)} \le Ce^{CV(t)}\left(2^{-N_0}+e^{CV(t)}-1\right)\|f_j\|_{L_t^\rho(L^p)}. \quad (3.48)$$

We now bound the last term in (3.47). Using the fact that $\dot{\Delta}_{j'} f_{0,j} = 0$ for $|j' - j| > 1$ and summing the inequality (3.46) for $j' \geq j - N_0$, we obtain

$$\sum_{j' \geq j - N_0} \nu^{\frac{1}{\rho}} 2^{j\left(s + \frac{2}{\rho}\right)} \|\dot{\Delta}_{j'} \tilde{f}_j\|_{L_t^\rho(L^p)} \leq C \bigg(2^{js} \|f_{0,j}\|_{L^p}$$

$$+ e^{CV(t)} 2^{3N_0} \nu^{-\frac{1}{\rho_1}} 2^{j\left(s - \frac{2}{\rho_1}\right)} \|g_j\|_{L_t^{\rho_1}(L^p)}$$

$$+ 2^{2N_0} \left(e^{CV(t)} - 1 \right) \nu^{\frac{1}{\rho}} 2^{j\left(s + \frac{2}{\rho}\right)} \|f_j\|_{L_t^\rho(L^p)} + 2^{3N_0} \int_0^t c_j V'_{p_1} e^{CV} \|f\|_{\dot{B}_{p,r}^s} \, dt' \bigg).$$

Plugging this and the inequality (3.48) into (3.47), we discover that, up to a change of C, we have

$$\nu^{\frac{1}{\rho}} 2^{j\left(s + \frac{2}{\rho}\right)} \|f_j\|_{L_t^\rho(L^p)} \leq C e^{CV(t)} \bigg(2^{js} \|f_{0,j}\|_{L^p} + 2^{3N_0} \nu^{-\frac{1}{\rho_1}} 2^{j\left(s - \frac{2}{\rho_1}\right)} \|g_j\|_{L_t^{\rho_1}(L^p)}$$

$$+ \left(2^{-N_0} + 2^{2N_0} \left(e^{CV(t)} - 1 \right) \right) \nu^{\frac{1}{\rho}} 2^{j\left(s + \frac{2}{\rho}\right)} \|f_j\|_{L_t^\rho(L^p)} + 2^{3N_0} \int_0^t c_j V'_{p_1} \|f\|_{\dot{B}_{p,r}^s} \, dt' \bigg).$$

Choose N_0 to be the unique integer such that $2C 2^{-N_0} \in \,] \frac{1}{8}, \frac{1}{4}]$ and T_1 to be the largest real number such that

$$T_1 \leq T \quad \text{and} \quad CV(T_1) \leq \varepsilon \quad \text{with} \quad \varepsilon = \min \left\{ \log 2, \frac{2^{-2N_0}}{16C} \right\}. \tag{3.49}$$

With this choice of T_1 and N_0, the third term of the right-hand side of the above inequality may be absorbed by the left-hand side whenever t is in $[0, T_1]$. This yields, for some positive constant C_1,

$$\nu^{\frac{1}{\rho}} 2^{j\left(s + \frac{2}{\rho}\right)} \|f_j\|_{L_t^\rho(L^p)} \leq C_1 \bigg(2^{js} \|f_{0,j}\|_{L^p}$$

$$+ \nu^{-\frac{1}{\rho_1}} 2^{j\left(s - \frac{2}{\rho_1}\right)} \|g_j\|_{L_t^{\rho_1}(L^p)} + \int_0^t c_j(t') V'_{p_1}(t') \|f(t')\|_{\dot{B}_{p,r}^s} \, dt' \bigg).$$

Finally, performing an ℓ^r summation gives, for all $t \in [0, T_1]$ and $\rho \in [\rho_1, \infty]$,

$$\nu^{\frac{1}{\rho}} \|f\|_{\tilde{L}_t^\rho(\dot{B}_{p,r}^{s + \frac{2}{\rho}})} \leq C_1 \bigg(\|f_0\|_{B_{p,r}^s} + \nu^{-\frac{1}{\rho_1}} \|g\|_{\tilde{L}_t^{\rho_1}(\dot{B}_{p,r}^{s - \frac{2}{\rho_1}})}$$

$$+ \int_0^t V'_{p_1}(t') \|f(t')\|_{\dot{B}_{p,r}^s} \, dt' \bigg). \tag{3.50}$$

It is now easy to complete the proof. Indeed, it is only a matter of splitting the interval $[0, T]$ into a finite number m of subintervals $[0, T_1], [T_1, T_2]$, and so on, such that

$$\frac{\varepsilon}{2} \leq \int_{T_k}^{T_{k+1}} \|\nabla v(t)\|_{L^\infty} \, dt \leq \varepsilon.$$

By arguing as was done to prove (3.50), we get, for all $t \in [T_k, T_{k+1}]$,

$$\nu^{\frac{1}{\rho}}\|f\|_{\widetilde{L}^{\rho}_{[T_k,t]}(\dot{B}^{s+\frac{2}{\rho}}_{p,r})} \leq C_1\Bigg(\|f(T_k)\|_{B^s_{p,r}}$$
$$+ \nu^{-\frac{1}{\rho_1}}\|g\|_{\widetilde{L}^{\rho_1}_{[T_k,t]}(\dot{B}^{s-\frac{2}{\rho_1}}_{p,r})} + \int_{T_k}^t V'_{p_1}(t')\|f(t')\|_{\dot{B}^s_{p,r}}\,dt'\Bigg).$$

Note that if $k = 1$, then the first term on the right-hand side may be bounded according to (3.50) with $\rho = \infty$ and $t = T_1$. Hence, after an obvious induction, we get

$$\nu^{\frac{1}{\rho}}\|f\|_{\widetilde{L}^{\rho}_t(\dot{B}^{s+\frac{2}{\rho}}_{p,r})} \leq C_1^{k+1}\Bigg(\|f_0\|_{B^s_{p,r}}$$
$$+ \nu^{-\frac{1}{\rho_1}}\|g\|_{\widetilde{L}^{\rho_1}_t(\dot{B}^{s-\frac{2}{\rho_1}}_{p,r})} + \int_0^t V'_{p_1}(t')\|f(t')\|_{\dot{B}^s_{p,r}}\,dt'\Bigg).$$

Since the number of such subintervals is $m \approx CV(T)\varepsilon^{-1}$, we can readily conclude that up to a change of C, we have, for all ρ in $[\rho_1, \infty]$,

$$\nu^{\frac{1}{\rho}}\|f\|_{\widetilde{L}^{\rho}_T(\dot{B}^{s+\frac{2}{\rho}}_{p,r})} \leq Ce^{CV(T)}\Bigg(\|f_0\|_{B^s_{p,r}} + \nu^{-\frac{1}{\rho_1}}\|g\|_{\widetilde{L}^{\rho_1}_T(\dot{B}^{s-\frac{2}{\rho_1}}_{p,r})}$$
$$+ \int_0^T V'_{p_1}(t)\|f(t)\|_{\dot{B}^s_{p,r}}\,dt\Bigg). \quad (3.51)$$

Taking $\rho = \infty$ and using the fact that $V'(t) \leq CV'_{p_1}(t)$, the Gronwall lemma gives

$$\|f\|_{\widetilde{L}^{\infty}_T(\dot{B}^s_{p,r})} \leq Ce^{CV_{p_1}(T)}\Bigg(\|f_0\|_{B^s_{p,r}} + \nu^{-\frac{1}{\rho_1}}\|g\|_{\widetilde{L}^{\rho_1}_T(\dot{B}^{s-\frac{2}{\rho_1}}_{p,r})}\Bigg).$$

Plugging this estimate into (3.51), we get

$$\nu^{\frac{1}{\rho}}\|f\|_{\widetilde{L}^{\rho}_t(\dot{B}^{s+\frac{2}{\rho}}_{p,r})} \leq C\Bigg(\|f_0\|_{B^s_{p,r}} + \nu^{-\frac{1}{\rho_1}}\|g\|_{\widetilde{L}^{\rho_1}_t(\dot{B}^{s-\frac{2}{\rho_1}}_{p,r})}\Bigg)$$
$$\times e^{CV_{p_1}(T)}\Bigg(1 + C\int_0^t V'_{p_1}(t')dt'\Bigg).$$

This completes the proof for general $\rho \in [\rho_1, \infty]$.   □

By treating the low frequencies separately, we can state the following a priori estimates for (TD_ν) in *nonhomogeneous* Besov spaces.

Theorem 3.38. *Let $1 \leq p_1 \leq p \leq \infty$, $1 \leq r \leq \infty$, $s \in \mathbb{R}$ satisfy (3.12), and V_{p_1} be defined as in Theorem 3.14.*

There exists a constant C which depends only on d, r, s, and $s - 1 - \frac{d}{p_1}$ and is such that for any smooth solution f of (TD_ν) and $1 \leq \rho_1 \leq \rho \leq \infty$, we have

$$\nu^{\frac{1}{p}}\|f\|_{\widetilde{L}_T^p(B_{p,r}^{s+\frac{2}{p}})} \leq Ce^{C(1+\nu T)^{\frac{1}{p}}V_{p_1}(T)}\left((1+\nu T)^{\frac{1}{p}}\|f_0\|_{B_{p,r}^s}\right.$$

$$\left.+ (1+\nu T)^{1+\frac{1}{p}-\frac{1}{p_1}}\nu^{\frac{1}{p_1}-1}\|g\|_{\widetilde{L}_T^{p_1}(B_{p,r}^{s-2+\frac{2}{p_1}})}\right).$$

Remark 3.39. If $r = \infty$, then both Theorems 3.37 and 3.38 hold true with

$$V_{p_1}(T) \overset{\text{def}}{=} \int_0^T \|\nabla v(t)\|_{\dot{B}_{p_1,1}^{\frac{d}{p_1}}}\, dt \quad \text{and} \quad V_{p_1}(T) \overset{\text{def}}{=} \int_0^T \|\nabla v(t)\|_{B_{p_1,1}^{\frac{d}{p_1}}}\, dt,$$

respectively, in the limit case

$$s = -d\min(1/p_1, 1/p') \quad \text{or} \quad s = -d\min(1/p_1, 1/p') - 1 \quad \text{if} \quad \text{div}\, v = 0.$$

This is a consequence of the inequality (2.55) page 112 and Remark 2.102.

Finally, we point out that similar estimates may be proven for the nonstationary Stokes equation with convection:

$$(S_\nu) \qquad \begin{cases} \partial_t u + v \cdot \nabla u - \nu \Delta u + \nabla \varPi = g \\ \text{div}\, u = 0, \qquad u_{|t=0} = u_0. \end{cases}$$

Indeed, we shall see in Chapter 5 that the Leray projector on divergence-free vector fields is a homogeneous Fourier multiplier of order 0. Thanks to Lemma 2.2 page 53, such operators are continuous self-maps on $\widetilde{L}_t^p(\dot{B}_{p,r}^s)$.

3.4.2 Exponential Decay

In this final subsection, we study the effect of diffusion in (TD_ν) on compactly supported data. Our main result is the following.

Theorem 3.40. *A constant C exists which satisfies the following properties. Let v be a divergence-free vector field which belongs to $L_{loc}^1(\mathbb{R}^+; C^{0,1})$, f_0 be a compactly supported function in L^2, and ν be a positive real number. Consider a solution f of the equation (TD_ν) with right-hand side 0 and initial data f_0. We denote by ψ the flow of the vector field v and define*

$$F_t \overset{\text{def}}{=} \psi(t, \text{Supp}\,(f_0)),$$
$$(F_t)_h^c \overset{\text{def}}{=} \{x \in \mathbb{R}^2 \,/\, d(x, F_t) > h\},$$
$$(F_t^c)_h \overset{\text{def}}{=} \{x \in F_t \,/\, d(x, \partial F_t) > h\}.$$

Let $V(t) \overset{\text{def}}{=} \int_0^t \|\nabla v(t')\|_{L^\infty}\, dt'$. We then have, for all $(t, h) \in \mathbb{R}^+ \times \mathbb{R}^+$,

$$\|f(t)\|_{L^2((F_t)_h^c)} \leq \|f_0\|_{L^2}e^{-\frac{h^2}{4\nu t}}\exp(-4V(t)). \tag{3.52}$$

Moreover, if f_0 is the characteristic function of a bounded domain F_0, then we have

$$\|f(t)-\mathbb{1}_{F_t}\|_{L^2((F_t^c)_h)} \le \|f_0\|_{L^2} \min\left\{1, C\left(\frac{\nu t}{h^2}\right)^{\frac{1}{2}} e^{2V(t)-\frac{h^2}{32\nu t}} \exp(-4V(t))\right\}. \quad (3.53)$$

Proof. Proving this theorem relies on energy estimates. Using regularization arguments, we may assume that the vector field v and the function f are smooth. We consider a smooth function Φ_0, denote by ψ the flow of v, and define

$$\Phi(t,x) \stackrel{\text{def}}{=} \Phi_0(\psi^{-1}(t,x)).$$

It is obvious that

$$\partial_t(\Phi f) + v \cdot \nabla(\Phi f) - \nu \Delta(\Phi f) = -\nu f \Delta \Phi - 2\nu \nabla \Phi \cdot \nabla f.$$

Taking the L^2 inner product with Φf and performing integrations by parts gives

$$\frac{1}{2}\frac{d}{dt}\|\Phi f\|_{L^2}^2 + \nu\|\nabla(\Phi f)\|_{L^2}^2 = \nu\|f\nabla\Phi\|_{L^2}^2.$$

We choose $\Phi(t,x) = \exp(\phi(t,x))$ with $\phi(t,x) = \phi_0(\psi^{-1}(t,x))$. From the above relation, we get that

$$\frac{d}{dt}\|\Phi f\|_{L^2}^2 \le 2\nu\|\nabla\phi\|_{L^\infty}^2\|\Phi f\|_{L^2}^2.$$

From the Gronwall lemma, we thus infer that

$$\|(\Phi f)(t)\|_{L^2} \le \|(\Phi f)(0)\|_{L^2} \exp\left(\nu \int_0^t \|\nabla\phi(t')\|_{L^\infty}^2 \, dt'\right).$$

We define

$$\phi_0(x) \stackrel{\text{def}}{=} \alpha \min\{R, d(x, \text{Supp}(f_0))\} \star \chi_\varepsilon,$$

where $\chi_\varepsilon(x) \stackrel{\text{def}}{=} \varepsilon^{-d}\chi(\varepsilon^{-1}x)$ for some function χ of $\mathcal{D}(\mathbb{R}^d)$ with integral 1. Note that with this choice, the function $(\Phi f)(0)$ tends to the function f_0 a.e. when ε goes to 0. Using the fact that $\|\nabla\phi_\varepsilon(t)\|_{L^\infty} \le \alpha \exp V(t)$, we get, by the Gronwall lemma, that

$$\|\Phi f(t)\|_{L^2} \le \|\Phi f(0)\|_{L^2} e^{\nu\alpha^2 t \exp(2V(t))}.$$

Taking the limit when ε tends to 0, it turns out, by the definition of Φ, that if $0 < \eta \le R$, we have

$$e^{\alpha\eta}\|f(t)\|_{L^2\left(\psi_t((F_0)_\eta^c)\right)} \le \|f_0\|_{L^2} e^{\nu\alpha^2 t \exp(2V(t))}. \quad (3.54)$$

But, obviously,

$$(F_t)_h^c \subset \psi_t\big((F_0)_{\delta(t,h)}^c\big) \quad \text{with} \quad \delta(t,h) \overset{\text{def}}{=} \frac{h}{\|\nabla\psi_t\|_{L^\infty}}. \tag{3.55}$$

Thus, taking $\eta = \delta(t,h)$ in (3.54) and assuming that $\delta(t,h) \leq R$, we obtain

$$\|f(t)\|_{L^2((F_t)_h^c)} \leq \|f_0\|_{L^2} e^{\nu\alpha^2 t \exp(2V(t)) - \alpha h \exp(-V(t))}.$$

As the above inequality is independent of R, it is true for any (t,h). The best choice for α then gives the inequality (3.52).

The proof of (3.53) follows essentially the same lines. Let $w(t,x) = f(t,x) - \mathbb{1}_{F_t}(x)$ and $\Phi(t,x) = \Phi_0(\psi^{-1}(t,x))$ with Φ_0 in $\mathcal{D}(F_0)$. Then, due to

$$\Delta\big(\Phi_t \mathbb{1}_{F_t}\big) = \mathbb{1}_{F_t}\Delta\Phi_t \quad \text{and} \quad \nabla\phi_t \cdot \nabla\mathbb{1}_{F_t} = 0,$$

we have

$$(\partial_t + v \cdot \nabla - \nu\Delta)(\Phi w) = -\nu w\Delta\Phi - 2\nu\nabla\Phi \cdot \nabla w.$$

As above, by an energy estimate, we get

$$\frac{1}{2}\frac{d}{dt}\|\Phi w\|_{L^2}^2 + \nu\|\nabla(\Phi w)\|_{L^2}^2 = \nu\|w\nabla\Phi\|_{L^2}^2.$$

Fix a constant C such that for any positive h_0, a function χ exists in $\mathcal{D}(F_0)$ such that χ is identically 1 on $(F_0^c)_{h_0}$ and $\|\nabla\chi\|_{L^\infty} \leq Ch_0^{-1}$. Then, choosing $\Phi_0 = \chi e^{\phi_0}$, where ϕ_0 is equal to (a regularization of) the function $x \mapsto d(x, F_0^c)$, we get that

$$\frac{1}{2}\frac{d}{dt}\|\Phi w\|_{L^2}^2 \leq 2\nu\left\|\nabla\psi_t^{-1}\right\|_{L^\infty}^2 \big(\|\Phi w\|_{L^2}^2 \|\nabla\phi_0\|_{L^\infty}^2 + \|(w \circ \psi_t)e^{\phi_0}\nabla\chi\|_{L^2}^2\big),$$

from which it follows, since $\|w \circ \psi_t\|_{L^2} \leq 2\|f_0\|_{L^2}$, that

$$\frac{d}{dt}\|\Phi w\|_{L^2}^2 \leq \nu e^{2V(t)}\left(4\alpha^2\|\Phi w\|_{L^2}^2 + \frac{Ce^{2\alpha h_0}}{h_0^2}\|f_0\|_{L^2}^2\right).$$

Using (3.55) (with F_t^c instead of F_t) and the Gronwall lemma, we get, for any t and h such that $he^{-V(t)} \geq h_0$,

$$\|w(t)\|_{L^2((F_t^c)_h)}^2 \leq C\|f_0\|_{L^2}^2 \frac{e^{2\alpha(h_0 - h\exp(-V(t)))}}{\alpha^2 h_0}\big(e^{4\alpha^2 t \exp(2V(t))} - 1\big).$$

Now, using the fact that $e^{-x}(e^{\frac{x}{2}} - 1) \leq e^{-\frac{x}{2}}$ and choosing

$$h_0 = \frac{he^{-V(t)}}{2} \quad \text{and} \quad \alpha = \frac{he^{-3V(t)}}{8\nu t}$$

gives the result. $\qquad\qquad\qquad\qquad\qquad\qquad\qquad\qquad\qquad\qquad\square$

3.5 References and Remarks

Most of the material in Section 3.1 belongs to the mathematical folklore. It may somewhat extended to non-smooth vector fields (see e.g. [12]). Here, we chose to extend some of the results stated in Chapter 5 of [69].

The study of transport equations under minimal regularity assumptions on the vector field is currently very active. See, in particular, the recent works by L. Ambrosio and P. Bernard [13], F. Colombini and N. Lerner [83], and N. Depauw [111]. In this book, we chose to focus on the study of a priori estimates in the case where the vector field is at least quasi-Lipschitz. The a priori estimates and existence results for the transport equation which were stated in Section 3.2 are well known in the framework of Hölder spaces or Sobolev spaces with positive exponent. Their extension to Hölder spaces with *negative* indices of regularity (i.e., in $B^r_{\infty,\infty}$ with $-1 < r < 0$) in the case where the vector field v is divergence-free has been carried out in [69, Chapter 4]. The a priori estimates and the existence statement in general Besov spaces essentially come from works by the second and third authors (see, in particular, [102]). That estimates for (T) improve in Besov spaces with regularity index 0 was discovered by M. Vishik in [296]. For proving Theorem 3.18, we instead followed T. Hmidi and S. Keraani's approach, which turns out to be more robust. In particular, it also works (with no changes) for transport-diffusion equations (see [158]).

The so-called Camassa–Holm equation (3.25) was derived independently by A. Fokas and B. Fuchssteiner in [126], and by R. Camassa and D. Holm in [56]. Its systematic mathematical study was initiated in a series of papers by A. Constantin and J. Escher (see, e.g., [84]). It has infinitely many conservation laws, the most obvious ones being the conservation of the average over $\mathbb{R}$ and of the H^1 norm for smooth solutions with sufficient decay at infinity. By taking advantage of this latter property, Z. Xin and P. Zhang proved that (3.25) has global weak solutions for any data in H^1 (see [301]). The results stated in Section 3.2.4 are borrowed from [96]. Note that for proving uniqueness for data in $B^s_{p,r}$, we are led to estimate the difference between two solutions in $B^{s-1}_{p,r}$. Owing to the term $(\partial_x u)^2$, the additional condition $s > \max(\frac{3}{2}, 1 + \frac{1}{p})$ is thus required. In fact, uniqueness is also in true in $B^{\frac{3}{2}}_{2,1}$; see [96]. Further improvements were recently obtained in [108].

Losing estimates for transport equations associated with a log-Lipschitz vector field have been stated by a number of authors. The statement of Theorem 3.28 pertaining to loss of regularity in general Besov spaces comes from [102]. The phenomenon of exponential loss has been pointed out by the first two authors in [17]. Theorem 3.33 has been stated in [102], and a related result in Sobolev space has been proven by B. Desjardins in [113]. Theorem 3.36 may be seen as a borderline case of the results of Di Perna and Lions in [117] and of B. Desjardins in [112]. More details concerning the proof of Theorem 3.41 may be found in [100].

We give an application of Theorem 3.33 concerning the density-dependent incompressible Navier–Stokes equations:

$$\begin{cases} \partial_t \rho + u \cdot \nabla \rho = 0 \\ \rho(\partial_t u + u \cdot \nabla u) - \mu \Delta u + \nabla \Pi = 0 \\ \operatorname{div} u = 0. \end{cases} \tag{3.56}$$

Theorem 3.41. *Let $u_0 \in H^1(\mathbb{R}^2)$ with div $u_0 = 0$. Assume that $\rho_0 = 1/(1+a_0)$ with $a_0 \in H^{1+\beta}(\mathbb{R}^2)$ for some $\beta \in \,]0,1[$. Further, assume that $1+a_0 > 0$. Then, the system (3.56), supplemented with initial data (ρ_0, u_0), has a global unique solution (ρ, u) which satisfies*

$$a \overset{def}{=} \frac{1}{\rho} - 1 \in \mathcal{C}(\mathbb{R}^+; H^{1+\beta}), \quad \rho^{\pm 1} \in L^\infty, \quad \text{and} \quad u \in \mathcal{C}(\mathbb{R}^+; H^1) \cap \widetilde{L}^1_{loc}(\mathbb{R}^+; H^3).$$

Proof. We only sketch the proof, emphasizing how Theorem 3.33 is used. A more detailed proof is available in [100].

On the one hand, in dimension two, under the assumptions that $\rho_0^{\pm 1} \in L^\infty$ and $u_0 \in H^1$, it is well established (see, e.g., [14]) that (3.56) has a global weak solution (ρ, u) with $\rho^{\pm 1}$ bounded and

$$u \in \left(L^\infty_{loc}(\mathbb{R}^+; H^1) \cap L^2_{loc}(\mathbb{R}^+; H^2) \right)^2.$$

Now, because $\nabla u \in L^1_{loc}(\mathbb{R}^+; H^1)$ and, by assumption, $a_0 \in H^{1+\beta}$, Theorem 3.33 with $\alpha = 1/2$, $p_1 = p = 2$ ensures that a belongs to $\mathcal{C}(\mathbb{R}^+; H^{1+\beta'})$ for all $\beta' < \beta$.

On the other hand, the local well-posedness theory for density-dependent Navier–Stokes equations provides a unique local maximal solution $(\widetilde{a}, \widetilde{u})$ such that

$$\widetilde{a} \in \mathcal{C}([0, T^*[; H^{1+\beta}) \quad \text{and} \quad \widetilde{u} \in \mathcal{C}([0, T^*[; H^1) \cap \widetilde{L}^1_{loc}([0, T^*[; H^3).$$

Since ∇a remains for all time in some Sobolev space *with positive index*, and, by virtue of Sobolev embeddings, the vector field $\widetilde{u}$ belongs to $L^1_{loc}([0, T^*[; C^{0,1})$, it is not difficult to prove a *weak-strong* uniqueness statement. It is only a matter of writing the equation satisfied by $(a - \widetilde{a}, u - \widetilde{u})$ and applying Theorem 3.14 and the inequality (3.39). Therefore, we actually have $(a, u) \equiv (\widetilde{a}, \widetilde{u})$ on $[0, T^*[$. Now, if one assumes that T^* is finite, then we have $\|a(t)\|_{H^{1+\beta}}$ and $\|u(t)\|_{H^1}$ uniformly bounded on $[0, T^*[$ so that the local existence theory enables us to continue (a, u) beyond T^*. Hence, we must have $T^* = \infty$. $\qquad\square$

Remark 3.42. A similar statement may be proven under the weaker assumption that $u_0 \in H^\gamma(\mathbb{R}^2)$ for arbitrarily small $\gamma > 0$.

The proof of a priori estimates for transport-diffusion equations has a long history. The case of Sobolev spaces H^s is classical. The extension to more general Besov spaces was initiated in [90], then improved in [95] under the restrictions that $1 < p < \infty$ and that div $v = 0$. The proof was based on a slight generalization of Lemma 2.8 page 58 (see [90, 251, 95]), which fails in the limit cases $p = 1, \infty$. The extension to all $p \in [1, \infty]$ in the case div $v = 0$ is due to T. Hmidi in [156]. This is based on the Lagrangian approach that was used in the present chapter and on the smoothing property of the heat equation stated in (3.39) that was first observed in [72]. Finally, the whole statement of Theorems 3.37 and 3.38 was proven in [103]. Different types of estimates have been obtained by a number of authors (see, in particular, the work by E. Carlen and M. Loss in [59]).

The exponential decay results for transport-diffusion equations were been proven in [90]. Some generalizations have been obtained by J. Ben Ameur and the third author in [32], and by T. Hmidi in [156].

4

Quasilinear Symmetric Systems

Quasilinear and linear symmetric systems appear in a number of physical systems such as wave equations, systems of conservation laws, compressible Euler equations, and so on (some examples are given in the first section below).

In this chapter, we state a few elementary and classical facts concerning these systems. The first section is devoted to a short presentation on linear and quasilinear symmetric systems. In the second section, we focus on the linear case with suitably smooth coefficients. We demonstrate global well-posedness in Sobolev spaces H^s for any $s \geq 0$. We also establish that linear symmetric systems have the finite propagation speed property. In Section 4.3 we focus on quasilinear symmetric systems. We prove that they may be solved locally in any Sobolev space embedded in the set of Lipschitz functions and exhibit a blow-up criterion involving the $L^1(\mathrm{Lip})$ norm of the solution. Section 4.4 is dedicated to the study of the Cauchy problem for quasilinear symmetric systems under minimal regularity assumptions, as well as to refined blow-up criteria. In the last section, we investigate the regularity of the associated flow map.

4.1 Definition and Examples

We shall begin by explaining what is meant by a *linear symmetric system*. Let I be an interval of $\mathbb{R}$ and $(\mathcal{A}_k)_{0 \leq k \leq d}$ be a family of smooth bounded functions from $I \times \mathbb{R}^d$ into the space of $N \times N$ matrices with real coefficients. Let $t_0 \in I$. We want to solve the following initial boundary value problem for any suitably smooth functions $U_0 : \mathbb{R}^d \to \mathbb{R}^N$ and $F : I \times \mathbb{R}^d \to \mathbb{R}^N$:

$$(LS) : \begin{cases} \partial_t U + \displaystyle\sum_{k=1}^{d} \mathcal{A}_k \partial_k U + \mathcal{A}_0 U = F \\ \qquad\qquad\qquad U_{|t=t_0} = U_0. \end{cases}$$

We will first explain what it means to *solve* (LS).

H. Bahouri et al., *Fourier Analysis and Nonlinear Partial Differential Equations*, Grundlehren der mathematischen Wissenschaften 343, DOI 10.1007/978-3-642-16830-7_4, © Springer-Verlag Berlin Heidelberg 2011

Definition 4.1. *A function* $U \in \mathcal{C}\big(I;(\mathcal{S}'(\mathbb{R}^d))^N\big)$ *is called a* weak solution *of* (LS) *on* $I \times \mathbb{R}^d$ *if:*

(i) Functions $\sum_j U^j \mathcal{A}_{k,i,j}$ *and* $\sum_j U^j (\mathrm{div}\,\mathcal{A})_{i,j}$ *with* $(\mathrm{div}\,\mathcal{A})_{i,j} \overset{\mathrm{def}}{=} \sum_k \partial_k \mathcal{A}_{k,i,j}$
  are in $L^1(I;\mathcal{S}'(\mathbb{R}^d))$ *for all* $i \in \{1,\dots,N\}$ *and* $k \in \{1,\dots,d\}$.
(ii) For all $t \in I$ *and* $\varphi \in \mathcal{C}^1(I;(\mathcal{S}(\mathbb{R}^d))^N)$, *it holds that*

$$\sum_i \int_0^t \langle U^i, \partial_t \varphi_i \rangle_{\mathcal{S}' \times \mathcal{S}}\, d\tau + \sum_{i,j} \int_0^t \langle F^i + U^j\big((\mathrm{div}\,\mathcal{A})_{i,j} - \mathcal{A}_{0,i,j}\big), \varphi_i \rangle_{\mathcal{S}' \times \mathcal{S}}\, d\tau$$

$$+ \sum_{i,j,k} \int_0^t \langle U^j \mathcal{A}_{k,i,j}, \partial_k \varphi_i \rangle_{\mathcal{S}' \times \mathcal{S}}\, d\tau = \sum_i \Big(\langle U^i(t), \varphi_i(t) \rangle_{\mathcal{S}' \times \mathcal{S}} - \langle U_0^i, \varphi_i(0) \rangle_{\mathcal{S}' \times \mathcal{S}} \Big).$$

Formally, in order to control the energy of a solution U of (LS), we can proceed as follows. First, we take the $L^2(\mathbb{R}^d;\mathbb{R}^N)$ inner product of (LS) with U. We find that

$$\frac{1}{2}\frac{d}{dt}\|U(t)\|_{L^2}^2 = -\sum_{k=1}^d \Big(\mathcal{A}_k \partial_k U \,|\, U\Big)_{L^2} - (\mathcal{A}_0 U \,|\, U)_{L^2} + (F \,|\, U)_{L^2}.$$

If we further assume that the first order space derivatives of the functions $\mathcal{A}_k$ $(1 \le k \le d)$ are bounded, then we can next perform an integration by parts. This gives

$$-\Big(\mathcal{A}_k \partial_k U \,|\, U\Big)_{L^2} = -\sum_{i,j} \int_{\mathbb{R}^d} \mathcal{A}_{k,i,j} \partial_k U^j \, U^i \, dx$$

$$= \sum_{i,j} \int_{\mathbb{R}^d} \mathcal{A}_{k,i,j} U^j \partial_k U^i \, dx + \sum_{i,j} \int_{\mathbb{R}^d} \partial_k \mathcal{A}_{k,i,j} U^i \, U^j \, dx.$$

In general, due to the first term on the right-hand side, estimating the term $\big(\mathcal{A}_k \partial_k U \,|\, U\big)_{L^2}$ (and thus $\|U\|_{L^2}$) requires a bound on $\|\partial_k U\|_{L^2}$. This loss of one derivative precludes our closing the estimates and motivates the following definition.

Definition 4.2. *The above system* (LS) *is said to be* symmetric *if for any* k *in* $\{1,\dots,d\}$ *and any* $(t,x) \in I \times \mathbb{R}^d$, *the matrices* $\mathcal{A}_k(t,x)$ *are symmetric, that is, for any* i, j, *and* k, *we have* $\mathcal{A}_{k,i,j}(t,x) = \mathcal{A}_{k,j,i}(t,x)$.

We now resume the above computation under the additional assumption that (LS) is *symmetric*. We get

$$-\sum_{k=1}^d \Big(\mathcal{A}_k \partial_k U \,|\, U\Big)_{L^2} = \frac{1}{2}((\mathrm{div}\,\mathcal{A})U \,|\, U)_{L^2}.$$

This implies that

$$\left| \sum_{k=1}^{d} \left(A_k \partial_k U | U \right)_{L^2} \right| \leq \frac{1}{2} \| \operatorname{div} \mathcal{A} \|_{L^\infty} \| U \|_{L^2}^2.$$

Thus, we get that

$$\frac{d}{dt} \| U(t) \|_{L^2}^2 \leq a_0(t) \| U(t) \|_{L^2}^2 + 2(F(t) | U(t))_{L^2} \tag{4.1}$$

with $a_0(t) \stackrel{\text{def}}{=} \| \operatorname{div} \mathcal{A}(t) \|_{L^\infty} + 2 \| \mathcal{A}_0(t) \|_{L^\infty}$, so we may now control the energy of the solution in terms of the data by means of the Gronwall lemma.

We next define a *quasilinear symmetric system*. A "general" quasilinear system is of the form

$$(QS) : \begin{cases} \partial_t U + \sum_{k=1}^{d} A_k(U) \partial_k U + A_0(U) = F \\ \qquad\qquad\qquad\qquad U_{|t=t_0} = U_0, \end{cases}$$

where $A = (A_k)_{0 \leq k \leq d}$ is a family of $d+1$ smooth functions from $\mathbb{R}^N$ to the space of $N \times N$ matrices with real coefficients. Motivated by the linear case, we define symmetric quasilinear systems as follows.

Definition 4.3. *The system (QS) is said to be* symmetric *if for any k in $\{1, \ldots, d\}$, the function A_k is valued in the space of symmetric $N \times N$ matrices.*

As an example, we will consider the Euler system for a perfect gas in the whole space $\mathbb{R}^d$. Denoting by ρ the density of the particles of the gas and by v the velocity field of the particles, the system to be considered is

$$\begin{cases} \partial_t \rho + v \cdot \nabla \rho + \rho \operatorname{div} v = 0 \\ \partial_t v + v \cdot \nabla v + \rho^{-1} \nabla p = 0 \quad \text{with} \quad p = A \rho^\gamma. \end{cases}$$

The above system is *not* quasilinear symmetric. However, if we introduce the new unknown function c defined by

$$c \stackrel{\text{def}}{=} \frac{2}{\gamma - 1} \left(\frac{\partial p}{\partial \rho} \right)^{\frac{1}{2}} = \frac{(4 \gamma A)^{\frac{1}{2}}}{\gamma - 1} \rho^{\frac{\gamma - 1}{2}}$$

and define $\tilde{\gamma} \stackrel{\text{def}}{=} (\gamma - 1)/2$, then the system becomes

$$\begin{cases} \partial_t c + v \cdot \nabla c + \tilde{\gamma} c \operatorname{div} v = 0 \\ \partial_t v + v \cdot \nabla v + \tilde{\gamma} c \nabla c = 0. \end{cases}$$

This system *is* symmetric. For instance, if $d = 3$ and we write $U = (c, v)$, it is of the form (QS) with

$$A_1(U) = \begin{pmatrix} v_1 & \widetilde{\gamma}c & 0 & 0 \\ \widetilde{\gamma}c & v_1 & 0 & 0 \\ 0 & 0 & v_1 & 0 \\ 0 & 0 & 0 & v_1 \end{pmatrix}, \quad A_2(U) = \begin{pmatrix} v_2 & 0 & \widetilde{\gamma}c & 0 \\ 0 & v_2 & 0 & 0 \\ \widetilde{\gamma}c & 0 & v_2 & 0 \\ 0 & 0 & 0 & v_2 \end{pmatrix}, \quad A_3(U) = \begin{pmatrix} v_3 & 0 & 0 & \widetilde{\gamma}c \\ 0 & v_3 & 0 & 0 \\ 0 & 0 & v_3 & 0 \\ \widetilde{\gamma}c & 0 & 0 & v_3 \end{pmatrix}.$$

We shall temporarily suppose that the solution $U = (c, v)$ is a perturbation of order ε of the steady state $(\overline{c}, 0)$, where $\overline{c}$ is a given positive constant. By identification of powers of ε, we get, for the first order term,

$$\begin{cases} \partial_t c + \widetilde{\gamma}\,\overline{c}\,\mathrm{div}\, v = 0 \\ \partial_t v + \widetilde{\gamma}\,\overline{c}\,\nabla c = 0. \end{cases}$$

This is a symmetric linear system, called an *acoustic wave system*. In fact, an immediate computation shows that c satisfies the wave equation

$$\partial_t^2 c - \widetilde{\gamma}^2\,\overline{c}^2\,\Delta c = 0$$

so that c has a finite speed of propagation, namely $\widetilde{\gamma}\overline{c}$. We shall see in Section 4.2.2 that any linear first order symmetric system has the finite propagation speed property.

4.2 Linear Symmetric Systems

In this section we investigate linear symmetric systems. First, we want to solve them and then study a few basic properties of their solutions.

In all that follows, for s in $\mathbb{N}$, we define

$$|U(t)|_s^2 \overset{\mathrm{def}}{=} \sum_{\substack{1 \leq j \leq N \\ 0 \leq |\alpha| \leq s}} \|\partial_x^\alpha U^j(t)\|_{L^2}^2.$$

To simplify the presentation, we shall assume throughout this chapter that $I = [0, T]$ and $t_0 = 0$. Due to the time-reversibility and translational invariance of the systems that we here consider, however, similar results are true for any interval I and t_0 in I.

4.2.1 The Well-posedness of Linear Symmetric Systems

This subsection is devoted to the proof of the following well-posedness result.

Theorem 4.4. *Let* (LS) *be a linear symmetric system with smooth, bounded, and Lipschitz (with respect to the space variable) coefficients and let* s *be an integer. Let* U_0 *be in* H^s *and* F *be in* $\mathcal{C}(I; H^s)$. *Then,* (LS) *has a unique solution in the space* $\mathcal{C}(I; H^s) \cap \mathcal{C}^1(I; H^{s-1})$.

Proving this theorem requires four steps:

- First, we prove a priori estimates for sufficiently smooth solutions of the system (LS).
- Second, we apply the Friedrichs method so as to solve a sequence of ordinary differential equations which approximate (LS).
- Third, we pass to the limit in the case of sufficiently smooth initial data and get existence in any case by smoothing out the initial data.
- Finally, we get uniqueness using existence of the adjoint system.

We begin by stating a priori estimates for smooth solutions (the symmetry hypothesis is crucial here).

Lemma 4.5. *For any nonnegative integer s, a locally bounded nonnegative function a_s exists such that for any function U in $\mathcal{C}(I; H^{s+1}) \cap \mathcal{C}^1(I; H^s)$ and t in I, we have*

$$|U(t)|_s \leq |U_0|_s \exp\left(\frac{1}{2} \int_0^t a_s(t')\, dt'\right) + \int_0^t |F(t')|_s \exp\left(\frac{1}{2} \int_{t'}^t a_s(t'')\, dt''\right) dt'$$

with

$$F = \partial_t U + \sum_{k=1}^d \mathcal{A}_k \partial_k U + \mathcal{A}_0 U .$$

Proof. To begin, we prove this lemma for $s = 0$. Consider a function U in the space $\mathcal{C}(I; H^1) \cap \mathcal{C}^1(I; L^2)$. By the definition of F, we have

$$\frac{1}{2}\frac{d}{dt} |U(t)|_0^2 = (\partial_t U | U)_0$$

$$= (F|U)_0 - (\mathcal{A}_0 U | U)_0 - \sum_{k=1}^d (\mathcal{A}_k \partial_k U | U)_0 .$$

As the system (LS) is symmetric and U belongs to $\mathcal{C}(I; H^1) \cap \mathcal{C}^1(I; L^2)$, the computations carried out on page 171, leading to (4.1), are rigorous. Thus, we have

$$\frac{d}{dt} |U(t)|_0^2 \leq a_0(t)|U(t)|_0^2 + 2|F(t)|_0|U(t)|_0 \tag{4.2}$$

with $a_0(t) \stackrel{\text{def}}{=} \|\operatorname{div}\mathcal{A}(t,\cdot)\|_{L^\infty} + 2\|\mathcal{A}_0(t,\cdot)\|_{L^\infty}$. By the Gronwall lemma, we get

$$|U(t)|_0 \leq |U_0|_0\, e^{\frac{1}{2}\int_0^t a_0(t')\, dt'} + \int_0^t |F(t')|_0\, e^{\frac{1}{2}\int_{t'}^t a_0(t'')\, dt''} dt'. \tag{4.3}$$

In order to prove the lemma for any nonnegative integer, we shall proceed by induction. Assume that Lemma 4.5 is proved for some integer s. Let U be a function in $\mathcal{C}(I; H^{s+2}) \cap \mathcal{C}^1(I; H^{s+1})$ and introduce the function [with $N(d+1)$ components] $\widetilde{U}$ defined by

$$\widetilde{U} = (U, \partial_1 U, \dots, \partial_d U) .$$

As

$$F = \partial_t U + \sum_{k=1}^{d} \mathcal{A}_k \partial_k U + \mathcal{A}_0 U,$$

we obtain, for any j in $\{1, \ldots, d\}$, by differentiation of the equation,

$$\partial_t (\partial_j U) = -\sum_{k=1}^{d} \mathcal{A}_k \partial_k \partial_j U - \sum_{k=1}^{d} (\partial_j \mathcal{A}_k) \cdot \partial_k U - \partial_j (\mathcal{A}_0 U) + \partial_j F.$$

Let $\widetilde{F} \stackrel{\text{def}}{=} (F, \partial_1 F, \ldots, \partial_d F)$ and

$$\mathcal{B}_0 \widetilde{U} \stackrel{\text{def}}{=} \left(\mathcal{A}_0 U, \sum_{k=1}^{d} (\partial_1 \mathcal{A}_k) \cdot \partial_k U + \partial_1 (\mathcal{A}_0 U), \ldots, \sum_{k=1}^{d} (\partial_d \mathcal{A}_k) \cdot \partial_k U + \partial_d (\mathcal{A}_0 U) \right).$$

We may write

$$\partial_t \widetilde{U} + \sum_{k=1}^{d} \mathcal{B}_k \, \partial_k \widetilde{U} + \mathcal{B}_0 \widetilde{U} = \widetilde{F} \quad \text{with} \quad \mathcal{B}_k \stackrel{\text{def}}{=} \begin{pmatrix} \mathcal{A}_k & 0 & \cdots & 0 \\ 0 & \ddots & \ddots & \vdots \\ \vdots & \ddots & \mathcal{A}_k & 0 \\ 0 & \cdots & 0 & \mathcal{A}_k \end{pmatrix}.$$

The induction hypothesis then allows us to complete the proof of Lemma 4.5.
□

Remark 4.6. In the case $s = 0, 1$, the above computations are still valid when the matrices $\mathcal{A}_0, \ldots, \mathcal{A}_d$ are only continuous, bounded, and have bounded first order space derivatives.

We should point out that proving the inequalities of Lemma 4.5 requires one more derivative than in the statement of Theorem 4.4. Hence, existence does not follow from basic contraction mapping arguments. This leads us to smooth out both the system and the data. To do so, we shall use the Friedrichs method. More precisely, we consider the system (LS_n) defined by

$$(LS_n) : \begin{cases} \partial_t U_n + \displaystyle\sum_{k=1}^{d} \mathbb{E}_n \left(\mathcal{A}_k \partial_k U_n \right) + \mathbb{E}_n (\mathcal{A}_0 U_n) = \mathbb{E}_n F \\ \qquad\qquad\qquad\qquad\qquad \mathbb{E}_n U_{|t=0} = \mathbb{E}_n U_0, \end{cases}$$

where $\mathbb{E}_n$ is the cut-off operator defined on L^2 by

$$\mathbb{E}_n u \stackrel{\text{def}}{=} \mathcal{F}^{-1}(\mathbf{1}_{B(0,n)} \widehat{u}). \tag{4.4}$$

In other words $\mathbb{E}_n$ is the L^2 orthogonal projector over the closed space L_n^2 of L^2 functions with Fourier transforms supported in the ball with center 0 and radius n. Lemma 2.1 tells us, in particular, that the operator ∂_k is continuous on L_n^2. As the functions $\mathcal{A}_k$ are bounded, it turns out that the linear operator

$$V \longmapsto \sum_{k=1}^{d} \mathbb{E}_n \left(\mathcal{A}_k \partial_k V \right) + \mathbb{E}_n (\mathcal{A}_0 V)$$

is continuous on L_n^2. Thus, the system (LS_n) is a linear system of ordinary differential equations on L_n^2. This implies the existence of a unique function U_n in $\mathcal{C}^1(I; L_n^2)$ which is a solution of (LS_n). Of course, due to the definition of L_n^2, the function U_n is also in any space $\mathcal{C}^1(I; H^s)$ with $s \in \mathbb{N}$.

We claim that the functions U_n still satisfy the energy estimates of Lemma 4.5. More precisely, we have the following lemma.

Lemma 4.7. *For any nonnegative integer s, a locally bounded function a_s exists such that for any $n \in \mathbb{N}$ and any t in I, we have,*

$$|U_n(t)|_s \leq |\mathbb{E}_n U_0|_s \exp \int_0^t a_s(t')\,dt' + \int_0^t |\mathbb{E}_n F(t')|_s \exp \left(\int_{t'}^t a_s(t'')\,dt'' \right) dt'.$$

Proof. Taking the scalar product of (LS_n) with U_n in L^2 and using the facts that the operator $\mathbb{E}_n$ is self-adjoint on L^2 and $\mathbb{E}_n U_n = U_n$, we get

$$\frac{d}{dt}|U_n(t)|_0^2 = -2\sum_{k=1}^{d} (\mathcal{A}_k \partial_k U_n | U_n)_0 - 2(\mathcal{A}_0 U_n)|U_n)_0 + 2(\mathbb{E}_n F|U_n)_0.$$

We proceed exactly as in the proof of Lemma 4.5. As the system (LS) is symmetric and U_n belongs to $\mathcal{C}(I; H^1) \cap \mathcal{C}^1(I; L^2)$, the computations carried out on page 171 are rigorous. Thus, we have

$$\frac{d}{dt}|U_n(t)|_0^2 \leq a_0(t)|U_n(t)|_0^2 + 2|\mathbb{E}_n F(t)|_0|U_n(t)|_0 \tag{4.5}$$

with $a_0(t) \overset{\text{def}}{=} \|\operatorname{div} \mathcal{A}(t,\cdot)\|_{L^\infty} + 2\|\mathcal{A}_0(t,\cdot)\|_{L^\infty}$. The Gronwall lemma implies that

$$|U_n(t)|_0 \leq |\mathbb{E}_n U_0|_0 \, e^{\frac{1}{2}\int_0^t a_0(t')\,dt'} + \int_0^t |\mathbb{E}_n F(t')|_0 \, e^{\frac{1}{2}\int_{t'}^t a_0(t'')\,dt''} \, dt'.$$

Proving the lemma for any integer s works exactly the same as for Lemma 4.5 and is thus omitted.   □

The third step amounts to proving the following well-posedness result.

Proposition 4.8. *Let $s \geq 3$. Consider the linear symmetric system (LS) with F in $\mathcal{C}(I; H^s)$ and U_0 in H^s. A unique solution U exists in*

$$L^\infty(I; H^s) \cap \mathcal{C}(I; H^{s-2}) \cap \mathcal{C}^1(I; H^{s-3})$$

which, moreover, satisfies

$$|U(t)|_\sigma \leq |U_0|_\sigma \, \exp \int_0^t a_s(t') \, dt' + \int_0^t |F(t')|_\sigma \, \exp \left(\int_{t'}^t a_s(t'') \, dt'' \right) dt'$$

for all integers $\sigma \leq s$ and $t \in I$.

Proof. Consider the sequence $(U_n)_{n\in\mathbb{N}}$ of solutions of (LS_n). We shall prove that $(U_n)_{n\in\mathbb{N}}$ is a Cauchy sequence in $L^\infty(I; H^{s-2})$. In order to do so, we define $V_{n,p} \stackrel{\text{def}}{=} U_{n+p} - U_n$. We have

$$\begin{cases} \partial_t V_{n,p} + \sum_{k=1}^d \mathbb{E}_{n+p} \left(\mathcal{A}_k \, \partial_k V_{n,p} \right) + \mathbb{E}_{n+p}(\mathcal{A}_0 V_{n,p}) = F_{n,p} \\ V_{n,p|t=0} = (\mathbb{E}_{n+p} - \mathbb{E}_n) U_0 \end{cases} \qquad (4.6)$$

with

$$F_{n,p} \stackrel{\text{def}}{=} -\sum_{k=1}^d (\mathbb{E}_{n+p} - \mathbb{E}_n) \left(\mathcal{A}_k \, \partial_k U_n \right) - (\mathbb{E}_{n+p} - \mathbb{E}_n)(\mathcal{A}_0 U_n) + (\mathbb{E}_{n+p} - \mathbb{E}_n) F.$$

Lemma 4.7 tells us that the sequence $(U_n)_{n\in\mathbb{N}}$ is bounded in $L^\infty(I; H^s)$. Moreover, we have, for any real σ and any a in H^σ,

$$|(\mathbb{E}_{n+p} - \mathbb{E}_n) a|_{\sigma-1} \leq \frac{C}{n} |a|_\sigma.$$

Thus, we have

$$|(\mathbb{E}_{n+p} - \mathbb{E}_n) \left(\mathcal{A}_k \, \partial_k U_n(t) \right)|_{s-2} \leq \frac{C}{n} \sup_k |(\mathbb{E}_{n+p} - \mathbb{E}_n) \left(\mathcal{A}_k \, \partial_k U_n(t) \right)|_{s-1}$$
$$\leq \frac{C}{n} |U_n(t)|_s.$$

The same arguments give

$$\left| (\mathbb{E}_{n+p} - \mathbb{E}_n)(\mathcal{A}_0 U_n(t)) + (\mathbb{E}_{n+p} - \mathbb{E}_n) F(t) \right|_{s-2} \leq \frac{C}{n^2} \left(|U_n(t)|_s + |F(t)|_s \right). \qquad (4.7)$$

By using the energy estimate for (4.6) and Lemma 4.7, we get

$$|V_{n,p}(t)|_{s-2} \leq \frac{C}{n} (1+t) \exp \int_0^t a_s(t') \, dt'.$$

Thus, $(U_n)_{n\in\mathbb{N}}$ is a Cauchy sequence in $L^\infty(I; H^{s-2})$. Moreover, using (4.6) and (4.7), we infer that $(\partial_t U_n)_{n\in\mathbb{N}}$ is a Cauchy sequence in $L^\infty(I; H^{s-3})$. We denote by U the limit of $(U_n)_{n\in\mathbb{N}}$. Of course, U belongs to the space

$$\mathcal{C}(I; H^{s-2}) \cap \mathcal{C}^1(I; H^{s-3}).$$

We now check that this function U is a solution of (LS). As U_0 is in H^s and F belongs to $\mathcal{C}(I; H^s)$, we have that

$$\lim_{n\to\infty} \mathbb{E}_n U_0 = U_0 \text{ in } H^s \quad \text{and} \quad \lim_{n\to\infty} \mathbb{E}_n F = F \text{ in } L^\infty(I; H^s). \quad (4.8)$$

As the sequence $(U_n)_{n\in\mathbb{N}}$ is bounded in $L^\infty(I; H^s)$, we have

$$\|(\mathbb{E}_n - \mathrm{Id})\mathcal{A}_k \partial_k U_n\|_{L^\infty(I;H^{s-2})} \leq \frac{C}{n}.$$

Thus, U is a solution of (LS). To complete the proof of Proposition 4.8, we use the fact that $(U_n)_{n\in\mathbb{N}}$ is bounded in $L^\infty(I; H^s)$. Hence, for all t in I, the sequence $(U_n(t))_{n\in\mathbb{N}}$ weakly converges (up to extraction) in H^s. Thus, $U(t)$ belongs to H^s and

$$\|U(t)\|_{H^s} \leq \liminf_{n\to\infty} \|U_n(t)\|_{H^s}.$$

Now, combining the uniform bounds for $(U_n)_{n\in\mathbb{N}}$ in $L^\infty(I; H^s)$ with the above result on convergence in $L^\infty(I; H^{s-2})$ and using the interpolation inequality stated in Proposition 1.52, we get that for any $s' < s$, the sequence $(U_n)_{n\in\mathbb{N}}$ converges in $\mathcal{C}(I; H^{s'})$. Thus, U belongs to $\mathcal{C}(I; H^{s'})$. Using the fact that U is a solution of (LS), we get that U belongs to $\mathcal{C}(I; H^{s'}) \cap \mathcal{C}^1(I; H^{s'-1})$. So, finally, passing to the limit in Lemma 4.7, we find that

$$|U(t)|_\sigma \leq |U_0|_\sigma \exp\int_0^t a_\sigma(t')\,dt' + \int_0^t |F(t')|_\sigma \exp\left(\int_{t'}^t a_\sigma(t'')\,dt''\right)dt'$$

for all integers $\sigma \leq s$. Proposition 4.8 is thus proved. $\qquad\qquad\square$

In order to prove the existence part of Theorem 4.4, we now have to solve (LS) for general data $U_0 \in H^s$ and $F \in \mathcal{C}(I; H^s)$. We therefore consider the sequence $(\widetilde{U}_n)_{n\in\mathbb{N}}$ of solutions of

$$\begin{cases} \dfrac{\partial \widetilde{U}_n}{\partial t} + \displaystyle\sum_{k=1}^d \mathcal{A}_k \partial_k \widetilde{U}_n + \mathcal{A}_0 \widetilde{U}_n = \mathbb{E}_n F \\ \qquad\qquad\qquad \widetilde{U}_{n|t=0} = \mathbb{E}_n U_0. \end{cases}$$

Thanks to Proposition 4.8, $\widetilde{U}_n$ is well defined on I and belongs to $\mathcal{C}^1(I, H^s)$ for any positive real number s. Further, the function $V_{n,p} \overset{\mathrm{def}}{=} \widetilde{U}_{n+p} - \widetilde{U}_n$ satisfies

$$\begin{cases} \partial_t \widetilde{V}_{n,p} + \displaystyle\sum_{k=1}^d \mathcal{A}_k \partial_k \widetilde{V}_{n,p} + \mathcal{A}_0 \widetilde{V}_{n,p} = (\mathbb{E}_{n+p} - \mathbb{E}_n)F \\ \qquad\qquad\qquad \widetilde{V}_{n,p|t=0} = (\mathbb{E}_{n+p} - \mathbb{E}_n)U_0. \end{cases}$$

Lemma 4.5 implies that

$$|\widetilde{V}_{n,p}(t)|_s \leq |(\mathbb{E}_{n+p} - \mathbb{E}_n)U_0|_s \exp \int_0^t a_s(t')\,dt'$$

$$+ \int_0^t |(\mathbb{E}_{n+p} - \mathbb{E}_n)F(t')|_s \exp \left(\int_{t'}^t a_s(t')\,dt' \right) dt.$$

As the function F is continuous from I into H^s, the sequence $(\mathbb{E}_n F)_{n\in\mathbb{N}}$ converges to F in the space $L^\infty(I; H^s)$. This is a consequence of Dini's theorem applied to the nonincreasing sequence of continuous functions $t \mapsto \|(F - \mathbb{E}_n F)(t)\|_s$ on the compact interval I.

As U_0 belongs to H^s, the sequence $(\mathbb{E}_n U_0)_{n\in\mathbb{N}}$ converges to U_0 in H^s. Thus, the sequence $(\widetilde{U}_n)_{n\in\mathbb{N}}$ is Cauchy in $L^\infty(I; H^s)$ and therefore converges to some function U in $\mathcal{C}(I; H^s)$ which is, of course, a solution of the system (LS). The fact that $\partial_t U$ belongs to $\mathcal{C}(I; H^{s-1})$ comes immediately from the fact that U is a solution of the system (LS).

Remark 4.9. Assume that the matrices $\mathcal{A}_0, \ldots, \mathcal{A}_d$ are only continuous and bounded with bounded first order space derivatives. By taking advantage of Remark 4.6 and compactness arguments, it is possible to prove that for any data U_0 in H^1 and F in $L^\infty(I; H^1)$, the system (LS) has a solution U in the space $L^\infty(I; H^1) \cap C^{0,1}(I; L^2)$.

Finally, uniqueness in the case $s \geq 1$ is merely a consequence of Lemma 4.5. This completes the proof of Theorem 4.4 when $s \geq 1$.

Uniqueness in the case $s = 0$ follows from the following proposition.

Proposition 4.10. *Under the assumptions of Remark 4.9, let U be a solution in the space $\mathcal{C}(I; L^2)$ of the symmetric system (LS) with initial data $U_0 = 0$ and external force $F = 0$. Then, $U \equiv 0$.*

Proof. In order to prove this proposition, we shall use a duality method. Let ψ be a function in $\mathcal{D}(\,]0, T[\, \times \mathbb{R}^d)$ and consider the solution of the system

$$(^tLS) : \begin{cases} -\partial_t\varphi - \displaystyle\sum_{k=1}^d \partial_k(\mathcal{A}_k\varphi) + {}^t\mathcal{A}_0\varphi = \psi \\ \\ \qquad\qquad\qquad\qquad \varphi_{|t=T} = 0. \end{cases}$$

The system (^tLS) can be understood as the *adjoint system* of the system (LS). As we have $\partial_k(\mathcal{A}_k\varphi) = \mathcal{A}_k\partial_k\varphi + (\partial_k\mathcal{A}_k)\varphi$, it may be rewritten as

$$\begin{cases} -\partial_t\varphi - \displaystyle\sum_{k=1}^d \mathcal{A}_k\partial_k\varphi + \widetilde{\mathcal{A}}_0\varphi = \psi \\ \\ \qquad\qquad\qquad\qquad \varphi_{|t=T} = 0 \end{cases}$$

with $\widetilde{\mathcal{A}}_0 \stackrel{\text{def}}{=} {}^t\mathcal{A}_0 - \operatorname{div}\mathcal{A}$.

This is obviously a linear symmetric system. Since ψ belongs, in particular, to H^1, Remark 4.9 provides a solution φ for (^tLS) in $L^\infty(I, H^1) \cap C^{0,1}(I; L^2)$. Thus, we have

$$
\begin{aligned}
\langle U, \psi \rangle &= \Big\langle U, -\partial_t \varphi - \sum_{k=1}^d \mathcal{A}_k \partial_k \varphi + \tilde{\mathcal{A}}_0 \varphi \Big\rangle \\
&= -\int_I \big(U(t) \mid \partial_t \varphi(t) \big)_0 dt - \sum_{k=1}^d \int_I \big(U(t) \mid \partial_k (\mathcal{A}_k \varphi)(t) \big)_0 dt \\
&\qquad\qquad + \int_I \big(U(t) \mid {}^t\mathcal{A}_0 \varphi(t) \big)_0 dt.
\end{aligned}
$$

Owing to the weak regularity of U, the integrations by parts must be justified. Because each $\mathcal{A}_k$ is continuous and bounded with bounded gradient, $\partial_k(\mathcal{A}_k \varphi)$ is in $L^\infty(I; L^2)$. Therefore, we can write that

$$
\begin{aligned}
\big(U(t) \mid \partial_k (\mathcal{A}_k \varphi)(t) \big)_0 &= \sum_{i,j} \big(U^i(t) \mid \partial_k (\mathcal{A}_{k,i,j} \varphi^j)(t) \big)_{L^2} \\
&= -\sum_{i,j} \big\langle \partial_k U^i(t), \mathcal{A}_{k,i,j} \varphi^j(t) \big\rangle_{H^{-1} \times H^1}.
\end{aligned}
$$

Observe that $\mathcal{A}_k \partial_k U$ is in $L^\infty(I; H^{-1})$. Indeed, for any smooth function V, we have

$$
\begin{aligned}
\langle \mathcal{A}_k \partial_k V, \varphi \rangle &= -\langle V, (\partial_k {}^t\mathcal{A}_k) \varphi \rangle - \langle V, {}^t\mathcal{A}_k \partial_k \varphi \rangle \\
&\le \big(\|\mathcal{A}_k\|_{L^\infty} + \|\partial_k \mathcal{A}_k\|_{L^\infty} \big) \|V\|_{L^2} \|\varphi\|_{H^1}.
\end{aligned}
$$

Because the matrices $\mathcal{A}_k$ are symmetric, we therefore have, for any t in I,

$$
-\big(U(t) \mid \partial_k (\mathcal{A}_k \varphi)(t) \big)_0 = \big\langle \mathcal{A}_k \partial_k U(t), \varphi(t) \big\rangle_{H^{-1} \times H^1},
$$

from which it follows that

$$
(U \mid \psi)_0 = -(U \mid \partial_t \varphi)_0 - \Big\langle \sum_{k=1}^d \mathcal{A}_k \partial_k U + \mathcal{A}_0 U, \varphi \Big\rangle_{H^{-1} \times H^1}.
$$

In order to justify the time integration by parts, we observe that $\partial_t U$ belongs to $L^\infty(I; H^{-1})$. We now use the smoothing operator $\mathbb{E}_n$ defined by (4.4). The function $\mathbb{E}_n U$ belongs to $C^1(I; H^s)$ for any nonnegative integer s. Using this with s greater than $d/2 + 1$ implies that for any x, the function

$$
(t, x) \longmapsto \mathbb{E}_n U(t, x)
$$

is C^1 on $I \times \mathbb{R}^d$. Likewise, the function $\mathbb{E}_n \varphi$ is C^1 on $I \times \mathbb{R}^d$. This implies that

$$- \int_I \mathbb{E}_n U(t,x) \partial_t \mathbb{E}_n \varphi(t,x) dt = - \mathbb{E}_n U(T,x) \mathbb{E}_n \varphi(T,x)$$

$$+ \mathbb{E}_n U_0(x) \mathbb{E}_n \varphi(0,x) + \int_I \partial_t \mathbb{E}_n U(t,x) \mathbb{E}_n \varphi(t,x) dt.$$

Using the facts that $U_0 = 0$ and $\varphi(T, \cdot) = 0$, we get that

$$- \int_I \mathbb{E}_n U(t,x) \partial_t \mathbb{E}_n \varphi(t,x) dt = \int_I \partial_t (\mathbb{E}_n U)(t,x) \mathbb{E}_n \varphi(t,x) dt.$$

Integrating with respect to the variable x and interchanging the time and space integrations, we get that

$$- \int_I \left(\mathbb{E}_n U(t) \mid \partial_t \mathbb{E}_n \varphi(t) \right)_0 dt = \int_I \langle \partial_t (\mathbb{E}_n U)(t), \mathbb{E}_n \varphi(t) \rangle_{H^{-1} \times H^1} dt. \quad (4.9)$$

As U is a function of $\mathcal{C}(I; L^2) \cap \mathcal{C}^1(I; H^{-1})$, we have

$$\lim_{n \to \infty} \mathbb{E}_n U = U \quad \text{in} \quad L^\infty(I; L^2) \quad \text{and} \quad \lim_{n \to \infty} \mathbb{E}_n \partial_t U = \partial_t U \quad \text{in} \quad L^\infty(I; H^{-1}).$$

Similarly, as φ belongs to $L^\infty(I; H^1) \cap C^{0,1}(I; L^2)$, we have

$$\lim_{n \to \infty} \mathbb{E}_n \varphi = \varphi \quad \text{in} \quad L^\infty(I; H^1) \quad \text{and} \quad \lim_{n \to \infty} \mathbb{E}_n \partial_t \varphi = \partial_t \varphi \quad \text{in} \quad L^\infty(I; L^2).$$

Passing to the limit in (4.9) thus gives

$$- \int_I \left(U(t) \mid \partial_t \varphi(t) \right)_0 dt = \int_I \langle \partial_t U(t), \varphi(t) \rangle_{H^{-1} \times H^1} dt$$

and thus

$$\int_I \left(U(t) \mid \psi(t) \right)_0 dt = \int_I \left\langle \partial_t U(t) + \sum_{k=1}^d A_k \partial_k U(t) + A_0 U(t), \varphi(t) \right\rangle_{H^{-1} \times H^1} dt.$$

As U is a solution of (LS) with $F = 0$, we conclude that $U \equiv 0$. $\qquad \square$

4.2.2 Finite Propagation Speed

Linear symmetric systems have the *finite propagation speed property*. This means that there exists some positive constant C_0 (the maximal speed of propagation) such that the value of the solution U at some point (x_0, t_0) determines $U(t,x)$ only for those (t,x) such that $|x - x_0| \le C_0 |t - t_0|$.

This phenomenon is described in the following theorem.

Theorem 4.11. *Let (LS) be a symmetric system. A constant C_0 exists such that for any $R > 0$, x_0 in $\mathbb{R}^d$, F in $\mathcal{C}(I; L^2)$, and $U_0 \in L^2$ such that*

$$F(t,x) = 0 \ \text{for} \ |x - x_0| < R - C_0 t \quad \text{and} \quad U_0(x) = 0 \ \text{for} \ |x - x_0| < R, \quad (4.10)$$

the unique solution U of the system (LS) in $\mathcal{C}(I; L^2)$ with data F and U_0 satisfies

$$U(t,x) = 0 \quad \text{for} \quad |x - x_0| < R - C_0 t.$$

Another form of this statement is given by the following corollary.

Corollary 4.12. *If the data F and U_0 satisfy*

$$F(t,x) \equiv 0 \quad for \quad |x - x_0| > R + C_0 t \quad and \quad U_0(x) \equiv 0 \quad for \quad |x - x_0| > R,$$

then the solution U satisfies

$$U(t,x) \equiv 0 \quad when \quad |x - x_0| > R + C_0 t.$$

Proof. Of course, it suffices to consider the case $x_0 = 0$. To begin, we smooth out the data U_0 and F, perturbing their support as little as possible. Let χ be a function in $\mathcal{D}(B(0,1))$ with integral 1. For any positive ε, we define

$$\chi_\varepsilon(x) \overset{\text{def}}{=} \frac{1}{\varepsilon^d} \chi\left(\frac{x}{\varepsilon}\right)$$

and consider the data

$$U_{0,\varepsilon} \overset{\text{def}}{=} \chi_\varepsilon \star U_0 \quad and \quad F_\varepsilon(t,\cdot) \overset{\text{def}}{=} \chi_\varepsilon \star F(t,\cdot).$$

Of course, we have

$$\text{Supp } U_{0,\varepsilon} \subset \text{Supp } U_0 + B(0,\varepsilon) \quad and \quad \text{Supp } F_\varepsilon(t,\cdot) \subset \text{Supp } F(t,\cdot) + B(0,\varepsilon).$$

Hence, the support hypothesis is satisfied for $U_{0,\varepsilon}$ and F_ε with $R + \varepsilon$ instead of R, and the associated solution U_ε is in $\mathcal{C}^1(I; H^s)$ for any $s \in \mathbb{N}$ and tends to U in $\mathcal{C}(I; L^2)$. It is thus enough to prove Theorem 4.11 for those regular solutions, namely, the following statement.

Theorem 4.13. *Let (LS) be a symmetric system. A constant C_0 exists such that for any positive real number R and any data F in $\mathcal{C}(I; H^1)$ and U_0 in H^1 such that*

$$F(t,x) \equiv 0 \quad for \quad |x| < R - C_0 t \quad and \quad U_0(x) \equiv 0 \quad for \quad |x| < R, \quad (4.11)$$

the unique solution U of the system (LS) in $\mathcal{C}(I; H^1) \cap \mathcal{C}^1(I; L^2)$ with data F and U_0 satisfies

$$U(t,x) \equiv 0 \quad when \quad |x| < R - C_0 t.$$

Proof. The key to the proof is a weighted energy estimate. More precisely, for τ greater than 1, we introduce

$$U_\tau(t,x) \overset{\text{def}}{=} e^{\tau\phi(t,x)} U(t,x) \quad with \quad \phi(t,x) \overset{\text{def}}{=} -t + \psi(x).$$

Above, ψ stands for a smooth real-valued function on $\mathbb{R}^d$ which will be chosen later. We have

$$\partial_t U_\tau + \sum_{k=1}^d \mathcal{A}_k\, \partial_k U_\tau + \mathcal{B}_\tau U_\tau = F_\tau \quad \text{with}$$

$$F_\tau(t,x) \overset{\text{def}}{=} e^{\tau\phi(t,x)} F(t,x) \quad \text{and} \quad \mathcal{B}_\tau = \mathcal{A}_0 + \tau \left(\mathrm{Id} - \sum_{k=1}^d \partial_k \psi \mathcal{A}_k \right).$$

Thus, a constant $K > 0$ exists such that for any $(t,x) \in I \times \mathbb{R}^d$, any vector $W \in \mathbb{R}^N$, and any positive real number τ, we have

$$\|\nabla\psi\|_{L^\infty} \le K \Rightarrow (\mathcal{B}_\tau(t,x)W\,|\,W) \ge (\mathcal{A}_0(t,x)W\,|\,W).$$

Next, we write the energy estimate and use the above inequality and the relation (4.1) to obtain

$$\frac{d}{dt}|U_\tau(t)|_0^2 = -2\sum_{k=1}^d (\mathcal{A}_k\partial_k U_\tau | U_\tau)_{L^2} - 2(\mathcal{B}_\tau U_\tau | U_\tau)_{L^2} + 2(F_\tau | U_\tau)_{L^2}$$
$$\le a_0(t)|U_\tau(t)|_0^2 + 2(F_\tau(t)|U_\tau(t))_{L^2}.$$

Using the Gronwall lemma, we get

$$|U_\tau(t)|_0 \le |U_\tau(0)|_0 e^{\int_0^t a_0(t')\,dt'} + \int_0^t |F_\tau(t')|_0 e^{\int_{t'}^t a_0(t'')\,dt''}\,dt'. \tag{4.12}$$

Note that the above inequality is independent of τ. We now define

$$C_0 \overset{\text{def}}{=} \left(\sum_{k=1}^d \|A_k\|_{L^\infty}^2 \right)^{1/2} \quad \text{and} \quad K \overset{\text{def}}{=} 1/C_0,$$

and choose a smooth function $\psi = \psi(|x|)$ such that

$$-2\varepsilon + K(R - |x|) \le \psi(x) \le -\varepsilon + K(R - |x|) \quad \text{and} \quad \|\nabla\psi\|_{L^\infty} \le K. \tag{4.13}$$

We then have, for any (t,x) in $I \times \mathbb{R}^d$,

$$|x| \ge R - C_0 t \implies -t + \psi(x) \le -\varepsilon.$$

When τ tends to $+\infty$ in the inequality (4.12), we get, for any t in I,

$$\lim_{\tau \to \infty} \int_{\mathbb{R}^d} e^{2\tau\phi(t,x)} |U(t,x)|^2\,dx = 0.$$

Thus, $U(t,x) \equiv 0$ on the open set $t < \psi(x)$. If (t_0, x_0) satisfies $|x_0| < R - C_0 t_0$, then it is possible to choose a function ψ satisfying (4.13) and such that $t_0 < \psi(x_0)$. This proves the theorem. $\qquad\square$

4.2.3 Further Well-posedness Results for Linear Symmetric Systems

In this section, we are concerned with a priori estimates and existence results for (LS) in more general spaces: Sobolev spaces with noninteger indices or Besov spaces of type $B_{2,r}^s$. These results will be needed for proving existence results in general Sobolev spaces or in $B_{2,1}^{d/2+1}$ for symmetric quasilinear systems, and also for stating the continuity of the flow map.

For simplicity, we drop the 0 order term in (LS) (i.e., $\mathcal{A}_0 \equiv 0$ is assumed). Throughout this section, r is given in $[1, \infty]$ and $(c_j)_{j \geq -1}$ denotes a generic sequence of nonnegative locally integrable functions over I such that $\|(c_j(t))\|_{\ell^r} = 1$ for any t in I.

Lemma 4.14. Let $s > 0$, $r \in [1, \infty]$, and V satisfy

$$\partial_t V + \sum_{k=1}^{d} \mathcal{A}_k \partial_k V = F.$$

Let $V_j \stackrel{def}{=} \Delta_j V$, $\overline{S}_j \stackrel{def}{=} S_j$ if $j \geq 0$, and[1] $\overline{S}_j \stackrel{def}{=} \Delta_{-1}$ if $p \in \{-2, -1\}$. We have

$$\partial_t V_j + \sum_{k=1}^{d} (\overline{S}_{j-1} \mathcal{A}_k)\, \partial_k V_j = \Delta_j F + R_j \quad \text{for all} \quad j \geq -1,$$

where R_j satisfies, for all $t \in I$,

$$2^{js} \|R_j(t)\|_{L^2} \leq C c_j(t) \sum_{k=1}^{d} \Big(\|\nabla \mathcal{A}_k(t)\|_{L^\infty} \|\nabla V(t)\|_{B_{2,r}^{s-1}}$$

$$+ \|\nabla V(t)\|_{L^\infty} \|\nabla \mathcal{A}_k(t)\|_{B_{2,r}^{s-1}} \Big). \quad (4.14)$$

If $0 < s < d/2 + 1$, then we also have

$$2^{js} \|R_j(t)\|_{L^2} \leq C c_j(t) \|\nabla V(t)\|_{B_{2,r}^{s-1}} \sum_{k=1}^{d} \|\nabla \mathcal{A}_k(t)\|_{L^\infty \cap B_{2,\infty}^{\frac{d}{2}}}, \quad (4.15)$$

and if $s = d/2 + 1$, then for all $\varepsilon > 0$,

$$2^{j(\frac{d}{2}+1)} \|R_j(t)\|_{L^2} \leq C c_j(t) \|\nabla V(t)\|_{B_{2,r}^{\frac{d}{2}}} \sum_{k=1}^{d} \|\nabla \mathcal{A}_k(t)\|_{B_{2,\infty}^{\frac{d}{2}+\varepsilon}}. \quad (4.16)$$

---

[1] This unusual choice for the low-frequency cut-off is motivated by the wish to have only the gradient of $\mathcal{A}_k$ involved in the estimates of R_j. This refinement turns out to be important in the next section for functions $\mathcal{A}_k$ which need not tend to 0 at infinity.

Proof. First, we write

$$\partial_t V_j + \Delta_j \sum_{k=1}^d \mathcal{A}_k \partial_k V = \Delta_j F.$$

Recall that

$$\left(\mathcal{A}_k \partial_k V \right)^i = \sum_\ell A_{k,i,\ell} \partial_k V^\ell.$$

To simplify the notation, we shall drop the indices i and ℓ in the following computations.

In order to better describe the commutation between the multiplication operator and Δ_j, we shall use a simplified version of the Bony decomposition defined in Section 2.8. We write

$$\mathcal{A}_k \partial_k V = \overline{T}_{\mathcal{A}_k} \partial_k V + \overline{T}'_{\partial_k V} \mathcal{A}_k \quad \text{with}$$

$$\overline{T}_{\mathcal{A}_k} \partial_k V = \sum_{j' \geq -1} \overline{S}_{j'-1} \mathcal{A}_k \, \Delta_{j'} \partial_k V \quad \text{and} \quad \overline{T}'_{\partial_k V} \mathcal{A}_k = \sum_{j' \geq 0} S_{j'+2} \partial_k V \, \Delta_{j'} \mathcal{A}_k.$$

As the support of the Fourier transform of $\overline{S}_{j'-1} \mathcal{A}_k \Delta_{j'} \partial_k V$ is included in an annulus of the type $\{ \xi \in \mathbb{R}^d \,/\, c_1 2^{j'} \leq |\xi| \leq c_2 2^{j'} \}$, and $\Delta_j \Delta_{j'} = 0$ for $|j - j'| \geq 2$ (see Proposition 2.10), we have, for some fixed integer N_1,

$$\Delta_j \sum_{j'} \overline{S}_{j'-1} \mathcal{A}_k \, \Delta_{j'} \partial_k V = \Delta_j \sum_{|j'-j| \leq N_1} \overline{S}_{j'-1} \mathcal{A}_k \, \Delta_{j'} \partial_k V$$

$$= R_{j,k}^1 + R_{j,k}^2 + \overline{S}_{j-1} \mathcal{A}_k \, \partial_k V_j$$

$$\text{with} \quad \begin{cases} R_{j,k}^1 \overset{\text{def}}{=} \displaystyle\sum_{|j'-j| \leq N_1} \left[\Delta_j, \overline{S}_{j'-1} \mathcal{A}_k \right] \Delta_{j'} \partial_k V \\[2mm] R_{j,k}^2 \overset{\text{def}}{=} \displaystyle\sum_{|j'-j| \leq 1} (\overline{S}_{j'-1} \mathcal{A}_k - \overline{S}_{j-1} \mathcal{A}_k) \Delta_j \Delta_{j'} \partial_k V. \end{cases}$$

Finally, then, the commutation between the operator Δ_j and the equation can be described by the following formula:

$$\partial_t V_j + \sum_{k=1}^d \overline{S}_{j-1} \mathcal{A}_k \, \partial_k V_j = \Delta_j F + \sum_{m=1}^3 R_j^m \quad \text{with} \qquad (4.17)$$

$$R_j^1 \overset{\text{def}}{=} \sum_{1 \leq k \leq d} R_{j,k}^1,$$

$$R_j^2 \overset{\text{def}}{=} \sum_{1 \leq k \leq d} R_{j,k}^2,$$

$$R_j^3 \overset{\text{def}}{=} \Delta_j \sum_{1 \leq k \leq d} \overline{T}'_{\partial_k V} \mathcal{A}_k.$$

Lemma 2.97 page 110 implies that

$$2^j \|R_j^1\|_{L^2} \leq C \sum_{\substack{|j'-j| \leq N_1 \\ 1 \leq k \leq d}} \|\nabla \overline{S}_{j'-1} \mathcal{A}_k\|_{L^\infty} \|\Delta_{j'} \partial_k V\|_{L^2} .$$

Hence, because $\|\nabla \overline{S}_{j'-1} \mathcal{A}_k\|_{L^\infty} \leq C \|\nabla \mathcal{A}_k\|_{L^\infty}$, we get that

$$2^{js} \|R_j^1\|_{L^2} \leq C \sum_{\substack{|j'-j| \leq N_1 \\ 1 \leq k \leq d}} 2^{(j-j')(s-1)} \|\nabla \mathcal{A}_k\|_{L^\infty} 2^{j'(s-1)} \|\Delta_{j'} \partial_k V\|_{L^2} .$$

We thus get, according to the definition of the $B_{2,r}^s$ norm,

$$2^{js} \|R_j^1\|_{L^2} \leq C c_j \|\nabla \mathcal{A}\|_{L^\infty} \|\nabla V\|_{B_{2,r}^{s-1}}. \tag{4.18}$$

In order to estimate R_j^2, we observe that, due to the fact that $|j'-j| \leq 1$, the block $\Delta_{-1} \mathcal{A}_k$ does not play any role. Now, Bernstein's inequality ensures that

$$\|\Delta_\ell \mathcal{A}_k\|_{L^\infty} \leq C 2^{-\ell} \|\nabla \mathcal{A}_k\|_{L^\infty} \quad \text{for } \ell \in \mathbb{N}.$$

This implies that

$$2^{js} \|R_j^2\|_{L^2} \leq C c_j \|\nabla \mathcal{A}\|_{L^\infty} \|\nabla V\|_{B_{2,r}^{s-1}}. \tag{4.19}$$

Finally, as $s > 0$ and $\Delta_{-1} \mathcal{A}_k$ is not involved in $\overline{T}'_{\partial_k V} \mathcal{A}_k$ either, arguing as in Remark 2.83 page 103 enables us to get

$$\|\overline{T}'_{\partial_k V} \mathcal{A}_k\|_{B_{2,r}^s} \leq C \|\nabla V\|_{L^\infty} \|\nabla \mathcal{A}_k\|_{B_{2,r}^{s-1}}$$

whenever s is positive, hence

$$2^{js} \|R_j^3\|_{L^2} \leq C c_j \|\nabla V\|_{L^\infty} \|\nabla \mathcal{A}\|_{B_{2,r}^{s-1}}. \tag{4.20}$$

Combining the three estimates (4.18)–(4.20), we get the inequality (4.14).

Proving the other two inequalities follows along the same lines. It is only a matter of using appropriate continuity results for the paraproduct and remainder when bounding the term R_j^3 (see Propositions 2.82 and 2.85). The details are left to the reader. $\qquad \square$

Theorem 4.15. *Let $r \in [1,\infty]$, $s > 0$, U_0 be in $B_{2,r}^s$, and F be in $\mathcal{C}(I; B_{2,r}^s)$. Assume that the matrices $\mathcal{A}_k$ are symmetric and continuous with respect to (t,x), and that*

- $\nabla \mathcal{A}_k \in \mathcal{C}(I; B_{2,r}^{s-1})$ *if $s > d/2 + 1$, or $s = d/2 + 1$ and $r = 1$,*
- $\nabla \mathcal{A}_k \in \mathcal{C}(I; B_{2,\infty}^{\frac{d}{2}+\varepsilon})$ *for some $\varepsilon > 0$ if $s = d/2 + 1$ and $r > 1$,*
- $\nabla \mathcal{A}_k \in \mathcal{C}(I; B_{2,\infty}^{\frac{d}{2}} \cap L^\infty)$ *if $0 < s < d/2 + 1$.*

The system

$$(LS_0) : \begin{cases} \partial_t U + \sum_{k=1}^d \mathcal{A}_k \partial_k U = F \\ U_{|t=0} = U_0 \end{cases}$$

then has a unique solution U in the space $\mathcal{C}(I; B_{2,r}^s) \cap \mathcal{C}^1(I; B_{2,r}^{s-1})$ if $r > 1$ and in the space $L^\infty(I; B_{2,r}^s) \cap C^{0,1}(I; B_{2,r}^{s-1})$ if $r = \infty$. Moreover, for all $t \in I$ and some constant C depending only on d and s, we have

$$|U(t)|_{B_{2,r}^s} \leq |U_0|_{B_{2,r}^s} \exp\left(\int_0^t C a_s(t')\, dt'\right)$$
$$+ \int_0^t |f(t')|_{B_{2,r}^s} \exp\left(\int_{t'}^t C a_s(t'')\, dt''\right) dt' \qquad (4.21)$$

with $|U|_{B_{2,r}^s} \overset{def}{=} \|2^{qs}|\Delta_j U|_0\|_{\ell^r}$ and

$$a_s(t) \overset{def}{=} \begin{cases} \sum_k \|\nabla \mathcal{A}_k(t)\|_{B_{2,r}^{s-1}}, & \text{if } s > d/2+1, \text{ or } s = d/2+1 \text{ and } r = 1, \\ \sum_k \|\nabla \mathcal{A}_k(t)\|_{B_{2,\infty}^{\frac{d}{2}+\varepsilon}}, & \text{if } s = d/2+1 \text{ and } r > 1, \\ \sum_k \|\nabla \mathcal{A}_k(t)\|_{B_{2,\infty}^{\frac{d}{2}} \cap L^\infty}, & \text{if } 0 < s < d/2+1. \end{cases}$$

Proof. We first prove (4.21) for smooth solutions U of (LS_0). Defining $U_j \overset{def}{=} \Delta_j U$, we have

$$\partial_t U_j + \sum_{k=1}^d (\overline{S}_{j-1} \mathcal{A}_k) \partial_k U_j = \Delta_j F + R_j \qquad (4.22)$$

with, according to Lemma 4.14 and the embedding $B_{2,r}^{s-1} \hookrightarrow L^\infty$ if $s > 1+d/2$ (or if $s \geq 1+d/2$ and $r = 1$),

$$\|R_j\|_{L^2} \leq C c_j 2^{-js} a_s |U|_{B_{2,r}^s}. \qquad (4.23)$$

Now, applying the usual energy method to the equation (4.22) yields

$$\frac{1}{2}\frac{d}{dt}|U_j|_0^2 \leq \frac{1}{2}\|\text{div}\,\mathcal{A}\|_{L^\infty}|U_j|_0^2 + \left(|R_j|_0 + |\Delta_j F|_0\right)|U_j|_0.$$

Inserting the inequality (4.23), we get, for all positive α,

$$\frac{d}{dt}\sqrt{|U_j|_0^2 + \alpha} \leq |\Delta_j F|_0 + \frac{1}{2}\|\text{div}\,\mathcal{A}\|_{L^\infty}|U_j|_0 + C c_j 2^{-js} a_s |U|_{B_{2,r}^s}.$$

Integrating over $[0, t]$ and letting α tend to 0, we end up with

$$|U_j(t)|_0 \leq |U_j(0)|_0 + \int_0^t |\Delta_j F(t')|_0\, dt' + C 2^{-js} \int_0^t a_s(t') c_j(t') |U(t')|_{B_{2,r}^s}\, dt'.$$

Next, we multiply both sides by 2^{qs} and take the ℓ^r norm to obtain

$$|U(t)|_{B_{2,r}^s} \leq |U_0|_{B_{2,r}^s} + \int_0^t |F(\tau)|_{B_{2,r}^s} \, d\tau + C \int_0^t a_s(\tau)|U(\tau)|_{B_{2,r}^s} \, d\tau.$$

Applying the Gronwall lemma then leads to the inequality (4.21).

In order to prove the existence of a solution of (LS_0) under the assumption of Theorem 4.15, we can use exactly the same Friedrichs method as on page 174: We consider the ordinary differential equation

$$\begin{cases} \partial_t U^n + \sum_{k=1}^d \mathbb{E}_n\big(A_k \, \partial_k U^n\big) = \mathbb{E}_n \, F \\ U^n_{|t=0} = \mathbb{E}_n \, U_0, \end{cases}$$

which admits a unique solution U^n in $\mathcal{C}^1(I; L_n^2)$, thus in $\mathcal{C}^1(I; B_{2,r}^\sigma)$ for any $r \in [1, \infty]$ and $\sigma \in \mathbb{R}$, owing to the spectral localization. As $\mathbb{E}_n^2 = \mathbb{E}_n$ and $\mathbb{E}_n U^n = U^n$, the above estimates remain unchanged, so (4.21) is satisfied.

Mimicking the proof of Theorem 4.4, it is now easy to complete the proof of existence. Note, however, that in the case $r = \infty$, the sequence $(\mathbb{E}_n U_0)_{n \in \mathbb{N}}$ does not converge to U_0 in $B_{2,\infty}^s$, so time continuity does not hold up to index s.

Finally, if $s > 1$, then uniqueness is a consequence of Lemma 4.5. In the case where $0 < s \leq 1$, we still have $U \in \mathcal{C}(I; L^2)$, and the functions A_k are continuous with bounded first order space derivatives. Hence, Proposition 4.10 yields uniqueness.   □

4.3 The Resolution of Quasilinear Symmetric Systems

The purpose of this section is to prove local well-posedness for the following quasilinear symmetric system:

$$(S) : \begin{cases} \partial_t U + \sum_{k=1}^d A_k(U)\partial_k U = 0 \\ U_{|t=0} = U_0. \end{cases}$$

For the sake of simplicity, we do not consider any 0-order term or source term in the system. Further, we assume that the functions A_k are of the type

$$A_k(U) = A_k^{(0)} + \sum_{\ell=1}^N A_k^\ell U^\ell$$

for some constant real matrices $A_k^{(0)}$ and A_k^ℓ $(1 \leq k \leq d$ and $1 \leq \ell \leq N)$.

We aim to prove the following statement.

Theorem 4.16. *Let U_0 belong to H^s for some $s > d/2+1$. There then exists a positive time T such that a unique solution U of (S) exists in*

$$\mathcal{C}([0,T]; H^s) \cap \mathcal{C}^1([0,T]; H^{s-1}).$$

Moreover, T can be bounded from below by $c\|U_0\|_{H^s}^{-1}$, where c depends only on the family $A = (A_k)_{1 \le k \le d}$. Finally, the maximal time of existence T^* of such a solution does not depend on s and satisfies

$$T^* < \infty \implies \int_0^{T^*} \|\nabla U(t, \cdot)\|_{L^\infty} \, dt = \infty.$$

Remark 4.17. Note that, due to Sobolev embedding (see Theorem 1.50), the solution U is C^1 and therefore it is a solution of (S) in the classical sense.

Remark 4.18. The above blow-up criterion implies that the maximum time of existence does not depend on s.

Indeed, let U_0 be in H^s for some $s > 1 + d/2$ and consider some s' in $]1 + d/2, s[$. Denote by U_s (resp., $U_{s'}$) the corresponding maximal H^s (resp., $H^{s'}$) solution given by the above theorem. Denote by T_s^* (resp., $T_{s'}^*$) the lifespan of U_s (resp., $U_{s'}$). Because $H^s \subset H^{s'}$, uniqueness entails that $T_s^* \le T_{s'}^*$ and that $U_s \equiv U_{s'}$ on $[0, T_s^*[$. Now, if $T_s^* < T_{s'}^*$, then we must have $U_{s'}$ in $\mathcal{C}([0, T_s^*]; H^{s'})$ so that, due to Sobolev embedding, $\nabla U_{s'} \in L^1([0, T_s^*]; L^\infty)$. This stands in contradiction to the above blow-up criterion. Hence, $T_s^* = T_{s'}^*$.

Proof of Theorem 4.16.

To prove existence, we shall use the following iterative scheme: Consider the sequence $(U^n)_{n \in \mathbb{N}}$ defined by $U^0 = 0$ and

$$\begin{cases} \partial_t U^{n+1} + \sum_{k=1}^d A_k(U^n)\partial_k U^{n+1} = 0 \\ \qquad\qquad\qquad U^{n+1}_{|t=0} = S_{n+1}U_0. \end{cases}$$

Theorem 4.4 ensures that this sequence is well defined and that U^n belongs to $\mathcal{C}^1(\mathbb{R}; H^s)$ for any s. The proof of Theorem 4.16 proceeds in three steps:

- First, we prove that for T sufficiently small, the sequence $(U^n)_{n \in \mathbb{N}}$ is bounded in $L^\infty([0,T]; H^s)$.
- Second, we establish that for T sufficiently small, $(U^n)_{n \in \mathbb{N}}$ is a Cauchy sequence in $L^\infty([0,T]; H^{s'})$ for any $s' < s$.
- Finally, we check that the limit of this sequence is a solution of (S) and that it belongs to $\mathcal{C}([0,T]; H^s) \cap \mathcal{C}^1([0,T]; H^{s-1})$.

As we shall see, the proof relies on Littlewood–Paley theory and paradifferential calculus.

4.3.1 Paralinearization and Energy Estimates

We aim to prove uniform estimates in H^s for the approximate solution U^n. We claim that some constant C_0 can be found such that

$$C_0 T\|U_0\|_{H^s} < 1 \Longrightarrow \forall n \in \mathbb{N},\ \|U^n\|_{L^\infty([0,T];H^s)} \le \sqrt{2}\,\|U_0\|_{H^s}. \qquad (4.24)$$

We shall proceed by induction. The above assertion is of course true for $n = 0$. We assume that it is satisfied for some n. In order to bound U^{n+1}, we shall perform a paralinearization of the system satisfied by U^{n+1}, according to Lemma 4.14. For all $j \ge -1$, we get

$$\partial_t \Delta_j U^{n+1} + \sum_{k=1}^d (\overline{S}_{j-1} A_k(U^n)) \partial_k \Delta_j U^{n+1} = R_j^n$$

for some remainder term R_j^n satisfying, for all $t \in I$,

$$\|R_j^n(t)\|_{L^2} \le C c_j(t) 2^{-js} \Big(\|\nabla U^n(t)\|_{L^\infty} \|\nabla U^{n+1}(t)\|_{H^{s-1}}$$
$$+ \|\nabla U^{n+1}(t)\|_{L^\infty} \|\nabla U^n(t)\|_{H^{s-1}} \Big) \quad \text{with } \|(c_j^n(t))\|_{\ell^2} \le 1.$$

The L^2 energy estimate (4.2) and the fact that

$$\left\|\nabla \overline{S}_{j-1} A_k(U^n)\right\|_{L^\infty} \le C \|\nabla U^n\|_{L^\infty}$$

together imply that

$$\frac{1}{2}\frac{d}{dt}\|U_j^{n+1}\|_{L^2}^2 \le C\|\nabla U^n\|_{L^\infty}\|U_j^{n+1}\|_{L^2}^2 + C\|R_j^n\|_{L^2}\|U_j^{n+1}\|_{L^2}.$$

As $s - 1 > d/2$, the space H^{s-1} is continuously embedded in L^∞. Hence, thanks to the induction hypothesis, for any $t \in [0,T]$, we get

$$\frac{d}{dt}\|U_j^{n+1}\|_{L^2}^2 \le C\|U_0\|_{H^s}\|U_j^{n+1}\|_{L^2}\Big(\|U_j^{n+1}\|_{L^2} + c_j 2^{-js}\|U^{n+1}\|_{H^s}\Big).$$

By definition of the Sobolev norm, we thus get

$$\frac{d}{dt}\|U_j^{n+1}\|_{L^2}^2 \le C\|U_0\|_{H^s} c_j^2 2^{-2js}\|U^{n+1}\|_{H^s}^2.$$

By time integration, we obtain that

$$\|U_j^{n+1}\|_{L_T^\infty(L^2)}^2 \le \|\Delta_j U_0\|_{L^2}^2 + C\|U_0\|_{H^s}\|U^{n+1}\|_{L_T^\infty(H^s)}^2 2^{-2js}\int_0^T c_j^2(t)\,dt.$$

Recall that for any t, we have $\sum_j c_j^2(t) = 1$. Multiplying by 2^{2js} and taking the sum over j thus gives

$$\sum_j 2^{2js} \|U_j^{n+1}\|_{L_T^\infty(L^2)}^2 \leq \|U_0\|_{H^s}^2 + C\|U_0\|_{H^s} T\|U^{n+1}\|_{L_T^\infty(H^s)}^2. \tag{4.25}$$

Now, by virtue of Minkowski's inequality, we have

$$\|U^{n+1}\|_{L_T^\infty(H^s)}^2 \leq \sum_j 2^{2js} \|U_j^{n+1}\|_{L_T^\infty(L^2)}^2$$

so that choosing $C_0 \geq 2C$, where C is the constant that appears in the above inequality, we get that

$$\|U^{n+1}\|_{L_T^\infty(H^s)}^2 \leq 2\|U_0\|_{H^s}^2. \tag{4.26}$$

This is the conclusion of the first step of the proof.

Remark 4.19. We should point out that we have proven slightly more than what was originally suggested. In fact, plugging (4.26) into (4.25) gives

$$\sum_j 2^{2js} \|U_j^{n+1}\|_{L_T^\infty(L^2)}^2 \leq 2\|U_0\|_{H^s}^2. \tag{4.27}$$

This will be the key to proving the continuity of the solution with values in H^s.

4.3.2 Convergence of the Scheme

We first prove that $(U^n)_{n\in\mathbb{N}}$ is a Cauchy sequence in $L^\infty(([0,T]; L^2)$. We have

$$\partial_t(U^{n+1} - U^n) + \sum_{k=1}^d A_k(U^n)\partial_k(U^{n+1} - U^n)$$

$$= -\sum_{k=1}^d \Big(A_k(U^n) - A_k(U^{n-1})\Big)\partial_k U^n.$$

Using the energy estimate (4.2), we then get, for any $\varepsilon > 0$,

$$\frac{d}{dt}\big(\|U^{n+1} - U^n\|_{L^2}^2 + \varepsilon^2\big) \leq C\|\nabla U^n\|_{L^\infty}\|U^{n+1} - U^n\|_{L^2}$$

$$\times \Big(\|U^{n+1} - U^n\|_{L^2} + \|U^n - U^{n-1}\|_{L^2}\Big).$$

Define $v_n \overset{\text{def}}{=} \|U^n - U^{n-1}\|_{L_T^\infty(L^2)}$. From the above inequality and the fact that for any positive x and any positive ε, we have $x \leq (x^2 + \varepsilon^2)^{\frac{1}{2}}$, we deduce that for all $t \in [0, T]$,

$$\frac{d}{dt}\big(\|(U^{n+1} - U^n)(t)\|_{L^2}^2 + \varepsilon^2\big)^{\frac{1}{2}} \leq C\|\nabla U^n(t)\|_{L^\infty}(v_{n+1} + v_n).$$

Integrating and using the estimate (4.26) together with the Sobolev embedding $H^{s-1} \hookrightarrow L^\infty$ gives

$$(v_{n+1}^2 + \varepsilon^2)^{\frac{1}{2}} \leq (\|\Delta_n U_0\|_{L^2}^2 + \varepsilon^2)^{\frac{1}{2}} + C\|U_0\|_{H^s} T(v_{n+1} + v_n).$$

Passing to the limit when ε tends to 0 gives

$$v_{n+1} \leq \|\Delta_n U_0\|_{L^2} + C\|U_0\|_{H^s} T(v_{n+1} + v_n).$$

Assuming that $4CT\|U_0\|_{H^s} \leq 1$, we then have

$$v_{n+1} \leq \frac{4}{3}\|\Delta_n U_0\|_{L^2} + \frac{1}{3}v_n.$$

As $\|\Delta_n U_0\|_{L^2} \leq C2^{-ns}$, the series $\sum v_n$ converges. Hence, $(U^n)_{n\in\mathbb{N}}$ is a Cauchy sequence in $L^\infty(([0,T]; L^2)$.

Now, using Proposition 1.52 page 38 and (4.26), we get, for any s' in $[0, s[$,

$$\|U^{n+p} - U^n\|_{L_T^\infty(H^{s'})} \leq C\|U^{n+p} - U^n\|_{L_T^\infty(L^2)}^{1-\frac{s'}{s}d} \|U_0\|_{H^s}^{\frac{s'}{s}},$$

and hence convergence also holds true in $L^\infty([0, T]; H^{s'})$. Therefore, as the product continuously maps $H^{s'} \times H^{s'-1}$ into $H^{s'-1}$ when s' is greater than $d/2$, we may pass to the limit in (S). In addition, from the weak compactness properties of Sobolev spaces and the fact that the sequence $(U^n)_{n\in\mathbb{N}}$ is bounded in $L^\infty([0, T]; H^s)$, we deduce that U belongs to $L^\infty([0, T]; H^s)$.

4.3.3 Completion of the Proof of Existence

To summarize, the whole existence part of Theorem 4.16 is now proved, except for the fact that U is continuous in time with values in H^s. This may be achieved by passing to the limit in (4.27). However, we shall proceed slightly differently. In fact, we shall instead state a new estimate for the solution which will be most useful for proving the continuation criterion.

We therefore consider a solution U of (S) belonging to

$$L^\infty([0, T]; H^s) \cap \mathcal{C}([0, T]; H^1) \cap \mathcal{C}^1([0, T]; L^2).$$

By Lemma 4.14, $\Delta_j U$ satisfies

$$\begin{cases} \partial_t \Delta_j U + \sum_{k=1}^d (\overline{S}_{j-1} A_k(U)) \partial_k \Delta_j U = R_j \\ \\ \Delta_j U_{|t=0} = \Delta_j U_0 \end{cases}$$

with

$$\|R_j(t)\|_{L^2} \leq Cc_j(t)2^{-js}\|\nabla U(t)\|_{L^\infty}\|U(t)\|_{H^s}.$$

By an L^2 energy estimate and time integration, this leads to

$$\|\Delta_j U(t)\|_{L^2}^2 \le \|\Delta_j U_0\|_{L^2}^2 + C2^{-2js} \int_0^t c_j^2(t') \|\nabla U(t')\|_{L^\infty} \|U(t')\|_{H^s}^2 \, dt'.$$

After multiplication by 2^{2js} and summation in j, we find that for all $t \in [0, T]$,

$$\sum_j 2^{2js} \|\Delta_j U\|_{L_t^\infty(L^2)}^2 \le \|U_0\|_{H^s}^2 + C \int_0^t \|\nabla U\|_{L^\infty} \|U\|_{H^s}^2 \, dt'. \qquad (4.28)$$

Minkowski's inequality and the Gronwall lemma then finally imply that

$$\|U\|_{L_t^\infty(H^s)}^2 \le \sum_j 2^{2js} \|\Delta_j U\|_{L_t^\infty(L^2)}^2 \le \|U_0\|_{H^s}^2 \exp\left(C \int_0^t \|\nabla U\|_{L^\infty} \, dt'\right). \qquad (4.29)$$

Because H^{s-1} is continuously embedded in L^∞ and $U \in L^\infty([0, T]; H^s)$, we can thus conclude that

$$\sum_j 2^{2js} \|\Delta_j U\|_{L_T^\infty(L^2)}^2 < \infty.$$

We now consider any positive ε. The above inequality implies that an integer j_0 exists such that

$$\sum_{j \ge j_0} 2^{2js} \|\Delta_j U\|_{L_T^\infty(L^2)}^2 \le \frac{\varepsilon^2}{4}.$$

Thus, we have

$$\|U(t) - U(t')\|_{H^s}^2 \le \sum_{j < j_0} 2^{2js} \|\Delta_j(U(t) - U(t'))\|_{L^2}^2$$

$$+ 2 \sum_{j \ge j_0} 2^{2js} \|\Delta_j U\|_{L_T^\infty(L^2)}^2$$

$$\le \sum_{j < j_0} 2^{2js} \|\Delta_j(U(t) - U(t'))\|_{L^2}^2 + \frac{\varepsilon^2}{2}$$

$$\le C 2^{2j_0 s} \|U(t) - U(t')\|_{L^2}^2 + \frac{\varepsilon^2}{2}.$$

As U is in $\mathcal{C}([0, T]; L^2)$, we can now conclude that $U \in \mathcal{C}([0, T]; H^s)$.

4.3.4 Uniqueness and Continuation Criterion

The uniqueness is an obvious consequence of the following proposition.

Proposition 4.20. *Let U and V be two solutions of (S) in the space*

$$\mathcal{C}([0, T]; H^1) \cap \mathcal{C}^1(([0, T]; L^2)$$

with continuous and bounded gradients on $[0, T] \times \mathbb{R}^d$. We then have

$$\|U(t) - V(t)\|_{L^2} \le \|U_0 - V_0\|_{L^2} \exp\left(C \int_0^t (\|\nabla U(t')\|_{L^\infty} + \|\nabla V(t')\|_{L^\infty}) \, dt'\right).$$

Proof. We have

$$\partial_t(U - V) + \sum_{k=1}^{d} A_k(U)\partial_k(U - V) = \sum_{k=1}^{d} A_k(V - U)\partial_k V.$$

Using (4.3), which is valid under the assumptions of the proposition, we get the result.   □

In order to prove the blow-up condition, we first observe that, according to (4.24), the maximal time of existence $T^\star$ satisfies

$$T^\star \geq \frac{c}{\|U_0\|_{H^s}}.$$

Let $\widetilde{U}$ be the solution of the Cauchy problem for (S) with data $U(t)$ at time t. By virtue of uniqueness, we must have $\widetilde{U}(\tau) = U(t + \tau)$ for $0 \leq t + \tau < T^\star$ so that the maximal time of existence for $\widetilde{U}$ is $T^\star - t$. Thus, we have

$$T^\star - t \geq \frac{c}{\|U(t)\|_{H^s}},$$

which can be written

$$\|U(t)\|_{H^s} \geq \frac{C}{(T^\star - t)}. \tag{4.30}$$

This implies that $\|U(t)\|_{H^s}$ does not remain bounded when t tends to $T^\star$.

Now, if $\nabla U \in L^1([0, T^\star[; L^\infty)$, then the inequality (4.29) obviously implies that U is in $L^\infty(0, T^\star[; H^s)$. Combining this with the inequality (4.30) completes the proof of the whole of Theorem 4.16.   □

4.4 Data with Critical Regularity and Blow-up Criteria

In this section we give a generalization and refinements of Theorem 4.16. This involves two directions: First, we consider more general spaces for the initial data, and second, we give a refined blow-up criterion.

4.4.1 Critical Besov Regularity

The following theorem can be understood as a borderline case for well-posedness.

Theorem 4.21. *Let U_0 be in $B_{2,1}^{\frac{d}{2}+1}$. Then, (S) has a unique maximal solution U in $\mathcal{C}([0, T^\star[; B_{2,1}^{\frac{d}{2}+1}) \cap \mathcal{C}^1([0, T^\star[; B_{2,1}^{\frac{d}{2}})$. Moreover, there exists a positive constant c, depending only on the functions A_k, such that*

$$T^\star \geq \frac{c}{\|U_0\|_{B_{2,1}^{\frac{d}{2}+1}}}.$$

Finally, if $T^\star$ is finite, then

$$\int_0^{T^\star} \|\nabla U(t)\|_{L^\infty} \, dt = \infty.$$

Proof. The first step is to prove an a priori estimate in $L^\infty([0,T]; B_{2,1}^{\frac{d}{2}+1})$ of any solution given by Theorem 4.16. To achieve this, we paralinearize the system (S). Let U be a suitably smooth solution of (S) defined on some time interval $[0, T^\star[$ and define $U_j \overset{\text{def}}{=} \Delta_j U$. We have

$$\partial_t U_j + \sum_{k=1}^d (\overline{S}_{j-1} A_k(U)) \, \partial_k U_j = R_j \quad \text{for all} \quad j \geq -1$$

with, according to Lemma 4.14,

$$\|R_j\|_{L^2} \leq C c_j 2^{-j(1+\frac{d}{2})} \|\nabla U\|_{L^\infty} \|\nabla U\|_{B_{2,1}^{\frac{d}{2}}}.$$

Throughout this proof, we agree that $\|(c_j)\|_{\ell^1} = 1$.

Next, applying the usual energy method to the above paralinearized system yields, for any time t in $[0, T^\star[$,

$$\frac{d}{dt} \|U_j\|_{L^2}^2 \leq C 2^{-j(\frac{d}{2}+1)} c_j \|\nabla U\|_{L^\infty} \|U_j\|_{L^2} \|\nabla U\|_{B_{2,1}^{\frac{d}{2}}}.$$

Let ε be a positive number. From the previous inequality, we infer that

$$\frac{d}{dt} \left(\|U_j\|_{L^2}^2 + \varepsilon \right)^{\frac{1}{2}} \leq C 2^{-j(\frac{d}{2}+1)} c_j \|\nabla U\|_{L^\infty} \|\nabla U\|_{B_{2,1}^{\frac{d}{2}}}.$$

A time integration yields

$$\left(\|U_j(t)\|_{L^2}^2 + \varepsilon \right)^{\frac{1}{2}} \leq \left(\|\Delta_j U_0\|_{L^2}^2 + \varepsilon \right)^{\frac{1}{2}}$$
$$+ C 2^{-j(\frac{d}{2}+1)} \int_0^t c_j(t') \|\nabla U(t')\|_{L^\infty} \|\nabla U(t')\|_{B_{2,1}^{\frac{d}{2}}} \, dt'.$$

Taking the limit when ε tends to 0 and then summing over j, we get that

$$\|U\|_{L_T^\infty(B_{2,1}^{\frac{d}{2}+1})} \leq \sum_j 2^{j(\frac{d}{2}+1)} \|U_j\|_{L_T^\infty(L^2)}$$

$$\leq \|U_0\|_{B_{2,1}^{\frac{d}{2}+1}} + C \int_0^T \|\nabla U(t)\|_{L^\infty} \|\nabla U(t)\|_{B_{2,1}^{\frac{d}{2}}} \, dt. \quad (4.31)$$

Using the Gronwall lemma, we get that

$$\|U(t)\|_{B_{2,1}^{\frac{d}{2}+1}} \le \|U_0\|_{B_{2,1}^{\frac{d}{2}+1}} \exp\left(C \int_0^t \|\nabla U(t')\|_{L^\infty}\, dt'\right). \qquad (4.32)$$

From (4.31) and the fact that the space $B_{2,1}^{\frac{d}{2}}$ is continuously included in L^∞, we infer that

$$\|U(t)\|_{B_{2,1}^{\frac{d}{2}+1}} \le \|U_0\|_{B_{2,1}^{\frac{d}{2}+1}} \exp\left(C \int_0^t \|U(t')\|_{B_{2,1}^{\frac{d}{2}+1}}\, dt'\right).$$

Therefore, if

$$T < \min\{T^\star, T_0\} \quad \text{with} \quad T_0 \overset{\text{def}}{=} \frac{1}{2C\|U_0\|_{B_{2,1}^{\frac{d}{2}+1}}},$$

then

$$\|U\|_{L_T^\infty(B_{2,1}^{\frac{d}{2}+1})} \le 2\|U_0\|_{B_{2,1}^{\frac{d}{2}+1}}. \qquad (4.33)$$

Because $B_{2,1}^{\frac{d}{2}} \hookrightarrow L^\infty$, the blow-up condition of Theorem 4.16 thus implies that

$$T^\star \ge T_0 = \frac{1}{2C\|U_0\|_{B_{2,1}^{\frac{d}{2}+1}}}. \qquad (4.34)$$

We now consider the sequence $(U^n)_{n\in\mathbb{N}}$ of solutions to (S) with the initial data $S_n U_0$ for some fixed U_0 in the nonhomogeneous Besov space $B_{2,1}^{\frac{d}{2}+1}$. Using (4.34), we see that the lifespan of U^n is bounded from below by T_0. Therefore, according to Proposition 4.20, for any time $t \le T_0$, we have

$$\|(U^n - U^m)(t)\|_{L^2} \le \|S_n U_0 - S_m U_0\|_{L^2}$$
$$\times \exp\left(C \int_0^t (\|\nabla U^n(t')\|_{L^\infty} + \|\nabla U^m(t')\|_{L^\infty})\, dt'\right).$$

By the inequality (4.33) and thanks to the fact that $B_{2,1}^{\frac{d}{2}} \hookrightarrow L^\infty$, we get

$$\int_0^t (\|\nabla U^n(t')\|_{L^\infty} + \|\nabla U^m(t')\|_{L^\infty})\, dt' \le Ct\|U_0\|_{B_{2,1}^{\frac{d}{2}+1}}.$$

Thus, $(U^n)_{n\in\mathbb{N}}$ is a Cauchy sequence in $L^\infty([0,T_0]; L^2)$ and, by interpolation (see Theorem 2.80 page 102), in $L^\infty([0,T_0]; B_{2,1}^{s'})$ for any $s' < d/2 + 1$. The limit U of $(U^n)_{n\in\mathbb{N}}$ is obviously a solution of (S). Using the Fatou property for the Besov space $B_{2,1}^{\frac{d}{2}+1}$ (see Theorem 2.72), we conclude that U belongs to

$$L^\infty([0,T_0]; B_{2,1}^{\frac{d}{2}+1}) \cap \mathcal{C}([0,T_0]; B_{2,1}^{s'}) \cap \mathcal{C}^1([0,T_0]; B_{2,1}^{s'-1}) \quad \text{for any} \ \ s' < d/2 + 1.$$

In order to prove that U belongs to $\mathcal{C}([0,T_0]; B_{2,1}^{\frac{d}{2}+1})$, we pass to the limit in the inequality (4.31) (for $U^{(n)}$), thereby obtaining

$$\sum_{j\geq -1} 2^{j(\frac{d}{2}+1)}\|U_j\|_{L^\infty_{T_0}(L^2)} \leq 2\|U_0\|_{B^{\frac{d}{2}+1}_{2,1}}.$$

We can now conclude as in the Sobolev case: consider a positive ε, then, owing to the above inequality, we can find some integer j_0 such that

$$\sum_{j\geq j_0} 2^{j(\frac{d}{2}+1)}\|\Delta_j U\|_{L^\infty_{T_0}(L^2)} \leq \frac{\varepsilon}{4}.$$

Therefore, we have

$$\|U(t)-U(t')\|_{B^{\frac{d}{2}+1}_{2,1}} \leq \sum_{j<j_0} 2^{j(\frac{d}{2}+1)}\|\Delta_j(U(t)-U(t'))\|_{L^2}$$

$$+ 2\sum_{j\geq j_0} 2^{j(\frac{d}{2}+1)}\|\Delta_j U\|_{L^\infty_{T_0}(L^2)}$$

$$\leq C2^{j_0(\frac{d}{2}+1)}\|U(t)-U(t')\|_{L^2} + \frac{\varepsilon}{2}.$$

Because U is in $\mathcal{C}([0,T_0];L^2)$, the first term on the right-hand side tends to 0 when t' goes to t. This implies that U is continuous in time with values in $B^{\frac{d}{2}+1}_{2,1}$. □

4.4.2 A Refined Blow-up Condition

Here, we prove a more accurate blow-up condition than the (classical) one given in Theorem 4.16. We are going to substitute for the Lipschitz norm any norm associated with an admissible Osgood modulus of continuity (see Definition 2.108 page 117 and Definition 3.1 page 124).

Theorem 4.22. *Let* $s > d/2 + 1$ *and* U *be a maximal solution of* (S) *in* $\mathcal{C}([0,T^\star[;H^s)$. *If* $T^\star$ *is finite, then for any admissible Osgood modulus of continuity, we have*

$$\int_0^{T^\star} \|U(t)\|_{C_\mu}\, dt = \infty.$$

Proof. In order to prove this theorem, we define the C^1 nondecreasing function R_s as follows:

$$R_s(t) \overset{\text{def}}{=} \left(\|U_0\|^2_{H^s} + C\int_0^t \|\nabla U(t')\|_{L^\infty}\|U(t')\|^2_{H^s}\, dt'\right)^{\frac{1}{2}}.$$

Note that the inequality (4.28) guarantees that if the constant C has been chosen sufficiently large, then we have $R_s(t) \geq \sup_{0\leq t'\leq t}\|U(t')\|_{H^s}$. Therefore,

$$R_s(t) \leq \|U_0\|_{H^s} + C\int_0^t \|\nabla U(t')\|_{L^\infty} R_s(t')\, dt'.$$

Let $\varepsilon = \min\left(1, s - \frac{d}{2} - 1\right)$ and Γ be the function associated with the modulus of continuity $\mu : [0, a] \to \mathbb{R}^+$ introduced in Definition 2.108 page 117. Using Proposition 2.112 page 119 with $\Lambda = \|U_0\|_{H^s}$, we see that for some constant depending only on ε and on a, we have

$$\|\nabla U\|_{L^\infty} \leq C\big(\|U\|_{C_\mu} + \|U_0\|_{H^s}\big)\left(1 + \Gamma\left(\left(\frac{\|\nabla U\|_{C^{0,\varepsilon}}}{\|U\|_{C_\mu} + \|U_0\|_{H^s}}\right)^{\frac{1}{\varepsilon}}\right)\right)$$

whenever the argument of Γ is greater than or equal to $1/a$.

Note that if this latter condition is not satisfied, then the above inequality is trivially satisfied (if we agree that Γ is continued by the constant function $\Gamma(1/a)$ on the interval $[0, 1/a]$). Therefore, taking advantage of the continuous embedding of H^s in $C^{0,\varepsilon}$ and of the fact that Γ is nondecreasing (which, in particular, enables us to drop the exponent $1/\varepsilon$), we infer that

$$R_s(t) \leq \|U_0\|_{H^s} + C \int_0^t \gamma(t')\left(1 + \Gamma\left(\left(\frac{CR_s(t')}{\|U_0\|_{H^s}}\right)^{\frac{1}{\varepsilon}}\right)\right) R_s(t')\, dt'$$

with $\gamma(t) \overset{\text{def}}{=} \|U(t)\|_{C_\mu} + \|U_0\|_{H^s}$.

Thus, defining $\rho_s(t) \overset{\text{def}}{=} \dfrac{CR_s(t)}{\|U_0\|_{H^s}}$ and $\Gamma_\varepsilon(y) \overset{\text{def}}{=} \Gamma\big(y^{\frac{1}{\varepsilon}}\big)$, we get

$$\rho_s(t) \leq C\left(1 + \int_0^t \gamma(t')\big(1 + \Gamma_\varepsilon(\rho_s(t'))\big)\rho_s(t')\, dt'\right).$$

Let $a_\varepsilon \overset{\text{def}}{=} a^{\frac{1}{\varepsilon}}$. Given that μ is an Osgood modulus of continuity, it is easy to check that the function $\mathcal{G}_\varepsilon(y) \overset{\text{def}}{=} \displaystyle\int_{a_\varepsilon^{-1}}^y \frac{dy'}{y'\Gamma_\varepsilon(y')}$ maps $[a_\varepsilon^{-1}, +\infty[$ onto and one-to-one $[0, +\infty[$. Therefore, arguing as in Lemma 3.8 page 128, we infer that[2]

$$\rho_s(t) \leq \mathcal{G}_\varepsilon^{-1}\left(C + C \int_0^t \gamma(t')\, dt'\right).$$

By the definition of ρ_s, this implies that

$$\|U(t)\|_{H^s} \leq \frac{1}{C}\|U_0\|_{H^s}\mathcal{G}_\varepsilon^{-1}\left(C + C \int_0^t \gamma(t')\, dt'\right).$$

This means that if the solution U belongs to $\mathcal{C}([0, T[; H^s) \cap L^1([0, T[; C_\mu)$ for some finite T, then $\|U(t)\|_{H^s}$ stays bounded on $[0, T[$. Thus, the inequality (4.30) ensures that Theorem 4.22 holds. $\qquad\square$

---

[2] Note that we can assume with no loss of generality that $C \geq 1/a_\varepsilon$.

Example 4.23. Recall that the function $r \mapsto r(1 - \log r)$ is an Osgood modulus of continuity and that $B^1_{\infty,\infty}$ is embedded in the set LL of log-Lipschitz functions (see Proposition 2.107 page 116). Therefore, Theorem 4.22 implies that no blow-up may occur at time T unless

$$\int_0^T \|U(t)\|_{B^1_{\infty,\infty}}\, dt = \infty.$$

4.5 Continuity of the Flow Map

Let $s > 1 + d/2$. According to Theorem 4.16, for any data U_0 in H^s, the system (S) has a unique solution U on some nontrivial time interval $[0, T]$. Moreover, by taking advantage of the lower bound that we have stated for the lifespan of the solution of (S) and using the inequality (4.26), we can find some H^s-neighborhood $\mathcal{V}_{U_0}$ of U_0 and some positive constant K such that for any $V_0 \in \mathcal{V}_{U_0}$, the system (S) with data V_0 has a solution V in $\mathcal{C}([0, T]; H^s) \cap \mathcal{C}^1([0, T]; H^{s-1})$ which satisfies

$$\|V\|_{L^\infty_T(H^s)} + \|\partial_t V\|_{L^\infty_T(H^{s-1})} \le K. \tag{4.35}$$

In the present section, we address the question of continuity of the flow map

$$\Phi : \begin{cases} \mathcal{V}_{U_0} \longrightarrow \mathcal{C}([0, T]; H^s) \cap \mathcal{C}^1([0, T]; H^{s-1}) \\ V_0 \longmapsto V. \end{cases}$$

To begin, we observe that by combining the inequality (4.35) with the stability result stated in Proposition 4.20, we can deduce that the flow map Φ is continuous on $\mathcal{V}_{U_0}$, in the sense of the norm $L^\infty([0, T]; L^2)$. Also, note that by interpolating with the H^s bound given by (4.35), we find that continuity holds for the $L^\infty([0, T]; H^{s'})$ norm whenever $s' < s$.

We claim that continuity holds true *up to index* s. In other words, the system (S) is locally well posed in the sense of Hadamard.

Theorem 4.24. *Let U_0 be any data in H^s with $s > 1 + d/2$. There exists a neighborhood $\mathcal{V}_{U_0}$ of U_0 and a positive time T such that the flow map Φ defined above is continuous.*

Remark 4.25. A similar result holds true in the critical Besov space $B^{1+d/2}_{2,1}$. To simplify the presentation, however, we shall focus on the Sobolev case.

The following stability result for *linear* symmetric systems is the cornerstone of the proof of Theorem 4.24.

Lemma 4.26. *Define $\overline{\mathbb{N}} \stackrel{def}{=} \mathbb{N} \cup \{\infty\}$. For k in $\{1, \ldots, d\}$, we consider a sequence $(\mathcal{A}^n_k)_{n \in \overline{\mathbb{N}}}$ of continuous bounded functions on $I \times \mathbb{R}$ with values in the set of symmetric $N \times N$ matrices. Assume, in addition, that there exists a* real number $s > 1 + d/2$ such that for all k in $\{1, \ldots, d\}$ and $n \in \overline{\mathbb{N}}$, the function $\nabla \mathcal{A}_k^n$ belongs to $\mathcal{C}(I; H^{s-1})$, that there exists a nonnegative integrable function α over I such that

$$\|\nabla \mathcal{A}_k^n(t)\|_{H^{s-1}} \leq \alpha(t) \quad \text{for all } t \in I, \ k \in \{1, \ldots, d\}, \ n \in \overline{\mathbb{N}}, \tag{4.36}$$

and that

$$\mathcal{A}_k^n - \mathcal{A}_k^\infty \xrightarrow{n \to \infty} 0 \quad \text{in } L^1(I; H^{s-1}). \tag{4.37}$$

Let $F \in \mathcal{C}(I; H^{s-1})$ and $V_0 \in H^{s-1}$. For $n \in \overline{\mathbb{N}}$, denote by V^n the solution of

$$\begin{cases} \partial_t V^n + \sum_k \mathcal{A}_k^n \partial_k V^n = F \\ V^n_{|t=0} = V_0. \end{cases}$$

The sequence $(V^n)_{n \in \mathbb{N}}$ then converges to V^∞ in $\mathcal{C}(I; H^{s-1})$.

Proof. We first consider the smooth case: $V_0 \in H^s$ and $F \in \mathcal{C}(I; H^s)$.

By virtue of Theorem 4.15 and the assumption (4.36), the sequence $(V^n)_{n \in \overline{\mathbb{N}}}$ is bounded in $\mathcal{C}(I; H^s)$. In order to prove that V^n tends to V^∞ in $\mathcal{C}(I; H^{s-1})$, we shall use the fact that

$$\partial_t (V^n - V^\infty) + \sum_k \mathcal{A}_k^n \partial_k (V^n - V^\infty) = \sum_k (\mathcal{A}_k^\infty - \mathcal{A}_k^n) \partial_k V^\infty.$$

Indeed, because $V^n(0) = V^\infty(0)$, Theorem 4.15 and the assumption (4.36) together yield

$$\|(V^n - V^\infty)(t)\|_{H^{s-1}} \leq \int_0^t e^{C \int_\tau^t \alpha(\tau') d\tau'} \|(\mathcal{A}_k^\infty - \mathcal{A}_k^n) \partial_k V^\infty\|_{H^{s-1}} \, d\tau.$$

Because $s - 1 > d/2$, the Sobolev space H^{s-1} is an algebra. Therefore,

$$\|(V^n - V^\infty)(t)\|_{H^{s-1}} \leq C \int_0^t e^{C \int_\tau^t \alpha(\tau') d\tau'} \|\mathcal{A}_k^\infty - \mathcal{A}_k^n\|_{H^{s-1}} \|\partial_k V^\infty\|_{H^{s-1}} \, d\tau.$$

Taking advantage of (4.37), it is now easy to conclude that V^n tends to V^∞ in $\mathcal{C}(I; H^{s-1})$.

Consider now the rough case $V_0 \in H^{s-1}$ and $F \in \mathcal{C}(I; H^{s-1})$. For all $n \in \overline{\mathbb{N}}$ and $j \in \mathbb{N}$, we introduce the solution V_j^n to

$$\begin{cases} \partial_t V_j^n + \sum_k \mathcal{A}_k^n \partial_k V_j^n = \mathbb{E}_j F \\ (V_j^n)_{|t=0} = \mathbb{E}_j V_0. \end{cases}$$

Since

$$\begin{cases} \partial_t (V^n - V_j^n) + \sum_k \mathcal{A}_k^n \partial_k (V^n - V_j^n) = F - \mathbb{E}_j F \\ V^n_{|t=0} = V_0 - \mathbb{E}_j V_0, \end{cases}$$

Theorem 4.15 and the assumption (4.36) guarantee that for all $t \in I$,

$$\|(V^n - V_j^n)(t)\|_{H^{s-1}} \leq e^{C \int_0^t \alpha(\tau) \, d\tau} \left(\|V_0 - \mathbb{E}_j V_0\|_{H^{s-1}} \right.$$
$$\left. + \int_0^t \|F - \mathbb{E}_j F\|_{H^{s-1}} \, d\tau \right). \quad (4.38)$$

We are now ready to prove that V^n tends to V^∞ in $\mathcal{C}(I; H^{s-1})$. Indeed, fix an arbitrary $\varepsilon > 0$ and write

$$\|V^n - V^\infty\|_{L^\infty(I; H^{s-1})} \leq \|V^n - V_j^n\|_{L^\infty(I; H^{s-1})}$$
$$+ \|V_j^n - V_j^\infty\|_{L^\infty(I; H^{s-1})} + \|V_j^\infty - V^\infty\|_{L^\infty(I; H^{s-1})}. \quad (4.39)$$

On the one hand, because $\mathbb{E}_j V_0$ tends to V_0 in H^{s-1} and $\mathbb{E}_j F$ tends to F in the space $\mathcal{C}(I; H^{s-1})$, we can, according to (4.38), find some $j \in \mathbb{N}$ such that

$$\|V^n - V_j^n\|_{L^\infty(I; H^{s-1})} \leq \varepsilon/3 \quad \text{for all} \quad n \in \overline{\mathbb{N}}.$$

On the other hand, since the data $\mathbb{E}_j V_0$ and $\mathbb{E}_j F$ are smooth, we can, according to the first part of the proof, find some integer n_0 such that the second term in the right-hand side of (4.39) is less than $\varepsilon/3$ for all $n \geq n_0$. This completes the proof of the lemma. $\square$

Proof of Theorem 4.24. In the introductory part of this section, we stated the existence of some H^s-neighborhood $\mathcal{V}_{U_0}$ of U_0 and some positive T such that for all $V_0 \in \mathcal{V}_{U_0}$, the system (S) has a unique H^s solution $\Phi(V_0)$ over $[0, T]$ which is bounded independently of V_0 and such that $\Phi(V_0)$ tends to $\Phi(U_0)$ in $\mathcal{C}(I; H^{s-1})$ with $I \overset{\text{def}}{=} [0, T]$.

We claim that convergence holds true in $\mathcal{C}(I; H^s)$. To prove this fact, consider a sequence of data U_0^n converging to $U_0^\infty \overset{\text{def}}{=} U_0$ in H^s. Of course, with no loss of generality, we can assume that all the terms of the sequence belong to $\mathcal{V}_{U_0}$. For $n \in \overline{\mathbb{N}}$, denote by U^n the solution of (S) with initial data U_0^n. Given that $U^n \to U^\infty$ in $\mathcal{C}(I; H^{s-1})$, it suffices to prove that, in addition, $V^n \overset{\text{def}}{=} \nabla U^n$ tends to $V^\infty \overset{\text{def}}{=} \nabla U^\infty$ in $\mathcal{C}(I; H^{s-1})$. This latter task may be achieved by splitting V_n into $W^n + Z^n$ with (W^n, Z^n) satisfying

$$\begin{cases} \partial_t W^n + \sum_k \mathcal{A}_k^n \partial_k W^n = F^\infty \\ W_{|t=0}^n = V_0^\infty \end{cases} \quad \text{and} \quad \begin{cases} \partial_t Z^n + \sum_k \mathcal{A}_k^n \partial_k Z^n = F^n - F^\infty \\ Z_{|t=0}^n = V_0^n - V_0^\infty \end{cases}$$

$$\text{with} \quad \mathcal{A}_k^n \overset{\text{def}}{=} A_k(U^n) \quad \text{and} \quad F^n \overset{\text{def}}{=} -\sum_k \nabla \mathcal{A}_k^n \partial_k U^n.$$

Because $(U^n)_{n \in \overline{\mathbb{N}}}$ is bounded in $\mathcal{C}(I; H^s)$, it is obvious that $(\nabla \mathcal{A}_k^n)_{n \in \overline{\mathbb{N}}}$ is bounded in $\mathcal{C}(I; H^{s-1})$. Further,

$$\mathcal{A}_k^n - \mathcal{A}_k^\infty = \sum_{j=1}^{N} \mathcal{A}_k^j (U^n - U^\infty)$$

and therefore, owing to the fact that $(U^n - U^\infty)$ goes to 0 in $\mathcal{C}(I; H^{s-1})$, the sequence $(\mathcal{A}_k^n - \mathcal{A}_k^\infty)_{n \in \mathbb{N}}$ converges to 0 in $\mathcal{C}(I; H^{s-1})$. Lemma 4.26 thus ensures that W^n tends to W^∞ (i.e., V^∞) in $\mathcal{C}(I; H^{s-1})$.

Next, according to Theorem 4.15, we have, for all $n \in \mathbb{N}$ and $t \in [0, T]$,

$$\|Z^n(t)\|_{H^{s-1}} \le e^{C \int_0^t \|\nabla \mathcal{A}_k^n\|_{H^{s-1}} \, d\tau} \left(\|V_0^n - V_0^\infty\|_{H^{s-1}} + \int_0^t \|F^n - F^\infty\|_{H^{s-1}} \, d\tau \right).$$

Using the definition of $\mathcal{A}_k^n$ and the fact that H^{s-1} is an algebra, we deduce that

$$\|F^n - F^\infty\|_{H^{s-1}} \le C \left(\|V^n\|_{H^{s-1}} + \|V^\infty\|_{H^{s-1}} \right) \|V^n - V^\infty\|_{H^{s-1}}$$
$$\le C \left(\|V^n\|_{H^{s-1}} + \|V^\infty\|_{H^{s-1}} \right) \left(\|Z^n\|_{H^{s-1}} + \|W^n - W^\infty\|_{H^{s-1}} \right).$$

Denoting by K a bound in $\mathcal{C}(I; H^{s-1})$ for $(\nabla \mathcal{A}_k^n)_{n \in \mathbb{N}}$, we thus get

$$\|Z^n(t)\|_{H^{s-1}} \le e^{CKt} \left(\|V_0^n - V_0^\infty\|_{H^{s-1}} + C \int_0^t \left(\|V^n\|_{H^{s-1}} + \|V^\infty\|_{H^{s-1}} \right) \right.$$
$$\left. \times \left(\|Z^n\|_{H^{s-1}} + \|W^n - W^\infty\|_{H^{s-1}} \, d\tau \right) \right).$$

Applying the Gronwall lemma and using the facts that

- $(V^n)_{n \in \overline{\mathbb{N}}}$ is bounded in $\mathcal{C}([0, T]; H^{s-1})$,
- V_0^n tends to V_0^∞ in H^{s-1},
- W^n goes to W^∞ in $\mathcal{C}([0, T]; H^{s-1})$,

it is now easy to conclude that Z^n tends to 0 in $\mathcal{C}([0, T]; H^{s-1})$.   □

4.6 References and Remarks

There are a number of references concerning the study of more general linear or quasilinear systems. Results related to the well-posedness theory in H^s and finite propagation speed for (LS), (QS), or more general systems may be found in the monographs by T. Kato [177], S. Alinhac and P. Gérard [11], L. Hörmander [168], D. Serre [262], or S. Benzoni-Gavage and D. Serre [33]. For results concerning the particular case of the compressible Euler system introduced at the end of Section 5.1, one may refer to e.g. [63, 261]. The concept of paralinearization was introduced by J.-M. Bony in his pioneering paper [39]. The standard blow-up criterion involving the $L^1([0, T[; \mathrm{Lip})$ norm of the solution is part of mathematical folklore.

The well-posedness for data with critical regularity was first stated by D. Iftimie in the Appendix of [172]. We mention in passing that a slightly more accurate

lower bound for the lifespan T^* of the solution to (S) may be proven, namely,
$$T \geq c\|\nabla U_0\|_{\dot{B}_{2,1}^{\frac{d}{2}}}^{-1} .$$

To the best of our knowledge, the fact that the $L^1([0,T[;\text{Lip})$ assumption in Theorem 4.16 may be replaced by a slightly weaker condition goes back to the pioneering paper [31] by J. Beale, T. Kato, and A. Majda for the incompressible Euler equations.

The continuity of the flow map up to index s belongs to the mathematical folklore. In Section 4.5 the method introduced by T. Kato in [177] (in the framework of *abstract quasilinear evolution equations*) has been applied. We should mention that an alternative method combining viscous regularization of the system and regularization of the data may be used (see, e.g., [38] for the KdV equation).

5

The Incompressible Navier–Stokes System

This chapter is devoted to the mathematical study of the Navier–Stokes system for incompressible fluids evolving in the whole space[1] $\mathbb{R}^d$, where $d = 2$ or 3. Denoting by $u \in \mathbb{R}^d$ the velocity field, by $P \in \mathbb{R}$ the pressure function, and by $\nu > 0$ the kinematic viscosity, the Cauchy problem for the incompressible Navier–Stokes system can be written as follows:

$$\begin{cases} \partial_t u + u \cdot \nabla u - \nu \Delta u = -\nabla P \\ \operatorname{div} u = 0 \\ u_{|t=0} = u_0, \end{cases}$$

where

$$\operatorname{div} u = \sum_{j=1}^{d} \partial_j u^j, \quad u \cdot \nabla = \sum_{j=1}^{d} u^j \partial_j, \quad \text{and} \quad \Delta = \sum_{j=1}^{d} \partial_j^2.$$

The first section of this chapter is devoted to the presentation of a few basic results concerning the Navier–Stokes system. There, we introduce the weak formulation of the system, state Leray's theorem, and prove a fixed point theorem which will be of constant use in the sections which follow.

In the second section, we solve a generalized Navier–Stokes system locally in time for general data in $\dot{H}^{\frac{d}{2}-1}$, or globally in time for small data in $\dot{H}^{\frac{d}{2}-1}$. In the third section, we present results which use the special structure of the nonlinearity in the Navier–Stokes system. First, we prove the uniqueness of finite energy solutions in dimension two. Next, in dimension three, we establish a result concerning the asymptotics of possible large global solutions. As a consequence, we show that the set of initial data which give rise to global solutions in $L_{loc}^4(\mathbb{R}^+; \dot{H}^1)$ is an open subset of $\dot{H}^{\frac{1}{2}}$.

In the fourth section, we prove local well-posedness for general data in $L^3(\mathbb{R}^3)$ and global well-posedness for small data. This result is a by-product of a more general result where Besov spaces embedded in $\dot{B}_{\infty,\infty}^{-1}$ arise naturally. The next section is devoted to the study of the well-posedness issue in

---

[1] This means that boundary effects are neglected.

H. Bahouri et al., *Fourier Analysis and Nonlinear Partial Differential Equations*, Grundlehren der mathematischen Wissenschaften 343, DOI 10.1007/978-3-642-16830-7_5, © Springer-Verlag Berlin Heidelberg 2011

the so-called endpoint space for the Picard scheme. There, we consider data which are scarcely better than $\dot{B}^{-1}_{\infty,\infty}$.

Up to this point, all results concerning the Navier–Stokes system are obtained by means of elementary methods: nothing more than the classical Sobolev embedding and Young's and Hölder's inequalities. The last section, however, is more demanding. There, we present a result concerning well-posedness in the context of Besov spaces which uses the smoothing effect of the heat flow described by the inequality (3.39) page 157. Next, we take advantage of that approach in order to study the problem of the existence of a flow for the velocity field in a scaling invariant framework.

5.1 Basic Facts Concerning the Navier–Stokes System

We begin by introducing the *weak formulation* of the Navier–Stokes system. From Leibniz's formula it is clear that when the vector field u is smooth and divergence-free, we have

$$u \cdot \nabla u = \operatorname{div}(u \otimes u), \quad \text{where} \quad \operatorname{div}(u \otimes u)^j \overset{\text{def}}{=} \sum_{k=1}^{d} \partial_k(u^j u^k) = \operatorname{div}(u^j u),$$

so that the Navier–Stokes system may be written as

$$(NS_\nu) \qquad \begin{cases} \partial_t u + \operatorname{div}(u \otimes u) - \nu \Delta u = -\nabla P \\ \operatorname{div} u = 0 \\ u_{|t=0} = u_0. \end{cases}$$

The advantage of this formulation is that it makes sense for more singular vector fields than the previous formulation, a fact which will be used extensively in what follows.

Based on this observation, we now define a *weak solution* of (NS). The following definition may be seen, in the nonlinear framework, as the analog of Definition 3.13 page 132.

Definition 5.1. *A time-dependent vector field u with components in the space $L^2_{loc}(0,T]\times\mathbb{R}^d)$ is a weak solution of (NS_ν) if, for any smooth, compactly supported, time-dependent, divergence-free vector field Ψ, we have*

$$\int_{\mathbb{R}^d} u(t,x) \cdot \Psi(t,x)\,dx = \int_0^t \int_{\mathbb{R}^d} \big(\nu u \cdot \Delta\Psi + u \otimes u : \nabla\Psi + u \cdot \partial_t \Psi\big)(t',x)\,dx\,dt'$$

$$+ \int_{\mathbb{R}^d} u_0(x) \cdot \Psi(0,x)\,dx. \tag{5.1}$$

We now formally[2] derive the well-known *energy estimate*. First, taking the $(L^2(\mathbb{R}^d))^d$ scalar product of the system with the solution u gives

---

[2] These computations will be made rigorous in the next sections.

$$\frac{1}{2}\frac{d}{dt}\|u\|_{L^2}^2 + (u \cdot \nabla u|u)_{L^2} - \nu(\Delta u|u)_{L^2} = -(\nabla P|u)_{L^2}.$$

Using formal integration by parts, we may write

$$
\begin{aligned}
(u \cdot \nabla u|u)_{L^2} &= \sum_{1 \le j,k \le d} \int_{\mathbb{R}^d} u^j (\partial_j u^k) u^k \, dx \\
&= \frac{1}{2} \sum_{1 \le j \le d} \int_{\mathbb{R}^d} u^j \partial_j (|u|^2) \, dx \\
&= -\frac{1}{2} \int_{\mathbb{R}^d} (\operatorname{div} u)|u|^2 \, dx \\
&= 0.
\end{aligned}
$$

Moreover, we obviously have

$$-\nu(\Delta u|u)_{L^2} = \nu\|\nabla u\|_{L^2}^2.$$

Again, (formal) integration by parts yield

$$
\begin{aligned}
-(\nabla P|u)_{L^2} &= -\sum_{j=1}^{d} \int_{\mathbb{R}^d} u^j \partial_j P \, dx \\
&= \int_{\mathbb{R}^d} P \operatorname{div} u \, dx \\
&= 0.
\end{aligned}
$$

It therefore turns out that

$$\frac{1}{2}\frac{d}{dt}\|u(t)\|_{L^2}^2 + \nu\|\nabla u(t)\|_{L^2}^2 = 0,$$

from which it follows, by time integration, that

$$\|u(t)\|_{L^2}^2 + 2\nu \int_0^t \|\nabla u(t')\|_{L^2}^2 \, dt' = \|u_0\|_{L^2}^2. \tag{5.2}$$

It follows that the natural assumption for the initial data u_0 is that it is square integrable and divergence-free. This leads to the following statement, first proven by J. Leray in 1934.

Theorem 5.2 (Leray). *Let u_0 be a divergence-free vector field in $L^2(\mathbb{R}^d)$. Then, (NS_ν) has a weak solution u in the energy space*

$$L^\infty(\mathbb{R}^+; L^2) \cap L^2(\mathbb{R}^+; \dot{H}^1)$$

such that the energy inequality holds, namely,

$$\|u(t)\|_{L^2}^2 + 2\nu \int_0^t \|\nabla u(t')\|_{L^2}^2 \, dt' \le \|u_0\|_{L^2}^2. \tag{5.3}$$

Remark 5.3. The Leray solutions satisfy the Navier–Stokes system in a stronger sense than that of Definition 5.1: For any smooth, compactly supported, time-dependent, divergence-free vector field Ψ, we have

$$
\int_{\mathbb{R}^d} u(t,x) \cdot \Psi(t,x) \, dx + \int_0^t \int_{\mathbb{R}^d} \Big(\nu \nabla u : \nabla \Psi - u \otimes u : \nabla \Psi - u \cdot \partial_t \Psi \Big)(t',x) \, dx \, dt'
$$
$$
= \int_{\mathbb{R}^d} u_0(x) \cdot \Psi(0,x) \, dx.
$$

Proving Leray's theorem relies on a compactness method analogous to that of the first section of Chapter 6:

- First, approximate solutions with compactly supported Fourier transforms satisfying (5.3) are built. This may be done by solving an appropriate sequence of ordinary differential equations in L^2-type spaces.
- Next, a time compactness result is derived.
- Finally, the solution is obtained by passing to the limit in the weak formulation.[3]

In dimension two, the Leray weak solutions are unique. More precisely, we have the following theorem, which we shall prove in Section 5.3.1.

Theorem 5.4. *If $d = 2$, then the solutions given by the above theorem are unique, continuous with values in $L^2(\mathbb{R}^2)$, and satisfy the energy equality*

$$
\|u(t)\|_{L^2}^2 + 2\nu \int_0^t \|\nabla u(t')\|_{L^2}^2 \, dt' = \|u_0\|_{L^2}^2.
$$

Another important feature of the Navier–Stokes system in the whole space $\mathbb{R}^d$ is that there is an explicit formula giving the pressure in terms of the velocity field. Indeed, in Fourier variables, the Leray projector $\mathbb{P}$ on divergence-free vector fields is as follows:

$$
\mathcal{F}(\mathbb{P} f)^j(\xi) = \widehat{f^j}(\xi) - \frac{1}{|\xi|^2} \sum_{k=1}^d \xi_j \xi_k \widehat{f^k}(\xi)
$$
$$
= \sum_{k=1}^d (\delta_{j,k} - 1) \frac{\xi_j \xi_k}{|\xi|^2} \widehat{f^k}(\xi), \tag{5.4}
$$

where $\delta_{jk} = 1$ if $j = k$ and 0 if $j \neq k$.

Therefore, applying the Leray projector to the Navier–Stokes system and denoting by Q_{NS} the bilinear operator defined by

$$
Q_{NS}(v,w) \overset{\text{def}}{=} -\frac{1}{2} \mathbb{P}\big(\mathrm{div}(v \otimes w) + (\mathrm{div}\, w \otimes v) \big)
$$

---

[3] For the proof of Theorem 5.2, the reader is referred to the magnificent original paper by J. Leray (see [207]). For a modern proof, see, for instance, [75] or [86].

yields

$$\begin{cases} \partial_t u - \nu \Delta u = Q_{NS}(u, u) \\ \quad u_{|t=0} = u_0. \end{cases}$$

Note that the divergence-free condition is satisfied by u whenever $\operatorname{div} u_0 = 0$. Hence, u satisfies the "original" system (NS_ν).

Throughout this chapter, we shall denote by Q any bilinear map of the form

$$Q^j(u, v) \overset{\text{def}}{=} \sum_{k,\ell,m} q_{k,\ell}^{j,m} \partial_m (u^k v^\ell),$$

where $q_{k,\ell}^{j,m}$ are Fourier multipliers of the form

$$q_{k,\ell}^{j,m} a \overset{\text{def}}{=} \sum_{n,p} \alpha_{k,\ell}^{j,m,n,p} \mathcal{F}^{-1}\left(\frac{\xi_n \xi_p}{|\xi|^2} \widehat{a}(\xi)\right),$$

and $\alpha_{k,\ell}^{j,m,n,p}$ are real numbers.

As pointed out above, the incompressible Navier–Stokes system is a particular case of the system

$$(GNS_\nu) : \begin{cases} \partial_t u - \nu \Delta u = Q(u, u) \\ \quad u_{|t=0} = u_0 \end{cases}$$

with the operator Q defined as above.

Let $B(u, v)$ [resp., $B_{NS}(u, v)$] be the solution to the heat equation

$$\begin{cases} \partial_t B(u, v) - \nu \Delta B(u, v) = Q(u, v) \quad [\text{resp.,} \ Q_{NS}(u, v)] \\ \quad B(u, v)_{|t=0} = 0. \end{cases}$$

Solving (GNS_ν) [resp., (NS_ν)] amounts to finding a fixed point for the map

$$u \longmapsto e^{\nu t \Delta} u_0 + B(u, u) \quad [\text{resp.,} \ B_{NS}(u, u)].$$

Throughout this chapter, we shall solve (GNS_ν) or (NS_ν) by means of a contraction mapping argument in a suitable Banach space. This is based on a classical lemma that we recall (and prove) here.

Lemma 5.5. *Let E be a Banach space, $\mathcal{B}$ a continuous bilinear map from $E \times E$ to E, and α a positive real number such that*

$$\alpha < \frac{1}{4\|\mathcal{B}\|} \quad \text{with} \quad \|\mathcal{B}\| \overset{\text{def}}{=} \sup_{\|u\|, \|v\| \leq 1} \|\mathcal{B}(u, v)\|. \tag{5.5}$$

For any a in the ball $B(0, \alpha)$ (i.e., with center 0 and radius α) in E, a unique x then exists in $B(0, 2\alpha)$ such that

$$x = a + \mathcal{B}(x, x).$$

Proof. The proof involves an application of the classical iterative scheme defined by

$$x_0 = a \quad \text{and} \quad x_{n+1} = a + \mathcal{B}(x_n, x_n).$$

By induction we may prove that $\|x_n\| \leq 2\alpha$. Indeed, using (5.5) and the definition of x_{n+1}, we get

$$\|x_{n+1}\| \leq \alpha(1 + 4\alpha\|\mathcal{B}\|) \leq 2\alpha.$$

Thus, the sequence $(x_n)_{n \in \mathbb{N}}$ remains in the ball $B(0, 2\alpha)$. Now,

$$\begin{aligned} x_{n+1} - x_n &= \mathcal{B}(x_n, x_n) - \mathcal{B}(x_{n-1}, x_{n-1}) \\ &= \mathcal{B}(x_n - x_{n-1}, x_n) + \mathcal{B}(x_{n-1}, x_n - x_{n-1}). \end{aligned}$$

Therefore, we obtain

$$\|x_{n+1} - x_n\| \leq 4\alpha\|\mathcal{B}\| \, \|x_n - x_{n-1}\|.$$

Hence, by virtue of (5.5), $(x_n)_{n \in \mathbb{N}}$ is a Cauchy sequence in E, the limit of which is a fixed point of $x \mapsto a + \mathcal{B}(x, x)$ in the ball $B(0, 2\alpha)$. This fixed point is unique because if x and y are two such fixed points, then

$$\|x - y\| \leq \|\mathcal{B}(x - y, y) + \mathcal{B}(x, x - y)\| \leq 4\alpha\|\mathcal{B}\| \, \|x - y\|.$$

The lemma is thus proved. □

Proving the existence of global solutions for (GNS_ν) or (NS_ν) by means of Lemma 5.5 requires a Banach space X with a norm invariant under the transformations that preserve the set of global solutions. This set contains the translations with respect to the space variable and, more importantly for our purposes, the so-called *scaling transformations* defined by

$$u_\lambda(t, x) \stackrel{\text{def}}{=} \lambda u(\lambda^2 t, \lambda x).$$

The following spaces obviously meet these conditions:

$$L^\infty(\mathbb{R}^+; L^d), \ L^\infty(\mathbb{R}^+; \dot{H}^{\frac{d}{2}-1}), \ L^4(\mathbb{R}^+; \dot{H}^{\frac{d-1}{2}}),$$
$$L^\infty(\mathbb{R}^+; \dot{H}^{\frac{d}{2}-1}) \cap L^2(\mathbb{R}^+; \dot{H}^{\frac{d}{2}}).$$

When $d = 2$, the energy space itself, $L^\infty(\mathbb{R}^+; L^2) \cap L^2(\mathbb{R}^+; \dot{H}^1)$, is scaling invariant. This is the key to the proof of Theorem 5.4. In the case where $d = 3$, however, the regularity of the energy space is *below* that of the scaling invariant space $\dot{H}^{\frac{1}{2}}$. In other words, in dimension $d = 2$, demonstrating the global existence of regular solutions of the Navier–Stokes system is a critical problem, whereas in dimension $d = 3$, this can be interpreted as a supercritical problem. This is the core of the difficulty. As we shall see, being able to use the special structure of the equation *in a scaling invariant framework* is one of the challenges involved in resolving the global well-posedness issue.

5.2 Well-posedness in Sobolev Spaces

In this section, we investigate the local and global well-posedness issues for the generalized Navier–Stokes system (GNS_ν).

5.2.1 A General Result

The main theorem of this subsection is the following one.

Theorem 5.6. *Let u_0 be in $\dot{H}^{\frac{d}{2}-1}(\mathbb{R}^d)$. There exists a positive time T such that the system (GNS_ν) has a unique solution u in $L^4([0,T];\dot{H}^{\frac{d-1}{2}})$ which also belongs to*

$$\mathcal{C}([0,T];\dot{H}^{\frac{d}{2}-1}) \cap L^2([0,T];\dot{H}^{\frac{d}{2}}).$$

Let T_{u_0} denote the maximal time of existence of such a solution. Then:

- *There exists a constant c such that*

$$\|u_0\|_{\dot{H}^{\frac{d}{2}-1}} \le c\nu \Longrightarrow T_{u_0} = \infty.$$

- *If T_{u_0} is finite, then*

$$\int_0^{T_{u_0}} \|u(t)\|_{\dot{H}^{\frac{d-1}{2}}}^4 \, dt = \infty. \tag{5.6}$$

Moreover, the solutions are stable in the following sense: If u and v are solutions, then

$$\|u(t) - v(t)\|_{\dot{H}^{\frac{d}{2}-1}}^2 + \nu \int_0^t \|u(t') - v(t')\|_{\dot{H}^{\frac{d}{2}}}^2 \, dt' \le \|u_0 - v_0\|_{\dot{H}^{\frac{d}{2}-1}}^2$$
$$\times \exp\left(\frac{C}{\nu^3} \int_0^t \left(\|u(t')\|_{\dot{H}^{\frac{d-1}{2}}}^4 + \|v(t')\|_{\dot{H}^{\frac{d-1}{2}}}^4 \right) dt' \right).$$

Remark 5.7. We note that for any small data, the corresponding solution u belongs to $L^4(\mathbb{R}^+;\dot{H}^{\frac{d-1}{2}})$. In fact, we shall see in Theorem 5.17 that *any* global solution of (GNS_ν) belongs to $L^4(\mathbb{R}^+;\dot{H}^{\frac{d-1}{2}})$.

Remark 5.8. As a by-product of the proof, under the condition $\|u_0\|_{\dot{H}^{\frac{d}{2}-1}} \le c\nu$, we actually get $\|u(t)\|_{\dot{H}^{\frac{d}{2}-1}} \le 2c\nu$ for any time t.

Proof of Theorem 5.6. We shall prove that the map

$$u \longmapsto e^{\nu t \Delta} u_0 + B(u,u)$$

has a unique fixed point in the space $L^4([0,T];\dot{H}^{\frac{d-1}{2}})$ for an appropriate T. This basically relies on the following two lemmas, the first of which is simply a variation on Sobolev embedding.

Lemma 5.9. *A constant C exists such that*

$$\|Q(a,b)\|_{\dot{H}^{\frac{d}{2}-2}} \leq C \|a\|_{\dot{H}^{\frac{d-1}{2}}} \|b\|_{\dot{H}^{\frac{d-1}{2}}}.$$

Proof. We focus on the cases $d = 2, 3$, where the result may be proven by elementary arguments. For $d \geq 4$ the result follows from Corollary 2.55 page 90 (the proof of which requires more elaborate techniques).

Beginning with the case $d = 2$ we can use Sobolev embedding (see Theorem 1.38 page 29) to write

$$
\begin{aligned}
\|Q(a,b)\|_{\dot{H}^{-1}} &\leq C \|ab\|_{L^2} \\
&\leq C \|a\|_{L^4} \|b\|_{L^4} \\
&\leq C \|a\|_{\dot{H}^{\frac{1}{2}}} \|b\|_{\dot{H}^{\frac{1}{2}}}.
\end{aligned}
$$

Next, if $d = 3$, then we have, by the definition of Q,

$$\|Q(a,b)\|_{\dot{H}^{-\frac{1}{2}}} \leq C \sup_{k,\ell} \left(\|a^k \partial b^\ell\|_{\dot{H}^{-\frac{1}{2}}} + \|b^\ell \partial a^k\|_{\dot{H}^{-\frac{1}{2}}} \right).$$

Thanks to the dual Sobolev embedding (see Corollary 1.39 page 29) and to the Sobolev embedding itself, we have

$$
\begin{aligned}
\|Q(a,b)\|_{\dot{H}^{-\frac{1}{2}}} &\leq C \sup_{k,\ell} \left(\|a^k \partial b^\ell\|_{L^{\frac{3}{2}}} + \|b^\ell \partial a^k\|_{L^{\frac{3}{2}}} \right) \\
&\leq C \left(\|a\|_{L^6} \|\nabla b\|_{L^2} + \|\nabla a\|_{L^2} \|b\|_{L^6} \right) \\
&\leq C \|a\|_{\dot{H}^1} \|b\|_{\dot{H}^1}.
\end{aligned}
$$

This proves the lemma.   □

The second lemma describes an aspect of the smoothing effect of the heat flow and may be seen as a particular case of the inequality (3.39) page 157. Here, we provide an elementary self-contained proof which does not require Littlewood–Paley decomposition.

Lemma 5.10. *Let v be the solution in $\mathcal{C}([0,T]; \mathcal{S}'(\mathbb{R}^d))$ of the Cauchy problem*

$$
\begin{cases}
\partial_t v - \nu \Delta v = f \\
v_{|t=0} = v_0
\end{cases}
$$

with f in $L^2([0,T]; \dot{H}^{s-1})$ and v_0 in $\dot{H}^s(\mathbb{R}^d)$. Then,

$$v \in \left(\bigcap_{p=2}^{\infty} L^p([0,T]; \dot{H}^{s+\frac{2}{p}}) \right) \bigcap \mathcal{C}([0,T]; \dot{H}^s).$$

Moreover, we have the following estimates:

$$\|v(t)\|_{\dot{H}^s}^2 + 2\nu \int_0^t \|\nabla v(t')\|_{\dot{H}^s}^2 \, dt' = \|v_0\|_{\dot{H}^s}^2 + 2 \int_0^t \langle f(t'), v(t')\rangle_s \, dt',$$

$$\left(\int_{\mathbb{R}^d} |\xi|^{2s} \left(\sup_{0 \le t' \le t} |\widehat{v}(t', \xi)| \right)^2 d\xi \right)^{\frac{1}{2}} \le \|v_0\|_{\dot{H}^s} + \frac{1}{(2\nu)^{\frac{1}{2}}} \|f\|_{L_T^2(\dot{H}^{s-1})},$$

$$\|v(t)\|_{L_T^p(\dot{H}^{s+\frac{2}{p}})} \le \frac{1}{\nu^{\frac{1}{p}}} \left(\|v_0\|_{\dot{H}^s} + \frac{1}{\nu^{\frac{1}{2}}} \|f\|_{L_T^2(\dot{H}^{s-1})} \right)$$

with $\langle a, b\rangle_s \overset{def}{=} \int |\xi|^{2s} \widehat{a}(\xi) \overline{\widehat{b}(\xi)} \, d\xi.$

Proof. The first estimate is just the energy estimate. The proof of the second one is based around writing Duhamel's formula in Fourier space, namely,

$$\widehat{v}(t, \xi) = e^{-\nu t|\xi|^2} \widehat{v}_0(\xi) + \int_0^t e^{-\nu(t-t')|\xi|^2} \widehat{f}(t', \xi) \, dt'.$$

The Cauchy–Schwarz inequality implies that

$$\sup_{0 \le t' \le t} |\widehat{v}(t', \xi)| \le |\widehat{v}_0(\xi)| + \frac{1}{\sqrt{2\nu|\xi|^2}} \|\widehat{f}(\cdot, \xi)\|_{L^2([0,t])}.$$

Taking the L^2 norm with respect to $|\xi|^{2s} \, d\xi$ then allows us to conclude that

$$V(t) \overset{def}{=} \left(\int_{\mathbb{R}^d} \left(\sup_{0 \le t' \le t} |\widehat{v}(t', \xi)| \right)^2 |\xi|^{2s} \, d\xi \right)^{\frac{1}{2}}$$

$$\le \|v_0\|_{\dot{H}^s} + \frac{1}{(2\nu)^{\frac{1}{2}}} \left(\int_{\mathbb{R}^d} \|\widehat{f}(\cdot, \xi)\|_{L^2([0,t])}^2 |\xi|^{2s-2} \, d\xi \right)^{\frac{1}{2}}$$

$$\le \|v_0\|_{\dot{H}^s} + \frac{1}{(2\nu)^{\frac{1}{2}}} \left(\int_{[0,t] \times \mathbb{R}^d} |\widehat{f}(t', \xi)|^2 |\xi|^{2s-2} \, d\xi \, dt' \right)^{\frac{1}{2}}$$

$$\le \|v_0\|_{\dot{H}^s} + \frac{1}{(2\nu)^{\frac{1}{2}}} \|f\|_{L^2([0,t]; \dot{H}^{s-1})}.$$

Since, for almost all fixed $\xi \in \mathbb{R}^d$, the map $t \mapsto \widehat{v}(t, \xi)$ is continuous over $[0, T]$, the Lebesgue dominated convergence theorem ensures that $v \in \mathcal{C}([0, T]; \dot{H}^s)$.

Finally, the last inequality follows by interpolation. $\qquad\square$

Combining Lemmas 5.9 and 5.10, we get the following result.

Corollary 5.11. *A constant C exists such that*

$$\|B(u, v)\|_{L_T^4(\dot{H}^{\frac{d-1}{2}})} \le \frac{C}{\nu^{\frac{3}{4}}} \|u\|_{L_T^4(\dot{H}^{\frac{d-1}{2}})} \|v\|_{L_T^4(\dot{H}^{\frac{d-1}{2}})}.$$

Proof of Theorem 5.6 (continued). To prove the first part of Theorem 5.6, we shall use Lemma 5.5. We know that if

$$\|e^{\nu t \Delta} u_0\|_{L_T^4(\dot{H}^{\frac{d-1}{2}})} \leq \frac{\nu^{\frac{3}{4}}}{4C_0} \tag{5.7}$$

with $C_0 > C$, then there exists a unique solution of (GNS_ν) in the ball with center 0 and radius $(\frac{\nu^{\frac{3}{4}}}{2C_0})$ in the space $L^4([0,T]; \dot{H}^{\frac{d-1}{2}})$.

Next, we investigate when the condition (5.7) is satisfied. Applying the last inequality of Lemma 5.10 with $s = d/2 - 1$ and $p = 4$ yields, for any positive time T,

$$\|e^{\nu t \Delta} u_0\|_{L_T^4(\dot{H}^{\frac{d-1}{2}})} \leq \frac{1}{\nu^{\frac{1}{4}}} \|u_0\|_{\dot{H}^{\frac{d}{2}-1}}. \tag{5.8}$$

Thus, if $\|u_0\|_{\dot{H}^{\frac{d}{2}-1}} \leq (4C_0)^{-1}\nu$, then the smallness condition (5.7) is satisfied and we have a global solution.

We now consider the case of a *large* initial data u_0 in $\dot{H}^{\frac{d}{2}-1}$. We shall split u_0 into a small part in $\dot{H}^{\frac{d}{2}-1}$ and a large part with compactly supported Fourier transform. For that, we fix some positive real number ρ_{u_0} such that

$$\left(\int_{|\xi| \geq \rho_{u_0}} |\xi|^{d-2} |\hat{u}_0(\xi)|^2 \, d\xi \right)^{\frac{1}{2}} \leq \frac{\nu}{8C_0}.$$

Using (5.8) and defining $u_0^\flat \stackrel{\text{def}}{=} \mathcal{F}^{-1}(\mathbf{1}_{B(0,\rho_{u_0})} \hat{u}_0)$, we get

$$\|e^{\nu t \Delta} u_0\|_{L_T^4(\dot{H}^{\frac{d-1}{2}})} \leq \frac{\nu^{\frac{3}{4}}}{8C_0} + \|e^{\nu t \Delta} u_0^\flat\|_{L_T^4(\dot{H}^{\frac{d-1}{2}})}.$$

We note that

$$\|e^{\nu t \Delta} u_0^\flat\|_{L_T^4(\dot{H}^{\frac{d-1}{2}})} \leq \rho_{u_0}^{\frac{1}{2}} \|e^{\nu t \Delta} u_0^\flat\|_{L_T^4(\dot{H}^{\frac{d}{2}-1})}$$
$$\leq (\rho_{u_0}^2 T)^{\frac{1}{4}} \|u_0\|_{\dot{H}^{\frac{d}{2}-1}}.$$

Thus, if

$$T \leq \left(\frac{\nu^{\frac{3}{4}}}{8C_0 \rho_{u_0}^{\frac{1}{2}} \|u_0\|_{\dot{H}^{\frac{d}{2}-1}}} \right)^4, \tag{5.9}$$

then we have the existence of a unique solution in the ball with center 0 and radius $\frac{\nu^{\frac{3}{4}}}{2C_0}$ in the space $L^4([0,T]; \dot{H}^{\frac{d-1}{2}})$.

Finally, we observe that if u is a solution of (GNS_ν) in $L^4([0,T]; \dot{H}^{\frac{d-1}{2}})$, then, by Lemma 5.9, $Q(u,u)$ belongs to $L^2([0,T]; \dot{H}^{\frac{d}{2}-2})$. Hence, Lemma 5.10 implies that the solution u belongs to

$$\mathcal{C}([0,T]; \dot{H}^{\frac{d}{2}-1}) \cap L^2([0,T]; \dot{H}^{\frac{d}{2}}).$$

In order to prove the stability estimate, consider the difference w between two solutions u and v. We note that w satisfies

$$\begin{cases} \partial_t w - \nu \Delta w = Q(w, u + v) \\ \qquad w_{|t=0} = w_0 \overset{\text{def}}{=} u_0 - v_0. \end{cases}$$

Thus, by the energy estimate in $\dot{H}^{\frac{d}{2}-1}$ (see Lemma 5.10), we have

$$\Delta_{w(t)} \overset{\text{def}}{=} \|w(t)\|_{\dot{H}^{\frac{d}{2}-1}}^2 + 2\nu \int_0^t \|\nabla w(t')\|_{\dot{H}^{\frac{d}{2}-1}}^2 \, dt'$$

$$\leq \|w_0\|_{\dot{H}^{\frac{d}{2}-1}}^2 + 2 \int_0^t \langle Q(w(t'), u(t') + v(t')), w(t') \rangle_{\frac{d}{2}-1} \, dt'.$$

The nonlinear term is treated by means of the following lemma.

Lemma 5.12. *A constant C exists such that*

$$\langle Q(a, b), c \rangle_{\frac{d}{2}-1} \leq C \|a\|_{\dot{H}^{\frac{d-1}{2}}} \|b\|_{\dot{H}^{\frac{d-1}{2}}} \|\nabla c\|_{\dot{H}^{\frac{d}{2}-1}}.$$

Proof. Let $\alpha = Q(a, b)$. By definition of the $\dot{H}^{\frac{d}{2}-1}$ scalar product, we have, thanks to the Cauchy–Schwarz inequality,

$$\langle \alpha, c \rangle_{\frac{d}{2}-1} = \int \widehat{\alpha}(\xi) \overline{\widehat{c}}(\xi) |\xi|^{d-2} \, d\xi$$

$$= \int |\xi|^{\frac{d}{2}-2} \widehat{\alpha}(\xi) \, |\xi|^{\frac{d}{2}-1} |\xi| \overline{\widehat{c}}(\xi) \, d\xi$$

$$\leq \|\alpha\|_{\dot{H}^{\frac{d}{2}-2}} \|\nabla c\|_{\dot{H}^{\frac{d}{2}-1}},$$

which, by virtue of Lemma 5.9, leads to the result. $\qquad\square$

Completion of the proof of Theorem 5.6. We now resume the proof of the stability. We deduce from the above lemma that

$$\Delta_w(t) \leq \|w_0\|_{\dot{H}^{\frac{d}{2}-1}}^2 + C \int_0^t \|w(t')\|_{\dot{H}^{\frac{d-1}{2}}} N(t') \|\nabla w(t')\|_{\dot{H}^{\frac{d}{2}-1}} \, dt'$$

with $N(t) \overset{\text{def}}{=} \|u(t)\|_{\dot{H}^{\frac{d-1}{2}}} + \|v(t)\|_{\dot{H}^{\frac{d-1}{2}}}$. By the interpolation inequality between $\dot{H}^{\frac{d}{2}-1}$ and $\dot{H}^{\frac{d}{2}}$, we infer that

$$\Delta_w(t) \leq \|w_0\|_{\dot{H}^{\frac{d}{2}-1}}^2 + C \int_0^t \|w(t')\|_{\dot{H}^{\frac{d}{2}-1}}^{\frac{1}{2}} N(t') \|\nabla w(t')\|_{\dot{H}^{\frac{d}{2}-1}}^{\frac{3}{2}} \, dt'.$$

Using the convexity inequality $ab \leq \frac{1}{4} a^4 + \frac{3}{4} b^{\frac{4}{3}}$, we deduce that

$$\Delta_w(t) \leq \|w_0\|_{\dot{H}^{\frac{d}{2}-1}}^2 + \frac{C}{\nu^3} \int_0^t \|w(t')\|_{\dot{H}^{\frac{d}{2}-1}}^2 N^4(t') \, dt' + \nu \int_0^t \|\nabla w(t')\|_{\dot{H}^{\frac{d}{2}-1}}^2 \, dt'.$$

By definition of Δ_w, this can be written

$$\|w(t)\|^2_{\dot{H}^{\frac{d}{2}-1}} + \nu \int_0^t \|\nabla w(t')\|^2_{\dot{H}^{\frac{d}{2}-1}} \, dt'$$

$$\leq \|w_0\|^2_{\dot{H}^{\frac{d}{2}-1}} + \frac{C}{\nu^3} \int_0^t \|w(t')\|^2_{\dot{H}^{\frac{d}{2}-1}} N^4(t') \, dt'.$$

Using the Gronwall lemma, we infer that

$$\|w(t)\|^2_{\dot{H}^{\frac{d}{2}-1}} + \nu \int_0^t \|\nabla w(t')\|^2_{\dot{H}^{\frac{d}{2}-1}} \, dt' \leq \|w_0\|^2_{\dot{H}^{\frac{d}{2}-1}} \exp\left(\frac{C}{\nu^3} \int_0^t N^4(t') \, dt'\right).$$

The theorem is thus proved up to the blow-up criterion. Assume that we have a solution u of (GNS_ν) on a time interval $[0, T[$ such that

$$\int_0^T \|u(t)\|^4_{\dot{H}^{\frac{d-1}{2}}} \, dt < \infty.$$

We claim that the lifespan T_{u_0} of u is greater than T. Indeed, thanks to Lemmas 5.9 and 5.10, we have

$$\int_{\mathbb{R}^d} |\xi|^{d-2} \left(\sup_{t \in [0,T[} |\hat{u}(t,\xi)|\right)^2 d\xi < \infty.$$

Thus, a positive number ρ exists such that

$$\forall t \in [0, T[, \ \int_{|\xi| \geq \rho} |\xi|^{d-2} |\hat{u}(t,\xi)|^2 \, d\xi < \frac{c\nu}{2}.$$

The condition (5.9) now implies that for any $t \in [0, T[$, the lifespan for a solution of (GNS_ν) with initial data $u(t)$ is bounded from below by a positive real number τ which is *independent of* t. Thus, $T_{u_0} > T$, and the whole of Theorem 5.6 is now proved. □

5.2.2 The Behavior of the $\dot{H}^{\frac{d}{2}-1}$ Norm Near 0

In this subsection, we show that for small solutions, the $\dot{H}^{\frac{d}{2}-1}$ norm behaves as a Lyapunov function near 0.

Proposition 5.13. *Let u_0 be in the ball with center 0 and radius $c\nu$ in the space $\dot{H}^{\frac{d}{2}-1}(\mathbb{R}^d)$. The function*

$$t \longmapsto \|u(t)\|_{\dot{H}^{\frac{d}{2}-1}}$$

is then nonincreasing.

Proof. We shall again use the fact that the function u is a solution of the equation

$$\partial_t u - \nu \Delta u = Q(u,u) \quad \text{with} \quad Q(u,u) \in L^2(\mathbb{R}^+; \dot{H}^{\frac{d}{2}-2}).$$

Thus, thanks to Lemma 5.10, we infer that

$$\|u(t)\|^2_{\dot{H}^{\frac{d}{2}-1}} + 2\nu \int_0^t \|\nabla u(t')\|^2_{\dot{H}^{\frac{d}{2}-1}} \, dt'$$

$$= \|u_0\|^2_{\dot{H}^{\frac{d}{2}-1}} + 2 \int_0^t \langle Q(u(t'), u(t')), u(t') \rangle_{\frac{d}{2}-1} \, dt'.$$

Using Lemma 5.12 and an interpolation inequality, we get, for any $0 \le t_1 \le t_2$,

$$U(t_1, t_2) \overset{\text{def}}{=} \|u(t_2)\|^2_{\dot{H}^{\frac{d}{2}-1}} + 2\nu \int_{t_1}^{t_2} \|\nabla u(t')\|^2_{\dot{H}^{\frac{d}{2}-1}} \, dt'$$

$$\le \|u(t_1)\|^2_{\dot{H}^{\frac{d}{2}-1}} + C \int_{t_1}^{t_2} \|u(t')\|^2_{\dot{H}^{\frac{d-1}{2}}} \|\nabla u(t')\|_{\dot{H}^{\frac{d}{2}-1}} \, dt'$$

$$\le \|u(t_1)\|^2_{\dot{H}^{\frac{d}{2}-1}} + C \int_{t_1}^{t_2} \|u(t')\|_{\dot{H}^{\frac{d}{2}-1}} \|\nabla u(t')\|^2_{\dot{H}^{\frac{d}{2}-1}} \, dt'.$$

By Theorem 5.6, we know that $u(t)$ remains in the ball with center 0 and radius $2c\nu$ in the space $\dot{H}^{\frac{d}{2}-1}(\mathbb{R}^d)$. Thus, if c is small enough, we get that

$$\|u(t_2)\|^2_{\dot{H}^{\frac{d}{2}-1}} + \nu \int_{t_1}^{t_2} \|\nabla u(t')\|^2_{\dot{H}^{\frac{d}{2}-1}} \, dt' \le \|u(t_1)\|^2_{\dot{H}^{\frac{d}{2}-1}}.$$

This proves the proposition.   □

5.3 Results Related to the Structure of the System

In this section we present results which are related to the very structure of the Navier–Stokes system. Here, the energy estimate will play a fundamental role.

5.3.1 The Particular Case of Dimension Two

As explained above, in dimension two the energy estimate turns out to be scaling invariant for the Navier–Stokes system. This fact will enable us to prove that (NS_ν) is globally well posed for *any* initial data in $L^2(\mathbb{R}^2)$, as follows.

Theorem 5.14. *Let u_0 be a divergence-free vector field in $L^2(\mathbb{R}^2)$. A unique solution then exists in the space $L^4(\mathbb{R}^+; \dot{H}^{\frac{1}{2}})$ which also belongs to*

$$\mathcal{C}(\mathbb{R}^+; L^2) \cap L^\infty(\mathbb{R}^+; L^2) \cap L^2(\mathbb{R}^+; \dot{H}^1)$$

and satisfies the energy equality

$$\|u(t)\|_{L^2}^2 + 2\nu \int_0^t \|\nabla u(t')\|_{L^2}^2 \, dt' = \|u_0\|_{L^2}^2.$$

Proof. Let u be the solution given by Theorem 5.6. Thanks to Lemma 5.9, we know that $Q_{NS}(u, u)$ belongs to $L^2_{loc}([0, T_{u_0}[; \dot{H}^{-1})$. Therefore, Lemma 5.10 implies that u is continuous with values in $L^2(\mathbb{R}^2)$ and satisfies

$$\|u(t)\|_{L^2}^2 + 2\nu \int_0^t \|\nabla u(t')\|_{L^2}^2 \, dt' = \|u_0\|_{L^2}^2 + 2 \int_0^t \langle Q_{NS}(u(t'), u(t')), u(t') \rangle_0 \, dt'.$$

We temporarily assume the following lemma.

Lemma 5.15. *Let u and v be time-dependent, divergence-free vector fields over $\mathbb{R}^d$. If u and v belong to $L^4([0, T]; L^4) \cap L^2([0, T]; H^1)$, then we have*

$$\int_0^t \langle Q_{NS}(u(t'), v(t')), v(t') \rangle_0 \, dt' = 0.$$

Combining interpolation and the Sobolev embedding $\dot{H}^{\frac{1}{2}}(\mathbb{R}^2) \hookrightarrow L^4(\mathbb{R}^2)$, we see that u is in $L^4([0, T] \times \mathbb{R}^2)$. Therefore, we deduce that for any $t < T_{u_0}$,

$$\|u(t)\|_{L^2}^2 + 2\nu \int_0^t \|\nabla u(t')\|_{L^2}^2 \, dt' = \|u_0\|_{L^2}^2.$$

Thanks to the above energy estimate and using an interpolation inequality between L^2 and $\dot{H}^1$, we obtain, for any $T < T_{u_0}$,

$$\int_0^T \|u(t)\|_{\dot{H}^{\frac{1}{2}}}^4 \, dt \leq \|u_0\|_{L^2}^2 \int_0^T \|\nabla u(t)\|_{L^2}^2 \, dt$$

$$\leq \frac{1}{2\nu} \|u_0\|_{L^2}^4.$$

The blow-up condition (5.6) then implies the theorem.   □

For the sake of completeness, we now prove Lemma 5.15. We know that

$$Q^j_{NS}(u, v) \overset{\text{def}}{=} -\operatorname{div}(v^j u) - \sum_{1 \leq k, \ell \leq d} \partial_j (-\Delta)^{-1} \partial_k \partial_\ell (u^k v^\ell).$$

Note that all the terms on the right-hand side are in $L^2([0, T]; \dot{H}^{-1})$. Therefore,

$$\langle Q_{NS}(u,v), v\rangle = -\sum_{1\le j\le d}\int_{\mathbb{R}^d} v^j\,\mathrm{div}(v^j u)\,dx$$

$$+\sum_{1\le j,k,\ell\le d}\int_{\mathbb{R}^d} v^j\partial_j(\Delta^{-1}\partial_k\partial_\ell(u^k v^\ell))\,dx$$

$$= -\frac{1}{2}\int_{\mathbb{R}^d}(\mathrm{div}\,u)|v|^2\,dx$$

$$-\sum_{1\le k,\ell\le d}\int_{\mathbb{R}^d}(\mathrm{div}\,v)\Delta^{-1}\partial_k\partial_\ell(u^k v^\ell)\,dx. \qquad (5.10)$$

As $\mathrm{div}\,u = \mathrm{div}\,v = 0$, this completes the proof of the lemma. $\qquad\square$

5.3.2 The Case of Dimension Three

The case of dimension three is much more involved. The question of whether or not (NS_ν) is globally well posed for large data in $\dot{H}^{\frac{1}{2}}(\mathbb{R}^3)$ is still open. The purpose of this section is first to prove the energy equality for solutions of (NS_ν) given by Theorem 5.6 and then to show that any global solution is stable.

Proposition 5.16. *Consider an initial data u_0 in $H^{\frac{1}{2}}(\mathbb{R}^3)$ with $\mathrm{div}\,u_0 = 0$. If u denotes the solution given by Theorem 5.6, then u is continuous with values in $L^2(\mathbb{R}^3)$ and satisfies the energy equality*

$$\|u(t)\|_{L^2}^2 + 2\nu\int_0^t \|\nabla u(t')\|_{L^2}^2\,dt' = \|u_0\|_{L^2}^2.$$

Proof. As the solution u belongs to

$$L_{loc}^\infty([0, T_{u_0}[; \dot{H}^{\frac{1}{2}}) \cap L_{loc}^4([0, T_{u_0}[; \dot{H}^1),$$

the interpolation inequality between Sobolev norms (see Proposition 1.32 page 25) implies that u belongs to the space $L_{loc}^8([0, T_{u_0}[; \dot{H}^{\frac{3}{4}})$, which, in view of Sobolev embedding, is a subspace of $L_{loc}^4([0, T_{u_0}[; L^4)$. Therefore, we may apply Lemma 5.15, and the energy equality is thus satisfied. Now, because $u \in L_{loc}^4([0, T_{u_0}[; L^4)$, we have $Q_{NS}(u,u) \in L_{loc}^2([0, T_{u_0}[; \dot{H}^{-1})$, so applying Lemma 5.10 yields the desired continuity result. $\qquad\square$

Next, we shall investigate qualitative properties of global solutions. In fact, any global solution is stable, even if associated with large initial data. More precisely, we have the following statement.

Theorem 5.17. *Let u be a global solution of (NS_ν) in $L_{loc}^4(\mathbb{R}^+; \dot{H}^1)$. We then have*

$$\lim_{t\to\infty}\|u(t)\|_{\dot{H}^{\frac{1}{2}}} = 0 \quad and \quad \int_0^\infty \|u(t)\|_{\dot{H}^1}^4\,dt < \infty.$$

Remark 5.18. We know that if $\|u_0\|_{\dot{H}^{\frac{1}{2}}}$ satisfies the smallness condition of Theorem 5.6, then the global solution associated with the Cauchy data u_0 belongs to the space $L^4(\mathbb{R}^+; \dot{H}^1)$. Hence, it suffices to prove that $\lim\limits_{t\to\infty} \|u(t)\|_{\dot{H}^{\frac{1}{2}}} = 0$.

Remark 5.19. If u_0 also belongs to $L^2(\mathbb{R}^3)$, then this theorem is an immediate consequence of Proposition 5.16. Indeed, interpolating between L^2 and $\dot{H}^1$ yields

$$\int_{\mathbb{R}^+} \|u(t)\|_{\dot{H}^{\frac{1}{2}}}^4 \, dt \leq \frac{1}{2\nu} \|u_0\|_{L^2}^4,$$

from which the result follows since the $\dot{H}^{\frac{1}{2}}$ norm is a Lyapunov function near 0.

Proof of Theorem 5.17. For fixed, given $\rho > 0$, we decompose the initial data u_0 as

$$u_0 = u_{0,h} + u_{0,\ell} \quad \text{with} \quad u_{0,\ell} \stackrel{\text{def}}{=} \mathcal{F}^{-1}(\mathbf{1}_{B(0,\rho)} \widehat{u}_0).$$

Let ε be any positive real number. We can choose ρ such that

$$\|u_{0,\ell}\|_{\dot{H}^{\frac{1}{2}}} \leq \min\left\{ c\nu, \frac{\varepsilon}{2} \right\}.$$

Denote by u_ℓ the global solution of (NS_ν) given by Theorem 5.6 for the initial data $u_{0,\ell}$. Thanks to Proposition 5.13, we have

$$\forall t \in \mathbb{R}^+, \ \|u_\ell(t)\|_{\dot{H}^{\frac{1}{2}}} \leq \frac{\varepsilon}{2}. \tag{5.11}$$

Define $u_h \stackrel{\text{def}}{=} u - u_\ell$. This satisfies

$$\begin{cases} \partial_t u_h - \nu \Delta u_h = Q_{NS}(u, u_h) + Q_{NS}(u_h, u_\ell) \\ u_{h|t=0} = u_{0,h}. \end{cases}$$

Obviously, $u_{0,h}$ belongs to L^2 (with an L^2 norm which depends on ρ and thus on ε). Moreover, both $Q_{NS}(u, u_h)$ and $Q_{NS}(u_h, u_\ell)$ belong to the space $L^2_{loc}(\mathbb{R}^+; \dot{H}^{-1})$. Applying Lemma 5.10 and Lemma 5.15, we get

$$\|u_h(t)\|_{L^2}^2 + 2\nu \int_0^t \|\nabla u_h(t')\|_{L^2}^2 \, dt' = \|u_{0,h}\|_{L^2}^2$$

$$+ 2 \int_0^t \langle Q_{NS}(u_h(t'), u_\ell(t')), u_h(t') \rangle_{\dot{H}^{-1} \times \dot{H}^1} \, dt'.$$

From Sobolev embedding, we infer that

$$\left| \langle Q_{NS}(u_h(t), u_\ell(t)), u_h(t) \rangle_{\dot{H}^{-1} \times \dot{H}^1} \right| \leq C \|u_h(t) u_\ell(t)\|_{L^2} \|\nabla u_h(t)\|_{L^2}$$

$$\leq C \|u_h(t)\|_{L^6} \|u_\ell(t)\|_{L^3} \|\nabla u_h(t)\|_{L^2}$$

$$\leq C \|u_\ell(t)\|_{\dot{H}^{\frac{1}{2}}} \|\nabla u_h(t)\|_{L^2}^2.$$

We then deduce that

$$\|u_h(t)\|_{L^2}^2 + 2\nu \int_0^t \|\nabla u_h(t')\|_{L^2}^2 \, dt' \leq \|u_{0,h}\|_{L^2}^2 + C\varepsilon \int_0^t \|\nabla u_h(t')\|_{L^2}^2 \, dt'.$$

Choosing ε small enough ensures that

$$\|u_h(t)\|_{L^2}^2 + \nu \int_0^t \|\nabla u_h(t')\|_{L^2}^2 \, dt' \leq \|u_{0,h}\|_{L^2}^2.$$

This implies that a positive time t_ε exists such that $\|u_h(t_\varepsilon)\|_{\dot{H}^{\frac{1}{2}}} < \varepsilon/2$. Thus, $\|u(t_\varepsilon)\|_{\dot{H}^{\frac{1}{2}}} \leq \varepsilon$. Theorem 5.6 and Proposition 5.13 then allow us to complete the proof. $\qquad\square$

Theorem 5.17 has the following interesting consequence.

Corollary 5.20. *The set of initial data u_0 such that the solution u given by Theorem 5.6 is global is an open subset of $\dot{H}^{\frac{1}{2}}$.*

Proof. Let u_0 in $\dot{H}^{\frac{1}{2}}$ be such that the associated solution is global. Let w_0 be in $\dot{H}^{\frac{1}{2}}$. Denote by v the maximal local solution associated with the initial data $v_0 \overset{\text{def}}{=} u_0 + w_0$. The function $w \overset{\text{def}}{=} v - u$ is solution of

$$\begin{cases} \partial_t w - \nu \Delta w = Q_{NS}(u,w) + Q_{NS}(w,u) + Q_{NS}(w,w) \\ w_{|t=0} = w_0. \end{cases}$$

Lemma 5.12, together with an interpolation inequality, gives

$$\langle Q_{NS}(u,w) + Q_{NS}(w,u), w \rangle_{\dot{H}^{\frac{1}{2}}} \leq C\|u\|_{\dot{H}^1} \|w\|_{\dot{H}^{\frac{1}{2}}}^{\frac{1}{2}} \|\nabla w\|_{\dot{H}^{\frac{1}{2}}}^{\frac{3}{2}},$$

$$\langle Q_{NS}(w,w), w \rangle_{\dot{H}^{\frac{1}{2}}} \leq C\|w\|_{\dot{H}^{\frac{1}{2}}} \|\nabla w\|_{\dot{H}^{\frac{1}{2}}}^2.$$

Assume that $\|w_0\|_{\dot{H}^{\frac{1}{2}}} \leq \dfrac{\nu}{8C}$ and define

$$T_{w_0} \overset{\text{def}}{=} \sup\Big\{ t \;/\; \max_{0 \leq t' \leq t} \|w(t')\|_{\dot{H}^{\frac{1}{2}}} \leq \frac{\nu}{4C} \Big\}.$$

From Lemma 5.10 and the convexity inequality $ab \leq \dfrac{1}{4}a^4 + \dfrac{3}{4}b^{\frac{4}{3}}$, we then infer that for any $t < T_{w_0}$,

$$\|w(t)\|_{\dot{H}^{\frac{1}{2}}}^2 + \nu \int_0^t \|\nabla w(t')\|_{\dot{H}^{\frac{1}{2}}}^2 \, dt' \leq \|w_0\|_{\dot{H}^{\frac{1}{2}}}^2 + \frac{C}{\nu^3} \int_0^t \|u(t')\|_{\dot{H}^1}^4 \|w(t')\|_{\dot{H}^{\frac{1}{2}}}^2 \, dt'.$$

The Gronwall lemma and Theorem 5.17 together imply that for any $t < T_{w_0}$,

$$\|w(t)\|_{\dot{H}^{\frac{1}{2}}}^2 + \nu \int_0^t \|\nabla w(t')\|_{\dot{H}^{\frac{1}{2}}}^2 \, dt' \leq \|w_0\|_{\dot{H}^{\frac{1}{2}}}^2 \exp\Big(\frac{C}{\nu^3} \int_0^t \|u(t')\|_{\dot{H}^1}^4 \, dt' \Big).$$

Now, according to Theorem 5.17, u is in $L^4(\mathbb{R}^+; \dot{H}^1)$. Hence, we can conclude that if the smallness condition

$$\|w_0\|^2_{\dot{H}^{\frac{1}{2}}} \exp\left(\frac{C}{\nu^3}\int_0^\infty \|u(t)\|^4_{\dot{H}^1}\, dt\right) \leq \frac{\nu^2}{16C^2}$$

is satisfied, then the blow-up condition for v is never satisfied. Corollary 5.20 is thus proved. □

5.4 An Elementary L^p Approach

As announced in the introduction of this chapter, we here prove local well-posedness for initial data in $L^3(\mathbb{R}^3)$. The main result is the following theorem.

Theorem 5.21. *Let u_0 be in $L^3(\mathbb{R}^3)$. A positive time T then exists such that (GSN_ν) has a unique solution u in the space $\mathcal{C}([0,T]; L^3)$. Moreover, there exists a positive constant c such that T can be chosen equal to infinity if $\|u_0\|_{L^3} \leq c\nu$.*

Proving this theorem cannot be achieved by means of a fixed point argument in the space $L^\infty([0,T]; L^3)$. Indeed, as discovered by F. Oru in [243], the bilinear functional B_{NS} does not map $L^\infty([0,T]; L^3) \times L^\infty([0,T]; L^3)$ into $L^\infty([0,T]; L^3)$.

As in the preceding section, we shall use the smoothing effect of the heat equation to define a space in which the fixed point method applies. This motivates the introduction of the following *Kato spaces*.

Definition 5.22. *If p is in $[3,\infty]$ and T is in $]0,\infty[$, then we define $K_p(T)$ by*

$$K_p(T) \overset{def}{=} \left\{ u \in \mathcal{C}(]0,T]; L^p) \,/\, \|u\|_{K_p(T)} \overset{def}{=} \sup_{t\in]0,T]} (\nu t)^{\frac{1}{2}\left(1-\frac{3}{p}\right)}\|u(t)\|_{L^p} < \infty \right\}.$$

If $p \in [1,3[$, then we define $K_p(T)$ by

$$K_p(T) \overset{def}{=} \left\{ u \in \mathcal{C}([0,T]; L^p) \,/\, \|u\|_{K_p(T)} \overset{def}{=} \sup_{t\in]0,T]} (\nu t)^{\frac{1}{2}\left(1-\frac{3}{p}\right)}\|u(t)\|_{L^p} < \infty \right\}.$$

We denote by $K_p(\infty)$ the space defined as above with $]0,\infty[$ (resp., $[0,\infty[$) instead of $]0,T]$ (resp., $[0,T]$).

Remark 5.23. Kato spaces are Banach spaces. Moreover, $K_p(\infty)$ is invariant under the scaling of the Navier–Stokes system.

Remark 5.24. Consider some u_0 in L^3 and $p \geq 3$. As

$$e^{\nu t\Delta}u_0 = \frac{1}{(4\pi\nu t)^{\frac{3}{2}}}e^{-\frac{|\cdot|^2}{4\nu t}} \star u_0,$$

we have, thanks to Young's inequality,

$$\|e^{\nu t \Delta} u_0\|_{L^p} \leq \frac{1}{(4\pi\nu t)^{\frac{3}{2}}} \|e^{-\frac{|\cdot|^2}{4\nu t}}\|_{L^r} \|u_0\|_{L^3} \quad \text{with} \quad \frac{1}{r} = \frac{2}{3} + \frac{1}{p}.$$

This gives $\|e^{\nu t \Delta} u_0\|_{L^p} \leq c(\nu t)^{-\frac{1}{2}\left(1 - \frac{3}{p}\right)} \|u_0\|_{L^3}$ and thus

$$\|e^{\nu t \Delta} u_0\|_{K_p(\infty)} \leq C\|u_0\|_{L^3}. \tag{5.12}$$

We note that if u_0 belongs to L^3, then, for any positive ε, a function ϕ can be found in $\mathcal{S}$ such that $\|u_0 - \phi\|_{L^3} \leq \varepsilon$. This implies, in particular, that

$$\|e^{\nu t \Delta}(u_0 - \phi)\|_{K_p(\infty)} \leq C\varepsilon.$$

Observing that $\|e^{\nu t \Delta} \phi\|_{L^p} \leq \|\phi\|_{L^p}$, we then get, for $p > 3$,

$$\|e^{\nu t \Delta} u_0\|_{K_p(T)} \leq C\varepsilon + (\nu T)^{\frac{1}{2}\left(1 - \frac{3}{p}\right)} \|\phi\|_{L^p}. \tag{5.13}$$

We can thus conclude that $\|e^{\nu t \Delta} u_0\|_{K_p(T)}$ tends to 0 when T goes to 0.

Remark 5.25. We now give an example of a sequence $(\phi_n)_{n \in \mathbb{N}}$ such that the L^3 norm is constant, the $\dot{H}^{\frac{1}{2}}$ norm tends to infinity, and the $K_p(\infty)$ norm of $e^{\nu t \Delta} \phi_n$ tends to 0 for any $p > 3$. Consider, for some ω in the unit sphere, the sequence

$$\phi_n(x) \stackrel{\text{def}}{=} e^{in(x|\omega)} \phi(x),$$

where ϕ is a function in $\mathcal{S}$ with a compactly supported Fourier transform.

On the one hand, since $\widehat{\phi}_n(\xi) = \widehat{\phi}(\xi - n\omega)$, straightforward computations give

$$\lim_{n \to \infty} n^{-1/2} \|\phi_n\|_{\dot{H}^{1/2}} = \|\widehat{\phi}\|_{L^2}.$$

On the other hand, we have

$$e^{\nu t \Delta} \phi_n(x) = (2\pi)^{-3} e^{in(x|\omega)} \int_{\mathbb{R}^3} e^{i(x|\eta)} e^{-\nu t|\eta + n\omega|^2} \widehat{\phi}(\eta) \, d\eta.$$

Hence, because $\widehat{\phi}$ is compactly supported, we find that for large enough n,

$$\|e^{\nu t \Delta} \phi_n\|_{L^\infty} \leq C e^{-\frac{\nu}{2} t n^2} \|\widehat{\phi}\|_{L^1} \quad \text{and} \quad \|e^{\nu t \Delta} \phi_n\|_{L^2} \leq C e^{-\frac{\nu}{2} t n^2} \|\widehat{\phi}\|_{L^2},$$

from which it follows, by Hölder's inequality, that

$$\|e^{\nu t \Delta} \phi_n\|_{L^p} \leq C(\nu t n^2)^{-\frac{1}{2}\left(1 - \frac{3}{p}\right)}.$$

Thus,

$$\|e^{\nu t \Delta} \phi_n\|_{K_p(\infty)} \leq \frac{C}{n^{1 - \frac{3}{p}}},$$

which implies that the $K_p(\infty)$ norm of $e^{\nu t \Delta} \phi_n$ tends to 0 when n goes to infinity. Finally, as $\|\phi_n\|_{L^3} = \|\phi\|_{L^3}$, this example has the announced properties.

Remark 5.26. We emphasize that when $p > 3$, $\|e^{\nu t \Delta} u_0\|_{K_p(\infty)}$ is equivalent to the norm of the homogeneous Besov space $\dot{B}_{p,\infty}^{-1+\frac{3}{p}}$ (see Theorem 2.34 page 72).

In fact, Theorem 5.21 turns out to be a corollary of the following theorem.

Theorem 5.27. *For any p in $]3, \infty[$, a constant c exists which satisfies the following property. Let u_0 be an initial data in S' such that for some positive T,*

$$\|e^{\nu t \Delta} u_0\|_{K_p(T)} \leq c\nu. \tag{5.14}$$

A unique solution u of (GNS_ν) then exists in the ball with center 0 and radius $2c\nu$ in the Banach space $K_p(T)$.

Remark 5.28. Thanks to the inequality (5.13), this theorem implies that for any initial data in L^3 we have a local solution. Thanks to the inequality (5.12) this solution is global if $\|u_0\|_{L^3}$ is small enough.

Proof of Theorem 5.27. The proof relies on Lemma 5.5 applied in $K_p(T)$. It therefore suffices to state the following result.

Lemma 5.29. *Let p, q, and r be such that*

$$0 < \frac{1}{p} + \frac{1}{q} \leq 1 \quad and \quad \frac{1}{r} \leq \frac{1}{p} + \frac{1}{q} < \frac{1}{3} + \frac{1}{r}.$$

For any positive T, the bilinear functional B maps $K_p(T) \times K_q(T)$ into $K_r(T)$. Moreover, a constant C (independent of T) exists such that

$$\|B(u,v)\|_{K_r(T)} \leq \frac{C}{\nu} \|u\|_{K_p(T)} \|v\|_{K_q(T)}.$$

Proof. This will involve writing B as a convolution operator. More precisely, we have the following lemma.

Lemma 5.30. *Define the operator L_m by*

$$\begin{cases} \partial_t L_m f - \nu \Delta L_m f + \nabla P = \partial_m f \\ \operatorname{div} L_m f = 0 \\ L_m f_{|t=0} = 0. \end{cases}$$

We have

$$L_m f^j(t,x) = \sum_k \int_0^t \int_{\mathbb{R}^3} \Gamma_{m,k}^j(t-t',y) \star f^k(t',x-y) \Big) dt',$$

where the functions $\Gamma_{k,\ell}^j$ belong to $\mathcal{C}(]0, \infty[; L^s)$ for any s in $[1, \infty[$ and satisfy

$$|\Gamma_{m,k}^j(t,x)| \leq \frac{C}{(\sqrt{\nu t} + |x|)^4}. \tag{5.15}$$

Proof. In Fourier space, we have

$$\mathcal{F}L_m f^j(t, \xi) = i \int_0^t e^{-\nu(t-t')|\xi|^2} \sum_{k=1}^3 \xi_m \mathcal{F}(\mathbb{P}f)^j(t', \xi) \, dt'.$$

Using the computation of the pressure (5.4), we get

$$\mathcal{F}L_m f^j(t, \xi) = i \int_0^t e^{-\nu(t-t')|\xi|^2} \sum_{k=1}^3 \alpha_{k,j} \frac{\xi_j \xi_k \xi_m}{|\xi|^2} \widehat{f^k}(t', \xi) \, dt'.$$

Thus, defining

$$\Gamma_{m,k}^j(t, \cdot) \overset{\text{def}}{=} i\alpha_{k,j} \mathcal{F}^{-1}\left(e^{-\nu t|\xi|^2} \frac{\xi_j \xi_k \xi_m}{|\xi|^2}\right)$$

gives the lemma, provided that we have the pointwise estimate (5.15). Define, for $\beta \in \mathbb{N}^3$ with length 3,

$$\Gamma_\beta(t, \cdot) \overset{\text{def}}{=} i\mathcal{F}^{-1}\left(e^{-\nu t|\xi|^2} \frac{\xi^\beta}{|\xi|^2}\right).$$

Using the fact that

$$e^{-\nu t|\xi|^2} |\xi|^{-2} = \nu \int_t^\infty e^{-\nu t'|\xi|^2} \, dt',$$

we get

$$\Gamma_\beta(t, x) = (2\pi)^{-3} \nu i \int_t^\infty \int_{\mathbb{R}^3} \xi^\beta e^{i(x|\xi) - \nu t'|\xi|^2} \, dt' \, d\xi$$

$$= -(2\pi)^{-3} \nu \partial^\beta \int_t^\infty \int_{\mathbb{R}^3} e^{i(x|\xi) - \nu t'|\xi|^2} \, dt' \, d\xi.$$

Using the formula (1.20) page 18 for the Fourier transform of Gaussian functions, we obtain

$$\Gamma_\beta(t, x) = -\nu \partial^\beta \int_t^\infty \frac{1}{(4\pi\nu t')^{\frac{3}{2}}} e^{-\frac{|x|^2}{4\nu t'}} \, dt'$$

$$= -\frac{\nu}{\pi^{\frac{3}{2}}} \int_t^\infty \frac{1}{(4\nu t')^3} \Psi_\beta\left(\frac{x}{\sqrt{4\nu t'}}\right) \, dt' \quad \text{with} \quad \Psi_\beta(z) \overset{\text{def}}{=} \partial^\beta e^{-|z|^2}.$$

The change of variable $r = (4\nu t')^{-1}|x|^2$ leads to

$$|\Gamma_\beta(t, x)| \leq \frac{C}{|x|^4} \int_0^{\frac{|x|^2}{4\nu t}} r \Psi_\beta\left(\frac{x}{|x|} r^{\frac{1}{2}}\right) \, dr.$$

This implies that

$$|\Gamma_\beta(t,x)| \le c\min\left\{\frac{1}{(\nu t)^2}, \frac{1}{|x|^4}\right\} \tag{5.16}$$

and thus that

$$\|\Gamma_{k,\ell}^j(t,\cdot)\|_{L^s} \le \frac{C}{(\nu t)^{2-\frac{3}{2s}}}.$$

In order to prove the continuity, we observe that there exists some $\delta > 0$ such that for $0 < c \le t_1 \le t_2$, we have

$$|\Gamma_{k,\ell}^j(t_2,x) - \Gamma_{k,\ell}^j(t_1,x)| \le \frac{C}{|x|^4}\int_{\frac{|x|^2}{4\nu t_2}}^{\frac{|x|^2}{4\nu t_1}} re^{-\delta r}\,dr.$$

This implies that

$$|\Gamma_{k,\ell}^j(t_2,x) - \Gamma_{k,\ell}^j(t_1,x)| \le C\min\left\{\frac{t_2^2-t_1^2}{(\nu t_1 t_2)^2}, \frac{1}{|x|^4}\right\}.$$

The lemma is thus proved. □

Completion of the proof of Lemma 5.29. Thanks to Young's and Hölder's inequalities, and to the condition

$$\frac{1}{r} \le \frac{1}{p} + \frac{1}{q} \le 1,$$

we have, according to Lemma 5.30 with s defined by $1 + \dfrac{1}{r} = \dfrac{1}{s} + \dfrac{1}{p} + \dfrac{1}{q}$,

$$\|B(u,v)(t)\|_{L^r} \le C\int_0^t \frac{1}{\sqrt{\nu(t-t')}^{\,4-3\left(1+\frac{1}{r}-\frac{1}{p}-\frac{1}{q}\right)}}\|u(t')\|_{L^p}\|v(t')\|_{L^q}\,dt'.$$

By the definition of the $K_p(T)$ norms, we thus get that

$$\|B(u,v)(t)\|_{L^r} \le C\|u\|_{K_p(T)}\|v\|_{K_q(T)}$$
$$\times \int_0^t \frac{1}{\sqrt{\nu(t-t')}^{\,1-3\left(\frac{1}{r}-\frac{1}{p}-\frac{1}{q}\right)}}\frac{1}{\sqrt{\nu t'}^{\,2-3\left(\frac{1}{p}+\frac{1}{q}\right)}}\,dt'$$
$$\le \frac{C}{\nu}\frac{1}{\sqrt{\nu t}^{\,1-\frac{3}{r}}}\|u\|_{K_p(T)}\|v\|_{K_q(T)}.$$

Lemma 5.29 is proved, and thus Lemma 5.5 implies Theorem 5.27. □

Completion of the proof of Theorem 5.21. According to Remark 5.24 we may apply Theorem 5.27 with $p = 6$ and T suitably small. Note that if the initial data is small in L^3, then the inequality (5.12) enables us to take $T = \infty$. Hence, it remains only to check the following two points:

– The solution u is continuous with values in L^3.
– The solution u is unique among all continuous functions with values in L^3.

These two problems are solved using a method which turns out to be important in the study of the (generalized) Navier–Stokes system: It consists in the consideration of the new unknown

$$w \overset{\text{def}}{=} u - e^{\nu t \Delta} u_0.$$

The idea is that w is *smoother* than u. Obviously, we have $w = B(u, u)$. Lemma 5.29 applied with $p = q = 6$ and $r = 3$ implies that w belongs to $\mathcal{C}(]0, T]; L^3(\mathbb{R}^3))$. The continuity of w at the origin will follow from the fact that, still using Lemma 5.29, we have

$$\|w\|_{L^\infty([0,t];L^3)} \leq \frac{C}{\nu} \|u\|_{K_6(t)}^2.$$

However, the solution u given by Lemma 5.5 satisfies

$$\|u\|_{K_6(t)} \leq 2\|e^{\nu t \Delta} u_0\|_{K_6(t)}.$$

Remark 5.24 thus implies that $\lim_{t \to 0} \|w\|_{L^\infty([0,t];L^3)} = 0$. As the heat flow is continuous with values in L^3, we have proven that the solution u is continuous with values in L^3.

We will now prove that there is at most one solution in $\mathcal{C}([0, T]; L^3)$. Observe that by applying Lemma 5.29 with $p = q = 3$ and $r = 2$, we get

$$w = B(u, u) \in K_2(T).$$

In particular, w belongs to $\mathcal{C}([0, T]; L^2)$. Consider two solutions u_1 and u_2 of (GNS_ν) in the space $\mathcal{C}([0, T]; L^3)$ associated with the same initial data and denote by u_{21} the difference $u_2 - u_1$. Because $u_{21} = w_2 - w_1$, it belongs to $\mathcal{C}([0, T]; L^2)$ and satisfies

$$\begin{cases} \partial_t u_{21} - \nu \Delta u_{21} = f_{21} \\ u_{21|t=0} = 0 \end{cases} \quad \text{with}$$

$$f_{21} = Q(e^{\nu t \Delta} u_0, u_{21}) + Q(u_{21}, e^{\nu t \Delta} u_0) + Q(w_2, u_{21}) + Q(u_{21}, w_1).$$

Via Sobolev embeddings, we have

$$\|Q(a, b)\|_{\dot{H}^{-\frac{3}{2}}} \leq C \sup_{1 \leq k, \ell \leq 3} \|a^k b^\ell\|_{\dot{H}^{-\frac{1}{2}}}$$

$$\leq C \sup_{1 \leq k, \ell \leq 3} \|a^k b^\ell\|_{L^{\frac{3}{2}}} \tag{5.17}$$

$$\leq C \|a\|_{L^3} \|b\|_{L^3}. \tag{5.18}$$

Thus, the external force f_{21} belongs to $L^2([0, T]; \dot{H}^{-\frac{3}{2}})$. As u_{21} is the unique solution in the space of continuous functions with values in $\mathcal{S}'$, we infer that u_{21} belongs to

$$L^\infty([0,T]; \dot{H}^{-\frac{1}{2}}) \cap L^2([0,T]; \dot{H}^{\frac{1}{2}})$$

and satisfies, thanks to Lemma 5.10,

$$
\begin{aligned}
U_{21}(t) &\overset{\text{def}}{=} \|u_{21}(t)\|_{\dot{H}^{-\frac{1}{2}}}^2 + 2\nu \int_0^t \|u_{21}(t')\|_{\dot{H}^{\frac{1}{2}}}^2 \, dt' \\
&= 2 \int_0^t \langle f_{21}(t'), u_{21}(t') \rangle_{-\frac{1}{2}} \, dt' \\
&\leq 2 \int_0^t \|f_{21}(t')\|_{\dot{H}^{-\frac{3}{2}}} \|u_{21}(t')\|_{\dot{H}^{\frac{1}{2}}} \, dt'. \qquad (5.19)
\end{aligned}
$$

As the space of continuous and compactly supported functions is dense in L^3, we may decompose u_0 into the sum of a small function in L^3 norm and a (possibly large) function of L^6:

$$u_0 = u_0^\sharp + u_0^\flat \quad \text{with} \quad \|u_0^\sharp\|_{L^3} \leq c\nu \quad \text{and} \quad u_0^\flat \in L^6. \qquad (5.20)$$

Defining $g_{21} \overset{\text{def}}{=} f_{21} - Q(e^{\nu t \Delta} u_0^\flat, u_{21}) - Q(u_{21}, e^{\nu t \Delta} u_0^\flat)$ and applying (5.18) gives, again via Sobolev embeddings,

$$
\begin{aligned}
A_{21}(t) &\overset{\text{def}}{=} \|g_{21}(t)\|_{\dot{H}^{-\frac{3}{2}}} \\
&\leq C\Big(\|e^{\nu t \Delta} u_0^\sharp\|_{L^3} + \|w_1\|_{K_3(t)} + \|w_2\|_{K_3(t)} \Big) \|u_{21}(t)\|_{L^3} \\
&\leq C\Big(\|u_0^\sharp\|_{L^3} + \|w_1\|_{K_3(t)} + \|w_2\|_{K_3(t)} \Big) \|u_{21}(t)\|_{\dot{H}^{\frac{1}{2}}}.
\end{aligned}
$$

If t is sufficiently small, and c is chosen sufficiently small in (5.20), we get

$$A_{21}(t) \leq \frac{\nu}{4} \|u_{21}(t)\|_{\dot{H}^{\frac{1}{2}}}. \qquad (5.21)$$

Still using Sobolev embeddings and Hölder inequality, we can write

$$
\begin{aligned}
B_{21}(t) &\overset{\text{def}}{=} \Big\| Q(e^{\nu t \Delta} u_0^\flat, u_{21}) + Q(u_{21}, e^{\nu t \Delta} u_0^\flat) \Big\|_{\dot{H}^{-\frac{3}{2}}} \\
&\leq C \sup_{1 \leq k,\ell \leq d} \|(e^{\nu t \Delta} u_0^{\flat,k}) u_{21}^\ell\|_{L^{\frac{3}{2}}} \\
&\leq C \|e^{\nu t \Delta} u_0^\flat\|_{L^6} \|u_{21}(t)\|_{L^2}.
\end{aligned}
$$

Using the fact that the heat flow is a contraction over the L^p spaces and then the interpolation inequality between $\dot{H}^{-\frac{1}{2}}$ and $\dot{H}^{\frac{1}{2}}$, we get

$$B_{21}(t) \leq C \|u_0^\flat\|_{L^6} \|u_{21}(t)\|_{\dot{H}^{-\frac{1}{2}}}^{\frac{1}{2}} \|u_{21}(t)\|_{\dot{H}^{\frac{1}{2}}}^{\frac{1}{2}}.$$

Using (5.19) and (5.21), we then deduce that

$$\|u_{21}(t)\|^2_{\dot{H}^{-\frac{1}{2}}} + \frac{3}{2}\nu \int_0^t \|u_{21}(t')\|^2_{\dot{H}^{\frac{1}{2}}}\, dt'$$

$$\leq C\|u_0^\flat\|_{L^6} \int_0^t \|u_{21}(t')\|^{\frac{1}{2}}_{\dot{H}^{-\frac{1}{2}}} \|u_{21}(t')\|^{\frac{3}{2}}_{\dot{H}^{\frac{1}{2}}}\, dt'.$$

Using the classical convexity inequality $ab \leq \frac{1}{4}a^4 + \frac{3}{4}b^{\frac{4}{3}}$, we then get

$$\|u_{21}(t)\|^2_{\dot{H}^{-\frac{1}{2}}} + \nu \int_0^t \|u_{21}(t')\|^2_{\dot{H}^{\frac{1}{2}}}\, dt' \leq \frac{C}{\nu^3}\|u_0^\flat\|^4_{L^6} \int_0^t \|u_{21}(t')\|^2_{\dot{H}^{-\frac{1}{2}}}\, dt'.$$

The Gronwall lemma implies that $u_{21} \equiv 0$ on a sufficiently small time interval. Basic connectivity arguments then yield uniqueness on $[0, T]$. This completes the proof of Theorem 5.21. □

5.5 The Endpoint Space for Picard's Scheme

According to Theorems 2.34 and 5.27, the generalized Navier–Stokes system (GNS_ν) is globally well posed whenever the initial data u_0 is small with respect to ν in the homogeneous Besov space $\dot{B}_{p,\infty}^{-1+\frac{3}{p}}$ with $3 < p < \infty$. In this section, we seek to find the largest space for solving (GNS_ν) by means of an iterative scheme. Since the spaces $\dot{B}_{p,\infty}^{-1+\frac{3}{p}}$ are increasing with p, a good candidate would be the space $\dot{B}_{\infty,\infty}^{-1}$. In fact, the following proposition guarantees that it is pointless to go beyond that space.

Proposition 5.31. *Let B be a Banach space continuously embedded in the set $\mathcal{S}'(\mathbb{R}^3)$. Assume that for any (λ, a) in $\mathbb{R}_\star^+ \times \mathbb{R}^3$,*

$$\|f(\lambda(\cdot - a))\|_B = \lambda^{-1}\|f\|_B.$$

B is then continuously embedded in $\dot{B}_{\infty,\infty}^{-1}$.

Proof. As B is continuously included in $\mathcal{S}'$, we have that $|\langle f, e^{-|\cdot|^2}\rangle| \leq C\|f\|_B$. By dilation and translation, we then deduce that

$$\|f\|_{\dot{B}_{\infty,\infty}^{-1}} = \sup_{t>0} t^{\frac{1}{2}}\|e^{t\Delta}f\|_{L^\infty} \leq C\|f\|_B.$$

This proves the proposition. □

It turns out, however, that $\dot{B}_{\infty,\infty}^{-1}$ is too large a space. The main reason why is that if we want to solve the problem using an iterative scheme, then we need $e^{t\Delta}u_0$ to belong to $L^2_{loc}(\mathbb{R}^+ \times \mathbb{R}^3)$ so that $B(e^{t\Delta}u_0, e^{t\Delta}u_0)$ makes sense. Taking into consideration the scaling and translation invariance thus leads to the following definition.

Definition 5.32. *We denote by X_0 the space[4] of tempered distributions u such that*

$$\|u\|_{X_0} \stackrel{def}{=} \|u\|_{\dot{B}_{\infty,\infty}^{-1}} + \sup_{\substack{x\in\mathbb{R}^3 \\ R>0}} R^{-\frac{3}{2}} \Big(\int_{P(x,R)} |e^{t\Delta}u(y)|^2 \, dy \, dt \Big)^{\frac{1}{2}} < \infty,$$

where $P^\nu(x,R) \stackrel{def}{=} [0,\nu^{-1}R^2] \times B(x,R)$ and $B(x,R)$ denotes the ball in $\mathbb{R}^3$ with center x and radius R.

We denote by X^ν the space of functions f on $\mathbb{R}_\star^+ \times \mathbb{R}^3$ such that

$$\|f\|_{X^\nu} \stackrel{def}{=} \sup_{t>0}(\nu t)^{\frac{1}{2}}\|f(t)\|_{L^\infty} + \sup_{\substack{x\in\mathbb{R}^3 \\ R>0}} \nu^{\frac{1}{2}}R^{-\frac{3}{2}} \Big(\int_{P^\nu(x,R)} |f(t,y)|^2 \, dy \, dt \Big)^{\frac{1}{2}} < \infty.$$

We denote by Y^ν the space of functions on $\mathbb{R}_\star^+ \times \mathbb{R}^3$ such that

$$\|f\|_Y \stackrel{def}{=} \sup_{t>0}\nu t\|f(t)\|_{L^\infty} + \sup_{\substack{x\in\mathbb{R}^3 \\ R>0}} \nu R^{-3} \int_{P^\nu(x,R)} |f(t,y)| \, dy \, dt < \infty.$$

Remark 5.33. The spaces X_0 and X^1 are related by the fact that $\|u_0\|_{X_0}$ is equal to $\|e^{t\Delta}u_0\|_{X^1}$. We also emphasize that any space $\dot{B}_{p,\infty}^{-1+\frac{3}{p}}$ with $1 \le p < \infty$ is continuously embedded in X_0. Indeed, since we can assume with no loss of generality that $p \ge 3$, it suffices to note that for any $x \in \mathbb{R}^3$ and $R > 0$, we have

$$\int_0^{R^2} \int_{B(x,R)} |e^{t\Delta}u_0|^2 \, dx \, dt \le |B(x,R)|^{1-\frac{2}{p}} \int_0^{R^2} \Big(\int_{B(x,R)} |e^{t\Delta}u_0|^p \, dx \Big)^{\frac{2}{p}} dt.$$

Now, according to Theorem 2.34, we have, for some constant C,

$$\|e^{t\Delta}u_0\|_{L^p}^2 \le Ct^{-1+\frac{3}{p}}\|u_0\|_{\dot{B}_{p,\infty}^{-1+\frac{3}{p}}}^2,$$

which obviously entails the announced embedding.

We now show that the space Y^ν is stable under mollifiers.

Proposition 5.34. *Let θ be in $\mathcal{S}(\mathbb{R}^3)$. There exists some $C > 0$ such that for all $t > 0$, $\widetilde{f}_\theta \stackrel{def}{=} t^{-\frac{3}{2}}\theta(t^{-\frac{1}{2}}\cdot) \star f(t,\cdot)$ satisfies $\|\widetilde{f}_\theta\|_{Y^\nu} \le C\|f\|_{Y^\nu}$.*

Proof. To simplify the notation, we will just consider the case $\nu = 1$. Observe that for any x in the ball with center 0 and radius R, we have

---

[4] In the original work by H. Koch and D. Tataru in [196], this space is denoted by BMO^{-1}.

$$|\widetilde{f}_\theta(t,x)| \le t^{-\frac{3}{2}} \int_{\mathbb{R}^3} \left|\theta\left(\frac{x-y}{\sqrt{t}}\right)\right| \mathbf{1}_{B(0,2R)}(y)|f(t,y)|\,dy$$

$$+ Ct^{-\frac{3}{2}} \int_{\mathbb{R}^3} \frac{1}{\left(1+\frac{|x-y|}{\sqrt{t}}\right)^4} \frac{t}{R^2}|f(t,y)|\,dy$$

$$\le t^{-\frac{3}{2}} \left(|\theta(t^{-\frac{1}{2}}\cdot)| \star \mathbf{1}_{B(0,2R)}|f(t,\cdot)|\right)(x) + \frac{C}{R^2}\sup_{t>0} t\|f(t,\cdot)\|_{L^\infty}.$$

Hence,

$$\frac{1}{R^3}\|\widetilde{f}_\theta\|_{L^1(P(0,R))} \le \frac{C}{R^3} \int_{P(0,R)} |f(t,y)|\,dt\,dy + C\sup_{t>0} t\|f(t,\cdot)\|_{L^\infty}.$$

This proves the proposition. □

The following theorem tells us that the space X^ν is suitable for solving the generalized Navier–Stokes system.

Theorem 5.35. *A constant c exists such that if u_0 is in X_0 and $\|u_0\|_{X_0} \le c\nu$, then (GNS_ν) has a unique solution u in X^ν such that $\|u\|_{X^\nu} \le 2\|u_0\|_{X_0}$.*

Proof. Using the change of functions

$$u(t,x) = \nu v(\nu t, x) \quad \text{and} \quad u_0(x) = \nu v_0(x),$$

we see that it suffices to treat the case $\nu = 1$. Indeed, we have

$$\|u\|_{X^\nu} = \nu\|v\|_{X^1} \quad \text{and} \quad \|u_0\|_{X^0} = \nu\|v_0\|_{X^0}.$$

Therefore, we assume from now on that $\nu = 1$ and define $X \stackrel{\text{def}}{=} X^1$, $Y \stackrel{\text{def}}{=} Y^1$, and $P(x,R) \stackrel{\text{def}}{=} P^1(x,R)$. According to Lemma 5.5, it suffices to prove that there exists some constant C such that

$$\|B(u,v)\|_X \le C\|u\|_X\|v\|_X. \tag{5.22}$$

Observing that $\|fg\|_Y \le \|f\|_X\|g\|_X$, we see that the above inequality is implied by the following lemma.

Lemma 5.36. *If $\nu = 1$, then the operator L_j defined in Lemma 5.30 maps Y continuously into X.*

Proof. Using Lemma 5.30, we get that

$$(L_j f)^k(t,x) = \sum_{\ell=1}^{3} \Gamma_{j,\ell}^k(t-t', x-y)f^\ell(t',y)\,dt'\,dy$$

with, for all positive real numbers R,

$$|\Gamma_{j,\ell}^k(\tau,\varsigma)| \le \frac{C}{(\sqrt{\tau}+|\varsigma|)^4} \le C'\left(\Gamma_R^{(1)}(\tau,\varsigma) + \Gamma_R^{(2)}(\tau,\varsigma)\right)$$

with $\Gamma_R^{(1)}(\tau,\varsigma) \stackrel{\text{def}}{=} \mathbf{1}_{|\varsigma|\ge R}\frac{1}{|\varsigma|^4}$ and $\Gamma_R^{(2)}(\tau,\varsigma) \stackrel{\text{def}}{=} \mathbf{1}_{|\varsigma|\le R}\frac{1}{(\sqrt{\tau}+|\varsigma|)^4}$.

The functions $\Gamma_R^{(1)}$ and $\Gamma_R^{(2)}$ may be bounded according to the following proposition.

Proposition 5.37. *There exists a constant C such that, for any $R > 0$,*

$$\|\Gamma_R^{(1)} \star f\|_{L^\infty([0,R^2] \times \mathbb{R}^3)} \le \frac{C}{R} \|f\|_Y, \tag{5.23}$$

$$\|\Gamma_R^{(2)} \star f\|_{L^\infty([R^2,\infty[\times \mathbb{R}^3)} \le \frac{C}{R} \|f\|_Y. \tag{5.24}$$

Proof. Splitting $\Gamma_R^{(1)} \star f$ into a sum of integrals over the annuli $C(0, 2^p R, 2^{p+1} R)$ yields

$$|(\Gamma_R^{(1)} \star f)(t, x)| \le \sum_{p=0}^{\infty} \int_0^t \int_{C(0,2^p R, 2^{p+1} R)} \frac{1}{|y|^4} |f(t', x - y)| \, dy \, dt'$$

$$\le \frac{1}{R} \sum_{p=0}^{\infty} 2^{-p+3} (2^{p+1} R)^{-3} \int_0^t \int_{B(0, 2^{p+1} R)} |f(t', x - y)| \, dy \, dt'.$$

As p is nonnegative, we have, for $t \le R^2$,

$$|\Gamma_R^{(1)} \star f(t, x)| \le \frac{C}{R} \sum_{p=0}^{\infty} 2^{-p} (2^{p+1} R)^{-3} \int_{P(x, 2^{p+1} R)} |f(t, z)| \, dt \, dz$$

$$\le \frac{C}{R} \sum_{p=0}^{\infty} 2^{-p} \sup_{R'>0} \frac{1}{R'^3} \int_{P(x, R')} |f(t, z)| \, dt \, dz.$$

By the definition of $\| \cdot \|_Y$, the inequality (5.23) is proved.

In order to prove the second inequality, we observe that for all $x \in \mathbb{R}^3$ and $t \ge R^2$, we have

$$|(\Gamma_R^{(2)} \star f)(t, x)| \le \Gamma_R^{(21)}(t, x) + \Gamma_R^{(22)}(t, x) \quad \text{with}$$

$$\Gamma_R^{(21)}(t, x) \stackrel{\text{def}}{=} \int_0^{\min(R^2, \frac{t}{2})} \int_{B(0,R)} \frac{1}{(\sqrt{t - t'} + |y|)^4} |f(t', x - y)| \, dy \, dt',$$

$$\Gamma_R^{(22)}(t, x) \stackrel{\text{def}}{=} \int_{\min(R^2, \frac{t}{2})}^t \int_{B(0,R)} \frac{1}{(\sqrt{t - t'} + |y|)^4} |f(t', x - y)| \, dy \, dt'.$$

To bound $\Gamma_R^{(21)}(t, x)$, we use the fact that $t \le 2(t - t')$. We get

$$\Gamma_R^{(21)}(t, x) \le C \frac{R^3}{t^2} \left(\frac{1}{R^3} \int_0^{R^2} \int_{B(0,R)} |f(t', x - y)| \, dt' \, dy \right)$$

so that, for any $t \ge R^2$ and x in $\mathbb{R}^3$,

$$\Gamma_R^{(21)}(t, x) \le \frac{C}{t^{\frac{1}{2}}} \|f\|_Y. \tag{5.25}$$

In order to estimate $\Gamma_R^{(22)}$, we use the facts that $t \leq 2t'$ and, for any $a > 0$,

$$\int_{B(0,R)} \frac{dy}{(a+|y|)^4} \leq \frac{1}{a} \int_{\mathbb{R}^3} \frac{dz}{(1+|z|)^4}.$$

This enables us to write that

$$\Gamma_R^{(22)}(t,x) \leq \int_{\min(R^2,\frac{t}{2})}^t \int_{B(0,R)} \frac{1}{(\sqrt{t-t'}+|y|)^4} \|f(t',\cdot)\|_{L^\infty} \, dy \, dt'$$

$$\leq C\|f\|_Y \left(\int_{t/2}^t \frac{1}{\sqrt{t-t'}} \frac{dt'}{t'} + \int_{R^2}^t \frac{|B(0,R)|}{t^2} \frac{dt'}{t'} \right)$$

$$\leq C\|f\|_Y \left(\frac{1}{t^{\frac{1}{2}}} + \frac{1}{R^2} \frac{tR^3}{t^2} \right).$$

As $R \leq \sqrt{t}$ this completes the proof of the proposition.   □

Completion of the proof of Lemma 5.36. Note that applying the above proposition with $R = \sqrt{t}$ yields

$$\|(L_j f)(t,\cdot)\|_{L^\infty} \leq \frac{C}{t^{\frac{1}{2}}} \|f\|_Y. \tag{5.26}$$

Hence, it suffices to estimate $\|L_j f\|_{L^2(P(x,R))}$ for an arbitrary $x \in \mathbb{R}^3$. Using translations and dilations, we can assume that $x = 0$ and $R = 1$. We write

$$L_j f = L_j(\mathbf{1}_{{}^c B(0,2)} f) + L_j(\mathbf{1}_{B(0,2)} f).$$

Observing that for any $y \in B(0,1)$ we have

$$|L_j(\mathbf{1}_{{}^c B(0,2)} f)(t,y)| \leq C K_1^{(1)} \star (\mathbf{1}_{{}^c B(0,2)} |f|)(t,y)$$

and using the inequality (5.23), we get

$$\|L_j(\mathbf{1}_{{}^c B(0,2)} f)\|_{L^\infty(P(0,1))} \leq C\|f\|_Y.$$

As the volume of $P(0,1)$ is finite we infer that

$$\|L_j(\mathbf{1}_{{}^c B(0,2)} f)\|_{L^2(P(0,1))} \leq C\|f\|_Y. \tag{5.27}$$

The proof of Lemma 5.36 is now reduced to the proof of the following proposition.

Proposition 5.38. *For any function* $f : [0,1] \times \mathbb{R}^3 \to \mathbb{R}$ *such that* $f(t,\cdot)$ *is supported in* $B(0,2)$ *for all* $t \in [0,1]$, *we have*

$$\|(L_j f)(t,\cdot)\|_{L^\infty} \leq C\|f\|_Y \quad \text{for all} \quad t \in [0,1].$$

Proof. We decompose f into low and high frequencies, in the sense of the heat flow:

$$f = f^\flat + f^\sharp \quad \text{with} \quad f^\flat(t, \cdot) \overset{\text{def}}{=} \mathcal{F}^{-1}(\widehat{\theta}(t^{\frac{1}{2}}\xi)\widehat{f}(t, \xi)),$$

where θ denotes a function such that $\widehat{\theta}$ is compactly supported and with value 1 near the origin. We write

$$\|f^\sharp\|_{L^2([0,1];\dot{H}^{-1})}^2 = (2\pi)^{-3} \int_{[0,1]\times\mathbb{R}^3} \frac{|1 - \widehat{\theta}(t^{\frac{1}{2}}\xi)|^2}{t|\xi|^2} \, t|\widehat{f}(t, \xi)|^2 \, dt \, d\xi$$

$$\leq C \int_{[0,1]\times\mathbb{R}^3} t\|f(t, \cdot)\|_{L^2}^2 \, dt$$

$$\leq C\|f\|_{L^1([0,1]\times\mathbb{R}^3)} \sup_{t>0} t\|f(t, \cdot)\|_{L^\infty}.$$

Using the energy estimate for the heat equation, we thus end up with

$$\|L_j f^\sharp\|_{L^2([0,1]\times\mathbb{R}^3)} \leq C\|f\|_Y. \tag{5.28}$$

We now estimate $\|L_j f^\flat\|_{L^2([0,1]\times\mathbb{R}^3)}$. First, observe that by the definitions of L_j and $f^\flat$, we have

$$L_j f^\flat = \partial_j \int_0^t e^{(t-t')\Delta} f^\flat(t') \, dt'$$

$$= \partial_j e^{t\Delta} \int_0^t \widetilde{f}^\flat(t') \, dt' \quad \text{with} \quad \mathcal{F}\widetilde{f}^\flat(t', \xi) \overset{\text{def}}{=} e^{t'|\xi|^2}\widehat{\theta}(t'|\xi|^2)\widehat{f}(t, \xi).$$

Note that, by the definition of θ, we have

$$\widetilde{f}^\flat(t, \cdot) = t^{-\frac{3}{2}}\widetilde{\theta}\left(\frac{\cdot}{\sqrt{t}}\right) \star f(t, \cdot) \quad \text{with} \quad \widetilde{\theta} \in \mathcal{S}(\mathbb{R}^3). \tag{5.29}$$

Thus,

$$\mathcal{L}f \overset{\text{def}}{=} \sum_{j=1}^3 \|L_j f^\flat\|_{L^2([0,1]\times\mathbb{R}^3)}^2$$

$$= \int_0^1 \left\|\nabla e^{t\Delta} \int_0^t \widetilde{f}^\flat(t') \, dt'\right\|_{L^2}^2 \, dt.$$

By symmetry, we have

$$\mathcal{L}f = 2 \int_0^1 \int_0^t \int_0^{t'} \left(\nabla e^{t\Delta}\widetilde{f}^\flat(t'') \middle| \nabla e^{t\Delta}\widetilde{f}^\flat(t')\right)_{L^2} dt'' \, dt' \, dt.$$

By integration by parts and because $e^{t\Delta}$ is self-adjoint on L^2, we get

$$\left(\nabla e^{t\Delta}\widetilde{f}^\flat(t'') \middle| \nabla e^{t\Delta}\widetilde{f}^\flat(t')\right)_{L^2} = -\langle \Delta e^{2t\Delta}\widetilde{f}^\flat(t''), \widetilde{f}^\flat(t')\rangle.$$

Moreover, as $2\Delta e^{2t\Delta} = \partial_t e^{2t\Delta}$, we infer that

$$\left(\nabla e^{t\Delta}\widetilde{f^\flat}(t'')\,\Big|\,\nabla e^{t\Delta}\widetilde{f^\flat}(t')\right)_{L^2} = -\frac{1}{2}\frac{d}{dt}\langle e^{2t\Delta}\widetilde{f^\flat}(t''), \widetilde{f^\flat}(t')\rangle.$$

We then deduce that

$$\mathcal{L}f = -\int_0^1\int_0^{t'}\left\langle\left(\int_{t'}^1\frac{d}{dt}e^{2t\Delta}\,dt\right)\widetilde{f^\flat}(t''), \widetilde{f^\flat}(t')\right\rangle dt''\,dt'$$

$$= \int_0^1\left\langle(e^{2t'\Delta} - e^{2\Delta})\int_0^{t'}\widetilde{f^\flat}(t'')\,dt'', \widetilde{f^\flat}(t')\right\rangle dt'$$

$$\leq \|\widetilde{f^\flat}\|_{L^1([0,1]\times\mathbb{R}^3)}\sup_{t'\in[0,1]}\left\|(e^{2t'\Delta} - e^{2\Delta})\int_0^{t'}\widetilde{f^\flat}(t'')\,dt''\right\|_{L^\infty}.$$

First, note that using (5.29) and the fact that the operator $e^{2\Delta}$ maps $L^1(\mathbb{R}^3)$ into $L^\infty(\mathbb{R}^3)$, we have

$$\left\|e^{2\Delta}\int_0^{t'}\widetilde{f^\flat}(t'')\,dt''\right\|_{L^\infty} \leq C\|f\|_{L^1([0,1]\times\mathbb{R}^3)}. \tag{5.30}$$

Thanks to Proposition 5.34, $\widetilde{f^\flat}$ belongs to Y. We write

$$\left|e^{2t'\Delta}\int_0^{t'}\widetilde{f^\flat}(t'',x)\,dt''\right| \leq \sum_{n\in\mathbb{Z}^3}\frac{1}{(4\pi t')^{\frac{3}{2}}}\int_{\mathbb{R}^3}\int_0^{t'}e^{-\frac{|x-y|^2}{4t'}}\mathbf{1}_{B_{n,t'}}(y)\big|\widetilde{f^\flat}(t'',y)\big|\,dt''dy,$$

where $B_{n,t'}$ denotes the ball with center $n\sqrt{t'}$ and radius $\sqrt{t'}$. Using translation invariance, it is enough to estimate the above integral at the point $x = 0$. We write, thanks to Proposition 5.34,

$$\left|\left(e^{2t'\Delta}\int_0^{t'}\widetilde{f^\flat}(t'')\,dt''\right)(0)\right| \leq \sum_{|n|>2}e^{-\frac{|n|^2}{4}}\left(\frac{1}{|n|^3}\int_{P(n,t')}|\widetilde{f^\flat}(t'',y)|\,dt''dy\right)$$

$$+ \sum_{|n|\leq 2}\frac{1}{(4\pi t')^{\frac{3}{2}}}\int_{\mathbb{R}^3}\int_0^{t'}e^{-\frac{|x-y|^2}{4t'}}\mathbf{1}_{B_{n,t'}}(y)\big|\widetilde{f^\flat}(t'',y)\big|\,dt''dy$$

$$\leq C\|f\|_Y.$$

Thanks to the inequality (5.28), this completes the proof of the proposition. □

As explained above, this completes the proof of Lemma 5.36 and thus the proof of Theorem 5.35. □

5.6 The Use of the L^1-smoothing Effect of the Heat Flow

According to Theorem 2.34 page 72, the smallness condition (5.14) in the case where $T = \infty$ satisfies the smallness condition for the $\dot{B}_{p,\infty}^{-1+\frac{3}{p}}$ norm. The purpose of this section is to provide another approach to Theorem 5.27, one which relies on Littlewood–Paley theory and on the smoothing effect of the heat flow described in Corollary 2.5 page 55.

5.6.1 The Cannone–Meyer–Planchon Theorem Revisited

We assume that u_0 belongs to $\dot{B}_{p,\infty}^{-1+\frac{3}{p}}$. We deduce from Lemma 2.4 page 54 that $\|\dot{\Delta}_j e^{\nu t \Delta} u_0\|_{L^p} \leq C e^{-c\nu t 2^{2j}} \|\dot{\Delta}_j u_0\|_{L^p}$. By time integration, we get

$$\|\dot{\Delta}_j e^{\nu t \Delta} u_0\|_{L^1(L^p)} \leq \frac{C}{\nu 2^{2j}} 2^{-j(-1+\frac{3}{p})} \|u_0\|_{\dot{B}_{p,\infty}^{-1+\frac{3}{p}}}. \qquad (5.31)$$

This leads to the following definition.

Definition 5.39. *For p in $[1,\infty]$, we denote by E_p the space of functions u in $L^\infty(\mathbb{R}^+; \dot{B}_{p,\infty}^{-1+\frac{3}{p}})$ such that*

$$\|u\|_{E_p} \overset{def}{=} \sup_j 2^{j\left(-1+\frac{3}{p}\right)} \|\dot{\Delta}_j u\|_{L^\infty(L^p)} + \sup_j \nu 2^{2j} 2^{j\left(-1+\frac{3}{p}\right)} \|\dot{\Delta}_j u\|_{L^1(L^p)}$$

is finite.

We note that the estimate (5.31) implies that

$$\|e^{\nu t \Delta} u_0\|_{E_p} \leq C \|u_0\|_{\dot{B}_{p,\infty}^{-1+\frac{3}{p}}}.$$

This motivates the following statement (which should be compared with the global existence result stated in Theorem 5.27).

Theorem 5.40. *Let $p \in [1,\infty[$. There exists a constant c such that the system (GNS_ν) has a unique solution u in the ball with center 0 and radius $2c\nu$ in E_p whenever $\|u_0\|_{\dot{B}_{p,\infty}^{-1+\frac{3}{p}}} \leq c\nu$.*

Proof. Since the proof relies on Lemma 5.5, it suffices to prove the following.

Lemma 5.41. *There exists a constant C such that for any p in $[1,\infty[$,*

$$\|B(u,v)\|_{E_p} \leq \frac{Cp}{\nu} \|u\|_{E_p} \|v\|_{E_p}. \qquad (5.32)$$

Proof. We recall that the nonlinear term $Q(u,v)$ can be written as

$$Q^m(u,v) = \sum_{k,\ell} A_{k,\ell}^m(D)(u^k v^\ell),$$

where the $A_{k,\ell}^m(D)$ are homogeneous Fourier multipliers of degree 1. With the notation of Chapter 2 page 61, we may write

$$u^k v^\ell = \sum_j \dot{S}_j u^k \dot{\Delta}_j v^\ell + \sum_j \dot{\Delta}_j u^k \dot{S}_{j+1} v^\ell.$$

As the supports of the Fourier transforms of $\dot{S}_j u^k \dot{\Delta}_j v^\ell$ and $\dot{\Delta}_j u^k \dot{S}_{j+1} v^\ell$ are included in $2^j \mathcal{B}$ for some ball $\mathcal{B}$ in $\mathbb{R}^3$, an integer N_0 exists such that if j' is less than $j - N_0$, then

$$\dot{\Delta}_j Q(\dot{S}_{j'} u, \dot{\Delta}_{j'} v) = \dot{\Delta}_j Q(\dot{\Delta}_{j'} u, \dot{S}_{j'+1} v) = 0. \tag{5.33}$$

We now decompose B as

$$B(u, v) = B_1(u, v) + B_2(u, v) \quad \text{with}$$
$$B_1(u, v) \overset{\text{def}}{=} \sum_j B(\dot{S}_j u, \dot{\Delta}_j v) \quad \text{and} \quad B_2(u, v) \overset{\text{def}}{=} \sum_j B(\dot{\Delta}_j u, \dot{S}_{j+1} v).$$

According to (5.33) and the definition of B in Fourier space, we have

$$\dot{\Delta}_j B_1(u, v) \overset{\text{def}}{=} \sum_{j' \geq j - N_0} \dot{\Delta}_j B(\dot{S}_{j'} u, \dot{\Delta}_{j'} v), \tag{5.34}$$

$$\dot{\Delta}_j B_2(u, v) \overset{\text{def}}{=} \sum_{j' \geq j - N_0} \dot{\Delta}_j B(\dot{\Delta}_{j'} u, \dot{S}_{j'+1} v). \tag{5.35}$$

We shall treat only B_1 since B_2 is similar. Using Lemma 2.1 page 52, we infer that

$$\|\dot{\Delta}_j Q(\dot{S}_{j'} u, \dot{\Delta}_{j'} v)\|_{L^p} \leq C 2^j \sup_{k,\ell} \|\dot{S}_{j'} u^k \dot{\Delta}_{j'} v^\ell\|_{L^p}.$$

Hence, using Lemma 2.4 page 54, we get

$$\|\dot{\Delta}_j B(\dot{S}_{j'} u, \dot{\Delta}_{j'} v)(t)\|_{L^p} \leq C \int_0^t e^{-c\nu(t-t')2^{2j}} \|\dot{\Delta}_j Q(\dot{S}_{j'} u(t'), \dot{\Delta}_{j'} v(t'))\|_{L^p} dt'$$

$$\leq C 2^j \int_0^t e^{-c\nu(t-t')2^{2j}} \sup_{k,\ell} \|\dot{S}_{j'} u^k(t') \dot{\Delta}_{j'} v^\ell(t')\|_{L^p} dt'$$

$$\leq C 2^j \int_0^t e^{-c\nu(t-t')2^{2j}} \|\dot{S}_{j'} u(t')\|_{L^\infty} \|\dot{\Delta}_{j'} v(t')\|_{L^p} dt'.$$

By the definitions of the operators $\dot{S}_j$ and of the E_p norm, we get, thanks to Lemma 2.1,

$$\|\dot{S}_{j'} u(t')\|_{L^\infty} \leq \sum_{j'' < j'} \|\dot{\Delta}_{j''} u(t')\|_{L^\infty}$$

$$\leq \sum_{j'' < j'} 2^{j'' \frac{3}{p}} \|\dot{\Delta}_{j''} u(t')\|_{L^p}$$

$$\leq C 2^{j'} \|u\|_{E_p}.$$

Thus, we deduce that

$$\|\dot{\Delta}_j B(\dot{S}_{j'}u, \dot{\Delta}_{j'}v)(t)\|_{L^p} \le C\|u\|_{E_p} 2^j 2^{j'} \int_0^t e^{-c\nu(t-t')2^{2j}} \|\dot{\Delta}_{j'}v(t')\|_{L^p} \, dt'.$$

Using Young's inequality for the time integral, we obtain, by the definition of the E_p norm, that

$$
\begin{aligned}
\mathcal{B}_{j,j'}(u,v) &\stackrel{\text{def}}{=} \|\dot{\Delta}_j B(\dot{S}_{j'}u, \dot{\Delta}_{j'}v)\|_{L^\infty(L^p)} + \nu 2^{2j}\|\dot{\Delta}_j B(\dot{S}_{j'}u, \dot{\Delta}_{j'}v)\|_{L^1(L^p)} \\
&\le C\|u\|_{E_p} 2^j 2^{j'}\|\dot{\Delta}_{j'}v\|_{L^1(L^p)} \\
&\le \frac{C}{\nu}\|u\|_{E_p}\|v\|_{E_p} 2^j 2^{-j'\frac{3}{p}}.
\end{aligned}
$$

Thanks to (5.34) and (5.35), we thus get

$$\sup_j 2^{j\left(-1+\frac{3}{p}\right)}\left(\|\dot{\Delta}_j B_1(u,v)\|_{L_T^\infty(L^p)} + \nu 2^{2j}\|\dot{\Delta}_j B_1(u,v)\|_{L_T^1(L^p)}\right)$$

$$\le \frac{C}{\nu}\|u\|_{E_p}\|v\|_{E_p} \sum_{j'\ge j-N_0} 2^{-(j'-j)\frac{3}{p}}.$$

The lemma, and thus Theorem 5.40, is proved. $\qquad\square$

Remark 5.42. For any divergence-free data u_0 in $\dot{B}_{p,\infty}^{-1+\frac{3}{p}}$ (with $p \in [1,\infty[$), we can construct a *local* solution which belongs to the space E_p restricted to $[0,T]$.

5.6.2 The Flow of the Solutions of the Navier–Stokes System

In this final section, we seek to determine whether the solutions constructed in the previous sections have flows. We first consider the solutions of (GNS_ν) associated with initial data in the space $H^{\frac{d}{2}-1}$. In what follows, we write $\omega_\eta(r) = r(-\log r)^{1-\eta}$ for η in $]0,1[$ and r in $]0,1]$.

Theorem 5.43. *Let $u \in \mathcal{C}([0,T]; H^{\frac{d}{2}-1}) \cap L^2([0,T]; H^{\frac{d}{2}})$ satisfy (GNS_ν) on the time interval $[0,T]$. Then, u belongs to $L^1([0,T]; C_{\omega_\eta}(\mathbb{R}^d; \mathbb{R}^d))$ for all η in $]0,1/2[$ and there exists a unique continuous map $\psi : [0,T] \times \mathbb{R}^d \to \mathbb{R}^d$ such that*

$$\psi(t,x) = x + \int_0^t u(t', \psi(t',x)) \, dt'.$$

Moreover, ψ belongs to $L^\infty([0,T]; C^{0,1-\varepsilon})$ for any positive ε.

The proof of this theorem relies on the following two lemmas.

Lemma 5.44. *Under the hypotheses of Theorem 5.43, the fluctuation $B(u,u)$ belongs to $L^1([0,T]; B_{2,1}^{\frac{d}{2}+1})$.*

Lemma 5.45. *Let E be a Banach space, $\eta \in {]}0, 1{[}$, and v be a vector field with coefficients in the space $L^1([0,T]; C_{\omega_\eta}(E; E))$. Let*

$$V_\eta(t) = \int_0^t \|v(\tau)\|_{\omega_\eta}\, d\tau \quad \text{and} \quad \omega_{\eta,t}(r) = \exp\left(-\left((\log \tfrac{1}{r})^\eta - \eta V_\eta(t)\right)^{\frac{1}{\eta}}\right).$$

There exists a unique continuous map $\psi : [0,T] \times \mathbb{R}^d \to \mathbb{R}^d$ such that

$$\psi(t,x) = x + \int_0^t v(t', \psi(t',x))\, dt'.$$

Moreover, ψ is such that for any time $t \in [0,T]$, we have $\psi(t, \cdot) \in C_{\omega_{\eta,t}}$ and

$$t \longmapsto \|\psi(t, \cdot)\|_{\omega_{\eta,t}} \in L^\infty([0,T]).$$

In particular, $\psi \in L^\infty([0,T]; C^{0,1-\varepsilon})$ for any positive ε.

Proof of Theorem 5.43. We first introduce some notation. For T positive, s in $\mathbb{R}$, and ρ in $[1,\infty]$, we denote by $\widetilde{L}_T^\rho(H^s)$ the set of tempered distributions u over $[0,T] \times \mathbb{R}^d$ such that

$$\sum_{j \geq -1} 2^{2js} \|\Delta_j u\|_{L_T^\rho(L^2)}^2 < \infty.$$

Since $u = e^{t\nu\Delta} u_0 + B(u,u)$, combining Corollary 2.5 and Lemma 5.44 shows that the solution u belongs to $\widetilde{L}_T^1(H^{\frac{d}{2}+1})$. We claim that $\widetilde{L}_T^1(H^{\frac{d}{2}+1})$ is embedded in the space $L^1([0,T]; C_{\omega_\eta})$ for all $\eta \in {]}0, \tfrac{1}{2}{[}$. Indeed, if x, y are distinct elements of $\mathbb{R}^d$ such that $|x-y| \leq 1$ and $t \in [0,T]$, we may write that

$$
\begin{aligned}
\left|u(t,y) - u(t,x)\right| &\leq |x-y| \sum_{j \leq N} \|\nabla \Delta_j u(t)\|_{L^\infty} + 2\sum_{j > N} \|\Delta_j u(t)\|_{L^\infty} \\
&\leq (2+N)^{1-\eta}|x-y| \sum_{j \leq N} \frac{\|\nabla \Delta_j u(t)\|_{L^\infty}}{(2+j)^{1-\eta}} \\
&\quad + C \sum_{j > N} 2^{-j}(2+j)^{1-\eta} \frac{\|\Delta_j \nabla u(t)\|_{L^\infty}}{(2+j)^{1-\eta}}.
\end{aligned}
$$

Choosing $N = \left[1 - \log|x-y|\right] - 2$ and defining $\alpha_\eta(t) \overset{\text{def}}{=} \sum_j \frac{\|\nabla \Delta_j u(t)\|_{L^\infty}}{(2+j)^{1-\eta}}$,

we deduce that

$$\left|u(t,y) - u(t,x)\right| \leq C\alpha_\eta(t)|x-y|\left(1 - \log|x-y|\right)^{1-\eta}.$$

Bernstein's lemma now ensures that

$$\int_0^T \alpha_\eta(t)\, dt \leq C \sum_j (2+j)^{\eta-1} \int_0^T 2^{j(1+\frac{d}{2})} \|\Delta_j u(t)\|_{L^2}\, dt,$$

from which it follows, according to the Cauchy–Schwarz inequality for series, that

$$\int_0^T \alpha_\eta(t)\, dt \leq C\|u\|_{\widetilde{L}_T^1(H^{\frac{d}{2}+1})}.$$

Therefore, the solution u belongs to the space $L^1([0,T]; C_{\omega_\eta})$. Applying Lemma 5.45 completes the proof of Theorem 5.43.   □

Proof of Lemma 5.44. Since $u \in \mathcal{C}([0,T]; H^{\frac{d}{2}-1}) \cap L^2([0,T]; H^{\frac{d}{2}})$, a straightforward interpolation argument ensures that $u \in L^3([0,T]; H^{\frac{d}{2}-\frac{1}{3}})$. By taking advantage of Hölder's inequality and the continuity results stated in Section 2.8 page 102, we thus find that

$$Q(u,u) \in L^{\frac{3}{2}}([0,T]; H^{\frac{d}{2}-\frac{5}{3}}).$$

Using the smoothing properties of the heat flow (namely Proposition 2.5) and the fact that

$$\partial_t B(u,u) - \nu \Delta B(u,u) = Q(u,u), \qquad B(u,u)(0) = 0,$$

we deduce that

$$B(u,u) \in \widetilde{L}_T^{\frac{3}{2}}(H^{\frac{d}{2}+\frac{1}{3}}) \cap \widetilde{L}_T^3(H^{\frac{d}{2}-\frac{1}{3}}).$$

Of course, as $u_0 \in H^{\frac{d}{2}-1}$, Corollary 2.5 also ensures that $e^{t\nu\Delta}u_0$ belongs to the above space.

In order to complete the proof, it suffices to note that the operator $(a,b) \longmapsto ab$ maps $\left(\widetilde{L}_T^{\frac{3}{2}}(H^{\frac{d}{2}+\frac{1}{3}}) \cap \widetilde{L}_T^3(H^{\frac{d}{2}-\frac{1}{3}})\right)^2$ into $L^1([0,T]; B_{2,1}^{\frac{d}{2}})$. This may be easily proven by taking advantage of Bony's decomposition for ab and the continuity results for the paraproduct and remainder [generalized to the spaces $\widetilde{L}_T^\rho(B_{p,r}^s)$].

The above continuity result now entails that $Q(u,u)$ belongs to the space $L^1([0,T]; B_{2,1}^{\frac{d}{2}-1})$, so once again applying Corollary 2.5 leads to $B(u,u) \in L^1([0,T]; B_{2,1}^{\frac{d}{2}+1})$.   □

Proof of Lemma 5.45. The fact that for any Cauchy data, we have a unique, global, continuous integral curve follows immediately from Theorem 3.2 and the fact that the vector field v belongs to $L^1([0,T]; C_{\omega_\eta}(E;E))$.

In order to prove the regularity of the flow, consider two integral curves, γ_1 and γ_2, of the vector field v, coming, respectively, from x_1 and x_2 such that $\|x_1 - x_2\| < e^{-1}$. By the definition of the space C_{ω_η}, we have

$$\|\gamma_1(t) - \gamma_2(t)\| \leq \|x_1 - x_2\| + \int_0^t \|v(\tau, \gamma_1(\tau)) - v(\tau, \gamma_2(\tau))\|\, d\tau$$

$$\leq \|x_1 - x_2\| + \int_0^t \|v(\tau)\|_{\omega_\eta} \times \omega_\eta(\|\gamma_1(\tau) - \gamma_2(\tau)\|)\, d\tau.$$

We now apply Lemma 3.4 with $\rho(t) = \|\gamma_1(t) - \gamma_2(t)\|$, $\mu = \omega_\eta$, $c = \|x_1 - x_2\|$, and $\gamma(\tau) = \|v(\tau)\|_{\omega_\eta}$. We find that

$$(-\log \|\gamma_1(t) - \gamma_2(t)\|)^\eta \geq (-\log \|x_1 - x_2\|)^\eta - \eta V_\eta(t). \tag{5.36}$$

Assume that $1 + \eta V_\eta(t) \leq (-\log \|x_1 - x_2\|)^\eta$, which means that

$$\|x_1 - x_2\| \leq \exp\left(-(1 + \eta V_\eta(t))^{1/\eta}\right). \tag{5.37}$$

We deduce from the inequality (5.36) that if $\|x_1 - x_2\| \leq \exp(-(1 + \eta V_\eta(t))^{1/\eta})$, then we have

$$\|\gamma_1(t) - \gamma_2(t)\| \leq \exp\left(-\left((-\log \|x_1 - x_2\|)^\eta - \eta V_\eta(t)\right)^{1/\eta}\right).$$

This proves the lemma.   □

To conclude, we consider whether the solutions constructed in Theorem 5.40 have flows. In the following proposition, we establish that constructing such flows *cannot* be done according to Osgood's theorem.

Proposition 5.46. *Let u_0 be a nonzero homogeneous distribution of degree -1 which is smooth outside the origin. Let μ be any admissible modulus of continuity such that $e^{t\Delta}u_0$ belongs to $L^1([0,T]; C_\mu)$ for some positive T. Then, μ does not satisfy the Osgood condition.*

Proof. As u_0 is homogeneous of degree -1, we have $\nabla \dot{S}_j u_0 = 2^{2j}(\nabla \dot{S}_0 u_0)(2^j \cdot)$, hence

$$\|e^{t\Delta}\nabla \dot{S}_j u_0\|_{L^\infty} = 2^{2j}\|e^{t2^{2j}\Delta}\nabla \dot{S}_0 u_0\|_{L^\infty}.$$

Let j_t denote the greatest integer j such that $2^{-2j} \geq t$. According to Definition 2.108, the function Γ given by $\Gamma(y) \overset{\text{def}}{=} y\mu\left(\frac{1}{y}\right)$ is nondecreasing. Since $(e^{\tau\Delta})_{\tau>0}$ is a semigroup of contractions over L^∞, we deduce that

$$\sup_j \frac{\|e^{t\Delta}\nabla \dot{S}_j u_0\|_{L^\infty}}{\Gamma(2^j)} \geq 2^{2j_t}\frac{\|e^{t2^{2j_t}\Delta}\nabla \dot{S}_0 u_0\|_{L^\infty}}{\Gamma(2^{j_t})} \geq \frac{\left\|e^{\Delta}\nabla \dot{S}_0 u_0\right\|_{L^\infty}}{2t\Gamma\left(1/\sqrt{t}\right)}.$$

Note that since u_0 is nonzero and homogeneous of degree -1, we must have $\nabla \dot{S}_0 u_0 \neq 0$, hence also $\left\|e^{\Delta}\nabla \dot{S}_0 u_0\right\|_{L^\infty} \neq 0$. Thus, if $e^{t\Delta}u_0$ belongs to $L^1([0,T]; C_\mu)$, then we have, by definition of Γ and under Proposition 2.111 page 118,

$$\int_0^{\sqrt{T}} \frac{dr}{\mu(r)} = \frac{1}{2}\int_0^T \frac{dt}{t\,\Gamma\left(\frac{1}{\sqrt{t}}\right)} \leq c\int_0^T \|e^{t\Delta}u_0\|_{C_\mu}\,dt.$$

The proposition is thus proved.   □

Even though the Osgood lemma cannot be used, the following theorem states that *small* elements of E_p have a flow.

Theorem 5.47. *A constant C exists such that for any positive r, and any v in $L^1([0,T]; \dot{B}_{\infty,\infty}^{-r})$ such that for some positive integer j_0,*

$$N_{j_0}(T,v) \stackrel{def}{=} \sup_{j \geq j_0} 2^j \|\Delta_j v\|_{L_T^1(L^\infty)} < \frac{1}{C},$$

a unique continuous map ψ of $[0,T] \times \mathbb{R}^d$ to $\mathbb{R}^d$ exists such that

$$\psi(t,x) = x + \int_0^t v(t', \psi(t',x)) \, dt' \quad and \quad \psi(t,\cdot) - \mathrm{Id} \in C^{1-CN_{j_0}(t,v)}.$$

Proof. Uniqueness is an immediate consequence of the following lemma.

Lemma 5.48. *Under the hypothesis of the above theorem, if γ_1 and γ_2 are continuous functions such that*

$$\gamma_j(t) = x_j + \int_0^t v(t', \gamma_j(t')) \, dt' \quad for \quad j = 1, 2,$$

and if, in addition, $|x_1 - x_2| \leq 2^{-j_0}$, then we have, for all $t_0 \leq [0,T]$,

$$|\gamma_1(t_0) - \gamma_2(t_0)| \leq C|x_1 - x_2|^{1-CN_{j_0}(t_0,v)} \exp\left(2^{j_0(r+1)} \int_0^{t_0} \|v(t,\cdot)\|_{\dot{B}_{\infty,\infty}^{-r}} \, dt\right).$$

Proof. Splitting the vector field v into low and high frequencies yields

$$|\gamma_1(t) - \gamma_2(t)| \leq |x_1 - x_2| + \int_0^t |\dot{S}_j v(t', \gamma_1(t')) - \dot{S}_j v(t', \gamma_2(t'))| \, dt'$$

$$+ 2\int_0^t \sum_{j' \geq j} \|\Delta_{j'} v(t')\|_{L^\infty} \, dt'$$

$$\leq |x_1 - x_2| + \int_0^t \|\nabla \dot{S}_j v(t')\|_{L^\infty} |\gamma_1(t') - \gamma_2(t')| \, dt'$$

$$+ 2^{1-j} \sum_{j' \geq j} 2^{j-j'} 2^{j'} \int_0^t \|\Delta_{j'} v(t')\|_{L^\infty} \, dt'.$$

For $0 \leq t \leq t_0 \leq T$, we define $\rho(t) \stackrel{def}{=} \sup_{t' \leq t} |\gamma_1(t') - \gamma_2(t')|$ and

$$D_j(t) \stackrel{def}{=} |x_1 - x_2| + 2^{2-j} N_{j_0}(t_0, v) + \int_0^t \|\nabla \dot{S}_j v(t')\|_{L^\infty} |\gamma_1(t') - \gamma_2(t')| \, dt'.$$

By the definition of $N_{j_0}(t,v)$, we have $\rho(t) \leq D_j(t)$ for any $j \geq j_0$. Therefore, for any $t \leq t_0$,

$$D_j(t) \leq |x_1 - x_2| + 2^{2-j} N_{j_0}(t_0, v) + \int_0^t \|\nabla \dot{S}_j v(t')\|_{L^\infty} D_j(t') \, dt'.$$

The Gronwall lemma implies that, for any $t \leq t_0$,

$$D_j(t) \leq \left(|x_1 - x_2| + 2^{2-j} N_{j_0}(t_0, v) \right) \exp \left(\int_0^t \|\nabla \dot{S}_j v(t')\|_{L^\infty} \, dt' \right).$$

Using Lemma 2.1 page 52, we deduce, for any $t \leq t_0$, that

$$\int_0^t \|\nabla \dot{S}_j v(t')\|_{L^\infty} \, dt' \leq \int_0^t \sum_{j' < j_0} 2^{j'} \|\dot{\Delta}_{j'} v(t')\|_{L^\infty} \, dt'$$

$$+ \sum_{j'=j_0}^{j} \int_0^t 2^{j'} \|\dot{\Delta}_{j'} v(t')\|_{L^\infty} \, dt'$$

$$\leq 2^{j_0(r+1)} \int_0^t \|v(t')\|_{\dot{B}_{\infty,\infty}^{-r}} \, dt' + j N_{j_0}(t, v). \qquad (5.38)$$

Thus, for any integer $j \geq j_0$ and any $t \leq t_0$, we have

$$D_j(t) \leq \left((|x_1 - x_2| + 2^{2-j} N_{j_0}(t_0, v) \right)$$

$$\times \exp \left(2^{j_0(r+1)} \int_0^t \|v(t')\|_{\dot{B}_{\infty,\infty}^{-r}} \, dt' + j N_{j_0}(t, v) \right).$$

Choose j such that $1 \leq 2^j |x_1 - x_2| < 2$. We then infer that

$$\rho(t_0) \leq C |x_1 - x_2|^{1 - C N_{j_0}(t_0, v)} \exp \left(2^{j_0(r+1)} \int_0^{t_0} \|v(t')\|_{\dot{B}_{\infty,\infty}^{-r}} \, dt' \right)$$

and the lemma is proved.   □

In order to prove the existence, we shall establish the convergence of the classical Picard scheme,

$$x_{k+1}(t) = x_0 + \int_0^t v(t', x_k(t')) \, dt'.$$

We define

$$\rho_k(t) \overset{\text{def}}{=} \sup_{\substack{t' \leq t \\ n \geq 0}} |x_{k+n}(t') - x_k(t')|.$$

Along the same lines as the proof of Lemma 5.48, separately treating the high and low frequencies, we get, for any $j \geq j_0$,

$$\rho_{k+1}(t) \leq \int_0^t |v(t', x_{k+n}(t')) - v(t', x_k(t'))| \, dt'$$

$$\leq 2^{2-j} N_{j_0}(T, v) + \int_0^t \|\nabla \dot{S}_j v(t')\|_{L^\infty} \rho_k(t') \, dt'.$$

Now, setting $\rho(t) \overset{\text{def}}{=} \limsup\limits_{k \to \infty} \rho_k(t)$ and

$$D_j(t) \overset{\text{def}}{=} 2^{2-j} N_{j_0}(T, v) + \int_0^t \|\nabla \dot{S}_j v(t')\|_{L^\infty} \rho(t')\, dt',$$

we obtain, passing to the limit,

$$D_j(t) \le 2^{2-j} N_{j_0}(T, v) + \int_0^t \|\nabla \dot{S}_j v(t')\|_{L^\infty} D_j(t')\, dt'.$$

The Gronwall lemma ensures that

$$D_j(T) \le C 2^{-j} \exp\left(\int_0^T \|\nabla \dot{S}_j v(t')\|_{L^\infty}\, dt' \right). \tag{5.39}$$

Appealing to (5.38), this leads to

$$D_j(T) \le C 2^{-j(1 - C N_{j_0}(T, v))} \exp\left(2^{j_0(r+1)} \int_0^T \|v(t')\|_{\dot{B}_{\infty,\infty}^{-r}}\, dt' \right)$$

for any $j \ge j_0$, which completes the proof of the theorem. $\qquad\square$

5.7 References and Remarks

For a much more detailed introduction to the incompressible Navier–Stokes system, the reader can consult [57, 86, 214, 286, 299]. For a complete and up-to-date bibliography, see [205].

The mathematical theory of the incompressible Navier–Stokes system originates with J. Leray's celebrated 1934 paper (see [207] and also the work by E. Hopf in [165]). There, the concept of weak solutions was introduced and the existence of such solutions was proven. The regularity properties of those weak solutions have been studied by a number of authors (see, in particular, [54]). In this seminal paper [207], J. Leray also proved that if the initial data satisfies a smallness condition of the type

$$\|u_0\|_{L^2} \|\nabla u_0\|_{L^2} \le c\nu^2 \quad \text{or} \quad \|u_0\|_{L^2}^2 \|\nabla u_0\|_{L^\infty} \le c\nu^3,$$

then the solution exists in a space in which the uniqueness of such a solution holds. The smallness condition was improved by H. Fujita and T. Kato in 1964: In [129], they essentially proved Theorem 5.6 (see also [144, 145]). The proof presented here relies mainly on Sobolev inequalities and our proof of these classical inequalities comes from [67].

The proof of uniqueness in dimension two (Theorem 5.14) is contained in the works by J. Leray [206, 207], J.-L. Lions and G. Prodi [211], and O. Ladyzhenskaya [201]. It has been extended in [106] to the Boussinesq system with partial viscosity (a coupling between the Navier–Stokes system and some transport equation). The global stability result Theorem 5.17 was proven by I. Gallagher, D. Iftimie, and F. Planchon in [134], and the idea of Corollary 5.20 can be found in [252]. The existence part of Theorem 5.21 is close to T. Kato's theorem of 1972, proven in [176]. The uniqueness of continuous solutions with values in L^3 was proven by G. Furioli, P.-G. Lemarié-Rieusset, and E. Terraneo in [130] (see also [82]). The proof of Lemma 5.30 follows the computations carried out by, for instance, F. Vigneron in [295].

In dimension three, the question of global solvability for general large data has remained unsolved. Let us emphasize that on the one hand it has been proved in [239] that the self-similar solutions introduced by J. Leray in [207] as models of blow-up solutions cannot have finite energy, and that, on the other hand, solutions blowing-up in finite time have been constructed in [234] for a Navier–Stokes like system that enters in the class (GNS_ν). For more results concerning the lifespan of solutions to the three-dimensional Navier–Stokes system, the reader may refer to [132, 143].

Theorem 5.40 was proven by M. Cannone, Y. Meyer, and F. Planchon in [58], by a different method. A local version and various extensions of Theorem 5.40 can be found in [72] and [198]. The endpoint case (Theorem 5.35) was first studied by H. Koch and D. Tataru in [196].

The rest of this chapter comes essentially from [76] and [73]. We mention in passing that in dimension three, the Leray solutions have a (possibly nonunique) flow (see the work by C. Foias, C. Guillopé and R. Temam in [127]).

For an extensive study of the Navier–Stokes equations by means of Fourier analysis techniques, the reader may refer to the books [57] by M. Cannone and [205] by P.-G. Lemarié-Rieusset.

6

Anisotropic Viscosity

The purpose of this chapter is to study a modified version of the incompressible Navier–Stokes system in $\mathbb{R}^3$, where the usual Laplace operator Δ is replaced by the Laplace operator Δ_h in the horizontal variables, namely $\Delta_h \overset{\text{def}}{=} \partial_1^2 + \partial_2^2$. The system we will consider is thus of the form

$$(ANS_\nu) \qquad \begin{cases} \partial_t u + u \cdot \nabla u - \nu \Delta_h u = -\nabla P \\ \operatorname{div} u = 0 \\ u_{|t=0} = u_0. \end{cases}$$

Systems of this type appear in geophysical fluids. In fact, in order to model turbulent diffusion, physicists often consider a diffusion term of the form $-\nu_h \Delta_h - \nu_3 \partial_3^2$, where ν_h and ν_3 are empirical constants. In most applications, it turns out that ν_3 is much smaller than ν_h.

Obviously, the system (ANS_ν) has the same scaling invariance as the standard Navier–Stokes system studied in Chapter 5. That is, (u, P) satisfies (ANS_ν) with data u_0 if and only if for all $\lambda > 0$,

$$(u_\lambda, P_\lambda)(t, x) \overset{\text{def}}{=} \left(\lambda u(\lambda^2 t, \lambda x), \lambda^3 P(\lambda^2 t, \lambda x) \right)$$

satisfies (ANS_ν) with data $\lambda u_0(\lambda \cdot)$.

In contrast with the system (NS_ν), however, the system (ANS_ν) is of mixed type: parabolic in the horizontal variables and hyperbolic in the vertical variable so that the classical approach for the Navier–Stokes system (which strongly relies on parabolicity) is bound to fail. Nevertheless, we shall see in this chapter that some global well-posedness results for small data in suitable scaling invariant spaces may be proven.

This chapter is structured as follows. In order to make the basic ideas clear, we first prove a theorem which is not optimal (i.e., not at the scaling), but requires only elementary tools. More precisely, in Section 6.1 we prove an existence and uniqueness result for L^2 data with one vertical derivative in L^2. The rest of the chapter is devoted to the study of the well-posedness

H. Bahouri et al., *Fourier Analysis and Nonlinear Partial Differential Equations*, Grundlehren der mathematischen Wissenschaften 343, DOI 10.1007/978-3-642-16830-7_6, © Springer-Verlag Berlin Heidelberg 2011

issue in a function space with the right scaling. Roughly speaking, we shall consider three-dimensional data which have horizontal derivative $-\frac{1}{2}$ in L^4 and vertical derivative $\frac{1}{2}$ in L^2. The corresponding function spaces are introduced in Section 6.2, together with some technical tools (nonisotropic paradifferential calculus in particular). Global existence is proved in Section 6.3, and the last section is devoted to the proof of uniqueness.

6.1 The Case of L^2 Data with One Vertical Derivative in L^2

In this section, we will show that the system (ANS_ν) is well posed for any divergence-free data in L^2 with one vertical derivative in L^2.

Since the horizontal variable $x_h \overset{\text{def}}{=} (x_1, x_2)$ does not play the same role as the vertical variable x_3, it is natural to introduce the following *anisotropic Sobolev spaces.*

Definition 6.1. *Let s and s' be real numbers. We define the Banach space $H^{s,s'}$ as the set of tempered distributions u such that $\widehat{u}$ belongs to $L^2_{loc}(\mathbb{R}^3)$ and*

$$\|u\|^2_{H^{s,s'}} \overset{\text{def}}{=} \int_{\mathbb{R}^3} (1 + |\xi_h|^2)^s (1 + |\xi_3|^2)^{s'} |\widehat{u}(\xi)|^2 \, d\xi < \infty.$$

Before stating the main result of this section, we shall introduce some more notation. Throughout this chapter, we write $\mathbb{R}^3 = \mathbb{R}^2_h \times \mathbb{R}_v$. The components of the three-dimensional vector field v are denoted (v^h, v^3), and it is understood that $\nabla_h \overset{\text{def}}{=} (\partial_1, \partial_2)$ and $\mathrm{div}_h \, v = \partial_1 v^1 + \partial_2 v^2$. Finally, the notation X_h (resp., X_v) means that X_h is a function space over $\mathbb{R}^2_h$ (resp., $\mathbb{R}_v$). A function space over $\mathbb{R}^3$ is simply denoted by X. For instance, $L^p \overset{\text{def}}{=} L^p(\mathbb{R}^3)$, $L^p_h \overset{\text{def}}{=} L^p(\mathbb{R}^2_h)$, and $L^p_v \overset{\text{def}}{=} L^p(\mathbb{R}_v)$.

We can now state the main result of this section.

Theorem 6.2. *Let u_0 be a divergence-free vector field with coefficients in $H^{0,1}$. There exists a positive time T such that the system (ANS_ν) has a unique solution u in the space*

$$L^\infty([0,T]; H^{0,1}) \cap L^2([0,T]; H^{1,1}).$$

Moreover, the solution u is in $\mathcal{C}([0,T]; L^2)$ and satisfies the energy equality

$$\|u(t)\|^2_{L^2} + 2\nu \int_0^t \|\nabla_h u(t')\|^2_{L^2} \, dt' = \|u_0\|^2_{L^2} \quad \text{for all } t \in [0,T]. \tag{6.1}$$

Furthermore, if we have

$$\|u_0\|^{\frac{1}{2}}_{L^2} \|\partial_3 u_0\|^{\frac{1}{2}}_{L^2} \le c\nu \tag{6.2}$$

for some small enough constant c, then the solution is global.

Proof. The lack of smoothing effect in the vertical variable x_3 prevents us from solving the system by a fixed point method (as in Section 5.2) and from using compactness methods based on the L^2 energy estimate. The structure of the proof is as follows:

– First, we define a family of approximate problems with global smooth solutions.
– Second, we prove uniform bounds for this family on some fixed time interval.
– Third, we show that the sequence defined by this procedure converges to some solution of (ANS_ν) with the desired properties.
– Finally, we establish a stability estimate in L^2 which implies uniqueness.

Step 1: The family of approximate solutions. We shall use the Friedrichs method introduced in Chapter 4. We wish to solve

$$(ANS_{\nu,n}) \qquad \begin{cases} \partial_t u_n - \nu \Delta_h u_n + \mathbb{E}_n(u_n \cdot \nabla u_n) + \nabla P_n = 0 \\ P_n = \mathbb{E}_n \sum_{j,k} (-\Delta)^{-1} \partial_j \partial_k (u_n^j u_n^k) \\ u_{n|t=0} = \mathbb{E}_n u_0, \end{cases}$$

where $(-\Delta)^{-1} \partial_j \partial_k$ stands for the Fourier multiplier with symbol $|\xi|^{-2} \xi_j \xi_k$, and $\mathbb{E}_n$ denotes the Fourier multiplier defined by (4.4) page 174. As in Chapter 4, the system $(ANS_{\nu,n})$ turns out to be an ordinary differential equation on the space

$$L_n^{2,\sigma} \overset{\text{def}}{=} \left\{ v \in L^2(\mathbb{R}^3) / \ \text{div } v = 0 \quad \text{and} \quad \text{Supp } \widehat{v} \subset B(0,n) \right\}$$

endowed with the L^2 norm. Indeed, we have, thanks to Lemma 2.1, for any u and v in $L_n^{2,\sigma}$,

$$Q_n(u,v) \overset{\text{def}}{=} \left\| \mathbb{E}_n(u \cdot \nabla v) + \mathbb{E}_n \nabla \sum_{1 \leq j,k \leq 3} (-\Delta)^{-1} \partial_j \partial_k (u^j v^k) \right\|_{L^2}$$

$$\leq C n^{\frac{3}{2}+1} \|u\|_{L^2} \|v\|_{L^2}.$$

Thus, for any n, there exists a $T_n > 0$ such that the system $(ANS_{\nu,n})$ has a maximal solution u_n in $\mathcal{C}^\infty([0, T_n[; L_n^{2,\sigma})$.

Step 2: A priori bounds. Arguing as on page 205 (which is rigorous since u_n is smooth), we get, for all $t \in [0, T_n[$,

$$\|u_n(t)\|_{L^2}^2 + 2\nu \int_0^t \|\nabla_h u_n(t')\|_{L^2}^2 \, dt' = \| \mathbb{E}_n u_0 \|_{L^2}^2 \leq \|u_0\|_{L^2}^2. \qquad (6.3)$$

Thanks to the blow-up condition for ordinary differential equations given by Corollary 3.12 page 131, this implies that for any n, the solution u_n of $(ANS_{\nu,n})$ is global and belongs to $\mathcal{C}^\infty(\mathbb{R}_+; L_n^{2,\sigma})$.

Bounding u_n in $L^\infty([0,T]; H^{0,1}) \cap L^2([0,T]; H^{1,1})$ for some T independent of n is more involved. We differentiate the system $(ANS_{\nu,n})$ with respect to ∂_3. This gives, dropping the index n in order to ease notation,

$$\|\partial_3 u(t)\|_{L^2}^2 + 2\nu \int_0^t \|\nabla_h \partial_3 u(t')\|_{L^2}^2 \, dt'$$

$$= \|\partial_3 \mathbb{E}_n u_0\|_{L^2}^2 - 2 \sum_{1 \le k, \ell \le 3} \int_0^t I_{k,\ell}(t') \, dt' \qquad (6.4)$$

with $I_{k,\ell}(t) \stackrel{\text{def}}{=} \int_{\mathbb{R}^3} \partial_3 u^k(t,x) \partial_k u^\ell(t,x) \partial_3 u^\ell(t,x) \, dx$.

We will start with the terms $I_{k,\ell}$ where $k \ne 3$, namely, the terms which contain only two vertical derivatives. The following proposition will be useful.

Proposition 6.3. *A constant C exists such that*

$$\left(\int_{\mathbb{R}^3} a(x)b(x)c(x) \, dx \right)^2 \le C \|a\|_{L^\infty(\mathbb{R}_v; L_h^2)} \|b\|_{L^2} \|\nabla_h c\|_{L^2} \|c\|_{L^2}$$

$$\times \min \left\{ \|a\|_{L^\infty(\mathbb{R}_v; L_h^2)} \|\nabla_h b\|_{L^2}, \|\nabla_h a\|_{L^\infty(\mathbb{R}_v; L_h^2)} \|b\|_{L^2} \right\}.$$

Proof. Define

$$J(a,b,c) \stackrel{\text{def}}{=} \int_{\mathbb{R}^3} a(x)b(x)c(x) \, dx$$

$$= \int_{\mathbb{R}} dx_3 \int_{\mathbb{R}^2} a(x_h, x_3) b(x_h, x_3) c(x_h, x_3) \, dx_h.$$

Hölder's inequality implies that

$$J(a,b,c) \le \int_{\mathbb{R}} \|a(\cdot, x_3)\|_{L_h^2} \|b(\cdot, x_3)\|_{L_h^4} \|c(\cdot, x_3)\|_{L_h^4} \, dx_3$$

$$\le \|a\|_{L^\infty(\mathbb{R}_v; L_h^2)} \|b\|_{L^2(\mathbb{R}_v; L_h^4)} \|c\|_{L^2(\mathbb{R}_v; L_h^4)}.$$

Using the Sobolev embedding $\dot{H}_h^{\frac{1}{2}} \hookrightarrow L_h^4$, the interpolation inequality between $\dot{H}_h^1$ and L_h^2, and the Cauchy–Schwarz inequality, we then get that

$$\|b\|_{L^2(\mathbb{R}_v; L_h^4)}^2 \le C \int_{\mathbb{R}} \|b(x_3, \cdot)\|_{\dot{H}_h^{\frac{1}{2}}}^2 \, dx_3$$

$$\le C \int_{\mathbb{R}} \|\nabla_h b(\cdot, x_3)\|_{L_h^2} \|b(\cdot, x_3)\|_{L_h^2} \, dx_3$$

$$\le C \|\nabla_h b\|_{L^2} \|b\|_{L^2}.$$

The proof of the other inequality is similar. □

We shall also use the following corollary of Proposition 6.3.

Corollary 6.4. *A constant C exists such that*

$$\left(\int_{\mathbb{R}^3} a(x)b(x)c(x)\,dx\right)^2 \le C\|\partial_3 a\|_{L^2}\|a\|_{L^2}\|\nabla_h b\|_{L^2}\|b\|_{L^2}\|\nabla_h c\|_{L^2}\|c\|_{L^2}.$$

Proof. According to the previous proposition, we have

$$\left(\int_{\mathbb{R}^3} a(x)b(x)c(x)\,dx\right)^2 \le C\|a\|_{L^\infty(\mathbb{R}_v;L^2_h)}^2 \|\nabla_h b\|_{L^2}\|b\|_{L^2}\|\nabla_h c\|_{L^2}\|c\|_{L^2}.$$

Noting that

$$
\begin{aligned}
\|a(\cdot, x_3)\|_{L^2_h}^2 &= \int_{-\infty}^{x_3} \frac{d}{dy_3}\left(\int_{\mathbb{R}^2} |a(x_h, y_3)|^2\,dx_h\right)dy_3 \\
&= 2\int_{-\infty}^{x_3}\int_{\mathbb{R}^2} a(x_h, y_3)\partial_{y_3} a(x_h, y_3)\,dx_h\,dy_3,
\end{aligned}
$$

the Cauchy–Schwarz inequality then implies that

$$\forall x_3 \in \mathbb{R}, \ \|a(\cdot, x_3)\|_{L^2_h}^2 \le 2\|\partial_3 a\|_{L^2}\|a\|_{L^2}.$$

The corollary is thus proved. $\square$

Proof of Theorem 6.2 (continued). Applying the above corollary for $a = \partial_k u^\ell$, $b = \partial_3 u^k$, and $c = \partial_3 u^\ell$ gives, for $k \ne 3$,

$$I_{k,\ell}(t) \le C\|\nabla_h \partial_3 u(t)\|_{L^2}^{\frac{3}{2}} \|\partial_3 u(t)\|_{L^2}\|\nabla_h u(t)\|_{L^2}^{\frac{1}{2}}.$$

Bounding the terms $I_{3,\ell}$ relies on the special structure of the system: We use the fact that the nonlinear term is $u \cdot \nabla u$ and that $\operatorname{div} u = 0$. Indeed, the divergence-free condition implies that

$$
\begin{aligned}
I_{3,\ell}(t) &= \int_{\mathbb{R}^3} \partial_3 u^3(t,x)\,\partial_3 u^\ell(t,x)\,\partial_3 u^\ell(t,x)\,dx \\
&= -\int_{\mathbb{R}^3} \operatorname{div}_h u^h(t,x)\,\partial_3 u^\ell(t,x)\,\partial_3 u^\ell(t,x)\,dx.
\end{aligned}
$$

This term is strictly analogous to the preceding ones. Thus, we have, for any k and ℓ, that

$$I_{k,\ell}(t) \le C\|\nabla_h \partial_3 u(t)\|_{L^2}^{\frac{3}{2}} \|\partial_3 u(t)\|_{L^2}\|\nabla_h u(t)\|_{L^2}^{\frac{1}{2}}.$$

Plugging this into the energy estimate (6.4) gives

$$
\begin{aligned}
\|\partial_3 u(t)\|_{L^2}^2 + 2\nu \int_0^t \|\nabla_h \partial_3 u(t')\|_{L^2}^2\,dt' &\le \|\partial_3 u_0\|_{L^2}^2 \\
&+ C\int_0^t \|\nabla_h \partial_3 u(t')\|_{L^2}^{\frac{3}{2}} \|\partial_3 u(t')\|_{L^2}\|\nabla_h u(t')\|_{L^2}^{\frac{1}{2}}\,dt'.
\end{aligned}
$$

Using the convexity inequality $ab \leq \dfrac{1}{4}a^4 + \dfrac{3}{4}b^{\frac{4}{3}}$, we obtain

$$\|\partial_3 u(t)\|_{L^2}^2 + \nu \int_0^t \|\nabla_h \partial_3 u(t')\|_{L^2}^2 \, dt' \leq \|\partial_3 u_0\|_{L^2}^2$$

$$+ \frac{C}{\nu^3} \int_0^t \|\partial_3 u(t')\|_{L^2}^4 \|\nabla_h u(t')\|_{L^2}^2 \, dt'. \quad (6.5)$$

We now reintroduce the index n and define

$$T_n \overset{\text{def}}{=} \sup\left\{ t > 0 \,/\, \|\partial_3 u_n\|_{L_t^\infty(L^2)}^2 + \nu \|\nabla_h \partial_3 u_n\|_{L_t^2(L^2)}^2 \leq 2\|\partial_3 u_0\|_{L^2}^2 \right\}.$$

The function u_n is continuous with values in H^s for any s, and $\|\partial_3 \mathbb{E}_n u_0\|_{L^2}$ is less than or equal to $\|\partial_3 u_0\|_{L^2}$. Thus, the time T_n is positive and, for any $t < T_n$, we have

$$\|\partial_3 u_n(t)\|_{L^2}^2 + \nu \int_0^t \|\nabla_h \partial_3 u_n(t')\|_{L^2}^2 \, dt' \leq \|\partial_3 u_0\|_{L^2}^2$$

$$\times \left(1 + \frac{C}{\nu^3} \|\partial_3 u_0\|_{L^2}^2 \int_0^t \|\nabla_h u_n(t')\|_{L^2}^2 \, dt' \right). \quad (6.6)$$

Thanks to the energy estimate (6.3), we have, for any $t < T_n$,

$$\|\partial_3 u_n(t)\|_{L^2}^2 + \nu \int_0^t \|\nabla_h \partial_3 u_n(t')\|_{L^2}^2 \, dt' \leq \|\partial_3 u_0\|_{L^2}^2 \left(1 + \frac{C}{\nu^4} \|\partial_3 u_0\|_{L^2}^2 \|u_0\|_{L^2}^2 \right).$$

Thus, under the smallness condition (6.2), we have that $T_n = +\infty$ and thus

$$\forall t \geq 0, \, \forall n \in \mathbb{N}, \, \|\partial_3 u_n(t)\|_{L^2}^2 + \nu \int_0^t \|\nabla_h \partial_3 u_n(t')\|_{L^2}^2 \, dt' \leq 2\|\partial_3 u_0\|_{L^2}^2.$$

We now investigate the case where the initial data *does not* satisfy the smallness condition. We write u_n as a perturbation of the free solution $u_{N_0,F} \overset{\text{def}}{=} e^{\nu t \Delta_h} \mathbb{E}_{N_0} u_0$. Let

$$w_n \overset{\text{def}}{=} u_n - u_{N_0,F}$$

for some integer N_0 to be chosen later. The inequality (6.6) becomes

$$\|\partial_3 u_n(t)\|_{L^2}^2 + \nu \int_0^t \|\nabla_h \partial_3 u_n(t')\|_{L^2}^2 \, dt' \leq \|\partial_3 u_0\|_{L^2}^2$$

$$\times \left(1 + \frac{C}{\nu^3} \|\partial_3 u_0\|_{L^2}^2 \left(\int_0^t \|\nabla_h u_{N_0,F}(t')\|_{L^2}^2 \, dt' + \int_0^t \|\nabla_h w_n(t')\|_{L^2}^2 \, dt' \right) \right).$$

From the definition of $u_{N_0,F}$, we infer that

$$\|\partial_3 u_n(t)\|_{L^2}^2 + \nu \int_0^t \|\nabla_h \partial_3 u_n(t')\|_{L^2}^2 \, dt' \leq \|\partial_3 u_0\|_{L^2}^2$$

$$\times \left(1 + \frac{C}{\nu^3} \|\partial_3 u_0\|_{L^2}^2 \left(t N_0^2 \|u_0\|_{L^2}^2 + \int_0^t \|\nabla_h w_n(t')\|_{L^2}^2 \, dt' \right) \right).$$

We now estimate the last integral. By the definition of w_n, we have

$$\begin{cases} \partial_t w_n - \nu \Delta_h w_n + \mathbb{E}_n(u_n \cdot \nabla w_n) + \mathbb{E}_n(u_n \cdot \nabla u_{N_0,F}) = -\nabla P_n \\ \operatorname{div} w_n = 0 \\ w_{n|t=0} = (\mathrm{Id} - \mathbb{E}_{N_0}) \mathbb{E}_n u_0. \end{cases}$$

Using the divergence-free condition, we get, by the energy estimate, that

$$\nu \int_0^t \|\nabla_h w_n(t')\|_{L^2}^2 \, dt' \leq \|(\mathrm{Id} - \mathbb{E}_{N_0}) u_0\|_{L^2}^2 - 2 \int_0^t \langle u_n(t') \cdot \nabla u_{N_0,F}(t'), w_n(t') \rangle \, dt'.$$

Note that using Lemma 2.1 page 52 and (6.3) yields

$$\begin{aligned} |\langle u_n(t') \cdot \nabla u_{N_0,F}(t'), w_n(t') \rangle| &\leq \|\nabla u_{N_0,F}(t')\|_{L^\infty} \|u_n(t')\|_{L^2} \|w_n(t')\|_{L^2} \\ &\leq C\|u_0\|_{L^2}^2 \|\nabla u_{N_0,F}(t')\|_{L^\infty} \\ &\leq C N_0^{\frac{5}{2}} \|u_0\|_{L^2}^3. \end{aligned}$$

Thus, for any $n \in \mathbb{N}$,

$$\nu \int_0^t \|\nabla_h w_n(t')\|_{L^2}^2 \, dt' \leq \|(\mathrm{Id} - \mathbb{E}_{N_0}) u_0\|_{L^2}^2 + C t N_0^{\frac{5}{2}} \|u_0\|_{L^2}^3.$$

We infer that for all $T > 0$,

$$\|\partial_3 u_n(T)\|_{L^2}^2 + \nu \int_0^T \|\nabla_h \partial_3 u_n(t')\|_{L^2}^2 \, dt' \leq \|\partial_3 u_0\|_{L^2}^2$$

$$\times \left(1 + \frac{C}{\nu^3} \|\partial_3 u_0\|_{L^2}^2 \left(T N_0^2 \|u_0\|_{L^2}^2 + \frac{1}{\nu} \|(\mathrm{Id} - \mathbb{E}_{N_0}) u_0\|_{L^2}^2 + \frac{1}{\nu} T N_0^{\frac{5}{2}} \|u_0\|_{L^2}^3 \right) \right).$$

First choosing N_0 sufficiently large and then T sufficiently small so that the above quantity is small enough ensures that for all $t \in [0, T]$ and $n \in \mathbb{N}$,

$$\|\partial_3 u_n(t)\|_{L^2}^2 + \nu \int_0^t \|\nabla_h \partial_3 u_n(t')\|_{L^2}^2 \, dt' \leq 2\|\partial_3 u_0\|_{L^2}^2. \qquad (6.7)$$

Step 3: Convergence. To simplify the presentation, we only consider the case where T is finite. Since $(u_n)_{n \in \mathbb{N}}$ is bounded in $L^\infty([0, T]; H^{0,1}) \cap L^2([0, T]; H^{1,1})$, we also have $(u_n)_{n \in \mathbb{N}}$ bounded in $L^4([0, T]; H^{\frac{1}{2},1})$ by interpolation.[1] Assume, temporarily, that

$$H^{\frac{1}{2},1} \hookrightarrow L_v^2(L_h^4) \cap L_v^\infty(L_h^4). \qquad (6.8)$$

We then deduce that the convection and pressure terms of $(ANS_{\nu,n})$ are bounded in $L^2([0, T]; H^{-1})$. Therefore,

---

[1] In fact, this may be proven directly using the definition of $H^{s,s'}$ and Hölder's inequality.

$$\partial_t u_n \quad \text{is bounded in} \quad L^2([0,T];H^{-1}). \tag{6.9}$$

Since the embedding of H^{-1} in L^2 is locally compact (see Theorem 1.68 page 45), we can now conclude, by combining Ascoli's theorem and the Cantor diagonal process, that up to extraction, $(u_n)_{n\in\mathbb{N}}$ converges to some u in $\mathcal{C}([0,T];\mathcal{S}')$. Because $(u_n)_{n\in\mathbb{N}}$ is bounded in $L^\infty([0,T];H^{0,1}) \cap L^2([0,T]; H^{1,1})$, we actually have $u \in L^\infty([0,T];H^{0,1}) \cap L^2([0,T];H^{1,1})$ (use the weak compactness properties of the Hilbert spaces $H^{0,1}$ and $H^{1,1}$), and it is possible to pass to the limit in $(ANS_{\nu,n})$. Hence, u is a solution of (ANS_ν).

We now prove that $u \in \mathcal{C}([0,T];L^2)$. Since u satisfies (ANS), it is not difficult to show that $\partial_t u$ is bounded in $L^2([0,T];H^{-1})$ [just proceed as in the proof of (6.9)]. Since, in addition, u is bounded in $L^2([0,T];H^1)$, a classical interpolation argument ensures that u belongs to $\mathcal{C}([0,T];L^2)$.

Finally, we note that Lemma 5.15 page 216, combined with the fact that $u \in L^4([0,T];L^4) \cap L^2([0,T];H^1)$, implies that the energy equality (6.1) is satisfied.

For the sake of completeness, we shall justify (6.8). Note that $H^{\frac{1}{2},1}$ is embedded in $L_v^2(H_h^{\frac{1}{2}})$, and $H_h^{\frac{1}{2}}$ is embedded in L_h^4. Hence, $H^{\frac{1}{2},1} \hookrightarrow L_v^2(L_h^4)$. In order to prove the embedding in $L_v^\infty(L_h^4)$, consider some function a in $\mathcal{S}$. For all x_3 in $\mathbb{R}_v$, we may write

$$4\int_{\mathbb{R}_h^2} a^4(x_h,x_3)\,dx_h = \int_{\mathbb{R}_h^2} \left(\int_{-\infty}^{x_3} (a\partial_3 a)(x_h,y_3)\,dy_3 \right)^2 dx_h.$$

Therefore, by virtue of the Cauchy–Schwarz inequality,

$$4\int_{\mathbb{R}_h^2} a^4(x_h,x_3)\,dx_h \le \|a\|_{L_h^4(L_v^2)}^2 \|\partial_3 a\|_{L_h^4(L_v^2)}^2.$$

Applying the Minkowski inequality then completes the proof of (6.8).

Step 4: Uniqueness. This is obviously implied by the following lemma.

Lemma 6.5. *Let u_j, $j \in \{1,2\}$, be solutions of (ANS_ν) in the space*

$$L^\infty([0,T];H^{0,1}) \cap L^2([0,T];H^{1,1}).$$

We then have

$$\|u_2(t)-u_1(t)\|_{L^2}^2 + \nu \int_0^t \|\nabla_h(u_2-u_1)(t')\|_{L^2}^2\,dt' \le \|(u_2-u_1)(0)\|_{L^2}^2 \exp M_{u_1}(t)$$

$$\text{with} \quad M_{u_1(t)} \overset{def}{=} \frac{C}{\nu} \int_0^t \|\partial_3 \nabla_h u_1(t')\|_{L^2} \|\nabla u_1(t')\|_{L^2}\,dt'.$$

Remark 6.6. As u_1 belongs to $L^\infty([0,T];H^{0,1}) \cap L^2([0,T];H^{1,1})$, we have

$$M_{u_1}(T) \le \frac{C}{\nu} \|\partial_3 \nabla_h u_1\|_{L_T^2(L^2)} \left(\frac{1}{\sqrt{2\nu}} \|u_1(0)\|_{L^2} + T^{\frac{1}{2}} \|\partial_3 u_1\|_{L_T^\infty(L^2)} \right) < \infty.$$

Remark 6.7. We note that this lemma is a stability result for initial data in $H^{0,1}$. We should point out that the stability is proved in $L^\infty_t(L^2) \cap L^2_t(H^{1,0})$, which corresponds to the loss of one vertical derivative with respect to the regularity of the initial data.

Proof of Lemma 6.5. Defining $u_{21} \stackrel{\text{def}}{=} u_2 - u_1$, we get, by an L^2 energy estimate,

$$\|u_{21}(t)\|^2_{L^2} + 2\nu \int_0^t \|\nabla u_{21}(t')\|^2_{L^2}\,dt' = -2I^h(t) - 2I^v(t)$$

with

$$I^h(t) \stackrel{\text{def}}{=} \sum_{\substack{1 \le k \le 2 \\ 1 \le \ell \le 3}} \int_0^t \int_{\mathbb{R}^3} u^k_{21}(t',x)\,\partial_k u^\ell_1(t',x)\,u^\ell_{21}(t',x)\,dt'\,dx,$$

$$I^v(t) \stackrel{\text{def}}{=} \sum_{1 \le \ell \le 3} \int_0^t \int_{\mathbb{R}^3} u^3_{21}(t',x)\,\partial_3 u^\ell_1(t',x)\,u^\ell_{21}(t',x)\,dt'\,dx.$$

Corollary 6.4 applied with $a = \partial_k u^\ell_1$, $b = u^k_{21}$, and $c = u^\ell_{21}$ implies that

$$I^h(t) \le C \int_0^t \|\partial_3 \nabla_h u_1(t')\|^{\frac{1}{2}}_{L^2} \|\nabla_h u_1(t')\|^{\frac{1}{2}}_{L^2} \|\nabla_h u_{21}(t')\|_{L^2} \|u_{21}(t')\|_{L^2}\,dt'$$

$$\le \frac{\nu}{2} \int_0^t \|\nabla_h u_{21}(t')\|^2_{L^2}\,dt'$$

$$+ \frac{C}{\nu} \int_0^t \|\partial_3 \nabla_h u_1(t')\|_{L^2} \|\nabla_h u_1(t')\|_{L^2} \|u_{21}(t')\|^2_{L^2}\,dt'.$$

Proposition 6.3 applied with $a = u^3_{21}$, $b = \partial_3 u^\ell_1$, and $c = u^\ell_{21}$ gives

$$I^v(t) \le \int_0^t \|u^3_{21}(t')\|_{L^\infty(\mathbb{R}_v;L^2_h)} \|\partial_3 \nabla_h u_1(t')\|^{\frac{1}{2}}_{L^2} \|\partial_3 u_1(t')\|^{\frac{1}{2}}_{L^2}$$

$$\times \|\nabla_h u_{21}(t')\|^{\frac{1}{2}}_{L^2} \|u_{21}(t')\|^{\frac{1}{2}}_{L^2}\,dt'.$$

We shall temporarily assume the following result.

Lemma 6.8. *Let v be a divergence-free vector field. We then have*

$$\|v^3\|^2_{L^\infty(\mathbb{R}_v;L^2_h)} \le 2\|\operatorname{div}_h v^h\|_{L^2}\|v^3\|_{L^2}.$$

We now have

$$I^v(t) \le \int_0^t \|\nabla_h u_{21}(t')\|_{L^2} \|u_{21}(t')\|_{L^2} \|\partial_3 \nabla_h u_1(t')\|^{\frac{1}{2}}_{L^2} \|\partial_3 u_1(t')\|^{\frac{1}{2}}_{L^2}\,dt'$$

$$\le \frac{\nu}{2} \int_0^t \|\nabla_h u_{21}(t')\|^2_{L^2}\,dt$$

$$+ \frac{C}{\nu} \int_0^t \|\partial_3 \nabla_h u_1(t')\|_{L^2} \|\partial_3 u_1(t')\|_{L^2} \|u_{21}(t')\|^2_{L^2}\,dt'.$$

Applying the Gronwall lemma then completes the proof.   □

Proof of Lemma 6.8. Write

$$\|v^3(\cdot, x_3)\|_{L_h^2}^2 = 2 \int_{-\infty}^{x_3} \left(\int_{\mathbb{R}^2} \partial_3 v^3(x_h, y_3) v^3(x_h, y_3) \, dx_h \right) dx_3$$

$$= -2 \int_{-\infty}^{x_3} \left(\int_{\mathbb{R}^2} \mathrm{div}_h \, v^h(x_h, y_3) v^3(x_h, y_3) \, dx_h \right) dx_3.$$

Applying the Cauchy–Schwarz inequality then completes the proof.   □

6.2 A Global Existence Result in Anisotropic Besov Spaces

Theorem 6.2 asserts global well-posedness under the smallness condition (6.2). On the one hand, this smallness condition is scaling invariant. On the other hand, the $H^{0,1}$ regularity which was needed in Theorem 6.2 is not scaling invariant. The rest of this chapter is devoted to the proof of a *global* existence statement for *small* data in some suitable scaling invariant function space. Motivated by the results presented in the previous chapter, we seek a functional framework in which a suitable class of highly oscillating data generates global solutions.

6.2.1 Anisotropic Localization in Fourier Space

In order to define the spaces we shall work with, we first have to construct an *anisotropic* version of the dyadic decomposition of the Fourier space introduced in Proposition 2.10 page 59.

For (k, ℓ) in $\mathbb{Z}^2$, we define

$$\Delta_k^h a = \mathcal{F}^{-1}(\varphi(2^{-k}|\xi_h|)\widehat{a}), \qquad \Delta_\ell^v a = \mathcal{F}^{-1}(\varphi(2^{-\ell}|\xi_3|)\widehat{a}),$$

$$S_k^h a = \sum_{k' \leq k-1} \Delta_{k'}^h a, \quad \text{and} \quad S_\ell^v a = \sum_{\ell' \leq \ell-1} \Delta_{\ell'}^v a, \qquad (6.10)$$

where $\widehat{a}$ denotes the Fourier transform of the tempered distribution a over $\mathbb{R}^3$, and φ denotes a function in $\mathcal{D}([3/4, 8/3])$ such that, for any positive τ,

$$\sum_{j \in \mathbb{Z}} \varphi(2^{-j}\tau) = 1.$$

Remark 6.9. Note that if we define

$$\chi(\tau) \overset{\text{def}}{=} 1 - \sum_{j \in \mathbb{N}} \varphi(2^{-j}\tau), \qquad (6.11)$$

then we have, for all $a \in \mathcal{S}(\mathbb{R}^3)$,

$$\mathcal{F}(S_\ell^h a)(\xi) = \chi(2^{-\ell}|\xi_h|)\mathcal{F}a(\xi) \quad \text{and} \quad \mathcal{F}(S_\ell^v a)(\xi) = \chi(2^{-\ell}|\xi_3|)\mathcal{F}a(\xi).$$

In what follows, we shall always consider functions a for which $\left\|S_\ell^h a\right\|_{L^\infty}$ and $\left\|S_\ell^v a\right\|_{L^\infty}$ converge to 0 when k goes to $-\infty$ so that we may write $S_\ell^h a = \chi(2^{-\ell}D_h)a$ and $S_\ell^v a = \chi(2^{-\ell}D_3)a$.

The following lemma can be understood as an anisotropic version of Lemma 2.1 page 52.

Lemma 6.10. *Let $\mathcal{B}_h$ (resp., $\mathcal{B}_v$) be a ball in $\mathbb{R}_h^2$ (resp., $\mathbb{R}_v$) and $\mathcal{C}_h$ (resp., $\mathcal{C}_v$) be an annulus in $\mathbb{R}_h^2$ (resp., $\mathbb{R}_v$). Let $1 \leq p_2 \leq p_1 \leq \infty$ and $1 \leq q_2 \leq q_1 \leq \infty$. We then have the following results:*

- *If the support of $\widehat{a}$ is included in $2^k \mathcal{B}_h$, then*

$$\left\|\partial_{x_h}^\alpha a\right\|_{L_h^{p_1}(L_v^{q_1})} \leq C2^{k\left(|\alpha|+2\left(\frac{1}{p_2}-\frac{1}{p_1}\right)\right)}\|a\|_{L_h^{p_2}(L_v^{q_1})}.$$

- *If the support of $\widehat{a}$ is included in $2^\ell \mathcal{B}_v$, then*

$$\left\|\partial_3^\beta a\right\|_{L_h^{p_1}(L_v^{q_1})} \leq C2^{\ell\left(|\beta|+\left(\frac{1}{q_2}-\frac{1}{q_1}\right)\right)}\|a\|_{L_h^{p_1}(L_v^{q_2})}.$$

- *If the support of $\widehat{a}$ is included in $2^k \mathcal{C}_h$, then*

$$\|a\|_{L_h^{p_1}(L_v^{q_1})} \leq C2^{-kN} \sup_{|\alpha|=N} \left\|\partial_h^\alpha a\right\|_{L_h^{p_1}(L_v^{q_1})}.$$

- *If the support of $\widehat{a}$ is included in $2^\ell \mathcal{C}_v$, then*

$$\|a\|_{L_h^{p_1}(L_v^{q_1})} \leq C2^{-\ell N}\left\|\partial_3^N a\right\|_{L_h^{p_1}(L_v^{q_1})}$$

Proof. This is analogous to the proof of Lemma 2.1. As an example, we prove the last inequality. As usual, using dilations, we can assume without loss of generality that $\ell = 0$. Let $\widetilde{\varphi}$ be a function in $\mathcal{D}(\mathbb{R}\backslash\{0\})$ with value 1 near $\mathcal{C}_v$. We have

$$\widehat{a}(\xi_h, \xi_3) = \frac{\widetilde{\varphi}(\xi_3)}{(i\xi_3)^N}\mathcal{F}(\partial_3^N a). \tag{6.12}$$

Defining $h_N \overset{\text{def}}{=} \mathcal{F}^{-1}\left(\widetilde{\varphi}(\xi_3)(i\xi_3)^{-N}\right)$, we may write

$$a(x_h, x_3) = \int_\mathbb{R} h_N(x_3 - y_3)a(x_h, y_3)\, dy_3.$$

Young's inequality then gives the result. $\qquad\qquad\square$

6.2.2 The Functional Framework

This subsection is devoted to the presentation of the function spaces we shall work with when globally solving the anisotropic Navier–Stokes equations.

In the following definition, we introduce two scaling invariant spaces in which (ANS_ν) turns out to be well posed.

Definition 6.11. *We denote by* $\mathcal{B}^{0,\frac{1}{2}}$ *and* $\mathcal{B}_4^{-\frac{1}{2},\frac{1}{2}}$ *the respective completions of* $\mathcal{S}(\mathbb{R}^3)$ *for the norms*

$$\|a\|_{\mathcal{B}^{0,\frac{1}{2}}} \overset{def}{=} \sum_{\ell \in \mathbb{Z}} 2^{\frac{\ell}{2}} \|\Delta_\ell^v a\|_{L^2(\mathbb{R}^3)} \quad and$$

$$\|a\|_{\mathcal{B}_4^{-\frac{1}{2},\frac{1}{2}}} \overset{def}{=} \sum_{\ell \in \mathbb{Z}} 2^{\frac{\ell}{2}} \left(\sum_{k=\ell-1}^{\infty} 2^{-k} \|\Delta_k^h \Delta_\ell^v a\|_{L_h^4(L_v^2)}^2 \right)^{\frac{1}{2}} + \sum_{j \in \mathbb{Z}} 2^{\frac{j}{2}} \|S_{j-1}^h \Delta_j^v a\|_{L^2}.$$

Remark 6.12. The definition of $\mathcal{B}^{0,\frac{1}{2}}$ is "natural". Indeed, the functions of $\mathcal{B}^{0,\frac{1}{2}}$ are L^2 in the horizontal variable and have vertical derivative $1/2$ in L^2. The choice of an ℓ^1 summation in the vertical variable allows us to get for free an $L_h^2(L_v^\infty)$ control which turns out to be of paramount importance for treating the nonlinear terms. Note, in passing, that this control would not be given if we used the (slightly smaller) $H^{0,\frac{1}{2}}$ norm instead.

The reason for the choice of the space $\mathcal{B}_4^{-\frac{1}{2},\frac{1}{2}}$ is probably less obvious. Of course, it has the required scaling (roughly $-1/2$ horizontal derivative in L^4 and $1/2$ vertical derivative in L^2), and Lemma 6.10 ensures that $\mathcal{B}^{0,\frac{1}{2}}$ is continuously included in $\mathcal{B}_4^{-\frac{1}{2},\frac{1}{2}}$. Having a negative regularity index for the horizontal variables will enable us to show global existence for highly oscillating data in the horizontal variable. The choice of the norm is also motivated by the following consideration: If we consider the linear equation

$$\partial_t u - \nu \Delta_h u = f \quad on \quad \mathbb{R}^+ \times \mathbb{R}^3,$$

then the terms $\Delta_k^h \Delta_\ell^v u$ satisfy

$$\partial_t \left(\Delta_k^h \Delta_\ell^v \right) u - \nu \Delta_h \left(\Delta_k^h \Delta_\ell^v \right) u = \Delta_k^h \Delta_\ell^v f.$$

It is now clear (from Lemma 6.10) that whenever $k \geq \ell - 1$, the action of the operator Δ_h over $\Delta_k^h \Delta_\ell^v u$ is equivalent to that of the operator Δ [indeed, we have $|\xi_h|^2 \approx |\xi|^2$ for all ξ in the support of $\mathcal{F}(\Delta_k^h \Delta_\ell^v u)$]. Therefore, those terms will be treated by means of parabolic techniques. On the other hand, no smoothing effect is expected on the remaining terms $S_{j-1}^h \Delta_j^v u$, which should dealt with as solutions of a hyperbolic equation.

To study the evolution of (ANS_ν) with initial data in $\mathcal{B}^{0,\frac{1}{2}}$ (resp., $\mathcal{B}_4^{-\frac{1}{2},\frac{1}{2}}$), we also need to introduce the following subspace of the space $L^2([0,T]; \mathcal{B}_4^{-\frac{1}{2},\frac{1}{2}})$ (resp., $L^2([0,T]; \mathcal{B}^{0,\frac{1}{2}})$).

Definition 6.13. *We denote by $\mathcal{B}^{0,\frac{1}{2}}(T)$ and $\mathcal{B}_4^{-\frac{1}{2},\frac{1}{2}}(T)$ the respective completions of the space $\mathcal{C}^\infty([0,T], \mathcal{S}(\mathbb{R}^3))$ for the norms*

$$\|a\|_{\mathcal{B}^{0,\frac{1}{2}}(T)} \stackrel{def}{=} \sum_{\ell\in\mathbb{Z}} 2^{\frac{\ell}{2}} \left(\|\Delta_\ell^v a\|_{L_T^\infty(L^2(\mathbb{R}^3))} + \nu^{\frac{1}{2}} \|\nabla_h \Delta_\ell^v a\|_{L_T^2(L^2(\mathbb{R}^3))} \right),$$

$$\|a\|_{\mathcal{B}_4^{-\frac{1}{2},\frac{1}{2}}(T)} \stackrel{def}{=} \sum_{\ell\in\mathbb{Z}} 2^{\frac{\ell}{2}} \left(\left(\sum_{k=\ell-1}^\infty 2^{-k} \|\Delta_k^h \Delta_\ell^v a\|_{L_T^\infty(L_h^4(L_v^2))}^2 \right)^{\frac{1}{2}} \right.$$

$$\left. + \nu^{\frac{1}{2}} \left(\sum_{k=\ell-1}^\infty 2^k \|\Delta_k^h \Delta_\ell^v a\|_{L_T^2(L_h^4(L_v^2))}^2 \right)^{\frac{1}{2}} \right)$$

$$+ \sum_{j\in\mathbb{Z}} 2^{\frac{j}{2}} \left(\|S_{j-1}^h \Delta_j^v a\|_{L_T^\infty(L^2(\mathbb{R}^3))} + \nu^{\frac{1}{2}} \|\nabla_h S_{j-1}^h \Delta_j^v a\|_{L_T^2(L^2(\mathbb{R}^3))} \right).$$

Lemma 6.10 obviously implies the following result.

Corollary 6.14. *For all $T \in \,]0,\infty]$, the space $\mathcal{B}^{0,\frac{1}{2}}(T)$ is continuously embedded in $\mathcal{B}_4^{-\frac{1}{2},\frac{1}{2}}(T)$ and in $L_T^\infty(L_h^2(L_v^\infty))$. Moreover, the norm of the embedding is independent of T.*

We shall also make use of the fact that the space $\mathcal{B}_4^{-\frac{1}{2},\frac{1}{2}}$ is embedded in the space of distributions which are $\mathcal{B}_{4,2}^{-\frac{1}{2}}$ in the horizontal variable and $\mathcal{B}_{2,1}^{\frac{1}{2}}$ in the vertical variable. More precisely, we have the following.

Corollary 6.15. *There exists a constant C such that for all $a \in \mathcal{B}_4^{-\frac{1}{2},\frac{1}{2}}(T)$, we have*

$$\sum_{\ell\in\mathbb{Z}} 2^{\frac{\ell}{2}} \left(\sum_{k\in\mathbb{Z}} 2^{-k} \|\Delta_k^h \Delta_\ell^v a(0)\|_{L_h^4(L_v^2)}^2 \right)^{\frac{1}{2}} \leq C\|a(0)\|_{\mathcal{B}_4^{-\frac{1}{2},\frac{1}{2}}},$$

$$\sum_{\ell\in\mathbb{Z}} 2^{\frac{\ell}{2}} \left(\sum_{k\in\mathbb{Z}} \left(2^{-k} \|\Delta_k^h \Delta_\ell^v a\|_{L_T^\infty(L_h^4(L_v^2))}^2 + \nu 2^k \|\Delta_k^h \Delta_\ell^v a\|_{L_T^2(L_h^4(L_v^2))}^2 \right) \right)^{\frac{1}{2}}$$

$$\leq C\|a\|_{\mathcal{B}_4^{-\frac{1}{2},\frac{1}{2}}(T)}.$$

Proof. We only treat the first inequality, the proof of the second being similar. Obviously, it suffices to show that

$$I \stackrel{def}{=} \sum_{\ell\in\mathbb{Z}} 2^{\frac{\ell}{2}} \left(\sum_{k\leq\ell-2} 2^{-k} \|\Delta_k^h \Delta_\ell^v a(0)\|_{L_h^4(L_v^2)}^2 \right)^{\frac{1}{2}} \leq C\|a(0)\|_{\mathcal{B}_4^{-\frac{1}{2},\frac{1}{2}}}.$$

According to the second inequality of Lemma 6.10, we have

$$I \leq C \sum_{\ell\in\mathbb{Z}} 2^{\frac{\ell}{2}} \left(\sum_{k\leq\ell-2} \|\Delta_k^h \Delta_\ell^v a(0)\|_{L^2}^2 \right)^{\frac{1}{2}}.$$

Now, since (horizontal) Littlewood–Paley decomposition is almost orthogonal in L^2, we get, arguing as in the proof of (2.11),

$$\sum_{k\leq\ell-2} \|\Delta_k^h\Delta_\ell^v a\|_{L^2}^2 \leq 2\|S_{\ell-1}^h\Delta_\ell^v a\|_{L^2}^2,$$

from which the desired inequality follows.   □

6.2.3 Statement of the Main Result

We now explain briefly how we may proceed in order to show that the system (ANS_ν) is globally well posed for small data in $\mathcal{B}_4^{-\frac{1}{2},\frac{1}{2}}$. We shall search for a solution of the form $u = u_F + w$ with

$$u_F \overset{\text{def}}{=} e^{\nu t\Delta_h}u_{hh} \quad\text{and}\quad u_{hh} \overset{\text{def}}{=} \sum_{k\geq\ell-1}\Delta_k^h\Delta_\ell^v u_0. \tag{6.13}$$

Note that $u_{\ell h} \overset{\text{def}}{=} u_0 - u_{hh}$ satisfies

$$u_{\ell h} = \sum_{j\in\mathbb{Z}}S_{j-1}^h\Delta_j^v u_0. \tag{6.14}$$

It turns out that $u_{\ell h}$ is *smoother* than u_0. Indeed,

$$\Delta_j^v u_{\ell h} = \sum_{|j-j'|\leq 1}S_{j'-1}^h\Delta_{j'}^v\Delta_j^v u_0,$$

and thus

$$\|\Delta_j^v u_{\ell h}\|_{L^2} \leq C\sum_{|j-j'|\leq 1}\|S_{j'-1}^h\Delta_{j'}^v u_0\|_{L^2}.$$

This implies that if u_0 belongs to $\mathcal{B}_4^{-\frac{1}{2},\frac{1}{2}}$, then $u_{\ell h}$ belongs to $\mathcal{B}^{0,\frac{1}{2}}$ and

$$\|u_{\ell h}\|_{\mathcal{B}^{0,\frac{1}{2}}} \leq C\|u_0\|_{\mathcal{B}_4^{-\frac{1}{2},\frac{1}{2}}}. \tag{6.15}$$

In turn, this implies that w is also more regular than the free solution u_F.

We can now state the main result of this chapter.

Theorem 6.16. *There exists a constant c such that for all divergence-free initial data u_0 in $\mathcal{B}_4^{-\frac{1}{2},\frac{1}{2}}$ satisfying $\|u_0\|_{\mathcal{B}_4^{-\frac{1}{2},\frac{1}{2}}} \leq c\nu$, the system (ANS_ν) has a unique global solution u in $\mathcal{B}_4^{-\frac{1}{2},\frac{1}{2}}(\infty)$. Moreover, the vector field $u - u_F$ belongs to $\mathcal{B}^{0,\frac{1}{2}}(\infty)$.*

The above theorem will be proven in the next two sections. For the time being, we will show that the $\mathcal{B}_4^{-\frac{1}{2},\frac{1}{2}}$ norm may be made small by fast *horizontal* oscillations.

Proposition 6.17. *Let ϕ be in $\mathcal{S}(\mathbb{R}^3)$ and define $\phi_\varepsilon(x) \stackrel{def}{=} e^{ix_1/\varepsilon}\phi(x)$. A constant C_ϕ exists such that for any positive ε,*

$$\|\phi_\varepsilon\|_{\mathcal{B}_4^{-\frac{1}{2},\frac{1}{2}}} \leq C_\phi \varepsilon^{\frac{1}{2}}.$$

Proof. By definition of the norm $\|\cdot\|_{\mathcal{B}_4^{-\frac{1}{2},\frac{1}{2}}}$ and because the $\|\cdot\|_{\ell^2}$ norm is less than or equal to the $\|\cdot\|_{\ell^1}$ norm, we have

$$\|\phi_\varepsilon\|_{\mathcal{B}_4^{-\frac{1}{2},\frac{1}{2}}} \leq \sum_{j=1}^{4} \Phi_\varepsilon^{(j)} \qquad \text{with}$$

$$\Phi_\varepsilon^{(1)} \stackrel{def}{=} \sum_{\substack{\varepsilon 2^k > 1 \\ k \geq \ell-1}} 2^{-\frac{k-\ell}{2}} \|\Delta_k^h \Delta_\ell^v \phi_\varepsilon\|_{L_h^4(L_v^2)},$$

$$\Phi_\varepsilon^{(2)} \stackrel{def}{=} \sum_{\substack{\varepsilon 2^k \leq 1 \\ k \geq \ell-1}} 2^{-\frac{k-\ell}{2}} \|\Delta_k^h \Delta_\ell^v \phi_\varepsilon\|_{L_h^4(L_v^2)},$$

$$\Phi_\varepsilon^{(3)} \stackrel{def}{=} \sum_{\varepsilon 2^j > 1} 2^{\frac{j}{2}} \|S_{j-1}^h \Delta_j^v \phi_\varepsilon\|_{L^2},$$

$$\Phi_\varepsilon^{(4)} \stackrel{def}{=} \sum_{\varepsilon 2^j \leq 1} 2^{\frac{j}{2}} \|S_{j-1}^h \Delta_j^v \phi_\varepsilon\|_{L^2}.$$

In order to estimate $\Phi_\varepsilon^{(1)}$, we note that

$$\Phi_\varepsilon^{(1)} \leq \left(\sum_{\varepsilon 2^k > 1} 2^{-\frac{k}{2}} \right) \sum_{\ell \in \mathbb{Z}} 2^{\frac{\ell}{2}} \sup_{k \in \mathbb{Z}} \|\Delta_k^h \Delta_\ell^v \phi_\varepsilon\|_{L_h^4(L_v^2)}$$

$$\leq \varepsilon^{\frac{1}{2}} \sum_{\ell \in \mathbb{Z}} 2^{\frac{\ell}{2}} \sup_{k \in \mathbb{Z}} \|\Delta_k^h \Delta_\ell^v \phi_\varepsilon\|_{L_h^4(L_v^2)}.$$

Using Lemma 6.10 and the definition of ϕ_ε, we get

$$\sup_{k \in \mathbb{Z}} \|\Delta_k^h \Delta_\ell^v \phi_\varepsilon\|_{L_h^4(L_v^2)} \leq C\|\phi_\varepsilon\|_{L_h^4(L_v^2)} \leq C\|\phi\|_{L_h^4(L_v^2)}$$

and also

$$\sup_{k \in \mathbb{Z}} \|\Delta_k^h \Delta_\ell^v \phi_\varepsilon\|_{L_h^4(L_v^2)} \leq C2^{-\ell} \|\partial_3 \phi_\varepsilon\|_{L_h^4(L_v^2)} \leq C2^{-\ell} \|\partial_3 \phi\|_{L_h^4(L_v^2)}.$$

Thus, taking the sum over $\ell \leq N$ and $\ell > N$ and choosing the best N gives

$$\Phi_\varepsilon^{(1)} \leq \varepsilon^{\frac{1}{2}} \sum_{\ell \in \mathbb{Z}} 2^{\frac{\ell}{2}} \sup_{k \in \mathbb{Z}} \|\Delta_k^h \Delta_\ell^v \phi_\varepsilon\|_{L_h^4(L_v^2)} \leq \varepsilon^{\frac{1}{2}} \|\phi\|_{L_h^4(L_v^2)}^{\frac{1}{2}} \|\partial_3 \phi\|_{L_h^4(L_v^2)}^{\frac{1}{2}}.$$

Estimating $\Phi_\varepsilon^{(2)}$ demands the use of oscillations. Let

$$\widetilde{\phi}_{k,\ell}^{2,\varepsilon}(x) \overset{\text{def}}{=} 2^{2k}2^{\ell} \int_{\mathbb{R}^3} (\partial_1 \widetilde{g})(2^k(x_h - y_h))\widetilde{h}(2^{\ell}(x_3 - y_3))e^{i\frac{y_1}{\varepsilon}} \phi(y)\, dy$$

with $\mathcal{F}\widetilde{g}(\xi_h) = \widetilde{\varphi}(|\xi_h|)$ and $\mathcal{F}\widetilde{h}(\xi_3) = \widetilde{\varphi}(\xi_3)$. Integration by parts gives

$$\Delta_k^h \Delta_\ell^v \phi_\varepsilon = \phi_{k,\ell}^{1,\varepsilon} + \phi_{k,\ell}^{2,\varepsilon} \text{ with } \phi_{k,\ell}^{1,\varepsilon} \overset{\text{def}}{=} i\varepsilon \Delta_k^h \Delta_\ell^v (e^{i\frac{y_1}{\varepsilon}} \partial_1 \phi) \text{ and } \phi_{k,\ell}^{2,\varepsilon} \overset{\text{def}}{=} -i\varepsilon 2^k \widetilde{\phi}_{k,\ell}^{2,\varepsilon}.$$

Using Lemma 6.10, we get

$$2^{-\frac{k}{2}} \sum_{\ell \leq k+1} 2^{\frac{\ell}{2}} \|\phi_{k,\ell}^{1,\varepsilon}\|_{L_h^4(L_v^2)} \leq C\varepsilon \sup_{\ell \in \mathbb{Z}} \|\Delta_k^h \Delta_\ell^v (e^{i\frac{y_1}{\varepsilon}} \partial_1 \phi)\|_{L_h^4(L_v^2)}$$

$$\leq C\varepsilon 2^{\frac{k}{2}} \|\partial_1 \phi\|_{L^2}.$$

Moreover, we have

$$2^{-\frac{k}{2}} \sum_{\ell \leq k+1} 2^{\frac{\ell}{2}} \|\phi_{k,\ell}^{2,\varepsilon}\|_{L_h^4(L_v^2)} \leq \varepsilon 2^{\frac{k}{2}} \sum_{\ell \in \mathbb{Z}} 2^{\frac{\ell}{2}} \|\widetilde{\phi}_{k,\ell}^{2,\varepsilon}\|_{L_h^4(L_v^2)}.$$

Using Lemma 6.10, we get

$$\|\widetilde{\phi}_{k,\ell}^{2,\varepsilon}\|_{L_h^4(L_v^2)} \leq C\|\phi\|_{L_h^4(L_v^2)} \quad \text{and} \quad \|\widetilde{\phi}_{k,\ell}^{2,\varepsilon}\|_{L_h^4(L_v^2)} \leq C2^{-\ell}\|\partial_3 \phi\|_{L_h^4(L_v^2)}.$$

Again, taking the sum over $\ell \leq N$ and $\ell > N$ and choosing the best N, we get

$$\sum_{\ell \in \mathbb{Z}} 2^{\frac{\ell}{2}} \|\widetilde{\phi}_{k,\ell}^{2,\varepsilon}\|_{L_h^4(L_v^2)} \leq C\|\phi\|_{L_h^4(L_v^2)}^{\frac{1}{2}} \|\partial_3 \phi\|_{L_h^4(L_v^2)}^{\frac{1}{2}}.$$

Therefore,

$$\Phi_\varepsilon^{(2)} \leq C_\phi \varepsilon \sum_{\varepsilon 2^k \leq 1} 2^{\frac{k}{2}} \leq C_\phi \varepsilon^{\frac{1}{2}}.$$

In order to estimate $\Phi_\varepsilon^{(3)}$, we note that, thanks to Lemma 6.10, we have

$$\Phi_\varepsilon^{(3)} \leq C \sum_{\varepsilon 2^j > 1} 2^{-\frac{j}{2}} \|S_{j-1}^h \Delta_j^v \partial_3 \phi_\varepsilon\|_{L^2}$$

$$\leq C\|\partial_3 \phi_\varepsilon\|_{L^2} \sum_{\varepsilon 2^j > 1} 2^{-\frac{j}{2}}$$

$$\leq C\varepsilon^{\frac{1}{2}} \|\partial_3 \phi\|_{L^2}.$$

Estimating $\Phi_\varepsilon^{(4)}$ requires use of the oscillations. Integrating by parts, we get

$$S_{j-1}^h \Delta_j^v \phi_\varepsilon = \phi_j^{1,\varepsilon} + \phi_j^{2,\varepsilon} \text{ with } \phi_j^{1,\varepsilon} \overset{\text{def}}{=} i\varepsilon S_{j-1}^h \Delta_j^v (e^{i\frac{y_1}{\varepsilon}} \partial_1 \phi) \text{ and } \phi_j^{2,\varepsilon} \overset{\text{def}}{=} -i\varepsilon 2^j \widetilde{\phi}_j^{2,\varepsilon}$$

with $\widetilde{\phi}_j^{2,\varepsilon}(x) \overset{\text{def}}{=} 2^{3j} \int (\partial_1 \overline{g})(2^j(x_h - y_h))\widetilde{h}(2^j(x_3 - y_3))e^{i\frac{y_1}{\varepsilon}} \phi(y)\, dy$ for some function $\overline{g}$ in $\mathcal{S}(\mathbb{R}^2)$.

Using Lemma 6.10, we get

$$\sum_{\varepsilon 2^j \leq 1} 2^{\frac{j}{2}} \|\phi_j^{1,\varepsilon}\|_{L^2} \leq C\varepsilon \|\partial_1 \phi\|_{L^2} \sum_{\varepsilon 2^j \leq 1} 2^{\frac{j}{2}} \leq C\varepsilon^{\frac{1}{2}} \|\partial_1 \phi\|_{L^2}.$$

Using Lemma 6.10 again, we get $2^j \|\widetilde{\phi}_j^{2,\varepsilon}\|_{L^2} \leq \|\partial_3 \phi\|_{L^2}$. Thus, we infer that

$$\sum_{\varepsilon 2^j \leq 1} 2^{\frac{j}{2}} \|\phi_j^{2,\varepsilon}\|_{L^2} \leq C\varepsilon \|\partial_3 \phi\|_{L^2} \sum_{\varepsilon 2^j \leq 1} 2^{\frac{j}{2}} \leq C\varepsilon^{\frac{1}{2}} \|\partial_3 \phi\|_{L^2}.$$

This completes the proof of Proposition 6.17. □

Combining Theorem 6.16 with the above result, we deduce that data with high oscillations with respect to the horizontal variable generate global solutions of the system (ANS_ν).

Corollary 6.18. *For any ϕ in $\mathcal{S}(\mathbb{R}^3)$, there exists some $\varepsilon_0 > 0$ such that for all ε in $]0, \varepsilon_0[$, the system (ANS_ν) has a global unique solution with data*

$$u_0^\varepsilon(x) = \sin\left(\frac{x_1}{\varepsilon}\right)(0, -\partial_3 \phi, \partial_2 \phi). \tag{6.16}$$

6.2.4 Some Technical Lemmas

For the remainder of this chapter, it will be understood that $(c_k)_{k \in \mathbb{Z}}$ [resp., $(d_j)_{j \in \mathbb{Z}}$] denotes a generic element of the sphere of $\ell^2(\mathbb{Z})$ [resp., $\ell^1(\mathbb{Z})$]. Furthermore, $(c_{k,\ell})_{(k,\ell) \in \mathbb{Z}^2}$ will denote a generic element of the sphere of $\ell^2(\mathbb{Z}^2)$ and $(d_{k,\ell})_{(k,\ell) \in \mathbb{Z}^2}$ a generic sequence such that

$$\sum_{\ell \in \mathbb{Z}} \left(\sum_{k \in \mathbb{Z}} d_{k,\ell}^2\right)^{\frac{1}{2}} = 1.$$

We shall often use the following property, the proof of which is omitted.

Lemma 6.19. *Let α be in $]0, \infty[$ and N_0 be in $\mathbb{Z}$. We then have*

$$\sum_{\substack{(k,\ell) \in \mathbb{Z}^2 \\ \ell \geq j - N_0}} 2^{-\alpha(\ell - j)} d_{k,\ell} c_k \leq \frac{2^{\alpha N_0}}{1 - 2^{-\alpha}} d_j.$$

The following lemma will be of frequent use in this chapter. It describes some estimates of dyadic parts of functions in $\mathcal{B}_4^{-\frac{1}{2},\frac{1}{2}}(T)$.

Lemma 6.20. *For any $a \in \mathcal{B}_4^{-\frac{1}{2},\frac{1}{2}}(T)$, we have*

$$\|S_k^h \Delta_\ell^v a\|_{L_T^\infty(L_h^4(L_v^2))} + \nu^{\frac{1}{2}} \|\nabla_h S_k^h \Delta_\ell^v a\|_{L_T^2(L_h^4(L_v^2))} \leq C d_{k,\ell} 2^{\frac{k}{2}} 2^{-\frac{\ell}{2}} \|a\|_{\mathcal{B}_4^{-\frac{1}{2},\frac{1}{2}}(T)},$$

$$\|S_k^h a\|_{L_T^\infty(L_h^4(L_v^\infty))} + \nu^{\frac{1}{2}} \|\nabla_h S_k^h a\|_{L_T^2(L_h^4(L_v^\infty))} \leq C c_k 2^{\frac{k}{2}} \|a\|_{\mathcal{B}_4^{-\frac{1}{2},\frac{1}{2}}(T)}.$$

Proof. By definition of S_k^h, we have

$$\mathcal{S}_{k,\ell}(a) \overset{\text{def}}{=} \|S_k^h \Delta_\ell^v a\|_{L_T^\infty(L_h^4(L_v^2))} + \nu^{\frac{1}{2}} \|\nabla_h S_k^h \Delta_\ell^v a\|_{L_T^2(L_h^4(L_v^2))}$$

$$\leq \sum_{k' \leq k-1} \left(\|\Delta_{k'}^h \Delta_\ell^v a\|_{L_T^\infty(L_h^4(L_v^2))} + \nu^{\frac{1}{2}} \|\nabla_h \Delta_{k'}^h \Delta_\ell^v a\|_{L_T^2(L_h^4(L_v^2))} \right).$$

Noting that

$$2^{\frac{\ell}{2}} 2^{-\frac{k}{2}} \mathcal{S}_{k,\ell}(a) \leq 2^{\frac{\ell}{2}} \sum_{k' \leq k-1} 2^{-\frac{k-k'}{2}} 2^{-\frac{k'}{2}}$$

$$\times \left(\|\Delta_{k'}^h \Delta_\ell^v a\|_{L_T^\infty(L_h^4(L_v^2))} + \nu^{\frac{1}{2}} \|\nabla_h \Delta_{k'}^h \Delta_\ell^v a\|_{L_T^2(L_h^4(L_v^2))} \right),$$

we get, by applying the Cauchy–Schwarz inequality, that

$$2^{\frac{\ell}{2}} \left(\sum_{k \in \mathbb{Z}} 2^{-k} \mathcal{S}_{k,\ell}(a)^2 \right)^{\frac{1}{2}} \leq 2^{\frac{\ell}{2}} \left(\sum_{k' \in \mathbb{Z}} 2^{-k'} \left(\|\Delta_{k'}^h \Delta_\ell^v a\|_{L_T^\infty(L_h^4(L_v^2))} \right. \right.$$

$$\left. \left. + \nu^{\frac{1}{2}} \|\nabla_h \Delta_{k'}^h \Delta_\ell^v a\|_{L_T^2(L_h^4(L_v^2))} \right)^2 \right)^{\frac{1}{2}}.$$

By Corollary 6.15, this proves the first inequality.

In order to establish the second inequality, we shall prove that for any sequence $(c_k)_{k \in \mathbb{Z}}$ in the unit ball of $\ell^2(\mathbb{Z})$, we have

$$I(a) \overset{\text{def}}{=} \sum_{k \in \mathbb{Z}} 2^{-\frac{k}{2}} \mathcal{S}_k c_k \leq C \|a\|_{\mathcal{B}_4^{-\frac{1}{2},\frac{1}{2}}(T)} \quad \text{with} \tag{6.17}$$

$$\mathcal{S}_k \overset{\text{def}}{=} \|S_k^h a\|_{L_T^\infty(L_h^4(L_v^\infty))} + \nu^{\frac{1}{2}} \|\nabla_h S_k^h a\|_{L_T^2(L_h^4(L_v^\infty))}. \tag{6.18}$$

Again using Lemma 6.10, we have

$$\mathcal{S}_k \leq C \sum_{k' \leq k-1} \sum_{\ell \in \mathbb{Z}} 2^{\frac{\ell}{2}} \left(\|\Delta_{k'}^h \Delta_\ell^v a\|_{L_T^\infty(L_h^4(L_v^2))} + \nu^{\frac{1}{2}} \|\Delta_{k'}^h \Delta_\ell^v \nabla_h a\|_{L_T^2(L_h^4(L_v^2))} \right).$$

We deduce that

$$I(a) \leq C \sum_{\ell \in \mathbb{Z}} 2^{\frac{\ell}{2}} \sum_{\substack{(k,k') \in \mathbb{Z}^2 \\ k' \leq k-1}} 2^{-\frac{k-k'}{2}} 2^{-\frac{k'}{2}} c_k \left(\|\Delta_{k'}^h \Delta_\ell^v a\|_{L_T^\infty(L_h^4(L_v^2))} \right.$$

$$\left. + \nu^{\frac{1}{2}} \|\Delta_{k'}^h \Delta_\ell^v \nabla_h a\|_{L_T^2(L_h^4(L_v^2))} \right).$$

From the Cauchy–Schwarz inequality with the weight $2^{-\frac{k-k'}{2}} \mathbf{1}_{k' \leq k-1}$, we infer that

$$I(a) \le C \left(\sum_{\substack{(k,k')\in\mathbb{Z}^2 \\ k'\le k-1}} 2^{-\frac{k-k'}{2}} c_k^2 \right)^{\frac{1}{2}} \sum_{\ell\in\mathbb{Z}} 2^{\frac{\ell}{2}} \left(\sum_{\substack{(k,k')\in\mathbb{Z}^2 \\ k'\le k-1}} 2^{-\frac{k-k'}{2}} 2^{-k'} \right.$$

$$\left. \times \left(\|\Delta_{k'}^h \Delta_\ell^v a\|_{L_T^\infty(L_h^4(L_v^2))} + \nu^{\frac{1}{2}} \|\Delta_{k'}^h \Delta_\ell^v \nabla_h a\|_{L_T^2(L_h^4(L_v^2))} \right)^2 \right)^{\frac{1}{2}}.$$

From this, we deduce that

$$I(a) \le C \sum_{\ell\in\mathbb{Z}} 2^{\frac{\ell}{2}} \left(\sum_{\substack{(k,k')\in\mathbb{Z}^2 \\ k'\le k-1}} 2^{-\frac{k-k'}{2}} 2^{-k'} \left(\|\Delta_{k'}^h \Delta_\ell^v a\|_{L_T^\infty(L_h^4(L_v^2))} \right. \right.$$

$$\left. \left. + \nu^{\frac{1}{2}} \|\Delta_{k'}^h \Delta_\ell^v \nabla_h a\|_{L_T^2(L_h^4(L_v^2))} \right)^2 \right)^{\frac{1}{2}}$$

$$\le C \sum_{\ell\in\mathbb{Z}} 2^{\frac{\ell}{2}} \left(\sum_{k'\in\mathbb{Z}} 2^{-k'} \left(\|\Delta_{k'}^h \Delta_\ell^v a\|_{L_T^\infty(L_h^4(L_v^2))} \right. \right.$$

$$\left. \left. + \nu^{\frac{1}{2}} \|\Delta_{k'}^h \Delta_\ell^v \nabla_h a\|_{L_T^2(L_h^4(L_v^2))} \right)^2 \right)^{\frac{1}{2}}$$

$$\le C \|a\|_{\mathcal{B}_4^{-\frac{1}{2},\frac{1}{2}}(T)},$$

which proves (6.17) and thus the whole Lemma 6.20. $\square$

With Lemma 6.20 at our disposal, we will now establish a result which is very close to Sobolev embedding and which will be of constant use in proving the existence part of Theorem 6.16.

Lemma 6.21. *The space* $\mathcal{B}_4^{-\frac{1}{2},\frac{1}{2}}(T)$ *is embedded in* $L_T^4(L_h^4(L_v^\infty))$. *More precisely, for any function* a *in* $\mathcal{B}_4^{-\frac{1}{2},\frac{1}{2}}(T)$, *we have*

$$\|\Delta_j^v a\|_{L_T^4(L_h^4(L_v^2))} \le C \frac{d_j}{\nu^{\frac{1}{4}}} 2^{-\frac{j}{2}} \|a\|_{\mathcal{B}_4^{-\frac{1}{2},\frac{1}{2}}(T)}$$

$$\|a\|_{L_T^4(L_h^4(L_v^\infty))} \le \frac{C}{\nu^{\frac{1}{4}}} \|a\|_{\mathcal{B}_4^{-\frac{1}{2},\frac{1}{2}}(T)}.$$

Proof. First, note that

$$\|\Delta_j^v a\|_{L_T^4(L_h^4(L_v^2))}^2 = \|(\Delta_j^v a)^2\|_{L_T^2(L_h^2(L_v^1))}.$$

Then, according to Bony's decomposition in the horizontal variables, we may write

$$(\Delta_j^v a)^2 = \sum_{k\in\mathbb{Z}} S_{k-1}^h \Delta_j^v a \, \Delta_k^h \Delta_j^v a + \sum_{k\in\mathbb{Z}} S_{k+2}^h \Delta_j^v a \, \Delta_k^h \Delta_j^v a.$$

The two terms on the right-hand side may be estimated exactly in the same way, so we first focus on the first term. Applying Hölder's inequality, we get

$$\|S^h_{k-1}\Delta^v_j a\, \Delta^h_k\Delta^v_j a\|_{L^2_T(L^2_h(L^1_v))}$$

$$\leq 2^{-\frac{k}{2}}\|S^h_{k-1}\Delta^v_j a\|_{L^\infty_T(L^4_h(L^2_v))}2^{\frac{k}{2}}\|\Delta^h_k\Delta^v_j a\|_{L^2_T(L^4_h(L^2_v))}.$$

Using the first inequality of Lemma 6.20 and Corollary 6.15, we infer that

$$\|S^h_{k-1}\Delta^v_j a\, \Delta^h_k\Delta^v_j a\|_{L^2_T(L^2_h(L^1_v))} \leq C\frac{d^2_{k,j}}{\nu^{\frac{1}{2}}}2^{-j}\|a\|^2_{\mathcal{B}^{-\frac{1}{2},\frac{1}{2}}_4(T)}.$$

Taking the sum over k, we thus deduce that

$$\|(\Delta^v_j a)^2\|_{L^2_T(L^2_h(L^1_v))} \leq C\frac{d^2_j}{\nu^{\frac{1}{2}}}2^{-j}\|a\|^2_{\mathcal{B}^{-\frac{1}{2},\frac{1}{2}}_4(T)},$$

which is exactly the first inequality of the lemma. Now, using Lemma 6.10, we have

$$\|\Delta^v_j a\|_{L^4_T(L^4_h(L^\infty_v))} \leq C2^{\frac{j}{2}}\|\Delta^v_j a\|_{L^4_T(L^4_h(L^2_v))}.$$

This proves the whole lemma. $\qquad\square$

We will now use Lemma 6.10 to study the free evolution u_F of the high horizontal frequency part of the initial data u_0, as defined in (6.13). In order to do this, we first recall a result, in the spirit of Corollary 2.5 page 55, which describes the action of the semigroup of the heat equation on distributions with Fourier transforms supported in a fixed annulus.

Lemma 6.22. Let $u_0 \in \mathcal{B}^{-\frac{1}{2},\frac{1}{2}}_4$, u_F be as in (6.13), $\alpha \in \mathbb{N}^3$, and $p \in [1,\infty]$. Then, $\Delta^h_k\Delta^v_\ell u_F = 0$ if $k \leq \ell - 3$, and

$$\|\Delta^h_k\Delta^v_\ell u_F\|_{L^p_T(L^4_h(L^2_v))} \leq C\frac{d_{k,\ell}}{\nu^{\frac{1}{p}}}2^{k(\frac{1}{2}-\frac{2}{p})}2^{-\frac{\ell}{2}}\|u_0\|_{\mathcal{B}^{-\frac{1}{2},\frac{1}{2}}_4} \quad \text{if } k \geq \ell-2. \qquad (6.19)$$

Moreover, u_F belongs to $\mathcal{B}^{-\frac{1}{2},\frac{1}{2}}_4(\infty)$ and satisfies

$$\|u_F\|_{\mathcal{B}^{-\frac{1}{2},\frac{1}{2}}_4(\infty)} \leq C\|u_0\|_{\mathcal{B}^{-\frac{1}{2},\frac{1}{2}}_4}. \qquad (6.20)$$

Proof. From the relations (2.2) and (2.3) page 54, we deduce that

$$\Delta^h_k\Delta^v_\ell u_F(t) = 2^{2k}g(t,2^k\cdot)\star\Delta^h_k\Delta^v_\ell u_0 \quad \text{with } \|g(t,\cdot)\|_{L^1(\mathbb{R}^2)} \leq Ce^{-c\nu t2^{2k}}. \qquad (6.21)$$

Here, the convolution must be understood as the convolution on $\mathbb{R}^2$. Thus,

$$\|\Delta^h_k\Delta^v_\ell u_F(t,x_h,\cdot)\|_{L^2_v} \leq 2^{2k}|g(t,2^k\cdot)|\star\|\Delta^h_k\Delta^v_\ell u_0(x_h,\cdot)\|_{L^2_v}.$$

Using (6.21) and Lemma 6.10, we get

$$\|\Delta^h_k\Delta^v_\ell u_F(t)\|_{L^4_h(L^2_v)} \leq Ce^{-c\nu t2^{2k}}\|\Delta^h_k\Delta^v_\ell u_0\|_{L^4_h(L^2_v)}$$

$$\leq Ce^{-c\nu t2^{2k}}d_{k,\ell}2^{\frac{k}{2}}2^{-\frac{\ell}{2}}\|u_0\|_{\mathcal{B}^{-\frac{1}{2},\frac{1}{2}}_4}.$$

By time integration, the lemma then follows. $\qquad\square$

From Lemma 6.22, we immediately deduce the following corollary.

Corollary 6.23. *For any (p,q) in $[1,\infty] \times [4,\infty]$, we have*

$$\|\Delta_k^h u_F\|_{L^p(\mathbb{R}^+;L_h^q(L_v^\infty))} \leq C \frac{1}{\nu^{\frac{1}{p}}} c_k 2^{-k\left(2\left(\frac{1}{p}+\frac{1}{q}\right)-1\right)} \|u_0\|_{\mathcal{B}_4^{-\frac{1}{2},\frac{1}{2}}}.$$

If, in addition, $\dfrac{1}{p}+\dfrac{1}{q} > \dfrac{1}{2}$, then we have

$$\|\Delta_j^v u_F\|_{L^p(\mathbb{R}^+;L_h^q(L_v^2))} \leq C \frac{1}{\nu^{\frac{1}{p}}} d_j 2^{-j\left(2\left(\frac{1}{p}+\frac{1}{q}\right)-\frac{1}{2}\right)} \|u_0\|_{\mathcal{B}_4^{-\frac{1}{2},\frac{1}{2}}}.$$

The following lemma corresponds to the endpoint of the second estimate of Corollary 6.23.

Lemma 6.24. *Under the assumptions of Lemma 6.22, we have*

$$\|\Delta_j^v u_F\|_{L^2(\mathbb{R}^+;L_h^\infty(L_v^2))} \leq C \frac{d_j}{\sqrt{\nu}} 2^{-\frac{j}{2}} \|u_0\|_{\mathcal{B}_4^{-\frac{1}{2},\frac{1}{2}}} \quad and$$

$$\|u_F\|_{L^2(\mathbb{R}^+;L^\infty)} \leq C \frac{1}{\sqrt{\nu}} \|u_0\|_{\mathcal{B}_4^{-\frac{1}{2},\frac{1}{2}}}.$$

Proof. Trivially, we have

$$\|\Delta_j^v u_F\|_{L_T^2(L_h^\infty(L_v^2))}^2 = \|(\Delta_j^v u_F)^2\|_{L_T^1(L_h^\infty(L_v^1))}.$$

Using Bony's decomposition in the horizontal variables, we obtain

$$(\Delta_j^v u_F)^2 = \sum_{k\in\mathbb{Z}} S_{k-1}^h \Delta_j^v u_F \, \Delta_k^h \Delta_j^v u_F + \sum_{k\in\mathbb{Z}} \Delta_k^h \Delta_j^v u_F \, S_{k+2}^h \Delta_j^v u_F. \quad (6.22)$$

Now, the idea is to take advantage of the smoothing effect on the highest possible horizontal frequencies of u_F. Applying Hölder's inequality and Lemma 6.10, we get

$$\|S_{k-1}^h \Delta_j^v u_F \, \Delta_k^h \Delta_j^v u_F\|_{L_T^1(L_h^\infty(L_v^1))}$$
$$\leq C 2^k \|S_{k-1}^h \Delta_j^v u_F\|_{L_T^\infty(L_h^4(L_v^2))} \|\Delta_k^h \Delta_j^v u_F\|_{L_T^1(L_h^4(L_v^2))}.$$

Note that by (6.19) and the fact that $S_{k-1}^h = \sum\limits_{k' \leq k-2} \Delta_{k'}^h$, we have

$$\|S_{k-1}^h \Delta_j^v u_F\|_{L_T^\infty(L_h^4(L_v^2))} \leq C \Big(\sum_{k' \leq k-2} d_{k,j}^2 \Big)^{\frac{1}{2}} 2^{\frac{k}{2}} 2^{-\frac{j}{2}} \|u_0\|_{\mathcal{B}_4^{-\frac{1}{2},\frac{1}{2}}}.$$

Therefore, by using (6.19) once again, we arrive at

$$\Big\| \sum_{k\in\mathbb{Z}} S_{k-1}^h \Delta_j^v u_F \Delta_k^h \Delta_j^v u_F \Big\|_{L_T^1(L_h^\infty(L_v^1))} \leq C \frac{2^{-j}}{\nu} \Big(\sum_{k'\in\mathbb{Z}} d_{k',j}^2 \Big) \|u_0\|_{\mathcal{B}_4^{-\frac{1}{2},\frac{1}{2}}}^2.$$

Estimating the other term in (6.22) follows along the same lines. Therefore,

$$\|\Delta_j^v u_F\|_{L_T^2(L_h^\infty(L_v^2))}^2 \le C \frac{d_j^2}{\nu} 2^{-j} \|u_0\|_{\mathcal{B}_4^{-\frac{1}{2},\frac{1}{2}}}^2.$$

From Lemma 6.10 we then conclude that

$$\|S_j^v u_F\|_{L_T^2(L^\infty)} \le C \sum_{j' \le j-1} 2^{\frac{j'}{2}} \|\Delta_{j'}^v u_F\|_{L_T^2(L_h^\infty(L_v^2))} \le C \frac{1}{\sqrt{\nu}} \|u_0\|_{\mathcal{B}_4^{-\frac{1}{2},\frac{1}{2}}}.$$

This completes the proof of the lemma.   □

6.3 The Proof of Existence

As announced in the previous section, we seek a solution of the form

$$u = u_F + w.$$

By substituting the above formula into (ANS_ν), we find that w must satisfy

$$(\widetilde{ANS_\nu}) \begin{cases} \partial_t w + w \cdot \nabla w - \nu \Delta_h w + w \cdot \nabla u_F + u_F \cdot \nabla w = -u_F \cdot \nabla u_F - \nabla P \\ \operatorname{div} w = 0 \\ w|_{t=0} = u_{\ell h} \overset{\text{def}}{=} u_0 - u_{hh}. \end{cases}$$

Recall that, according to (6.15), if u_0 belongs to $\mathcal{B}_4^{-\frac{1}{2},\frac{1}{2}}$, then $u_{\ell h}$ belongs to $\mathcal{B}^{0,\frac{1}{2}}$.

As in the proof of Theorem 6.2, we shall use the Friedrichs regularization method to construct the approximate solutions to $(\widetilde{ANS_\nu})$. Define $u_{F,n} \overset{\text{def}}{=} (\operatorname{Id} - \mathbb{E}_n) u_F$. The approximate system $(\widetilde{ANS_{\nu,n}})$ we consider is of the form

$$\begin{cases} \partial_t w_n - \nu \Delta_h w_n + \mathbb{E}_n(w_n \cdot \nabla w_n) + \mathbb{E}_n(w_n \cdot \nabla u_{F,n}) + \mathbb{E}_n(u_{F,n} \cdot \nabla w_n) \\ \qquad = -\mathbb{E}_n(u_{F,n} \cdot \nabla u_{F,n}) - \mathbb{E}_n \nabla(-\Delta)^{-1} \partial_j \partial_k \left((u_{F,n}^j + w_n^j)(u_{F,n}^k + w_n^k) \right) \\ \operatorname{div} w_n = 0, \\ w_n|_{t=0} = \mathbb{E}_n(u_{\ell h}) \overset{\text{def}}{=} \mathbb{E}_n(u_0 - u_{hh}). \end{cases}$$

Arguing as in the first section of this chapter, we can prove that the system $(\widetilde{ANS_{\nu,n}})$ is an ordinary differential equation in the space $L_n^{2,\sigma}$. Thanks to Theorem 3.11 page 131, this ordinary differential equation is globally well posed because

$$\frac{d}{dt} \|w_n(t)\|_{L^2}^2 \le C_n \|u_{F,n}(t)\|_{L^\infty} \|w_n\|_{L^2}^2 + C_n \|u_{F,n}(t)\|_{L_h^4(L_v^2)}^2 \|w_n(t)\|_{L^2},$$

and, according to Corollary 6.23 and Lemma 6.24, the function $u_{F,n}$ belongs to $L^2(\mathbb{R}^+; L^\infty \cap L_h^4(L_v^2))$.

The proof of Theorem 6.16 now reduces to the following three propositions, which we shall assume for the time being.[2]

Proposition 6.25. Let u_0 be in $\mathcal{B}_4^{-\frac{1}{2},\frac{1}{2}}$ and a be in $\mathcal{B}^{0,\frac{1}{2}}(T)$. Define

$$I_j(T) \stackrel{def}{=} \int_0^T \left(\Delta_j^v(u_F \cdot \nabla u_F)|\Delta_j^v a\right) dt.$$

Then, for any j in $\mathbb{Z}$, we have

$$|I_j(T)| \leq Cd_j^2 \nu^{-1} 2^{-j} \|u_0\|^2_{\mathcal{B}_4^{-\frac{1}{2},\frac{1}{2}}} \|a\|_{\mathcal{B}^{0,\frac{1}{2}}(T)}.$$

Proposition 6.26. Let a and b be vector fields in $\mathcal{B}^{0,\frac{1}{2}}(T)$. Define

$$J_j(T) \stackrel{def}{=} \int_0^T \left(\Delta_j^v(a \cdot \nabla u_F)|\Delta_j^v b\right) dt.$$

If $\operatorname{div} a = 0$, then, for any j in $\mathbb{Z}$,

$$|J_j(T)| \leq Cd_j^2 \nu^{-1} 2^{-j} \|a\|_{\mathcal{B}^{0,\frac{1}{2}}(T)} \|u_0\|_{\mathcal{B}_4^{-\frac{1}{2},\frac{1}{2}}} \|b\|_{\mathcal{B}^{0,\frac{1}{2}}(T)}.$$

Proposition 6.27. Let a be a divergence-free vector field in $\mathcal{B}_4^{-\frac{1}{2},\frac{1}{2}}(T)$ and b a vector field in $\mathcal{B}^{0,\frac{1}{2}}(T)$. Define

$$F_j(T) \stackrel{def}{=} \int_0^T \left(\Delta_j^v(a \cdot \nabla b)|\Delta_j^v b\right) dt.$$

Then, for any $j \in \mathbb{Z}$, we have

$$|F_j(T)| \leq Cd_j^2 \nu^{-1} 2^{-j} \|a\|_{\mathcal{B}_4^{-\frac{1}{2},\frac{1}{2}}(T)} \|b\|^2_{\mathcal{B}^{0,\frac{1}{2}}(T)}.$$

Completion of the proof of Theorem 6.16. Apply the operator Δ_j^v to $(\widetilde{ANS}_{\nu,n})$ and take the L^2 inner product of the resulting equation with $\Delta_j^v w_n$. Because $\mathbb{E}_n w_n = w_n$, we get

$$D_n(t) \stackrel{def}{=} \frac{d}{dt} \|\Delta_j^v w_n(t)\|^2_{L^2} + 2\nu \|\nabla_h \Delta_j^v w_n(t)\|^2_{L^2}$$
$$= -2(\Delta_j^v(w_n \cdot \nabla w_n)|\Delta_j^v w_n) - 2(\Delta_j^v(u_{F,n} \cdot \nabla w_n)|\Delta_j^v w_n)$$
$$\quad - 2(\Delta_j^v(w_n \cdot \nabla u_{F,n})|\Delta_j^v w_n) - 2(\Delta_j^v(u_{F,n} \cdot \nabla u_{F,n})|\Delta_j^v w_n).$$

By integrating the above equation over $[0, T]$, we get

$$2^j \|\Delta_j^v w_n\|^2_{L^\infty_T(L^2)} + 2^{j+1}\nu \|\nabla_h \Delta_j^v w_n\|^2_{L^2_T(L^2)}$$

---

[2] In the following three statements, we drop the index n from $u_{F,n}$ to simplify notation.

$$\leq 2^j \|\Delta_j^v w_n(0)\|_{L^2}^2 + 2\sum_{k=1}^{4}|W_j^k(T)| \quad (6.23)$$

with
$$W_j^1(T) \stackrel{\text{def}}{=} 2^j \int_0^T \left(\Delta_j^v(w_n(t)\cdot\nabla w_n(t))|\Delta_j^v w_n(t)\right) dt,$$
$$W_j^2(T) \stackrel{\text{def}}{=} 2^j \int_0^T \left(\Delta_j^v(u_{F,n}(t)\cdot\nabla w_n(t))|\Delta_j^v w_n(t)\right) dt,$$
$$W_j^3(T) \stackrel{\text{def}}{=} 2^j \int_0^T \left(\Delta_j^v(w_n(t)\cdot\nabla u_{F,n}(t))|\Delta_j^v w_n(t)\right) dt,$$
$$W_j^4(T) \stackrel{\text{def}}{=} 2^j \int_0^T \left(\Delta_j^v(u_{F,n}(t)\cdot\nabla u_{F,n}(t))|\Delta_j^v w_n(t)\right) dt.$$

Applying Proposition 6.27 with $a = b = w_n$, together with Corollary 6.14, gives

$$\left|W_j^1(T)\right| \leq C\nu^{-1}d_j^2\|w_n\|_{\mathcal{B}^{0,\frac{1}{2}}(T)}^3. \quad (6.24)$$

Thanks to Lemma 6.22, Proposition 6.27 applied with $a = u_{F,n}$ and $b = w_n$ implies, in particular, that

$$\left|W_j^2(T)\right| \leq C\nu^{-1}d_j^2\|u_0\|_{\mathcal{B}_4^{-\frac{1}{2},\frac{1}{2}}}\|w_n\|_{\mathcal{B}^{0,\frac{1}{2}}(T)}^2. \quad (6.25)$$

Proposition 6.26 applied with $a = b = w_n$ yields

$$\left|W_j^3(T)\right| \leq C\nu^{-1}d_j^2\|u_0\|_{\mathcal{B}_4^{-\frac{1}{2},\frac{1}{2}}}\|w_n\|_{\mathcal{B}^{0,\frac{1}{2}}(T)}^2. \quad (6.26)$$

Finally, Proposition 6.25 guarantees that

$$\left|W_j^4(T)\right| \leq C\nu^{-1}d_j^2\|u_0\|_{\mathcal{B}_4^{-\frac{1}{2},\frac{1}{2}}}^2\|w_n\|_{\mathcal{B}^{0,\frac{1}{2}}(T)}. \quad (6.27)$$

Plugging the estimates (6.24)–(6.27) into (6.23) gives

$$2^j\left(\|\Delta_j^v w_n\|_{L_T^\infty(L^2)} + \sqrt{2\nu}\|\nabla_h\Delta_j^v w_n\|_{L_T^2(L^2)}\right)^2 \leq 2^j\|\Delta_j^v w_n(0)\|_{L^2}^2$$
$$+ \frac{C}{\nu}d_j^2\left(\|w_n\|_{\mathcal{B}^{0,\frac{1}{2}}(T)}^2 + \|u_0\|_{\mathcal{B}_4^{-\frac{1}{2},\frac{1}{2}}}^2\right)\|w_n\|_{\mathcal{B}^{0,\frac{1}{2}}(T)}.$$

Using (6.15), we get, by the definition of $\mathcal{B}^{0,\frac{1}{2}}(T)$,

$$\|w_n\|_{\mathcal{B}^{0,\frac{1}{2}}(T)} \leq C\|u_0\|_{\mathcal{B}_4^{-\frac{1}{2},\frac{1}{2}}} + \frac{C}{\sqrt{\nu}}\left(\|w_n\|_{\mathcal{B}^{0,\frac{1}{2}}(T)} + \|u_0\|_{\mathcal{B}_4^{-\frac{1}{2},\frac{1}{2}}}\right)\|w_n\|_{\mathcal{B}^{0,\frac{1}{2}}(T)}^{\frac{1}{2}}.$$

Define

$$T_n \stackrel{\text{def}}{=} \sup\left\{T > 0 \,/\, \|w_n\|_{\mathcal{B}^{0,\frac{1}{2}}(T)} \leq 2C\|u_0\|_{\mathcal{B}_4^{-\frac{1}{2},\frac{1}{2}}}\right\}.$$

The fact that w_n is continuous with values in H^N for any integer N ensures that T_n is positive. The above inequality then implies that, for any n and any $T < T_n$, we have

$$\|w_n\|_{\mathcal{B}^{0,\frac{1}{2}}(T)} \leq C\|u_0\|_{\mathcal{B}_4^{-\frac{1}{2},\frac{1}{2}}} + \frac{\sqrt{2}\,C(2C+1)\sqrt{C}}{\sqrt{\nu}}\|u_0\|_{\mathcal{B}_4^{-\frac{1}{2},\frac{1}{2}}}^{\frac{3}{2}}.$$

Thus, if $2C(1+2C)^2\|u_0\|_{\mathcal{B}_4^{-\frac{1}{2},\frac{1}{2}}} < \nu$, then we get, for any n and any $T < T_n$,

$$\|w_n\|_{\mathcal{B}^{0,\frac{1}{2}}(T)} < 2C\|u_0\|_{\mathcal{B}_4^{-\frac{1}{2},\frac{1}{2}}}.$$

Thus, $T_n = +\infty$ for any n. Existence then follows from classical compactness methods, the details of which are omitted. Theorem 6.16 is then proved, provided, of course, that we have proven the three propositions 6.25–6.27. $\qquad\square$

Proof of Propositions 6.25–6.27. We shall proceed differently for terms involving a horizontal derivative and terms involving a vertical derivative. For the former, the following two lemmas will be crucial.

Lemma 6.28. *Let a be in $\mathcal{B}_4^{-\frac{1}{2},\frac{1}{2}}(T)$ and b be in $\mathcal{B}^{0,\frac{1}{2}}(T)$. We have, for $h = 1,2$,*

$$\|\Delta_j^v(a\partial_h b)\|_{L_T^{\frac{4}{3}}(L_h^{\frac{4}{3}}(L_v^2))} \leq C\frac{d_j}{\nu^{\frac{3}{4}}}2^{-\frac{j}{2}}\|a\|_{\mathcal{B}_4^{-\frac{1}{2},\frac{1}{2}}(T)}\|b\|_{\mathcal{B}^{0,\frac{1}{2}}(T)}.$$

Lemma 6.29. *Let a and b be in $\mathcal{B}_4^{-\frac{1}{2},\frac{1}{2}}(T)$. We have*

$$\|\Delta_j^v(ab)\|_{L_T^2(L^2)} \leq C\frac{d_j}{\nu^{\frac{1}{2}}}2^{-\frac{j}{2}}\|a\|_{\mathcal{B}_4^{-\frac{1}{2},\frac{1}{2}}(T)}\|b\|_{\mathcal{B}_4^{-\frac{1}{2},\frac{1}{2}}(T)}.$$

Proof of Lemma 6.28. Using Bony's decomposition in the vertical variable gives

$$\Delta_j^v(a\partial_h b) = \sum_{|j-j'|\leq 5}\Delta_j^v(S_{j'-1}^v a\Delta_{j'}^v\partial_h b) + \sum_{j'\geq j-3}\Delta_j^v(S_{j'+2}^v(\partial_h b)\Delta_{j'}^v a).$$

Using Hölder's inequality and then Lemma 6.21, we have

$$\|\Delta_j^v(S_{j'-1}^v a\Delta_{j'}^v\partial_h b)\|_{L_T^{\frac{4}{3}}(L_h^{\frac{4}{3}}(L_v^2))} \leq C\|S_{j'-1}^v a\|_{L_T^4(L_h^4(L_v^\infty))}\|\Delta_{j'}^v\partial_h b\|_{L_T^2(L^2)}$$

$$\leq C\frac{d_{j'}}{\nu^{\frac{3}{4}}}2^{-\frac{j'}{2}}\|a\|_{\mathcal{B}_4^{-\frac{1}{2},\frac{1}{2}}(T)}\|b\|_{\mathcal{B}^{0,\frac{1}{2}}(T)}.$$

Similarly, we have

$$\|\Delta_j^v(S_{j'+2}^v(\partial_h b)\Delta_{j'}^v a)\|_{L_T^{\frac{4}{3}}(L_h^{\frac{4}{3}}(L_v^2))} \leq C\|S_{j'+2}^v\partial_h b\|_{L_T^2(L_h^2(L_v^\infty))}\|\Delta_{j'}^v a\|_{L_T^4(L_h^4(L_v^2))}$$

$$\leq C\frac{d_{j'}}{\nu^{\frac{3}{4}}}2^{-\frac{j'}{2}}\|a\|_{\mathcal{B}_4^{-\frac{1}{2},\frac{1}{2}}(T)}\|b\|_{\mathcal{B}^{0,\frac{1}{2}}(T)}.$$

It then turns out that

$$2^{\frac{j}{2}}\|\Delta_j^v(a\partial_h b)\|_{L_T^{\frac{4}{3}}(L_h^{\frac{4}{3}}(L_v^2))} \le \frac{C}{\nu^{\frac{3}{4}}}\|a\|_{\mathcal{B}_4^{-\frac{1}{2},\frac{1}{2}}(T)}\|b\|_{\mathcal{B}^{0,\frac{1}{2}}(T)} \sum_{j'\ge j-5} 2^{-\frac{j'-j}{2}}d_{j'},$$

which implies the lemma.   □

Proof of Lemma 6.29. We write

$$\Delta_j^v(ab) = \sum_{|j'-j|\le 5} \Delta_j^v(S_{j'-1}^v a \Delta_{j'}^v b) + \sum_{j'\ge j-3} \Delta_j^v(S_{j'+2}^v b \Delta_{j'}^v a).$$

Again using Hölder's inequality and Lemma 6.21, we get

$$\|\Delta_j^v(S_{j'-1}^v a \Delta_{j'}^v b)\|_{L_T^2(L_h^4(L_v^2))} \le C\|S_{j'-1}^v a\|_{L_T^4(L_h^4(L_v^\infty))}\|\Delta_{j'}^v b\|_{L_T^4(L_h^4(L_v^2))}$$

$$\le C\frac{d_{j'}}{\nu^{\frac{1}{2}}}2^{-\frac{j'}{2}}\|a\|_{\mathcal{B}_4^{-\frac{1}{2},\frac{1}{2}}(T)}\|b\|_{\mathcal{B}_4^{-\frac{1}{2},\frac{1}{2}}(T)}.$$

We can now conclude as in the previous lemma.   □

Proof of Proposition 6.25. Note that, thanks to the fact that u_F is divergence-free, we have

$$I_j(T) = \int_0^T \left(\Delta_j^v(u_F\cdot\nabla u_F)|\Delta_j^v a\right) dt = I_j^h(T) + I_j^v(T) \qquad \text{with}$$

$$I_j^h(T) \stackrel{\text{def}}{=} \int_0^T \left(\Delta_j^v(u_F^h\otimes u_F)|\Delta_j^v\nabla_h a\right) dt \quad \text{and}$$

$$I_j^v(T) \stackrel{\text{def}}{=} \int_0^T \left(\partial_3\Delta_j^v(u_F^3 u_F)|\Delta_j^v a\right) dt.$$

Using Lemma 6.29 and the definition of $\mathcal{B}^{0,\frac{1}{2}}(T)$, we get

$$|I_j^h(T)| \le \|\Delta_j^v(u_F^h\otimes u_F)\|_{L_T^2(L^2)}\|\Delta_j^v(\nabla_h a)\|_{L_T^2(L^2)}$$

$$\le C\frac{d_j^2}{\nu}2^{-j}\|u_0\|_{\mathcal{B}_4^{-\frac{1}{2},\frac{1}{2}}}^2\|a\|_{\mathcal{B}^{0,\frac{1}{2}}(T)}.$$

For the term with the vertical derivative, we write, using Lemma 6.10,

$$|I_j^v(T)| \le C2^j\|\Delta_j^v(u_F^3 u_F)\|_{L_T^1(L^2)}\|\Delta_j^v b\|_{L_T^\infty(L^2)}.$$

Again using Bony's decomposition in the vertical variable, we infer that

$$\Delta_j^v(u_F^3 u_F) = \sum_{|j'-j|\le 5} \Delta_j^v(S_{j'-1}^v u_F^3 \Delta_{j'}^v u_F) + \sum_{j'\ge j-3} \Delta_j^v(\Delta_{j'}^v u_F^3 S_{j'+2}^v u_F).$$

Using Bony's decomposition in the horizontal variables, we get

$$S^v_{j'-1} u_F \Delta^v_{j'} u_F = \sum_{k \geq j'-4} \left\{ S^h_{k-1} S^v_{j'-1} u_F^3 \Delta^h_k \Delta^v_{j'} u_F + \Delta^h_k S^v_{j'-1} u_F^3 S^h_{k+2} \Delta^v_{j'} u_F \right\}.$$

The two terms in the above sum are estimated along exactly the same lines. As in the proof of Lemma 6.24, we use the smoothing effect on the highest possible horizontal frequencies of u_F. Using Hölder's inequality, this gives

$$\|S^h_{k-1} S^v_{j'-1} u_F^3 \Delta^h_k \Delta^v_{j'} u_F\|_{L^1_T(L^2)}$$
$$\leq 2^{-\frac{k}{2}} \|S^h_{k-1} S^v_{j'-1} u_F\|_{L^\infty_T(L^4_h(L^\infty_v))} 2^{\frac{k}{2}} \|\Delta^h_k \Delta^v_{j'} u_F\|_{L^1_T(L^4_h(L^2_v))}.$$

Lemma 6.22 guarantees that

$$2^{\frac{k}{2}} \|\Delta^h_k \Delta^v_{j'} u_F\|_{L^1_T(L^4_h(L^2_v))} \leq \frac{C}{\nu} d_{k,j} 2^{-\frac{j'}{2}} 2^{-k} \|u_0\|_{\mathcal{B}^{-\frac{1}{2},\frac{1}{2}}_4}.$$

Lemma 6.20 states, in particular, that

$$2^{-\frac{k}{2}} \|S^h_{k-1} S^v_{j'-1} u_F\|_{L^\infty_T(L^4_h(L^\infty_v))} \leq C c_k \|u_0\|_{\mathcal{B}^{-\frac{1}{2},\frac{1}{2}}_4}.$$

Using Lemma 6.19, it then turns out that

$$\|S^v_{j'-1} u_F \Delta^v_{j'} u_F\|_{L^1_T(L^2)} \leq \frac{C}{\nu} \left(\sum_{k \geq j'-2} c_k d_{k,j'} 2^{-k} \right) 2^{-\frac{j'}{2}} \|u_0\|^2_{\mathcal{B}^{-\frac{1}{2},\frac{1}{2}}_4}$$
$$\leq C \frac{d_{j'}}{\nu} 2^{-\frac{3j'}{2}} \|u_0\|^2_{\mathcal{B}^{-\frac{1}{2},\frac{1}{2}}_4}.$$

We deduce that

$$2^{\frac{3j}{2}} \|\Delta^v_j(u_F^3 u_F)\|_{L^1_T(L^2)} \leq \frac{C}{\nu} \|u_0\|^2_{\mathcal{B}^{-\frac{1}{2},\frac{1}{2}}_4} \sum_{j' \geq j-5} d_{j'} 2^{-\frac{3(j'-j)}{2}}.$$

This completes the proof of Proposition 6.25.   □

Proof of Proposition 6.26. Again, we distinguish the terms with horizontal derivatives from the terms with vertical ones, writing

$$J_j(T) = \int_0^T \left(\Delta^v_j(a \cdot \nabla u_F) | \Delta^v_j b \right) dt = J^h_j(T) + J^v_j(T) \quad \text{with}$$
$$J^h_j(T) \stackrel{\text{def}}{=} \int_0^T \left(\Delta^v_j(a^h \cdot \nabla_h u_F) | \Delta^v_j b \right) dt \quad \text{and}$$
$$J^v_j(T) \stackrel{\text{def}}{=} \int_0^T \left(\Delta^v_j(a^3 \partial_3 u_F) | \Delta^v_j b \right) dt.$$

Using integration by parts gives

$$\left(\Delta^v_j(a^h \cdot \nabla_h u_F) | \Delta^v_j b \right) = - \left(\Delta^v_j(u_F \operatorname{div}_h a^h) | \Delta^v_j b \right) - \left(\Delta^v_j(a^h \otimes u_F) | \nabla_h \Delta^v_j b \right).$$

From Lemma 6.21 and Lemma 6.28, we have

$$\int_0^T \left|\left(\Delta_j^v(u_F \operatorname{div}_h a^h)|\Delta_j^v b\right)\right| dt \leq \|\Delta_j^v(u_F \operatorname{div}_h a^h)\|_{L_T^{\frac{4}{3}}(L_h^{\frac{4}{3}}(L_v^2))} \|\Delta_j^v b\|_{L_T^4(L_h^4(L_v^2))}$$

$$\leq C\frac{d_j^2}{\nu} 2^{-j} \|u_0\|_{\mathcal{B}_4^{-\frac{1}{2},\frac{1}{2}}} \|a^h\|_{\mathcal{B}^{0,\frac{1}{2}}(T)} \|b\|_{\mathcal{B}^{0,\frac{1}{2}}(T)}.$$

Lemma 6.29 gives

$$\int_0^T \left|\left(\Delta_j^v(a^h \otimes u_F)|\nabla_h \Delta_j^v b\right)\right| dt \leq \|\Delta_j^v(a^h \otimes u_F)\|_{L_T^2(L^2)} \|\Delta_j^v \nabla_h b\|_{L_T^2(L^2)}$$

$$\leq C\frac{d_j^2}{\nu} 2^{-j} \|u_0\|_{\mathcal{B}_4^{-\frac{1}{2},\frac{1}{2}}} \|a^h\|_{\mathcal{B}^{0,\frac{1}{2}}(T)} \|b\|_{\mathcal{B}^{0,\frac{1}{2}}(T)}.$$

Therefore,

$$|J_j^h(T)| \leq C\frac{d_j^2}{\nu} 2^{-j} \|u_0\|_{\mathcal{B}_4^{-\frac{1}{2},\frac{1}{2}}} \|a^h\|_{\mathcal{B}^{0,\frac{1}{2}}(T)} \|b\|_{\mathcal{B}^{0,\frac{1}{2}}(T)}.$$

On the other hand, using Bony's decomposition in the vertical variables, we get

$$\Delta_j^v(a^3 \partial_3 u_F) = \sum_{|j'-j|\leq 5} \Delta_j^v(S_{j'-1}^v a^3 \partial_3 \Delta_{j'}^v u_F) \qquad (6.28)$$

$$+ \sum_{j'\geq j-3} \Delta_j^v(\Delta_{j'}^v a^3 S_{j'+2}^v \partial_3 u_F).$$

To deal with the first term, we use Hölder's inequality to get

$$\|S_{j'-1}^v a^3 \partial_3 \Delta_{j'}^v u_F\|_{L_T^1(L^2)} \leq C 2^{j'} \|S_{j'-1}^v a^3\|_{L_T^\infty(L_h^2(L_v^\infty))} \|\Delta_{j'}^v u_F\|_{L_T^1(L_h^\infty(L_v^2))}.$$

Corollary 6.14 and Corollary 6.23 applied with $p = 1$ and $q = \infty$ together imply that

$$\|S_{j'-1}^v a^3 \partial_3 \Delta_{j'}^v u_F\|_{L_T^1(L^2)} \leq C\frac{d_{j'}}{\nu} 2^{-\frac{j'}{2}} \|u_0\|_{\mathcal{B}_4^{-\frac{1}{2},\frac{1}{2}}} \|a\|_{\mathcal{B}^{0,\frac{1}{2}}(T)},$$

from which we infer that

$$\sum_{|j'-j|\leq 5} \|\Delta_j^v(S_{j'-1}^v a^3 \partial_3 \Delta_{j'}^v u_F)\|_{L_T^1(L^2)} \leq C\frac{d_j}{\nu} 2^{-\frac{j}{2}} \|u_0\|_{\mathcal{B}_4^{-\frac{1}{2},\frac{1}{2}}} \|a\|_{\mathcal{B}^{0,\frac{1}{2}}(T)}.$$

We now estimate the second term of (6.28). Hölder's inequality gives

$$\|\Delta_{j'}^v a^3 S_{j'+2}^v \partial_3 u_F\|_{L_T^1(L^2)} \leq C 2^{j'} \|\Delta_{j'}^v a^3\|_{L_T^2(L^2)} \|S_{j'+2}^v u_F\|_{L_T^2(L^\infty)}.$$

From Lemma 6.10, we get

$$\|\Delta^v_{j'}a^3\|_{L^2_T(L^2(\mathbb{R}^3))} \leq C2^{-j'}\|\Delta^v_{j'}\partial_3 a^3\|_{L^2_T(L^2)}.$$

Using the fact that $\mathrm{div}\, a = 0$, we have

$$\|\Delta^v_{j'}a^3\|_{L^2_T(L^2)} \leq C2^{-j'}\|\Delta^v_{j'}\,\mathrm{div}_h\, a^h\|_{L^2_T(L^2)} \leq C\frac{d_{j'}}{\sqrt{\nu}}2^{-\frac{3j'}{2}}\|a^h\|_{\mathcal{B}^{0,\frac{1}{2}}(T)}.$$

Together with Lemma 6.24, this implies that

$$\sum_{j'\geq j-3}\|\Delta^v_j(\Delta^v_{j'}a^3 S^v_{j'+2}\partial_3 u_F)\|_{L^1_T(L^2)} \leq C\frac{d_j}{\nu}2^{-\frac{j}{2}}\|u_0\|_{\mathcal{B}^{-\frac{1}{2},\frac{1}{2}}_4}\|a^h\|_{\mathcal{B}^{0,\frac{1}{2}}(T)}.$$

This completes the proof of Proposition 6.26.   □

Proof of Proposition 6.27. We decompose $F_j(T)$ into

$$F_j(T) = \int_0^T \left(\Delta^v_j(a\cdot\nabla b)|\Delta^v_j b\right)\,dt = F^h_j(T) + F^v_j(T) \quad \text{with}$$

$$F^h_j(T) \stackrel{\mathrm{def}}{=} \int_0^T \left(\Delta^v_j(a^h\cdot\nabla_h b)|\Delta^v_j b\right)\,dt \quad \text{and}$$

$$F^v_j(T) \stackrel{\mathrm{def}}{=} \int_0^T \left(\Delta^v_j(a^3\partial_3 b)|\Delta^v_j b\right)\,dt.$$

On the one hand, according to Hölder's inequality, we have

$$\left|F^h_j(T)\right| \leq \|\Delta^v_j(a^h\cdot\nabla_h b)\|_{L^{\frac{4}{3}}_T(L^{\frac{4}{3}}_h(L^2_v))}\|\Delta^v_j b\|_{L^4_T(L^4_h(L^2_v))},$$

so combining Lemma 6.28 with Corollary 6.14 and Lemma 6.21 yields

$$\left|F^h_j(T)\right| \leq C\frac{d^2_j}{\nu}2^{-j}\|a\|_{\mathcal{B}^{-\frac{1}{2},\frac{1}{2}}_4(T)}\|b\|^2_{\mathcal{B}^{0,\frac{1}{2}}(T)}.$$

On the other hand, the norms $\mathcal{B}^{0,\frac{1}{2}}(T)$ or $\mathcal{B}^{-\frac{1}{2},\frac{1}{2}}_4(T)$ do not have any gain of vertical derivative. This difficulty may be bypassed by taking advantage of the fact that $\mathrm{div}\, a = 0$. More precisely, the vertical Bony decomposition, combined with a straightforward commutator process, enables us to write

$$\Delta^v_j(a^3\partial_3 b) = S^v_{j-1}a^3\partial_3\Delta^v_j b + \sum_{|j-\ell|\leq 5}[\Delta^v_j, S^v_{\ell-1}a^3]\partial_3\Delta^v_\ell b$$

$$+ \sum_{|j-\ell|\leq 1}(S^v_{\ell-1}a^3 - S^v_{j-1}a^3)\partial_3\Delta^v_j\Delta^v_\ell b + \sum_{\ell\geq j-3}\Delta^v_j(\Delta^v_\ell a^3\partial_3 S^v_{\ell+2}b).$$

From this we may decompose $F^v_j(T)$ into

$$F^v_j(T) \stackrel{\mathrm{def}}{=} F^{1,v}_j + F^{2,v}_j + F^{3,v}_j + F^{4,v}_j \quad \text{with} \quad F^{1,v}_j \stackrel{\mathrm{def}}{=} \int_0^T \left(S^v_{j-1}a^3\partial_3\Delta^v_j b|\Delta^v_j b\right)\,dt,$$

and obvious definitions for $F^{2,v}_j$, $F^{3,v}_j$, and $F^{4,v}_j$.

In order to bound $F_j^{1,v}$ we use the fact that

$$\partial_3 a^3 = -\operatorname{div}_h a^h. \tag{6.29}$$

Integrating twice by parts we thus get

$$F_j^{1,v} = \frac{1}{2} \int_0^T \int_{\mathbb{R}^3} S_{j-1}^v \operatorname{div}_h a^h |\Delta_j^v b|^2 \, dx \, dt$$

$$= -\int_0^T \int_{\mathbb{R}^3} S_{j-1}^v a^h \cdot \nabla_h \Delta_j^v b \, \Delta_j^v b \, dx \, dt.$$

Applying Lemma 6.21, together with Corollary 6.14, yields

$$|F_j^{1,v}| \leq \|S_{j-1}^v a^h\|_{L_T^4(L_h^4(L_v^\infty))} \|\nabla_h \Delta_j^v b\|_{L_T^2(L^2)} \|\Delta_j^v b\|_{L_T^4(L_h^4(L_v^2))}$$

$$\leq C \frac{d_j^2}{\nu} 2^{-j} \|a^h\|_{\mathcal{B}_4^{-\frac{1}{2},\frac{1}{2}}(T)} \|b\|_{\mathcal{B}^{0,\frac{1}{2}}(T)}^2.$$

To deal with the commutator in $F_j^{2,v}$, we first use Taylor's formula. Writing $\bar{h}(x_3) = x_3 h(x_3)$ and integrating by parts, we find that

$$F_j^{2,v} = -\sum_{|j-\ell|\leq 5} \int_0^T \left(\int_{\mathbb{R}} \bar{h}(2^j(x_3 - y_3)) \right.$$

$$\times \left(\int_0^1 S_{\ell-1}^v \partial_3 a^3(x_h, \tau y_3 + (1-\tau)x_3) \, d\tau \right) \Delta_\ell^v b(x_h, y_3) \, dy_3 \left| \Delta_j^v b \right) dt.$$

Next, using (6.29) and integration by parts, we rewrite $F_j^{2,v}$ as

$$F_j^{2,v} = \sum_{|j-\ell|\leq 5} \int_0^T \left(\int_{\mathbb{R}} \bar{h}(2^j(x_3 - y_3)) \right.$$

$$\times \int_0^1 S_{\ell-1}^v a^h(x_h, \tau y_3 + (1-\tau)x_3) \, d\tau \cdot \nabla_h \Delta_\ell^v b(x_h, y_3) \, dy_3 | \Delta_j^v b \right) dt$$

$$+ \sum_{|j-\ell|\leq 5} \int_0^T \left(\int_{\mathbb{R}} \bar{h}(2^j(x_3 - y_3)) \right.$$

$$\times \left(\int_0^1 S_{\ell-1}^v a^h(x_h, \tau y_3 + (1-\tau)x_3) \, d\tau \right) \Delta_\ell^v b(x_h, y_3) \, dy_3 | \nabla_h \Delta_j^v b \right) dt.$$

Young's inequality, together with Corollary 6.14 and Lemma 6.21, then yields

$$|F_j^{2,v}| \leq C \sum_{|j-\ell|\leq 5} \|S_{\ell-1}^v a^h\|_{L_T^4(L_h^4(L_v^\infty))} \left(\|\nabla_h \Delta_\ell^v b\|_{L_T^2(L^2)} \|\Delta_j^v b\|_{L_T^4(L_h^4(L_v^2))} \right.$$

$$\left. + \|\Delta_\ell^v b\|_{L_T^4(L_h^4(L_v^2))} \|\nabla_h \Delta_j^v b\|_{L_T^2(L^2)} \right)$$

$$\leq C \frac{d_j^2}{\nu} 2^{-j} \|a^h\|_{\mathcal{B}_4^{-\frac{1}{2},\frac{1}{2}}(T)} \|b\|_{\mathcal{B}^{0,\frac{1}{2}}(T)}^2.$$

Note that

$$|F_j^{3,v}| \leq \sum_{\substack{|j-\ell'|\leq 1 \\ |j-\ell|\leq 1}} \int_0^T \left|(\Delta_{\ell'}^v a^3 \partial_3 \Delta_j^v \Delta_\ell^v b | \Delta_j^v b)\right| dt.$$

To estimate $F_j^{3,v}$, we then need to gain two derivatives from $\Delta_{\ell'}^v a^3$. In order to do this, we need to use (6.12) with $N = 1$, which implies that

$$\Delta_{\ell'}^v a^3(x) = \int_{\mathbb{R}} g^v(2^{\ell'}(x_3 - y_3))\partial_3 \Delta_{\ell'}^v a^3(x_h, y_3)\, dy_3, \qquad (6.30)$$

where $g^v \in \mathcal{S}(\mathbb{R})$ is defined via $\mathcal{F}(g^v)(\xi_3) = \dfrac{\widetilde{\varphi}(|\xi_3|)}{i\xi_3}$.

Plugging (6.30) into $F_j^{3,v}$, using (6.29), and then integrating by parts in the horizontal variables, we find that, up to an irrelevant multiplicative constant, the quantity $F_j^{3,v}$ is less than

$$\sum_{\substack{|j-\ell'|\leq 1 \\ |j-\ell|\leq 1}} \int_0^T \left|\left(\left(\int_{\mathbb{R}} g^v(2^{\ell'}(x_3 - y_3))\Delta_{\ell'}^v a^h(x_h, y_3)\, dy_3\right) \cdot \nabla_h \partial_3 \Delta_j^v \Delta_\ell^v b \Big| \Delta_j^v b\right)\right| dt$$

$$+ \sum_{\substack{|j-\ell'|\leq 1 \\ |j-\ell|\leq 1}} \int_0^T \left|\left(\left(\int_{\mathbb{R}} g^v(2^{\ell'}(x_3 - y_3))\Delta_{\ell'}^v a^h(x_h, y_3)\, dy_3\right)\partial_3 \Delta_j^v \Delta_\ell^v b \Big| \nabla_h \Delta_j^v b\right)\right| dt.$$

Together with Young's inequality, Corollary 6.14, and Lemma 6.21, this implies that

$$|F_j^{3,v}| \leq C \sum_{\substack{|j-\ell'|\leq 1 \\ |j-\ell|\leq 1}} 2^{\ell-\ell'} \|\Delta_{\ell'}^v a^h\|_{L_T^4(L_h^4(L_v^\infty))} \|\nabla_h \Delta_j^v b\|_{L_T^2(L^2)} \|\Delta_j^v b\|_{L_T^4(L_h^4(L_v^2))}$$

$$\leq C \frac{d_j^2}{\nu} 2^{-j} \|a^h\|_{\mathcal{B}_4^{-\frac{1}{2},\frac{1}{2}}(T)} \|b\|_{\mathcal{B}^{0,\frac{1}{2}}(T)}^2.$$

Finally, using (6.30) once again, we can write that $F_j^{4,v}$ is equal to

$$\sum_{\ell \geq j-3} \left(\int_0^T \left(\Delta_j^v \left(\int_{\mathbb{R}} g^v(2^\ell(x_3 - y_3))\Delta_\ell^v a^h(x_h, y_3)\, dy_3 \cdot \nabla_h \partial_3 S_{\ell+2}^v b\right) \Big| \Delta_j^v b\right) dt \right.$$

$$\left. + \int_0^T \left(\Delta_j^v \left(\int_{\mathbb{R}} g^v(2^\ell(x_3 - y_3))\Delta_\ell^v a^h(x_h, y_3)\, dy_3 \partial_3 S_{\ell+2}^v b\right) \Big| \nabla_h \Delta_j^v b\right) dt \right).$$

From Young's inequality, we deduce that

$$|F_j^{4,v}| \leq C \sum_{\ell \geq j-3} \|\Delta_\ell^v a^h\|_{L_T^4(L_h^4(L_v^2))} \left(\|\nabla_h S_{\ell+2}^v b\|_{L_T^2(L_h^2(L_v^\infty))} \|\Delta_j^v b\|_{L_T^4(L_h^4(L_v^2))} \right.$$

$$\left. + \|S_{\ell+2}^v b\|_{L_T^4(L_h^4(L_v^\infty))} \|\nabla_h \Delta_j^v b\|_{L_T^2(L^2)} \right),$$

which, together with Corollary 6.14 and Lemma 6.21, implies that

$$|F_j^{4,v}| \leq C \frac{d_j^2}{\nu} 2^{-j} \|a^h\|_{\mathcal{B}_4^{-\frac{1}{2},\frac{1}{2}}(T)} \|b\|_{\mathcal{B}^{0,\frac{1}{2}}(T)}^2.$$

This completes the proof of Proposition 6.27.   □

6.4 The Proof of Uniqueness

In the previous section we showed that any small divergence-free data in $\mathcal{B}_4^{-\frac{1}{2},\frac{1}{2}}$ generates a global solution u in $\mathcal{B}_4^{-\frac{1}{2},\frac{1}{2}}(\infty)$ such that, in addition, $(u - u_F) \in \mathcal{B}^{0,\frac{1}{2}}(\infty)$. In this section we want to prove uniqueness in the space $\mathcal{B}_4^{-\frac{1}{2},\frac{1}{2}}(\infty)$. As a first step we prove the following regularity theorem.

Theorem 6.30. *Let* $u \in \mathcal{B}_4^{-\frac{1}{2},\frac{1}{2}}(T)$ *be a solution of* (ANS_ν) *with initial data* u_0 *in* $\mathcal{B}_4^{-\frac{1}{2},\frac{1}{2}}$. *We then have*

$$w \overset{def}{=} u - u_F \in \mathcal{B}^{0,\frac{1}{2}}(T).$$

Proof. We have already observed (at the beginning of Section 6.3) that the vector field w is the solution of the linear system

$$(\widetilde{ANS_\nu}) \begin{cases} \partial_t w - \nu \Delta_h w = -u \cdot \nabla u - \nabla P \\ \operatorname{div} w = 0 \\ w_{|t=0} = u_{\ell h}, \end{cases}$$

where $u_{\ell h}$ is defined as in (6.14). As stated in Lemma 6.22, u_F belongs to the space $\mathcal{B}_4^{-\frac{1}{2},\frac{1}{2}}(T)$ and thus so does w. Hence, it is only a matter of proving that

$$\|(\operatorname{Id} - S_{j-1}^h)\Delta_j^v w\|_{L_T^\infty(L^2)} + \nu^{\frac{1}{2}} \|(\operatorname{Id} - S_{j-1}^h)\Delta_j^v \nabla_h w\|_{L_T^2(L^2)} \leq C d_j 2^{-\frac{j}{2}}.$$

In order to do so, we apply the operator $(\operatorname{Id} - S_{j-1}^h)\Delta_j^v$ to the system $(\widetilde{ANS_\nu})$ and define

$$w_j \overset{def}{=} (\operatorname{Id} - S_{j-1}^h)\Delta_j^v w.$$

This gives, by virtue of the L^2 energy estimate,

$$\|w_j(t)\|_{L^2}^2 + 2\nu \int_0^t \|\nabla_h w_j(t')\|_{L^2}^2 \, dt' \leq \|\Delta_j^v u_{\ell h}\|_{L^2}^2$$

$$+ 2 \int_0^t \left|\langle (\operatorname{Id} - S_{j-1}^h)\Delta_j^v (u(t') \cdot \nabla u(t')), w_j(t') \rangle\right| \, dt'.$$

From the Fourier–Plancherel theorem, we then infer that

$$\|w_j(t)\|_{L^2}^2 + \nu \int_0^t \|\nabla_h w_j(t')\|_{L^2}^2 \, dt' + c\nu 2^{2j} \int_0^t \|w_j(t')\|_{L^2}^2 \, dt'$$

$$\leq \|\Delta_j^v u_{\ell h}\|_{L^2}^2 + 2 \int_0^t \left| \langle (\mathrm{Id} - S_{j-1}^h) \Delta_j^v (u(t') \cdot \nabla u(t')), w_j(t') \rangle \right| \, dt'.$$

Observe that, thanks to the divergence-free condition, we have

$$u \cdot \nabla u^m = \mathrm{div}_h(u^m u^h) + \partial_3(u^m u^3).$$

Integrating by parts, we get

$$\left| \langle (\mathrm{Id} - S_{j-1}^h) \Delta_j^v \mathrm{div}_h(u^m u^h), w_j \rangle \right| \leq \left| ((\mathrm{Id} - S_{j-1}^h) \Delta_j^v (u^m u^h), \nabla_h w_j) \right|$$

$$\leq \|\Delta_j^v (u^m u^h)\|_{L^2} \|\nabla_h w_j\|_{L^2}$$

$$\leq \frac{\nu}{2} \|\nabla_h w_j\|_{L^2}^2 + \frac{C}{\nu} \|\Delta_j^v (u^m u^h)\|_{L^2}^2,$$

while, by using Lemma 6.10, we have

$$\left| \langle (\mathrm{Id} - S_{j-1}^h) \Delta_j^v \partial_3(u^m u^3), w_j \rangle \right| \leq 2^j \|\Delta_j^v (u^m u^3)\|_{L^2} \|w_j\|_{L^2}$$

$$\leq \frac{c\nu}{2} 2^{2j} \|w_j\|_{L^2}^2 + \frac{C}{\nu} \|\Delta_j^v (u^m u^3)\|_{L^2}^2.$$

Using the inequality (6.15) and Lemma 6.29, we deduce that

$$\|w_j\|_{L_T^\infty(L^2)} + \sqrt{\nu} \|\nabla_h w_j\|_{L_T^2(L^2)} \leq C d_j 2^{-\frac{j}{2}} \left(\|u_0\|_{\mathcal{B}_4^{-\frac{1}{2},\frac{1}{2}}} + \nu^{-1} \|u\|_{\mathcal{B}_4^{-\frac{1}{2},\frac{1}{2}}(T)}^2 \right).$$

This completes the proof of Theorem 6.30. □

The above theorem implies that if u_1 and u_2 are two solutions of (ANS_ν) in the space $\mathcal{B}_4^{-\frac{1}{2},\frac{1}{2}}(T)$ associated with the same initial data, then the difference $\delta \overset{\text{def}}{=} u_2 - u_1$ belongs to $\mathcal{B}^{0,\frac{1}{2}}(T)$. Moreover, it satisfies the system

$$(ANS_\nu') \qquad \begin{cases} \partial_t \delta - \nu \Delta_h \delta = L\delta - \nabla P \\ \mathrm{div}\,\delta = 0 \\ \delta_{|t=0} = 0, \end{cases}$$

where L is the linear operator defined as follows:

$$L\delta \overset{\text{def}}{=} -\delta \cdot \nabla u_1 - u_2 \cdot \nabla \delta.$$

In order to prove uniqueness, it suffices to establish that $\delta \equiv 0$. Because existence in Theorem 6.16 is not proved by using Picard's fixed point method, this is not obvious. The main reason why is that the system (ANS_ν) is *hyperbolic* in the vertical direction. Roughly speaking, we thus expect that the contraction argument may be realized with one less vertical derivative than for the existence space.

Before proceeding to the heart of the proof of uniqueness, we have to introduce more notation: Let $\Delta_j^{vi} \overset{\text{def}}{=} \Delta_j^v$, $S_j^{vi} = S_j^v$ if $j \geq 0$, $\Delta_{-1}^{vi} \overset{\text{def}}{=} S_0^{vi} = S_0^v$, and $\Delta_j^{vi} = S_{j+1}^{vi} = 0$ if $j \leq -2$.

Definition 6.31. *We denote by* $\mathcal{H}$ *the space of tempered distribution such that*

$$\|a\|_{\mathcal{H}}^2 \overset{def}{=} \sum_{j \in \mathbb{Z}} 2^{-j} \|\Delta_j^{vi} a\|_{L^2}^2 < \infty.$$

The corresponding inner product is denoted by $(\cdot \mid \cdot)_{\mathcal{H}}$.

Because the space $\mathcal{H}$ is nonhomogeneous, it is not true (owing to the low vertical frequencies) that $\mathcal{B}^{0,\frac{1}{2}}(T)$ is embedded in $L_T^{\infty}(\mathcal{H})$. Since δ satisfies (ANS_ν'), however, we have the following result.

Lemma 6.32. *The difference* δ *is in* $L_T^{\infty}(\mathcal{H})$ *and satisfies* $\nabla_h \delta \in L_T^2(\mathcal{H})$.

Proof. Let $S_0^v \delta$ be a solution (with initial value 0) of

$$\partial_t S_0^v \delta - \nu \Delta_h S_0^v \delta = g_1 + g_2 + g_3 \quad \text{with}$$

$$g_1 \overset{def}{=} \sum_{\lambda \in \Lambda} S_0^v \partial_3 (a_\lambda b_\lambda),$$

$$g_2 \overset{def}{=} \sum_{\lambda \in \Lambda} S_0^v \operatorname{div}_h (c_\lambda (\operatorname{Id} - S_0^v) \delta),$$

$$g_3 \overset{def}{=} \sum_{\lambda \in \Lambda} d_\lambda S_0^v \operatorname{div}_h (S_0^v \delta),$$

where Λ is a finite set of indices and a_λ, b_λ, c_λ, and d_λ belong to $\mathcal{B}_4^{-\frac{1}{2},\frac{1}{2}}(T)$. Using Lemmas 6.10 and 6.29, we get that

$$\|S_0^v \partial_3 (a_\lambda b_\lambda)\|_{L_T^2(L^2)} \leq C \sum_{j \leq -1} 2^j \|\Delta_j^v (a_\lambda b_\lambda)\|_{L_T^2(L^2)}$$

$$\leq \frac{C}{\nu^{\frac{1}{2}}} \|a_\lambda\|_{\mathcal{B}_4^{-\frac{1}{2},\frac{1}{2}}(T)} \|b_\lambda\|_{\mathcal{B}_4^{-\frac{1}{2},\frac{1}{2}}(T)}.$$

Defining $C_{12}(T) \overset{def}{=} \|u_1\|_{\mathcal{B}_4^{-\frac{1}{2},\frac{1}{2}}(T)} + \|u_2\|_{\mathcal{B}_4^{-\frac{1}{2},\frac{1}{2}}(T)}$, we thus have

$$\|g_1\|_{L_T^2(L^2)} \leq \frac{C}{\nu^{\frac{1}{2}}} C_{12}^2(T). \tag{6.31}$$

Estimating g_2 relies on Lemma 6.21. We get

$$\|(\operatorname{Id} - S_0^v)\delta\|_{L_T^4(L_h^4(L_v^2))} \leq C\nu^{-\frac{1}{4}} \|\delta\|_{\mathcal{B}_4^{-\frac{1}{2},\frac{1}{2}}(T)},$$

$$\|c_\lambda\|_{L_T^4(L_h^4(L_v^\infty))} \leq C\nu^{-\frac{1}{4}} \|c_\lambda\|_{\mathcal{B}_4^{-\frac{1}{2},\frac{1}{2}}(T)},$$

from which it follows that

$$\|c_\lambda (\operatorname{Id} - S_0^v)\delta\|_{L_T^2(L^2)} \leq \|c_\lambda\|_{L_T^4(L_h^4(L_v^\infty))} \|(\operatorname{Id} - S_0^v)\delta\|_{L_T^4(L_h^4(L_v^2))}$$

$$\leq C\nu^{-\frac{1}{2}} \|c_\lambda\|_{\mathcal{B}_4^{-\frac{1}{2},\frac{1}{2}}(T)} \|\delta\|_{\mathcal{B}_4^{-\frac{1}{2},\frac{1}{2}}(T)}.$$

This gives that

$$g_2 = \operatorname{div}_h \widetilde{g}_2 \quad \text{with} \quad \|\widetilde{g}_2\|_{L_T^2(L^2)} \le C\nu^{-\frac{1}{2}} C_{12}^2(T). \tag{6.32}$$

The term g_3 must be treated with a commutator argument based on the following lemma.

Lemma 6.33. *Let χ be a function of $\mathcal{S}(\mathbb{R})$. A constant C exists such that, for any function a in $L_h^2(L_v^\infty)$, we have*

$$\|[\chi(\varepsilon x_3), S_0^v]a\|_{L^2} \le C\varepsilon^{\frac{1}{2}} \|a\|_{L_h^2(L_v^\infty)}.$$

Proof. The first order Taylor formula gives

$$C_\varepsilon(a)(x_h, x_3) \stackrel{\text{def}}{=} [\chi(\varepsilon x_3), S_0^v]a(x_h, x_3)$$

$$= \varepsilon \int_{\mathbb{R} \times [0,1]} h(x_3 - y_3)\chi'(\varepsilon((1-\tau)x_3 + \tau y_3)) a(x_h, y_3) \, dy_3 \, d\tau.$$

Using the Cauchy–Schwarz inequality for the measure $|h(x_3 - y_3)| \, dx_3 \, dy_3 \, d\tau$ on $\mathbb{R}^2 \times [0,1]$, we may write that

$$\|C_\varepsilon(a)(x_h, \cdot)\|_{L_v^2}^2 \le \varepsilon^2 \|a(x_h, \cdot)\|_{L_v^\infty}^2$$

$$\times \sup_{\|\varphi\|_{L^2(\mathbb{R})} \le 1} \left(\int_{\mathbb{R}^2} |h(x_3 - y_3)| \varphi^2(x_3) \, dx_3 \, dy_3 \right) (H_1^\varepsilon + H_2^\varepsilon)$$

$$\le C\varepsilon^2 \|a(x_h, \cdot)\|_{L_v^\infty}^2 (H_1^\varepsilon + H_2^\varepsilon),$$

where we define H_1^ε and H_2^ε as follows:

$$H_1^\varepsilon \stackrel{\text{def}}{=} \int_{\mathbb{R}^2 \times [0,\frac{1}{2}]} (\chi')^2 (\varepsilon((1-\tau)x_3 + \tau y_3)) |h(x_3 - y_3)| \, dx_3 \, dy_3 \, d\tau,$$

$$H_2^\varepsilon \stackrel{\text{def}}{=} \int_{\mathbb{R}^2 \times [\frac{1}{2},1]} (\chi')^2 (\varepsilon((1-\tau)x_3 + \tau y_3)) |h(x_3 - y_3)| \, dx_3 \, dy_3 \, d\tau.$$

Changing variables

$$\begin{cases} x_\tau = (1-\tau)x_3 + \tau y_3 \\ y_\tau = y_3 \end{cases} \text{ in } H_1^\varepsilon \quad \text{and} \quad \begin{cases} x_\tau = x_3 \\ y_\tau = \tau y_3 + (1-\tau)x_3 \end{cases} \text{ in } H_2^\varepsilon$$

gives

$$H_1^\varepsilon = \int_{\mathbb{R}^2 \times [0,\frac{1}{2}]} \frac{1}{1-\tau} (\chi')^2 (\varepsilon x_\tau) \left| h\left(\frac{x_\tau - y_\tau}{1-\tau}\right) \right| dx_\tau \, dy_\tau \, d\tau,$$

$$H_2^\varepsilon = \int_{\mathbb{R}^2 \times [\frac{1}{2},1]} \frac{1}{\tau} (\chi')^2 (\varepsilon y_\tau) \left| h\left(\frac{x_\tau - y_\tau}{\tau}\right) \right| dx_\tau \, dy_\tau \, d\tau.$$

We immediately infer that $\|C_\varepsilon(a)(x_h, \cdot)\|_{L_v^2} \le C\varepsilon^{\frac{1}{2}} \|a(x_h, \cdot)\|_{L_v^\infty}$ and the lemma is thus proved. □

Completion of the proof of Lemma 6.32. Choose $\chi \in \mathcal{D}(\mathbb{R})$ with value 1 near 0 and define $S_{0,\varepsilon}^v a \overset{\text{def}}{=} \chi(\varepsilon \cdot) S_0^v a$. We get, via a classical L^2 energy estimate and a convexity inequality, that

$$\|S_{0,\varepsilon}^v \delta(t)\|_{L^2}^2 + \nu \int_0^t \|\nabla_h S_{0,\varepsilon}^v \delta(t')\|_{L^2}^2 \, dt' \le 2 \int_0^t \|g_1(t')\|_{L^2} \|S_{0,\varepsilon}^v \delta(t')\|_{L^2} \, dt'$$

$$+ \frac{1}{\nu} \int_0^t \|\widetilde{g}_2(t')\|_{L^2}^2 \, dt' + 2 \int_0^t \langle \chi(\varepsilon \cdot) g_3(t'), S_{0,\varepsilon}^v \delta(t') \rangle \, dt'.$$

By the definition of g_3, the integrand in the last term of the above equality is a finite sum of terms of the type

$$D_\lambda \overset{\text{def}}{=} \langle \chi(\varepsilon \cdot) S_0^v (d_\lambda S_0^v \delta), \partial_h S_{0,\varepsilon}^v \delta \rangle$$

with $h \in \{1,2\}$ and $d_\lambda \in \mathcal{B}_4^{-\frac{1}{2},\frac{1}{2}}(T)$. Writing $D_\lambda = D_\lambda^1 + D_\lambda^2$ with

$$D_\lambda^1 \overset{\text{def}}{=} \langle [\chi(\varepsilon \cdot), S_0^v](d_\lambda S_0^v \delta), \partial_h S_{0,\varepsilon}^v \delta \rangle \quad \text{and} \quad D_\lambda^2 \overset{\text{def}}{=} \langle S_0^v (d_\lambda S_0^v \delta), \partial_h S_{0,\varepsilon}^v \delta \rangle,$$

Lemmas 6.21 and 6.33 imply that

$$\int_0^t |D_\lambda^1(t')| \, dt' \le C \varepsilon^{\frac{1}{2}} C_{12}^2(t) \|\nabla_h S_{0,\varepsilon} \delta\|_{L_t^2(L^2)}$$

$$\le \frac{\nu}{4} \|\nabla_h S_{0,\varepsilon}^v \delta\|_{L_t^2(L^2)}^2 + \frac{C}{\nu} \varepsilon C_{12}^4(t).$$

Next, we write

$$|D_\lambda^2(t)| \le C \|d_\lambda(t)\|_{L_h^4(L_v^\infty)} \|S_{0,\varepsilon}^v \delta(t)\|_{L^2}^{\frac{1}{2}} \|\nabla_h S_{0,\varepsilon}^v \delta(t)\|_{L^2}^{\frac{3}{2}}$$

$$\le \frac{\nu}{4} \|\nabla_h S_{0,\varepsilon}^v(t)\|_{L^2}^2 + \frac{C}{\nu^3} \|d_\lambda(t)\|_{L_h^4(L_v^\infty)}^4 \|S_{0,\varepsilon}^v(t)\|_{L^2}^2.$$

Using (6.31) we get, for $\varepsilon \in \,]0,1[$,

$$\|S_{0,\varepsilon}^v \delta(t)\|_{L^2}^2 + \frac{\nu}{2} \int_0^t \|\nabla_h S_{0,\varepsilon}^v \delta(t')\|_{L^2}^2 \, dt' \le C(\nu^{-1} + \nu^{-2}) C_{12}^4(T)$$

$$+ C \int_0^t \left(1 + \frac{1}{\nu^3} \left(\|u_1\|_{L_h^4(L_v^\infty)}^4 + \|u_2\|_{L_h^4(L_v^\infty)}^4 \right) \right) \|S_{0,\varepsilon}^v \delta(t')\|_{L^2}^2 \, dt'.$$

The Gronwall lemma, together with (6.31), gives

$$\|S_{0,\varepsilon}^v \delta(t)\|_{L^2}^2 + \frac{\nu}{2} \int_0^t \|\nabla_h S_{0,\varepsilon}^v \delta(t')\|_{L^2}^2 \, dt' \le C(\nu^{-1} + \nu^{-2}) C_{12}^4(T)$$

$$\times \exp \left(C \int_0^t \left(1 + \frac{1}{\nu^3} \left(\|u_1\|_{L_h^4(L_v^\infty)}^4 + \|u_2\|_{L_h^4(L_v^\infty)}^4 \right) \right) dt' \right)$$

and thus, by Lemma 6.21,

$$\|S_{0,\varepsilon}^v \delta(t)\|_{L^2}^2 + \frac{\nu}{2} \int_0^t \|\nabla_h S_{0,\varepsilon}^v \delta(t')\|_{L^2}^2 \, dt'$$
$$\leq C(\nu^{-1} + \nu^{-2})C_{12}^4(T) \exp\left(C\left(1 + \frac{1}{\nu^3}C_{12}^4(T)\right)\right).$$

Passing to the limit when ε tends to 0 then allows us to complete the proof of Lemma 6.32. □

Proof of Theorem 6.16 (continued). Let us first point out the main difficulty we shall encounter here. Roughly speaking, a function in $\mathcal{B}_4^{-\frac{1}{2},\frac{1}{2}}(T)$ must be $B_{2,1}^{\frac{1}{2}}$ in the vertical direction, while a function in $\mathcal{H}$ is $H^{-\frac{1}{2}}$ in the vertical direction. Hence, we have to deal with products of distributions in $B_{2,1}^{\frac{1}{2}} \times H^{-\frac{1}{2}}$, which is known to be the "bad" critical case for product laws (see, e.g., Theorem 2.52 page 88). In order to bypass this ultimate difficulty, we introduce the seminorms

$$\|a\|_{H^{0,\frac{1}{2}}} \overset{\text{def}}{=} \left(\sum_{j \in \mathbb{Z}} 2^{js} \|\Delta_j^v a\|_{L^2}^2 \right)^{\frac{1}{2}} \quad \text{and} \quad \|b\|_{\mathcal{B}_u}^2 \overset{\text{def}}{=} \sum_{\substack{k \in \mathbb{Z} \\ j \in \mathbb{N}}} 2^{j-k} \|\Delta_k^h \Delta_j^v a\|_{L_h^4(L_v^2)}^2.$$

We note that as $\ell^1(\mathbb{Z})$ is included in $\ell^2(\mathbb{Z})$, we have

$$\|a\|_{L_T^\infty(H^{0,\frac{1}{2}})}^2 + \nu \|\nabla_h a\|_{L_T^2(H^{0,\frac{1}{2}})}^2 \leq C\|a\|_{\mathcal{B}^{0,\frac{1}{2}}(T)}^2. \tag{6.33}$$

$$\|b\|_{L_T^\infty(\mathcal{B}_u)}^2 + \nu \|\nabla_h b\|_{L_T^2(\mathcal{B}_u)}^2 \leq C\|b\|_{\mathcal{B}_4^{-\frac{1}{2},\frac{1}{2}}(T)}^2. \tag{6.34}$$

The key to the proof is the following lemma, which we will temporarily assume to hold.

Lemma 6.34. *Let a and b be two divergence-free vector fields such that a and $\nabla_h a$ are in $H^{0,\frac{1}{2}} \cap \mathcal{H}$, and b is in $\mathcal{B}_u \cap L_h^4(L_v^\infty)$ with $\nabla_h b \in \mathcal{B}_u$. We assume, in addition, that $\|a\|_{\mathcal{H}}^2 \leq 2^{-16}$. We then have*

$$|(b \cdot \nabla a | a)_{\mathcal{H}}| + |(a \cdot \nabla b | a)_{\mathcal{H}}| \leq \frac{\nu}{10} \|\nabla_h a\|_{\mathcal{H}}^2 + C(a,b)\mu(\|a\|_{\mathcal{H}}^2)$$

with $\mu(r) \overset{\text{def}}{=} r(1 - \log_2 r) \log_2(1 - \log_2 r)$ and

$$C(a,b) \overset{\text{def}}{=} \frac{C}{\nu} \|b\|_{L_h^4(L_v^\infty)}^2 \left(1 + \frac{\|b\|_{L_h^4(L_v^\infty)}^2}{\nu^2} \right) + \frac{C}{\nu}(1 + \|b\|_{\mathcal{B}_u}^2)$$
$$\times \left(1 + \frac{\|b\|_{\mathcal{B}_u}^4}{\nu^2} \right) \left(\|b\|_{\mathcal{B}_u}^2 \|\nabla_h b\|_{\mathcal{B}_u}^2 + \|a\|_{H^{0,\frac{1}{2}}}^2 \|\nabla_h a\|_{H^{0,\frac{1}{2}}}^2 \right).$$

Thus, we have

$$\|\delta(t)\|_{\mathcal{H}}^2 \leq \int_0^t f(t')\mu(\|\delta(t')\|_{\mathcal{H}}^2)\,dt' \text{ with } f(t) \overset{\text{def}}{=} C(u_1(t),\delta(t)) + C(u_2(t),\delta(t)).$$

Lemma 6.21 and assertions (6.33) and (6.34) collectively imply that $f \in L^1([0,T])$. The uniqueness then follows from Lemma 3.4 page 125. $\qquad\square$

Proof of Lemma 6.34. As both terms may be treated similarly, we focus on $(b \cdot \nabla a | a)_{\mathcal{H}}$. Using a nonhomogeneous Bony decomposition in the vertical variable, we may write

$$\Delta_j^{vi}(b \cdot \nabla a) = T_b^{vi}\nabla a + R^{vi}(b,\nabla a) \quad \text{with}$$

$$T_b^{vi}\nabla a \overset{\text{def}}{=} \sum_\ell S_{\ell-1}^{vi} b \cdot \nabla \Delta_\ell^{vi} a \quad \text{and} \quad R^{vi}(b,\nabla a) \overset{\text{def}}{=} \sum_\ell \Delta_\ell^{vi} b \cdot \nabla S_{\ell+2}^{vi} a.$$

As usual, we shall treat the terms involving vertical derivatives in a different way than the terms involving horizontal derivatives. This leads to

$$\Delta_j^{vi}(T_b^{vi}\nabla a) = \mathcal{T}_j^h + \mathcal{T}_j^v \quad \text{with}$$

$$\mathcal{T}_j^h \overset{\text{def}}{=} \Delta_j^{vi} \sum_{|j-\ell|\leq 5} S_{\ell-1}^{vi} b^h \cdot \nabla_h \Delta_\ell^{vi} a \quad \text{and} \quad \mathcal{T}_j^v \overset{\text{def}}{=} \Delta_j^{vi} \sum_{|j-\ell|\leq 5} S_{\ell-1}^{vi} b^3 \partial_3 \Delta_\ell^{vi} a.$$

By the definition of the space $\mathcal{H}$ and using the anisotropic Hölder inequality, we get

$$\|\mathcal{T}_j^h\|_{L_h^{\frac{4}{3}}(L_v^2)} \leq C\|b\|_{L_h^4(L_v^\infty)} \sum_{|j-\ell|\leq 5} \|\nabla_h \Delta_\ell^{vi} a\|_{L^2}$$

$$\leq Cc_j 2^{\frac{j}{2}} \|b\|_{L_h^4(L_v^\infty)} \|\nabla_h a\|_{\mathcal{H}}.$$

We immediately infer that

$$|(\mathcal{T}_j^h|\Delta_j^{vi} a)_{L^2}| \leq Cc_j 2^{\frac{j}{2}} \|b\|_{L_h^4(L_v^\infty)} \|\Delta_j^{vi} a\|_{L_h^4(L_v^2)} \|\nabla_h a\|_{\mathcal{H}}.$$

As we have

$$\|\Delta_j^{vi} a\|_{L_h^4(L_v^2)}^2 \leq C\|\Delta_j^{vi} a\|_{L^2} \|\nabla_h \Delta_j^{vi} a\|_{L^2}, \tag{6.35}$$

we get

$$\sum_j 2^{-j} |(\mathcal{T}_j^h|\Delta_j^{vi} a)_{L^2}| \leq C\|b\|_{L_h^4(L_v^\infty)} \|\nabla_h a\|_{\mathcal{H}}^{\frac{3}{2}} \|a\|_{\mathcal{H}}^{\frac{1}{2}}. \tag{6.36}$$

Estimating $(\mathcal{T}_j^v|\Delta_j^{vi} a)_{L^2}$ is more involved. We write

$$\mathcal{T}_j^v = \sum_{n=1}^{3} \mathcal{T}_j^{v,n} \quad \text{with}$$

$$\mathcal{T}_j^{v,1} \overset{\text{def}}{=} S_{j-1}^{vi} b^3 \partial_3 \Delta_j^{vi} a,$$

$$\mathcal{T}_j^{v,2} \overset{\text{def}}{=} \sum_{|j-\ell| \le 5} [\Delta_j^{vi}, S_{\ell-1}^{vi} b^3] \partial_3 \Delta_\ell^{vi} a, \quad \text{and}$$

$$\mathcal{T}_j^{v,3} \overset{\text{def}}{=} \sum_{|j-\ell| \le 1} (S_{\ell-1}^{vi} b^3 - S_{j-1}^{vi} b^3) \partial_3 \Delta_j^{vi} \Delta_\ell^{vi} a.$$

In order to estimate $\mathcal{T}_j^{v,1}$, we perform an integration by parts and obtain

$$(\mathcal{T}_j^{v,1} | \Delta_j^{vi} a)_{L^2} = -\frac{1}{2} \int_{\mathbb{R}^3} S_{j-1}^{vi} \partial_3 b^3 \left(\Delta_j^{vi} a\right)^2 dx.$$

Using the fact that $\partial_3 b^3 = -\operatorname{div}_h b^h$ and integrating by parts in the horizontal variables, we get

$$(\mathcal{T}_j^{v,1} | \Delta_j^{vi} a)_{L^2} = -\int_{\mathbb{R}^3} S_{j-1}^{vi} b^h \cdot \nabla_h \Delta_j^{vi} a \, \Delta_j^{vi} a \, dx.$$

Now, arguing as we did in proving (6.36), we end up with

$$\sum_j 2^{-j} |(\mathcal{T}_j^{v,1} | \Delta_j^{vi} a)_{L^2}| \le C \|b\|_{L_h^4(L_v^2)} \|\nabla_h a\|_{\mathcal{H}}^{\frac{3}{2}} \|a\|_{\mathcal{H}}^{\frac{1}{2}}. \tag{6.37}$$

In order to estimate the commutator, we use Taylor's formula. For a function f on $\mathbb{R}^3$, we define the function $\widetilde{f}$ on $\mathbb{R}^4$ by

$$\widetilde{f}(x, y_3) \overset{\text{def}}{=} \int_0^1 f(x_h, x_3 + \tau(y_3 - x_3)) \, d\tau.$$

Then, defining $\overline{h}(x_3) \overset{\text{def}}{=} x_3 h(x_3)$, we have

$$\mathcal{T}_j^{v,2} = \sum_{|j-\ell| \le 5} \int_{\mathbb{R}} \overline{h}(2^j(x_3 - y_3))(\widetilde{S_{\ell-1}^{vi} \partial_3 b^3})(x, y_3) \partial_3 \Delta_\ell^{vi} a(x_h, y_3) \, dy_3.$$

Using the fact that b is divergence-free and the fact that $\nabla_h \widetilde{f} = \widetilde{\nabla_h f}$, we infer that

$$\mathcal{T}_j^{v,2} = -\sum_{|j-\ell| \le 5} \int_{\mathbb{R}} \overline{h}(2^j(x_3 - y_3)) \operatorname{div}_h (\widetilde{S_{\ell-1}^{vi} b^h})(x, y_3) \partial_3 \Delta_\ell^{vi} a(x_h, y_3) \, dy_3.$$

Integrating by parts with respect to the horizontal variable, we then get that $(\mathcal{T}_j^{v,2} | \Delta_j^{vi} a)_{L^2}$ is equal to the sum over $\ell \in \{j - 5, \dots, j + 5\}$ of

$$\int_{\mathbb{R}^4} \overline{h}(2^j(x_3 - y_3))(\widetilde{S^{vi}_{\ell-1}b^h})(x, y_3)\partial_3\nabla_h\Delta^{vi}_\ell a(x_h, y_3)\Delta^{vi}_j a(x)\,dx\,dy_3$$

$$+ \int_{\mathbb{R}^4} \overline{h}(2^j(x_3 - y_3))(\widetilde{S^{vi}_{\ell-1}b^h})(x, y_3)\partial_3\Delta^{vi}_\ell a(x_h, y_3)\nabla_h\Delta^{vi}_j a(x)\,dx\,dy_3.$$

As we have $\|\widetilde{b}(x_h, \cdot, y_3)\|_{L^\infty_v} \leq \|b(x_h, \cdot)\|_{L^\infty_v}$, we infer that

$$\left|(\mathcal{T}^{v,2}_j|\Delta^{vi}_j a)_{L^2}\right| \leq C2^{-j}\|b\|_{L^4_h(L^\infty_v)} \sum_{|\ell-j|\leq 5} \left(\|\partial_3\nabla_h\Delta^{vi}_\ell a\|_{L^2}\|\Delta^{vi}_j a\|_{L^4_h(L^2_v)}\right.$$

$$\left. + \|\partial_3\Delta^{vi}_\ell a\|_{L^4_h(L^2_v)}\|\nabla_h\Delta^{vi}_j a\|_{L^2}\right)$$

$$\leq C\|b\|_{L^4_h(L^\infty_v)} \sum_{|\ell-j|\leq 5} \|\nabla_h\Delta^{vi}_\ell a\|_{L^2}\|\Delta^{vi}_j a\|_{L^4_h(L^2_v)}.$$

Using (6.35), we get that

$$\sum_j 2^{-j}|(\mathcal{T}^{v,2}_j|\Delta^{vi}_j a)_{L^2}| \leq C\|b\|_{L^4_h(L^\infty_v)}\|\nabla_h a\|_{\mathcal{H}}^{\frac{3}{2}}\|a\|_{\mathcal{H}}^{\frac{1}{2}}. \tag{6.38}$$

The estimation of $\mathcal{T}^{v,3}_j$ is based on the following observation. For any divergence-free vector field u, we have, from (6.30),

$$\Delta^v_\ell u^3(x) = \int_{\mathbb{R}} g^v(2^\ell(x_3 - y_3))\Delta^v_\ell\partial_3 u^3(x_h, y_3)\,dy_3$$

$$= -\operatorname{div}_h \int_{\mathbb{R}} g^v(2^\ell(x_3 - y_3))\Delta^v_\ell u^h(x_h, y_3)\,dy_3$$

$$= -2^{-\ell}\operatorname{div}_h\widetilde{\Delta}^v_\ell u^h \tag{6.39}$$

with $\widetilde{\Delta}^v_\ell \overset{\text{def}}{=} \widetilde{\varphi}(2^{-\ell}D_3)$ for some suitable smooth function $\widetilde{\varphi}$ supported in an annulus.

Note that if $j \geq 2$, then the term $S^{vi}_{\ell-1}b^3 - S^{vi}_{j-1}b^3$ which appears in $\mathcal{T}^{v,3}_j$ reduces to just $\Delta^{vi}_j b^3$ or $\Delta^{vi}_{j-2}b^3$. Thus, using (6.39) and integrating by parts in the horizontal variable, we get

$$(\mathcal{T}^{v,3}_j|\Delta^{vi}_j a)_{L^2} = \sum_{\substack{\ell'\in\{j-2,j\}\\|\ell-j|\leq 1}} 2^{-\ell'}\left(\left(\Delta^v_j\left(\widetilde{\Delta}^v_\ell b^h\nabla_h\Delta^v_j\Delta^v_\ell\partial_3 a\right)\Big|\Delta^{vi}_j a\right)\right.$$

$$\left. + \left(\Delta^v_j\left(\widetilde{\Delta}^v_\ell b^h\Delta^v_j\Delta^v_\ell\partial_3 a\right)\Big|\nabla_h\Delta^{vi}_j a\right)\right).$$

Now, following the lines of reasoning which led to (6.38), we get

$$\sum_{j\geq 2} 2^{-j}|(\mathcal{T}^{v,3}_j|\Delta^{vi}_j a)_{L^2}| \leq C\|b\|_{L^4_h(L^\infty_v)}\|\nabla_h a\|_{\mathcal{H}}^{\frac{3}{2}}\|a\|_{\mathcal{H}}^{\frac{1}{2}}. \tag{6.40}$$

If $j \leq 1$, we observe that

$$|(\mathcal{T}_j^{v,3}|\Delta_j^{vi}a)_{L^2}| \leq C\|b\|_{L_h^4(L_v^\infty)}\|\nabla_h a\|_\mathcal{H}\|a\|_\mathcal{H}.$$

Combining this with the inequalities (6.36)–(6.38) and (6.40), we end up with

$$|(T_b^{vi}\nabla a|a)_\mathcal{H}| \leq C\|b\|_{L_h^4(L_v^\infty)}\|\nabla_h a\|_\mathcal{H}^{\frac{3}{2}}\|a\|_\mathcal{H}^{\frac{1}{2}} + \|b\|_{L_h^4(L_v^\infty)}\|\nabla_h a\|_\mathcal{H}\|a\|_\mathcal{H}.$$

From the convexity inequality

$$\alpha\beta \leq \theta\alpha^{\frac{1}{\theta}} + (1-\theta)\beta^{\frac{1}{1-\theta}} \tag{6.41}$$

for $\theta = 1/4$ and $\theta = 1/2$, we infer that

$$|(T_b^{vi}\nabla a|a)_\mathcal{H}| \leq \frac{\nu}{100}\|\nabla_h a\|_\mathcal{H}^2 + \frac{C}{\nu}\|b\|_{L_h^4(L_v^\infty)}^2\left(1 + \frac{1}{\nu^2}\|b\|_{L_h^4(L_v^\infty)}^2\right)\|a\|_\mathcal{H}^2.$$

To bound $(R^{vi}(b,\nabla a)|a)_\mathcal{H}$, we have to deal with the fact that the sum of the indices of the vertical regularity is 0. Again, we separate the terms involving vertical derivatives from the terms involving horizontal derivatives. This leads to

$$\Delta_j^{vi}R^{vi}(b,\nabla a) = \mathcal{R}_j^h + \mathcal{R}_j^v + \mathcal{R}_j^0 \quad \text{with}$$

$$\mathcal{R}_j^h \overset{\text{def}}{=} \Delta_j^{vi}\sum_{\ell \geq (j-3)^+}\Delta_\ell^{vi}b^h \cdot \nabla_h S_{\ell+2}^{vi}a,$$

$$\mathcal{R}_j^v \overset{\text{def}}{=} \Delta_j^{vi}\sum_{\ell \geq (j-3)^+}\Delta_\ell^{vi}b^3 S_{\ell+2}^{vi}\partial_3 a,$$

$$\mathcal{R}_j^0 \overset{\text{def}}{=} \Delta_j^{vi}(S_0^v b \cdot \nabla S_2^v a).$$

We first estimate $\mathcal{R}_j^0$. It is obvious that if j is large enough, then $\mathcal{R}_j^0 \equiv 0$. We thus have

$$|(\mathcal{R}_j^0|\Delta_j^{vi}a)_{L^2}| \leq C\|b\|_{L_h^4(L_v^\infty)}\|\nabla_h a\|_\mathcal{H}\|a\|_\mathcal{H}$$

$$\leq \frac{\nu}{100}\|\nabla_h a\|_\mathcal{H}^2 + \frac{C}{\nu}\|b\|_{L_h^4(L_v^\infty)}^2\|a\|_\mathcal{H}^2.$$

Bounding $\mathcal{R}_j^h$ relies on the following lemma.

Lemma 6.35. *A constant C exists such that for any $p \in [4,\infty[$, we have*

$$\|\Delta_j^v b\|_{L_h^p(L_v^2)} \leq Cc_j\sqrt{p}\,2^{-\frac{j}{2}}\|b\|_{\mathcal{B}_u}^{\frac{2}{p}}\|\nabla_h b\|_{\mathcal{B}_u}^{1-\frac{2}{p}} \quad \text{for all} \quad j \geq 0.$$

Proof. By the definition of $\|\cdot\|_{\mathcal{B}_u}$ and using Lemma 6.10, we have, for any p in $[4,\infty[$,

$$2^{\frac{j}{2}}\|\Delta_j^v b\|_{L_h^p(L_v^2)} \leq C \sum_{k \leq N} 2^{k\left(1-\frac{2}{p}\right)} 2^{\frac{j-k}{2}} \|\Delta_k^h \Delta_j^v b\|_{L_h^4(L_v^2)}$$

$$+ C \sum_{k > N} 2^{-\frac{2k}{p}} 2^{\frac{j-k}{2}} \|\Delta_k^h \Delta_j^v \nabla_h b\|_{L_h^4(L_v^2)}$$

$$\leq C\|b\|_{\mathcal{B}_u} \sum_{k \leq N} 2^{k\left(1-\frac{2}{p}\right)} c_{k,j} + C\|\nabla_h b\|_{\mathcal{B}_u} \sum_{k > N} 2^{-\frac{2k}{p}} c_{k,j}.$$

Using the Cauchy–Schwarz inequality, we deduce that

$$2^{\frac{j}{2}}\|\Delta_j^v b\|_{L_h^p(L_v^2)} \leq C\left(\sum_k c_{k,j}^2\right)^{\frac{1}{2}} \left(\|b\|_{\mathcal{B}_u}\left(\sum_{k \leq N} 2^{2k\left(1-\frac{2}{p}\right)}\right)^{\frac{1}{2}}\right.$$

$$\left. + \|\nabla_h b\|_{\mathcal{B}_u}\left(\sum_{k > N} 2^{-\frac{4k}{p}}\right)^{\frac{1}{2}}\right)$$

$$\leq C\left(\sum_k c_{k,j}^2\right)^{\frac{1}{2}} \left(\|b\|_{\mathcal{B}_u} 2^{N\left(1-\frac{2}{p}\right)} + \|\nabla_h b\|_{\mathcal{B}_u} \sqrt{p}\, 2^{-\frac{2N}{p}}\right)$$

$$\leq C c_j \left(\|b\|_{\mathcal{B}_u} 2^{N\left(1-\frac{2}{p}\right)} + \|\nabla_h b\|_{\mathcal{B}_u} \sqrt{p}\, 2^{-\frac{2N}{p}}\right).$$

Choosing $2^N \approx \dfrac{\|\nabla_h b\|_{\mathcal{B}_u}}{\|b\|_{\mathcal{B}_u}}$ then gives the lemma.   □

We now derive a first estimate for $\mathcal{R}_j^h$ which takes care of the high vertical regularity of a. Using Lemmas 6.10 and 6.35 we get[3]

$$\|\mathcal{R}_j^h\|_{L_h^{\frac{4}{3}}(L_v^2)} \leq C 2^{\frac{j}{2}} \sum_{\ell \geq (j-3)^+} \|\Delta_\ell^v b^h \cdot \nabla_h S_{\ell+2}^v a\|_{L_h^{\frac{4}{3}}(L_v^1)}$$

$$\leq C 2^{\frac{j}{2}} \sum_{\ell \geq (j-3)^+} \|\Delta_\ell^v b^h\|_{L_h^4(L_v^2)} \|\nabla_h S_{\ell+2}^v a\|_{L^2}$$

$$\leq C 2^{\frac{j}{2}} \left(\sum_\ell c_\ell^2\right) \|b\|_{\mathcal{B}_u}^{\frac{1}{2}} \|\nabla_h b\|_{\mathcal{B}_u}^{\frac{1}{2}} \|\nabla_h a\|_{\mathcal{H}}.$$

Using (6.35) we then infer that

$$|(\mathcal{R}_j^h|\Delta_j^{vi} a)_{L^2}| \leq C\|b\|_{\mathcal{B}_u}^{\frac{1}{2}} \|\nabla_h b\|_{\mathcal{B}_u}^{\frac{1}{2}} \|\nabla_h a\|_{\mathcal{H}} 2^{\frac{j}{2}} \|\Delta_j^{vi} a\|_{L_h^4(L_v^2)}$$

$$\leq C\|b\|_{\mathcal{B}_u}^{\frac{1}{2}} \|\nabla_h b\|_{\mathcal{B}_u}^{\frac{1}{2}} \|\nabla_h a\|_{\mathcal{H}} \|a\|_{H^{0,\frac{1}{2}}}^{\frac{1}{2}} \|\nabla_h a\|_{H^{0,\frac{1}{2}}}^{\frac{1}{2}}. \quad (6.42)$$

We shall now estimate $\mathcal{R}_j^h$ using only the fact that a and $\nabla_h a$ belong to $\mathcal{H}$. This may be done by taking advantage of Lemmas 6.10 and 6.35. For any p in $[4, \infty[$ we get

---

[3] Below, $(j-3)^+$ means $\max(0, j-3)$.

$$\|\mathcal{R}_j^h\|_{L_h^{\frac{2p}{p+2}}(L_v^2)} \leq C2^{\frac{j}{2}} \sum_{\ell \geq (j-3)^+} \|\Delta_\ell^v b^h \cdot \nabla_h S_{\ell+2}^v a\|_{L_h^{\frac{2p}{p+2}}(L_v^1)}$$

$$\leq C2^{\frac{j}{2}} \sum_{\ell \geq (j-3)^+} \|\Delta_\ell^v b^h\|_{L_h^p(L_v^2)} \|\nabla_h S_{\ell+2}^v a\|_{L^2}$$

$$\leq C2^{\frac{j}{2}} \left(\sum_\ell c_\ell^2\right) \sqrt{p} \|b\|_{\mathcal{B}_u}^{\frac{2}{p}} \|\nabla_h b\|_{\mathcal{B}_u}^{1-\frac{2}{p}} \|\nabla_h a\|_{\mathcal{H}}.$$

By interpolation, a constant C exists (independent of p) such that, for any p in $[4, \infty[$, we have

$$\|\Delta_j^{vi} a\|_{L_h^{\frac{2p}{p-2}}(L_v^2)} \leq C\|\Delta_j^{vi} a\|_{L^2}^{1-\frac{2}{p}} \|\Delta_j^{vi} \nabla_h a\|_{L^2}^{\frac{2}{p}}.$$

Thus, we get

$$|(\mathcal{R}_j^h|\Delta_j^{vi} a)_{L^2}| \leq C2^{\frac{j}{2}} \sqrt{p} \|b\|_{\mathcal{B}_u}^{\frac{2}{p}} \|\nabla_h b\|_{\mathcal{B}_u}^{1-\frac{2}{p}} \|\nabla_h a\|_{\mathcal{H}} \|\Delta_j^{vi} a\|_{L^2}^{1-\frac{2}{p}} \|\Delta_j^{vi} \nabla_h a\|_{L^2}^{\frac{2}{p}}$$

$$\leq Cc_j 2^j \sqrt{p} \|b\|_{\mathcal{B}_u}^{\frac{2}{p}} \|\nabla_h b\|_{\mathcal{B}_u}^{1-\frac{2}{p}} \|a\|_{\mathcal{H}}^{1-\frac{2}{p}} \|\nabla_h a\|_{\mathcal{H}}^{1+\frac{2}{p}}. \tag{6.43}$$

Using the estimates (6.42) and (6.43), we infer that for any positive integer M and any p in $[4, \infty[$,

$$\sum_j 2^{-j} |(\mathcal{R}_j^h|\Delta_j^{vi} a)_{L^2}| = \sum_{0 \leq j \leq M} 2^{-j} |(\mathcal{R}_j^h|\Delta_j^{vi} a)_{L^2}| + \sum_{j > M} 2^{-j} |(\mathcal{R}_j^h|\Delta_j^{vi} a)_{L^2}|$$

$$\leq C\left(\sum_{j > M} 2^{-j}\right) \|b\|_{\mathcal{B}_u}^{\frac{1}{2}} \|\nabla_h b\|_{\mathcal{B}_u}^{\frac{1}{2}} \|\nabla_h a\|_{\mathcal{H}} \|a\|_{H^{0,\frac{1}{2}}}^{\frac{1}{2}} \|\nabla_h a\|_{H^{0,\frac{1}{2}}}^{\frac{1}{2}}$$

$$+ \left(\sum_{0 \leq j \leq M} c_j\right) \sqrt{p} \|b\|_{\mathcal{B}_u}^{\frac{2}{p}} \|\nabla_h b\|_{\mathcal{B}_u}^{1-\frac{2}{p}} \|a\|_{\mathcal{H}}^{1-\frac{2}{p}} \|\nabla_h a\|_{\mathcal{H}}^{1+\frac{2}{p}}.$$

Using the Cauchy–Schwarz inequality, we obtain

$$\sum_j 2^{-j} |(\mathcal{R}_j^h|\Delta_j^{vi} a)_{L^2}| \leq C2^{-M} \|b\|_{\mathcal{B}_u}^{\frac{1}{2}} \|\nabla_h b\|_{\mathcal{B}_u}^{\frac{1}{2}} \|\nabla_h a\|_{\mathcal{H}} \|a\|_{H^{0,\frac{1}{2}}}^{\frac{1}{2}} \|\nabla_h a\|_{H^{0,\frac{1}{2}}}^{\frac{1}{2}}$$

$$+ (pM)^{\frac{1}{2}} \|b\|_{\mathcal{B}_u}^{\frac{2}{p}} \|\nabla_h b\|_{\mathcal{B}_u}^{1-\frac{2}{p}} \|a\|_{\mathcal{H}}^{1-\frac{2}{p}} \|\nabla_h a\|_{\mathcal{H}}^{1+\frac{2}{p}}.$$

Using the convexity inequality (6.41) with $\theta = \frac{1}{2}$ and with $\theta = \frac{p+2}{2p}$, we deduce that

$$\sum_j 2^{-j} |(\mathcal{R}_j^h|\Delta_j^{vi} a)_{L^2}| \leq \frac{\nu}{10} \|\nabla_h a\|_{\mathcal{H}}^2 + \frac{C}{\nu^{\frac{p+2}{p-2}}} (pM)^{\frac{p}{p-2}} \|b\|_{\mathcal{B}_u}^{\frac{4}{p-2}} \|\nabla_h b\|_{\mathcal{B}_u}^2 \|a\|_{\mathcal{H}}^2$$

$$+ \frac{C}{\nu} 2^{-2M} \|b\|_{\mathcal{B}_u} \|\nabla_h b\|_{\mathcal{B}_u} \|\nabla_h a\|_{H^{0,\frac{1}{2}}} \|a\|_{H^{0,\frac{1}{2}}}.$$

Assume that $M \geq 16$. As p is in $[4, \infty[$, we can choose $p = \log_2 M$. We infer that for any $M \geq 16$, the sum $\sum_j 2^{-j} |(\mathcal{R}_j^h|\Delta_j^{vi} a)_{L^2}|$ is less than

$$\frac{\nu}{10}\|\nabla_h a\|_{\mathcal{H}}^2 + \frac{C}{\nu}2^{-2M}\|b\|_{\mathcal{B}_u}\|\nabla_h b\|_{\mathcal{B}_u}\|\nabla_h a\|_{H^{0,\frac{1}{2}}}\|a\|_{H^{0,\frac{1}{2}}}$$

$$+ \frac{\|b\|_{\mathcal{B}_u}^4}{\nu}\left(1 + \frac{\|b\|_{\mathcal{B}_u}^4}{\nu^2}\right)\|\nabla_h b\|_{\mathcal{B}_u}^2\|a\|_{\mathcal{H}}^2 M\log_2 M.$$

If $\|a\|_{\mathcal{H}} \leq 2^{-16}$, then we can choose M such that $2^{-M} \approx \|a\|_{\mathcal{H}}$. This gives

$$\sum_j 2^{-j}|(\mathcal{R}_j^h|\Delta_j^{vi}a)_{L^2}| \leq \frac{\nu}{10}\|\nabla_h a\|_{\mathcal{H}}^2 + C_1(a,b)\mu(\|a\|_{\mathcal{H}}^2) \qquad (6.44)$$

with

$$C_1(a,b) \stackrel{\text{def}}{=} \frac{C}{\nu}\|b\|_{\mathcal{B}_u}\|\nabla_h b\|_{\mathcal{B}_u}\|\nabla_h a\|_{H^{0,\frac{1}{2}}}\|a\|_{H^{0,\frac{1}{2}}}$$

$$+ \frac{\|b\|_{\mathcal{B}_u}^4}{\nu}\left(1 + \frac{\|b\|_{\mathcal{B}_u}^4}{\nu^2}\right)\|\nabla_h b\|_{\mathcal{B}_u}^2.$$

We now estimate $(\mathcal{R}_j^v|\Delta_j^{vi}a)_{L^2}$. First, we use (6.39). Together with integration by parts in the horizontal variable, this gives

$$(\mathcal{R}_j^v|\Delta_j^{vi}a)_{L^2} = \mathcal{R}_j^{v,1}(a) + \mathcal{R}_j^{v,2}(a) \quad \text{with}$$

$$\mathcal{R}_j^{v,1}(a) \stackrel{\text{def}}{=} \sum_{\ell \geq (j-3)^+} 2^{-\ell}\left(\Delta_j^{vi}(\widetilde{\Delta}_\ell^v b^h \cdot \nabla_h \partial_3 S_{\ell+2}^v a)|\Delta_j^{vi}a\right)_{L^2} \quad \text{and}$$

$$\mathcal{R}_j^{v,2}(a) \stackrel{\text{def}}{=} \sum_{\ell \geq (j-3)^+} 2^{-\ell}\left(\Delta_j^{vi}(\widetilde{\Delta}_\ell^v b^h \partial_3 S_{\ell+2}^v a)|\nabla_h\Delta_j^{vi}a\right)_{L^2}.$$

Having observed that for any $u \in H^{0,\frac{1}{2}} \cap \mathcal{H}$, we have

$$\|\partial_3 S_\ell^v u\|_{L^2} \leq Cc_\ell 2^{\frac{3\ell}{2}}\|u\|_{\mathcal{H}} \quad \text{and} \quad \|\partial_3 S_\ell^v u\|_{L^2} \leq Cc_\ell 2^{\frac{\ell}{2}}\|u\|_{H^{0,\frac{1}{2}}}, \quad (6.45)$$

by following exactly the lines of reasoning which led to (6.44), we find that

$$\sum_j 2^{-j}|\mathcal{R}_j^{v,1}(a)| \leq \frac{\nu}{10}\|\nabla_h a\|_{\mathcal{H}}^2 + C_1(a,b)\mu(\|a\|_{\mathcal{H}}^2). \qquad (6.46)$$

We now estimate $\mathcal{R}_j^{v,2}(a)$ by using the fact that a and $\nabla_h a$ are in $H^{0,\frac{1}{2}}$. Using Lemma 6.10, we get

$$\|\Delta_j^{vi}(\widetilde{\Delta}_\ell^v b^h \partial_3 S_{\ell+2}^v a)\|_{L^2} \leq C2^{\frac{j}{2}}\|\widetilde{\Delta}_\ell^v b^h \partial_3 S_{\ell+2}^v a\|_{L_h^2(L_v^1)}$$

$$\leq C2^{\frac{j}{2}}\|\widetilde{\Delta}_\ell^v b^h\|_{L_h^4(L_v^2)}\|\partial_3 S_{\ell+2}^v a\|_{L_h^4(L_v^2)}.$$

From (6.45), we infer that

$$\|\partial_3 S_{\ell+2}^v a\|_{L_h^4(L_v^2)} \leq Cc_\ell 2^{\frac{\ell}{2}}\|\nabla_h a\|_{H^{0,\frac{1}{2}}}^{\frac{1}{2}}\|a\|_{H^{0,\frac{1}{2}}}^{\frac{1}{2}}.$$

Lemma 6.35 applied with $p = 4$ then leads to

$$|\mathcal{R}_j^{v,2}(a)| \leq C2^j \Big(\sum_{\ell \geq j-3} 2^{-\ell} \Big) \|b\|_{\mathcal{B}_u}^{\frac{1}{2}} \|\nabla_h b\|_{\mathcal{B}_u}^{\frac{1}{2}} \|a\|_{H^{0,\frac{1}{2}}}^{\frac{1}{2}} \|\nabla_h a\|_{H^{0,\frac{1}{2}}}^{\frac{1}{2}} 2^{-\frac{j}{2}} \|\Delta_j^{vi} \nabla_h a\|_{L^2}$$

$$\leq C \|b\|_{\mathcal{B}_u}^{\frac{1}{2}} \|\nabla_h b\|_{\mathcal{B}_u}^{\frac{1}{2}} \|a\|_{H^{0,\frac{1}{2}}}^{\frac{1}{2}} \|\nabla_h a\|_{H^{0,\frac{1}{2}}}^{\frac{1}{2}} \|\nabla_h a\|_{\mathcal{H}}. \tag{6.47}$$

Finally, we estimate $|\mathcal{R}_j^{v,2}(a)|$ by using the fact that a and $\nabla_h a$ belong to $\mathcal{H}$. Lemma 6.35, applied for any $p \in [4, \infty[$, together with (6.45), gives

$$\|\Delta_j^{vi}(\Delta_\ell^v b^h \partial_3 S_{\ell+2}^v a)\|_{L^2} \leq C2^{\frac{j}{2}} \|\widetilde{\Delta}_\ell^v b\|_{L_h^p(L_v^2)} \|\partial_3 S_{\ell+2}^v a\|_{L_h^{\frac{2p}{p-2}}(L_v^2)}$$

$$\leq C2^{\frac{j}{2}} d_\ell \sqrt{p} \|b\|_{\mathcal{B}_u}^{\frac{2}{p}} \|\nabla_h b\|_{\mathcal{B}_u}^{1-\frac{2}{p}} \|a\|_{\mathcal{H}}^{1-\frac{2}{p}} \|\nabla_h a\|_{\mathcal{H}}^{\frac{2}{p}}.$$

Thus, we deduce that

$$|\mathcal{R}_j^{v,2}(a)| \leq Cc_j 2^j \sqrt{p} \|b\|_{\mathcal{B}_u}^{\frac{2}{p}} \|\nabla_h b\|_{\mathcal{B}_u}^{1-\frac{2}{p}} \|a\|_{\mathcal{H}}^{1-\frac{2}{p}} \|\nabla_h a\|_{\mathcal{H}}^{1+\frac{2}{p}}.$$

Using (6.47) and following exactly the same lines of reasoning which led to (6.44), we get that

$$\sum_j 2^{-j} |(\mathcal{R}_j^v | \Delta_j^{vi} a)_{L^2}| \leq \frac{\nu}{10} \|\nabla_h a\|_{\mathcal{H}}^2 + C_1(a, b)\mu(\|a\|_{\mathcal{H}}^2). \tag{6.48}$$

This proves Lemma 6.34. □

6.5 References and Remarks

For a more complete discussion of the geophysical considerations leading to the anisotropic Navier–Stokes system, the reader is referred to the book by J. Pedlosky [248] or the introduction of the book by J.-Y. Chemin et al. [75].

The use of anisotropic Sobolev spaces is not recent in the study of partial differential equations (if we have in mind boundary value problems); see, for example, the book by L. Hörmander [166]. An anisotropic paradifferential calculus was constructed by M. Sablé-Tougeron in [255]. Anisotropic Sobolev spaces were introduced in the context of the incompressible Navier–Stokes system by D. Iftimie in [172]. The study of the anisotropic incompressible Navier–Stokes system was initiated by J.-Y. Chemin et al. in [74] and D. Iftimie in [173]. The first sharp scaling invariant result was obtained by M. Paicu in [244], wherein he proved local existence for any divergence-free data in $\mathcal{B}^{0,\frac{1}{2}}$ and global existence for small data. Uniqueness was obtained in the class of solutions which belong to $L^\infty([0,T]; H^{0,\frac{1}{2}}) \cap L^2([0,T]; H^{1,\frac{1}{2}})$ [which is not comparable to our space $\mathcal{B}_4^{-\frac{1}{2},\frac{1}{2}}(T)$].

Except for the first section, all the material in the present chapter is borrowed from the paper by J.-Y. Chemin and P. Zhang [78]. The key Lemma 6.34, however, was first proven by M. Paicu in [244].

Finally, we note that it is possible to prove a *local* version of Theorem 6.16 for any (divergence-free) large data in $\mathcal{B}_4^{-\frac{1}{2},\frac{1}{2}}$.

7

Euler System for Perfect Incompressible Fluids

This chapter is devoted to the mathematical study of the Euler system for incompressible inviscid fluids with constant density:

$$(E) \qquad \begin{cases} \partial_t v + v \cdot \nabla v = -\nabla P \\ \operatorname{div} v = 0 \\ v_{|t=0} = v_0. \end{cases}$$

Here, $v = v(t,x)$ is a time-dependent divergence-free vector field on $\mathbb{R}^d$ ($d \geq 2$). The scalar function $P = P(t,x)$ may be interpreted as the Lagrange multiplier associated with the divergence-free constraint. From a physical viewpoint, v is the *speed* of a particle of the fluid located at x at time t, and P is the *pressure* field.

The choice of $\mathbb{R}^d$ instead of the more physical case of a bounded domain is for the purposes of simplicity (since we shall mainly use tools coming from Fourier analysis). Of course, the results that we shall present here carry over to the case of periodic boundary conditions.

The *vorticity* $\Omega \overset{\text{def}}{=} Dv - \nabla v$ (where Dv stands for the Jacobian matrix of v, and ∇v stands for its transposed matrix) plays a fundamental role in incompressible fluid mechanics. Indeed, on the one hand, Ω satisfies the following linear transport-like equation:

$$\partial_t \Omega + v \cdot \nabla \Omega + \Omega \cdot Dv + \nabla v \cdot \Omega = 0. \tag{7.1}$$

On the other hand, owing to the fact that $\operatorname{div} v = 0$ and $\sum_{j=1}^d \partial_j \Omega_j^i = \Delta v^i$, the vector field v may be computed in terms of Ω by the formula

$$v^i = -\sum_j \partial_j E_d * \Omega_j^i,$$

where E_d stands for the fundamental solution of $-\Delta$. In other words, we have

$$v^i(x) = c_d \sum_j \int_{\mathbb{R}^d} \frac{x^j - y^j}{|x-y|^d} \Omega_j^i(y)\,dy \tag{7.2}$$

H. Bahouri et al., *Fourier Analysis and Nonlinear Partial Differential Equations*, Grundlehren der mathematischen Wissenschaften 343, DOI 10.1007/978-3-642-16830-7_7, © Springer-Verlag Berlin Heidelberg 2011

with $c_d \overset{\text{def}}{=} \dfrac{\Gamma(1 + d/2)}{d \, \pi^{d/2}}$ and $\Gamma(s) \overset{\text{def}}{=} \displaystyle\int_0^{+\infty} t^{s-1} e^{-t} \, dt$ for $s > 0$.

The above relation is sometimes called the *Biot–Savart law*. The coupling between (7.1) and (7.2) is called the *vorticity formulation* of the Euler system and is formally equivalent to (E).

In dimension three, the skew-symmetric matrix Ω may be identified with the vector field $\omega = \nabla \times v$ and the vorticity formulation becomes

$$\partial_t \omega + v \cdot \nabla \omega = \omega \cdot \nabla v \quad \text{with} \tag{7.3}$$

$$v(x) = \frac{1}{4\pi} \int_{\mathbb{R}^3} \frac{(x - y) \times \omega(y)}{|x - y|^3} \, dy. \tag{7.4}$$

In dimension two, the vorticity may be identified with the scalar function $\omega \overset{\text{def}}{=} \partial_1 v^2 - \partial_2 v^1$ so that the vorticity formulation reduces to[1]

$$\partial_t \omega + v \cdot \nabla \omega = 0 \quad \text{with} \quad v(x) = \frac{1}{2\pi} \int_{\mathbb{R}^2} \frac{(x - y)^\perp}{|x - y|^2} \omega(y) \, dy. \tag{7.5}$$

Due to the fact that $\operatorname{div} v = 0$, this implies that all the L^p norms of the vorticity are conserved by the flow. As we shall see below, this is the main ingredient for proving the global existence of the two-dimensional Euler system. In dimension $d \geq 3$, however, the vorticity equation has an extra term (the so-called *stretching term*) so that one cannot expect any global control for the L^p norms of the vorticity. This is one of the reasons why, until now, no global results have been known for general data in dimension $d \geq 3$.

This chapter unfolds as follows. In the first section we prove local existence and uniqueness for the Euler system in general nonhomogeneous Besov spaces. Global existence in dimension two is addressed in Section 7.2. Section 7.3 is devoted to the study of the inviscid limit for incompressible fluids. The more specific case of vortex-patch-like structures in dimension two is postponed to Section 7.4.

7.1 Local Well-posedness Results for Inviscid Fluids

In this section we are concerned with the initial value problem for the Euler system in dimension $d \geq 2$. Before stating our main result, we introduce the set L_L^∞ of measurable functions u over $\mathbb{R}^d$ such that

$$\|u\|_{L_L^\infty} \overset{\text{def}}{=} \sup_{x \in \mathbb{R}^d} \frac{|u(x)|}{1 + \log\langle x \rangle} < \infty \quad \text{with} \quad \langle x \rangle \overset{\text{def}}{=} \sqrt{1 + |x|^2}.$$

The set L_L^∞ endowed with the norm $\|\cdot\|_{L_L^\infty}$ is obviously a Banach space.

---

[1] In what follows, it is understood that $z^\perp = (-z^2, z^1)$ if $z = (z^1, z^2)$.

We can now state the main result of this section.

Theorem 7.1. *Let* $1 \leq p, r \leq \infty$ *and* $s \in \mathbb{R}$ *be such that*[2] $B_{p,r}^s \hookrightarrow C^{0,1}$.
There exists a constant c, *depending only on* s, p, r, *and* d, *such that for all divergence-free data* $v_0 \in B_{p,r}^s(\mathbb{R}^d)$, *there exists a time* $T \geq c/\|v_0\|_{B_{p,r}^s}$ *such that* (E) *has a solution* (v, P) *on* $[-T, T] \times \mathbb{R}^d$ *satisfying*

$$v, \nabla P \in L^\infty([-T, T]; B_{p,r}^s) \quad and \quad P \in L^\infty([-T, T]; L^1 + L_L^\infty).$$

Moreover, if $(\widetilde{v}, \widetilde{P})$ *also satisfies* (E) *with the same data and belongs to the above class, then* $\widetilde{v} \equiv v$ *and* $\nabla \widetilde{P} = \nabla P$.

Finally, if $r < \infty$ *(resp.,* $r = \infty$*), then* v *and* ∇P *are continuous (resp., weakly continuous) in time with values in* $B_{p,r}^s$.

Remark 7.2. In Sections 7.1.5 and 7.1.6 we shall state a more accurate uniqueness result and a blow-up criterion. Global results in the two-dimensional case will be proven in the next section.

Remark 7.3. We should also point out that in the case $1 < p < \infty$, we can define the pressure P such that $P \in L^\infty([-T, T]; B_{p,r}^{s+1})$.

We shall first provide some guidance concerning the reading of this section. As explained in the introduction, the vorticity and the way the velocity can be computed from the vorticity (the Biot–Savart law) play a fundamental role in the study of the Euler system. For that reason, the first part of this section will be devoted to the Biot–Savart law. It is well known that dealing with the pressure term is one of the main difficulties involved in solving the Euler system. However, it turns out that for sufficiently smooth solutions with reasonable growth at infinity, the pressure may be computed in terms of the velocity field, leading to the study of a *modified* Euler system. Estimates for the pressure will be given in the second part of this section, whereas the modified Euler system will be solved in the fourth part. In the third part, we give conditions under which the standard Euler system and the modified Euler system are equivalent. The study of uniqueness is postponed to the fifth part. In the final part, we give continuation criteria for the standard Euler system.

7.1.1 The Biot–Savart Law

In dimension two, the vorticity is preserved along the trajectories so that the way we can deduce information about the vector field from information about the vorticity is obviously fundamental. In fact, even in dimension $d \geq 3$ the question of global existence of the Euler system is intimately entangled with the control of the vorticity.

---

[2] That is, $s > 1 + d/p$, or $s = 1 + d/p$ and $r = 1$.

Throughout this section, it is assumed that the divergence-free vector field v over $\mathbb{R}^d$ is computed from the vorticity Ω according to the formula (7.2). We aim to prove various estimates for the velocity in terms of the vorticity.

We begin with a straightforward estimate.

Proposition 7.4. *If $1 < a < d < b < \infty$, then*

$$\|v\|_{L^\infty} \le C\|\Omega\|_{L^a \cap L^b}.$$

Proof. We can split $\mathbb{R}^d$ as $\{y \in \mathbb{R}^d \ / \ |x - y| \le 1\} \cup \{y \in \mathbb{R}^d \ / \ |x - y| > 1\}$ and use convolution inequalities to bound the integral in (7.2). $\square$

The next estimate that we shall give is much harder to prove. It relies on the fact that the map $\Omega \mapsto \nabla v$ is a *Calderon–Zygmund operator*. As a consequence, we get the following fundamental estimate that we shall assume throughout this book.

Proposition 7.5. *There exists a constant C, depending only on the dimension d, such that for any $1 < p < \infty$ and any divergence-free vector field v with gradient in L^p, we have*

$$\|\nabla v\|_{L^p} \le C\frac{p^2}{p-1}\|\Omega\|_{L^p}.$$

The above inequality turns out to be false in the limit cases $p = 1$ and $p = \infty$. In particular, even in dimension two, we cannot find a constant C such that the inequality

$$\|\nabla v\|_{L^\infty} \le C\|\omega\|_{L^1 \cap L^\infty}$$

is true for all divergence-free vector fields v satisfying (7.2).[3] However, v is quasi-Lipschitz in the sense of Definition 2.106 page 116: For any finite a, there exists a constant C such that

$$\|v\|_{LL} \le C\|\Omega\|_{L^a \cap L^\infty}. \tag{7.6}$$

This is a consequence of Proposition 2.107 combined with the decomposition

$$\nabla v = \Delta_{-1}\nabla v + \left(\text{Id} - \Delta_{-1}\right)\nabla v$$

and the following lemma.

Lemma 7.6. *For any $a \in [1, \infty[$ and $b \in [1, \infty]$, we have*

$$\|\Delta_{-1}\nabla v\|_{L^\infty} \le C_1\|\Omega\|_{L^a} \quad and \quad \|\Delta_{-1}\nabla v\|_{L^\infty} \le C_2\|v\|_{L^b}$$

with C_1 depending only on a and d, and C_2 depending only on d.

---

[3] For example, if we take for ω the characteristic function of the square $[0, 1]^2$, then v is *not* Lipschitz. In fact, ∇v blows up as the logarithm of the distance to the corners of the square. See [69] for more details.

For all $s \in \mathbb{R}$ and $1 \le p, r \le \infty$, there exists a constant C' such that

$$\|(\mathrm{Id} - \Delta_{-1})\nabla v\|_{B^s_{p,r}} \le C' \|\Omega\|_{\dot{B}^s_{p,r}}.$$

Proof. That $\|\Delta_{-1}\nabla v\|_{L^\infty} \le C\|v\|_{L^b}$ follows from Bernstein's lemma. Further, in the case $1 < a < \infty$, Proposition 7.5 yields $\|\Delta_{-1}\nabla v\|_{L^a} \le C\|\Delta_{-1}\Omega\|_{L^a}$, from which follows the desired bound for $\|\Delta_{-1}\nabla v\|_{L^\infty}$, according to Bernstein's lemma. In the case $a = 1$, we can still write

$$\|\Delta_{-1}\nabla v\|_{L^\infty} \le C\|\Delta_{-1}\nabla v\|_{L^2} \le C\|\Delta_{-1}\Omega\|_{L^2} \le C\|\Omega\|_{L^1}.$$

To prove the last inequality, we may write[4]

$$(\mathrm{Id} - \Delta_{-1})\nabla v^i = \sum_j B_j(D)\Omega^i_j \quad \text{with} \quad B_j(D) \overset{\text{def}}{=} -(\mathrm{Id} - \Delta_{-1})|D|^{-2}\nabla \partial_j.$$

Because the operator $B_j(D)$ is an S^0-multiplier, the desired inequality is a consequence of Proposition 2.78 page 101. □

Finally, if the vorticity has enough regularity, then v has to be Lipschitz. More precisely, we have the following result.

Proposition 7.7. *Let $s \in \mathbb{R}$ and $1 \le p, r \le \infty$ satisfy $s > 1 + d/p$. If, in addition, $v \in L^b$ for some $b \in [1, \infty]$ or $\Omega \in L^a$ for some $a \in [1, \infty[$, then there exists a constant C such that*

$$\|\nabla v\|_{L^\infty} \le C\left(\min(\|v\|_{L^b}, \|\Omega\|_{L^a}) + \|\Omega\|_{L^\infty} \log\left(e + \frac{\|\Omega\|_{B^{s-1}_{p,r}}}{\|\Omega\|_{L^\infty}}\right)\right).$$

Proof. We decompose ∇v into low and high frequencies:

$$\nabla v = \nabla \Delta_{-1} v + (\mathrm{Id} - \Delta_{-1})\nabla v.$$

The first term may be bounded according to Lemma 7.6. For the second term, we use Proposition 2.104 page 116. As $\varepsilon \overset{\text{def}}{=} s - d/p - 1 > 0$, we can write

$$\|(\mathrm{Id} - \Delta_{-1})\nabla v\|_{L^\infty} \le C\|(\mathrm{Id} - \Delta_{-1})\nabla v\|_{B^0_{\infty,\infty}} \log\left(e + \frac{\|(\mathrm{Id} - \Delta_{-1})\nabla v\|_{B^\varepsilon_{\infty,\infty}}}{\|(\mathrm{Id} - \Delta_{-1})\nabla v\|_{B^0_{\infty,\infty}}}\right).$$

Next, by virtue of Lemma 7.6 and the embedding $L^\infty \hookrightarrow \dot{B}^0_{\infty,\infty}$, we may write

$$\|(\mathrm{Id} - \Delta_{-1})\nabla v\|_{B^0_{\infty,\infty}} \le C\|\Omega\|_{L^\infty} \quad \text{and} \quad \|(\mathrm{Id} - \Delta_{-1})\nabla v\|_{B^\varepsilon_{\infty,\infty}} \le C\|\Omega\|_{B^\varepsilon_{\infty,\infty}}.$$

Since, in addition, $B^{s-1}_{p,r} \hookrightarrow B^\varepsilon_{\infty,\infty}$, we get the desired estimate. □

---

[4] Here, $|D|^{-2}$ stands for the Fourier multiplier with symbol $|\xi|^{-2}$.

7.1.2 Estimates for the Pressure

We first explain formally how the pressure may be computed from the velocity field. First, we apply div to (E) and get, as the vector field v is divergence-free,

$$-\Delta P = \operatorname{div}(v \cdot \nabla v) = \operatorname{tr}(Dv)^2.$$

Therefore, we must have

$$\nabla P = \nabla \operatorname{div} E_d * (v \cdot \nabla v) = \nabla E_d * \big(\operatorname{tr}(Dv)^2\big).$$

This induces us to set $\nabla P = \Pi(v,v)$ with

$$\Pi(v,w) = \Pi_1(v,w) + \Pi_2(v,w) + \Pi_3(v,w) + \Pi_4(v,w) + \Pi_5(v,w) \qquad (7.7)$$

and, denoting by θ some function of $\mathcal{D}(B(0,2))$ with value 1 on $B(0,1)$, we have

$$\Pi_1(v,w) = \nabla |D|^{-2} T_{\partial_i v^j} \partial_j w^i,$$
$$\Pi_2(v,w) = \nabla |D|^{-2} T_{\partial_j w^i} \partial_i v^j,$$
$$\Pi_3(v,w) = \nabla |D|^{-2} \partial_i \partial_j (\operatorname{Id} - \Delta_{-1}) R(v^i, w^j),$$
$$\Pi_4(v,w) = \theta E_d * \nabla \partial_i \partial_j \Delta_{-1} R(v^i, w^j),$$
$$\Pi_5(v,w) = \nabla \partial_i \partial_j \widetilde{E}_d * \Delta_{-1} R(v^i, w^j) \quad \text{with} \quad \widetilde{E}_d \stackrel{\text{def}}{=} (1-\theta)E_d.$$

In the above formulas, as in the rest of this chapter, the summation convention over repeated indices is used.

This subsection is devoted to estimating the bilinear operator Π in various function spaces. We first state L^p bounds.

Lemma 7.8. *Let $1 < p < \infty$. Assume that v is divergence-free. There exists a constant C, depending only on d and p, such that*

$$\|\Pi(v,v)\|_{L^p} \le C \min\big(\|v\|_{L^\infty} \|\Omega\|_{L^p}, \|v\|_{L^p} \|\nabla v\|_{L^\infty}\big).$$

Proof. It suffices to note that if $\operatorname{div} v = 0$, then

$$\Pi(v,v) = \nabla \operatorname{div} |D|^{-2}(v \cdot \nabla v)$$

so that, according to the Marcinkiewicz theorem,

$$\|\Pi(v,v)\|_{L^p} \le C \|v \cdot \nabla v\|_{L^p}.$$

Applying Hölder's inequality and Proposition 7.5 then completes the proof.

$\square$

Lemma 7.9. *For all $s > -1$ and $1 \le p, r \le \infty$, there exists a constant C such that*

$$\|\Pi(v,w)\|_{B_{p,r}^s} \le C\big(\|v\|_{C^{0,1}} \|w\|_{B_{p,r}^s} + \|w\|_{C^{0,1}} \|v\|_{B_{p,r}^s}\big).$$

Proof. We first note that the first three terms of $\Pi(v, w)$ are spectrally supported away from the origin. Hence, in the definitions of Π_1, Π_2, and Π_3, the operator $\nabla |D|^{-2}$ may be replaced by an S^{-1}-multiplier, in the sense of Proposition 2.78. Further, by virtue of Theorems 2.82 and 2.85, if $s > -1$, then we have

$$\|T_{\partial_i v^j}\partial_j w^i\|_{B_{p,r}^{s-1}} \leq C\|\nabla v\|_{L^\infty}\|\nabla w\|_{B_{p,r}^{s-1}},$$
$$\|T_{\partial_j w^i}\partial_i v^j\|_{B_{p,r}^{s-1}} \leq C\|\nabla w\|_{L^\infty}\|\nabla v\|_{B_{p,r}^{s-1}},$$
$$\|R(v, w)\|_{B_{p,r}^{s+1}} \leq C\|v\|_{B_{\infty,\infty}^1}\|w\|_{B_{p,r}^s}.$$

Hence, Π_1, Π_2, and Π_3 satisfy the desired inequality.

Next, since $\Pi_4(v, w)$ and $\Pi_5(v, w)$ are spectrally supported in a ball, it suffices to bound their L^p norm. Because $\theta E_d \in L^1$, we have, by virtue of Young's inequalities and Bernstein's lemma page 52,

$$\|\Pi_4(v, w)\|_{L^p} \leq \|\theta E_d\|_{L^1}\|\nabla\partial_i\partial_j\Delta_{-1}R(v^i, w^j)\|_{L^p}$$
$$\leq C\|\theta E_d\|_{L^1}\|\Delta_{-1}R(v, w)\|_{L^p}$$
$$\leq C\|\theta E_d\|_{L^1}\|R(v, w)\|_{B_{p,r}^{s+1}}$$
$$\leq C\|\theta E_d\|_{L^1}\|v\|_{B_{\infty,\infty}^1}\|w\|_{B_{p,r}^s}.$$

As $\nabla\partial_i\partial_j\widetilde{E}_d$ is in L^1, similar computations yield the desired inequality for $\Pi_5(v, w)$. □

Lemma 7.10. *Let $1 \leq p, r \leq \infty$ and $0 < \varepsilon < 2 + d/p$. We have*

$$\|\Pi(v, w)\|_{B_{p,r}^{\frac{d}{p}+1-\varepsilon}} \leq C\big(\|v\|_{C^{0,1}}\|w\|_{B_{p,r}^{\frac{d}{p}+1-\varepsilon}} + \|w\|_{B_{\infty,\infty}^{1-\varepsilon}}\|\nabla v\|_{B_{p,r}^{\frac{d}{p}}}\big).$$

Proof. The proof is very similar to that of the previous lemma. First, owing to the spectral properties of $\Pi_i(v, w)$ ($i = 1, 2, 3$) and the continuity results for the paraproduct and remainder, we have, if $0 < \varepsilon < 2 + d/p$,

$$\|\Pi_1(v, w)\|_{B_{p,r}^{1-\varepsilon+\frac{d}{p}}} \leq C\|\nabla v\|_{L^\infty}\|\nabla w\|_{B_{p,r}^{\frac{d}{p}-\varepsilon}},$$
$$\|\Pi_2(v, w)\|_{B_{p,r}^{1-\varepsilon+\frac{d}{p}}} \leq C\|\nabla w\|_{B_{\infty,\infty}^{-\varepsilon}}\|\nabla v\|_{B_{p,r}^{\frac{d}{p}}},$$
$$\|\Pi_3(v, w)\|_{B_{p,r}^{1-\varepsilon+\frac{d}{p}}} \leq C\|v\|_{B_{\infty,\infty}^1}\|w\|_{B_{p,r}^{1-\varepsilon+\frac{d}{p}}}.$$

The last terms, $\Pi_4(v, w)$ and $\Pi_5(v, w)$, may be treated by arguing as in Lemma 7.9. □

Lemma 7.11. *Let $1 < p < \infty$. There exists a constant C, depending only on d and p, such that if $\operatorname{div} v = 0$, then*

$$\|\Pi(v, v)\|_{B_{\infty,\infty}^{1-\frac{d}{p}}} \leq C\big(\|\Omega\|_{L^\infty} + \|v\|_{L^\infty}\big)\|\Omega\|_{L^p}.$$

Proof. Owing to the fact that the low frequencies of b are not involved in the definition of the paraproduct $T_a b$ (see Remark 2.83 page 103) and that, according to Lemma 7.6,

$$\|(\mathrm{Id} - \Delta_{-1})\nabla v\|_{B^0_{\infty,\infty}} \le C \|\Omega\|_{L^\infty},$$

applying Proposition 2.82 yields

$$\|\Pi_1(v,v)\|_{B^{1-\frac{d}{p}}_{\infty,\infty}} + \|\Pi_2(v,v)\|_{B^{1-\frac{d}{p}}_{\infty,\infty}} \le \|\nabla v\|_{B^{-\frac{d}{p}}_{\infty,\infty}} \|\Omega\|_{L^\infty}. \tag{7.8}$$

Because

$$\Pi_3(v,v) = -\nabla |D|^{-2}\partial_i(\mathrm{Id} - \Delta_{-1})R(\partial_j v^i, v^j)$$

and $B^1_{p,\infty} \hookrightarrow B^{1-\frac{d}{p}}_{\infty,\infty}$, we have

$$\|\Pi_3(v,v)\|_{B^{1-\frac{d}{p}}_{\infty,\infty}} \le C\|\Pi_3(v,v)\|_{B^1_{p,\infty}} \le C\|v\|_{B^1_{\infty,\infty}} \|\nabla v\|_{L^p}.$$

Note that it is enough to bound the L^∞ norm of $\Pi_4(v,v)$ and of $\Pi_5(v,v)$. Hence, those two terms satisfy the same inequality as $\Pi_3(v,v)$.

Finally, Proposition 7.5 and Lemma 7.6 ensure that

$$\|\nabla v\|_{L^p} \le C\|\Omega\|_{L^p} \quad \text{and} \quad \|v\|_{B^1_{\infty,\infty}} \le C(\|v\|_{L^\infty} + \|\Omega\|_{L^\infty}). \tag{7.9}$$

This completes the proof of the lemma. □

In the case where v is divergence-free, we expect that $\mathrm{div}\, \Pi(v,v) = -\mathrm{tr}\,(Dv)^2$. This is a consequence of the following lemma.

Lemma 7.12. *Let $1 \le p, r \le \infty$ and $s > 1$. There exists a constant C such that*

$$\|\mathrm{div}\, \Pi(v,w) + \mathrm{tr}(Dv\, Dw)\|_{B^{s-1}_{p,r}}$$
$$\le C\Big(\|\mathrm{div}\, v\|_{B^0_{\infty,\infty}} \|w\|_{B^s_{p,r}} + \|\mathrm{div}\, w\|_{B^0_{\infty,\infty}} \|v\|_{B^s_{p,r}}\Big).$$

In the limit case $s = 1$ we have

$$\|\mathrm{div}\, \Pi(v,w) + \mathrm{tr}(Dv\, Dw)\|_{B^0_{p,\infty}}$$
$$\le C\Big(\|\mathrm{div}\, v\|_{B^0_{\infty,\infty}} \|w\|_{B^1_{p,1}} + \|\mathrm{div}\, w\|_{B^0_{\infty,\infty}} \|v\|_{B^1_{p,1}}\Big).$$

Proof. From the definition of Π we get

$$-\mathrm{div}\, \Pi(v,w) = T_{\partial_i v^j}\partial_j w^i + T_{\partial_j w^i}\partial_i v^j + \partial_i\partial_j R(v^j, w^i).$$

Hence, after a few calculations we get

$$-\mathrm{div}\, \Pi(v,w) = \mathrm{tr}\,(Dv\, Dw) + \partial_i R(\mathrm{div}\, v, w^i) + R(v^i, \partial_i\, \mathrm{div}\, w).$$

The desired inequalities thus follow from continuity results for the remainder operator (see Proposition 2.85). □

Note that by construction, if v and w are suitably smooth, then $\Pi(v, w)$ is the gradient of some tempered distribution. Indeed, for $i = 1, 2, 3, 4$ it is obvious that $\Pi_i(v, w) = \nabla P_i(v, w)$ with

$$P_1(v, w) = |D|^{-2} T_{\partial_i v^j} \partial_j w^i, \qquad P_3(v, w) = |D|^{-2} \partial_i \partial_j (\mathrm{Id} - \Delta_{-1}) R(v^i, w^j),$$

$$P_2(v, w) = |D|^{-2} T_{\partial_j w^i} \partial_i v^j, \qquad P_4(v, w) = \theta E_d * \partial_i \partial_j \Delta_{-1} R(v^i, w^j).$$

If, in addition, $\Delta_{-1} R(v, w)$ belongs to some L^p space (which is of course the case if, say, $v \in B_{p,r}^s$ with $s > -1$ and $w \in C^{0,1}$), then $\Pi_5(v, w)$ is the gradient of some smooth function $P_5(v, w)$. Moreover, if $1 < p < \infty$, as the operator of convolution by $\partial_i \partial_j \widetilde{E}_d$ is a Calderon–Zygmund operator, we may write

$$\Pi_5(v, w) = \nabla P_5(v, w) \quad \text{with} \quad P_5(v, w) = \partial_i \partial_j \widetilde{E}_d \star \Delta_{-1} R(v^i, w^j),$$

and, owing to the spectral localization, we find that $P_5(v, w)$ belongs to any space $B_{p,r}^\sigma$ with $\sigma \in \mathbb{R}$.

Since $D^2 \widetilde{E}_d$ is *not* an integrable function, however, in the case $p = 1$ or ∞, expressing $P_5(v, w)$ in terms of v and w requires some care. Therefore, we set

$$P_5(v, w) = \sum_{1 \le i,j \le d} \sum_{m=1}^{3} L_{ij}^m \big(\Delta_{-1} R(v^i, w^j)\big),$$

where the operators L_{ij}^1, L_{ij}^2, and L_{ij}^3 are defined by

$$L_{ij}^1(u)(x) \overset{\text{def}}{=} \int_{\mathbb{R}^d} \partial_i \partial_j \widetilde{E}_d(x - y) \theta\Big(\frac{x - y}{\langle x \rangle}\Big) u(y) \, dy,$$

$$L_{ij}^2(u)(x) \overset{\text{def}}{=} \int_0^1 \int_{\mathbb{R}^d} x^k \partial_i \partial_j \partial_k \widetilde{E}_d(tx - y) (1 - \theta)\Big(\frac{tx - y}{\langle tx \rangle}\Big) u(y) \, dy \, dt,$$

$$L_{ij}^3(u)(x) \overset{\text{def}}{=} -\int_0^1 \int_{\mathbb{R}^d} x^k \partial_i \partial_j \widetilde{E}_d(tx - y) \frac{\partial}{\partial x_k}\Big\{\theta\Big(\frac{tx - y}{\langle tx \rangle}\Big)\Big\} u(y) \, dy \, dt.$$

Obviously, if u is a continuous, bounded function, then

$$\nabla\big(L_{ij}^1(u) + L_{ij}^2(u) + L_{ij}^3(u)\big) = \nabla \partial_i \partial_j \widetilde{E}_d \star u.$$

Furthermore, the operators L_{ij}^m are continuous from L^∞ to L_L^∞, as the following result shows.

Lemma 7.13. *There exists a constant C such that for m in $\{1, 2, 3\}$ and (i, j) in $\{1, \ldots, d\}^2$, we have, for any bounded function u,*

$$\forall x \in \mathbb{R}^d, \ \big|L_{ij}^m(u)(x)\big| \le C\big(1 + \log\langle x \rangle\big) \|u\|_{L^\infty}.$$

Proof. We obviously have

$$\big|L_{ij}^1(u)(x)\big| \le C\|u\|_{L^\infty} \int_{1 \le |z| \le 2\langle x \rangle} |z|^{-d} \, dz,$$

hence $L_{ij}^1(u)(x)$ satisfies the desired inequality.

Next, using Fubini's theorem and an obvious change of variables, we get

$$\left|L_{ij}^2(u)(x)\right| \leq C\|u\|_{L^\infty} \int_0^1 \int_{|tx-y|\geq \max(1,\langle tx\rangle)} \frac{|x|}{|tx-y|^{d+1}}\, dy\, dt$$

$$\leq C\|u\|_{L^\infty} \int_{|z|\geq 1} \int_{t\in[0,1]\,/\,|z|\geq\langle tx\rangle} \frac{|x|}{|z|^{d+1}}\, dt\, dx$$

$$\leq C\|u\|_{L^\infty} \int_{|z|\geq 1} \left(\int_0^{\min\left(1,\frac{|z|}{|x|}\right)} dt\right) \frac{|x|\, dz}{|z|^{d+1}}$$

$$\leq C\|u\|_{L^\infty} \left(1+\log\langle x\rangle\right).$$

Finally, we note that due to the support properties of $\widetilde{E}_d$ and $\nabla\theta$, the integration in the definition of $L_{ij}^3(u)(x)$ may be restricted to those (t,y) for which

$$|tx-y| \geq 1 \quad \text{and} \quad 1 \leq \frac{|tx-y|}{\langle tx\rangle} \leq 2.$$

Since, for such (t,y), we have

$$\frac{\partial}{\partial x_k}\left(\frac{tx-y}{\langle tx\rangle}\right) \leq \frac{C}{\langle x\rangle},$$

the same argument as for $L_{ij}^1(u)(x)$ leads to the desired inequality. □

Setting $P(v,w) = \sum_{i=1}^5 P_i(v,w)$ and using continuity results for the paraproduct, remainder, and Lemma 7.13, we end up with the following statement.

Lemma 7.14. *For any $\sigma \in \mathbb{R}$, $1 \leq p, r \leq \infty$, define*

- $\widetilde{B}_{p,r}^\sigma \overset{def}{=} B_{p,r}^\sigma$ *if $1 < p < \infty$,*
- $\widetilde{B}_{\infty,r}^\sigma \overset{def}{=} \left\{u \in B_{p,r}^\sigma + L_L^\infty\,/\,\nabla u \in B_{p,r}^{\sigma-1}\right\},$
- $\widetilde{B}_{1,r}^\sigma \overset{def}{=} B_{1,r}^{\sigma+1} + \bigcap_{q>1}\bigcap_{s\in\mathbb{R}} B_{q,r}^s.$

There exists a bilinear operator P such that $\Pi(v,w) = \nabla P(v,w)$, and

- *if v,w are in $C^{0,1}\cap B_{p,r}^s$ for some $s > -1$, then*

$$\|P(v,w)\|_{\widetilde{B}_{p,r}^{s+1}} \leq C\big(\|v\|_{C^{0,1}}\|w\|_{B_{p,r}^s} + \|w\|_{C^{0,1}}\|v\|_{B_{p,r}^s}\big);$$

- *if v,w are in $B_{\infty,\infty}^0\cap B_{p,r}^s$ for some $s > 0$, then*

$$\|P(v,w)\|_{\widetilde{B}_{p,r}^s} \leq C\big(\|v\|_{B_{\infty,\infty}^0}\|w\|_{B_{p,r}^s} + \|w\|_{B_{\infty,\infty}^0}\|v\|_{B_{p,r}^s}\big).$$

7.1.3 Another Formulation of the Euler System

In the previous subsection, we gave conditions under which the gradient of the pressure may be computed from the velocity. This motivates our studying the following *modified Euler system*:

$$(\widetilde{E}) \qquad\qquad \partial_t v + v \cdot \nabla v + \Pi(v, v) = 0.$$

This new formulation is easier to deal with since only the vector field v has to be determined. Since we are ultimately interested in solving the Cauchy problem for the true Euler system (E), however, it is important to find conditions under which solving $(\widetilde{E})$ does provide a solution for (E). This is the purpose of the following proposition.

Proposition 7.15. *Let (v, P) satisfy (E) on $[T_1, T_2] \times \mathbb{R}^d$. Assume that for some $s > 0$ and $1 \leq p, r \leq \infty$, we have*

$$v \in L^1([T_1, T_2]; B^s_{p,r} \cap B^0_{\infty,\infty}) \quad and \quad P \in L^1([T_1, T_2]; L^1 + L^\infty_L). \qquad (7.10)$$

Then, v satisfies $(\widetilde{E})$ and $\nabla P = \Pi(v, v)$.

Conversely, assume that $v \in L^\infty([T_1, T_2]; B^s_{p,r})$ satisfies $(\widetilde{E})$ on $[T_1, T_2]$ with s sufficiently large enough that $B^s_{p,r} \hookrightarrow C^{0,1}$. If, in addition, $\operatorname{div} v(t_0) = 0$ for some t_0 in $[T_1, T_2]$, then $(v, P(v, v))$ satisfies (E).

Proof. We begin by proving the first statement. Applying the operator div to the Euler system (E), we get, for all $t \in [T_1, T_2]$,

$$-\Delta P(t) = \operatorname{div}(v(t) \cdot \nabla v(t)) = -\Delta P(v(t), v(t)).$$

Hence, $P(t) - P(v(t), v(t))$ is a harmonic polynomial. Note that the assumption (7.10) and Lemma 7.14 guarantee that $P(t) - P(v(t), v(t))$ is in $L^1 + L^\infty_L$ for almost every $t \in [T_1, T_2]$. Hence, $P(t) - P(v(t), v(t))$ depends only on t. This entails that $\nabla P = \Pi(v, v)$.

We now prove the second part of the proposition. Because $B^s_{p,r}$ is continuously included in $C^{0,1}$, Lemma 7.14 ensures that $\Pi(v, v)$ is the gradient of $P(v, v)$. In order to conclude that $(v, P(v, v))$ satisfies (E), however, we still have to check that the vector field v is divergence-free. This may be achieved by applying div to $(\widetilde{E})$. We get

$$(\partial_t + v \cdot \nabla) \operatorname{div} v = -\operatorname{div} \Pi(v, v) - \operatorname{tr}(Dv)^2.$$

Assume for simplicity that $[T_1, T_2] = [0, T]$ and $t_0 = 0$. If $s > 1$, we then deduce from Theorem 3.14 that for all $t \in [0, T]$,

$$\|\operatorname{div} v(t)\|_{B^{s-1}_{p,r}} \leq \int_0^t e^{C \int_{t'}^t \|v\|_{B^s_{p,r}} dt''} \|\operatorname{div} \Pi(v, v) + \operatorname{tr}(Dv)^2\|_{B^{s-1}_{p,r}} dt'. \quad (7.11)$$

Now, according to Lemma 7.12, we have

$$\| \operatorname{div} \Pi(v,v) + \operatorname{tr}(Dv)^2 \|_{B_{p,r}^{s-1}} \leq C \|\operatorname{div} v\|_{L^\infty} \|v\|_{B_{p,r}^s} \leq C\| \operatorname{div} v\|_{B_{p,r}^{s-1}} \|v\|_{B_{p,r}^s}.$$

Plugging this inequality into (7.11) and using Gronwall's inequality, we conclude that $\operatorname{div} v(t) = 0$ in $B_{p,r}^{s-1}$ for all $t \in [0,T]$.

In the limit case where $s = 1$, due to $B_{p,r}^s \hookrightarrow C^{0,1}$, we must have $p = \infty$ and $r = 1$. Then, using the last inequality of Lemma 7.12 and performing the estimates for $\operatorname{div} v$ in the space $L^\infty([0,T]; B_{\infty,\infty}^0)$, we still get $\operatorname{div} v \equiv 0$.   □

7.1.4 Local Existence of Smooth Solutions

This section is devoted to the proof of the existence part of Theorem 7.1. We first state the local existence for the modified Euler system.

Proposition 7.16. *Let* $1 \leq p, r \leq \infty$ *and* $s \in \mathbb{R}$ *be such that* $B_{p,r}^s \hookrightarrow C^{0,1}$. *There exists a constant* c, *depending only on* s, p, r, *and* d, *such that for all initial data* v_0 *in* $B_{p,r}^s(\mathbb{R}^d)$, *there exists a time* $T \geq c/\|v_0\|_{B_{p,r}^s}$ *such that* $(\widetilde{E})$ *has a solution* v *in* $L^\infty([-T,T]; B_{p,r}^s)$.

If $r < \infty$ *(resp.,* $r = \infty$*), then* v *is continuous (resp., weakly continuous) in time with values in* $B_{p,r}^s$.

The proof relies mainly on estimates for the transport equation and on Lemmas 7.9 and 7.10. It is structured as follows:

- First, we inductively solve linear transport equations so as to get a sequence of approximate solutions.
- Second, we prove local a priori estimates in large norm.
- Third, we prove the convergence in small norm.
- Finally, we pass to the limit in the equation.

First Step: Construction of Approximate Solutions

In order to define a sequence $(v^n)_{n\in\mathbb{N}}$ of (global) approximate solutions to $(\widetilde{E})$, we use an iterative scheme. First, we set $v^0 = v_0$, then, assuming that v^n belongs to $L_{loc}^\infty(\mathbb{R}; B_{p,r}^s)$, we solve the following *linear* transport equation:

$$\begin{cases} \partial_t v^{n+1} + v^n \cdot \nabla v^{n+1} = \Pi(v^n, v^n) \\ v^n_{|t=0} = v_0. \end{cases} \tag{7.12}$$

Since $v^n \in L_{loc}^\infty(\mathbb{R}; B_{p,r}^s)$ and $B_{p,r}^s \hookrightarrow C^{0,1}$, Lemma 7.9 ensures that $\Pi(v^n, v^n)$ belongs to $L_{loc}^\infty(\mathbb{R}; B_{p,r}^s)$. Therefore, Theorem 3.19 provides a global solution v^{n+1} to the equation (7.12) which belongs to $L_{loc}^\infty(\mathbb{R}; B_{p,r}^s)$.

Second Step: A Priori Estimates

Combining Lemma 7.9 with Theorem 3.19 yields, for all $n \in \mathbb{N}$ and $t \in \mathbb{R}^+$,

$$\|v^{n+1}(t)\|_{B^s_{p,r}} \le e^{CV_n(t)}\left(\|v_0\|_{B^s_{p,r}} + C\int_0^t e^{-CV_n(t')}(V_n'(t'))^2\, dt'\right)$$

$$\text{with }\ V_n(t) \stackrel{\text{def}}{=} \int_0^t \|v^n(t')\|_{B^s_{p,r}}\, dt'.$$

A similar inequality holds for negative times. Hence, arguing as in the proof of the existence for the Camassa–Holm equation in Chapter 3, we deduce that for all $n \in \mathbb{N}$,

$$\|v^n(t)\|_{B^s_{p,r}} \le \frac{\|v_0\|_{B^s_{p,r}}}{1 - 2C|t|\|v_0\|_{B^s_{p,r}}} \quad \text{whenever}\quad 2C|t|\|v_0\|_{B^s_{p,r}} < 1. \qquad (7.13)$$

Third Step: Convergence of the Sequence

Let us fix some T such that $2CT\|v_0\|_{B^s_{p,r}} < 1$. Let $(m,n) \in \mathbb{N}^2$. By taking the difference between the equations for v^{n+m+1} and v^{n+1}, we find that

$$\begin{aligned}(\partial_t + v^{n+m}\cdot\nabla)(v^{n+m+1} - v^{n+1})\\ = (v^n - v^{n+m})\cdot\nabla v^{n+1} + \Pi(v^{n+m} - v^n, v^{n+m} + v^n).\end{aligned} \qquad (7.14)$$

We first consider the case where $s > 1$. We claim that $(v^n)_{n\in\mathbb{N}}$ is a Cauchy sequence in $L^\infty([-T,T]; B^{s-1}_{p,r})$. Indeed, Lemma 7.10, combined with the fact that $B^s_{p,r} \hookrightarrow C^{0,1}$, yields[5]

$$\|\Pi(v^{n+m} - v^n, v^{n+m} + v^n)\|_{B^{s-1}_{p,r}} \le C\|v^{n+m} - v^n\|_{B^{s-1}_{p,r}}\|v^{n+m} + v^n\|_{B^s_{p,r}}.$$

By taking advantage of Bony's decomposition and of continuity results for the paraproduct and the remainder, it is not difficult to check that

$$\|(v^n - v^{n+m})\cdot\nabla v^{n+1}\|_{B^{s-1}_{p,r}} \le C\|v^{n+m} - v^n\|_{B^{s-1}_{p,r}}\|v^{n+1}\|_{B^s_{p,r}}.$$

Applying Theorem 3.14 to (7.14), we thus get, for all $t \in [0,T]$,

$$\begin{aligned}\|v^{n+m+1} - v^{n+1}\|_{B^{s-1}_{p,r}} \le Ce^{CV_{n+m}(t)}\int_0^t e^{-CV_{n+m}(t')}\\ \times\left(\|v^n\|_{B^s_{p,r}} + \|v^{n+1}\|_{B^s_{p,r}} + \|v^{n+m}\|_{B^s_{p,r}}\right)\|v^{n+m} - v^n\|_{B^{s-1}_{p,r}}\, dt'\end{aligned}$$

and a similar inequality for $t \in [-T, 0]$.

Using (7.13), we conclude by induction that for all $(n,m) \in \mathbb{N}^2$,

$$\|v^{n+m} - v^n\|_{L^\infty([-T,T]; B^{s-1}_{p,r})} \le \frac{1}{n!}\left(\frac{C\|v_0\|_{B^s_{p,r}}}{1 - 2CT\|v_0\|_{B^s_{p,r}}}\right)^n \|v^m - v^0\|_{L^\infty([-T,T]; B^{s-1}_{p,r})}.$$

Since $(v^m)_{m\in\mathbb{N}}$ is bounded in $L^\infty([-T,T]; B^s_{p,r})$, this ensures that $(v^n)_{n\in\mathbb{N}}$ is indeed a Cauchy sequence in $L^\infty([-T,T]; B^{s-1}_{p,r})$.

---

[5] Without loss of generality, we may assume that $s < 2 + d/p$.

Let us now consider the limit case $s = 1$. Due to the fact that $B^s_{p,r} \hookrightarrow C^{0,1}$, we must have $p = \infty$ and $r = 1$. Now, on the one hand, since the vector fields considered here need not be divergence-free, Theorem 3.14 does not provide any control for the norm of $v^{n+m+1} - v^{n+1}$ in $B^0_{\infty,1}$. On the other hand, that theorem may be used to bound the norm in $B^0_{\infty,\infty}$.

According to Lemma 7.10, the operator Π maps $B^0_{\infty,\infty} \times B^1_{\infty,1}$ into $B^0_{\infty,\infty}$, and it is not difficult to check (by combining Bony decomposition with the properties of continuity for the paraproduct and remainder) that

$$\|(v^n - v^{n+m}) \cdot \nabla v^{n+1}\|_{B^0_{\infty,\infty}} \leq C\|v^{n+m} - v^n\|_{B^0_{\infty,\infty}}\|v^{n+1}\|_{B^1_{\infty,1}}.$$

Therefore, arguing as above, we can conclude that $(v^n)_{n\in\mathbb{N}}$ is a Cauchy sequence in $L^\infty([-T,T]; B^0_{\infty,\infty})$.

Fourth Step: Passing to the Limit

Let v be the limit of the sequence $(v^n)_{n\in\mathbb{N}}$. Using the uniform bounds given by (7.13) and the Fatou property (see Theorem 2.72 page 100), we see that v belongs to $L^\infty([-T,T]; B^s_{p,r})$. Next, by interpolating with the convergence properties stated in the previous step, we discover that $(v^n)_{n\in\mathbb{N}}$ tends to v in every space $L^\infty([-T,T]; B^{s'}_{p,r})$ with $s' < s$, which suffices to pass to the limit in $(\widetilde{E})$.

Hence, v is a solution of the modified Euler system $(\widetilde{E})$. Note that $\Pi(v,v)$ is in $L^\infty([-T;T]; B^s_{p,r})$, so Theorem 3.19 ensures that v satisfies the desired properties of continuity with respect to time. This completes the proof of Proposition 7.16.   □

Taking advantage of Proposition 7.15 and Lemma 7.14, we can now conclude that if, in addition, the initial vector field v_0 is divergence-free, then $(v, P(v,v))$ satisfies the true Euler system (E) and has the required regularity. This completes the proof of the existence part of Theorem 7.1.

7.1.5 Uniqueness

Recall that LL stands for the set of log-Lipschitz functions defined on page 116 and that, according to Proposition 2.111 page 118, the (semi)norms

$$\|f\|_{LL} \quad \text{and} \quad \sup_{j\in\mathbb{N}} \frac{\|\nabla S_j f\|_{L^\infty}}{j+1}$$

are equivalent.

In this subsection, we establish a uniqueness result for the Euler system (E) under the sole assumptions that v belongs to $\mathcal{C}([0,T]; B^0_{\infty,\infty}) \cap L^1([0,T]; LL)$ and that the pressure is a measurable function with at most logarithmic growth at infinity.

We first state a uniqueness result for the modified Euler system $(\widetilde{E})$.

Theorem 7.17. *Let v^1 and v^2 solve $(\widetilde{E})$ on $[0, T]$. Assume that v^1 and v^2 belong to*

$$\mathcal{C}([0, T]; B^0_{\infty, \infty}) \cap L^1([0, T]; LL).$$

If, in addition, $v^1(0) = v^2(0)$, then $v^1 \equiv v^2$ on $[0, T] \times \mathbb{R}^d$.

Proof. The proof relies on the fact that $\delta v \overset{\text{def}}{=} v^2 - v^1$ satisfies a transport equation associated with a log-Lipschitz vector field, namely,

$$\partial_t \delta v + v^2 \cdot \nabla \delta v = \Pi(\delta v, v^1) + \Pi(v^2, \delta v) - \delta v \cdot \nabla v^1. \tag{7.15}$$

We claim that the bilinear operator Π satisfies the following estimate for all $\varepsilon \in \,]0, 1[$ and $k \geq -1$:

$$\|\Delta_k \Pi(v, w)\|_{L^\infty} \leq C(k+2) 2^{k\varepsilon} \min\left(\|v\|_{B^{-\varepsilon}_{\infty, \infty}} \|w\|_{\overline{LL}}, \|w\|_{B^{-\varepsilon}_{\infty, \infty}} \|v\|_{\overline{LL}}\right), \tag{7.16}$$

where we have used the notation $\|\cdot\|_{\overline{LL}} \overset{\text{def}}{=} \|\cdot\|_{L^\infty} + \|\cdot\|_{LL}$.

According to (7.7), it suffices to establish this inequality for $\Delta_k \Pi_i(v, w)$ with $i \in \{1, \ldots, 5\}$.

We begin with $\Delta_k \Pi_1(v, w)$. As $\nabla(-\Delta)^{-1}$ is a homogeneous operator of degree -1 and

$$\left(S_{k-1} \partial_i v^j \, \Delta_k \partial_j w^i\right)_{k \geq -1}$$

is spectrally supported in dyadic shells, we see that it suffices to establish that

$$\|S_{k-1} \partial_i v^j \Delta_k \partial_j w^i\|_{L^\infty} \leq C(k+2) 2^{k(1+\varepsilon)} \min\left(\|v\|_{B^{-\varepsilon}_{\infty, \infty}} \|w\|_{\overline{LL}}, \|w\|_{B^{-\varepsilon}_{\infty, \infty}} \|v\|_{\overline{LL}}\right).$$

We may now write

$$\left\|S_{k-1} \partial_i v^j \Delta_k \partial_j w^i\right\|_{L^\infty} \leq \|S_{k-1} \nabla v\|_{L^\infty} \|\Delta_k \nabla w\|_{L^\infty}.$$

Note that we obviously have

$$\|S_{k-1} \nabla v\|_{L^\infty} \leq C \min\left((k+2)\|v\|_{LL}, 2^{k(1+\varepsilon)} \|\nabla v\|_{B^{-1-\varepsilon}_{\infty, \infty}}\right),$$
$$\|\Delta_k \nabla w\|_{L^\infty} \leq C \min\left((k+2)\|w\|_{LL}, 2^{k(1+\varepsilon)} \|\nabla w\|_{B^{-1-\varepsilon}_{\infty, \infty}}\right).$$

Therefore, $\|\Delta_k \Pi_1(v, w)\|_{L^\infty}$ is bounded by the right-hand side of (7.16). As $\Pi_2(v, w) = \Pi_1(w, v)$, the same inequality holds for $\Delta_k \Pi_2(v, w)$.

As the roles of v and w may be exchanged, in order to bound the other terms of (7.7), it suffices to establish that

$$\|\Delta_k R(v, w)\|_{L^\infty} \leq C(k+2) 2^{k(\varepsilon-1)} \|v\|_{\overline{LL}} \|w\|_{B^{-\varepsilon}_{\infty, \infty}}.$$

By virtue of Proposition 2.10 page 59, we may write that

$$\Delta_k R(v^i, w^j) = \sum_{k' \geq k-3} \Delta_k (\Delta_{k'} v^i \, \widetilde{\Delta}_{k'} w^j).$$

Therefore,

$$\begin{aligned}
\left\| \Delta_k R(v^i, w^j) \right\|_{L^\infty} &\leq C \sum_{k' \geq k-3} \left\| \Delta_{k'} v \right\|_{L^\infty} \left\| \widetilde{\Delta}_{k'} w \right\|_{L^\infty} \\
&\leq C \sum_{k' \geq k-3} (k' + 2) 2^{k'(\varepsilon-1)} \|v\|_{\overline{LL}} \|w\|_{B_{\infty,\infty}^{-\varepsilon}}.
\end{aligned}$$

As $\varepsilon - 1 < 0$, we get the desired inequality for $\Delta_k R(v, w)$. This completes the proof of (7.16).

We now focus on the term $v \cdot \nabla w$. We claim that

$$\|\Delta_k (v \cdot \nabla w)\|_{L^\infty} \leq C(k+2) 2^{k\varepsilon} \|w\|_{\overline{LL}} \|v\|_{B_{\infty,\infty}^{-\varepsilon}}. \tag{7.17}$$

This is, in fact, a consequence of the following Bony decomposition (where we have used the fact that div $v = 0$):

$$v \cdot \nabla w^i = T_{v^j} \partial_j w^i + \partial_j R(v^j, w^i) + T_{\partial_j w^i} v^j.$$

By mimicking the computations leading to (7.16), it is easy to get (7.17). The details are left to the reader.

We can now resume the proof of uniqueness. Combining the inequalities (7.16) and (7.17), we see that (7.15) is a transport equation associated with a vector field with coefficients in $L^1([0,T]; LL)$ and a right-hand side δf which satisfies, for all $\varepsilon \in \,]0, 1[$,

$$\|\Delta_k \delta f\|_{L^\infty} \leq C(k+2) 2^{k\varepsilon} \left(\|v^1\|_{\overline{LL}} + \|v^2\|_{\overline{LL}} \right) \|\delta v\|_{B_{\infty,\infty}^{-\varepsilon}}. \tag{7.18}$$

Let $\varepsilon_t \stackrel{\text{def}}{=} C \int_0^t \left(\|v^1\|_{\overline{LL}} + \|v^2\|_{\overline{LL}} \right) dt'$. As (7.18) is satisfied, Theorem 3.28 ensures that if C is taken sufficiently large, then, for all $k \geq -1$,

$$2^{-k\varepsilon_t} \|\Delta_k \delta v(t)\|_{L^\infty} \leq \frac{1}{2} \sup_{t' \in [0,t]} \|\delta v(t')\|_{B_{\infty,\infty}^{-\varepsilon_{t'}}}$$

whenever t belongs to the time interval $[0, T_0]$ defined by

$$T_0 = \sup \left\{ t \in [0, T], \ C \int_0^t \left(\|v^1\|_{\overline{LL}} + \|v^2\|_{\overline{LL}} \right) dt' \leq \frac{1}{2} \right\}.$$

This yields uniqueness on $[0, T_0]$.

Because v^1 and v^2 are in $L^1([0,T]; LL) \cap L^\infty([0,T]; B_{\infty,\infty}^0)$, the argument may be repeated a finite number of times, yielding uniqueness on the whole interval $[0, T]$. $\qquad\square$

Corollary 7.18. *Let (v^1, P^1) and (v^2, P^2) satisfy the Euler system (E) with the same initial data. Assume, in addition, that for $i = 1, 2$,*

$$v^i \in \mathcal{C}([0, T]; B^0_{\infty, \infty}) \cap L^1([0, T]; LL) \quad and \quad P^i \in L^1([0, T]; L^1 + L^\infty_L).$$

We then have $v^1 \equiv v^2$ and $\nabla P^1 \equiv \nabla P^2$ on $[0, T] \times \mathbb{R}^d$.

Proof. Note that the assumptions on (v^1, P^1) and (v^2, P^2) guarantee that v^1 and v^2 both solve $(\widetilde{E})$ with the same data (see Proposition 7.15). Hence, the previous theorem implies that $v^1 \equiv v^2$ and that

$$\nabla P^1 = \Pi(v^1, v^1) = \Pi(v^2, v^2) = \nabla P^2.$$

This proves the corollary.   □

Remark 7.19. The logarithmic growth assumption on the pressure cannot be omitted. Indeed, let v_0 be a nonzero constant vector field and set

$$\bigl(v^1(t, x), P^1(t, x)\bigr) = (v_0, 0) \quad and \quad \bigl(v^2(t, x), P^2(t, x)\bigr) = (v_0 \cos t, (v_0 \cdot x) \sin t).$$

Then, (v^1, P^2) and (v^2, P^2) are two distinct smooth solutions of (E) pertaining to the same initial vector field.

7.1.6 Continuation Criteria

In this subsection, we state various continuation criteria for smooth solutions of the Euler system. We first explain what we mean by a *smooth solution*.

Definition 7.20. *Let $T_1 < T_2$. Let $s \in \mathbb{R}$ and $1 \le p, r \le \infty$. A divergence-free time-dependent vector field v is called a $B^s_{p,r}$ solution to the Euler system on $]T_1, T_2[$ if it belongs to $L^\infty_{loc}(]T_1, T_2[; B^s_{p,r})$ and satisfies (E) in the space $\mathcal{S}'(]T_1, T_2[\times \mathbb{R}^d)$ for some $P \in L^\infty_{loc}(]T_1, T_2[; L^1 + L^\infty_L)$.*

To simplify the presentation, we focus on continuation criteria for positive times. Due to the time reversibility of the Euler system, however, similar results hold for negative times. We begin with a very general statement.

Theorem 7.21. *Let $s \in \mathbb{R}$ and $1 \le p, r \le \infty$ satisfy $B^s_{p,r} \hookrightarrow C^{0,1}$. Assume that (E) has a $B^s_{p,r}$ solution over $[0, T[$. If*

$$\int_0^T \|v(t)\|_{C^{0,1}} \, dt < \infty, \tag{7.19}$$

then v may be continued beyond T to a $B^s_{p,r}$ solution of (E).

If, in addition, $v_0 \in L^a$ or $\nabla v_0 \in L^a$ for some finite a, then (7.19) may be replaced by the weaker condition

$$\int_0^T \|\nabla v(t)\|_{L^\infty} \, dt < \infty. \tag{7.20}$$

Proof. Our definition of a $B^s_{p,r}$ solution guarantees that v satisfies $(\widetilde{E})$. Hence, according to Theorem 3.19 page 136 and the fact that, according to Lemma 7.9,

$$\|\Pi(v,v)\|_{B^s_{p,r}} \leq C\|v\|_{C^{0,1}}\|v\|_{B^s_{p,r}},$$

we get, for all $t \in [0,T[$,

$$\|v(t)\|_{B^s_{p,r}} \leq \|v(0)\|_{B^s_{p,r}} + C\int_0^t \|v(t')\|_{C^{0,1}}\|v(t')\|_{B^s_{p,r}} \, dt'. \tag{7.21}$$

Hence, Gronwall's lemma implies that

$$\|v(t)\|_{B^s_{p,r}} \leq \|v(0)\|_{B^s_{p,r}} e^{\int_0^t \|v(t')\|_{C^{0,1}} \, dt'} \quad \text{for all } t \in [0,T[. \tag{7.22}$$

This ensures that $v \in L^\infty([0,T[;B^s_{p,r})$.

Let $\tau \overset{\text{def}}{=} c/\|v\|_{L^\infty_T(B^s_{p,r})}$ (where c stands for the constant defined in Theorem 7.1). The Euler system with data $v(T-\tau/2)$ then has a $B^s_{p,r}$ solution $\widetilde{v}$ over $[0,\tau]$. By virtue of uniqueness, we must have

$$\widetilde{v}(t) = v(T-\tau/2+t) \quad \text{for} \quad 0 \leq t < \tau/2.$$

Hence, $\widetilde{v}$ provides a continuation of v beyond T. This yields the first statement.

We now assume that $v_0 \in L^a$ for some finite a. As, of course, $v_0 \in L^\infty$, we can assume with no loss of generality that $1 < a < \infty$. On the one hand, according to Lemma 7.8,

$$\|\Pi(v,v)\|_{L^a} \leq C\|v\|_{L^a}\|\nabla v\|_{L^\infty}. \tag{7.23}$$

On the other hand, because v satisfies $(\widetilde{E})$, we have

$$\|v(t)\|_{L^a} \leq \|v_0\|_{L^a} + \int_0^t \|\Pi(v,v)\|_{L^a} \, dt'.$$

Inserting (7.23) into the above inequality and then using the Gronwall inequality, we thus conclude that $v \in L^\infty([0,T[;L^a)$. Now, by splitting v into low and high frequencies and using Bernstein's lemma, we see that

$$\|v\|_{L^\infty} \leq C\big(\|v\|_{L^a} + \|\nabla v\|_{L^\infty}\big).$$

Therefore, $v \in L^1([0,T[;L^\infty)$. Applying the first part of Theorem 7.21 thus shows that v may be continued beyond T.

Finally, we treat the case where $\nabla v_0 \in L^a$ for some finite a. Of course, we can assume that $a > d$ so that $1 - d/a > 0$. Hence, by virtue of Lemma 7.11,

$$\|\Pi(v,v)\|_{L^\infty} \leq C(\|\Omega\|_{L^\infty} + \|v\|_{L^\infty})\|\Omega\|_{L^a}.$$

Plugging this into the inequality

$$\|v(t)\|_{L^\infty} \leq \|v_0\|_{L^\infty} + \int_0^t \|\Pi(v,v)\|_{L^\infty}\, dt'$$

and applying Gronwall's lemma, we thus get

$$e^{-C\int_0^t \|\Omega\|_{L^a}\, dt'} \|v(t)\|_{L^\infty} \leq \|v_0\|_{L^\infty}$$
$$+ C\int_0^t e^{-C\int_0^\tau \|\Omega\|_{L^a}\, dt''} \|\Omega\|_{L^a} \|\Omega\|_{L^\infty}\, dt'. \quad (7.24)$$

We will temporarily assume that the following lemma holds.

Lemma 7.22. *For any $a \in \,]1, \infty[$, there exists a constant C such that the vorticity satisfies*

$$\forall t \in [0, T[, \ \|\Omega(t)\|_{L^a} \leq \|\Omega_0\|_{L^a} \exp\Big(C\int_0^t \|\Omega(t')\|_{L^\infty}\, dt'\Big).$$

Due to the fact that $\nabla v \in L^1([0,T[;L^\infty)$, we thus have $\Omega \in L^1([0,T[;L^a \cap L^\infty)$, so the inequality (7.24) entails that $v \in L^1([0,T[;L^\infty)$. This completes the proof of the theorem. □

Proof of Lemma 7.22. From equation (7.1) and Hölder's inequality, we get

$$\|\Omega(t)\|_{L^a} \leq \|\Omega_0\|_{L^a} + 2\int_0^t \|\Omega\|_{L^\infty} \|Dv\|_{L^a}\, dt.$$

Applying Proposition 7.5 for bounding $\|Dv\|_{L^a}$ and Gronwall's lemma completes the proof. □

For sufficiently smooth solutions, the above continuation criterion may be slightly refined, as follows.

Theorem 7.23. *Let s and $1 \leq p, r \leq \infty$ be such that $s > 1 + d/p$. Assume that (E) has a $B^s_{p,r}$ solution v on $[0,T[$ for some finite $T > 0$. If, in addition, there exists some admissible Osgood modulus of continuity μ such that*

$$\int_0^T \|v(t)\|_{C_\mu}\, dt < \infty,$$

then v may be continued beyond T to a $B^s_{p,r}$ solution of (E).

Proof. The proof, based on Proposition 2.112 page 119, is the same as for quasilinear systems (see Theorem 4.22 page 196). Indeed, let us set

$$R(t) \overset{\text{def}}{=} \|v_0\|_{B^s_{p,r}} + C\int_0^t \|v(t')\|_{C^{0,1}} \|v(t')\|_{B^s_{p,r}}\, dt'.$$

According to the inequality (7.21), if C has been chosen sufficiently large, then

$$\|v(t)\|_{B^s_{p,r}} \leq R(t) \quad \text{for all } t \in [0,T[.$$

Let $\varepsilon = \min\left(1, s - \frac{d}{p} - 1\right)$ and $\Gamma : [0, a] \to [0, +\infty[$ be the function associated with the modulus of continuity μ.

Using Proposition 2.112 page 119 with $\Lambda = \|v_0\|_{B_{p,r}^s}$ and the embedding $B_{p,r}^s \hookrightarrow B_{\infty,\infty}^{s-d/p}$, we get

$$R(t) \leq \|v_0\|_{B_{p,r}^s} + C \int_0^t \gamma(t') \left(1 + \Gamma\left(\left(\frac{CR(t')}{\|v_0\|_{B_{p,r}^s}}\right)^{\frac{1}{\varepsilon}}\right)\right) R(t') \, dt'$$

with $\gamma(t) \stackrel{\text{def}}{=} \|v(t)\|_{C_\mu} + \|v_0\|_{B_{p,r}^s}$.

Mimicking the proof of Proposition 2.112, we then get, after a few computations,

$$\|v(t)\|_{B_{p,r}^s} \leq \frac{1}{C} \|v_0\|_{B_{p,r}^s} \mathcal{G}_\varepsilon^{-1}\left(C + C \int_0^t \gamma(t') \, dt'\right)$$

with $\mathcal{G}_\varepsilon(y) \stackrel{\text{def}}{=} \int_{a_\varepsilon^{-1}}^y \frac{dy'}{y' \Gamma\left((y')^{\frac{1}{\varepsilon}}\right)}$ and $a_\varepsilon \stackrel{\text{def}}{=} a^{\frac{1}{\varepsilon}}$.

Therefore, $\|v(t)\|_{B_{p,r}^s}$ stays bounded on $[0, T[$, and the proof may be completed by arguing as in the proof of Theorem 7.21. $\square$

As a corollary, we get the following generalization of the celebrated *Beale–Kato–Majda continuation criterion*.

Corollary 7.24. *Let $s > 1 + d/p$ and v be a $B_{p,r}^s$ solution of the Euler system. Assume that $\nabla v_0 \in L^a$ for some finite a. If T is finite and*

$$\int_0^T \|\Omega(t)\|_{L^\infty} \, dt < \infty,$$

then v may be continued beyond T to a $B_{p,r}^s$ solution of (E).

Proof. As pointed out in Example 4.23 page 198, the space $B_{\infty,\infty}^1$ is continuously embedded in the space C_μ, where μ stands for the admissible Osgood modulus of continuity defined by $\mu(r) = r(1 - \log r)$. Now, by virtue of the second inequality in (7.9), we have

$$\|v\|_{B_{\infty,\infty}^1} \leq C\left(\|v\|_{L^\infty} + \|\Omega\|_{L^\infty}\right).$$

Because Ω_0 is in L^a, the inequality (7.24) and Lemma 7.22 imply that v belongs to $L^\infty([0, T[; L^\infty)$. Therefore, Theorem 7.23 applies. $\square$

7.2 Global Existence Results in Dimension Two

As explained in the introduction, in dimension two, the vorticity equation reduces to

$$\partial_t \omega + v \cdot \nabla \omega = 0. \tag{7.25}$$

So, at least formally, all the L^a norms of the vorticity are conserved by the flow. Based on Corollary 7.24, we thus expect the solution to be global.

In this section, we justify this heuristic in various contexts.

7.2.1 Smooth Solutions

In this subsection, we state a global result for two-dimensional data with high regularity in Besov spaces.

Theorem 7.25. *Let $v_0 \in B_{p,r}^s(\mathbb{R}^2)$ with $\operatorname{div} v_0 = 0$ and $s > 1 + 2/p$. Assume, in addition,[6] that $\nabla v_0 \in L^a$ for some finite a. The Euler system (E) then has a unique global $B_{p,r}^s$ solution v satisfying $\nabla v \in L^\infty(\mathbb{R}; L^a)$.*

Proof. Local existence in $B_{p,r}^s$ has already been proven, so we denote by $]T_*, T^*[$ the maximal interval of existence for v. Due to $\nabla v \in L_{loc}^\infty(]T_*, T^*[; L^\infty)$ and (7.25), it is clear that $\omega \in L^\infty(]T_*, T^*[; L^\infty)$. If T^* is finite, then Corollary 7.24 enables us to continue the solution beyond T^*, which stands in contradiction to the definition of T^*. Hence, $T^* = +\infty$. A similar argument leads to $T_* = -\infty$. □

7.2.2 The Borderline Case

Proving the global existence in the borderline case $s = 1 + 2/p$ and $r = 1$ is more involved. This is because no continuation criterion is known which is solely in terms of the vorticity (whether or not Corollary 7.24 is true in this case is an open question). Nevertheless, as stated in the following theorem, the global well-posedness in the borderline case is true.

Theorem 7.26. *Let $1 \le p \le \infty$ and $v_0 \in B_{p,1}^{1+2/p}$ with $\operatorname{div} v_0 = 0$. Assume, in addition, that $\nabla v_0 \in L^a$ for some finite a. The Euler system then has a unique global solution v in $\mathcal{C}(\mathbb{R}; B_{p,1}^{1+2/p})$ with $\nabla v \in L^\infty(\mathbb{R}; L^a)$.*

Proof. For the sake of conciseness, we treat only the case where $p = \infty$, the case where $p < \infty$ being easier. We therefore assume that $v_0 \in B_{\infty,1}^1$ and that $\nabla v_0 \in L^a$ for some finite a.

Stating global estimates for the vorticity ω in $B_{\infty,1}^0$ is the key to the proof. Theorem 7.1 provides a $B_{\infty,1}^1$ solution v defined on some maximal time interval $]T_*, T^*[$. Taking advantage of (7.25) and of Theorem 3.18 page 135, we deduce that

$$\forall t \in [0, T^*[, \quad \|\omega(t)\|_{B_{\infty,1}^0} \le \|\omega_0\|_{B_{\infty,1}^0}\left(1 + C\int_0^t \|\nabla v(t')\|_{L^\infty}\, dt'\right).$$

In order to bound $\|\nabla v\|_{L^\infty}$, we may combine Lemma 7.6 with the continuous embedding $B_{\infty,1}^0 \hookrightarrow L^\infty$ to get

$$\|\nabla v\|_{L^\infty} \le C\big(\|\omega\|_{L^a} + \|\omega\|_{B_{\infty,1}^0}\big).$$

Because $\|\omega(t)\|_{L^a} = \|\omega_0\|_{L^a}$, we thus have

---

[6] Of course, this assumption is relevant only if $p = \infty$.

$$\forall t \in [0, T^*[, \quad \|\omega(t)\|_{B^0_{\infty,1}} \leq \|\omega_0\|_{B^0_{\infty,1}} \left(1 + C\|\omega_0\|_{L^a} t + C \int_0^t \|\omega\|_{B^0_{\infty,1}} \, dt'\right),$$

from which it follows, by virtue of Gronwall's lemma, that

$$\|\omega(t)\|_{B^0_{\infty,1}} \leq e^{Ct\|\omega_0\|_{B^0_{\infty,1}}} \|\omega_0\|_{B^0_{\infty,1}} (1 + Ct\|\omega_0\|_{L^a}).$$

Therefore, if $T^* < \infty$, then Lemma 7.6 ensures that $\nabla v \in L^\infty([0, T^*[; L^\infty)$. So, according to Corollary 7.24, the solution may be continued beyond T^*, which contradicts the definition of T^*. Hence, $T^* = +\infty$. Proving that $T_* = -\infty$ relies on similar arguments.   □

7.2.3 The Yudovich Theorem

As the vorticity is constant along the trajectories, it is natural to wonder what happens if the initial vorticity is only bounded with no additional regularity assumption (note that in the global existence results stated thus far, the vorticity was at least continuous).

As pointed out before, even if the vorticity is compactly supported, the corresponding vector field need not be Lipschitz. Nevertheless, we shall prove the following result.

Theorem 7.27. *Let v_0 be a divergence-free vector field in $B^1_{\infty,\infty}(\mathbb{R}^2)$ with vorticity ω_0 in $L^a \cap L^\infty$ for some finite a. Then, (E) has a unique solution $(v, \nabla P)$ with*

$$v \in L^\infty_{loc}(\mathbb{R}; B^1_{\infty,\infty}), \quad \omega \in L^\infty(\mathbb{R}; L^a \cap L^\infty), \quad and \quad P \in L^\infty_{loc}(\mathbb{R}; (L^1 + L^\infty_L)).$$

Moreover, v has a generalized flow ψ, in the sense of Theorem 3.7 page 128, and there exists a constant C such that

$$\psi(t) - \text{Id} \in C^{\exp(-C|t| \|\omega_0\|_{L^a \cap L^\infty})} \quad for \; all \; t \in \mathbb{R}.$$

Proof. Uniqueness follows from Theorem 7.17. To prove existence, we may smooth out the data. Let v^n be the (global) solution of the Euler system with mollified initial velocity $n^2 \chi(n\cdot) \star v_0$ [where χ is in $\mathcal{S}(\mathbb{R}^2)$ and has integral 1 over $\mathbb{R}^2$]. According to Theorem 7.25, v^n is global and smooth. It is not difficult to prove uniform estimates for v^n and ω^n in the desired spaces. Indeed, we have

$$\|\omega^n(t)\|_{L^a \cap L^\infty} = n^2 \|\chi(n\cdot) \star \omega_0\|_{L^a \cap L^\infty} \leq \|\omega_0\|_{L^a \cap L^\infty} \quad \text{for all } t \in \mathbb{R}, \quad (7.26)$$

and hence, according to Lemma 7.6,

$$\|\nabla v^n(t)\|_{B^0_{\infty,\infty}} \leq C\|\omega_0\|_{L^a \cap L^\infty} \quad \text{for all } t \in \mathbb{R}.$$

Also, note that combining the inequalities (7.24) and (7.26) provides us with uniform bounds for v^n in $L^\infty_{loc}(\mathbb{R}^+; B^1_{\infty,\infty})$.

Now, from the boundedness of the time derivatives in convenient function spaces, we get some compactness, and it is then possible to pass to the limit in the equation. This yields a solution (v, P) with the desired regularity.

Finally, since

$$\|v\|_{LL} \le C\|\omega\|_{L^a \cap L^\infty},$$

we can conclude, thanks to Theorem 3.7 and Lemma 3.8, that v has a flow ψ such that $\psi(t) - \mathrm{Id}$ is in $C^{\exp(-C|t|\|\omega_0\|_{L^a \cap L^\infty})}$ for all real numbers t.   □

Remark 7.28. The regularity result for the flow given in the above theorem is essentially optimal. Indeed, it turns out that if the initial vorticity ω_0 is supported in the square $[-1, 1]^2$, is odd with respect to the two axes, and equal to 1 in $[0, 1]^2$, then the corresponding flow ψ at time $t > 0$ does not belong to any C^α for $\alpha > e^{-t}$.

Finally, if the vorticity has some positive regularity, then the following result is available [see the definitions of F_p^s and $\sigma(s, t)$ on page 151].

Theorem 7.29. *Let v_0 be a divergence-free vector field, the vorticity of which is in $L^\infty \cap F_p^s$ for some $s \in]0, 1[$ and $p \in [1, \infty]$. Let v be a solution of the two-dimensional incompressible Euler system with data v_0.*

Then, for any $t > 0$, the vorticity at time t belongs to the space $F_p^{\sigma(s,t)}$.

Proof. Note that this corollary is obvious if $s > 2/p$. Indeed, in this case, due to the fact that $F_p^s \hookrightarrow B_{p,\infty}^s \hookrightarrow B_{\infty,\infty}^{s-2/p}$, the vector field v_0 has Hölder regularity greater than 1 so that the standard existence theorem for smooth solutions applies and the initial regularity is globally preserved by the flow.

Now, in the more interesting case where $s \le 2/p$, we can apply Theorem 3.32 to the vorticity equation (7.25). This yields the result.   □

7.3 The Inviscid Limit

In this section, we investigate the *inviscid limit* for the incompressible Navier–Stokes system. More precisely, given some initial divergence-free vector field v_0, we want to obtain as much information as possible on the convergence of the solution v_ν to the Navier–Stokes equations

$$(NS_\nu) \qquad \begin{cases} \partial_t v_\nu + v_\nu \cdot \nabla v_\nu - \nu \Delta v_\nu = -\nabla P_\nu \\ \mathrm{div}\, v_\nu = 0 \\ (v_\nu)_{|t=0} = v_0 \end{cases}$$

when the viscosity ν goes to 0.

7.3.1 Regularity Results for the Navier–Stokes System

We emphasize the fact that all the existence and uniqueness results which have been stated thus far remain true in the viscous case *for positive times*. Further, all the estimates pertaining to the solutions of (NS_ν) for sufficiently small ν are the same as in the case $\nu = 0$.

This may be easily proven by taking advantage of the results of Section 3.4 (in particular, Theorem 3.38) and of the following lemma.

Lemma 7.30. *Let $\nu \geq 0$, $a \in [1, \infty]$, and $T > 0$. Assume that Ω satisfies the following vorticity equation on $[0, T] \times \mathbb{R}^d$:*

$$\partial_t \Omega + v \cdot \nabla \Omega + \Omega \cdot Dv + {}^T Dv \cdot \Omega - \nu \Delta \Omega = 0, \qquad \Omega_{|t=0} = \Omega_0 \in L^a(\mathbb{R}^d).$$

For all $t \in [0, T]$, we then have:

- $\|\Omega(t)\|_{L^a} \leq \|\Omega_0\|_{L^a} \, e^{2 \int_0^t \|\nabla v\|_{L^\infty} \, dt'}$.
- $\|\Omega(t)\|_{L^a} \leq \|\Omega_0\|_{L^a} \, e^{C \int_0^t \|\Omega\|_{L^\infty} \, dt'}$, *if $1 < a < \infty$.*
- $\|\Omega(t)\|_{L^a} \leq \|\Omega_0\|_{L^a}$, *if $d = 2$.*

Proof. In contrast with the case $\nu = 0$, which was treated earlier in Lemma 7.22, we have to take care of the term $-\nu \Delta \Omega$.

We first assume that $2 \leq a < \infty$. Arguing by density, we can assume that Ω is smooth and decays at infinity so that, integrating by parts, we get

$$-\int_{\mathbb{R}^2} \Omega |\Omega|^{a-2} \cdot \Delta \Omega \, dx = (a-1) \int_{\mathbb{R}^2} |\Omega|^{a-2} |\nabla \Omega|^2 \, dx \geq 0.$$

Hence, the inequalities satisfied by $\|\Omega\|_{L^a}$ are exactly the same as if $\nu = 0$. This yields the result in the case $2 \leq a < \infty$. The case $a = \infty$ follows by passing to the limit, and the case $1 \leq a < 2$ follows by duality. □

7.3.2 The Smooth Case

In what follows, we shall focus on the rate of convergence of v_ν toward v for the L^2 norm. Of course, due to the uniform estimates which are available in $B^s_{p,r}$, interpolating provides convergence in all intermediate spaces.

In this subsection, we shall state that for smooth solutions, the rate of convergence (in any dimension) for $\|v_\nu - v\|_{L^2}$ is at least of order ν. Our result will be based on the following lemma.

Lemma 7.31. *Let $\mathcal{A}$ be a measurable function defined on $[0, T]$ and valued in $\mathcal{L}(L^2)$. Assume that for some positive integrable function K and almost every $t \in [0, T]$, we have*

$$\forall w \in L^2, \ -(\mathcal{A}(t)w \mid w)_{L^2} \leq K(t) \|w\|_{L^2}^2.$$

Let v be a time-dependent, divergence-free vector field with coefficients in $L^\infty([0,T]; C^{0,1})$, f be in $L^1([0,T]; L^2)$, and w_0 be in L^2. The system

$$\begin{cases} \partial_t w + v \cdot \nabla w + \mathcal{A}(t)w - \nu \Delta w = f \\ w_{|t=0} = w_0 \end{cases}$$

then has a unique solution w in $\mathcal{C}([0,T]; L^2)$ which, moreover, satisfies

$$\|w(t)\|_{L^2} \leq e^{\int_0^t K(t')\,dt'} \left(\|w_0\|_{L^2} + \int_0^t e^{-\int_0^{t'} K(t'')\,dt''} \|f(t')\|_{L^2}\,dt' \right).$$

Proof. The proof relies on the following energy estimate:

$$\frac{1}{2} \frac{d}{dt} \|w\|_{L^2}^2 + \nu \|\nabla w\|_{L^2}^2 \leq \|f\|_{L^2} \|w\|_{L^2} + K(t)\|w\|_{L^2}^2. \tag{7.27}$$

Following the proof of Theorem 4.4 page 172 and taking advantage of Lemma 3.3 page 125 then yields the result. $\qquad\square$

Theorem 7.32. *Let* v *(resp.,* v_ν*) be a* $C^{0,1}$*-solution of* (E) *[resp.,* (NS_ν)*] over* $[0,T]$*. Assume that* Δv *belongs to* $L^1([0,T]; L^2)$ *and that* $w_\nu \overset{\text{def}}{=} v_\nu - v$ *belongs to* $\mathcal{C}([0,T]; L^2)$*. We then have, for all* $t \in [0,T]$*,*

$$\|w_\nu(t)\|_{L^2} \leq e^{V(t)} \left(\|w_\nu(0)\|_{L^2} + \nu \int_0^t e^{-V(t')} \|\Delta v(t')\|_{L^2}\,dt' \right)$$

with $V(t) \overset{\text{def}}{=} \int_0^t \|\nabla v(t')\|_{L^\infty}\,dt'.$

Proof. The equation satisfied by w_ν reads

$$\partial_t w_\nu + v_\nu \cdot \nabla w_\nu + w_\nu \cdot \nabla v + \Pi(w_\nu, v + v_\nu) - \nu \Delta w_\nu = \nu \Delta v.$$

Note that since $v + v_\nu$ is in $L^\infty([0,T]; C^{0,1})$, Lemma 7.8 ensures that

$$w_\nu \longmapsto w_\nu \cdot \nabla v + \Pi(w_\nu, v + v_\nu)$$

is a linear self-map on $L^1([0,T]; L^2)$. In addition, we have

$$-(w_\nu \cdot \nabla v \mid w_\nu)_{L^2} \leq \|\nabla v\|_{L^\infty} \|w_\nu\|_{L^2}^2,$$

and, because $\Pi(w_\nu, v + v_\nu)$ is a gradient and $\operatorname{div} w_\nu = 0$,

$$(\Pi(w_\nu, v + v_\nu) \mid w_\nu)_{L^2} = 0.$$

Applying Lemma 7.31 thus yields the desired inequality. $\qquad\square$

7.3.3 The Rough Case

Owing to Lemma 7.30, Theorem 7.27 also holds for the two-dimensional Navier–Stokes equation (NS_ν). More precisely, we can prove the following statement.

Theorem 7.33. *Let v_0 be a divergence-free vector field in $B^1_{\infty,\infty}(\mathbb{R}^2)$ with vorticity ω_0 in $L^2 \cap L^\infty$. Then, for all $\nu > 0$, the system (NS_ν) with data v_0 has a unique solution $(v_\nu, \nabla P_\nu)$ with (uniformly with respect to ν)*

$$v_\nu \in L^\infty_{loc}(\mathbb{R}^+; B^1_{\infty,\infty}), \quad \omega_\nu \in L^\infty(\mathbb{R}^+; L^2 \cap L^\infty), \quad and \quad P_\nu \in L^\infty_{loc}(\mathbb{R}^+; (L^1 + L^\infty_L)).$$

In this subsection, we investigate the rate of convergence of (NS_ν) toward (E) for (not necessarily two-dimensional) solutions having the above regularity. We shall see that the rate strongly depends on the regularity of the inviscid solution. We first establish that the rate is $\nu^{\frac{1}{2}}$ if the inviscid solution is Lipschitz.

Theorem 7.34. *Let v (resp., v_ν) be a $B^1_{\infty,\infty}$ solution of (E) [resp., (NS_ν)] over $[0,T]$, and let $w_\nu \overset{def}{=} v_\nu - v$. If, in addition,*

$$\nabla v \in L^1([0,T]; L^\infty) \cap L^2([0,T]; L^2)$$

and w_ν is in $\mathcal{C}([0,T]; L^2)$, then we have, for any t in $[0,T]$,

$$\|w_\nu(t)\|^2_{L^2} + \nu \int_0^t \|\nabla w_\nu\|^2_{L^2} \, dt' \le e^{2V(t)} \left(\|w_\nu(0)\|^2_{L^2} + \nu \int_0^t e^{-2V(t')} \|\nabla v\|^2_{L^2} \, dt' \right).$$

Proof. The starting point of the proof is the inequality (7.27) with $K(t) = V(t)$ and $f = -\nu\Delta v$. Now, integrating by parts and using Young's inequality, we note that

$$\nu \int_{\mathbb{R}^d} w_\nu \cdot \Delta v \, dx = -\nu \int_{\mathbb{R}^d} \nabla w_\nu \cdot \nabla v \, dx \le \frac{\nu}{2} \|\nabla v\|^2_{L^2} + \frac{\nu}{2} \|\nabla w_\nu\|^2_{L^2}.$$

Plugging this into the equality (7.27) and integrating, we thus obtain

$$\|w_\nu(t)\|^2_{L^2} + \nu \int_0^t \|\nabla w_\nu\|^2_{L^2} \, dt' \le \|w_\nu(0)\|^2_{L^2}$$
$$+ 2 \int_0^t \|\nabla v\|_{L^\infty} \|w_\nu\|^2_{L^2} \, dt' + \nu \int_0^t \|\nabla v\|^2_{L^2} \, dt'.$$

Using Gronwall's lemma then leads to the desired inequality. □

In the case where, in addition, the limit vorticity Ω belongs to the homogeneous Besov space $\dot{B}^\alpha_{2,\infty}$ for some $\alpha \in \,]0,1[$, we get a better rate of convergence, namely, $\nu^{\frac{1+\alpha}{2}}$.

Theorem 7.35. *Under the assumptions of Theorem 7.34, assume, in addition, that Ω is in $L^{\frac{2}{1+\alpha}}([0,T]; \dot{B}^{\alpha}_{2,\infty})$ for some $\alpha \in]0,1[$. We then have*

$$\|w_{\nu}(t)\|^{\frac{2}{1+\alpha}}_{L^2} \le e^{V(t)}\left(\|w_{\nu}(0)\|^{\frac{2}{1+\alpha}}_{L^2} + C\nu\int_0^t e^{-V(t')}\|\Omega\|^{\frac{2}{1+\alpha}}_{\dot{B}^{\alpha}_{2,\infty}}\,dt'\right).$$

Proof. The duality result stated in Proposition 2.29 page 70 ensures that

$$\nu\int_{\mathbb{R}^2} w_{\nu}\cdot\Delta v\,dx = -\nu\int_{\mathbb{R}^2}\nabla w_{\nu}\cdot\nabla v\,dx$$
$$\le C\nu\|\nabla v\|_{\dot{B}^{\alpha}_{2,\infty}}\|\nabla w_{\nu}\|_{\dot{B}^{-\alpha}_{2,1}}.$$

Using real interpolation (see Proposition 2.22 page 65) and the fact that the map $\Omega \to \nabla v$ is homogeneous of degree 0, we thus get

$$\nu\int_{\mathbb{R}^2} w_{\nu}\cdot\Delta v\,dx \le C\nu\|\Omega\|_{\dot{B}^{\alpha}_{2,\infty}}\|w_{\nu}\|^{\alpha}_{L^2}\|\nabla w_{\nu}\|^{1-\alpha}_{L^2}$$
$$\le C\nu\|\Omega\|^{\frac{2}{1+\alpha}}_{\dot{B}^{\alpha}_{2,\infty}}\|w_{\nu}\|^{\frac{2\alpha}{1+\alpha}}_{L^2} + \frac{\nu}{2}\|\nabla w_{\nu}\|^2_{L^2}.$$

Plugging this latter inequality into the inequality (7.27) and then applying Gronwall's lemma completes the proof of the theorem. $\qquad\square$

Remark 7.36. Appealing to the characterization of Besov spaces in terms of finite differences, it is not difficult to prove that the characteristic function of any bounded domain Ω_0 belongs to $\dot{B}^{\frac{1}{2}}_{2,\infty}$. In the next section, we shall state (in the two-dimensional case) that if the initial vorticity is the characteristic function of a $C^{1,r}$ domain, then the corresponding solution v is Lipschitz. Hence, the above theorem states that the rate of convergence for the L^2 norm is of order $\nu^{\frac{3}{4}}$. This rate proves to be optimal in the case of a circular domain.

If the limit vector field is no longer Lipschitz, then the $\nu^{\frac{1}{2}}$ rate of convergence is likely to coarsen, as we see in the following result.

Theorem 7.37. *Let v_0 be a two-dimensional divergence-free vector field in $B^1_{\infty,\infty}$ with vorticity ω_0 in $L^2 \cap L^{\infty}$. Denote by v (resp., v_{ν}) the corresponding global solution of (E) [resp., (NS_{ν})] and define $w_{\nu} \overset{def}{=} v_{\nu} - v$.*
   Then, w_{ν} is in $\mathcal{C}(\mathbb{R}^+; L^2)$ and satisfies, for some universal constant C,

$$\|w_{\nu}\|_{L^{\infty}_T(L^2)} \le (\nu T)^{\frac{1}{2}\exp(-C\|\omega^0\|_{L^2\cap L^{\infty}}T)}\|\omega_0\|_{L^2\cap L^{\infty}}e^{1-\exp(-C\|\omega_0\|_{L^2\cap L^{\infty}}T)}$$

whenever $(\nu T)^{\frac{1}{2}\exp(-C\|\omega^0\|_{L^2\cap L^{\infty}}T)}e^{1-\exp(-C\|\omega_0\|_{L^2\cap L^{\infty}}T)} \le 1$.

Proof. Let us bound the right-hand side of (7.27) as in the proof of Theorem 7.32. Because, in dimension two,

$$\|\nabla v\|^2_{L^2} = \|\omega\|^2_{L^2} \le \|\omega_0\|^2_{L^2},$$

we get

$$\frac{d}{dt}\|w_\nu\|_{L^2}^2 + \nu\|\nabla w_\nu\|_{L^2}^2 \leq \nu\|\omega_0\|_{L^2}^2 + \left|\int_{\mathbb{R}^2} (w_\nu \cdot \nabla v) \cdot w_\nu \, dx\right|. \qquad (7.28)$$

By combining Hölder's inequality and Proposition 7.5, we get, for all $a \in [2, \infty[$,

$$\left|\int_{\mathbb{R}^2} (w_\nu \cdot \nabla v) \cdot w_\nu \, dx\right| \leq Ca\|\omega\|_{L^a} \|w_\nu\|_{L^\infty}^{\frac{2}{a}} \|w_\nu\|_{L^2}^{2-\frac{2}{a}}.$$

Recall that $\|\omega\|_{L^a} \leq \|\omega_0\|_{L^a}$. Further, using the fact that $w_\nu = \Delta_{-1}w_\nu + (\mathrm{Id} - \Delta_1)w_\nu$ and Lemma 7.6, we easily get that

$$\|w_\nu\|_{L^\infty} \leq C(\|w_\nu\|_{L^2} + \|\omega_0\|_{L^\infty}).$$

So, finally, for all $a \in [2, \infty[$,

$$\frac{d}{dt}\|w_\nu\|_{L^2}^2 \leq \nu\|\omega_0\|_{L^2}^2 + Ca\|\omega_0\|_{L^2 \cap L^\infty}\|w_\nu\|_{L^2}^2 + Ca\|\omega_0\|_{L^2 \cap L^\infty}^{1+\frac{2}{a}}\|w_\nu\|_{L^2}^{2-\frac{2}{a}}.$$

Fix some small positive δ and define[7]

$$\delta_\nu(t) \overset{\text{def}}{=} \frac{\|w_\nu(t)\|_{L^2}^2}{\|\omega_0\|_{L^2 \cap L^\infty}^2} + \delta.$$

Assuming that $\delta_\nu \leq 1$ on $[0, T]$, the previous inequality yields

$$\delta_\nu'(t) \leq \nu + 2Ca\|\omega_0\|_{L^2 \cap L^\infty}(\delta_\nu(t))^{1-\frac{1}{a}}.$$

So, choosing $a = 2 - 2\log \delta_\nu(t)$, after performing a time integration (up to a change of C), we get

$$\delta_\nu(t) \leq \nu t + \delta + C\|\omega_0\|_{L^2 \cap L^\infty} \int_0^t \delta_\nu(t')(2 - \log \delta_\nu(t')) \, dt'.$$

We note that $\mu(r) \overset{\text{def}}{=} r(2 - \log r)$ is an Osgood modulus of continuity. Hence, applying Lemma 3.4 page 125 and having δ tend to 0 completes the proof. □

7.4 Viscous Vortex Patches

The original *vortex patch problem* has been addressed for the two-dimensional incompressible Euler system. Assuming that the initial vorticity ω_0 is a vortex patch (that is the characteristic function of some bounded domain D_0) Yudovich's theorem ensures that (E) has a global solution with bounded vorticity. Since, in addition, that solution has a flow ψ, and ω satisfies (7.5), we

---

[7] We rule out the trivial case where $\omega_0 \equiv 0$.

may deduce that $\omega(t)$ is the characteristic function of the domain transported by the flow:

$$\omega(t) = \mathbb{1}_{D_t} \quad \text{with} \quad D_t = \psi_t(D_0).$$

Note that having ω bounded does not imply that v is Lipschitz, so ψ need not be Lipschitz either. Hence, the above relation does not guarantee that the initial smoothness of the patch is preserved by the flow. Nevertheless, we shall see that if ∂D_0 is a simple $C^{1,r}$ curve for some $r \in]0,1[$, then ∂D_t remains so for all time.

The purpose of the present section is twofold. First, we shall study to what extent the global persistence of vortex patches remains true for viscous fluids, that is, when v solves the two-dimensional incompressible Navier–Stokes equation (NS_ν). Second, we shall study the inviscid limit for vortex-patch-like structures or, more generally, for data having *striated regularity* in a sense that we shall explain below.

7.4.1 Results Related to Striated Regularity

Note that if $\omega = \mathbb{1}_D$, where D is a $C^{1,r}$ simply connected bounded domain of $\mathbb{R}^2$, then ω is "more regular" in the direction which is tangent to ∂D. Indeed, for any smooth vector field X which is tangent to ∂D, we have

$$\partial_X \omega \stackrel{\text{def}}{=} X^1 \partial_1 \omega + X^2 \partial_2 \omega = 0.$$

Since

$$\operatorname{div}(X\omega) - \partial_X \omega = \omega \operatorname{div} X,$$

we can deduce that if X is sufficiently smooth and has bounded divergence, then $\operatorname{div}(X\omega)$ is in L^∞ (instead of being a linear combination of derivatives of L^∞ functions if ω is just bounded). This motivates the following definition.

Definition 7.38. *A family $(X_\lambda)_{\lambda \in \Lambda}$ of vector fields over $\mathbb{R}^2$ is said to be* nondegenerate *whenever*

$$I(X) \stackrel{\text{def}}{=} \inf_{x \in \mathbb{R}^d} \sup_{\lambda \in \Lambda} |X_\lambda(x)| > 0.$$

Let $r \in]0,1[$ and $(X_\lambda)_{\lambda \in \Lambda}$ be a nondegenerate family of $B^r_{\infty,\infty}$ vector fields over $\mathbb{R}^2$. A bounded function ω is said to be in the function space C^r_X if it satisfies

$$\|\omega\|_{C^r_X} \stackrel{\text{def}}{=} \sup_{\lambda \in \Lambda} \left(\frac{\|\omega\|_{L^\infty} \|X_\lambda\|_{B^r_{\infty,\infty}} + \|\operatorname{div}(X_\lambda \omega)\|_{B^{r-1}_{\infty,\infty}}}{I(X)} \right) < \infty.$$

Proving that C^r_X is a Banach space is left to the reader as an exercise.

Assuming that the initial vorticity ω_0 is in some space $C_{X_0}^r$, it seems reasonable (at least in the inviscid case, where ω is constant along the trajectories) that $\omega(t)$ remains in $C_{X(t)}^r$, where $X(t)$ is the family *transported by the flow of* v, that is, $X(t) = \left(X_\lambda(t)\right)_{\lambda \in \Lambda}$ with

$$X_\lambda(t) = \partial_{X_0}\psi(t) \quad \text{for all} \quad \lambda \in \Lambda.$$

The following theorem states that this is indeed the case (even in the viscous case), and that properties of striated regularity are conserved in the inviscid limit.

Theorem 7.39. *Let r be in $]0,1[$ and $(X_{0,\lambda})_{1\leq\lambda\leq m}$ be a nondegenerate family of $B_{\infty,\infty}^r$ vector fields over $\mathbb{R}^2$. Let v_0 be a $C^{0,1}$ divergence-free vector field with vorticity ω_0 in $C_{X_0}^r \cap L^2$. Then, for all positive ν, the system (NS_ν) [resp., the system (E)] has a unique global solution v_ν (resp., v) in $L_{loc}^\infty(\mathbb{R}^+; C^{0,1})$ with vorticity in $L^\infty(\mathbb{R}^+; L^2)$, and there exists a constant K depending only on the data and a universal constant C such that for all $t \geq 0$ we have*

$$\|\nabla v(t)\|_{L^\infty} \leq Ke^{C\|\omega_0\|_{L^\infty}t} \quad \text{and} \quad \|\nabla v_\nu(t)\|_{L^\infty} \leq K(1+\nu t)e^{C\|\omega_0\|_{L^\infty}t}.$$

Moreover, the family $(X_{\nu,\lambda})_{1\leq\lambda\leq m}$ [resp., $(X_\lambda)_{1\leq\lambda\leq m}$] of time-dependent vector fields transported by the flow ψ_ν (resp., ψ) of v_ν (resp., v) remains $B_{\infty,\infty}^r$ and nondegenerate for all $t \in \mathbb{R}^+$, and $\omega_\nu(t)$ [resp., $\omega(t)$] belongs to $C_{X_\nu(t)}^r$ (resp., $C_{X(t)}^r$). In addition, for any bounded subsets I and J of $[0,\infty[$ and $]0,\infty[$, there exists a constant C such that

$$\sup_{t\in I} \|\omega(t)\|_{C_{X(t)}^r} + \sup_{t\in I}\sup_{\nu\in J} \|\omega_\nu(t)\|_{C_{X_\nu(t)}^r} \leq C,$$

and the following convergence results hold true:

- $v_\nu \to v$ *and* $\psi_\nu - \psi \to 0$ *in* $L_{loc}^\infty(\mathbb{R}^+; B_{\infty,\infty}^{1-\varepsilon})$ *for all* $\varepsilon > 0$.
- $X_{\lambda,\nu} \to X_\lambda$ *and* $\partial_{X_{\lambda,\nu}}\psi_\nu \to \partial_{X_\lambda}\psi$ *in* $L_{loc}^\infty(\mathbb{R}^+; B_{\infty,\infty}^{r'})$ *for all* $r' < r$.
- $\partial_{X_{\lambda,\nu}}\omega_\nu \to \partial_{X_\lambda}\omega$ *in* $L_{loc}^\infty(\mathbb{R}^+; B_{\infty,\infty}^{r'-1})$ *for all* $r' < r$.

7.4.2 A Stationary Estimate for the Velocity Field

One of the keys to the proof of Theorem 7.39 is the following estimate, which states that any velocity field with striated vorticity is Lipschitz and may be bounded in terms of $\|\omega\|_{L^\infty}$ and of the logarithm of $\|\omega\|_{C_X^r}$.

Theorem 7.40. *Let r be in $]0,1[$ and $(X_\lambda)_{1\leq\lambda\leq m}$ be a nondegenerate family of $B_{\infty,\infty}^r$ vector fields over $\mathbb{R}^2$. Let v be a divergence-free vector field over $\mathbb{R}^2$ with vorticity ω in C_X^r. Assume, in addition, that $v \in L^q$ for some $q \in [1,\infty]$ or that $\nabla v \in L^p$ for some finite p. There then exists a constant C, depending only on m, p, and r, and such that*

$$\|\nabla v\|_{L^\infty} \leq C\left(\min\left(\|v\|_{L^q}, \|\omega\|_{L^p}\right) + \|\omega\|_{L^\infty} \log\left(e + \frac{\|\omega\|_{C_X^r}}{\|\omega\|_{L^\infty}}\right)\right).$$

Proof. We will first give a sketch of a proof in the flat case. We thus assume that the family X reduces to the unique vector field ∂_1. Having $\omega \in C_X^r$ then means that $\omega \in L^\infty$ and $\partial_1 \omega \in B_{\infty,\infty}^{r-1}$. This obviously entails that all the second order derivatives of ω except $\partial_2^2 \omega$ are in $B_{\infty,\infty}^{r-2}$. From the relation

$$\nabla v = (-\Delta)^{-1} \nabla \nabla^\perp \omega,$$

we thus discover that all the components of ∇v except $\partial_2 v^1$ are in $B_{\infty,\infty}^r$. Now, $\partial_2 v^1 = \partial_1 v^2 - \omega$, so, owing to the fact that $\omega \in L^\infty$, this last component is bounded.

We now turn to the proof of the theorem in the general case. According to the Biot–Savart law, we have

$$\nabla v = \Delta_{-1} \nabla v - \Lambda^{-2} \nabla \nabla^\perp \omega \quad \text{with} \quad \Lambda^{-2} \overset{\text{def}}{=} |D|^{-2} (\text{Id} - \Delta_{-1}).$$

Bounding the first term according to Lemma 7.6, we thus get

$$\|\nabla v\|_{L^\infty} \le C \Big(\min(\|v\|_{L^q}, \|\omega\|_{L^p}) + \sum_{i,j} \|\partial_i \partial_j \Lambda^{-2} \omega\|_{L^\infty} \Big).$$

We will temporarily assume that the following lemma holds.

Lemma 7.41. *There exist some functions a_{ij} and $b_{ij}^{k\lambda}$ ($1 \le i,j,k \le 2$ and $1 \le \lambda \le m$) in $B_{\infty,\infty}^r$ and a universal constant C such that*

$$\forall (x,\xi) \in \mathbb{R}^2 \times \mathbb{R}^2, \quad \xi_i \xi_j = a_{ij}(x)|\xi|^2 + \sum_{\lambda,k} b_{ij}^{k\lambda}(x) \xi_k (X_\lambda(x) \cdot \xi), \quad (7.29)$$

$$\|b_{ij}^{k\lambda}\|_r \le C \frac{m^2}{I(X)} \left(\frac{\sup_\lambda \|X_\lambda\|_{B_{\infty,\infty}^r}}{I(X)} \right)^8, \quad (7.30)$$

$$\|b_{ij}^{k\lambda} X_\lambda\|_{L^\infty} \le C, \quad \text{and} \quad \|a_{ij}\|_{L^\infty} \le 1. \quad (7.31)$$

We then have, for all $(x,\xi) \in \mathbb{R}^2 \times \mathbb{R}^2$ and $1 \le i,j \le 2$,

$$\frac{\xi_i \xi_j (1 - \chi(\xi))}{|\xi|^2} \widehat{\omega}(\xi) = a_{ij}(x)(1 - \chi(\xi))\widehat{\omega}(\xi)$$

$$+ \sum_{\lambda,k,\ell} \frac{\xi_k \xi_\ell (1 - \chi(\xi))}{|\xi|^2} b_{ij}^{k\lambda}(x) X_\lambda^\ell(x) \widehat{\omega}(\xi).$$

Evaluating the Fourier transform (with respect to x) of the above equality in ξ and then applying the inverse Fourier transform, we thus get

$$\partial_i \partial_j \Lambda^{-2} \omega = (\text{Id} - \Delta_{-1})(a_{ij} \omega) + \sum_{\lambda,k,\ell} \Lambda^{-2} \partial_k \partial_\ell (b_{ij}^{k\lambda} X_\lambda^\ell \omega).$$

On the one hand, according to the inequality (7.31), the first term on the right-hand side may be bounded by $C \|\omega\|_{L^\infty}$. On the other hand, the terms in the sum may be estimated by using the logarithmic interpolation inequality stated in Proposition 2.104 page 116:

$$\left\|\Lambda^{-2}\partial_k\partial_\ell\big(b_{ij}^{k\lambda}X_\lambda^\ell\omega\big)\right\|_{L^\infty} \le C\|\Lambda^{-2}\partial_k\partial_\ell\big(b_{ij}^{k\lambda}X_\lambda^\ell\omega\big)\|_{B_{\infty,\infty}^0}$$
$$\times \log\left(e + \frac{\|\Lambda^{-2}\partial_k\partial_\ell\big(b_{ij}^{k\lambda}X_\lambda^\ell\omega\big)\|_{B_{\infty,\infty}^r}}{\|\Lambda^{-2}\partial_k\partial_\ell\big(b_{ij}^{k\lambda}X_\lambda^\ell\omega\big)\|_{B_{\infty,\infty}^0}}\right).$$

Because the operator $\Lambda^{-2}\partial_k\partial_\ell$ is an S^0-multiplier in the sense of Proposition 2.78, we have, by virtue of (7.31),

$$\|\Lambda^{-2}\partial_k\partial_\ell\big(b_{ij}^{k\lambda}X_\lambda^\ell\omega\big)\|_{B_{\infty,\infty}^0} \le C\left\|b_{ij}^{k\lambda}X_\lambda^\ell\omega\right\|_{L^\infty} \le C\left\|\omega\right\|_{L^\infty}.$$

As the operator $\Lambda^{-2}\partial_k$ is an S^{-1}-multiplier and for any $\alpha > 0$, the function $t \mapsto t\log(e+\alpha/t)$ is increasing, we thus get

$$\left\|\partial_i\partial_j\Lambda^{-2}\omega\right\|_{L^\infty} \le C\left\|\omega\right\|_{L^\infty} \sum_{\lambda,k,\ell} \log\left(e + \frac{\|\partial_\ell\big(b_{ij}^{k\lambda}X_\lambda^\ell\omega\big)\|_{B_{\infty,\infty}^{r-1}}}{\|\omega\|_{L^\infty}}\right).$$

We can now write that [see the definition of T' in (2.42) page 103]

$$\partial_\ell(b_{ij}^{k\lambda}X_\lambda^\ell\omega) = \partial_\ell T'_{X_\lambda^\ell\omega}b_{ij}^{k\lambda} + T_{\partial_\ell(X_\lambda^\ell\omega)}b_{ij}^{k\lambda} + T_{X_\lambda^\ell\omega}\partial_\ell b_{ij}^{k\lambda},$$

so applying Theorems 2.82 and 2.85 gives

$$\|\partial_\ell\big(b_{ij}^{k\lambda}X_\lambda^\ell\omega\big)\|_{B_{\infty,\infty}^{r-1}} \le C\Big(\|X_\lambda\omega\|_{L^\infty}\|b_{ij}^{k\lambda}\|_{B_{\infty,\infty}^r} + \|\operatorname{div}(X_\lambda\omega)\|_{B_{\infty,\infty}^{r-1}}\|b_{ij}^{k\lambda}\|_{L^\infty}\Big).$$

As the functions $b_{ij}^{k\lambda}$ satisfy the inequality (7.30), we get the desired inequality. $\qquad\square$

Proof of Lemma 7.41. We first state a local version of the lemma pertaining to only one of the vector fields, X_λ. For that purpose, we introduce the open set

$$U_\lambda = \big\{x \in \mathbb{R}^2, \, |X_\lambda(x)| > I(X)/2\big\}.$$

We claim that for any $\lambda \in \{1,\dots,m\}$ and $(i,j) \in \{1,2\}^2$, there exist some functions $\widetilde{b}_{ij}^{k\lambda}$ which are homogeneous of degree 3 with respect to the components of X_λ and such that for all $x \in U_\lambda$ and $\xi \in \mathbb{R}^2$, we have

$$\xi_i\xi_j = \frac{Y_\lambda^i(x)Y_\lambda^j(x)}{|X_\lambda(x)|^2}|\xi|^2 + \sum_k \frac{\widetilde{b}_{ij}^{k\lambda}(x)}{|X_\lambda(x)|^4}\xi_k(X_\lambda(x)\cdot\xi) \quad \text{with} \quad Y_\lambda \overset{\text{def}}{=} X_\lambda^\perp. \quad (7.32)$$

To prove this identity, we may set, for $x \in U_\lambda$,

$$\widetilde{a}_{ij}(x) = |X_\lambda(x)|^{-2}q_{ij}(Y_\lambda(x)) \quad \text{with} \quad q_{ij}(\xi) = \xi_i\xi_j.$$

Of course, we have $q_{ij}(Y_\lambda(x)) - \widetilde{a}_{ij}(x)|X_\lambda(x)|^2 = 0$, so if we introduce the matrix Q_{ij} associated with the quadratic form q_{ij} and

$$A(x) \overset{\text{def}}{=} \frac{1}{|X_\lambda(x)|^2} \begin{pmatrix} X_\lambda^1(x) & -X_\lambda^2(x) \\ X_\lambda^2(x) & X_\lambda^1(x) \end{pmatrix},$$

then we discover that, owing to $^T A(x) A(x) = I_2/|X_\lambda(x)|^2$, we have

$$^T A(x) Q_{ij} A(x) - \tilde{a}_{ij}(x) \,{}^T A(x) A(x) = \frac{1}{|X_\lambda(x)|^4} \begin{pmatrix} m_{11}(x) & m_{12}(x) \\ m_{21}(x) & 0 \end{pmatrix}.$$

The coefficients $m_{11}(x)$, $m_{12}(x)$, and $m_{21}(x)$ are homogeneous of degree 2 with respect to the components of X_λ. So, applying the above equality to the vector $\eta = A^{-1}(x)\xi$, we find that for all $x \in U_\lambda$, we have

$$\xi_i \xi_j - \frac{q_{ij}(Y_\lambda(x))}{|X_\lambda(x)|^2}|\xi|^2 = \sum_k \frac{\tilde{b}_{ij}^{k\lambda}(x)}{|X_\lambda(x)|^4}\big(\xi_k(X_\lambda(x) \cdot \xi)\big)$$

with $\tilde{b}_{ij}^{k\lambda}$ homogeneous of degree 3 with respect to the components of X_λ. This yields (7.32).

In order to complete the proof of Lemma 7.41, it suffices to construct a family of smooth functions $(\phi_\lambda)_{1 \le \lambda \le m}$ satisfying:

(i) $\displaystyle\sum_{1 \le \lambda \le m} \phi_\lambda \equiv 1.$

(ii) Supp $\phi_\lambda \subset U_\lambda$.

(iii) $\|\phi_\lambda\|_{B^r_{\infty,\infty}} \le Cm\left(\dfrac{\sup_\lambda \|X_\lambda\|_{B^r_{\infty,\infty}}}{I(X)}\right).$

Indeed, we can set

$$b_{ij}^{k\lambda}(x) = \frac{\tilde{b}_{ij}^{k\lambda}}{|X_\lambda(x)|^4}\phi_\lambda \quad \text{and} \quad a_{ij}(x) = \sum_\lambda \frac{q_{ij}(Y_\lambda(x))}{|X_\lambda(x)|^2}\phi_\lambda.$$

We therefore construct the family $(\phi_\lambda)_{1 \le \lambda \le m}$. First, we introduce a family $(\chi_\varepsilon)_{\varepsilon > 0}$ of mollifiers, the sets

$$F_\lambda \overset{\text{def}}{=} \{x \in \mathbb{R}^2, |X_\lambda(x)| \ge I(X)\} \quad \text{and} \quad F_\lambda^\epsilon \overset{\text{def}}{=} \{x \in \mathbb{R}^2, d(x, F_\lambda) \le \epsilon\},$$

and the functions

$$\phi_\lambda^\epsilon \overset{\text{def}}{=} 1_{F_\lambda^{\epsilon/2}} \star \chi_{\epsilon/2} \quad \text{and} \quad \phi_\lambda = \phi_\lambda^\epsilon \prod_{j < \lambda}(1 - \phi_j^\epsilon).$$

Because $\mathbb{R}^2 = \bigcup_{1 \le \lambda \le m} F_\lambda$ and

$$1 - \sum_{1 \le \lambda \le m} \phi_\lambda = \prod_{1 \le \lambda \le m}(1 - \phi_\lambda^\epsilon),$$

it is not difficult to check that the family $(\phi_\lambda)_{1\leq\lambda\leq m}$ satisfies *(i)*.

Now, if we take $\epsilon = (2(I(X))^{-1} \sup_\lambda \|X_\lambda\|_{B^r_{\infty,\infty}})^{-\frac{1}{r}}$, then the property *(iii)* is also satisfied. In addition, for all $x \in F^\epsilon_\lambda$, we have

$$|X_\lambda|(x) \geq I(X) - \epsilon^r\|X_i\|_{B^r_{\infty,\infty}} \geq I(X)/2.$$

Hence, Supp $\phi_\lambda \subset F^\epsilon_\lambda \subset U_\lambda$ and *(ii)* is verified.   □

7.4.3 Uniform Estimates for Striated Regularity

The basic idea is that the conormal regularity of the vorticity also provides conormal regularity for the velocity. We shall make this more precise with the following lemma.

Lemma 7.42. *For any $r \in]0,1[$ there exists a constant C such that the following estimates hold true:*

$$\| \operatorname{div}(X\omega) - T_{X^j}\partial_j\omega\|_{B^{r-1}_{\infty,\infty}} \leq C\|X\|_{B^r_{\infty,\infty}}\|\omega\|_{L^\infty}, \tag{7.33}$$

$$\|\partial_X v\|_{B^r_{\infty,\infty}} \leq C\big(\|\nabla v\|_{L^\infty}\|X\|_{B^r_{\infty,\infty}} + \|\operatorname{div}(X\omega)\|_{B^{r-1}_{\infty,\infty}}\big). \tag{7.34}$$

Proof. To prove the inequality (7.33) it suffices to use the fact that

$$\operatorname{div}(X\omega) - T_{X^j}\partial_j\omega = \operatorname{div}\big(T'_\omega X\big) + [\partial_j, T_{X^j}]\omega$$

and to take advantage of continuity results stated in Chapter 2 for the paraproduct and the remainder, and of Lemma 2.99 page 111.

We now turn to the proof of (7.34). The Biot–Savart law states that

$$\forall (i,j) \in \{1,2\}^2, \ \partial_j v^i = -(-\Delta)^{-1}\partial^\perp_i \partial_j\omega \quad \text{with} \quad \partial^\perp_1 = \partial_2 \quad \text{and} \quad \partial^\perp_2 = -\partial_1.$$

Therefore, using Bony decomposition, we get

$$\partial_X v^i = T'_{\partial_j v^i} X^j - (-\Delta)^{-1}\partial^\perp_i T_{X^j}\partial_j\omega + [(-\Delta)^{-1}\partial^\perp_i, T_{X^j}]\partial_j\omega.$$

Since the multiplier $(-\Delta)^{-1}\partial^\perp_i$ is homogeneous of degree -1, combining Propositions 2.82 and 2.85, the inequality (7.33), and Lemma 2.99 yields the inequality (7.34).   □

Proposition 7.43. *Let $r \in]0,1[$. There exists a constant C such that for all $\nu \geq 0$, any smooth vector field v satisfying (NS_ν) on $[0,T]$, and any time-dependent vector field X transported by the flow of v, we have, for all t in $[0,T]$,*

$$\| \operatorname{div}(X\omega)(t)\|_{B^{r-1}_{\infty,\infty}} \leq Ce^{C\nu t^2\|\omega_0\|_{L^\infty}}e^{CV(t)}$$
$$\times \Big((1+\nu t)\|\omega_0\|_{L^\infty}\|X_0\|_{B^r_{\infty,\infty}} + \|\operatorname{div}(X_0\omega_0)\|_{B^{r-1}_{\infty,\infty}}\Big), \tag{7.35}$$

$$\|X(t)\|_{B^r_{\infty,\infty}} \leq e^{CV(t)}\Big(\|X_0\|_{B^r_{\infty,\infty}} + Cte^{C\nu t^2\|\omega_0\|_{L^\infty}}\big(\|\operatorname{div}(X_0\omega_0)\|_{B^{r-1}_{\infty,\infty}}$$
$$+ (1+\nu t)\|\omega_0\|_{L^\infty}\|X_0\|_{B^r_{\infty,\infty}}\big)\Big) \tag{7.36}$$

with $V(t) \overset{def}{=} \displaystyle\int_0^t \|\nabla v(t')\|_{L^\infty} \, dt'$.

Proof. We first consider the evolution equation for X. We have

$$D_t X = \partial_X v \quad \text{with} \quad D_t \overset{\text{def}}{=} \partial_t + v \cdot \nabla.$$

Hence, according to Proposition 3.14 page 133, we have, for some constant C depending only on r,

$$\|X(t)\|_{B^r_{\infty,\infty}} \leq e^{CV(t)} \left(\|X_0\|_{B^r_{\infty,\infty}} + \int_0^t e^{-CV(t')} \|\partial_X v\|_{B^r_{\infty,\infty}} \, dt' \right). \quad (7.37)$$

The right-hand side may be bounded by taking advantage of the inequality (7.34). Applying Gronwall's lemma, we conclude that

$$\|X(t)\|_{B^r_{\infty,\infty}} \leq e^{CV(t)} \left(\|X_0\|_{B^r_{\infty,\infty}} + C \int_0^t e^{-CV(t')} \|\operatorname{div}(X\omega)\|_{B^{r-1}_{\infty,\infty}} \, dt' \right). \quad (7.38)$$

In order to bound $\|\operatorname{div}(X\omega)\|_{B^{r-1}_{\infty,\infty}}$, we may write an evolution equation for $\operatorname{div}(X\omega)$. Since $\operatorname{div}(X\omega) = \partial_X \omega + \omega \operatorname{div} X$, we have

$$D_t \operatorname{div}(X\omega) = D_t \partial_X \omega + D_t \omega \operatorname{div} X + \omega D_t \operatorname{div} X.$$

Given that

- the vector fields D_t and ∂_X commute,
- due to $\operatorname{div} v = 0$, we have $D_t \operatorname{div} X = 0$,
- the vorticity satisfies $D_t \omega = \nu \Delta \omega$,

we discover, after a few computations, that

$$D_t \operatorname{div}(X\omega) - \nu \Delta \operatorname{div}(X\omega) = \nu \operatorname{div}\left(X \Delta \omega - \Delta(X\omega) \right).$$

We write $X \Delta \omega - \Delta(X\omega) = F + G$ with

$$F \overset{\text{def}}{=} [T_X, \Delta]\omega + T_{\Delta\omega} X - \Delta T'_\omega X \quad \text{and} \quad G \overset{\text{def}}{=} R(X, \Delta\omega).$$

On the one hand, applying Theorem 3.38 page 162 yields, for all $t \in [0, T]$,

$$\| \operatorname{div}(X\omega)\|_{L^\infty_t(B^{r-1}_{\infty,\infty})} \leq Ce^{CV(t)} \Big(\| \operatorname{div}(X_0\omega_0)\|_{B^{r-1}_{\infty,\infty}}$$
$$+ (1 + \nu t)\|F\|_{L^\infty_t(B^{r-2}_{\infty,\infty})} + \nu\|G\|_{\widetilde{L}^1_t(B^r_{\infty,\infty})} \Big) \quad (7.39)$$

for some constant C depending only on r. On the other hand, Propositions 2.82 and 2.85, together with Lemma 2.99 page 111, ensure that

$$\|F\|_{L^\infty_t(B^{r-2}_{\infty,\infty})} \leq C\|\omega\|_{L^\infty_t(L^\infty)}\|X\|_{L^\infty_t(B^r_{\infty,\infty})},$$
$$\|G\|_{\widetilde{L}^1_t(B^r_{\infty,\infty})} \leq C\|\omega\|_{\widetilde{L}^1_t(B^2_{\infty,\infty})}\|X\|_{L^\infty_t(B^r_{\infty,\infty})}$$

for some constant depending only on r.

As $D_t\omega = \nu\Delta\omega$, we have $\|\omega(t)\|_{L^\infty} \leq \|\omega_0\|_{L^\infty}$, according to Lemma 7.30. Furthermore, Theorem 3.38 implies that

$$\nu\|\omega\|_{\widetilde{L}^1_t(B^2_{\infty,\infty})} \leq Ce^{CV(t)}(1+\nu t)\,\|\omega_0\|_{L^\infty}\,.$$

Therefore, plugging the above inequalities into (7.39), we end up with

$$\|\operatorname{div}(X\omega)\|_{L^\infty_t(B^{r-1}_{\infty,\infty})} \leq Ce^{CV(t)}$$
$$\times \left(\|\operatorname{div}(X_0\omega_0)\|_{B^{r-1}_{\infty,\infty}} + (1+\nu t)\,\|\omega_0\|_{L^\infty}\,\|X\|_{L^\infty_t(B^r_{\infty,\infty})}\right).$$

In order to complete the proof, it suffices to insert the inequality (7.38) into the above inequality. We readily get

$$e^{-CV(t)}\|\operatorname{div}(X\omega)\|_{L^\infty_t(B^{r-1}_{\infty,\infty})} \leq C\Big(\|\operatorname{div}(X_0\omega_0)\|_{B^{r-1}_{\infty,\infty}}$$
$$+ (1+\nu t)\,\|\omega_0\|_{L^\infty}\left(\|X_0\|_{B^r_{\infty,\infty}} + \int_0^t e^{-CV(t')}\|\operatorname{div}(X\omega)\|_{B^{r-1}_{\infty,\infty}}\,dt'\right)\Big),$$

and hence Gronwall's lemma yields

$$e^{-CV(t)}\|\operatorname{div}(X\omega)\|_{L^\infty_t(B^{r-1}_{\infty,\infty})} \leq Ce^{Ct(1+\nu t)\|\omega_0\|_{L^\infty}}$$
$$\times \left(\|\operatorname{div}(X_0\omega_0)\|_{B^{r-1}_{\infty,\infty}} + (1+\nu t)\,\|\omega_0\|_{L^\infty}\,\|X_0\|_{B^r_{\infty,\infty}}\right).$$

As $t\,\|\omega_0\|_{L^\infty} \leq V(t)$, we get (7.35).

Finally, plugging the last inequality into (7.38), we get (7.36). This completes the proof. □

7.4.4 A Global Convergence Result for Striated Regularity

This subsection is devoted to the proof of Theorem 7.39. Note that since the initial vorticity is in $L^2 \cap L^\infty$, Yudovich's theorem page 312 provides a global solution with vorticity in $L^2 \cap L^\infty$. However, as explained before, this does *not* imply that v is in $C^{0,1}$. Hence, we shall proceed as follows:

– First, we smooth out the data. From the global existence theory of Section 7.2, we thus get a global smooth solution.
– Second, we prove uniform estimates for striated norms of those smooth solutions.
– Third, we prove convergence to a solution of (NS_ν) or (E) with initial data v_0.
– Finally, we let ν tend to 0.

For notational convenience, we shall drop the index ν in the first three steps of the proof and at the same time shall solve indistinctly (NS_ν) or (E).

First Step: Construction of Smooth Solutions

For all $n \in \mathbb{N}$, the vector field $S_n v^0$ has vorticity ω_0^n in L^2 and belongs to the set $\mathcal{C}_b^\infty$ of smooth bounded functions with bounded derivatives of all orders. Hence, Theorem 7.25 implies that (E) and (NS_ν) have a unique global smooth solution v^n with vorticity in L^2.

Second Step: Uniform Estimates for Striated Regularity

By definition, the time-dependent vector field X_λ^n transported by the flow ψ^n of v^n satisfies

$$X_\lambda^n(t) \circ \psi^n(t) = \partial_{X_{0,\lambda}} \psi^n(t).$$

Now, it is clear that

$$\partial_t \partial_{X_{0,\lambda}} \psi^n(t,x) = \partial_{X_{0,\lambda}} v^n(t, \psi^n(t,x)) = \nabla v^n(t, \psi^n(t,x)) \partial_{X_{0,\lambda}} \psi^n(t,x),$$

so Gronwall's lemma ensures that

$$\left| X_\lambda^n(t)(\psi^n(t,x)) \right|^{\pm 1} \leq |X_{0,\lambda}(x)| e^{V^n(t)}.$$

Therefore,

$$I(X^n(t)) \geq e^{-V^n(t)} I(X_0) \quad \text{with} \quad V^n(t) \stackrel{\text{def}}{=} \int_0^t \|\nabla v^n(t')\|_{L^\infty} \, dt'. \tag{7.40}$$

Let Y_0 be one of the vector fields of the family $(X_{0,\lambda})_{1 \leq \lambda \leq m}$. Since v^n is smooth, we have

$$\partial_t Y^n + v^n \cdot \nabla Y^n = \partial_{Y^n} v^n.$$

Hence, according to Proposition 7.43,

$$\| \operatorname{div}(Y^n \omega^n)(t) \|_{B_{\infty,\infty}^{r-1}} \leq C(1 + \nu t) e^{C\nu t^2 \|\omega_0^n\|_{L^\infty}} e^{CV^n(t)}$$
$$\times \left(\|\omega_0^n\|_{L^\infty} \|Y_0\|_{B_{\infty,\infty}^r} + \| \operatorname{div}(Y_0 \omega_0^n) \|_{B_{\infty,\infty}^{r-1}} \right),$$

$$\|Y^n(t)\|_{B_{\infty,\infty}^r} \leq C(1 + \nu t) e^{C\nu t^2 \|\omega_0^n\|_{L^\infty}} e^{CV^n(t)}$$
$$\times \left(\|Y_0\|_{B_{\infty,\infty}^r} + Ct \left(\| \operatorname{div}(Y_0 \omega_0^n) \|_{B_{\infty,\infty}^{r-1}} + \|\omega_0^n\|_{L^\infty} \|Y_0\|_{B_{\infty,\infty}^r} \right) \right).$$

We claim that the right-hand side of the above two inequalities may be bounded independently of n. First, it is clear that $\|S_n \omega_0\|_{L^\infty} \leq C \|\omega_0\|_{L^\infty}$. Next, we have

$$\| \operatorname{div}(Y_0 S_n \omega_0) \|_{B_{\infty,\infty}^{r-1}} \leq C \left(\| \operatorname{div}(Y_0 \omega_0) \|_{B_{\infty,\infty}^{r-1}} + \|\omega_0\|_{L^\infty} \|Y_0\|_{B_{\infty,\infty}^r} \right). \tag{7.41}$$

Indeed, according to the inequality (7.33),

$$\|\operatorname{div}(Y_0\,S_n\omega_0) - T_{Y_0^j}\partial_j S_n\omega_0\|_{B_{\infty,\infty}^{r-1}} \leq C\|Y_0\|_{B_{\infty,\infty}^r}\|S_n\omega_0\|_{L^\infty}.$$

We can now write that

$$T_{Y_0^j}\partial_j S_n\omega_0 = S_n T_{Y_0^j}\partial_j\omega_0 + [T_{Y_0^j}, S_n]\partial_j\omega_0.$$

Again using the inequality (7.33) and the fact that the operator S_n maps $B_{\infty,\infty}^{r-1}$ to itself with norm independent of n, we see that the first term satisfies

$$\|S_n T_{Y_0^j}\partial_j\omega_0\|_{B_{\infty,\infty}^{r-1}} \leq C\big(\|\operatorname{div}(Y_0\omega_0)\|_{B_{\infty,\infty}^{r-1}} + \|\omega_0\|_{L^\infty}\|Y_0\|_{B_{\infty,\infty}^r}\big).$$

Next, according to Proposition 2.10 page 59, we have, for some fixed integer N_0,

$$[T_{Y_0^j}, S_n]\partial_j\omega_0 = \sum_{k\leq n+N_0} [S_{k-1}Y_0^j, S_n]\Delta_k\partial_j\omega_0.$$

Resorting to Lemma 2.97 page 110, we discover that

$$\begin{aligned}
\big\|[S_{k-1}Y_0^j, S_n]\Delta_k\partial_j\omega_0\big\|_{L^\infty} &\leq C2^{-n}\|\nabla S_{k-1}Y_0\|_{L^\infty}\|\Delta_k\partial_j\omega_0\|_{L^\infty}\\
&\leq C2^{k-n}\,2^{k(1-r)}\|Y_0\|_{B_{\infty,\infty}^r}\|\omega_0\|_{L^\infty},
\end{aligned}$$

which completes the proof of (7.41).

So, finally, we find that for any vector X_λ^n and time $t\geq 0$, we have

$$\begin{aligned}
\|\operatorname{div}(X_\lambda^n\omega^n)(t)\|_{B_{\infty,\infty}^{r-1}} &\leq C(1+\nu t)e^{C\nu t^2\|\omega_0^n\|_{L^\infty}}e^{CV^n(t)}\\
&\times \Big(\|\omega_0\|_{L^\infty}\|X_{0,\lambda}\|_{B_{\infty,\infty}^r} + \|\operatorname{div}(X_{0,\lambda}\omega_0)\|_{B_{\infty,\infty}^{r-1}}\Big), \quad (7.42)
\end{aligned}$$

$$\begin{aligned}
\|X_\lambda^n(t)\|_{B_{\infty,\infty}^r} &\leq (1+\nu t)e^{C\nu t^2\|\omega_0^n\|_{L^\infty}}e^{CV^n(t)}\Big(\|X_{0,\lambda}\|_{B_{\infty,\infty}^r}\\
&+ Ct\big(\|\operatorname{div}(X_{0,\lambda}\omega_0)\|_{B_{\infty,\infty}^{r-1}} + \|\omega_0\|_{L^\infty}\|X_{0,\lambda}\|_{B_{\infty,\infty}^r}\big)\Big). \quad (7.43)
\end{aligned}$$

In order to complete the proof of our claim, we have to bound V^n (or, rather, $\|\nabla v^n\|_{L^\infty}$) independently of n. We have already proven that for any nonnegative t, the family $(X_\lambda^n(t))_{1\leq\lambda\leq m}$ is $B_{\infty,\infty}^r$ and nondegenerate. Therefore, Theorem 7.40 yields

$$\|\nabla v^n(t)\|_{L^\infty} \leq C\Big(\|\omega_0\|_{L^2} + \|\omega_0\|_{L^\infty}\log\Big(e + \frac{\|\omega^n(t)\|_{C_{X^n(t)}^r}}{\|\omega_0\|_{L^\infty}}\Big)\Big).$$

Taking advantage of the definition of the norm $\|\cdot\|_{C_{X^n(t)}^r}$ and of the inequalities (7.42) and (7.43), we get, after a few calculations,

$$\begin{aligned}
\|\nabla v^n(t)\|_{L^\infty} &\leq C\Big(\|\omega_0\|_{L^2} + \|\omega_0\|_{L^\infty}\log\Big(e + (1+\nu t)\frac{\|\omega_0\|_{C_{X_0}^r}}{\|\omega_0\|_{L^\infty}}\Big)\Big)\\
&+ C\big(\|\omega_0\|_{L^\infty}V^n(t) + \nu t^2\|\omega_0\|_{L^\infty}^2\big),
\end{aligned}$$

so applying Gronwall's lemma leads to

$$\|\nabla v^n(t)\|_{L^\infty} \leq C(1+\nu t)\left(\|\omega_0\|_{L^2\cap L^\infty}\log\left(e+\frac{\|\omega_0\|_{C_{X_0}^r}}{\|\omega_0\|_{L^\infty}}\right)\right)e^{C\|\omega_0\|_{L^\infty}t}. \quad (7.44)$$

Together with the inequalities (7.42) and (7.43), this completes the proof of the global uniform estimates for striated regularity and for $\|\nabla v^n\|_{L^\infty}$. Finally, using the fact that

$$\|\omega^n(t)\|_{L^2} \leq C\,\|\omega_0\|_{L^2} \quad \text{and} \quad \|\omega^n(t)\|_{L^\infty} \leq C\,\|\omega_0\|_{L^\infty},$$

together with the inequality (7.24) (which also holds true in the viscous case), we deduce that $(v^n)_{n\in\mathbb{N}}$ is bounded in $L^\infty_{loc}(\mathbb{R}^+;L^\infty)$. Therefore, the sequence $(v^n)_{n\in\mathbb{N}}$ is bounded in $L^\infty_{loc}(\mathbb{R}^+;C^{0,1})$.

Third Step: Convergence of Smooth Solutions

We claim that $(v^n)_{n\in\mathbb{N}}$ is a Cauchy sequence (and thus converges) in any space $L^\infty_{loc}(\mathbb{R}^+;B^{-\varepsilon}_{\infty,\infty})$ with $\varepsilon \in\,]0,1[$. To prove this, we write the evolution equation for $v^n - v^m$. For any $(n,m) \in \mathbb{N}^2$ we find that

$$\big(\partial_t + v^n\cdot\nabla - \nu\Delta\big)(v^n - v^m) = \Pi(v^n - v^m, v^n + v^m) + (v^m - v^n)\cdot\nabla v^m.$$

So, according to Theorem 3.14 page 133,

$$\|(v^n - v^m)(t)\|_{B^{-\varepsilon}_{\infty,\infty}} \leq e^{CV^n(t)}\left(\|(S_n - S_m)v_0\|_{B^{-\varepsilon}_{\infty,\infty}}\right.$$

$$\left.+ \int_0^t e^{-CV^n(t')}\left(\|\Pi(v^n - v^m, v^n + v^m)\|_{B^{-\varepsilon}_{\infty,\infty}} + \|(v^m - v^n)\cdot\nabla v^m\|_{B^{-\varepsilon}_{\infty,\infty}}\right)dt'\right).$$

Now, according to Lemma 7.10 we have

$$\|\Pi(v^n - v^m, v^n + v^m)\|_{B^{-\varepsilon}_{\infty,\infty}} \leq C\|v^n - v^m\|_{B^{-\varepsilon}_{\infty,\infty}}\|v^n + v^m\|_{C^{0,1}},$$

and using the (simplified) Bony decomposition and $\mathrm{div}(v^m - v^n) = 0$, we get

$$(v^m - v^n)\cdot\nabla v^m = T_{\nabla v^m}(v^m - v^n) + \mathrm{div}\,T'_{v^m - v^n}v^m.$$

We also find that

$$\|(v^m - v^n)\cdot\nabla v^m\|_{B^{-\varepsilon}_{\infty,\infty}} \leq C\|v^n - v^m\|_{B^{-\varepsilon}_{\infty,\infty}}\|v^m\|_{C^{0,1}}.$$

Using Gronwall's lemma and the uniform bounds of the previous step, it is now easy to conclude that $(v^n)_{n\in\mathbb{N}}$ converges to some vector field v in the space $\mathcal{C}(\mathbb{R}^+;B^{-\varepsilon}_{\infty,\infty})$. The details are left to the reader.

As usual, the uniform bounds of the previous step enable us to state that v satisfies (E) or (NS_ν), belongs to $L^\infty_{loc}(\mathbb{R}^+;C^{0,1})$, and satisfies (7.44). We can also deduce that the family $X(t)$ remains $B^r_{\infty,\infty}$ and nondegenerate for all $t \geq 0$, and that $\omega(t)$ is in $C^r_{X(t)}$.

Final Step: The Inviscid Limit

On the one hand, since $\nabla v \in L_{loc}^{\infty}(\mathbb{R}^{+}; L^{\infty})$, applying Theorem 7.34 yields

$$v_{\nu} - v \to 0 \quad \text{in} \quad L_{loc}^{\infty}(\mathbb{R}^{+}; L^{2}).$$

Because $L^{2}(\mathbb{R}^{2})$ is continuously included in $B_{\infty,\infty}^{-1}(\mathbb{R}^{2})$, and v_{ν} and v both belong to the space $L_{loc}^{\infty}(\mathbb{R}^{+}; B_{\infty,\infty}^{-1})$, we thus have

$$v_{\nu} \to v \quad \text{in} \quad L_{loc}^{\infty}(\mathbb{R}^{+}; B_{\infty,\infty}^{-1}).$$

On the other hand, all the bounds which have been stated in the previous steps are *independent of* ν for ν going to 0. Hence, we may interpolate with those bounds and complete the proof of Theorem 7.39.   □

7.4.5 Application to Smooth Vortex Patches

We want to apply Theorem 7.39 to the particular case of smooth vortex patches. We therefore consider a bounded simply connected domain D_0 such that ∂D_0 is a $C^{1,r}$ simple curve. We aim to solve (E) or (NS_{ν}) in the case where the initial velocity v_0 has vorticity $\omega_0 \overset{\text{def}}{=} \mathbb{1}_{D_0}$.

Let f_0 be a $C^{1,r}$ compactly supported function over $\mathbb{R}^2$ such that, for some neighborhood V of ∂D_0, we have

$$f_0^{-1}(\{0\}) \cap V = \partial D_0 \quad \text{and} \quad \nabla f_0 \text{ does not vanish on } V.$$

Because $\nabla^{\perp} f_0$ is C^r and has modulus bounded away from 0 on ∂D_0, given any x_0 in ∂D_0, solving the ordinary differential equation

$$\begin{cases} \partial_{\sigma}\gamma_0(\sigma) = \nabla^{\perp} f_0(\gamma_0(\sigma)) \\ \gamma_0(0) = x_0 \end{cases}$$

provides a $C^{1,r}$ parameterization of the curve ∂D_0.

Let W be a neighborhood of ∂D_0 such that $W \subset\subset V$. Introduce a smooth function α supported in V and with value 1 on W. We set

$$X_{0,1} = \nabla^{\perp} f_0, \qquad X_{0,2} = (1-\alpha)\partial_1, \quad \text{and} \quad X_{0,3} = (1-\alpha)\partial_2.$$

It is obvious that $(X_{0,1}, X_{0,2}, X_{0,3})$ is a nondegenerate family of $B_{\infty,\infty}^r$ vector fields and that ω_0 belongs to $C_{X_0}^r$. Therefore, Theorem 7.39 provides global Lipschitz solutions v and v_{ν} for (E) and (NS_{ν}) with initial data v_0, and uniform bounds in terms of striated vorticity with respect to the family $X(t)$ [resp., $X_{\nu}(t)$] transported by the flow ψ of v [resp., ψ_{ν} of v_{ν}].

Defining $D_t \overset{\text{def}}{=} \psi(t, D_0)$ and $D_{t,\nu} \overset{\text{def}}{=} \psi_{\nu}(t, D_0)$, we discover that $\gamma(t) \overset{\text{def}}{=} \psi_t \circ \gamma_0$ [resp., $\gamma_{\nu}(t) \overset{\text{def}}{=} \psi_{t,\nu} \circ \gamma_0$] is a parameterization for ∂D_t (resp., $\partial D_{t,\nu}$). Because $\partial_{\sigma}\gamma(t) = X_1(t, \gamma(t))$ and $\partial_{\sigma}\gamma_{\nu}(t) = X_{1,\nu}(t, \gamma_{\nu}(t))$, we can thus conclude that ∂D_t and $\partial D_{t,\nu}$ are $C^{1,r}$ simple curves and that $\partial_{\sigma}\gamma_{\nu}$ is in $L_{loc}^{\infty}(\mathbb{R}^{+}; C^{0,r})$, uniformly with respect to ν. So, we eventually get the following statement.

Theorem 7.44. *Let D_0 be a $C^{1,r}$ simply connected bounded open set of $\mathbb{R}^2$. Let $\omega_0 \overset{def}{=} \mathbb{1}_{D_0}$ and v_0 be given by the Biot–Savart law (7.2).*

For all $\nu > 0$, the system (NS_ν) (resp., E) with data v_0 then has a unique solution v_ν (resp., v) in $L^\infty_{loc}(\mathbb{R}^+; C^{0,1})$ and there exists a constant C depending only on D_0 and such that for all $t \in \mathbb{R}^+$,

$$\|\nabla v_\nu(t)\|_{L^\infty} \leq C(1 + \nu t)e^{Ct} \quad and \quad \|\nabla v(t)\|_{L^\infty} \leq Ce^{Ct}.$$

Further, for all time, the domain $D_{t,\nu}$ (resp., D_t) transported by the flow $\psi_\nu(t)$ [resp., $\psi(t)$] of v_ν (resp., v) remains $C^{1,r}$, and we can find $C^{1,r}$ parameterizations $\gamma(t)$ for ∂D_t and $\gamma_\nu(t)$ for $\partial D_{t,\nu}$ such that

$$\partial_\sigma \gamma_\nu \to \partial_\sigma \gamma \quad in \quad L^\infty_{loc}(\mathbb{R}^+; C^{0,r'}) \quad for\ all\ \ r' < r.$$

Remark 7.45. By taking advantage of Theorem 3.40 and of the uniform estimate for ∇v_ν, we deduce that at time t, the vorticity is equal to the characteristic function of the domain $\psi_\nu(t, D_0)$, up to an error term which decays as $e^{-ch^2/(\nu t)}$ at distance h from the boundary. More precisely, we have

$$\|\omega_\nu(t)\|_{L^2(d(x, D_{\nu,t}) > h)} \leq \|\omega_0\|_{L^2} e^{-\frac{h^2}{4\nu t} \exp(-4(e^{Ct} - 1))}$$

and

$$\|\omega_\nu(t) - \mathbb{1}_{D_{\nu,t}}\|_{L^2(d(x, \partial D_{\nu,t}) > h)}$$

$$\leq \|\omega_0\|_{L^2} \min\left\{1, C\left(\frac{\nu t}{h^2}\right)^{\frac{1}{2}} e^{2(e^{Ct} - 1) - \frac{h^2}{32\nu t} \exp(-4(e^{Ct} - 1))}\right\}.$$

7.5 References and Remarks

Most of the results which are presented here are generalizations to the viscous case (or to more general function spaces) of some results which may be found in a monograph by the second author (see [69]). Many other results on the incompressible Euler system are presented in the books by Bertozzi and Majda [36] and Marchioro and Pulvirenti [222]. For geometrical aspects of the Euler equation, the reader is referred to the works by V. Arnold in [16], Y. Brenier in [45], E. Ebin and J. Marsden in [121], and A. Shnirelman in [266].

The existence of $\mathcal{C}^\infty$ local-in-time solutions for the incompressible Euler system goes back to a series of papers by L. Lichtenstein in the 1920s (see [208]). Analytic data have been considered by C. Bardos and S. Benachour in [28, 30]. The existence theorem in the $W^{s,p}$ spaces was developed by T. Kato and G. Ponce in [178]. The local existence theorem in Besov spaces (namely Theorem 7.1) is a straightforward generalization of the work by J.-Y. Chemin in [69] devoted to Hölder spaces and has been extended by the third author to nonhomogeneous incompressible fluids in [106]. The endpoint case of data in $B^1_{\infty,1}(\mathbb{R}^d)$ has been studied in [245].

Similar regularity results have been obtained by A. Dutrifoy in smooth bounded domains (see [119]). The case of rough convex domains has been treated in [285].

The estimate for the Biot–Savart law given in Proposition 7.5 is a consequence of the well-known Marcinkiewicz theorem, the proof of which may be found in any book on harmonic analysis (see, e.g., [287, 273], or [150]).

The fact that having the vorticity in L^∞ implies uniqueness was first noted by V. Yudovich in [302]. Some recent improvements have been obtained by V. Yudovich, again in [303], and by M. Vishik in [297]. Theorem 7.17 and its corollary are a slight improvement of a result by the third author in [105].

The so-called *Beale–Kato–Majda* continuation criterion given in Corollary 7.24 was first proven in [31] in the three-dimensional case for H^s solutions with s greater than 5/2. The extension to Hölder spaces was achieved by H. Bahouri and B. Dehman in [25]. Some recent improvements have been obtained by a number of authors (see, e.g., [197]). To the best of our knowledge, Theorem 7.23 is new.

As explained above, the conservation of the L^∞ norm is the key to proving global existence in the two-dimensional case. Based on that insight, in 1933, W. Wolibner proved the global existence of smooth solutions (see [300]). Global well-posedness for data in the critical Besov space $B_{p,1}^{1+\frac{2}{p}}(\mathbb{R}^2)$ has been proven by M. Vishik if $p < \infty$. The endpoint index $p = \infty$ was treated by T. Hmidi and S. Keraani in [157].

In 1963, V. Yudovich proved the global existence and uniqueness of two-dimensional flows with bounded vorticity. Theorem 7.27 is in the same spirit as Yudovich's result. The proof of Remark 7.28 may be found in [69].

We mention in passing that there exist classes of definitely three-dimensional data for which global well-posedness is known. This is the case for *axisymmetric data without swirl*, that is, $v_0 := v_{0,r}(r,z)e_r + v_{0,z}(r,z)e_z$ in cylindrical coordinates (see the pioneering paper by [292] and also [256, 265, 105]). It was observed by A. Dutrifoy in [118] that data with *helicoidal* symmetry generate global solutions. The question of global solvability for general three-dimensional data is open. For related numerical or theoretical results, the reader may consult [151] and [260].

Even in the whole space (where no boundary layer is expected), the study of the inviscid limit for the Navier–Stokes equations has a long history. The fact that the rate of convergence in L^2 is of order ν for smooth solutions may be found in the works by T. Kato (see, e.g., [176]). The statement of Theorem 7.34 is essentially contained in a paper by P. Constantin and J. Wu [88], whereas Theorem 7.37 was proven by the second author in [70]. The fact that for a smooth vortex patch, the rate of convergence is of order at least $\nu^{\frac{3}{4}}$ was observed by H. Abidi and the third author in [1]. There, it was shown that the rate is optimal in the case of a circular patch. In this present chapter, to prove Theorem 7.35, we used the method introduced by N. Masmoudi in [224].

The study of the *vortex patch problem* for the two-dimensional Euler equations also has a long history. In a celebrated survey paper by A. Majda [221], it was conjectured that a singularity may appear in finite time in the boundary of an initially smooth vortex patch. Some theoretical results (see the work by S. Alinhac in [6], P. Constantin in [85] and Constantin and Titi in [87]) and numerical experiments corroborated the possible appearance of a singularity (see, in particular, the papers by T. Buttke [52, 53] and Hughes, Roberts and Zabusky in [170]). Nevertheless taking advantage of techniques from [62], it was shown by the second author in [68] (see also [64–66, 69, 136, 138]) that striated regularity is transported for all time by Eulerian flows. As a consequence, an initially $C^{1,r}$ vortex patch remains so for all time. We mention that other proofs have since been provided by A. Bertozzi and

P. Constantin in [35] and by P. Serfati in [258]. The case where the initial patch has a singularity has been studied both theoretically in [89, 92] and numerically in [81]. Generalizations to the three-dimensional case have been given by different people (see, in particular, [137, 259], and [169]). Vortex patches in bounded domains have been studied by N. Depauw in [110] (dimension two) and by A. Dutrifoy in [120] (dimension three). More singular solutions as the so-called *vortex sheet* have been sudied by e.g. J.-M. Delort in [109].

The study of the inviscid limit in the framework of two-dimensional striated regularity (thus for vortex patches in particular) was initiated by the third author in [90]. A simpler proof—the one presented in this chapter—was later proposed by T. Hmidi in [156]. We should mention that a local-in-time version of Theorem 7.39 may be proven in the d-dimensional case (see [91]) and that a global result may be proven for three-dimensional axisymmetric data with striated regularity (see [32]). Some very significant results on the localization properties of viscous patches have been obtained recently by F. Sueur in [277].

8

Strichartz Estimates and Applications to Semilinear Dispersive Equations

Dispersive phenomena often play a crucial role in the study of evolution partial differential equations. Mathematically, exhibiting dispersion often amounts to proving a decay estimate for the L^∞ norm of the solution at time t in terms of some (negative) power of t and of the L^1 norm of the data.

In many cases, proving these estimates relies on the stationary phase theorem and on a (possibly approximate) explicit representation of the solution. The basic idea is that fast oscillations induce a small average, as may be seen by performing suitable integrations by parts. As an example, in the case of the wave equation with constant coefficients, the geometric optics allow the solutions to be written in terms of oscillating functions, the frequencies of which grow linearly in time. As a consequence, a polynomial time decay may be exhibited for suitable norms.

It is now well established that these decay estimates, combined with an abstract functional analysis argument—the $TT^\star$ *argument*—yield a number of inequalities involving space-time Lebesgue norms. In the last two decades, these inequalities—the so-called *Strichartz estimates*—have proven to be of paramount importance in the study of semilinear or quasilinear Schrödinger and wave equations.

The purpose of this chapter is to give dispersive estimates for some linear partial differential equations and to provide a few examples of applications to solving semilinear problems. Although we shall focus mostly on Schrödinger and wave equations, the basic dispersive estimates that we derive apply to a much more general framework, whenever waves propagate in a physical medium.

The first section of this chapter is devoted to a few basic examples. First, we study the case of the free transport equation and the Schrödinger equation where decay inequalities may be proven by means of elementary tools. Next, we come to the study of oscillatory integrals and (a class of) Fourier integral operators. Oscillatory integrals arise naturally when proving dispersive estimates for the wave equation, while the L^2 boundedness of Fourier integral operators will be needed in the next chapter.

H. Bahouri et al., *Fourier Analysis and Nonlinear Partial Differential Equations*, Grundlehren der mathematischen Wissenschaften 343, DOI 10.1007/978-3-642-16830-7_8, © Springer-Verlag Berlin Heidelberg 2011

Section 8.2 is devoted to proving Strichartz inequalities for groups of operators satisfying a suitable decay inequality called a *dispersive inequality*. As an application, we prove a global well-posedness result for the cubic Schrödinger equation in $\mathbb{R}^2$. In the next section, we establish Strichartz estimates for the wave equation with data in Sobolev spaces. In the following two sections, we apply those Strichartz estimates to the investigation of some semilinear wave equations, namely, the quintic and cubic wave equations in $\mathbb{R}^3$. In the last section of this chapter, we establish local well-posedness in a suitable Besov space for a class of semilinear wave equations with quadratic nonlinearity with respect to the first order space derivatives of the solution. This result will help us to investigate some quasilinear wave equations in the next chapter.

8.1 Examples of Dispersive Estimates

In this section, we provide a few examples of linear equations, the solutions of which satisfy a dispersive estimate. We shall study three examples: the free transport equation, the Schrödinger equation, and the wave equation. In passing, we will establish decay estimates for oscillatory integrals and the boundedness in L^2 of a class of Fourier integral operators.

8.1.1 The Dispersive Estimate for the Free Transport Equation

In this subsection, we prove basic dispersive estimates for the free transport equation,

$$(T) \qquad \begin{cases} \partial_t f + v \cdot \nabla_x f = 0 \\ \quad f_{|t=0} = f_0, \end{cases}$$

which describes the evolution of the microscopic density $f(t, x, v) \in \mathbb{R}^+$ of free particles which, at time $t \in \mathbb{R}$, are located at $x \in \mathbb{R}^d$ and have speed $v \in \mathbb{R}^d$.

The dispersive estimates for (T) follow from the explicit formula for the solution, as the solution may be easily computed in terms of the Cauchy data f_0. It is only a matter of integrating along the characteristics.

Proposition 8.1. *The solution of the free transport equation (T) is given by*

$$f(t, x, v) = f_0(x - vt, v).$$

In fact, *even though the total mass is preserved*, the dispersive effect occurs for the macroscopic density

$$\rho(t, x) \overset{\text{def}}{=} \int_{\mathbb{R}^d} f(t, x, v) \, dv,$$

as stated in the following proposition.

Proposition 8.2. *If f is a solution of the transport equation (T), then we have*

$$\|\rho(t,\cdot)\|_{L^\infty} \leq \frac{1}{|t|^d} \|\sup_{v'} f_0(\cdot, v')\|_{L^1}.$$

Proof. For any x, we have, thanks to Proposition 8.1,

$$\int_{\mathbb{R}^d} f(t,x,v)\, dv = \int_{\mathbb{R}^d} f_0(x-vt,v)\, dv.$$

Now, the change of variable $y = x - vt$ leads to the inequalities

$$\int_{\mathbb{R}^d} f_0(x-vt,v)\, dv \leq \int_{\mathbb{R}^d} \sup_{v'} f_0(x-vt,v')\, dv$$

$$\leq \frac{1}{|t|^d} \int_{\mathbb{R}^d} \sup_{v'} f_0(y,v')\, dy,$$

which means that the macroscopic density ρ decays in L^∞, completing the proof of the proposition. □

8.1.2 The Dispersive Estimates for the Schrödinger Equation

The linear Schrödinger equation was introduced in the context of quantum mechanics and takes the form

$$(S) \qquad \begin{cases} i\partial_t u - \dfrac{1}{2}\Delta u = 0 \\ u_{|t=0} = u_0, \end{cases}$$

where the unknown complex-valued function u depends on $(t,x) \in \mathbb{R} \times \mathbb{R}^d$.

As we consider initial data (and thus solutions) which are *not* regular functions, solutions have to be understood in the *weak sense,* as introduced in Chapter 5. More precisely, a distribution $u \in \mathcal{C}(\mathbb{R}; \mathcal{S}'(\mathbb{R}^d))$ is a *weak solution* of (S) if it satisfies, for all φ in $C^\infty(\mathbb{R}; \mathcal{S}(\mathbb{R}^d))$,

$$\int_0^t \left\langle u(t'), \frac{1}{2}\Delta\varphi(t') + i\partial_t\varphi(t') \right\rangle dt' = \langle u(t), i\varphi(t) \rangle - \langle u_0, i\varphi(0) \rangle.$$

By using the Fourier transform, the solution may be expressed in terms of the Cauchy data. More precisely, we have the following result.

Proposition 8.3. *For any $u_0 \in \mathcal{S}'$, the Schrödinger equation (S) has a unique solution u in $\mathcal{S}(\mathbb{R}; \mathcal{S}')$. For $t \neq 0$, that solution is of the form*

$$u(t) = \mathcal{F}^{-1}\left(e^{it\frac{|\xi|^2}{2}} \widehat{u}_0 \right) = S_t \star u_0 \quad with \quad S_t(x) \overset{def}{=} \frac{e^{id\frac{t}{|t|}\frac{\pi}{4}}}{(2\pi|t|)^{\frac{d}{2}}} e^{-i\frac{|x|^2}{2t}}. \qquad (8.1)$$

Remark 8.4. The identity $\mathcal{F}u(t, \xi) = e^{it\frac{|\xi|^2}{2}}\widehat{u}_0(\xi)$ implies the conservation of the H^s norm. Also, under this identity, it is easy to see that $(U(t))_{t\in\mathbb{R}}$, where $U(t) : u_0 \longmapsto U(t)u_0$ and $U(t)u_0$ is the solution of (S) at time t, is a one-parameter group of unitary operators.

Remark 8.5. On the one hand, in the case where the Cauchy data u_0 is the Dirac mass δ_0, we get, thanks to (8.1), for any time $t \neq 0$,

$$u(t) = S_t,$$

and therefore for each fixed time $t \neq 0$, $u(t)$ is analytic.

On the other hand, if the Cauchy data is $u_0(x) \stackrel{\text{def}}{=} e^{ia|x|^2}$, then we get

$$u\left(\frac{1}{2a}\right) = \left(\frac{\pi}{ia}\right)^{\frac{d}{2}}\delta_0.$$

Hence, the support of the solution at time $1/(2a)$ collapses to a single point, even though its support is equal to the whole space $\mathbb{R}^d$ at time 0. This phenomenon is due to the infinite speed of the propagation for the Schrödinger equation. Also, note that the regularity of the solution depends on the behavior of the Cauchy data at infinity.

Proof of Proposition 8.3. Let $u(t) = \mathcal{F}^{-1}\left(e^{it\frac{|\xi|^2}{2}}\widehat{u}_0(\xi)\right)$. For $\varphi \in \mathcal{C}^\infty(\mathbb{R}; \mathcal{S})$, define

$$I_\varphi(t) \stackrel{\text{def}}{=} \int_0^t \left\langle u(t'), \frac{1}{2}\Delta\varphi(t') + i\partial_t\varphi(t') \right\rangle dt'.$$

By the definition of u, we have

$$
\begin{aligned}
I_\varphi(t) &= \int_0^t \left\langle \mathcal{F}^{-1}\left(e^{it'\frac{|\xi|^2}{2}}\widehat{u}_0(\xi)\right), \frac{1}{2}\Delta\varphi(t') + i\partial_t\varphi(t') \right\rangle dt' \\
&= \int_0^t \left\langle e^{it'\frac{|\xi|^2}{2}}\widehat{u}_0(\xi), \mathcal{F}^{-1}\left(\frac{1}{2}\Delta\varphi(t') + i\partial_t\varphi(t')\right) \right\rangle dt' \\
&= \int_0^t (2\pi)^{-d} \left\langle \widehat{u}_0(\xi), e^{it'\frac{|\xi|^2}{2}}\left(-\frac{1}{2}|\xi|^2\widehat{\varphi}(t', -\xi) + i\partial_t\widehat{\varphi}(t', -\xi)\right) \right\rangle dt'.
\end{aligned}
$$

Because the distribution $\widehat{u}_0$ may be interchanged with the integral, we get

$$I_\varphi(t) = (2\pi)^{-d} \left\langle \widehat{u}_0, \int_0^t e^{it'\frac{|\xi|^2}{2}}\left(-\frac{1}{2}|\xi|^2\widehat{\varphi}(t', -\xi) + i\partial_t\widehat{\varphi}(t', -\xi)\right) dt' \right\rangle.$$

As we have

$$\partial_{t'}\left(e^{it'\frac{|\xi|^2}{2}}i\widehat{\varphi}(t', -\xi)\right) = e^{it'\frac{|\xi|^2}{2}}\left(-\frac{1}{2}|\xi|^2\widehat{\varphi}(t', -\xi) + i\partial_{t'}\widehat{\varphi}(t', -\xi)\right),$$

we get that

$$\int_0^t e^{it'\frac{|\xi|^2}{2}} \left(-\frac{1}{2}|\xi|^2 \widehat{\varphi}(t',-\xi) + i\partial_{t'}\widehat{\varphi}(t',-\xi) \right) dt' = ie^{it\frac{|\xi|^2}{2}} \widehat{\varphi}(t,-\xi) - i\widehat{\varphi}(0,-\xi).$$

Thus,

$$\begin{aligned}
I_\varphi(t) &= i(2\pi)^{-d}\langle \widehat{u}_0, e^{it\frac{|\xi|^2}{2}} \widehat{\varphi}(t,-\xi)\rangle - i(2\pi)^{-d}\langle \widehat{u}_0, \widehat{\varphi}(0,-\xi)\rangle \\
&= i\langle \widehat{u}(t), \mathcal{F}^{-1}\varphi(t)\rangle - i\langle \widehat{u}_0, \mathcal{F}^{-1}\varphi(0)\rangle \\
&= i\langle u(t), \varphi(t)\rangle - i\langle u_0, \varphi(0)\rangle.
\end{aligned}$$

This proves that u is a weak solution of the Schrödinger equation. Uniqueness in $\mathcal{C}(\mathbb{R}; \mathcal{S}')$ may be proven by taking advantage of the duality method introduced in Chapter 4. Since its adaptation to the Schrödinger equation is straightforward, we leave the details to the reader.

To complete the proof, it remains to observe that, according to Proposition 1.28 page 23, we have

$$\mathcal{F}^{-1}\left(e^{it\frac{|\xi|^2}{2}} \right)(x) = \frac{1}{(-2i\pi t)^{\frac{d}{2}}} e^{-i\frac{|x|^2}{2t}},$$

from which follows the desired formula for S_t. □

From the above proposition and convolution inequalities, we readily get the following proposition.

Proposition 8.6. *If u is a solution of the linear Schrödinger equation (S), then we have, for $t \neq 0$,*

$$\|u(t)\|_{L^\infty} \leq \frac{1}{(2\pi|t|)^{\frac{d}{2}}} \|u_0\|_{L^1}.$$

Remark 8.7. Proposition 8.3, together with the conservation of the L^2 norm and the interpolation between L^p spaces (see Corollary 1.13 page 12), implies that

$$\forall t \in \mathbb{R}\setminus\{0\}, \ \forall p \in [2,\infty], \ \|u(t)\|_{L^p} \leq \frac{1}{(2\pi|t|)^{d(\frac{1}{2}-\frac{1}{p})}} \|u_0\|_{L^{p'}}.$$

8.1.3 Integral of Oscillating Functions

Proving dispersive estimates for the wave equation requires more elaborate techniques that we will now introduce. As we will see in the next subsection, we shall have to estimate integrals of the form

$$I_\psi(\tau) = \int_{\mathbb{R}^d} e^{i\tau\Phi(\xi)} \psi(\xi)\, d\xi,$$

where τ must be understood as a large parameter.

Notation. Throughout this section, ψ will denote a function in $\mathcal{D}(\mathbb{R}^d)$ and Φ a real-valued smooth function on a neighborhood of the support of ψ. Moreover, the constants which will appear will be generically denoted by C and will depend on a finite number of derivatives of ψ and on a finite number of derivatives of order greater than or equal to 2 of the phase function Φ.

We shall distinguish the case where $\nabla \Phi$ does not vanish (the *nonstationary phase* case) from the case where it may vanish (the *stationary phase* case).

Theorem 8.8. *Consider a compact set K of $\mathbb{R}^d$ and assume that a constant $c_0 \in]0,1]$ exists such that*

$$\forall \xi \in K, \ |\nabla \Phi(\xi)| \geq c_0.$$

Then, for any integer N and any function ψ in the set $\mathcal{D}_K$ of smooth functions supported in K, a constant C exists such that

$$|I_\psi(\tau)| \leq \frac{C_N}{(c_0\tau)^N}.$$

Proof. Note that changing Φ to Φ/c_0 and τ to $c_0\tau$ reduces the proof to the case $c_0 = 1$ (we leave the reader to check that after this change of function, the dependence with respect to c_0 is harmless since $c_0 \leq 1$). Assume, then, that $c_0 = 1$. It is then simply a matter of using the oscillations to produce decay. This will be achieved by means of suitable integrations by parts. Indeed, consider the following first order differential operator, defined for any function a in $\mathcal{D}_K$:

$$\mathcal{L}a \overset{\text{def}}{=} -i\sum_{j=1}^{d} \frac{\partial_j \Phi}{|\nabla \Phi|^2} \partial_j a.$$

This operator obviously satisfies

$$\mathcal{L}e^{i\tau\Phi} = \tau e^{i\tau\Phi},$$

hence, by repeated integrations by parts, we get that

$$I_\psi(\tau) = \frac{1}{\tau^N} \int_{\mathbb{R}^d} e^{i\tau\Phi} (({}^t\mathcal{L})^N \psi)(\xi) \, d\xi.$$

We now compute ${}^t\mathcal{L}$ for $a \in \mathcal{D}_K$. We have

$$^t\mathcal{L}a = -\mathcal{L}a + i\frac{\Delta\Phi}{|\nabla\Phi|^2}a - 2i \sum_{1\leq j,k\leq d} \frac{\partial_j\Phi\,\partial_k\Phi\,\partial_j\partial_k\Phi}{|\nabla\Phi|^4}a.$$

Thus, it is obvious that

$$({}^t\mathcal{L}\psi)(\xi) = f_{1,1}(\xi, \nabla\Phi(\xi)) + f_{1,2}(\xi, \nabla\Phi(\xi)),$$

where the function $f_{1,j}(\xi, \theta)$ belongs to $\mathcal{D}(K \times \mathbb{R}^d \setminus \{0\})$, is homogeneous of degree $-j$ with respect to θ, and satisfies

$$\forall (\alpha, \beta) \in (\mathbb{N}^d)^2, \quad \sup_{(\xi, \theta) \in K \times \mathbb{S}^{d-1}} \left| \partial_\xi^\alpha \partial_\theta^\beta f_{1,j}(\xi, \theta) \right| \leq C_{j,\alpha,\beta}.$$

As the coefficients of the differential operator $\mathcal{L}$ and all their derivatives are bounded on K, an obvious (and omitted) induction implies that

$$({}^t\mathcal{L})^N \psi(\xi) = \sum_{j=0}^N f_{N,j}(\xi, \nabla\Phi(\xi)),$$

where the function $f_{N,j}(\xi, \theta)$ belongs to $\mathcal{D}(K \times \mathbb{R}^d \setminus \{0\})$, is homogeneous of degree $-N - j$ in θ, and satisfies

$$\forall (\alpha, \beta) \in (\mathbb{N}^d)^2, \quad \sup_{(\xi, \theta) \in K \times \mathbb{S}^{d-1}} \left| \partial_\xi^\alpha \partial_\theta^\beta f_{N,j}(\xi, \theta) \right| \leq C_{N,j,\alpha,\beta}.$$

This proves the theorem. $\qquad\qquad\qquad\qquad\qquad\qquad\qquad\qquad\qquad\qquad\square$

We will now consider the case where the gradient of the phase function may vanish.

Theorem 8.9. *Consider a compact K of $\mathbb{R}^d$ and assume that a constant $c_0 \in \,]0,1]$ exists such that*

$$\forall \xi \in K, \quad |\nabla\Phi(\xi)| \leq c_0.$$

Then, for any integer N and any function ψ in $\mathcal{D}_K$, there exists a constant C_N such that

$$|I_\psi(\tau)| \leq C_N \int_K \frac{d\xi}{(1 + c_0\tau |\nabla\Phi(\xi)|^2)^N}.$$

Proof. As in the preceding theorem, it suffices to consider the case $c_0 = 1$, and we may perform suitable integrations by parts to pinpoint the decay with respect to τ. Consider the first order differential operator

$$\mathcal{L}_\tau \stackrel{\text{def}}{=} \frac{1}{1 + \tau |\nabla\Phi|^2} \left(\text{Id} - i\nabla\Phi \cdot \partial \right) \quad \text{with} \quad \nabla\Phi \cdot \partial = \sum_{j=1}^d \partial_j \Phi \, \partial_j. \qquad (8.2)$$

This operator obviously satisfies

$$\mathcal{L}_\tau e^{i\tau\Phi} = e^{i\tau\Phi}.$$

Now, by integration by parts, we get that

$$I_\psi(\tau) = \int_{\mathbb{R}^d} e^{i\tau\Phi} ({}^t\mathcal{L}_\tau)^N \psi(\xi) \, d\xi.$$

Hence, to complete the proof of the theorem, it suffices to demonstrate that for any integer N, a constant C exists such that

$$\left| ({}^t\mathcal{L}_\tau)^N \psi(\xi) \right| \leq \frac{C}{(1 + \tau |\nabla\Phi(\xi)|^2)^N}. \qquad (8.3)$$

In order to do this, we define the following class of functions.

Definition 8.10. *Given an integer N, we denote by S^N the set of smooth functions on $K \times \mathbb{R}^d$ such that*

$$\forall (\alpha, \beta) \in \mathbb{N}^d \times \mathbb{N}^d, \ \exists C \ / \ \forall (\xi, \theta) \in K \times \mathbb{R}^d, \ |\partial_\xi^\alpha \partial_\theta^\beta f(\xi, \theta)| \leq C(1 + |\theta|^2)^{\frac{N - |\beta|}{2}}.$$

It is obvious that the space S^N is increasing with N and that the product of a function in S^{N_1} by a function in S^{N_2} is a function in $S^{N_1 + N_2}$. Moreover, we have $\partial_\theta^\beta (S^N) \subset S^{N - |\beta|}$.

It is clear that the following lemma implies the inequality (8.3).

Lemma 8.11. *For any N in $\mathbb{N}$, a function f_N exists in S^{-2N} such that*

$$({}^t\mathcal{L}_\tau)^N \psi(\xi) = f_N \left(\xi, \tau^{\frac{1}{2}} \nabla \Phi(\xi) \right) \quad \text{for all} \quad \xi \in K.$$

Proof. By noting that S^0 contains the space $\mathcal{D}_K$ and by an immediate induction, it is enough to prove that if f belongs to S^M, then

$$ {}^t\mathcal{L}_\tau \left(f(\xi, \tau^{\frac{1}{2}} \nabla \Phi(\xi)) \right) = g(\xi, \tau^{\frac{1}{2}} \nabla \Phi(\xi)) \quad \text{with} \quad g \in S^{M-2}. \tag{8.4}$$

For any $a \in \mathcal{D}_K$, we have

$$ {}^t\mathcal{L}_\tau a(\xi) = i \frac{\nabla \Phi(\xi) \cdot \nabla a(\xi)}{1 + \tau |\nabla \Phi(\xi)|^2} + \sigma\left(\xi, \tau^{\frac{1}{2}} \nabla \Phi(\xi)\right) a(\xi) \tag{8.5}$$

$$\text{with} \quad \sigma(\xi, \theta) = \frac{i \Delta \Phi(\xi) + 1}{1 + |\theta|^2} - 2i \frac{D^2 \Phi(\theta, \theta)}{(1 + |\theta|^2)^2},$$

where, from now on, we agree that

$$D^2 \Phi(\theta_1, \theta_2) \overset{\text{def}}{=} \sum_{j,k} \theta_1^j \theta_2^k \partial_{jk}^2 \Phi.$$

It is obvious that $\sigma \in S^{-2}$. By using the chain rule, we get

$$\nabla \Phi \cdot \nabla f(\xi, \tau^{\frac{1}{2}} \nabla \Phi(\xi)) = \left(\nabla \Phi \cdot \nabla_\xi f + D^2 \Phi(\theta, \nabla_\theta f) \right)(\xi, \tau^{\frac{1}{2}} \nabla \Phi(\xi)).$$

Thus, we have the relation (8.4) with

$$g(\xi, \theta) = \frac{i}{1 + |\theta|^2} \left(\nabla \Phi(\xi) \cdot \nabla_\xi f(\xi, \theta) + D^2 \Phi(\theta, \nabla_\theta f(\xi, \theta)) \right) + (\sigma f)(\xi, \theta). \tag{8.6}$$

The lemma is proved and thus so is Theorem 8.9. □

Combining the above two theorems, we get the following statement.

Theorem 8.12. *Let ψ be in $\mathcal{D}(\mathbb{R}^d)$ and Φ be a real-valued smooth function defined on a neighborhood of the support of ψ. Fix some positive real number $c_0 \in \,]0, 1]$. Then, for any couple (N, N') of positive real numbers, there exist two constants, C_N and $C_{N'}$, such that*

$$|I_\psi(\tau)| \leq \frac{C_N}{(c_0\tau)^N} + C_{N'}\int \frac{\mathbf{1}_{\{\xi\in\mathbb{R}^d\,,\,|\nabla\Phi(\xi)|\leq c_0\}}}{(1+c_0\tau|\nabla\Phi|^2)^{N'}}\,d\xi.$$

Further, the constants C_N and $C_{N'}$ depend only on N, N', a finite number of derivatives of ψ, and a finite number of derivatives of order greater than or equal to 2 of Φ.

Proof. Let χ be a smooth function supported in the unit ball and with value 1 for $|x| \leq 1/2$. We may write

$$I_\psi(\tau) = I_1(\tau) + I_2(\tau) \quad \text{with} \quad \begin{cases} I_1(\tau) = \displaystyle\int e^{i\tau\Phi(\xi)}\left(1 - \chi\left(\frac{\nabla\Phi(\xi)}{c_0}\right)\right)\psi(\xi)\,d\xi \\[2mm] I_2(\tau) = \displaystyle\int e^{i\tau\Phi(\xi)}\chi\left(\frac{\nabla\Phi(\xi)}{c_0}\right)\psi(\xi)\,d\xi. \end{cases}$$

Applying Theorem 8.8 to I_1 and Theorem 8.9 to I_2 gives the result. $\qquad\square$

In the one-dimensional case, we can prove more accurate estimates. More precisely, we have the following theorem.

Theorem 8.13. *Let a be a function in the closure of a smooth compactly supported function of one real variable with respect to the norm $\|a'\|_{L^1(\mathbb{R})}$. Let Φ be a C^2 function on $\mathbb{R}$ such that a positive constant c_0 exists, where*

$$\forall x \in \text{Supp } a\,, \quad \Phi''(x) \geq c_0.$$

The integral defined by

$$I(t) \stackrel{def}{=} \int_{\mathbb{R}} e^{it\Phi(x)}a(x)dx$$

then satisfies

$$|I(t)| \leq \frac{C_0}{t^{\frac{1}{2}}}\|a'\|_{L^1} \quad \text{with} \quad C_0 \stackrel{def}{=} \frac{1}{2} + \frac{\pi}{2}\left(\frac{1}{c_0} + 3\right).$$

Proof. Using integration by parts with respect to the first order differential operator

$$(\mathcal{L}_t b)(x) \stackrel{def}{=} \frac{1}{1 + t(\Phi'(x))^2}\left(b(x) - i\Phi'(x)b'(x)\right),$$

we get

$$I(t) = I_1(t) + I_2(t) \quad \text{with}$$

$$I_1(t) \stackrel{def}{=} \int_{\mathbb{R}} e^{it\Phi(x)}\frac{i\Phi'(x)}{1 + t(\Phi'(x))^2}a'(x)\,dx \quad \text{and}$$

$$I_2(t) \stackrel{def}{=} \int_{\mathbb{R}} \frac{e^{it\Phi(x)}}{1 + t(\Phi'(x))^2}\left(1 + i\Phi''(x) - 2i\frac{t(\Phi'(x))^2\Phi''(x)}{1 + t(\Phi'(x))^2}\right)a(x)\,dx.$$

As $\left|\dfrac{i\Phi'(x)}{1+t(\Phi'(x))^2}\right| \le \dfrac{1}{2t^{\frac{1}{2}}}$, we get

$$I_1(t) \le \frac{1}{2t^{\frac{1}{2}}}\|a'\|_{L^1(\mathbb{R})}. \tag{8.7}$$

We now bound I_2. As $\Phi''(x) \ge c_0$, we have

$$\frac{1}{1+t(\Phi'(x))^2} \le \frac{1}{c_0}\frac{\Phi''(x)}{1+t(\Phi'(x))^2}.$$

Thus,

$$|I_2(t)| \le \left(\frac{1}{c_0}+3\right)\int_{\mathbb{R}}\frac{\Phi''(x)}{1+t(\Phi'(x))^2}|a(x)|\,dx.$$

For any positive ε, we have $|a(x)| \le \left(a(x)^2+\varepsilon^2\right)^{\frac{1}{2}}$. We infer that

$$|I_2(t)| \le \left(\frac{1}{c_0}+3\right)\int_{\mathbb{R}}\frac{\Phi''(x)}{1+t(\Phi'(x))^2}\left(a(x)^2+\varepsilon^2\right)^{\frac{1}{2}}\,dx.$$

By integration by parts, we deduce that

$$\begin{aligned}
|I_2(t)| &\le \left(\frac{1}{c_0}+3\right)\frac{1}{t^{\frac{1}{2}}}\int_{\mathbb{R}}\arctan\left(t^{\frac{1}{2}}\Phi'(x)\right)a'(x)\frac{a(x)}{(a(x)^2+\varepsilon^2)^{\frac{1}{2}}}\,dx\\
&\le \frac{C_0-1}{t^{\frac{1}{2}}}\int_{\mathbb{R}}|a'(x)|\,dx\\
&\le \frac{C_0-1}{t^{\frac{1}{2}}}\|a'\|_{L^1(\mathbb{R})}.
\end{aligned}$$

Together with (8.7), this completes the proof of the theorem.

8.1.4 Dispersive Estimates for the Wave Equation

The wave equation is a simplified model for the propagation of waves in a physical medium. In this subsection, we shall only consider the case of an isotropic medium so that the corresponding system reduces (after suitable normalization) to

$$(W) \qquad \begin{cases} \Box u = 0 \\ (u,\partial_t u)_{|t=0} = (u_0, u_1). \end{cases}$$

Here, $\Box$ denotes the wave operator $\partial_t^2 - \Delta$. The unknown function $u = u(t,x)$ is real-valued and depends only on $(t,x) \in \mathbb{R} \times \mathbb{R}^d$.

In the one-dimensional case $d = 1$, it may be easily shown that the solution of (W) is given (in the smooth case) by *d'Alembert's formula*,

$$u(t,x) = \frac{1}{2}\left(u_0(x+t) + u_0(x-t) + \int_{x-t}^{x+t}u_1(y)\,dy\right),$$

so we cannot expect the wave operator to have any (global) dispersive property or smoothing effect.

In the case of dimension $d \geq 2$ that we are going to study in the rest of this section, the situation is rather different. Easy computations in Fourier variables (similar to those which were carried out in the proof of Proposition 8.3) show that we have the following result.

Proposition 8.14. *If u_0 and u_1 are tempered distributions, then the unique solution of the linear wave equation (W) in $\mathcal{C}(\mathbb{R}; \mathcal{S}')$ is of the form*

$$u(t) = U^+(t)\gamma_+ + U^-(t)\gamma_- \quad \text{with}$$

$$\mathcal{F}\big(U^\pm(t)f\big)(\xi) \overset{def}{=} e^{\pm it|\xi|}\,\widehat{f}(\xi) \quad \text{and} \quad \widehat{\gamma}_\pm(\xi) \overset{def}{=} \frac{1}{2}\left(\widehat{u}_0(\xi) \pm \frac{1}{i|\xi|}\widehat{u}_1(\xi)\right).$$

Combining the above formula with Theorem 8.12 will enable us to prove the following dispersive estimate.

Proposition 8.15. *Assume that $d \geq 2$. Let $\mathcal{C} \overset{def}{=} \{\xi \in \mathbb{R}^d \ \ r \leq |\xi| \leq R\}$ for some positive r and R such that $r < R$. A constant C then exists such that if $\widehat{u}_0$ and $\widehat{u}_1$ are supported in the annulus $\mathcal{C}$, then u, the associate solution of the wave equation (W), satisfies*

$$\|u(t)\|_{L^\infty} \leq \frac{C}{|t|^{\frac{d-1}{2}}} \big(\|u_0\|_{L^1} + \|u_1\|_{L^1}\big) \quad \text{for all} \quad t \neq 0.$$

Remark 8.16. As the support of the Fourier transform is preserved by the flow of the constant coefficients wave equation (a property which is no longer true in the case of variable coefficients), the Fourier transform of the solution u is, at each time t, supported in the annulus $\mathcal{C}$.

Proof of Proposition 8.15. Due to the time reversibility of the wave equation, it suffices to prove the result for positive times. Let φ be a function in $\mathcal{D}(\mathbb{R}^d \setminus \{0\})$ with value 1 near $\mathcal{C}$. According to Proposition 8.14, we then have

$$u(t) = K^+(t, \cdot) \star \widetilde{\gamma}^+ + K^-(t, \cdot) \star \widetilde{\gamma}^- \quad \text{with}$$

$$\widetilde{\gamma}^\pm \overset{def}{=} \mathcal{F}^{-1}(\varphi\widehat{\gamma}^\pm) \quad \text{and} \quad K^\pm(t, x) \overset{def}{=} \int_{\mathbb{R}^d} e^{i(x|\xi)}e^{\pm it|\xi|}\varphi(\xi)\,d\xi.$$

We will temporarily assume the inequality

$$\|K^\pm(t, \cdot)\|_{L^\infty} \leq \frac{C}{t^{\frac{d-1}{2}}} \quad \text{for} \quad t > 0. \tag{8.8}$$

We then immediately get

$$\|u(t)\|_{L^\infty} \leq \frac{C}{t^{\frac{d-1}{2}}}\left(\|\widetilde{\gamma}^+\|_{L^1} + \|\widetilde{\gamma}^-\|_{L^1}\right).$$

Now, because

$$\tilde{\gamma}^{\pm} = \frac{1}{2}(u_0 \mp ih \star u_1),$$

where the L^1 function h stands for the inverse Fourier transform of $|\cdot|^{-1}\varphi$, we get the desired inequality for $\|u(t)\|_{L^\infty}$.

In order to complete the proof, we establish the inequality (8.8). As the L^∞ norm is invariant under dilation, it suffices to estimate $\|K(t,t\cdot)\|_{L^\infty}$. Now, Theorem 8.12 implies that

$$|K^{\pm}(t,tx)| \leq \frac{C}{t^{\frac{d-1}{2}}} + C \int_{\mathcal{C}_x} \left(1 + t\left|x \pm \frac{\xi}{|\xi|}\right|^2\right)^{-d} d\xi, \quad \text{where}$$

$$\mathcal{C}_x \stackrel{\text{def}}{=} \left\{\xi, \in \mathcal{C} \ / \ \left|x \pm \frac{\xi}{|\xi|}\right| \leq \frac{1}{2}\right\}.$$

If $\mathcal{C}_x$ is not empty, then $x \neq 0$. Hence, we can write the following orthogonal decomposition for any $\xi \in \mathcal{C}_x$:

$$\xi = \zeta_1 + \zeta' \quad \text{with} \quad \zeta_1 = \left(\xi\Big|\frac{x}{|x|}\right)\frac{x}{|x|} \quad \text{and} \quad \zeta' = \xi - \left(\xi\Big|\frac{x}{|x|}\right)\frac{x}{|x|}.$$

Knowing that ζ' is orthogonal to the vector x, we infer that

$$\left|x \pm \frac{\xi}{|\xi|}\right| \geq \frac{|\zeta'|}{|\xi|}.$$

Therefore, using the fact that $r \leq |\xi| \leq R$ for any $\xi \in \mathcal{C}$, we get

$$|K^{\pm}(t,tx)| \leq \frac{C}{t^{\frac{d-1}{2}}} + C \int_{\mathcal{C}} \frac{1}{(1 + t|\zeta'|^2)^d} \, d\zeta' \, d\zeta_1.$$

The change of variables $\tilde{\zeta} = t^{\frac{1}{2}}\zeta'$ gives (8.8). This completes the proof of the proposition. $\qquad\square$

8.1.5 The L^2 Boundedness of Some Fourier Integral Operators

In this subsection, we prove the L^2 boundedness of a particular case of Fourier integral operators. The proof relies on the techniques of Section 8.1.3 and will be useful in Chapter 9.

Consider a real-valued smooth function Φ over a neighborhood of $\mathbb{R}^d \times A$, where A is a compact subset of $\mathbb{R}^d$ such that for any ξ in A,

$$x \longmapsto \partial_\xi \Phi(x,\xi)$$

is a global 1-diffeomorphism of $\mathbb{R}^d$, in the sense given on page 41 (with a constant C independent of ξ), and is such that for any $\ell \geq 2$,

$$N_\ell(\Phi) \stackrel{\text{def}}{=} \sup_{\substack{(x,\xi)\in\mathbb{R}^d \times A \\ |\alpha|\leq\ell}} |\partial_\xi^\alpha \Phi(x,\xi)| < \infty.$$

Theorem 8.17. *Let Φ be a phase function satisfying the above hypotheses. Let σ be a smooth function supported in $\mathbb{R}^d \times A$. Consider the operator $\mathcal{I}$ defined on $\mathcal{S}(\mathbb{R}^d)$ by*

$$\mathcal{I}(\psi)(x) \overset{def}{=} \int_A e^{i\Phi(x,\xi)} \sigma(x,\xi) \widehat{\psi}(\xi) \, d\xi.$$

Then, $\mathcal{I}$ extends to a bounded linear operator on L^2, and there exists a constant C, depending only on $N_\ell(\Phi)$ and the supremum of a certain number of derivatives of σ, such that

$$\|\mathcal{I}(\psi)\|_{L^2} \leq C\|\psi\|_{L^2} \quad \text{for all } \psi \in L^2. \tag{8.9}$$

Proof. Arguing by density, it suffices to prove that (8.9) holds true for all ψ in $\mathcal{S}$. In that case we may write that

$$\mathcal{I}(\psi)(x) = \int_{\mathbb{R}^d} K(x,y)\psi(y) \, dy \quad \text{with} \quad K(x,y) \overset{def}{=} \int_A e^{i(\Phi(x,\xi)-(y|\xi))}\sigma(x,\xi) \, d\xi.$$

We now define the first order differential operator $\mathcal{L}$ by

$$\mathcal{L}a \overset{def}{=} \frac{a - i(\partial_\xi \Phi(x,\xi) - y)\cdot\partial_\xi a}{1 + |\partial_\xi \Phi(x,\xi) - y|^2}.$$

As $\mathcal{L}e^{i(\Phi(x,\xi)-(y|\xi))} = e^{i(\Phi(x,\xi)-(y|\xi))}$ we have, for any integer M,

$$|K(x,y)| = \left| \int e^{i(\Phi(x,\xi)-(y|\xi))} ({}^t\mathcal{L}^M \sigma)(x,y,\xi) \, d\xi \right|$$

$$\leq \int \left| ({}^t\mathcal{L}^M \sigma)(x,y,\xi) \right| d\xi.$$

We will temporarily assume the following inequality:

$$\left| ({}^t\mathcal{L}^M \sigma)(x,y,\xi) \right| \leq C_M(\Phi,\sigma) \frac{1}{(1 + |\partial_\xi \Phi(x,\xi) - y|^2)^{\frac{M}{2}}} \tag{8.10}$$

with $C_M(\Phi,\sigma) \overset{def}{=} C_M N_M(\Phi) \sup_{|\alpha| \leq M} \|\partial_\xi^\alpha \sigma\|_{L^\infty(\mathbb{R}^d \times A)}$.

Take $M = d + 1$ and define $C_{\Phi,\sigma} \overset{def}{=} C_{d+1}(\Phi,\sigma)$. For any φ in $L^2(\mathbb{R}^d)$ we then have

$$|(\mathcal{I}(\psi)|\varphi(x))| \leq \int_{\mathbb{R}^d \times \mathbb{R}^d} |K(x,y)| \, |\psi(y)| \, |\varphi(x)| \, dx \, dy$$

$$\leq C_{\Phi,\sigma} \int_{\mathbb{R}^d \times \mathbb{R}^d \times A} \frac{1}{(1 + |\partial_\xi \Phi(x,\xi) - y|^2)^{\frac{d+1}{2}}} |\psi(y)| \, |\varphi(x)| \, dx \, dy \, d\xi.$$

Applying the Cauchy–Schwarz inequality for the measure

$$(1 + |\partial_\xi \Phi(x,\xi) - y|^2)^{-(\frac{d+1}{2})} dx\, dy\, d\xi$$

gives

$$\left|\big(\mathcal{I}(\psi)|\varphi(x)\big)\right|^2 \leq C_{\Phi,\sigma}^2 \left(\int \frac{|\varphi(x)|^2}{(1 + |\partial_\xi \Phi(x,\xi) - y|^2)^{\frac{d+1}{2}}} dx\, dy\, d\xi\right)$$
$$\times \left(\int \frac{|\psi(y)|^2}{(1 + |\partial_\xi \Phi(x,\xi) - y|^2)^{\frac{d+1}{2}}} dx\, dy\, d\xi\right).$$

Integrating with respect to (y,ξ) (recall that integration with respect to ξ is performed over the compact set A) and then in x in the first integral gives

$$\left|\big(\mathcal{I}(\psi)|\varphi(x)\big)\right|^2 \leq C_{\Phi,\sigma}^2 \|\varphi\|_{L^2}^2 \left(\int \frac{|\psi(y)|^2}{(1 + |\partial_\xi \Phi(x,\xi) - y|^2)^{\frac{d+1}{2}}} dx\, dy\, d\xi\right).$$

Making the change of variable $x' = \widetilde{\Phi}(x,\xi) \overset{\text{def}}{=} \partial_\xi \Phi(x,\xi)$ and integrating first in x', we conclude that the last integral may be bounded by $C\|\psi\|_{L^2}^2$, which completes the proof of the theorem.

In order to prove the inequality (8.10), we may argue by induction. We claim that

$$(\mathcal{A}_M) \qquad \begin{aligned} &({}^t\mathcal{L}^M \sigma)(x,y,\xi) = f_M(x,\xi,\partial_x \Phi(x,\xi-y)) \quad \text{with} \\ &|\partial_\xi^\alpha \partial_\theta^\beta f_M(x,\xi,\theta)| \leq N_{M+|\alpha|}(\Phi) \sup_{|\alpha| \leq M+|\alpha|} \|\partial_\xi^\alpha \sigma\|_{L^\infty(\mathbb{R}^d \times A)}. \end{aligned}$$

We begin by proving $(\mathcal{A}_1)$. We have

$$ {}^t\mathcal{L}a = \frac{a + i(\partial_\xi \Phi(x,\xi) - y) \cdot \partial_\xi a}{1 + |\partial_\xi \Phi(x,\xi) - y|^2} - a \operatorname{div} \mathcal{L}. \qquad (8.11)$$

This implies that

$$({}^t\mathcal{L}\sigma)(x,y,\xi) = f_1(x,\xi,\partial_\xi \Phi(x,\xi) - y) \quad \text{with}$$
$$f_1(x,\xi,\theta) \overset{\text{def}}{=} \frac{1 + i\theta \cdot \partial_\xi \sigma}{1 + |\theta|^2} - \sigma\, d_\mathcal{L} \quad \text{and}$$
$$d_\mathcal{L} \overset{\text{def}}{=} i \frac{\Delta_\xi \Phi}{1 + |\theta|^2} - 2i \frac{D^2\Phi(\theta,\theta)}{(1 + |\theta|^2)^2}.$$

Now, assume $(\mathcal{A}_M)$. Observing that $\operatorname{div} \mathcal{L}(x,y,\xi) = d_\mathcal{L}(x,\xi,\partial_\xi \Phi(x,\xi) - y)$ and using (8.11), we get

$$({}^t\mathcal{L}^{M+1})\sigma(x,y,\xi) = \frac{1 + i(\partial_x \Phi(x,\xi) - y) \cdot \partial_\xi \big(f_M(x,\xi,\partial_\xi \Phi(x,\xi) - y)\big)}{1 + |\partial_\xi \Phi(x,\xi) - y|^2}$$
$$+ (f_M d_\mathcal{L})(x,\xi,\partial_\xi \Phi(x,\xi) - y).$$

Leibniz's formula then implies that

$$({}^t\mathcal{L}^{M+1})\sigma(x,y,\xi) = \frac{1 + i(\partial_x\Phi(x,\xi) - y)\cdot(\partial_\xi f_M)(x,\xi,\partial_\xi\Phi(x,\xi) - y)}{1 + |\partial_\xi\Phi(x,\xi) - y|^2}$$

$$+ \sum_{j,k} \frac{1 + i(\partial_{\xi_j}\Phi(x,\xi) - y_j)\partial_{\xi_j}\partial_{\xi_k}\Phi(\partial_{\theta_k} f_M)(x,\xi,\partial_x\Phi(x,\xi) - y)}{1 + |\partial_\xi\Phi(x,\xi) - y|^2}$$

$$+ (f_M d_\mathcal{L})(x,\xi,\partial_\xi\Phi(x,\xi) - y).$$

Thus, $(\mathcal{A}_M)$ is satisfied with

$$f_{M+1}(x,\xi,\theta) = \frac{1 + i\theta\cdot\partial_\xi f_M}{1 + |\theta|^2} + \sum_{j,k}\frac{\partial_{\xi_j}\partial_{\xi_k}\theta_j\partial_{\theta_k} f_M}{1 + |\theta|^2} - d_\mathcal{L} f_M.$$

This completes the proof of the theorem. $\qquad\square$

8.2 Bilinear Methods

This section describes the so-called $TT^\star$ *argument*, which is the standard method for converting the dispersive estimates (presented in the previous section) into inequalities involving suitable space-time Lebesgue norms of the solution. At the end of this section, those inequalities—the so-called *Strichartz estimates*—will be used to solve the cubic semilinear Schrödinger equation in dimension two. More applications will be given at the end of the chapter and in Chapter 10.

Throughout this section, we agree that the notation $\|\cdot\|_{L^q(L^r)}$ stands for the norm in $L^q(\mathbb{R}; L^r(\mathbb{R}^d))$. We now state the "abstract" Strichartz estimates.

Theorem 8.18. *Let $(U(t))_{t\in\mathbb{R}}$ be a bounded family of continuous operators on $L^2(\mathbb{R}^d)$ such that for some positive real numbers σ and C_0, we have*

$$\|U(t)U^\star(t')f\|_{L^\infty} \le \frac{C_0}{|t - t'|^\sigma}\|f\|_{L^1}. \tag{8.12}$$

Then, for any $(q,r)\in[2,\infty]^2$ such that

$$\frac{1}{q} + \frac{\sigma}{r} = \frac{\sigma}{2} \quad and \quad (q,r,\sigma)\neq(2,\infty,1), \tag{8.13}$$

we have, for some positive constant C,

$$\|U(t)u_0\|_{L^q(L^r)} \le C\|u_0\|_{L^2}, \tag{8.14}$$

$$\left\|\int_\mathbb{R} U^\star(t)f(t)\,dt\right\|_{L^2} \le C\|f\|_{L^{q'}(L^{r'})}. \tag{8.15}$$

Moreover, for any (q_1, r_1) and (q_2, r_2) satisfying (8.13), we have

$$\left\|\int_\mathbb{R} U(t)U^\star(t')f(t')\,dt'\right\|_{L^{q_1}(L^{r_1})} \le C\|f\|_{L^{q'_2}(L^{r'_2})}, \tag{8.16}$$

$$\left\|\int_{t'<t} U(t)U^\star(t')f(t')\,dt'\right\|_{L^{q_1}(L^{r_1})} \le C\|f\|_{L^{q'_2}(L^{r'_2})}. \tag{8.17}$$

8.2.1 The Duality Method and the $TT^\star$ Argument

The proof of Theorem 8.18 is based on a duality argument and on the Hardy–Littlewood–Sobolev inequality stated in Theorem 1.7 page 6.

We first note that

$$\|U(t)u_0\|_{L_t^q(L_x^r)} = \sup_{\varphi \in \mathcal{B}_{q,r}} \left| \int_{\mathbb{R} \times \mathbb{R}^d} U(t)u_0(x)\varphi(t,x)\, dt\, dx \right|$$

$$= \sup_{\varphi \in \mathcal{B}_{q,r}} \left| \int_{\mathbb{R}} (U(t)u_0 | \varphi(t))_{L^2}\, dt \right|,$$

where

$$\mathcal{B}_{q,r} \overset{\text{def}}{=} \{\varphi \in \mathcal{D}(\mathbb{R}^{1+d}; \mathbb{C}) \, / \, \|\varphi\|_{L^{q'}(L^{r'})} \le 1\}.$$

By the definition of the adjoint operator, we have

$$\|U(t)u_0\|_{L_t^q(L_x^r)} = \sup_{\varphi \in \mathcal{B}_{q,r}} \left| \left(u_0 \, \Big| \int_{\mathbb{R}} U^\star(t)\varphi(t)\, dt \right)_{L^2} \right|.$$

By virtue of the Cauchy–Schwarz inequality, we deduce that

$$\|U(t)u_0\|_{L_t^q(L_x^r)} \le \|u_0\|_{L^2} \sup_{\varphi \in \mathcal{B}_{q,r}} \left\| \int_{\mathbb{R}} U^\star(t)\varphi(t)\, dt \right\|_{L^2}. \tag{8.18}$$

Therefore, the inequality (8.15) implies the inequality (8.14). In order to prove (8.15), we write that

$$\left\| \int_{\mathbb{R}} U^\star(t)\varphi(t)\, dt \right\|_{L^2}^2 = \int_{\mathbb{R}^2} \left(U^\star(t')\varphi(t') | U^\star(t)\varphi(t) \right)_{L^2} dt'\, dt$$

$$= \int_{\mathbb{R}^2} \left(U(t)U^\star(t')\varphi(t') | \varphi(t) \right)_{L^2} dt'\, dt$$

$$= \int_{\mathbb{R}^2} \left\langle U(t)U^\star(t')\varphi(t'), \overline{\varphi}(t) \right\rangle dt'\, dt. \tag{8.19}$$

Observe that if we denote by T the solution operator

$$T : u_0 \longmapsto \left[t \mapsto U(t)u_0\right],$$

then

$$T^\star : \varphi \longmapsto \int U^\star(t)\varphi(t)\, dt.$$

Moreover, $TT^\star$ coincides with the operator

$$\varphi \longmapsto \left[t \mapsto \int_{\mathbb{R}} U(t)U^\star(t')\varphi(t')\, dt'\right].$$

The so-called $TT^\star$ argument is the observation that the inequality (8.15) [and thus (8.14)] is a consequence of the inequality (8.16): This is just a matter of taking $(q_1, r_1) = (q_2, r_2) = (q, r)$.

In fact, in order to prove both inequalities (8.16) and (8.17), it will be convenient to introduce the bilinear operator

$$T_\chi(f,g) \overset{\text{def}}{=} \int_{\mathbb{R}^2} \chi(t,t')\langle U(t)U^\star(t')f(t'), \overline{g}(t)\rangle \, dt' \, dt, \qquad (8.20)$$

where χ is a measurable function on $\mathbb{R}^2$ with values in the unit disc of $\mathbb{C}$.

In effect, taking appropriate functions χ and arguing by duality, we see that those two inequalities are consequences of the bilinear estimate

$$|T_\chi(f,g)| \leq C\|f\|_{L^{q_1'}(L^{r_1'})}\|g\|_{L^{q_2'}(L^{r_2'})} \qquad (8.21)$$

for all couples (q_j, r_j) in $[2, \infty]^2$ satisfying the relation (8.13). The rest of the this section is devoted to the proof of the inequality (8.21).

8.2.2 Strichartz Estimates: The Case $q > 2$

We first consider the case where $(q_1, r_1) = (q_2, r_2)$ and $q_1 > 2$. As $(U(t))_{t \in \mathbb{R}}$ is a bounded family of operators on L^2, we get, thanks to the dispersive estimate (8.12) and the linear interpolation result of Corollary 1.13 page 12,

$$\forall p \in [2, \infty], \ \|U(t)U^\star(t')f\|_{L^p} \leq \frac{C}{|t - t'|^{\sigma\left(1 - \frac{2}{p}\right)}}\|f\|_{L^{p'}}. \qquad (8.22)$$

Therefore, taking $p = r_1$, using relation (8.13), and applying the Hölder inequality gives

$$|T_\chi(f,g)| \leq C \int_{\mathbb{R}^2} \frac{1}{|t - t'|^{\frac{2}{q_1}}}\|f(t')\|_{L^{r_1'}}\|g(t)\|_{L^{r_1'}} \, dt' \, dt.$$

Because $q_1 > 2$, the Hardy–Littlewood–Sobolev inequality page 6 gives

$$|T_\chi(f,g)| \leq C\|f\|_{L^{q_1'}(L^{r_1'})}\|g\|_{L^{q_1'}(L^{r_1'})}, \qquad (8.23)$$

which is the inequality (8.21) in the case where $(q_1, r_1) = (q_2, r_2)$ and $q_1 > 2$. As pointed out above, this is enough to conclude that (8.15) holds in the case $q > 2$.

Next, writing that

$$T_\chi(f,g) = \int_{\mathbb{R}^2} \chi(t,t')\left(U^\star(t')f(t')|U^\star(t)g(t)\right)_{L^2} \, dt' \, dt$$

$$= \int_{\mathbb{R}} \left(\int_{\mathbb{R}} U^\star(t')f_t(t') \, dt' \Big| U^\star(t)g(t)\right)_{L^2} \, dt,$$

where $f_t(t') \overset{\text{def}}{=} \chi(t,t')f(t')$, we get, according to the Cauchy–Schwarz inequality and the fact that $U^\star(t)$ is uniformly bounded on L^2,

$$|T_\chi(f,g)| \leq \left(\sup_t \left\| \int_{\mathbb{R}} U^\star(t') f_t(t')\, dt' \right\|_{L^2} \right) \int_{\mathbb{R}} \|g(t)\|_{L^2}\, dt. \qquad (8.24)$$

From (8.15), we infer that for any admissible couple (q_1, r_1) with $q_1 > 2$,

$$|T_\chi(f,g)| \leq C\|f\|_{L^{q_1'}(L^{r_1'})} \|g\|_{L^1(L^2)}.$$

Interpolating between the above inequality and the inequality (8.23) [i.e., applying Corollary 1.13 page 12 with (q_1', r_1') and $(1,2)$], we get the inequality (8.21) for any pair of admissible couples (q_j, r_j) such that $2 < q_1 \leq q_2$. Now, in the above computations, it is clear that the roles of f and g may be exchanged. Hence, by the same token, we get the inequality in the case $2 < q_2 \leq q_1$.

8.2.3 Strichartz Estimates: The Endpoint Case $q = 2$

It suffices to prove that if $\sigma > 1$, then we have

$$\left\| \int_{\mathbb{R}} U^\star(t)\varphi(t)\, dt \right\|_{L^2}^2 \leq C\|\varphi\|_{L^2(L^{r'})}^2 \quad \text{with} \quad r = \frac{2\sigma}{\sigma - 1}. \qquad (8.25)$$

Indeed, the above inequality clearly implies the inequality (8.15) [and thus (8.14)]. Next, again using the inequality (8.24) and arguing exactly as in the case $q > 2$, it is easy to get the inequalities (8.16) and (8.17).

To prove the inequality (8.25), we shall show that the operator T_χ introduced in (8.20) is continuous on $\left(L^2(\mathbb{R}; L^{r'}(\mathbb{R}^d))\right)^2$. This result may be achieved by proceeding along the lines of the method that we used to prove the Hardy–Littlewood–Sobolev inequality. Indeed, let us decompose the bilinear functional T_χ into

$$T_\chi(f,g) = \sum_{j \in \mathbb{Z}} T_j(f,g) \quad \text{with}$$

$$T_j(f,g) \overset{\text{def}}{=} \int_{\mathbb{R}^2} \chi_j(t,t')\langle U(t)U^\star(t')f(t), \overline{g}(t')\rangle\, dt\, dt'$$

and $\chi_j(t,t') \overset{\text{def}}{=} \mathbb{1}_{2^j \leq |t-t'| < 2^{j+1}}(t)\chi(t,t')$.

The key to the proof is the following lemma.

Lemma 8.19. *There exists a neighborhood V of (r,r) such that for any (a,b) in V and any integer $j \in \mathbb{Z}$, we have*

$$|T_j(f,g)| \leq C 2^{-j\beta(a,b)}\|f\|_{L^2(L^{a'})}\|g\|_{L^2(L^{b'})} \quad \text{with} \quad \beta(a,b) = \sigma - 1 - \frac{\sigma}{a} - \frac{\sigma}{b}.$$

Proof. Using a dilation of size 2^j reduces the proof to the case $j = 0$. It thus suffices to show that

$$|T_0(f,g)| \leq C\|f\|_{L^2(L^{a'})}\|g\|_{L^2(L^{b'})}. \tag{8.26}$$

First, using (8.22), for any $a \geq 2$, we get

$$|T_0(f,g)| \leq C \int_{1 \leq |t-t'| \leq 2} \|U(t)U^\star(t')f(t')\|_{L^a}\|g(t)\|_{L^{a'}} \, dt' \, dt$$

$$\leq C \int_{1 \leq |t-t'| \leq 2} \|f(t')\|_{L^{a'}}\|g(t)\|_{L^{a'}} \, dt' \, dt,$$

which implies, thanks to Young's inequality (in time), that for any $a \geq 2$,

$$|T_0(f,g)| \leq C\|f\|_{L^2(L^{a'})}\|g\|_{L^2(L^{a'})}. \tag{8.27}$$

We now prove that for any $a \in [2, r[$, the following estimate holds:

$$|T_0(f,g)| \leq C\|f\|_{L^2(L^{a'})}\|g\|_{L^2(L^2)}. \tag{8.28}$$

Let $f_t(t') \overset{\text{def}}{=} \mathbb{1}_{1 \leq |t-t'| < 2}(t')\chi(t,t')f(t')$. By the definition of T_0, we have

$$T_0(f,g) = \int_{\mathbb{R}} \left(\int_{\mathbb{R}} U^\star(t')f_t(t') \, dt' \Big| U^\star(t)g(t) \right)_{L^2} dt.$$

From the Cauchy–Schwarz inequality, we then infer that

$$|T_0(f,g)| \leq C \int_{\mathbb{R}} \left\| \int_{\mathbb{R}} U^\star(t')f_t(t') \, dt' \right\|_{L^2} \|g(t)\|_{L^2} \, dt.$$

Applying the estimate (8.15) with $q(a)$ and $a \in [2, r[$ satisfying (8.13)[1] leads to

$$|T_0(f,g)| \leq C \int_{\mathbb{R}} \|f_t\|_{L^{q(a)'}(L^{a'})}\|g(t)\|_{L^2} \, dt.$$

We define $F_{a'}(t) \overset{\text{def}}{=} \|f(t)\|_{L^{a'}}$. Because $\chi \leq 1$, we get, by the definition of f_t,

$$|T_0(f,g)| \leq C \int_{\mathbb{R}} \left(\int_{1 \leq |t-t'| < 2} F_{a'}(t')^{q(a)'} \, dt' \right)^{\frac{1}{q(a)'}} \|g(t)\|_{L^2} \, dt$$

$$\leq C \int_{\mathbb{R}} \left(\mathbb{1}_{\{1 \leq |\tau| < 2\}} \star F_{a'}^{q(a)'} \right)^{\frac{1}{q(a)'}}(t)\|g(t)\|_{L^2} \, dt.$$

Thus, by the Cauchy–Schwarz inequality, we get

$$|T_0(f,g)| \leq C \left\| \mathbb{1}_{\{1 \leq |\tau| < 2\}} \star F_{a'}^{q(a)'} \right\|_{L_t^{\frac{2}{q(a)'}}}^{\frac{1}{q(a)'}} \|g\|_{L^2(L^2)}.$$

---

[1] Note that this implies that $q(a)$ is greater than 2, which is the case proved in Section 8.2.2.

As $q(a)' < 2$ and $\mathbb{1}_{\{1 \leq |\tau| < 2\}} \in L^1$, Young's inequality implies that

$$\left\| \mathbb{1}_{\{1 \leq |\tau| < 2\}} \star F_{a'}^{q(a)'} \right\|_{L_t^{\frac{2}{q(a)'}}} \leq C \|F_{a'}\|_{L_t^2}^{q(a)'}.$$

Thus, the inequality (8.28) is proved. Of course, as f and g play a symmetric role, similar arguments lead to

$$|T_0(f,g)| \leq C\|f\|_{L^2(L^2)} \|g\|_{L^2(L^{b'})} \quad \text{for any } b \in [2, r[. \tag{8.29}$$

Taking advantage of the bilinear interpolation result stated in Proposition 1.10 page 10, we conclude that

$$|T_0(f,g)| \leq C\|f\|_{L^2(L^{a'})} \|g\|_{L^2(L^{b'})}$$

whenever $(1/a', 1/b')$ is in the convex hull of

$$\left(\left[\frac{1}{2}, \frac{1}{r'} \right] \times \left\{ \frac{1}{2} \right\} \right) \cup \left(\left\{ \frac{1}{2} \right\} \times \left[\frac{1}{2}, \frac{1}{r'} \right[\right) \cup \left\{ (\gamma, \gamma) / \gamma \in \left[\frac{1}{2}, 1 \right] \right\},$$

which is obviously a neighborhood of $(1/r', 1/r')$. Lemma 8.19 is thus proved. $\square$

Completion of the proof of Theorem 8.18. We shall use the atomic decomposition of $f(t)$ and $g(t)$ defined in Section 1.1.2 page 7. Writing

$$f(t,x) = \sum_{k \in \mathbb{Z}} c_k(t) f_k(t,x) \quad \text{and} \quad g(t',x) = \sum_{\ell \in \mathbb{Z}} d_\ell(t') g_\ell(t',x),$$

and knowing that

$$\sigma - 1 = \frac{2\sigma}{r},$$

we infer that for any (a,b) in V,

$$|T_j(c_k f_k, d_\ell g_\ell)| \leq C\|c_k\|_{L^2(\mathbb{R})} \|d_\ell\|_{L^2(\mathbb{R})} 2^{-j\beta(a,b)} 2^{-k\left(\frac{1}{r'} - \frac{1}{a'}\right)} 2^{-\ell\left(\frac{1}{r'} - \frac{1}{b'}\right)}$$
$$\leq C\|c_k\|_{L^2(\mathbb{R})} \|d_\ell\|_{L^2(\mathbb{R})} 2^{(-j\sigma + k)\left(\frac{1}{r} - \frac{1}{a}\right)} 2^{(-j\sigma + \ell)\left(\frac{1}{r} - \frac{1}{b}\right)}.$$

Choosing a and b such that

$$\left| \frac{1}{r} - \frac{1}{a} \right| = \left| \frac{1}{r} - \frac{1}{b} \right| = 2\varepsilon,$$

$$(-j\sigma + k) \left(\frac{1}{r} - \frac{1}{a} \right) < 0, \quad \text{and} \quad (-j\sigma + \ell) \left(\frac{1}{r} - \frac{1}{b} \right) < 0$$

for some suitably small ε, we then get

$$|T_j(f_k, g_\ell)| \leq C\|c_k\|_{L^2(\mathbb{R})} \|d_\ell\|_{L^2(\mathbb{R})} 2^{-2\varepsilon|j\sigma - k|} 2^{-2\varepsilon|j\sigma - \ell|}$$
$$\leq C\|c_k\|_{L^2(\mathbb{R})} \|d_\ell\|_{L^2(\mathbb{R})} 2^{-\varepsilon|j\sigma - k|} 2^{-\varepsilon|k - \ell|}.$$

This gives

$$|T(f,g)| \leq C \sum_{j,k,\ell} \|c_k\|_{L^2(\mathbb{R})} \|d_\ell\|_{L^2(\mathbb{R})} 2^{-\varepsilon|j\sigma-k|} 2^{-\varepsilon|k-\ell|}$$

$$\leq C \sum_{k,\ell} \|c_k\|_{L^2(\mathbb{R})} \|d_\ell\|_{L^2(\mathbb{R})} 2^{-\varepsilon|k-\ell|}.$$

Using Young's inequality for series, we deduce that

$$|T(f,g)| \leq C \Big(\sum_k \|c_k\|_{L^2(\mathbb{R})}^2 \Big)^{\frac{1}{2}} \Big(\sum_\ell \|d_\ell\|_{L^2(\mathbb{R})}^2 \Big)^{\frac{1}{2}}$$

$$\leq C \Big(\int_{\mathbb{R}} \|(c_k(t))_k\|_{\ell^2}^2 \, dt \Big)^{\frac{1}{2}} \Big(\int_{\mathbb{R}} \|(d_\ell(t))_\ell\|_{\ell^2}^2 \, dt \Big)^{\frac{1}{2}}.$$

The fact that $r' < 2$ implies that $\|(c_k(t))_k\|_{\ell^2} \leq \|(c_k(t))_k\|_{\ell^{r'}}$. Owing to the properties of the atomic decomposition, we thus get

$$|T(f,g)| \leq C \Big(\int_{\mathbb{R}} \|(c_k(t))_k\|_{\ell^{r'}}^2 \, dt \Big)^{\frac{1}{2}} \Big(\int_{\mathbb{R}} \|(d_\ell(t))_k\|_{\ell^{r'}}^2 \, dt \Big)^{\frac{1}{2}}$$

$$\leq C \|f\|_{L^2(L^{r'})} \|g\|_{L^2(L^{r'})}. \tag{8.30}$$

Taking $f = g = \varphi$, we get the inequality (8.25), from which follows Theorem 8.18 in the endpoint case.

8.2.4 Application to the Cubic Semilinear Schrödinger Equation

As an application of the results of the previous section, we here solve the initial boundary value problem for the *cubic semilinear Schrödinger equation* in $\mathbb{R}^2$:

$$(NLS_3) \quad \begin{cases} i\partial_t u - \dfrac{1}{2} \Delta u = P_3(u, \overline{u}) \\ u_{|t=0} = u_0, \end{cases}$$

where P_3 is some given homogeneous polynomial of degree 3.

Theorem 8.20. *There exists a constant c such that for any initial data u_0 in $L^2(\mathbb{R}^2)$ satisfying $\|u_0\|_{L^2} \leq c$, the system (NLS_3) has a unique solution u in the space $L^\infty(\mathbb{R}; L^2(\mathbb{R}^2)) \cap L^3(\mathbb{R}; L^6(\mathbb{R}^2))$.*

Remark 8.21. We will first look at the scaling properties of the equation (NLS_3). If u is a solution of (NLS_3), then $u_\lambda(t,x) \overset{\text{def}}{=} \lambda u(\lambda^2 t, \lambda x)$ is also a solution of the same equation. In the family of Sobolev spaces, $L^2(\mathbb{R}^2)$ is the only invariant space.

Proof of Theorem 8.20. Let Q be the nonlinear functional defined by

$$\begin{cases} i\partial_t Q(u) - \dfrac{1}{2}\Delta Q(u) = P_3(u, \overline{u}) \\ Q(u)_{|t=0} = 0. \end{cases}$$

According to Theorem 8.18 and Proposition 8.6, this functional continuously maps $L^3(\mathbb{R}; L^6(\mathbb{R}^2))$ into $L^\infty(\mathbb{R}; L^2(\mathbb{R}^2)) \cap L^3(\mathbb{R}; L^6(\mathbb{R}^2))$. Indeed, using the fact that the group $(U(t))_{t\in\mathbb{R}}$ defined in Remark 8.4 is unitary, together with Duhamel's formula, we may write

$$Q(u)(t) = \int_0^t U(t-t') P_3(u(t'), \overline{u}(t')) \, dt'.$$

The inequality (8.17) leads to

$$\|Q(u)\|_{L^3(\mathbb{R};L^6(\mathbb{R}^2))} \le C \|P_3(u, \overline{u})\|_{L^1(\mathbb{R};L^2(\mathbb{R}^2))}$$
$$\le C \|u\|^3_{L^3(\mathbb{R};L^6(\mathbb{R}^2))}.$$

As $Q(u) - Q(v)$ satisfies

$$\left(i\partial_t + \dfrac{1}{2}\Delta\right)(Q(u) - Q(v)) = P_3(u, \overline{u}) - P_3(v, \overline{v}),$$

we get, again using the inequality (8.17),

$$\|Q(u) - Q(v)\|_{L^\infty(\mathbb{R};L^2(\mathbb{R}^2)) \cap L^3(\mathbb{R};L^6(\mathbb{R}^2))} \le C \|u - v\|_{L^3(\mathbb{R};L^6(\mathbb{R}^2))}$$
$$\times \left(\|u\|^2_{L^3(\mathbb{R};L^6(\mathbb{R}^2))} + \|v\|^2_{L^3(\mathbb{R};L^6(\mathbb{R}^2))}\right). \qquad (8.31)$$

It is now obvious that u is a solution of (NLS_3) if and only if it is a fixed point of the map

$$F(u) \overset{\text{def}}{=} U(t)u_0 + Q(u).$$

Applying Theorem 8.18 and the estimate (8.31) with $v = 0$, we get that

$$\|F(u)\|_{L^3(\mathbb{R};L^6)} \le C\|u_0\|_{L^2} + C\|u\|^3_{L^3(\mathbb{R};L^6)}.$$

Thus, if $8C^2\|u_0\|^2_{L^2} \le 1$, then the ball $B(0, 2C\|u_0\|_{L^2})$ with center 0 and radius $2C\|u_0\|_{L^2}$ in the Banach space $L^3(\mathbb{R}; L^6(\mathbb{R}^2))$ is invariant with respect to the map F. Again using the inequality (8.31), we get, for any u and v in $B(0, 2C\|u_0\|_{L^2})$,

$$\|F(u) - F(v)\|_{L^3(\mathbb{R};L^6)} \le 8C^3\|u_0\|^2_{L^2}\|u - v\|_{L^3(\mathbb{R};L^6)}.$$

Thus, if, in addition,

$$8C^3\|u_0\|^2_{L^2} \le \dfrac{1}{2},$$

then Picard's fixed point theorem implies that a unique solution u exists in some neighborhood of 0 in $L^3(\mathbb{R}; L^6)$. Clearly, the inequality (8.31) (localized on a sufficiently small time interval) implies that uniqueness holds true in $L^3(\mathbb{R}; L^6)$ without any smallness condition.

Finally, the energy estimate entails that this solution belongs to $L^\infty(\mathbb{R}; L^2)$. Indeed, multiplying the equation (NLS_3) by $\overline{u}$, integrating over $\mathbb{R}^2$, and then taking the real part, we discover that

$$\frac{1}{2}\frac{d}{dt}\|u\|_{L^2}^2 = \mathcal{I}m \int \overline{u}P_3(u,\overline{u})\,dx,$$

from which it follows, for all $t \in \mathbb{R}$, that

$$\|u(t)\|_{L^2} \le \|u_0\|_{L^2} + C\left|\int_0^t \|u\|_{L^6}^3\,d\tau\right|.$$

This completes the proof of the theorem.   □

We end this subsection with a contraction mapping lemma, a generalization of the one stated on page 207.

Lemma 8.22. *Let X and Y be two Banach spaces such that Y is continuously included in X, and let L be a continuous linear map from X to X which also continuously maps Y into Y and satisfies*

$$\|L\|_{\mathcal{L}(X)} < 1 \quad and \quad \|L\|_{\mathcal{L}(Y)} < 1.$$

Consider a finite, increasing family of integers $(m_j)_{1\le j\le N}$ such that $m_1 \ge 2$ and a family $(B_j)_{1\le j\le N}$ of operators from X^{m_j} into X which also map Y^{m_j} into Y and are such that for all $\ell \in \{1,\ldots,m_j\}$,

$$\begin{cases} X \longrightarrow X \\ x \longmapsto B_j(x_1,\ldots,x_{\ell-1},x,x_{\ell+1},\ldots,x_{m_j}) \end{cases}$$

is linear or antilinear. Assume, in addition, that for any $j \in \{1,\ldots,N\}$,

$$\|B_j(x_1,\ldots,x_{m_j})\|_X \le c_j \prod_{m=1}^{m_j} \|x_m\|_X \quad and$$

$$\|B_j(y_1,\ldots,y_{m_j})\|_Y \le c_j \min_{1\le m\le m_j} \|y_m\|_Y \prod_{m'\neq m} \|y_{m'}\|_X.$$

If x_0 belongs to Y and satisfies $\|x_0\|_Y \le \alpha_0$ with

$$\alpha_0 \overset{def}{=} \min_{1\le j\le N}\left(\frac{1-\max\{\|L\|_{\mathcal{L}(X)},\|L\|_{\mathcal{L}(Y)}\}}{4(N+1)m_jA^{m_j-1}c_j}\right)^{\frac{1}{m_j-1}} \quad and$$

$$A \overset{def}{=} \frac{N+1}{1-\max\{\|L\|_{\mathcal{L}(X)},\|L\|_{\mathcal{L}(Y)}\}},$$

then the equation

$$x = x_0 + Lx + \sum_{j=1}^{N} B_j(x,\ldots,x)$$

has a unique solution in the ball with center 0 and radius $2\alpha_0$ in X, which also belongs to Y.

Proof. Consider the classical iterative scheme

$$x_{n+1} = x_0 + L(x_n) + \sum_{j=1}^{N} B_j(x_n, \dots, x_n).$$

We have

$$\|x_{n+1}\|_X \leq \|x_0\|_X + \|L\|_{\mathcal{L}(X)}\|x_n\|_X + \sum_{j=1}^{N} c_j\|x_n\|_X^{m_j}.$$

Assume that for any $n' \leq n$, we have $\|x_{n'}\|_X \leq A\|x_0\|_X$. Then,

$$\|x_{n+1}\|_X \leq \|x_0\|_X \left(1 + A\|L\|_{\mathcal{L}(X)} + A\sum_{j=1}^{N} A^{m_j-1}c_j\|x_0\|_X^{m_j-1}\right).$$

Thus, if $\|x_0\|_X \leq \alpha$ with

$$\alpha \stackrel{\text{def}}{=} \min_{1 \leq j \leq N} \left(\frac{(1 - \|L\|_{\mathcal{L}(X)})}{c_j(N+1)A^{m_j-1}}\right)^{\frac{1}{m_j-1}},$$

then $\|x_{n+1}\|_X \leq A\|x_0\|_X$. We will now prove that if $\|x_0\|_X$ is sufficiently small, then $(x_n)_{n\in\mathbb{N}}$ is a Cauchy sequence in both X and Y. We have

$$x_{n+1} - x_n = L(x_n) - L(x_{n-1}) + \sum_{j=1}^{N} B_j(x_n, \dots, x_n) - B_j(x_{n-1}, \dots, x_{n-1}).$$

For each j, the difference $B_j(x_n, \dots, x_n) - B_j(x_{n-1}, \dots, x_{n-1})$ is the sum of m_j terms of the form $B_j(x_n, \dots, x_n, x_n - x_{n-1}, \dots, x_{n-1}, \dots, x_{n-1})$. This gives

$$\|x_{n+1} - x_n\|_X \leq \|x_n - x_{n-1}\|_X \left(\|L\|_{\mathcal{L}(X)} + \sum_{j=1}^{N} c_j m_j A^{m_j-1}\|x_0\|_X^{m_j-1}\right).$$

Thus, if

$$\|x_0\|_X \leq \alpha' \stackrel{\text{def}}{=} \min_{1 \leq j \leq N} \left(\frac{(1 - \|L\|_{\mathcal{L}(X)})}{(N+1)m_j A^{m_j-1}c_j}\right)^{\frac{1}{m_j-1}},$$

then $(x_n)_{n\in\mathbb{N}}$ is a Cauchy sequence in X. Now, using the estimate involving the space Y, we get that if

$$\|x_0\|_Y \leq \alpha_0 \stackrel{\text{def}}{=} \min_{1 \leq j \leq N} \left(\frac{(1 - \max\{\|L\|_{\mathcal{L}(X)}), \|L\|_{\mathcal{L}(Y)}\})}{(N+1)m_j 2^{m_j-1}c_j}\right)^{\frac{1}{m_j-1}},$$

then $(x_n)_{n\in\mathbb{N}}$ is also a Cauchy sequence in Y. The uniqueness is then obvious and the lemma is proved. □

8.3 Strichartz Estimates for the Wave Equation

We now come to some applications and refinements of the Strichartz estimates for the linear wave equation. Those estimates turn out to be of particular importance for the study of semilinear wave equations (see Sections 8.4 and 8.5). For simplicity, we shall focus on inequalities pertaining to the interval $[0, T[$ for some given T in $]0, \infty]$. It goes without saying that similar results may be proven for *any* time interval since *the generic constant C that we shall use below does not depend on T.*

In the rest of this chapter and in Chapter 9, we adopt the notation

$$\partial \overset{\text{def}}{=} (\partial_{x_1}, \dots, \partial_{x_d}), \quad \nabla \overset{\text{def}}{=} (\partial_t, \partial_{x_1}, \dots, \partial_{x_d}), \quad \text{and} \quad \partial_0 \overset{\text{def}}{=} \partial_t,$$

which is commonly used for semilinear and quasilinear wave equations (and, in particular, for the Einstein equations in relativity theory). Note that, here, the meaning of the operator ∇ is different from in the other parts of this book as it also involves the first order time derivative.

8.3.1 The Basic Strichartz Estimate

We first introduce the following definition.

Definition 8.23. *We will say that a pair (q, r) in $[2, \infty]^2$ is wave admissible if there exists some $\widetilde{r}$ in $[2, r]$ such that*

$$\frac{2}{q} + \frac{d-1}{\widetilde{r}} = \frac{d-1}{2} \quad \text{with} \quad (q, \widetilde{r}, d) \neq (2, \infty, 3). \tag{8.32}$$

The main result of this subsection is the following.

Theorem 8.24. *Assume that the space dimension d is greater than or equal to 2. For any wave admissible pairs (q_1, r_1) and (q_2, r_2), a constant C exists such that for any j in $\mathbb{Z}$,*

$$\|\nabla \dot{\Delta}_j u\|_{L_T^{q_1}(L^{r_1})} \leq C 2^{j\mu_1} \|\dot{\Delta}_j \nabla u(0)\|_{L^2} + C 2^{j\mu_{12}} \|\dot{\Delta}_j \Box u\|_{L_T^{q_2'}(L^{r_2'})} \tag{8.33}$$

with

$$\mu_1 \overset{\text{def}}{=} d\left(\frac{1}{2} - \frac{1}{r_1}\right) - \frac{1}{q_1} \quad \text{and} \quad \mu_{12} = d\left(1 - \frac{1}{r_1} - \frac{1}{r_2}\right) - \frac{1}{q_1} - \frac{1}{q_2}. \tag{8.34}$$

Proof. The solution u of the linear Cauchy problem (W) can be written as $u = v + w$, where v is the solution of the homogeneous wave equation

$$\begin{cases} \partial_t^2 v - \Delta v = 0 \\ (v, \partial_t v)_{|t=0} = (u_0, u_1), \end{cases}$$

and w is the solution of the nonhomogeneous wave equation

$$\begin{cases} \partial_t^2 w - \Delta w = f \stackrel{\text{def}}{=} \Box u \\ (w, \partial_t w)_{|t=0} = (0,0). \end{cases}$$

Using the notation introduced in Proposition 8.14 and Duhamel's principle, we can write, for all $t \in [0,T]$,

$$v(t) = U^+(t)\gamma_+ + U^-(t)\gamma_-,$$

$$w(t) = \int_0^t \left(U^+(t-t')f_+(t') + U^-(t-t')f_-(t') \right) dt'$$

with $\widehat{f_\pm}(t',\xi) = \pm\dfrac{1}{2i|\xi|}\widehat{f}(t',\xi)$.

From Bernstein's inequality, Proposition 8.15, and Theorem 8.18, we infer that for any couple $(q_j, \widetilde{r}_j)$ satisfying (8.32), we have

$$\|\dot{\Delta}_0 \nabla u\|_{L_T^{q_1}(L^{\widetilde{r}_1})} \leq C\left(\|\dot{\Delta}_0 \nabla u(0)\|_{L^2} + \|\dot{\Delta}_0 f\|_{L_T^{q_2'}(L^{\widetilde{r}_2'})} \right).$$

Since $r_1 \geq \widetilde{r}_1$ and $r_2' \leq \widetilde{r}_2'$, we deduce, using Bernstein's inequality, that

$$\|\dot{\Delta}_0 \nabla u\|_{L_T^{q_1}(L^{r_1})} \leq C\left(\|\dot{\Delta}_0 \nabla u(0)\|_{L^2} + \|\dot{\Delta}_0 f\|_{L_T^{q_2'}(L^{r_2'})} \right).$$

This gives the result for $j = 0$. The result for all $j \in \mathbb{Z}$ follows by means of an obvious rescaling. □

The two simple corollaries that we state next will prove to be very useful in the next sections.

Corollary 8.25. *For any wave admissible pairs (q_j, r_j) and any real σ, a constant C exists such that, using the notation of Theorem 8.24,*

$$\|\nabla u\|_{L_T^{q_1}(\dot{B}_{r_1,2}^{\sigma})} \leq C\left(\|\nabla u(0)\|_{\dot{H}^{\sigma+\mu_1}} + \|\Box u\|_{L_T^{q_2'}(\dot{B}_{r_2',2}^{\sigma+\mu_{12}})} \right). \quad (8.35)$$

Proof. Thanks to Theorem 8.24, we have, for any j in $\mathbb{Z}$,

$$2^{j\sigma}\|\dot{\Delta}_j \nabla u\|_{L_T^{q_1}(L^{r_1})} \leq C2^{j(\sigma+\mu_1)}\|\dot{\Delta}_j \nabla u(0)\|_{L^2} + C2^{j(\sigma+\mu_{12})}\|\dot{\Delta}_j \Box u\|_{L_T^{q_2'}(L^{r_2'})}.$$

Taking the $\ell^2(\mathbb{Z})$ norm of both sides, we get

$$\left(\sum_{j\in\mathbb{Z}} 2^{2j\sigma}\|\dot{\Delta}_j \nabla u\|_{L_T^{q_1}(L^{r_1})}^2 \right)^{\frac{1}{2}} \leq C\Bigg(\|\nabla u(0)\|_{\dot{H}^{\sigma+\mu_1}}$$

$$+ \left(\sum_{j\in\mathbb{Z}} 2^{2j(\sigma+\mu_{12})}\|\dot{\Delta}_j \Box u\|_{L_T^{q_2'}(L^{r_2'})}^2 \right)^{\frac{1}{2}} \Bigg).$$

As $q_1 \geq 2$ and $q_2' \leq 2$, the Minkowski inequality implies the theorem. □

Remark 8.26. Note that the "natural" norms which appear are the ones which were introduced in Definition 2.67 page 98. For instance, as a by-product of the proof of Corollary 8.25, we have the (slightly more accurate) inequality

$$\|\nabla u\|_{\widetilde{L}_T^{q_1}(\dot{B}_{r_1,2}^{\sigma})} \leq C\left(\|\nabla u(0)\|_{\dot{H}^{\sigma+\mu_1}} + \|\Box u\|_{\widetilde{L}_T^{q_2'}(\dot{B}_{r_2',2}^{\sigma+\mu_{12}})}\right) \tag{8.36}$$

whenever (q_1, r_1) and (q_2, r_2) are wave admissible pairs.

The following corollary is particularly useful.

Corollary 8.27. *For any wave admissible pair* (q, r), *a constant* C *exists such that*

$$\|u\|_{L_T^q(L^r)} \leq C\left(\|\nabla u(0)\|_{\dot{H}^{\mu-1}} + \|\Box u\|_{L_T^1(\dot{H}^{\mu-1})}\right) \quad \text{with} \quad \mu = d\left(\frac{1}{2} - \frac{1}{r}\right) - \frac{1}{q}.$$

Proof. Applying Corollary 8.25 with $(q_2, r_2) = (\infty, 2)$ and $\sigma = -1$, we get

$$\|u\|_{L_T^q(\dot{B}_{r,2}^0)} \leq C\left(\|\nabla u(0)\|_{\dot{H}^{\mu-1}} + \|\Box u\|_{L_T^1(\dot{H}^{\mu-1})}\right).$$

Theorem 2.40 page 79 implies the result.   □

Remark 8.28. The term $1/q$ in the definition of the index μ may be interpreted as a gain of $1/q$ derivative compared with the Sobolev embedding.

Corollary 8.29. *For any wave admissible pairs* (q_1, r_1) *and* (q_2, r_2), *and any* μ *such that*

$$\mu = d\left(\frac{1}{2} - \frac{1}{r_1}\right) - \frac{1}{q_1} \quad \text{and} \quad d\left(1 - \frac{1}{r_1} - \frac{1}{r_2}\right) = 1 + \frac{1}{q_1} + \frac{1}{q_2},$$

a constant C *exists such that*

$$\|u\|_{L_T^{q_1}(L^{r_1})} \leq C\left(\|\nabla u(0)\|_{\dot{H}^{\mu-1}} + \|\Box u\|_{L_T^{q_2'}(L^{r_2'})}\right). \tag{8.37}$$

Proof. Applying Corollary 8.25 with $\sigma = -1$, we get that

$$\|u\|_{L_T^{q_1}(\dot{B}_{r_1,2}^0)} \leq C\left(\|\nabla u(0)\|_{\dot{H}^{\mu-1}} + \|\Box u\|_{L_T^{q_2'}(\dot{B}_{r_2',2}^0)}\right).$$

Theorem 2.40 page 79 implies the result.   □

In dimension three, the endpoint estimate [i.e., the control of the $L^2(L^\infty)$ norm] for solutions of the Cauchy problem (W) turns out to be false. However, the following logarithmic estimate is available.

Theorem 8.30. *Assume that the dimension* d *is equal to 3. Let* $\mathcal{C}$ *be an annulus. There exists a constant* C *such that for any positive real numbers* λ *and* T, *and any function* u *such that for any* t, *the support of* $\mathcal{F}u(t, \cdot)$ *is included in* $\lambda\mathcal{C}$, *we have*

$$\|u\|_{L_T^2(L^\infty)} \leq C\left(\log(e + \lambda T)\right)^{\frac{1}{2}}\left(\|\nabla u(0)\|_{L^2} + \|\Box u\|_{L_T^1(L^2)}\right). \tag{8.38}$$

Proof. The proof relies on the TT^* argument in a rather simple way, starting as in the proof of Theorem 8.18. Defining $U(t)\gamma \overset{\text{def}}{=} \mathcal{F}^{-1}(e^{it|\xi|})\widehat{\gamma}(\xi)$, we have

$$\|U(t)\gamma\|_{L^2([0,T];L^\infty)} \leq \|\gamma\|_{L^2} \sup \left\| \int_0^T U(t)\varphi(t)\,dt \right\|_{L^2},$$

where the supremum is taken over the functions φ with $\|\varphi\|_{L^2_T(L^1)} \leq 1$ and such that for each t, the support of $\mathcal{F}\varphi(t,\cdot)$ is included in $\lambda\mathcal{C}$. Using a dilation of size λ, we then observe that it is enough to prove the inequality in the case where $\lambda = 1$. As previously, we write, using the fact that U is a unitary operator,

$$\left\| \int_0^T U(t)\varphi(t)\,dt \right\|_{L^2}^2 = \left(\int_0^T U(t')\varphi(t')\,dt' \middle| \int_0^T U(t)\varphi(t)\,dt \right)_{L^2}$$
$$= \int_{[0,T]^2} \left(U(t-t')\varphi(t') \middle| \varphi(t) \right)_{L^2} dt'\,dt.$$

Thanks to Proposition 8.15 and because $\mathrm{Supp}\,\widehat{\varphi}(t,\cdot) \subset \mathcal{C}$, we have

$$\|U(t-t')\varphi(t)\|_{L^\infty} \leq \frac{C}{|t-t'|}\|\varphi(t)\|_{L^1},$$
$$\|U(t-t')\varphi(t)\|_{L^\infty} \leq C\|U(t-t')\varphi(t)\|_{L^2} = C\|\varphi(t)\|_{L^2} \leq C'\|\varphi(t)\|_{L^1}.$$

Therefore,

$$\left\| \int_0^T U(t)\varphi(t)\,dt \right\|_{L^2}^2 \leq C \int_{[0,T]^2} \frac{1}{1+|t-t'|}\|\varphi(t)\|_{L^1}\|\varphi(t')\|_{L^1}\,dt'\,dt$$
$$\leq C\log(e+T)\|\varphi\|_{L^2([0,T];L^1)}^2.$$

Thus, Theorem 8.30 is proved. □

8.3.2 The Refined Strichartz Estimate

In some situations (which we shall encounter later in the study of nonlinear dispersive equations), the standard Strichartz estimates are not accurate enough to control the nonlinearity. In this subsection, we will give some refined Strichartz inequalities.

Theorem 8.31. *Let u be a function on $\mathbb{R} \times \mathbb{R}^d$ such that for any t, the support of the Fourier transform of $u(t,\cdot)$ is included in some ball $B(\xi_j,h)$ with $|\xi_j| \in [2^{j-2}, 2^{j+2}]$ and $h \leq |\xi_j|/2$. Then, for any wave admissible couple (q,r), we have*

$$\|\nabla u\|_{L^q(L^r)} \leq C2^{j\mu}h^{\frac{1}{q}}(\|\nabla u(0)\|_{L^2} + \|\Box u\|_{L^1(L^2)}) \quad \text{with} \quad \mu = d\left(\frac{1}{2} - \frac{1}{r}\right) - \frac{1}{q}.$$

Proof. By virtue of Theorem 8.18 page 349, dilation arguments, and Duhamel's principle, the theorem reduces to the following proposition.   □

Proposition 8.32. *Let ξ_0 in $\mathbb{R}^d$ be such that $|\xi_0| \in [1/4, 4]$. A constant C exists such that for any $h \in]0, |\xi_0|/2]$ and any $\gamma \in L^1$, the Fourier transform of which is supported in the ball with center ξ_0 and radius h, we have*

$$\forall t > 0, \quad \|\mathcal{F}^{-1}\big(e^{\pm it|\xi|}\widehat{\gamma}\big)\|_{L^\infty} \leq C \min\Big\{\frac{h}{t^{\frac{d-1}{2}}}, h^d\Big\}\|\gamma\|_{L^1}.$$

Proof. We shall follow the idea of the proof of Proposition 8.15. It is obvious that under the hypothesis of Proposition 8.32, we can write

$$u(t) = \sum_{\pm} K^{\pm}(t, h, \cdot) \star \widehat{\gamma}^{\pm} \quad \text{with}$$

$$\widehat{\gamma}^{\pm} \stackrel{\text{def}}{=} \mathcal{F}^{-1}(\varphi(\xi)\widehat{\gamma}^{\pm}(\xi)) \quad \text{and}$$

$$K^{\pm}(t, h, x) \stackrel{\text{def}}{=} \int_{\mathbb{R}^d} e^{i(x|\xi)\pm t|\xi|} \psi\Big(\frac{\xi - \xi_0}{h}\Big)\, d\xi,$$

where $\varphi \in \mathcal{D}(\mathbb{R}^d \setminus \{0\})$ has value 1 near the annulus $\{\xi \in \mathbb{R}^d \,/\, 1/4 \leq |\xi| \leq 4\}$, and ψ is a function in $\mathcal{D}(\mathbb{R}^d)$ with value 1 near the unit ball.

First, we note that we obviously have

$$\|K^{\pm}(t, h, \cdot)\|_{L^\infty} \leq Ch^d. \tag{8.39}$$

Therefore, along the same lines as the proof of Proposition 8.15, it is enough to prove that the kernel $K^{\pm}$ satisfies

$$\|K^{\pm}(t, h, \cdot)\|_{L^\infty} \leq \frac{Ch}{t^{\frac{d-1}{2}}}. \tag{8.40}$$

Note that the inequality (8.39) implies that

$$\|K^{\pm}(t, h, \cdot)\|_{L^\infty} \leq \frac{Ch}{t^{\frac{d-1}{2}}} \quad \text{for} \quad th^2 \leq 1. \tag{8.41}$$

In the case where $th^2 \geq 1$, we shall proceed as in the proof of Theorem 8.12, except that we will have to control the dependency with respect to h. In order to do so, we introduce the following definition.

Definition 8.33. *Let Ω be an open subset of $\mathbb{R}^d \times \mathbb{R}^\star$ such that*

$$\forall h \in]0, 1], \quad |\Omega_h| \leq 1 \quad \text{with} \quad \Omega_h \stackrel{\text{def}}{=} \{\xi \in \mathbb{R}^d \,/\, (\xi, h) \in \Omega\}.$$

Let $\widetilde{\mathcal{D}}(\Omega)$ be the set of functions ψ from Ω to $\mathbb{C}$ such that for any $h \in \Pi_{\mathbb{R}^\star}(\Omega)$, where $\Pi_{\mathbb{R}^\star}$ denotes the projection of Ω on $\mathbb{R}^\star$, the map

$$\xi \longmapsto \psi(\xi, h)$$

belongs to $\mathcal{D}(\Omega_h)$ and satisfies, for all $k \in \mathbb{N}$,

$$\|\psi\|_{k,\widetilde{\mathcal{D}}(\Omega)} \overset{def}{=} \sup_{\substack{|\alpha| \leq k \\ h \in]0,1]}} \|(h\partial_\xi)^\alpha \psi(\xi, h)\|_{L^\infty(\Omega)} < \infty.$$

We first consider the nonstationary part of the integral which is described by the following lemma.

Lemma 8.34. *Assume that*

$$\forall (\xi, h) \in \Omega, \ |\nabla\Phi(\xi)| \geq c_0 h \quad with \quad c_0 > 0.$$

Then, for any integer N, an integer k and a constant C exist such that

$$\left| \int_{\mathbb{R}^d} e^{it\Phi(\xi)} \psi(\xi, h)\, d\xi \right| \leq \frac{C}{(|t|h^2)^N} \|\psi\|_{k,\widetilde{\mathcal{D}}(\Omega)} |\Omega_h|$$

for any $t \neq 0$ and positive h.

Proof. Exactly as in the proof of Theorem 8.8 page 340, we shall consider the first order differential operator

$$\mathcal{L} \overset{def}{=} -i \sum_{j=1}^d \frac{\partial_j \Phi}{|\nabla\Phi|^2} \partial_j,$$

which obviously satisfies

$$\mathcal{L}(e^{it\Phi}) = t\mathcal{L}e^{it\Phi}.$$

From repeated integration by parts, we then infer that

$$I(t, h) \overset{def}{=} \int_{\mathbb{R}^d} e^{it\Phi(\xi)} \psi(\xi, h)\, d\xi$$

$$= \frac{1}{t^N} \int_{\mathbb{R}^d} e^{it\Phi(\xi)} (({}^t\mathcal{L})^N \psi)(\xi, h)\, d\xi.$$

We now observe that

$${}^t\mathcal{L}a = -\mathcal{L}a - (\operatorname{div} \mathcal{L})a \quad with \quad \operatorname{div} \mathcal{L} \overset{def}{=} -i\frac{\Delta\Phi}{|\nabla\Phi|^2} + 2i\frac{D^2\Phi(\nabla\Phi, \nabla\Phi)}{|\nabla\Phi|^4}.$$

For any $\psi \in \widetilde{\mathcal{D}}(\Omega)$, we write

$${}^t\mathcal{L}\psi = -\frac{1}{h^2} \left(\sum_{j=1}^d \frac{h^{-1}\partial_j\Phi\, h\partial_j\psi}{|h^{-1}\nabla\Phi|^2} - i\psi\left(\frac{\Delta\Phi}{|h^{-1}\nabla\Phi|^2} - \frac{2D^2\Phi(h^{-1}\nabla\Phi, h^{-1}\nabla\Phi)}{|h^{-1}\nabla\Phi|^4} \right) \right).$$

Recall that on the support of ψ, we have $|h^{-1}\nabla\Phi| \geq c_0$. Hence, for any integer k', there exist an integer k and a constant C such that

$$\|{}^t\mathcal{L}\psi\|_{\ell,\widetilde{\mathcal{D}}(\Omega)} \leq \frac{C\|\psi\|_{k,\widetilde{\mathcal{D}}(\Omega)}}{h^2}.$$

An obvious (and omitted) induction then implies that

$$\|({}^t\mathcal{L})^N\psi\|_{L^\infty(\Omega)} \leq \frac{C}{h^{2N}}.$$

This proves the lemma.   □

In the case where $|\nabla\Phi| \leq c_0 h$, we use the method of the proof of Theorem 8.9 page 341.

Lemma 8.35. *Assume that*

$$\forall(\xi,h) \in \Omega, \ |\nabla\Phi(\xi)| \leq c_0 h.$$

Then, for any integer N, a constant C and an integer k exist such that

$$\forall\psi \in \widetilde{\mathcal{D}}(\Omega), \ \left|\int_{\mathbb{R}^d} e^{it\Phi(\xi)}\psi(\xi,h)\,d\xi\right| \leq C\|\psi\|_{k,\widetilde{\mathcal{D}}(\Omega)} \int_{\Omega_h} \frac{1}{(1+t|\nabla\Phi(\xi)|^2)^N}\,d\xi.$$

Proof. We use the differential operator $\mathcal{L}_t$ introduced in (8.2) page 341:

$$\mathcal{L}_t a = \frac{a}{1+t|\nabla\Phi|^2} - i\sum_{j=1}^{d} \frac{\partial_j\Phi\partial_j a}{1+t|\nabla\Phi|^2}.$$

It is clear that $\mathcal{L}_t(e^{it\Phi}) = e^{it\Phi}$. Using repeated integration by parts, we get

$$\int_{\mathbb{R}^d} e^{it\Phi(\xi)}\psi(\xi,h)\,d\xi = \int_{\mathbb{R}^d} e^{it\Phi(\xi)}({}^t\mathcal{L}_t)^N\psi(\xi,h)\,d\xi.$$

We now state the equivalent of Definition 8.10 page 342 in the present context.

Definition 8.36. *Let Ω be a domain of $\mathbb{R}^d \times \mathbb{R}^\star$ and N a real number. We denote by S_{sc}^N the set of smooth functions f on $\Omega \times \mathbb{R}^d$ such that for any $(\alpha,\beta) \in \mathbb{N}^d \times \mathbb{N}^d$, a constant C exists such that*

$$\sup_{(\xi,h)\in\Omega} |(h\partial_\xi)^\alpha \partial_\theta^\beta f(\xi,h,\theta)| \leq C(1+|\theta|^2)^{\frac{N-|\beta|}{2}}.$$

We shall now prove that if f is in S_{sc}^M, then

$$ {}^t\mathcal{L}_t\left(f(\xi,h,t^{\frac{1}{2}}\nabla\Phi(\xi))\right) = g(\xi,h,t^{\frac{1}{2}}\nabla\Phi(\xi)) \quad \text{with} \quad g \in S_{sc}^{M-2}. \tag{8.42}$$

In order to do this, we recall the formula (8.5) page 342:

$$ {}^t\mathcal{L}_t a = i\frac{\nabla\Phi(\xi)\cdot\nabla a(\xi)}{1+t|\nabla\Phi|^2} + \sigma(\xi,t^{\frac{1}{2}}\nabla\Phi)a(\xi) \quad \text{with}$$

$$\sigma(\xi,\theta) = \frac{i\Delta\Phi(\xi)+1}{1+|\theta|^2} - \frac{2iD^2\Phi_\xi(\theta,\theta)}{(1+|\theta|^2)^2}.$$

The main point to check here is that if $f \in S^M$, then

$$\frac{\partial_{\xi_j} \Phi(\xi) \partial_{\xi_j} f(\xi, h, \theta)}{1 + |\theta|^2} \in S^{M-2}. \tag{8.43}$$

Leibniz's formula implies that

$$(h\partial_\xi)^\alpha \partial_\theta^\beta \left(\frac{\partial_{\xi_j} \Phi(\xi) \partial_{\xi_j} f(\xi, h, \theta)}{1 + |\theta|^2} \right) = \sum_{\substack{\alpha' \le \alpha \\ \beta' \le \beta}} C_\alpha^{\alpha'} C_\beta^{\beta'} (h\partial_\xi)^{\alpha'} \partial_{\xi_j} \Phi(\xi)$$

$$\times (h\partial_\xi)^{\alpha - \alpha'} \partial_{\xi_j} \partial_\theta^{\beta'} f(\xi, h, \theta) \partial_\theta^{\beta - \beta'} \left(\frac{1}{1 + |\theta|^2} \right).$$

Because we have $|\nabla \Phi(\xi)| \le c_0 h$ for any $(\xi, h) \in \Omega$, we get

$$\left| \partial_{\xi_j} \Phi(\xi) (h\partial_\xi)^\alpha \partial_{\xi_j} \partial_\theta^{\beta'} f(\xi, h, \theta) \partial_\theta^{\beta - \beta'} \left(\frac{1}{1 + |\theta|^2} \right) \right|$$

$$\le Ch \left| (h\partial_\xi)^\alpha \partial_{\xi_j} \partial_\theta^{\beta'} f(\xi, h, \theta) \partial_\theta^{\beta - \beta'} \left(\frac{1}{1 + |\theta|^2} \right) \right|.$$

Thus,

$$\left| \partial_{\xi_j} \Phi(\xi) (h\partial_\xi)^\alpha \partial_{\xi_j} \partial_\theta^{\beta'} f(\xi, h, \theta) \partial_\theta^{\beta - \beta'} \left(\frac{1}{1 + |\theta|^2} \right) \right| \le C(1 + |\theta|^2)^{\frac{M - 2 - |\beta|}{2}}.$$

If $\alpha' \ne 0$, then

$$\left| (h\partial_\xi)^{\alpha'} \partial_{\xi_j} \Phi(\xi) (h\partial_\xi)^{\alpha - \alpha'} \partial_{\xi_j} \partial_\theta^{\beta'} f(\xi, h, \theta) \partial_\theta^{\beta - \beta'} \left(\frac{1}{1 + |\theta|^2} \right) \right|$$

$$\le Ch \left(\sup_{2 \le |\alpha''| \le |\alpha| + 1} \|\partial_\xi^{\alpha''} \Phi\|_{L^\infty} \right) \left| (h\partial_\xi)^{\alpha - \alpha'} \partial_{\xi_j} \partial_\theta^{\beta'} f(\xi, h, \theta) \partial_\theta^{\beta - \beta'} \left(\frac{1}{1 + |\theta|^2} \right) \right|.$$

This completes the proof of the assertion (8.42) and thus of Lemma 8.35. $\square$

We can now give an analog of Theorem 8.12.

Lemma 8.37. *Let $\psi \in \tilde{\mathcal{D}}(\Omega)$ and*

$$I(t, h) = \int e^{it\Phi(\xi)} \psi(\xi, h) \, d\xi.$$

Then, for any couple (N, N') of positive real numbers, there exist two positive constants, C_N and $C_{N'}$, such that

$$|I(t, h)| \le \frac{C_N}{(th^2)^N} |\Omega_h| + \int_{\Omega_{h, \Phi}} \frac{C_{N'}}{(1 + t|\nabla \Phi(\xi)|^2)^{N'}} \, d\xi,$$

where $\Omega_{h, \Phi}$ denotes the set of points $\xi \in \Omega_h$ such that $|\nabla \Phi(\xi)| \le c_0 h$.

Proof. It is only a matter of decomposing $I(t,h)$ into

$$I(t,h) = \int e^{it\Phi(\xi)}(1-\chi)\left(\frac{\nabla\Phi(\xi)}{h}\right)\psi(\xi,h)\,d\xi + \int e^{it\Phi(\xi)}\chi\left(\frac{\nabla\Phi(\xi)}{h}\right)\psi(\xi,h)\,d\xi,$$

where χ is a function in $\mathcal{D}(\mathbb{R}^d)$ with value 1 near the unit ball, and applying Lemmas 8.34 and 8.35. □

Completion of the proof of Theorem 8.31. Applying Lemma 8.37 with

$$\Phi(\xi) = x \pm \frac{\xi}{|\xi|},$$

we get

$$|K^{\pm}(t,h,tx)| \leq \frac{C}{(th^2)^{\frac{d-1}{2}}}h^d + \int_{\Omega_{h,\Phi}} \frac{C'_N}{(1+t|\nabla\Phi(\xi)|^2)^N}\,d\xi.$$

As in the proof of Proposition 8.15, we decompose ξ into

$$\xi = \zeta_1 + \zeta' \quad \text{with} \quad \zeta_1 = \left(\xi\Big|\frac{x}{|x|}\right)\frac{x}{|x|} \quad \text{and} \quad \zeta' = \xi - \left(\xi\Big|\frac{x}{|x|}\right)\frac{x}{|x|}.$$

As $\Omega_{h,\Phi} \subset B(\xi_0,h)$, ζ_1 varies in an interval I_h of length $2h$. Thus,

$$\int_{\Omega_{h,\Phi}} \frac{1}{(1+t|\nabla\Phi(\xi)|^2)^N}\,d\xi \leq \int_{I_h}\left(\int_{\mathbb{R}^{d-1}}\frac{d\zeta'}{1+t|\zeta'|^2}\right)d\zeta_1$$

$$\leq \frac{Ch}{t^{\frac{d-1}{2}}}.$$

This amounts to proving Theorem 8.31. □

This refined Strichartz estimate leads to the following endpoint logarithmic Strichartz estimate in dimension $d \geq 3$.

Theorem 8.38. *A constant C exists such that for any T, any $h \leq 1$, and any function u such that for any time t, the support of $\hat{u}(t,\cdot)$ is included in a ball with radius h and in the annulus $\mathcal{C}$, we have*

$$\|u\|_{L^2_T(L^\infty)} \leq C\big(h^{d-2}\log(e+T)\big)^{\frac{1}{2}}\left(\|\nabla u(0)\|_{L^2} + \|\Box u\|_{L^1_T(L^2)}\right). \tag{8.44}$$

Proof. The proof is very close to that of Theorem 8.30. Indeed, if we define $U(t)\gamma \overset{\text{def}}{=} \mathcal{F}^{-1}(e^{it|\xi|})\hat{\gamma}(\xi)$, then

$$\|U(t)\gamma\|_{L^2([0,T];L^\infty)} = \|\gamma\|_{L^2}\sup\left\|\int_0^T U(t)\varphi(t)\,dt\right\|_{L^2},$$

where the supremum is taken over the functions φ with $\|\varphi\|_{L^2_T(L^1)} \leq 1$ and such that for each t, the support of $\mathcal{F}\varphi(t,\cdot)$ is included in a ball of radius h and in the annulus $\mathcal{C}$. As U is unitary, we have

$$\left\| \int_0^T U(t)\varphi(t)\, dt \right\|_{L^2}^2 = \left(\int_0^T U(t)\varphi(t)\, dt \,\Big|\, \int_0^T U(t')\varphi(t')\, dt' \right)_{L^2}$$

$$= \int_{[0,T]^2} (U(t-t')\varphi(t)\,|\,\varphi(t'))_{L^2}\, dt'\, dt.$$

From Proposition 8.32, we can easily prove that

$$\|U(t-t')\phi(t)\|_{L^\infty} \le C \frac{h^{d-2}}{1+|t-t'|}.$$

Therefore,

$$\left\| \int_0^T U(t)\varphi(t)\, dt \right\|_{L^2}^2 \le Ch^{d-2} \int_{[0,T]^2} \frac{1}{1+|t-t'|} \|\varphi(t)\|_{L^1} \|\varphi(t')\|_{L^1}\, dt'\, dt$$

$$\le Ch^{d-2} \log(e+T)\|\varphi\|_{L^2([0,T];L^1)}^2.$$

Thus, Theorem 8.38 is proved. □

8.4 The Quintic Wave Equation in $\mathbb{R}^3$

In this section, we investigate the quintic wave equation in $\mathbb{R}^3$:

$$(W_5^\pm) \quad \begin{cases} \Box u \pm u^5 = 0 \\ (u, \partial_t u)_{|t=0} = (u_0, u_1). \end{cases}$$

We shall prove that the equation $(W_5^\pm)$ is locally well posed in the scaling invariant space $\mathcal{C}([0,T];L^2) \cap L^5([0,T];L^{10})$.

Theorem 8.39. *If $\gamma \overset{def}{=} \nabla u_{|t=0}$ belongs to L^2, then a positive time T exists such that the Cauchy problem (W_5) has a unique solution u in*

$$E_T \overset{def}{=} \left\{ u \in L^5([0,T];L^{10}) \,/\, \nabla u \in \mathcal{C}([0,T];L^2) \right\}.$$

In addition, u satisfies the following continuation criterion. If $T^\star$ denotes the maximal time of existence of u in E_T, then:

- *There exists a constant c such that if $\|\gamma\|_{L^2} \le c$, then $T^\star = +\infty$ and the solution belongs to*

$$L^\infty(\mathbb{R}^+; \dot{H}^1) \cap L^5(\mathbb{R}^+; L^{10}).$$

- *If $T^\star$ is finite, then*

$$\int_0^{T^\star} \|u(t)\|_{L^{10}}^5\, dt = +\infty.$$

Proof. Denote by $B(u_1, \ldots, u_5)$ the solution of the wave equation

$$\begin{cases} \Box B(u_1, \ldots, u_5) = -\prod_{j=1}^{5} u_j \\ B(u_1, \ldots, u_5)_{|t=0} = \partial_t B(u_1, \ldots, u_5)_{|t=0} = 0, \end{cases}$$

and by u_F, the solution of the free wave equation $\Box u = 0$ satisfying $u_F(0) = u_0$ and $\partial_t u_F(0) = u_1$. A solution of (W_5) is a fixed point of the map

$$u \longmapsto u_F + B(u, \ldots, u).$$

The energy equality, Corollary 8.29, and Hölder's inequality together imply that for any T,

$$\|\nabla B(u_1, \ldots, u_5)\|_{L_T^\infty(L^2)} + \|B(u_1, \ldots, u_5)\|_{L_T^5(L^{10})} \leq C \prod_{j=1}^{5} \|u_j\|_{L_T^5(L^{10})}.$$

Provided that $\|u_F\|_{L_T^5(L^{10})}$ is sufficiently small, Lemma 8.22 page 357 ensures the existence of a solution with the desired properties on the interval $[0, T]$.

More precisely, in the case where $\|\gamma\|_{L^2}$ is small, we readily get *global* existence because, owing to Corollary 8.27,

$$\|u_F\|_{L^5(L^{10})} \leq C\|\gamma\|_{L^2}.$$

Now, if $\|\gamma\|_{L^2}$ is not small, we may decompose γ (as we often do in this book) into its high-frequency part and its low-frequency part, as follows:

$$\gamma = S_J \gamma + (\mathrm{Id} - S_J)\gamma.$$

Denote by $u_{F,J}^\ell$ and $u_{F,J}^h$ the respective solutions of the free wave equation $\Box u = 0$ associated with $S_J \gamma$ and $(\mathrm{Id} - S_J)\gamma$. As we know that

$$\lim_{j \to \infty} \|(\mathrm{Id} - S_j)\gamma\|_{L^2} = 0,$$

according to Corollary 8.27, for all positive ε there exists some $J \in \mathbb{Z}$ such that

$$\|u_{F,J}^h\|_{L^5(L^{10})} \leq \varepsilon. \tag{8.45}$$

For the low-frequency part we use Hölder's and Bernstein's inequalities, which imply that

$$\|u_{F,J}^\ell\|_{L_T^5(L^{10})} \leq T^{\frac{1}{5}} \|u_{F,J}^\ell\|_{L_T^\infty(L^{10})}$$

$$\leq CT^{\frac{1}{5}} 2^{\frac{J}{5}} \|u_{F,J}^\ell\|_{L_T^\infty(L^6)}.$$

Using Sobolev's inequality and the energy equality thus yields that

$$\|u_{F,J}^\ell\|_{L_T^5(L^{10})} \leq C 2^{\frac{J}{5}} T^{\frac{1}{5}} \|\gamma\|_{L^2}.$$

Together with (8.45), this gives that

$$\lim_{T \to 0} \|u_F\|_{L_T^5(L^{10})} = 0,$$

which leads to local well-posedness for any data in L^2.

Finally, we prove the blow-up criterion. Consider a solution u of $(W_5^\pm)$ on the interval $[0, T[$ such that

$$\int_0^T \|u(t)\|_{L^{10}}^5 \, dt < \infty.$$

Using the energy estimate between t' and t (with $t' \le t$) gives

$$\|\nabla u(t) - \nabla u(t')\|_{L^2} \le \int_{t'}^t \|u(t'')\|_{L^{10}}^5 \, dt''.$$

Thus, a function u_T exists in $\dot{H}^1$ such that

$$\lim_{t \to T} u(t) = u_T \quad \text{in} \quad \dot{H}^1.$$

The local well-posedness part of the theorem then implies that u can be continued beyond T. This completes the proof of Theorem 8.39.   □

8.5 The Cubic Wave Equation in $\mathbb{R}^3$

The cubic wave equation was introduced in the context of field theory and is of the form

$$(W_3^\pm) \quad \begin{cases} \Box u \pm u^3 = 0 \\ (u, \partial_t u)_{|t=0} = (u_0, u_1), \end{cases}$$

where the unknown function u has real values and depends on (t, x) in $\mathbb{R} \times \mathbb{R}^3$.

8.5.1 Solutions in $\dot{H}^1$

First, we shall prove that the equation $(W_3^\pm)$ is locally well posed for initial data (u_0, u_1) in $\dot{H}^1 \times L^2$.

Proposition 8.40. *Assume that $\gamma \overset{def}{=} \nabla u(0)$ belongs to L^2. There exists a positive time T such that $(W_3^\pm)$ has a unique solution u, where ∇u belongs to $\mathcal{C}([-T, T]; L^2)$. Moreover, if $]T_-^\star, T_+^\star[$ denotes the maximal interval of existence of the solution, then there exists some constant c such that $|T_\pm^\star| \ge c\|\gamma\|_{L^2}^{-2}$.*

Proof. Define the trilinear operator $B_\pm(a_1, a_2, a_3)$ as the solution of

$$\begin{cases} \Box B_\pm(a_1, a_2, a_3) = \mp a_1 a_2 a_3 \\ B_\pm(a_1, a_2, a_3)_{|t=0} = \partial_t B_\pm(a_1, a_2, a_3)_{|t=0} = 0. \end{cases}$$

Thanks to the energy estimate and the Sobolev embeddings (see Theorem 1.38 page 29), we get

$$\|\nabla B_{\pm}(a_1, a_2, a_3)\|_{L^\infty([-T,T];L^2)} \leq \|a_1 a_2 a_3\|_{L^1([-T,T];L^2)}$$

$$\leq T \prod_{\ell=1}^{3} \|a_\ell\|_{L^\infty([-T,T];L^6)}$$

$$\leq T \prod_{\ell=1}^{3} \|a_\ell\|_{L^\infty([-T,T];\dot{H}^1)}.$$

Now, the solution u_F of the free wave equation with data (u_0, u_1) satisfies

$$\|\nabla u_F\|_{L^\infty([-T,T];L^2)} \leq \|\gamma\|_{L^2}.$$

Hence, Lemma 8.22 yields the desired result. $\square$

In the defocusing case [namely, the case (W_3^+)], the equation is *globally* well posed, as stated by the following result.

Theorem 8.41. *If the initial data u_0 is in L^4 and such that γ belongs to L^2, then there exists a unique global solution u of (W_3^+) such that ∇u belongs to $\mathcal{C}(\mathbb{R}; L^2)$. Moreover, this solution belongs to $L^\infty(\mathbb{R}; L^4)$, and satisfies*

$$\frac{1}{2}\|\nabla u(t)\|_{L^2}^2 + \frac{1}{4}\|u(t)\|_{L^4}^4 \leq \frac{1}{2}\|\gamma\|_{L^2}^2 + \frac{1}{4}\|u(0)\|_{L^4}^4 \quad \text{for all } t \in \mathbb{R}.$$

Proof. Formally, this follows easily from the energy estimate. However, we have to justify that $u(t)$ belongs to L^4. Therefore, we consider a solution of (W_3^+) such that ∇u belongs to $\mathcal{C}([-T, T]; L^2)$ and $u(0)$ belongs to L^4, and a sequence $(u_n)_{n\in\mathbb{N}}$ of functions which are C^1 in time, and smooth and compactly supported in the space variable, such that

$$\lim_{n\to\infty} \nabla u_n = \nabla u \quad \text{in} \quad \mathcal{C}([-T, T]; L^2).$$

We can write that

$$\frac{1}{4}\int_{\mathbb{R}^3} u_n^4(t, x)\, dx = \frac{1}{4}\int_{\mathbb{R}^3} u_n^4(0, x)\, dx + \int_0^t \int_{\mathbb{R}^3} u_n^3(t', x)\partial_t u_n(t', x)\, dt'\, dx.$$

Thanks to Sobolev embeddings, we have

$$\lim_{n\to\infty} \int_0^t \int_{\mathbb{R}^3} u_n^3(t', x)\partial_t u_n(t', x)\, dt'\, dx = \int_0^t \int_{\mathbb{R}^3} u^3(t', x)\partial_t u(t', x)\, dt'\, dx.$$

This gives that $u(t)$ is in L^4 for any t and that

$$\frac{1}{4}\|u(t)\|_{L^4}^4 \leq \frac{1}{4}\int_{\mathbb{R}^3} u_0^4(x)\, dx + \int_0^t \int_{\mathbb{R}^3} u^3(t', x)\partial_t u(t', x)\, dt'\, dx.$$

As $u^3 = -\square u$, we have

$$\frac{1}{4}\|u(t)\|_{L^4}^4 \leq \frac{1}{4}\int_{\mathbb{R}^3} u_0^4(x)\,dx - \int_0^t\int_{\mathbb{R}^3}\Box u(t',x)\partial_t u(t',x)\,dt'\,dx.$$

Using an omitted density argument, we can write that

$$\int_0^t\int_{\mathbb{R}^3}\Box u(t',x)\partial_t u(t',x)\,dx\,dt' = -\frac{1}{2}\|\nabla u(t,\cdot)\|_{L^2}^2 + \frac{1}{2}\|\gamma\|_{L^2}^2.$$

This gives

$$\frac{1}{4}\|u(t)\|_{L^4}^4 \leq \frac{1}{4}\int_{\mathbb{R}^3} u_0^4(x)\,dx - \frac{1}{2}\|\nabla u(t)\|_{L^2}^2 + \frac{1}{2}\|\gamma\|_{L^2}^2.$$

Proposition 8.40 implies, in particular, that if, say, $T_+^\star$ is finite, then the norm $\|\nabla u(t)\|_{L^2}$ goes to infinity when t tends to $T_+^\star$. From the above inequality, we can thus deduce that the solution is global. $\qquad\square$

8.5.2 Local and Global Well-posedness for Rough Data

We first show that both equations (W_3^+) and (W_3^-) are locally well posed for data in the scaling invariant space $\dot{H}^{\frac{1}{2}} \times \dot{H}^{-\frac{1}{2}}$.

Theorem 8.42. *If γ belongs to $\dot{H}^{-\frac{1}{2}}$, then a positive time T exists such that a unique solution u exists in $L^4([-T,T]\times\mathbb{R}^3)$ which is, in addition, such that ∇u is in $\mathcal{C}([-T,T];\dot{H}^{-\frac{1}{2}})$. Moreover, there exist two positive constants, c and C, such that if $\|\gamma\|_{\dot{H}^{-\frac{1}{2}}} \leq c$, then the solution u is global and satisfies*

$$\|\nabla u\|_{L^\infty(\mathbb{R};\dot{H}^{-\frac{1}{2}})} + \|u\|_{L^4(\mathbb{R}^{1+3})} \leq C\|\gamma\|_{\dot{H}^{-\frac{1}{2}}}. \tag{8.46}$$

Proof. By Hölder's inequality and a Strichartz estimate (Corollary 8.29), we get

$$\|B(a_1,a_2,a_3)\|_{L^4([-T,T]\times\mathbb{R}^3)} \leq C\|a_1 a_2 a_3\|_{L^{\frac{4}{3}}([-T,T]\times\mathbb{R}^3)}$$

$$\leq C\prod_{\ell=1}^3 \|a_\ell\|_{L^4([-T,T]\times\mathbb{R}^3)} \tag{8.47}$$

and

$$\|u_F\|_{L^4(\mathbb{R}^{1+3})} \leq C\|\gamma\|_{\dot{H}^{-\frac{1}{2}}}.$$

Decomposing γ into

$$\gamma = \gamma_{1,R} + \gamma_{2,R} \quad \text{with} \quad \gamma_{1,R} \overset{\text{def}}{=} \mathcal{F}^{-1}(\mathbb{1}_{B(0,R)}\widehat{\gamma}),$$

we then get, by virtue of the Sobolev embedding $\dot{H}^{\frac{3}{4}} \hookrightarrow L^4$,

$$\|u_F\|_{L^4([-T,T]\times\mathbb{R}^3)} \leq CT^{\frac{1}{4}}R^{\frac{1}{4}}\|\gamma\|_{\dot{H}^{-\frac{1}{2}}} + \|\gamma_{2,R}\|_{\dot{H}^{-\frac{1}{2}}}.$$

As $\lim_{R\to\infty}\|\gamma_{2,R}\|_{\dot{H}^{-\frac{1}{2}}} = 0$, the whole theorem is proved using Lemma 8.22. $\qquad\square$

It turns out that the equation $(W_3^\pm)$ is also well posed in $\dot{H}^s \times \dot{H}^{s-1}$ for any $s \in \,]1/2, 1]$. Note that this result is *not* an obvious consequence of the previous local well-posedness statement since $\dot{H}^{\frac{1}{2}}$ and $\dot{H}^s$ are not included in one another.

Proposition 8.43. *Let s be in $]1/2, 1]$ and consider an initial data in $\dot{H}^s \times \dot{H}^{s-1}$. Define the wave admissible couple (q_1, r_1) by*

$$\left(\frac{1}{q_1}, \frac{1}{r_1}\right) \stackrel{def}{=} \left(\frac{1-s}{2}, \frac{2-s}{6}\right).$$

A positive time T then exists such that a unique solution u of $(W_3^\pm)$ exists in the space $L^{q_1}([-T, T]; L^{r_1}(\mathbb{R}^3))$ which is, in addition, such that ∇u belongs to the space $\mathcal{C}([-T, T]; \dot{H}^{s-1})$. Moreover, if $]T_-^\star, T_+^\star[$ denotes the maximal time interval of existence, then $|T_\pm^\star|$ is greater than $c\|\gamma\|_{\dot{H}^{s-1}}^{-\frac{2}{2s-1}}$.

Proof. We introduce the wave admissible couple

$$\left(\frac{1}{q_2}, \frac{1}{r_2}\right) \stackrel{def}{=} \left(\frac{1-s}{2}, \frac{s}{2}\right).$$

From Corollary 8.29, we infer that

$$\|u_F\|_{L^{q_1}(\mathbb{R}; L^{r_1}(\mathbb{R}^3))} \leq C\|\gamma\|_{\dot{H}^{s-1}},$$
$$\|B(a_1, a_2, a_3)\|_{L^{q_1}([-T,T]; L^{r_1}(\mathbb{R}^3))} \leq \|a_1 a_2 a_3\|_{L^{q_2'}([-T,T]; L^{r_2'}(\mathbb{R}^3))}.$$

Noting that

$$\frac{r_1}{3} = r_2' \quad \text{and} \quad \frac{1}{q_2'} - \frac{3}{q_1} = 2s - 1,$$

we get

$$\|B(a_1, a_2, a_3)\|_{L^{q_1}([-T,T]; L^{r_1}(\mathbb{R}^3))} \leq C \prod_{\ell=1}^{3} \|a_\ell\|_{L^{3q_2'}([-T,T]; L^{r_1}(\mathbb{R}^3))}$$

$$\leq T^{2s-1} \prod_{\ell=1}^{3} \|a_\ell\|_{L^{q_1}([-T,T]; L^{r_1}(\mathbb{R}^3))}.$$

Applying Lemma 8.22 then allows us to complete the proof. □

We now give a technical statement which will be useful in the next subsection.

Lemma 8.44. *Let s be in $]1/2, 1[$ and consider an initial data such that γ belongs to $\dot{H}^{s-1} \cap \dot{H}^{-\frac{1}{2}}$. Let u be the solution given by Theorem 8.42. There exists a constant C such that for any couple $(q, r) \neq (2, \infty)$ satisfying (8.32), we have*

$$\|u\|_{L^q([-T,T]; \dot{B}_{r,2}^{s-\frac{2}{q}}(\mathbb{R}^3))} \leq C\|\gamma\|_{\dot{H}^{s-1}}. \tag{8.48}$$

Proof. Observe that, according to Corollary 8.25 and Theorem 2.40 page 79,

$$\|\nabla B(a_1, a_2, a_3)\|_{L^q([-T,T];\dot{B}_{r,2}^{s-1-\frac{2}{q}})} \leq C\|a_1 a_2 a_3\|_{L^{\frac{2}{s+1}}([-T,T];L^{\frac{2}{2-s}})}$$

$$\leq C\prod_{\ell=1}^{2}\|a_\ell\|_{L^4([-T,T]\times\mathbb{R}^3)}\|a_3\|_{L^{\frac{2}{s}}([-T,T];L^{\frac{2}{1-s}})}. \qquad (8.49)$$

We first take $(q,r) = (2/s, 2/(1-s))$. Also using (8.47), Lemma 8.22 implies that u belongs to $L^{\frac{2}{s}}([-T,T];L^{\frac{2}{1-s}}) \cap L^4([-T,T];L^4)$ and satisfies

$$\|u\|_{L^{\frac{2}{s}}([-T,T];L^{\frac{2}{1-s}})} \leq C\|\gamma\|_{\dot{H}^{s-1}}.$$

Applying (8.49) and Corollary 8.29, we then get (8.48), and the lemma is proved. □

8.5.3 The Nonlinear Interpolation Method

In this subsection, we want to prove that in the defocusing case, the cubic wave equation is globally well posed for $(u_0, u_1) \in \dot{H}^{\frac{3}{4}} \times H^{-\frac{1}{4}}$, that is, at a level of regularity which is *less than* 1. For this, the very structure of the equation (namely, the defocusing assumption) has to be used, combined with an interpolation method between $\dot{H}^1$ and $\dot{H}^{\frac{1}{2}}$ (i.e., between spaces for which global well-posedness and local well-posedness, respectively, has been established).

We now state the main result of this subsection.

Theorem 8.45. *Assume that $\gamma \in \dot{H}^{-\frac{1}{4}}$. A unique global solution of (W_3^+) then exists in $L^4_{loc}(\mathbb{R};L^6)$ which is, in addition, such that ∇u is in $\mathcal{C}(\mathbb{R};\dot{H}^{-\frac{1}{4}})$.*

Proof. The proof relies on a nonlinear interpolation method: For any integer j, we decompose the initial data as

$$\gamma_j^\ell \overset{\text{def}}{=} (\partial\dot{S}_j u_0, \dot{S}_j u_1) \quad \text{and} \quad \gamma_j^h \overset{\text{def}}{=} (\partial(\text{Id}-\dot{S}_j)u_0, (\text{Id}-\dot{S}_j)u_1).$$

On the one hand, as (u_0, u_1) belongs to $\dot{H}^{\frac{3}{4}} \times \dot{H}^{-\frac{1}{4}}$, the high-frequency part will be small in $\dot{H}^{\frac{1}{2}} \times \dot{H}^{-\frac{1}{2}}$, giving rise to a global solution, according to Theorem 8.42. On the other hand, the low-frequency part satisfies a modified cubic wave equation for which the basic $\dot{H}^1$ energy estimate makes sense. For arbitrarily large time T, it will then be possible to choose j so that the solution exists on $[-T, T]$.

We will now be more specific. Denote by v_j the (global) solution of (W_3^+) associated with the Cauchy data γ_j^h given by Theorem 8.42, which exists provided we choose j such that

$$\|\gamma_j^h\|_{\dot{H}^{-\frac{1}{2}}} \leq 2^{-\frac{j}{4}}\|\gamma\|_{\dot{H}^{-\frac{1}{4}}} \leq c. \qquad (8.50)$$

As γ_j^h obviously belongs to $\dot{H}^{-\frac{1}{3}}$, applying Lemma 8.44 with $s = 2/3$ implies that $v_j \in L^3(\mathbb{R};L^6)$.

Next, we decompose the desired solution u into $u_j + v_j$, where u_j is the solution of the modified cubic equation

$$(W_{3,v}) \qquad \begin{cases} \Box w + w^3 + 3w^2 v + 3wv^2 = 0 \\ \qquad\qquad (\nabla w)_{|t=0} = \gamma \end{cases}$$

with $\gamma = \gamma_j^\ell$ and $v = v_j$. Note that γ_j^ℓ obviously belongs to $L^2 \cap \dot{H}^{-\frac{1}{4}}$.

The properties of the equation $(W_{3,v})$ are described by the following lemma.

Lemma 8.46. *Let $v \in L^3(\mathbb{R}; L^6)$ and $\gamma \in L^2 \cap \dot{H}^{-\frac{1}{4}}$. There exists a positive time T such that $(W_{3,v})$ has a unique solution w, where ∇w belongs to $\mathcal{C}([-T, T]; L^2)$. Moreover, T can be chosen greater than*

$$c \left(\|\gamma\|_{L^2} + \|v\|_{L^3(\mathbb{R};L^6)} \right)^{-2},$$

and ∇w belongs to $\mathcal{C}([-T, T]; \dot{H}^{-\frac{1}{4}})$.

Proof. Combining the Hölder inequality, the embedding $\dot{H}^1 \hookrightarrow L^6$, and the energy estimate yields

$$\|\nabla B(a_1, a_2, a_3)\|_{L^\infty([-T,T];L^2)} \leq T \prod_{\ell=1}^{3} \|\nabla a_\ell\|_{L^\infty([-T,T];L^2)},$$

$$\|\nabla B(v, a_2, a_3)\|_{L^\infty([-T,T];L^2)} \leq T^{\frac{2}{3}} \|v\|_{L^3(\mathbb{R};L^6)} \prod_{\ell=1}^{2} \|\nabla a_\ell\|_{L^\infty([-T,T];L^2)},$$

$$\|\nabla B(v, v, a)\|_{L^\infty([-T,T];L^2)} \leq T^{\frac{1}{3}} \|v\|_{L^3(\mathbb{R};L^6)}^2 \|\nabla a\|_{L^\infty([-T,T];L^2)}.$$

Lemma 8.22 then implies the first part of the lemma. In order to prove that ∇w belongs to $\mathcal{C}([-T, T]; \dot{H}^{-\frac{1}{4}})$, we can use the fact that, owing to the Sobolev embedding $L^{\frac{12}{7}}(\mathbb{R}^3) \hookrightarrow \dot{H}^{-\frac{1}{4}}(\mathbb{R}^3)$ (see Corollary 1.39 page 29), we have

$$\|\nabla w\|_{L^\infty([-T,T];\dot{H}^{-\frac{1}{4}})} \leq \|\nabla w(0)\|_{\dot{H}^{-\frac{1}{4}}} + C \|w^3 + 3w^2 v + 3wv^2\|_{L^1([-T,T];L^{\frac{12}{7}})}.$$

Now, using Hölder's inequality and the Sobolev embeddings $\dot{H}^{\frac{3}{4}}(\mathbb{R}^3) \hookrightarrow L^4(\mathbb{R}^3)$ and $\dot{H}^1(\mathbb{R}^3) \hookrightarrow L^6(\mathbb{R}^3)$ (see Theorem 1.38), we can write

$$\|w^3\|_{L^1([-T,T];L^{\frac{12}{7}})} \leq T \|\nabla w\|_{L^\infty([-T,T];L^2)}^2 \|\nabla w\|_{L^\infty([-T,T];\dot{H}^{-\frac{1}{4}})},$$

$$\|vw^2\|_{L^1([-T,T];L^{\frac{12}{7}})} \leq T^{\frac{2}{3}} \|\nabla w\|_{L^\infty([-T,T];\dot{H}^{-\frac{1}{4}})} \|\nabla w\|_{L^\infty([-T,T];L^2)} \|v\|_{L^3([-T,T];L^6)},$$

$$\|v^2 w\|_{L^1([-T,T];L^{\frac{12}{7}})} \leq T^{\frac{1}{3}} \|\nabla w\|_{L^\infty([-T,T];\dot{H}^{-\frac{1}{4}})} \|v\|_{L^3([-T,T];L^6)}^2.$$

The lemma is thus proved. $\qquad\qquad\qquad\qquad\qquad\qquad\qquad\qquad\qquad\quad\square$

Proof of Theorem 8.45 (continued). If we now denote by $T_j^\star$ the maximum time of existence of (W_{3,v_j}), the matter is reduced to proving that

$$\limsup_{j \to +\infty} T_j^\star = +\infty. \tag{8.51}$$

The last part of this section is devoted to the proof of (8.51). Because of Lemma 8.46, it suffices to prove an a priori bound on the energy of the solution u_j. This will be achieved via the energy inequality, provided that we can control the nonhomogeneous terms by the energy of u_j.

We introduce the notation

$$H_j \overset{\text{def}}{=} \frac{1}{2}\|\gamma_j^\ell\|_{L^2}^2 + \frac{1}{4}\|u_j(0)\|_{L^4}^4.$$

It will be useful to note that if j is nonnegative, then

$$H_j^{\frac{1}{2}} \leq C_\gamma c_j 2^{\frac{j}{4}}, \tag{8.52}$$

where, from now on, $(c_j)_{j \in \mathbb{N}}$ denotes a generic element of the unit sphere of $\ell^2(\mathbb{N})$ and $C_\gamma = f(\|\gamma\|_{\dot{H}^{-\frac{1}{4}}})$ for some locally bounded function f on $\mathbb{R}^+$. In fact, the quantity $\|u_j(0)\|_{L^4}^4$ is negligible compared to the energy of the initial data $\|\gamma_j^\ell\|_{L^2}^2$.

We now define T_j by

$$T_j \overset{\text{def}}{=} \sup\left\{t < T_j^\star \ / \ \frac{1}{2}\|\nabla u_j\|_{L_t^\infty(L^2)}^2 \leq 2H_j\right\}$$

and fix some $T > 0$ and $j_0 \in \mathbb{N}^*$.

We seek to prove that there exists some integer $j > j_0$ such that T_j is greater than T. As pointed out earlier, the key point is the control of the energy of u_j. Multiplying the equation by $\partial_t u_j$ and integrating over x and t, we get, for any T less than or equal to $T_j^\star$,

$$\frac{1}{2}\|\nabla u_j\|_{L_T^\infty(L^2)}^2 \leq H_j + 3\left|\int_0^T \int_{\mathbb{R}^3} v_j^2 u_j \partial_t u_j \, dx \, dt + \int_0^T \int_{\mathbb{R}^3} v_j u_j^2 \partial_t u_j \, dx \, dt\right|.$$

From now on, we denote by a_j a sequence such that

$$\liminf_{j \to \infty} a_j = 0.$$

Here, we easily see that the whole theorem is proved, provided that we can find some positive real number α such that

$$\left|\int_0^T \int_{\mathbb{R}^3} v_j^2 u_j \partial_t u_j \, dx \, dt + \int_0^T \int_{\mathbb{R}^3} v_j u_j^2 \partial_t u_j \, dx \, dt\right| \leq T^\alpha a_j \|\nabla u_j\|_{L_T^\infty(L^2)}^2. \tag{8.53}$$

By Hölder's inequality and Sobolev embedding, we have

$$\left|\int_0^T \int_{\mathbb{R}^3} v_j^2 u_j \partial_t u_j \, dx \, dt\right| \leq \int_0^T \|v_j(t)\|_{L^6}^2 \|u_j(t)\|_{L^6} \|\partial_t u_j(t)\|_{L^2} \, dt$$

$$\leq C\|\nabla u_j\|_{L_T^\infty(L^2)}^2 \int_0^T \|v_j(t)\|_{L^6}^2 \, dt.$$

Finally, using Lemma 8.44, we can write, for all $k \geq -2$,

$$\|v_j\|^2_{L^2_T(L^6)} \leq CT^{\frac{1}{3}}\|v_j\|^2_{L^3_T(L^6)}$$
$$\leq CT^{\frac{1}{3}}\|\gamma^h_j\|^2_{\dot{H}^{-\frac{1}{2}+\frac{1}{6}}}.$$

We deduce that

$$\left|\int_0^T \int_{\mathbb{R}^3} v_j^2 u_j \partial_t u_j \, dx \, dt\right| \leq CT^{\frac{1}{3}}\|\nabla u_j\|^2_{L^\infty_T(L^2)}\|\gamma^h_j\|^2_{\dot{H}^{-\frac{1}{2}+\frac{1}{6}}}$$
$$\leq CT^{\frac{1}{3}}2^{-\frac{j}{6}}\|\gamma\|^2_{\dot{H}^{-\frac{1}{4}}}\|\nabla u_j\|^2_{L^\infty_T(L^2)}.$$

Hence, the first term of (8.53) satisfies the desired inequality.

To handle the second term, we write (with obvious notation)

$$v_j = v_{j,F} + B(v_j, v_j, v_j).$$

Using Hölder's inequality and Corollary 8.29 with $(q_1, r_1) = (3, 6)$ and $(q_2, r_2) = (6, 3)$, we get

$$\|B(v_j, v_j, v_j)\|_{L^3_T(L^6)} \leq C\|v_j\|^2_{L^4([0,T]\times\mathbb{R}^3)}\|v_j\|_{L^3_T(L^6)}.$$

Using Lemma 8.44, we then get

$$\|B(v_j, v_j, v_j)\|_{L^3_T(L^6)} \leq C\|\gamma^h_j\|^2_{\dot{H}^{-\frac{1}{2}}}\|\gamma^h_j\|_{\dot{H}^{-\frac{1}{2}+\frac{1}{6}}}.$$

We will focus on the term

$$\int_0^T \int_{\mathbb{R}^3} B(v_j, v_j, v_j)\, u_j^2 \partial_t u_j \, dx \, dt,$$

which turns out to be the easiest one.

Again, thanks to Hölder's inequality and Sobolev embedding, we get

$$\left|\int_0^T \int_{\mathbb{R}^3} B(v_j, v_j, v_j)\, u_j^2 \partial_t u_j \, dx \, dt\right| \leq CT^{\frac{2}{3}}\|\nabla u_j\|^3_{L^\infty_T(L^2)}\|\gamma^h_j\|^2_{\dot{H}^{-\frac{1}{2}}}\|\gamma^h_j\|_{\dot{H}^{-\frac{1}{2}+\frac{1}{6}}}$$
$$\leq CT^{\frac{2}{3}}2^{-\frac{2j}{3}}\|\gamma\|^3_{\dot{H}^{-\frac{1}{4}}}\|\nabla u_j\|^3_{L^\infty_T(L^2)}.$$

Hence, this term is also bounded by the right-hand side of (8.53).

The term involving $v_{j,F}$ is more demanding and requires paradifferential calculus. Using Bony's decomposition, we can write

$$v_{j,F}u_j^2 = T'_{v_{j,F}}u_j^2 + T_{u_j^2}v_{j,F} \quad \text{with} \quad T'_a b \overset{\text{def}}{=} \sum_{k \geq -2} \dot{S}_{k+2}a \, \dot{\Delta}_k b.$$

As the support of the Fourier transform is preserved by the flow of the constant coefficients wave equation, the function $v_{j,F}$ has no low frequencies, and we can restrict the summation to those k such that $k \geq j - 2$.

Now, for any $k \geq j - 2$, we can write that

$$\|\dot{S}_{k+2}(v_{j,F})\dot{\Delta}_k(u_j^2)\|_{L_T^1(L^2)} \leq CT^{\frac{1}{2}}\|\dot{S}_{k+2}(v_{j,F})\|_{L_T^2(L^\infty)}\|\dot{\Delta}_k(u_j^2)\|_{L_T^\infty(L^2)}.$$

Note that, according to Bernstein's inequality, we may write

$$\begin{aligned}
\|\dot{\Delta}_k(u_j^2)\|_{L^2} &\leq C2^{-k}\|\dot{\Delta}_k\partial(u_j^2)\|_{L^2}\\
&\leq C2^{-k}\|\dot{\Delta}_k u_j\|_{L^4}\|\partial\dot{\Delta}_k u_j\|_{L^4}\\
&\leq C2^{-\frac{k}{2}}\|\partial\dot{\Delta}_k u_j\|_{L^2}.
\end{aligned}$$

Therefore,

$$\|\dot{S}_{k+2}(v_{j,F})\dot{\Delta}_k(u_j^2)\|_{L_T^1(L^2)} \leq CT^{\frac{1}{2}}\|\dot{S}_{k+2}(v_{j,F})\|_{L_T^2(L^\infty)}2^{-\frac{k}{2}}\|\nabla u_j\|_{L_T^\infty(L^2)}^2.$$

According to logarithmic Strichartz estimates (8.38), we have

$$\begin{aligned}
\|\dot{S}_{k+2}(v_{j,F})\|_{L_T^2(L^\infty)} &\leq \sum_{\ell \leq k+1}\|\dot{\Delta}_\ell(v_{j,F})\|_{L_T^2(L^\infty)}\\
&\leq C\sum_{\ell \leq k+1}\left(\log(e + 2^\ell T)\right)^{\frac{1}{2}}\|\dot{\Delta}_\ell\gamma_j^h\|_{L^2}.
\end{aligned}$$

We deduce that for any $\sigma < 1$,

$$\begin{aligned}
\|\dot{S}_{k+2}(v_{j,F})\|_{L_T^2(L^\infty)} &\leq \sum_{\ell \leq k+1}\left(\log(e + 2^\ell T)\right)^{\frac{1}{2}}2^{\ell(1-\sigma)}\|\gamma_j^h\|_{\dot{H}^{\sigma-1}}\\
&\leq C\left(\log(e + 2^k T)\right)^{\frac{1}{2}}2^{k(1-\sigma)}\|\gamma_j^h\|_{\dot{H}^{\sigma-1}}.
\end{aligned}$$

Finally, for any $\frac{1}{2} < \sigma < 1$,

$$\begin{aligned}
\|T'_{v_{j,F}}u_j^2\|_{L_T^1(L^2)} \leq{}& CT^{\frac{1}{2}}\|\gamma_j^h\|_{\dot{H}^{\sigma-1}}\|\nabla u_j\|_{L_T^\infty(L^2)}^2\\
&\times \sum_{k \geq j-2}2^{k(\frac{1}{2}-\sigma)}\left(\log(e + 2^k T)\right)^{\frac{1}{2}}.
\end{aligned}$$

Note that we can assume with no loss of generality that $T \leq 2^j$. Hence,

$$\begin{aligned}
\|T'_{v_{j,F}}u_j^2\|_{L_T^1(L^2)} \leq{}& CT^{\frac{1}{2}}\|\gamma_j^h\|_{\dot{H}^{\sigma-1}}\|\nabla u_j\|_{L_T^\infty(L^2)}^2\\
&\times \sum_{k \geq j-2}2^{k(\frac{1}{2}-\sigma)}\left(\log(e + 2^{2k})\right)^{\frac{1}{2}}\\
\leq{}& CT^{\frac{1}{2}}\|\gamma_j^h\|_{\dot{H}^{\sigma-1}}\|\nabla u_j\|_{L_T^\infty(L^2)}^2\sum_{k \geq j-2}2^{k(\frac{1}{2}-\sigma)}k^{\frac{1}{2}}.
\end{aligned}$$

Observing that, since $j \geq 2$,

$$\sum_{k \geq j-2} 2^{k(\frac{1}{2}-\sigma)} k^{\frac{1}{2}} = 2^{j(\frac{1}{2}-\sigma)} j^{\frac{1}{2}} \sum_{k \geq j-2} 2^{(k-j)(\frac{1}{2}-\sigma)} \left(\frac{k}{j}\right)^{\frac{1}{2}}$$

$$\leq 2^{j(\frac{1}{2}-\sigma)} j^{\frac{1}{2}} \sum_{k \geq j-2} 2^{(k-j)(\frac{1}{2}-\sigma)} (k-j+1)^{\frac{1}{2}},$$

we deduce that

$$\|T'_{v_{j,F}} u_j^2\|_{L_T^1(L^2)} \leq C T^{\frac{1}{2}} \|\gamma_j^h\|_{\dot{H}^{\sigma-1}} \|\nabla u_j\|_{L_T^\infty(L^2)}^2 2^{j(\frac{1}{2}-\sigma)} j^{\frac{1}{2}}.$$

Finally, taking $\sigma = \frac{3}{4}$, we get, thanks to the inequality (8.52),

$$\left| \int_0^T \int_{\mathbb{R}^3} T'_{v_{j,F}} u_j^2 \partial_t u_j \, dx \, dt \right| \leq C T^{\frac{1}{2}} \|\gamma_j^h\|_{\dot{H}^{-\frac{1}{4}}} j^{\frac{1}{2}} c_j \|\nabla u_j\|_{L_T^\infty(L^2)}^2.$$

Note that $(c_j) \in \ell^2$ implies that $\liminf_{j \to \infty} j^{\frac{1}{2}} c_j = 0$. Hence, this term also satisfies (8.53).

The last term, $T_{u_j^2} v_{j,F}$, is the most delicate to treat. We write that[2]

$$\int_0^T \int_{\mathbb{R}^3} T_{u_j^2} v_{j,F} \partial_t u_j \, dx \, dt = \sum_{k \geq j-2} \int_0^T \int_{\mathbb{R}^3} \dot{S}_{k-1}(u_j^2) \dot{\Delta}_k(v_{j,F}) \widetilde{\Delta}_k(\partial_t u_j) \, dx \, dt$$

$$= \sum_{k \geq j-2} \sum_{\ell \leq k-2} \int_0^T \int_{\mathbb{R}^3} \dot{\Delta}_\ell(u_j^2)$$

$$\times \widetilde{\Delta}_\ell(\dot{\Delta}_k(v_{j,F}) \widetilde{\Delta}_k(\partial_t u_j)) \, dx \, dt,$$

$\widetilde{\Delta}_k$ being the convolution operator by the inverse Fourier transform of $\widetilde{\varphi}(2^{-k} \cdot)$, where $\widetilde{\varphi}$ is a function in $\mathcal{D}(\mathbb{R}^d \setminus \{0\})$ with value 1 near the support of φ [see (2.5)].

According to the standard Strichartz estimates, the function $v_{j,F}$ fails to be controlled in $L_T^2(\dot{B}_{\infty,2}^0)$, whereas $\partial_t u_j$ belongs to $L_T^\infty(L^2)$. Therefore, the series with general term $\dot{\Delta}_k v_{j,F} \widetilde{\Delta}_k \partial_t u_j$ does not converge in any reasonable sense. In order to overcome this difficulty, we may use the logarithmic refined Strichartz estimate given by Theorem 8.38. For that purpose, we introduce a covering of $2^k \mathcal{C}$ by a family of balls of radius 2^ℓ centered at $(\xi_\nu^{k,\ell})_{\nu \in \Lambda_{k,\ell}}$. Let $\chi \in \mathcal{D}(B(0,1))$ be such that for all $\xi \in 2^k \mathcal{C}$,

$$\sum_{\nu \in \Lambda_{k,\ell}} \chi\left(\frac{\xi - \xi_\nu^{k,\ell}}{2^\ell}\right) = 1 \quad \text{and} \quad \frac{1}{C_0} \leq \sum_{\nu \in \Lambda_{k,\ell}} \chi^2\left(\frac{\xi - \xi_\nu^{k,\ell}}{2^\ell}\right) \leq C_0. \quad (8.54)$$

We write that

---

[2] As before, owing to the spectral properties of $v_{j,F}$, the summation may be restricted to those $k \in \mathbb{Z}$ such that $k \geq j - 2$.

$$I_{k,\ell}(\partial_t u_j) \stackrel{\text{def}}{=} \widetilde{\Delta}_\ell(\dot{\Delta}_k(v_{j,F})\widetilde{\Delta}_k(\partial_t u_j))$$

$$= \widetilde{\Delta}_\ell \sum_{\nu \in \Lambda_{k,\ell}} \Delta^\nu_{k,\ell}(v_{j,F})\widetilde{\Delta}_k(\partial_t u_j) \quad \text{with}$$

$$\Delta^\nu_{k,\ell}a \stackrel{\text{def}}{=} \mathcal{F}^{-1}\left((\varphi(2^{-k}\xi)\chi(2^{-\ell}(\xi - \xi^{k,\ell}_\nu)))\,\widehat{a}(\xi)\right).$$

As the support of the Fourier transform of a product is included in the sum of the support of each Fourier transform, we obtain

$$\int_0^T \int_{\mathbb{R}^3} T_{u_j^2}v_{j,F}\partial_t u_j \, dx \, dt$$

$$= \sum_{k \geq j-2} \sum_{\ell \leq k-2} \int_0^T \int_{\mathbb{R}^3} \sum_{\nu \in \Lambda_{k,\ell}} \dot{\Delta}_\ell(u_j^2)\widetilde{\Delta}_\ell(\Delta^\nu_{k,\ell}(v_{j,F})\widetilde{\Delta}^\nu_{k,\ell}(\partial_t u_j)) \, dx \, dt$$

with $\widetilde{\Delta}^\nu_{k,\ell}a \stackrel{\text{def}}{=} \mathcal{F}^{-1}\left(\widetilde{\varphi}(2^{-k}\xi)\mathbb{1}_{B(-\xi^{k,\ell}_\nu,C2^{-\ell})}\widehat{a}(\xi)\right).$

We deduce that

$$\int_0^T \int_{\mathbb{R}^3} T_{u^2}v_{j,F}\partial_t u_j \, dx \, dt \leq B_j \|\nabla u_j\|^2_{L^\infty_T(L^2)}, \quad \text{where}$$

$$B_j = \sum_{\substack{k \geq j-2 \\ \ell \leq k-2}} \int_0^T \sum_{\nu \in \Lambda_{k,\ell}} 2^{-\frac{\ell}{2}}c_{\ell,j}(t)\|\Delta^\nu_{k,\ell}(v_{j,F}(t))\|_{L^\infty}\|\widetilde{\Delta}^\nu_{k,\ell}(\partial_t u_j(t))\|_{L^2} \, dt,$$

and $(c_{\ell,j}(t))_{\ell \in \mathbb{Z}}$ denotes, as in all that follows, a generic element of the unit sphere of $\ell^2(\mathbb{Z})$ such that $c_{\ell,j}(t) = 0$ for $\ell \leq -2$.

For fixed k and ℓ, applying the Cauchy–Schwarz inequality with respect to ν and dt gives

$$B_j \leq \sum_{\substack{k \geq j-2 \\ \ell \leq k-2}} 2^{-\frac{\ell}{2}}\left(\int_0^T \sum_{\nu \in \Lambda_{k,\ell}} \|\Delta^\nu_{k,\ell}(v_{j,F}(t))\|^2_{L^\infty} \, dt\right)^{\frac{1}{2}}$$

$$\times \left(\int_0^T c^2_{\ell,j}(t) \sum_{\nu \in \Lambda_{k,\ell}} \|\widetilde{\Delta}^\nu_{k,\ell}(\partial_t u_j(t))\|^2_{L^2} \, dt\right)^{\frac{1}{2}}.$$

Applying the logarithmic refined Strichartz estimate (8.44), and using the quasi-orthogonality property stated in (8.54) and the fact that $T \leq 2^{k+2}$ for $k \geq j-2$, we get

$$\int_0^T \sum_{\nu \in \Lambda_{k,\ell}} \|\Delta^\nu_{k,\ell}(v_{j,F}(t))\|^2_{L^\infty} \, dt \leq C \sum_{\nu \in \Lambda_{k,\ell}} \log(e + 2^k)2^{\ell-k}\|\Delta^\nu_{k,\ell}\gamma^h_j\|^2_{L^2}$$

$$\leq Ck2^{\ell-k} \sum_{\nu \in \Lambda_{k,\ell}} \|\Delta^\nu_{k,\ell}\gamma^h_j\|^2_{L^2}$$

$$\leq Ck2^{\ell-k}\|\dot{\Delta}_k\gamma^h_j\|^2_{L^2}.$$

Therefore,

$$B_j \leq C \sum_{k \geq j-2} k^{\frac{1}{2}} 2^{-\frac{k}{2}} \|\dot{\Delta}_k \gamma_j^h\|_{L^2} \sum_{\ell \leq k-2} \left(\int_0^T c_{\ell,j}^2(t) \sum_{\nu \in \Lambda_{k,\ell}} \|\widetilde{\Delta}_{k,\ell}^\nu \partial_t u_j(t)\|_{L^2}^2 \, dt \right)^{\frac{1}{2}}.$$

Using the Cauchy–Schwarz inequality with respect to ℓ, the quasi-orthogonality properties, and the fact that the sequence $(c_{\ell,j}(t))_{\ell \in \mathbb{Z}}$ is an element of the unit sphere of $\ell^2(\mathbb{Z})$, we obtain

$$\sum_{\ell \leq k-2} \left(\int_0^T c_{\ell,j}^2 \sum_{\nu \in \Lambda_{k,\ell}} \|\widetilde{\Delta}_{k,\ell}^\nu \partial_t u_j\|_{L^2}^2 \, dt \right)^{\frac{1}{2}} \leq C \|\nabla u_j\|_{L_T^\infty(L^2)} \sum_{\ell \leq k} \left(\int_0^T c_{\ell,j}^2 \, dt \right)^{\frac{1}{2}}$$
$$\leq C T^{\frac{1}{2}} k^{\frac{1}{2}} \|\nabla u_j\|_{L_T^\infty(L^2)}.$$

This yields, for some sequence $(c_k')_{k \in \mathbb{N}}$ such that $\|(c_k')\|_{\ell^2} \leq 1$,

$$B_j \leq C T^{\frac{1}{2}} \sum_{k \geq j-2} k 2^{-\frac{k}{2}} \|\Delta_k \gamma_j^h\|_{L^2} \|\nabla u_j\|_{L_T^\infty(L^2)}$$
$$\leq C T^{\frac{1}{2}} \|\nabla u_j\|_{L_T^\infty(L^2)} \|\gamma_j^h\|_{\dot{H}^{\sigma-1}} \sum_{k \geq j-2} c_k' k 2^{-\frac{k}{2}} 2^{-k(\sigma-1)}.$$

Choosing $\sigma = 3/4$ and taking advantage of (8.52), we conclude that

$$B_j \leq C T^{\frac{1}{2}} c_j 2^{\frac{j}{4}} \|\gamma\|_{\dot{H}^{-\frac{1}{4}}} \sum_{k \geq j-2} c_k' k 2^{-\frac{k}{4}}$$
$$\leq C T^{\frac{1}{2}} j c_j \|\gamma\|_{\dot{H}^{-\frac{1}{4}}} \sum_{k \geq j-2} c_k' (k-j+1) 2^{-\frac{(k-j)}{4}}$$
$$\leq C T^{\frac{1}{2}} j c_j^2 \|\gamma\|_{\dot{H}^{-\frac{1}{4}}}.$$

Hence, the inequality (8.53) is satisfied, which completes the proof of the theorem. □

8.6 Application to a Class of Semilinear Wave Equations

This section is devoted to the study of a class of semilinear wave equations with quadratic nonlinearity with respect to ∇u. This type of nonlinearity arises naturally in different fields of mathematics and mathematical physics, and, in particular, for the so-called *wave maps equations*.

Here, we consider the semilinear wave equation

$$(SW) \qquad \begin{cases} \Box u = Q(t,u)(\nabla u, \nabla u) \\ (u, \partial_t u)_{|t=0} = (u_0, u_1), \end{cases}$$

where Q stands for a smooth function from $\mathbb{R}^2$ to the space of symmetric matrices on $\mathbb{R}^{1+d}$, which is bounded, as are all of its derivatives.

To simplify the presentation, we focus on the evolution for positive times and on the case $d \geq 4$. We now give the main statement of this section.

Theorem 8.47. *Assume that $d \geq 4$ and that $\gamma \overset{def}{=} (\partial u_0, u_1)$ belongs to $\dot{B}_{2,1}^{\frac{d}{2}-1} \cap \dot{B}_{2,1}^{\frac{d}{2}-\frac{1}{2}}$. There then exists a maximal positive time $T^\star$ such that (SW) has a unique solution u on $[0, T^\star[$ satisfying*

$$\nabla u \in \mathcal{C}\left([0, T^\star[; \dot{B}_{2,1}^{\frac{d}{2}-1} \cap \dot{B}_{2,1}^{\frac{d}{2}-\frac{1}{2}}\right).$$

Moreover, there exists a nonincreasing positive function c on $\mathbb{R}^+$ such that

$$T^\star \geq c\left(\|\gamma\|_{\dot{B}_{2,1}^{\frac{d}{2}-1}}\right)\|\gamma\|_{\dot{B}_{2,1}^{\frac{d}{2}-\frac{1}{2}}}^{-2}. \tag{8.55}$$

If $T^\star$ is finite, then

$$\limsup_{T \leq T^\star}\left(\|u(T)\|_{L^\infty} + \int_0^T \|\nabla u(t)\|_{L^\infty} dt\right) = \infty.$$

Proof. For a sufficiently regular function u, we introduce the solution $F(u)$ of the following linear wave equation:

$$\Box F(u) = Q(t, u)(\nabla u, \nabla u) \quad \text{with} \quad (F(u))_{|t=0} = 0 \quad \text{and} \quad (\partial_t F(u))_{|t=0} = 0.$$

Observe that, by virtue of Duhamel's formula, u is a solution of (SW) if and only if $u = u_F + F(u)$ with

$$\Box u_F = 0, \qquad u_{F|t=0} = u_0, \quad \text{and} \quad \partial_t u_{F|t=0} = u_1.$$

Therefore, the first part of Theorem 8.47 is a consequence of the following proposition and the Picard fixed point theorem.

Proposition 8.48. *For any positive T, define the norms*

$$\|a\|_T \overset{def}{=} \|\nabla a\|_{L_T^\infty(\dot{B}_{2,1}^{\frac{d}{2}-1})} + T^{\frac{1}{2}}\|a\|_{1,T} \quad \text{with}$$

$$\|a\|_{1,T} \overset{def}{=} \|\nabla a\|_{L_T^\infty(\dot{B}_{2,1}^{\frac{d}{2}-\frac{1}{2}})} + \|\nabla a\|_{L_T^2(L^\infty)}.$$

Given a couple of positive real numbers (M, r), define the set $X_T^{M,r}$ of functions a such that

$$\|\nabla a\|_{L_T^\infty(\dot{B}_{2,1}^{\frac{d}{2}-1})} \leq M \quad \text{and} \quad \|a\|_{1,T} \leq r.$$

There exists a positive constant C_d, depending only on d, and a nondecreasing continuous function $C : \mathbb{R}^+ \to \mathbb{R}^+$ such that if

$$M = 2\|\gamma\|_{\dot{B}^{\frac{d}{2}-1}_{2,1}}, \quad r = C_d\|\gamma\|_{\dot{B}^{\frac{d}{2}-\frac{1}{2}}_{2,1}}, \quad and \quad rT^{\frac{1}{2}}C(M) \le 1,$$

then $u \longmapsto u_F + F(u)$ maps $X_T^{M,r}$ into $X_T^{M,r}$, and for any u and v in $X_T^{M,r}$, we have

$$\|F(u) - F(v)\|_T \le \frac{1}{2}\|u - v\|_T. \tag{8.56}$$

Proof. We first establish (8.56). For u and v in $X_T^{M,r}$, we may write[3]

$$Q(u)(\nabla u, \nabla u) - Q(v)(\nabla v, \nabla v) = Q_1(u, v) + Q_2(u, v) \quad \text{with}$$
$$Q_1(u, v) \stackrel{\text{def}}{=} (Q(u) - Q(v))(\nabla u, \nabla u) \quad \text{and}$$
$$Q_2(u, v) \stackrel{\text{def}}{=} Q(v)(\nabla u + \nabla v, \nabla u - \nabla v).$$

Hence, $F(u) - F(v)$ satisfies

$$\begin{cases} \Box(F(u) - F(v)) = Q_1(u, v) + Q_2(u, v) \\ (F(u) - F(v))_{|t=0} = 0 \quad \text{and} \quad (\partial_t(F(u) - F(v)))_{|t=0} = 0. \end{cases}$$

On the one hand, applying the localization operator $\dot{\Delta}_j$ to the above equation and then using the basic energy estimate to bound each block $\dot{\Delta}_j(F(u) - F(v))$, we get, for $\alpha \in \{1/2, 1\}$,

$$\|\nabla(F(u) - F(v))\|_{L^\infty_T(\dot{B}^{\frac{d}{2}-\alpha}_{2,1})} \le \|Q_1(u, v) + Q_2(u, v)\|_{L^1_T(\dot{B}^{\frac{d}{2}-\alpha}_{2,1})}. \tag{8.57}$$

On the other hand, according to the Strichartz estimate stated in Corollary 8.27, we may write

$$\|\nabla(F(u) - F(v))\|_{L^2_T(L^\infty)} \le \|Q_1(u, v) + Q_2(u, v)\|_{L^1_T(\dot{B}^{\frac{d}{2}-\frac{1}{2}}_{2,1})}. \tag{8.58}$$

So, in order to prove (8.56), it is only a matter of exhibiting suitable bounds for $Q_1(u, v)$ and $Q_2(u, v)$ in $L^1_T(\dot{B}^{\frac{d}{2}-1}_{2,1})$ and in $L^1_T(\dot{B}^{\frac{d}{2}-\frac{1}{2}}_{2,1})$.

We first establish bounds in the space $L^1_T(\dot{B}^{\frac{d}{2}-1}_{2,1})$. Taking advantage of the product laws in Besov spaces (use Corollary 2.54 page 90) and Corollary 2.66 page 97, we get, for all u and v in $X_T^{M,r}$,

$$\|Q_1(u, v)\|_{\dot{B}^{\frac{d}{2}-1}_{2,1}} \le \|\nabla u \otimes \nabla u\|_{\dot{B}^{\frac{d}{2}-1}_{2,1}} \|Q(u) - Q(v)\|_{\dot{B}^{\frac{d}{2}}_{2,1}}$$
$$\le C(M)\|\nabla u\|_{L^\infty}\|\nabla u\|_{\dot{B}^{\frac{d}{2}-1}_{2,1}}\|\nabla u - \nabla v\|_{\dot{B}^{\frac{d}{2}-1}_{2,1}},$$

where C is a nondecreasing continuous function of M.

---

[3] For expository purposes, we omit the time dependency of the function Q.

Along the same lines, we get

$$\|Q_2(u,v)\|_{\dot{B}_{2,1}^{\frac{d}{2}-1}} \leq \big(|Q(0)|+\|Q(v)-Q(0)\|_{\dot{B}_{2,1}^{\frac{d}{2}}}\big)\|(\nabla u+\nabla v)\otimes(\nabla u-\nabla v)\|_{\dot{B}_{2,1}^{\frac{d}{2}-1}}$$
$$\leq C(M)\|\nabla u + \nabla v\|_{\dot{B}_{2,1}^{\frac{d}{2}-\frac{1}{2}}}\|\nabla u - \nabla v\|_{\dot{B}_{2,1}^{\frac{d}{2}-\frac{1}{2}}}.$$

Plugging these inequalities into the inequality (8.57) with $\alpha = 1$, we thus get

$$\|\nabla F(u) - \nabla F(v)\|_{L_T^\infty(\dot{B}_{2,1}^{\frac{d}{2}-1})} \leq C(M)\Big(T^{\frac{1}{2}}\|\nabla u\|_{L_T^2(L^\infty)}\|\nabla u\|_{L_T^\infty(\dot{B}_{2,1}^{\frac{d}{2}-1})}$$
$$\times\|\nabla u - \nabla v\|_{L_T^\infty(\dot{B}_{2,1}^{\frac{d}{2}-1})}+T\|\nabla u - \nabla v\|_{L_T^\infty(\dot{B}_{2,1}^{\frac{d}{2}-\frac{1}{2}})}\|\nabla u+\nabla v\|_{L_T^\infty(\dot{B}_{2,1}^{\frac{d}{2}-\frac{1}{2}})}\Big), \quad (8.59)$$

from which it follows, as u and v are in $X_T^{M,r}$, that

$$\|\nabla F(u) - \nabla F(v)\|_{L_T^\infty(\dot{B}_{2,1}^{\frac{d}{2}-1})} \leq C(M)T^{\frac{1}{2}}r\|u-v\|_T. \quad (8.60)$$

Next, again using the product laws in Besov spaces and Corollary 2.66, we may write

$$\|Q_1(u,v)\|_{\dot{B}_{2,1}^{\frac{d}{2}-\frac{1}{2}}} \leq \|Q(u) - Q(v)\|_{\dot{B}_{2,1}^{\frac{d}{2}}}\|\nabla u\otimes\nabla u\|_{\dot{B}_{2,1}^{\frac{d}{2}-\frac{1}{2}}}$$
$$\leq C(M)\|\nabla(u-v)\|_{\dot{B}_{2,1}^{\frac{d}{2}-1}}\|\nabla u\|_{L^\infty}\|\nabla u\|_{\dot{B}_{2,1}^{\frac{d}{2}-\frac{1}{2}}}. \quad (8.61)$$

Along the same lines, we have

$$\|Q_2(u,v)\|_{\dot{B}_{2,1}^{\frac{d}{2}-\frac{1}{2}}} \leq \big(|Q(0)|+\|Q(v)-Q(0)\|_{\dot{B}_{2,1}^{\frac{d}{2}}}\big)\|(\nabla u+\nabla v)\otimes(\nabla u-\nabla v)\|_{\dot{B}_{2,1}^{\frac{d}{2}-\frac{1}{2}}}$$
$$\leq C(M)\big(\|\nabla(u-v)\|_{L^\infty}\|\nabla(u+v)\|_{\dot{B}_{2,1}^{\frac{d}{2}-\frac{1}{2}}}$$
$$+ \|\nabla(u-v)\|_{\dot{B}_{2,1}^{\frac{d}{2}-\frac{1}{2}}}\|\nabla(u+v)\|_{L^\infty}\big).$$

Putting those inequalities together with the energy estimate (8.57) with $\alpha = 1/2$ and the Strichartz estimate (8.58), we thus get

$$\|F(u) - F(v)\|_{1,T} \leq C(M)\Big(T^{\frac{1}{2}}\|\nabla(u-v)\|_{L_T^2(L^\infty)}\|\nabla(u+v)\|_{L_T^\infty(\dot{B}_{2,1}^{\frac{d}{2}-\frac{1}{2}})}$$
$$+T^{\frac{1}{2}}\|\nabla(u-v)\|_{L_T^\infty(\dot{B}_{2,1}^{\frac{d}{2}-\frac{1}{2}})}\|\nabla(u+v)\|_{L_T^2(L^\infty)}$$
$$+T^{\frac{1}{2}}\|\nabla(u-v)\|_{L_T^\infty(\dot{B}_{2,1}^{\frac{d}{2}-1})}\|\nabla u\|_{L_T^2(L^\infty)}\|\nabla u\|_{L_T^\infty(\dot{B}_{2,1}^{\frac{d}{2}-\frac{1}{2}})}\Big), \quad (8.62)$$

from which it follows that

$$\|F(u) - F(v)\|_{1,T} \leq C(M)r(1 + rT^{\frac{1}{2}})\|u-v\|_T.$$

Combining this inequality with (8.60), we conclude that

$$\|F(u) - F(v)\|_T \leq C(M)rT^{\frac{1}{2}}(1 + rT^{\frac{1}{2}})\|u - v\|_T. \qquad (8.63)$$

Of course, we may assume with no loss of generality that $4C(M) \geq 1$. So, choosing T such that

$$4C(M)rT^{\frac{1}{2}} \leq 1,$$

we see that $rT^{\frac{1}{2}} \leq 1$. Hence, the inequality (8.63) implies that

$$\|F(u) - F(v)\|_T \leq \frac{1}{2}\|u - v\|_T.$$

We now establish that for an appropriate choice of r, M, and T, if u belongs to $X_T^{M,r}$, then so does $u_F + F(u)$.

First, we note that, thanks to the energy and Strichartz estimates, there exists some constant C_d, depending only on d, such that for all $j \in \mathbb{Z}$ and $t \in \mathbb{R}^+$, we have

$$\|\dot{\Delta}_j \nabla u_F(t)\|_{L^2} = \|\dot{\Delta}_j \gamma\|_{L^2} \quad \text{and} \quad \|\dot{\Delta}_j \nabla u_F\|_{L_t^2(L^\infty)} \leq C_d 2^{j\left(\frac{d}{2} - \frac{1}{2}\right)}\|\dot{\Delta}_j \gamma\|_{L^2}.$$

By summation over j, we thus get

$$\|\nabla u_F\|_{L_T^\infty(\dot{B}_{2,1}^{\frac{d}{2}-1})} = \|\gamma\|_{\dot{B}_{2,1}^{\frac{d}{2}-1}} \quad \text{and} \quad \|u_F\|_{1,T} \leq C_d\|\gamma\|_{\dot{B}_{2,1}^{\frac{d}{2}-\frac{1}{2}}}.$$

Second, applying the inequalities (8.59) and (8.62) with $v = 0$, we get, for all $u \in X_T^{M,r}$ [up to a harmless change of $C(M)$],

$$\|\nabla F(u)\|_{L_T^\infty(\dot{B}_{2,1}^{\frac{d}{2}-1})} \leq C(M)T^{\frac{1}{2}}r \quad \text{and} \quad \|F(u)\|_{1,T} \leq C(M)T^{\frac{1}{2}}r^2.$$

Taking $M = 2\|\gamma\|_{\dot{B}_{2,1}^{\frac{d}{2}-1}}$ and $r = 2C_d\|\gamma\|_{\dot{B}_{2,1}^{\frac{d}{2}-\frac{1}{2}}}$, we thus see that the above inequalities imply that $u_F + F(u)$ is in $X_T^{M,r}$ whenever T has been chosen sufficiently small so as to satisfy

$$2C(M)rT^{\frac{1}{2}} \leq 1.$$

This completes the proof of the proposition and thus of the existence of a solution of (SW). $\square$

The uniqueness is an easy consequence of the above computations. Indeed, if we consider two solutions, u and v, of (SW), then $v - u = F(v) - F(u)$. Hence, the inequality (8.63) reads

$$\|u - v\|_T \leq C(M)rT^{\frac{1}{2}}(1 + rT^{\frac{1}{2}})\|u - v\|_T$$

with $r = \max(\|u\|_{1,T}, \|v\|_{1,T})$ and $M = \max\left(\|\nabla u\|_{L_T^\infty(\dot{B}_{2,1}^{\frac{d}{2}-1})}, \|\nabla v\|_{L_T^\infty(\dot{B}_{2,1}^{\frac{d}{2}-1})}\right)$.

This implies uniqueness on a sufficiently small time interval $[0, T_0]$. Repeating the argument then yields uniqueness on $[T_0, 2T_0]$ and $[2T_0, 3T_0]$, and so on, until the whole interval $[0, T]$ is exhausted.

To prove the blow-up criterion, the starting point is the following energy estimate for $\alpha \in \{1/2, 1\}$:

$$\|\nabla u\|_{L_T^\infty(\dot{B}_{2,1}^{\frac{d}{2}-\alpha})} \leq \|\gamma\|_{\dot{B}_{2,1}^{\frac{d}{2}-\alpha}} + \|Q(u)(\nabla u, \nabla u)\|_{L_T^1(\dot{B}_{2,1}^{\frac{d}{2}-\alpha})}. \tag{8.64}$$

The term $Q(u)(\nabla u, \nabla u)$ is a linear combination of terms of the type

$$Q_{ij}(u)\partial_i u \partial_j u, \quad 0 \leq i, j \leq d.$$

Now, according to the (simplified) Bony decomposition, we have

$$Q_{ij}(u)\partial_i u \partial_j u = T'_{Q_{ij}(u)\partial_i u}\partial_j u + T_{\partial_j u}Q_{ij}(u)\partial_i u.$$

Since, for some smooth function $\mathcal{Q}_{ij}$ we have $Q_{ij}(u)\partial_i u = \partial_i(\mathcal{Q}_{ij}(u))$, the composition lemma, together with the paraproduct estimates, enables us to conclude that

$$\|Q(u)(\nabla u, \nabla u)\|_{\dot{B}_{2,1}^{\frac{d}{2}-\alpha}} \leq C(\|u\|_{L^\infty})\|\nabla u\|_{L^\infty}\|\nabla u\|_{\dot{B}_{2,1}^{\frac{d}{2}-\alpha}}.$$

Plugging this inequality into (8.64), we end up with

$$\|\nabla u\|_{L_T^\infty(\dot{B}_{2,1}^{\frac{d}{2}-\alpha})} \leq \|\gamma\|_{\dot{B}_{2,1}^{\frac{d}{2}-\alpha}} + C(\|u\|_{L_T^\infty(L^\infty)})\int_0^T \|\nabla u\|_{L^\infty}\|\nabla u\|_{\dot{B}_{2,1}^{\frac{d}{2}-\alpha}} dt.$$

Now, if u is in $L^\infty([0, T^\star[; L^\infty)$ and ∇u is in $L^1([0, T^\star[; L^\infty)$, then the Gronwall lemma ensures that ∇u is in $L^\infty([0, T^\star[; \dot{B}_{2,1}^{\frac{d}{2}-\frac{1}{2}} \cap \dot{B}_{2,1}^{\frac{d}{2}-1})$. From this, it is easy to conclude that the solution may be continued beyond $T^\star$. This is simply a matter of following the method that was used in the proof of Theorem 7.21 page 307. $\qquad\square$

8.7 References and Remarks

The study of the dispersive properties of linear equations has a long history. However, the idea of using them to achieve some gain of regularity (compared with Sobolev embedding) is rather recent. More general stationary or nonstationary phase arguments than the ones we used to prove the basic dispersive inequality for the wave equation (Proposition 8.15) may be found in, for example, the book [167] by L. Hörmander. The one-dimensional estimate may be found in [150] and [274].

The first global $L^q(L^r)$ estimate was stated in 1977 by R. Strichartz in [276] for the wave equation (see also the works by P. Brenner in [46, 47], and by H. Pecher in [246, 247]). The extension to the whole set of admissible indices was achieved by J. Ginibre and G. Velo in [147] for the Schrödinger equation, and in [149] for the wave equation, except for the endpoint case $q = 2$, $r = 2\sigma/(\sigma - 1)$ with $\sigma > 1$, which was established later by M. Keel and T. Tao in [180]. Let us emphasize that global $L^q(L^r)$ estimates are often the key to proving well-posedness results for semilinear wave of

Schrödinger equations (see for instance [26, 27, 60, 263, 264]). Such estimates are also available for other types of partial differential equations (see e.g. the works [48] and [148] concerning the Klein–Gordon equation).

The set of indices for which we proved the Strichartz inequality (8.14) is sharp. The so-called *Knapp wave* provides counterexamples away from the endpoint (see, e.g., [128]). On the one hand, it is also known that (8.14) fails for $(q, r, \sigma) = (2, \infty, 1)$ (see, e.g., [233] for the case of the wave or Schrödinger equations). On the other hand, it holds true for *radial functions* (see [278]).

Refined Strichartz inequalities were introduced by S. Klainerman and D. Tataru in [194] in order to prove a sharp result concerning the Yang–Mills equations. There is a huge literature concerning applications of Strichartz-type inequalities to nonlinear equations (see, e.g., [60] for the case of the Schrödinger equation and [253] for the semilinear wave equation).

Finally, that the defocusing cubic wave equation is globally well posed in the space $(\dot{H}^s \cap L^4)(\mathbb{R}^3) \times \dot{H}^{s-1}(\mathbb{R}^3)$ for $s \in \,]3/4, 1[$ was first proven by C. Kenig, G. Ponce, and L. Vega in [181], then by I. Gallagher and F. Planchon in [135] (see also [21]). The former proof follows the method introduced by J. Bourgain in [44], which amounts to first solving the equation for the low-frequency part of the data, then a modified cubic wave equation, while the latter work is based on a strategy introduced in the context of the Navier–Stokes equations by C. Calderón in [55]: The authors first solve the equation for the high-frequency part of the data. In this chapter, we adopted the latter approach. To the best our knowledge, our global well-posedness result in $\dot{H}^{\frac{3}{4}}(\mathbb{R}^3)$ is new. We should point out that since $\dot{H}^{\frac{3}{4}}(\mathbb{R}^3)$ is continuously embedded in $L^4(\mathbb{R}^3)$, we do not have any supplementary condition on the Cauchy data, in contrast with [181] and [135]. We also note out that in [209, 210], H. Lindblad and C. Sogge proved that the Cauchy problem for $(W_3^{\pm})$ in $\dot{H}^s$ is ill posed below $s = \frac{1}{2}$.

The local well-posedness result for the class of semilinear wave equations with quadratic nonlinearity considered in Section 8.6 is essentially contained in the work by G. Ponce and T. Sideris [253] (see also [267]). Here, we strived for a scaling invariant functional framework in which to apply the Picard fixed point theorem. Finally, we emphasize that if the nonlinearity satisfies the so-called *null condition*, then the best index of regularity for which local well-posedness holds true falls to $s = \frac{d}{2}$ (see, in particular, the works by S. Klainerman and S. Selberg in [193], and by D. Tataru in [279], dedicated to the wave maps equations).

9

Smoothing Effect in Quasilinear Wave Equations

This chapter is devoted to the local well-posedness issue for a class of quasilinear wave equations. The equations which we consider here may be seen as toy models for the Einstein equations in relativity theory. We shall see that the energy method presented in Chapter 4 allows to establish a local-in-time existence theorem for data in any H^s space embedded in the set of Lipschitz functions, or in the Besov space $B_{2,1}^{\frac{d}{2}+1}$. In this chapter, we aim to go beyond such classical results.

To be more specific, we now present the model that we are going to study here. As in the preceding chapter, we define

$$\partial_0 \overset{\text{def}}{=} \partial_t, \quad \partial \overset{\text{def}}{=} (\partial_{x_1}, \dots, \partial_{x_d}), \quad \text{and} \quad \nabla \overset{\text{def}}{=} (\partial_t, \partial_{x_1}, \dots, \partial_{x_d}).$$

Throughout the chapter, G will denote a smooth function, bounded on $\mathbb{R}^2$ (along with all of its derivatives) and valued in a compact subset K of the space of symmetric d-dimensional matrices. We assume, in addition, that $\text{Id} + K$ is included in the cone of positive definite matrices, a condition which ensures the ellipticity of the operator

$$\Delta + \partial \cdot (G(\cdot, u)\partial \cdot) \quad \text{with} \quad \partial \cdot (G(\cdot, u) \cdot \partial v) \overset{\text{def}}{=} \sum_{1 \leq j,k \leq d} \partial_j(G^{j,k}(\cdot, u)\partial_k v).$$

Let Q be a smooth function from $\mathbb{R}^2$ to the set of quadratic forms on $\mathbb{R}^{d+1}$, which is bounded as are all of its derivatives.

The quasilinear wave equations that we are going to consider in this chapter are of the form

$$(QW) \qquad \begin{cases} \partial_t^2 u - \Delta u - \partial \cdot (G(t,u) \cdot \partial u) = Q(t,u)(\nabla u, \nabla u) \\ \nabla u_{|t=0} = \gamma. \end{cases}$$

We point out that if $G \equiv 0$, then the equation (QW) reduces to the equation (SW) studied in Section 8.6. More generally, if G and Q are timeindependent, then it still has the following scaling invariance property: u is

H. Bahouri et al., *Fourier Analysis and Nonlinear Partial Differential Equations*, Grundlehren der mathematischen Wissenschaften 343, DOI 10.1007/978-3-642-16830-7_9, © Springer-Verlag Berlin Heidelberg 2011

a solution of (QW) on $[-T, T] \times \mathbb{R}^d$ if and only if $u_\lambda(t, x) \stackrel{\text{def}}{=} u(\lambda t, \lambda x)$ is a solution of (QW) on $[-\lambda^{-1}T, \lambda^{-1}T] \times \mathbb{R}^d$ [provided the second line of (QW) has been modified accordingly, of course]. Obviously, the Besov space $\dot{B}^s_{2,1}$ has the required invariance property if and only if $s = d/2$ or, in other words, if and only if γ belongs to $\dot{B}^{\frac{d}{2}-1}_{2,1}$. Therefore, we expect the quantity $\|\gamma\|_{\dot{B}^{\frac{d}{2}-1}_{2,1}}$ to play a decisive role in the study of (QW).

This chapter is structured as follows. The first section is devoted to the proof of the classical well-posedness result for initial data such that

$$\gamma \in \mathcal{B}^d \stackrel{\text{def}}{=} \dot{B}^{\frac{d}{2}-1}_{2,1} \cap \dot{B}^{\frac{d}{2}}_{2,1}. \tag{9.1}$$

We stress that this assumption is the weakest one (in the framework of Besov spaces related to L^2) for which $(\partial u_0, u_1)$ is bounded. Therefore, this space is somewhat critical, inasmuch as it is the largest one for which local well-posedness may be achieved by means of a basic energy method (which works in any dimension $d \geq 1$ as it is not related to any dispersive properties of the wave equation). In this section, we pay special attention to the scaling invariance of all the estimates as this will be important in the following sections.

The rest of the chapter is devoted to going weakening assumption (9.1). More precisely, in the second section of this chapter, we give our main statement, Theorem 9.5, and explain the strategy of its proof. As this will be based on geometrical optics, we need to regularize the metric $G(\cdot, u)$ both in time and space. As regards the time regularization, it turns out to be convenient to introduce a time cut-off so as to transform the initial quasilinear wave equation (QW) into a "truncated" linear wave equation (QW_T) with constant coefficients away from the time interval $[-T, T]$. If T is chosen suitably small, this will enable us to manipulate *globally defined* solutions only.

Still motivated by geometrical optics, in the third section, we introduce a refined paralinearization of the system (QW_T). This means that after localization about frequencies of size $\lambda_j = 2^j$, we regularize (in space-time variables) the metric $G_T(\cdot, u)$ by spectral truncation at frequency λ_j^δ *with δ less than 1* (instead of the λ_j used in the classical paralinearization procedure, such as in, e.g., Lemma 4.14 page 183). This refined procedure makes the method of geometrical optics more efficient. The price to be paid is that the remainder term is less regular (i.e., larger after frequency localization).

In the fourth section, we explain how to derive Theorem 9.5 from suitable *microlocal Strichartz estimates* (i.e., Strichartz estimates on small intervals, the lengths of which depend on the size of the frequency we are looking at). The key idea is to split the interval $[0, T]$ into sufficiently small intervals so that we may apply these microlocal estimates. Combining all these estimates leads to a Strichartz estimate with a loss, compared to the linear wave equation with constant coefficients.

The final section is devoted to the proof of rather general microlocal Strichartz estimates for a class of variable coefficients linear wave equations.

After suitable rescaling, we shall see that the general statement yields the estimates we are interested in for solving (QW_T). Our proof relies on the use of a geometrical optics method so as to build a sufficiently accurate approximation of the solution to the linear wave equation, and on the $TT^\star$ method.

9.1 A Well-posedness Result Based on an Energy Method

This section is devoted to the proof of a general local existence result for the quasilinear wave equation in $\mathbb{R}^d$ $(d \geq 1)$ with suitably smooth initial data. To begin, we recall some very basic facts about the variable coefficients wave equation. We fix some time-dependent *metric* $g = (g^{j,k})_{1 \leq j,k \leq d}$ on $\mathbb{R}^d$, that is, a function from $I \times \mathbb{R}^d$ (where I is a time interval) to the set $\mathcal{S}_d^+(\mathbb{R})$ of symmetric positive definite matrices on $\mathbb{R}^d$. Assume, in addition, that there exists some positive real number A_0 such that

$$A_0^{-1}|\eta|^2 \leq \sum_{j,k} g^{j,k}(t,x)\,\eta^j \eta^k \leq A_0|\eta|^2 \quad \text{for all} \quad (t,x,\eta) \in I \times \mathbb{R}^d \times \mathbb{R}^d. \quad (9.2)$$

Define

$$\partial \cdot (g \cdot \partial u) \overset{\text{def}}{=} \sum_{j,k} \partial_j \big(g^{j,k} \partial_k u\big).$$

We have the following lemma.

Lemma 9.1. *Consider a continuous function ϕ such that*

$$\phi(t) \geq A_0 \|\partial_t g(t,\cdot)\|_{L^\infty}.$$

For any function u such that ∇u is C^1 in time with values in L^2, and $f \overset{\text{def}}{=} \partial_t^2 u - \partial \cdot (g \cdot \partial u)$ is L^1 in time with values in L^2, we then have

$$\exp\Big(-\frac{1}{2}\int_0^t \phi(t')\,dt'\Big)\|\nabla u(t)\|_{L^2} \leq A_0\|\nabla u(0)\|_{L^2}$$

$$+ A_0^{\frac{1}{2}} \int_0^t \exp\Big(-\frac{1}{2}\int_0^{t'} \phi(t'')\,dt''\Big)\|f(t')\|_{L^2}\,dt'.$$

Proof. Taking the L^2 inner product of $\partial_t^2 u - \partial \cdot (g \cdot \partial u)$ with $\partial_t u$, we get

$$\frac{1}{2}\frac{d}{dt}\|\partial_t u\|_{L^2}^2 = (f|\partial_t u)_{L^2} + \sum_{j,k}\int_{\mathbb{R}^d} \partial_j\big(g^{j,k}\partial_k u\big)\partial_t u\,dx.$$

We now integrate by parts in the last term. As $g^{j,k} = g^{k,j}$ for all $1 \leq j,k \leq d$, we get

$$\sum_{j,k} \int_{\mathbb{R}^d} \partial_j \left(g^{j,k} \partial_k u\right) \partial_t u \, dx = -\frac{1}{2} \int_{\mathbb{R}^d} g^{j,k} \frac{\partial}{\partial t} \left(\partial_j u \partial_k u\right) dx.$$

Hence, we may conclude that

$$\frac{1}{2} \frac{d}{dt} \|\nabla u(t)\|_{L^2_{g(t)}}^2 = (f(t)|\partial_t u(t))_{L^2} + \frac{1}{2} \sum_{j,k} \int_{\mathbb{R}^d} \partial_t g^{j,k}(t,x) \partial_j u(t,x) \partial_k u(t,x) \, dx$$

with

$$\|\nabla u(t)\|_{L^2_{g(t)}}^2 \overset{\text{def}}{=} \|\partial_t u(t)\|_{L^2}^2 + \sum_{j,k} \int_{\mathbb{R}^d} g^{j,k}(t,x) \partial_j u(t,x) \, \partial_k u(t,x) \, dx.$$

As $\|\partial_t u\|_{L^2} \le \|\nabla u\|_{L^2_{g(t)}}$ and

$$A_0^{-1} \|\nabla u(t)\|_{L^2}^2 \le \|\nabla u(t)\|_{L^2_{g(t)}}^2 \le A_0 \|\nabla u(t)\|_{L^2}^2, \tag{9.3}$$

this gives

$$\frac{1}{2} \frac{d}{dt} \|\nabla u(t)\|_{L^2_{g(t)}}^2 \le \|f(t)\|_{L^2} \|\partial_t u(t)\|_{L^2} + \frac{1}{2} A_0 \|\partial_t g(t)\|_{L^\infty} \|\nabla u(t)\|_{L^2_{g(t)}}^2$$

$$\le \|f(t)\|_{L^2} \|\nabla u(t)\|_{L^2_{g(t)}} + \frac{1}{2} A_0 \|\partial_t g(t)\|_{L^\infty} \|\nabla u(t)\|_{L^2_{g(t)}}^2.$$

As $\phi(t) \ge A_0 \|\partial_t g(t,\cdot)\|_{L^\infty}$, Gronwall's lemma (see Lemma 3.3 page 125) thus gives

$$\exp\left(-\frac{1}{2} \int_0^t \phi(t') \, dt'\right) \|\nabla u(t)\|_{L^2_{g(t)}} \le \|\nabla u(0)\|_{L^2_{g(0)}}$$

$$+ \int_0^t \|f(t')\|_{L^2} \exp\left(-\frac{1}{2} \int_0^{t'} \phi(t'') \, dt''\right) dt'. \tag{9.4}$$

In order to conclude, we simply use the condition (9.2). $\qquad\qquad\square$

Before stating the basic existence result for the quasilinear wave equation (QW), we introduce an item of notation that will be used throughout this chapter.

Notation. We denote by C_γ a generic expression of the type $f(\|\gamma\|_{\dot{B}^{\frac{d}{2}-1}_{2,1}})$, where $f : \mathbb{R}^+ \to \mathbb{R}^+$ is a nondecreasing continuous function.

Theorem 9.2. *Assume that the metric* $\mathrm{Id} + G$ *satisfies the condition (9.2). If the initial data* (u_0, u_1) *is such that* $\gamma \overset{\text{def}}{=} (\partial u_0, u_1)$ *belongs to* $\mathcal{B}^d \overset{\text{def}}{=} \dot{B}^{\frac{d}{2}}_{2,1} \cap \dot{B}^{\frac{d}{2}-1}_{2,1}$, *then there exist two maximal positive times,* $T_\star$ *and* $T^\star$, *satisfying*

$$C_\gamma \min\{T_\star, T^\star\} \|\gamma\|_{\dot{B}^{\frac{d}{2}}_{2,1}} \ge 1$$

and such that (QW) *has a unique solution* u *in the space* $\mathcal{C}(]-T_\star, T^\star[; \mathcal{B}^d)$.

Moreover, if $T_\star$ (resp., $T^\star$) is finite, then

$$\limsup_{t \to T} \left(\|u(t)\|_{L^\infty} + \int_0^t \|\nabla u(t')\|_{L^\infty}\, dt' \right) = \infty$$

with $T = -T_\star$ (resp., $T = T^\star$).

Finally, if the initial data is such that γ belongs to $\dot B_{2,r}^{s-1}$ for some positive s and $r \in [1, \infty]$, then ∇u is continuous (or weakly continuous, if $r = \infty$) with values in the space $\dot B_{2,r}^{s-1}$.

Proof. As the equation (QW) is time-reversible, we shall focus (as in the whole of this chapter) on the proof for positive times. The proof has much in common with those of Theorems 4.16 page 188 and 4.21 page 193. Indeed, Section 4.2.1 can be effectively reproduced in the framework of linear wave equations with variable coefficients. Here, we define the sequence $(u_n)_{n \in \mathbb{N}}$ of approximate solutions by means of the following induction:

- The function u_0 is the solution of

$$\partial_t^2 u_0 - \Delta u_0 = 0 \quad \text{with} \quad \nabla u_0(0) = S_0 \gamma.$$

- Once u_n has been defined, u_{n+1} is the solution of[1]

$$\begin{cases} \partial_t^2 u_{n+1} - \Delta u_{n+1} - \partial \cdot (G(t, u_n) \cdot \partial u_{n+1}) = Q(t, u_n)(\nabla u_n, \nabla u_n) \\ \nabla u_{n+1}(0) = S_{n+1} \gamma. \end{cases}$$

In order to prove that the iterative scheme converges in $\mathcal{B}^d$, the following commutator lemma (in the spirit of Lemma 2.100 page 112) will be useful.

Lemma 9.3. *Let L be a compact subset of $]0, +\infty[$. A constant C exists such that for any s in L and any Lipschitz functions a and v which, in addition, belong to $\dot B_{2,r}^s$ for some $r \in [1, \infty]$, we have, for all $k \in \{1, \dots, d\}$,*

$$2^{j(s-1)} \|\dot\Delta_j(v \partial_k a) - v \partial_k \dot\Delta_j a\|_{\dot H^1} \leq c_j C \left(\|v\|_{\dot B_{2,r}^s} \|\partial a\|_{L^\infty} + \|a\|_{\dot B_{2,r}^s} \|\partial v\|_{L^\infty} \right),$$

and also

$$2^{j(s-1)} \|\dot\Delta_j(v \partial_k a) - v \partial_k \dot\Delta_j a\|_{\dot H^1} \leq c_j C \left(\|v\|_{\dot B_{2,r}^{s+1}} \|a\|_{L^\infty} + \|a\|_{\dot B_{2,r}^s} \|\partial v\|_{L^\infty} \right),$$

where $(c_j)_{j \in \mathbb{Z}}$ denotes an element of the unit sphere of $\ell^r(\mathbb{Z})$ which depends on v and a.

Proof. We have to prove that for all $k \in \{1, \dots, d\}$,

$$R_j(v, a) \overset{\text{def}}{=} \dot\Delta_j(v \partial_k a) - v \partial_k \dot\Delta_j a$$

---

[1] Recall that S_n is the low-frequency truncation operator defined on page 61.

satisfies the above estimate. We write that

$$v\partial_k a = \sum_{j'} \dot{\Delta}_{j'} v \, \partial_k \dot{S}_{j'+2} a + \sum_{j'} \dot{S}_{j'-1} v \, \partial_k \dot{\Delta}_{j'} a$$

and then, by virtue of the localization properties of the Littlewood–Paley decomposition, that

$$R_j(v,a) = \sum_{\ell=1}^{3} R_j^{(\ell)}(v,a) \quad \text{with}$$

$$R_j^{(1)}(v,a) \overset{\text{def}}{=} \sum_{j' \geq j-3} \dot{\Delta}_j(\dot{\Delta}_{j'} v \, \partial_k \dot{S}_{j'+2} a),$$

$$R_j^{(2)}(v,a) \overset{\text{def}}{=} \sum_{|j'-j| \leq 4} [\dot{\Delta}_j, \dot{S}_{j'-1} v] \partial_k \dot{\Delta}_{j'} a, \quad \text{and}$$

$$R_j^{(3)}(v,a) \overset{\text{def}}{=} \sum_{|j'-j| \leq 1} (\dot{S}_{j'-1} - \text{Id}) v \, \partial_k \dot{\Delta}_j \dot{\Delta}_{j'} a.$$

We now estimate each term. As $\|\partial_k \dot{S}_{j'+2} a\|_{L^\infty} \leq C\|\partial a\|_{L^\infty}$, we have

$$\|R_j^{(1)}(v,a)\|_{\dot{H}^1} \leq C2^j \|R_j^{(1)}(v,a)\|_{L^2}$$

$$\leq C2^j \sum_{j' \geq j-3} \|\partial a\|_{L^\infty} \|\dot{\Delta}_{j'} v\|_{L^2}.$$

Using Young's inequality for series, we get, for any positive s,

$$2^{j(s-1)} \|R_j^{(1)}(v,a)\|_{\dot{H}^1} \leq C2^{js} \|R_j^{(1)}(v,a)\|_{L^2}$$

$$\leq C\|\partial a\|_{L^\infty} \sum_{j' \geq j-3} 2^{-(j'-j)s} 2^{j's} \|\dot{\Delta}_{j'} v\|_{L^2}$$

$$\leq Cc_j \|\partial a\|_{L^\infty} \|v\|_{\dot{B}_{2,r}^s}. \tag{9.5}$$

As $\|\partial_k \dot{S}_{j'+2} a\|_{L^\infty} \leq C2^{j'} \|a\|_{L^\infty}$, we also have

$$2^{j(s-1)} \|R_j^{(1)}(v,a)\|_{\dot{H}^1} \leq C\|a\|_{L^\infty} \sum_{j' \geq j-3} 2^{-(j'-j)s} 2^{j'(s+1)} \|\dot{\Delta}_{j'} v\|_{L^2}$$

$$\leq Cc_j \|a\|_{L^\infty} \|v\|_{\dot{B}_{2,1}^{s+1}}. \tag{9.6}$$

We now estimate $\|R_j^{(2)}(v,a)\|_{\dot{H}^1}$. Using Lemma 2.97 page 110, we can write

$$2^{j(s-1)} \|R_j^{(2)}(v,a)\|_{\dot{H}^1} \leq C2^{js} \|R_j^{(2)}(v,a)\|_{L^2}$$

$$\leq C2^{js} \|\partial v\|_{L^\infty} \sum_{|j-j'| \leq 4} 2^{j'-j} \|\dot{\Delta}_{j'} a\|_{L^2}$$

$$\leq C\|\partial v\|_{L^\infty} \sum_{|j-j'| \leq 4} 2^{(j-j')(s-1)} 2^{j's} \|\dot{\Delta}_{j'} a\|_{L^2}.$$

Using the definition of the norm in $\dot{B}_{2,r}^s$, we thus get, for any s,

$$2^{j(s-1)}\|R_j^{(2)}(v,a)\|_{\dot{H}^1} \leq Cc_j\|\partial v\|_{L^\infty}\|a\|_{\dot{B}_{2,r}^s}. \tag{9.7}$$

Finally, we estimate $\|R_j^{(3)}(v,a)\|_{\dot{H}^1}$. We use the fact that, according to Leibniz's formula, we have

$$\|R_j^{(3)}(v,a)\|_{\dot{H}^1} \leq \|R_j^{(3)}(v,\partial a)\|_{L^2} + \|R_j^{(3)}(\partial v,a)\|_{L^2}.$$

From Lemma 2.1 page 52, we infer that a constant C exists such that for any integer j', we have

$$\|(\dot{S}_{j'-1} - \mathrm{Id})v\|_{L^\infty} \leq C \sum_{j'' \geq j'-1} 2^{-j''}\|\partial v\|_{L^\infty} \leq C2^{-j'}\|\partial v\|_{L^\infty}.$$

Thus, we deduce that for any s,

$$2^{j(s-1)}\|R_j^{(3)}(v,\partial a)\|_{L^2} \leq Cc_j\|\partial v\|_{L^\infty}\|a\|_{\dot{B}_{2,r}^s}. \tag{9.8}$$

Next, because $\|S_{j'-1}\partial v\|_{L^\infty} \leq C\|\partial v\|_{L^\infty}$, we have

$$2^{j(s-1)}\|R_j^{(3)}(\partial v,a)\|_{L^2} \leq Cc_j\|\partial v\|_{L^\infty}\|a\|_{\dot{B}_{2,r}^s}. \tag{9.9}$$

Combining (9.8) and (9.9) gives

$$2^{j(s-1)}\|R_j^{(3)}(v,a)\|_{\dot{H}^1} \leq Cc_j\|\partial v\|_{L^\infty}\|a\|_{\dot{B}_{2,r}^s}. \tag{9.10}$$

Combining the three estimates (9.5) [resp., (9.6)], (9.7), and (9.10) gives the first (resp., second) inequality. $\qquad\square$

Corollary 9.4. *Let s be a positive real number. There exists a continuous nondecreasing function $C_s : \mathbb{R}^+ \to \mathbb{R}^+$ which vanishes at 0 and satisfies the following properties. Consider a couple of functions (u,v), the space derivatives of which are locally integrable in time, with values in L^∞ and such that $v(t)$ is also locally integrable in time with values in L^∞. Assume, in addition, that u and v are locally integrable in time with values in $\dot{B}_{2,r}^s$. If*

$$\partial_t^2 u - \Delta u - \partial \cdot (G(t,v) \cdot \partial u) = f$$

with f in $L_{loc}^1(\dot{B}_{2,1}^{s-1})$, then we have, for any integer j,

$$\partial_t^2 \dot{\Delta}_j u - \Delta\dot{\Delta}_j u - \partial \cdot (G(t,v) \cdot \partial\dot{\Delta}_j u) = \dot{\Delta}_j f + R_j(u,v),$$

where the operator R_j is such that, for any t, there exists a sequence $(c_j(t))_{j\in\mathbb{Z}}$ in the unit sphere of $\ell^r(\mathbb{Z})$ such that

$$2^{j(s-1)}\|R_j(u,v)(t)\|_{L^2} \leq c_j(t)C(\|v(t)\|_{L^\infty})$$
$$\times \left(\|u(t)\|_{\dot{B}_{2,r}^s}\|\partial v(t)\|_{L^\infty} + \|v(t)\|_{\dot{B}_{2,r}^s}\|\partial u(t)\|_{L^\infty}\right) \tag{9.11}$$

and

$$2^{j(s-1)}\|R_j(u,v)(t)\|_{L^2} \leq c_j(t)C(\|v(t)\|_{L^\infty})$$
$$\times \left(\|u(t)\|_{\dot{B}_{2,r}^s}\|\partial v(t)\|_{L^\infty} + \|v(t)\|_{\dot{B}_{2,r}^{s+1}}\|u(t)\|_{L^\infty}\right). \quad (9.12)$$

Proof. Note that

$$R_j(u,v) \stackrel{\text{def}}{=} \partial \cdot \left(\dot{\Delta}_j(G(t,v)\cdot\partial u) - G(t,v)\cdot\dot{\Delta}_j\partial u\right)$$

satisfies

$$\|R_j(u,v)\|_{L^2} \leq \|\dot{\Delta}_j(G(t,v)\cdot\partial u) - G(t,v)\cdot\dot{\Delta}_j\partial u\|_{\dot{H}^1}.$$

The result is therefore a straightforward consequence of the previous lemma combined with Theorem 2.61 page 94. □

We now resume the proof of Theorem 9.2.

First Step: Uniform Bounds in Large Norm

We want to prove by induction that a positive constant B_0 and a positive time $\overline{T}$ exist such that for any n,

$$(\mathcal{P}_{n,\overline{T}}) \qquad \|\nabla u_n\|_{L_{\overline{T}}^\infty(\dot{B}_{2,1}^{s-1})} \leq B_0\|\gamma\|_{\dot{B}_{2,1}^{s-1}} \quad \text{for} \quad s = \frac{d}{2}, \frac{d}{2}+1.$$

Choosing $B_0 \geq 1$ makes the property $(\mathcal{P}_{0,\overline{T}})$ obvious for all $\overline{T} > 0$.

Assume that $(\mathcal{P}_{n,\overline{T}})$ is satisfied. Corollary 9.4 with $u = u_{n+1}$, $r = 1$, and s in $\{d/2, d/2+1\}$, hypothesis $(\mathcal{P}_{n,\overline{T}})$, and the embedding $\dot{B}_{2,1}^{\frac{d}{2}} \hookrightarrow L^\infty$ together give

$$\partial_t^2\dot{\Delta}_j u_{n+1} - \Delta\dot{\Delta}_j u_{n+1} - \partial \cdot (G(t,u_n)\cdot\partial\dot{\Delta}_j u_{n+1})$$
$$= \dot{\Delta}_j\left(Q(t,u_n)(\nabla u_n,\nabla u_n)\right) + R_j(u_{n+1},u_n)$$

with, for s in $\{d/2, d/2+1\}$,

$$2^{j(s-1)}\|R_j(u_{n+1},u_n)(t)\|_{L^2} \leq c_j(t)B_0 C_\gamma\|\gamma\|_{\dot{B}_{2,1}^{\frac{d}{2}}}\|\partial u_{n+1}(t)\|_{\dot{B}_{2,1}^{s-1}}.$$

More precisely, the case $s = d/2+1$ follows from the inequality (9.11), whereas the case $s = d/2$ is a consequence of the inequality (9.12).

Assume that

$$B_0 = 2A_0\exp(A_0A_1) \quad \text{with} \quad \int_0^{\overline{T}}\|\partial_t G(t,\cdot)\|_{L^\infty}\,dt \leq A_1. \quad (9.13)$$

Under the condition $(\mathcal{P}_{n,\overline{T}})$, we then have, owing to the chain rule and the embedding $\dot{B}_{2,1}^{\frac{d}{2}} \hookrightarrow L^\infty$,

$$A_0 \int_0^{\overline{T}} \|\partial_t(G(t,u_n(t)))\|_{L^\infty} \, dt \le A_0 \Big(A_1 + \|\partial_u G\|_{L^\infty(\mathbb{R}^2)} \int_0^{\overline{T}} \|\partial_t u_n(t)\|_{L^\infty} \, dt \Big)$$
$$\le A_0 \big(A_1 + C B_0 \|\partial_u G\|_{L^\infty(\mathbb{R}^2)} \overline{T} \|\gamma\|_{\dot{B}_{2,1}^{\frac{d}{2}}} \big).$$

Assume that

$$\overline{T} \le A_1 \big(C B_0 \|\partial_u G\|_{L^\infty(\mathbb{R}^2)} \|\gamma\|_{\dot{B}_{2,1}^{\frac{d}{2}}} \big)^{-1}. \tag{9.14}$$

We then get

$$A_0 \int_0^{\overline{T}} \|\partial_t(G(t,u_n(t,\cdot)))\|_{L^\infty} \, dt \le 2 A_0 A_1. \tag{9.15}$$

From Lemma 9.1 and the above inequalities, we infer that for any $t \le \overline{T}$ and j in $\mathbb{Z}$, we have

$$2^{j(s-1)} \|\nabla u_{n+1}\|_{L_t^\infty(L^2)} \le e^{A_0 A_1} \Big(A_0 2^{j(s-1)} \|\gamma\|_{L^2}$$
$$+ A_0^{\frac{1}{2}} B_0 C_\gamma \|\gamma\|_{\dot{B}_{2,1}^{\frac{d}{2}}} \int_0^t c_j(t') \|\partial u_{n+1}(t')\|_{\dot{B}_{2,1}^{s-1}} \, dt' \Big).$$

Hence, by summation over j, we get

$$\|\nabla u_{n+1}\|_{\widetilde{L}_t^\infty(\dot{B}_{2,1}^{s-1})} \le e^{A_0 A_1} \Big(A_0 \|\gamma\|_{\dot{B}_{2,1}^{s-1}} + A_0^{\frac{1}{2}} B_0 C_\gamma \|\gamma\|_{\dot{B}_{2,1}^{\frac{d}{2}}} \|\nabla u_{n+1}\|_{L_t^1(\dot{B}_{2,1}^{s-1})} \Big).$$

Choosing $\overline{T}$ such that

$$A_0^{\frac{1}{2}} B_0 e^{A_0 A_1} \overline{T} C_\gamma \|\gamma\|_{\dot{B}_{2,1}^{\frac{d}{2}}} \le \frac{1}{2}$$

implies that

$$\|\nabla u_{n+1}\|_{\widetilde{L}_{\overline{T}}^\infty(\dot{B}_{2,1}^{s-1})} \le 2 A_0 e^{A_0 A_1} \|\gamma\|_{\dot{B}_{2,1}^{s-1}} \quad \text{for} \ \ s \in \{d/2, d/2+1\}.$$

Together with (9.13), this gives $(\mathcal{P}_{n+1,\overline{T}})$.

Second Step: Convergence of the Approximate Sequence

We claim that if T_0 is sufficiently small, then $(\nabla u_n)_{n \in \mathbb{N}}$ is a Cauchy sequence in $L_{T_0}^\infty(\dot{B}_{2,1}^{\frac{d}{2}-1})$. Indeed, the difference $\widetilde{u}_n \overset{\text{def}}{=} u_{n+1} - u_n$ satisfies

$$(QW_n) \qquad \partial_t^2 \widetilde{u}_n - \Delta \widetilde{u}_n - \partial \cdot (G(t,u_n) \cdot \partial \widetilde{u}_n) = f_n$$

with

$$f_n \overset{\text{def}}{=} (Q(t, u_n) - Q(t, u_{n-1}))(\nabla u_n, \nabla u_n)$$
$$+Q(t, u_{n-1})(\nabla \widetilde{u}_{n-1}, \nabla u_n + \nabla u_{n-1}) + \partial \cdot ((G(t, u_n) - G(t, u_{n-1})) \cdot \partial u_n).$$

Applying Corollary 9.4 with $s = d/2$, $r = 1$, $u = \widetilde{u}_n$, and $v = u_n$ gives

$$\partial_t^2 \dot{\Delta}_j \widetilde{u}_n - \Delta \dot{\Delta}_j u - \partial \cdot (G(t, u_n) \cdot \partial \dot{\Delta}_j \widetilde{u}_n) = \dot{\Delta}_j f_n + R_j(\widetilde{u}_n, u_n).$$

To bound f_n, we may take advantage of the product laws (Corollary 2.54 page 90) and the composition estimates (Corollary 2.66 page 97). We deduce that for any $t \leq \overline{T}$,

$$2^{j(\frac{d}{2}-1)} \|\dot{\Delta}_j f_n(t)\|_{L^2} \leq c_j(t) C_\gamma \|\gamma\|_{\dot{B}_{2,1}^{\frac{d}{2}}} \|\nabla \widetilde{u}_{n-1}(t)\|_{\dot{B}_{2,1}^{\frac{d}{2}-1}}.$$

Observe that $\nabla \widetilde{u}_n(0) = \dot{\Delta}_n \gamma$. Hence, using Lemma 9.1, multiplying by $2^{j(\frac{d}{2}-1)}$, and summing over j, we get, for any $T \leq \overline{T}$,

$$\|\nabla \widetilde{u}_n\|_{L_T^\infty(\dot{B}_{2,1}^{\frac{d}{2}-1})} \leq B_0 \left(2^{-n} \|\gamma\|_{\dot{B}_{2,1}^{\frac{d}{2}}} + TC_\gamma \|\gamma\|_{\dot{B}_{2,1}^{\frac{d}{2}}} \|\nabla \widetilde{u}_{n-1}\|_{L_T^\infty(\dot{B}_{2,1}^{\frac{d}{2}-1})} \right).$$

Choosing $T \leq \overline{T}$ such that

$$B_0 T C_\gamma \|\gamma\|_{\dot{B}_{2,1}^{\frac{d}{2}}} \leq \frac{1}{2},$$

we then infer that

$$\|\nabla \widetilde{u}_n\|_{L_T^\infty(\dot{B}_{2,1}^{\frac{d}{2}-1})} \leq B_0 2^{-n} \|\gamma\|_{\dot{B}_{2,1}^{\frac{d}{2}}} + \frac{1}{2} \|\nabla \widetilde{u}_{n-1}\|_{L_T^\infty(\dot{B}_{2,1}^{\frac{d}{2}-1})}.$$

This readily implies that $(\nabla u_n)_{n \in \mathbb{N}}$ is a Cauchy sequence in $\mathcal{C}([0, T]; \dot{B}_{2,1}^{\frac{d}{2}-1})$. Hence, there exists some function u such that $(\nabla u_n)_{n \in \mathbb{N}}$ converges to ∇u in $\mathcal{C}([0, T]; \dot{B}_{2,1}^{\frac{d}{2}-1})$. Moreover, $(\nabla u_n)_{n \in \mathbb{N}}$ is a bounded sequence in $\widetilde{L}_T^\infty(\dot{B}_{2,1}^{\frac{d}{2}})$ and hence, by virtue of the Fatou property for Besov spaces (see Theorem 2.25 page 67), we have

$$\nabla u \in \mathcal{C}([0, T]; \dot{B}_{2,1}^{\frac{d}{2}-1}) \cap \widetilde{L}_T^\infty(\dot{B}_{2,1}^{\frac{d}{2}})$$

and may check that u is indeed a solution of (QW) on the time interval $[0, T]$.

Third Step: Time Continuity of the Solution

We must now check that ∇u belongs to $\mathcal{C}([0, T]; \dot{B}_{2,1}^{\frac{d}{2}})$. The argument of Section 4.3.2 page 190 can be repeated here. Indeed, if ε is a positive real number, then we have, for any integer j and $(t, t') \in [0, T]^2$,

$$\|\nabla u(t) - \nabla u(t')\|_{\dot{B}_{2,1}^{\frac{d}{2}}} = \sum_j 2^{j\frac{d}{2}}\|\dot{\Delta}_{j'}(\nabla u(t) - \nabla u(t'))\|_{L^2}$$

$$\leq 2^j \sum_{j'\leq j} 2^{j'(\frac{d}{2}-1)}\|\dot{\Delta}_{j'}(\nabla u(t) - \nabla u(t'))\|_{L^2}$$

$$+ 2\sum_{j'>j} 2^{j'\frac{d}{2}}\|\dot{\Delta}_{j'}\nabla u\|_{L_T^\infty(L^2)}$$

$$\leq 2^j \|\nabla u(t) - \nabla u(t')\|_{\dot{B}_{2,1}^{\frac{d}{2}-1}} + 2\sum_{j'>j} 2^{j'\frac{d}{2}}\|\dot{\Delta}_{j'}\nabla u\|_{L_T^\infty(L^2)}.$$

As $(2^{j\frac{d}{2}}\|\dot{\Delta}_j\nabla u\|_{L_T^\infty(L^2)})_{j\in\mathbb{Z}}$ is in $\ell^1(\mathbb{Z})$, an integer j_ε exists such that

$$\sum_{j'>j_\varepsilon} 2^{j\frac{d}{2}}\|\dot{\Delta}_{j'}\nabla u\|_{L_T^\infty(L^2)} < \frac{\varepsilon}{4}.$$

Thus, for all $(t, t') \in [0, T]^2$,

$$\|\nabla u(t) - \nabla u(t')\|_{\dot{B}_{2,1}^{\frac{d}{2}}} \leq 2^{j_\varepsilon}\|\nabla u(t) - \nabla u(t')\|_{\dot{B}_{2,1}^{\frac{d}{2}-1}} + \frac{\varepsilon}{2}.$$

As ∇u is continuous from $[0, T]$ to $\dot{B}_{2,1}^{\frac{d}{2}-1}$, the solution u is such that ∇u is also continuous from $[0, T]$ to $\dot{B}_{2,1}^{\frac{d}{2}}$.

Fourth Step: Uniqueness

This is a simple variation on the proof of the convergence of $(u_n)_{n\in\mathbb{N}}$. As in the previous step, it follows from stability estimates in the space $\mathcal{C}([0, T]; \dot{B}_{2,1}^{\frac{d}{2}-1})$.

Fifth Step: The Blow-up Criterion

We argue by contraposition. Let u be a solution of (QW) on $[0, T[$ such that ∇u belongs to $\mathcal{C}[0, T[; \mathcal{B}^d)$. Assume, in addition, that

$$\sup_{t\in[0,T[} \left(\|u(t)\|_{L^\infty} + \int_0^t \|\nabla u(t')\|_{L^\infty}\, dt'\right) < \infty. \tag{9.16}$$

We want to show that u may be continued beyond T to a solution of (QW).

We temporarily fix some $s > 0$. Applying Corollary 9.4 with $r = 1$ and $u = v$ then gives

$$\partial_t^2 \dot{\Delta}_j u - \Delta\dot{\Delta}_j u - \partial\cdot(G(t, u)\cdot\partial\dot{\Delta}_j u) = \dot{\Delta}_j\big(Q(t, u)(\nabla u, \nabla u)\big) + R_j(u, u)$$

with

$$2^{j(s-1)}\|R_j(u, u)(t)\|_{L^2} \leq c_j(t)C_s(\|u(t)\|_{L^\infty})\|\partial u(t)\|_{L^\infty}\|\partial u(t)\|_{\dot{B}_{2,1}^{s-1}}.$$

Thanks to the product laws in Besov spaces, we have

$$2^{j(s-1)}\|\dot\Delta_j\big(Q(t,u)(\nabla u,\nabla u)\big)\|_{L^2}\le c_j(t)C(\|u(t)\|_{L^\infty})\|\nabla u(t)\|_{L^\infty}\|\partial u(t)\|_{\dot B_{2,1}^{s-1}}.$$

We now define

$$\phi(t)\overset{\text{def}}{=}\|\partial_t(G(t,u(t)))\|_{L^\infty}\quad\text{and}$$

$$U_s(T)\overset{\text{def}}{=}\sum_j\sup_{t\le T}\exp\Big(-\frac12\int_0^t\phi(t')\,dt'\Big)2^{j(s-1)}\|\dot\Delta_j\nabla u(t)\|_{L^2}.$$

Using Lemma (9.1), we get, after multiplying by $2^{j(s-1)}$,

$$\exp\Big(-\frac12\int_0^t\phi\,dt'\Big)2^{j(s-1)}\|\dot\Delta_j\nabla u(t)\|_{L^2}\le A_0\|\dot\Delta_j\gamma\|_{L^2}$$

$$+A_0^{\frac12}\int_0^t c_jC(\|u(t')\|_{L^\infty})\|\nabla u\|_{L^\infty}\exp\Big(-\frac12\int_0^{t'}\phi\,dt''\Big)\|\nabla u\|_{\dot B_{2,1}^{s-1}}\,dt'.$$

Noting that for any $t'\le t$,

$$\exp\Big(-\frac12\int_0^{t'}\phi(t'')\,dt''\Big)\|\nabla u(t')\|_{\dot B_{2,1}^{s-1}}\le U_s(t),$$

we deduce, after summation over j, that

$$U_s(t)\le A_0\|\gamma\|_{\dot B_{2,1}^{s-1}}+A_0^{\frac12}\int_0^t C(\|u(t')\|_{L^\infty})\|\nabla u(t')\|_{L^\infty}U_s(t')\,dt'.$$

Gronwall's lemma then implies that

$$U_s(t)\le A_0^2\|\gamma\|_{\dot B_{2,1}^{s-1}}\exp\Big(A_0\int_0^t C(\|u(t')\|_{L^\infty})\|\nabla u(t')\|_{L^\infty}\,dt'\Big).$$

Hence, by the definition of U_s, we have

$$\|\nabla u\|_{\widetilde L_t^\infty(\dot B_{2,1}^{s-1})}\le A_0\|\gamma\|_{\dot B_{2,1}^{s-1}}$$

$$\times\exp\Big(\int_0^t\Big(\|\partial_t(G(t',u(t')))\|_{L^\infty}+A_0^{\frac12}C(\|u(t')\|_{L^\infty})\|\nabla u(t')\|_{L^\infty}\Big)\,dt'\Big).\tag{9.17}$$

Under the hypothesis (9.16), this ensures that $\|\nabla u\|_{\widetilde L_T^\infty(\dot B_{2,1}^{s-1})}$ is finite. Taking $s=d/2$, $s=d/2+1$ and using the lower bounds that were previously established for the lifespan, we can conclude that the solution may be continued beyond T.

Final Step: Additional Regularity

We have to establish that if, in addition, γ belongs to $\dot B_{2,r}^{\sigma-1}$ for some $\sigma>0$, then the solution u is such that ∇u belongs to $\mathcal C([0,T];\dot B_{2,r}^{\sigma-1})$. This follows from the fact that Corollary 9.4 holds for any positive index of regularity. As we may proceed exactly as in the first step, the details are left to the reader.

$\square$

9.2 The Main Statement and the Strategy of its Proof

In this section, we state the main result of this chapter—local well-posedness of (QW) with initial data which are not Lipschitz functions—and give an insight into the construction of its proof.

For expository purposes, we introduce the following notation:

$$\|a\|_\sigma \overset{\text{def}}{=} \|a\|_{\dot{B}^\sigma_{2,1}} \quad \text{and} \quad \|b\|_{T,\sigma} \overset{\text{def}}{=} \|b\|_{\widetilde{L}^\infty_T(\dot{B}^\sigma_{2,1})}. \tag{9.18}$$

Theorem 9.5. *Assume that the metric* $\mathrm{Id} + G$ *satisfies the condition (9.2) and that the initial data* (u_0, u_1) *are such that*

$$\gamma \in \dot{B}^{\frac{d}{2}-\frac{1}{4}}_{2,1} \cap \dot{B}^{\frac{d}{2}-\frac{5}{4}}_{2,1}, \quad \text{if } d \geq 4,$$

$$\gamma \in \dot{B}^{\frac{5}{4}+\varepsilon}_{2,1} \cap \dot{B}^{\frac{1}{4}+\varepsilon}_{2,1}, \quad \text{for some positive } \varepsilon \text{ if } d = 3,$$

$$\gamma \in \dot{B}^{\frac{7}{8}}_{2,1} \cap \dot{B}^{-\frac{1}{8}}_{2,1}, \quad \text{if } d = 2.$$

There then exist two maximal positive times, $T_\star$ *and* $T^\star$, *such that* (QW) *has a unique solution* u *with* $\nabla u \in L^2_{loc}(]-T_\star; T^\star[; L^\infty)$ *and*

$$\nabla u \in \mathcal{C}(]-T_\star, T^\star[; \dot{B}^{\frac{d}{2}-\frac{1}{4}}_{2,1} \cap \dot{B}^{\frac{d}{2}-\frac{5}{4}}_{2,1}), \text{ if } d \geq 4,$$

$$\nabla u \in \mathcal{C}(]-T_\star, T^\star[; \dot{B}^{\frac{5}{4}+\varepsilon}_{2,1} \cap \dot{B}^{\frac{1}{4}+\varepsilon}_{2,1}), \text{ if } d = 3,$$

$$\nabla u \in \mathcal{C}(]-T_\star, T^\star[; \dot{B}^{\frac{7}{8}}_{2,1} \cap \dot{B}^{-\frac{1}{8}}_{2,1}), \text{ if } d = 2.$$

Moreover, we have (using the notation of page 392),

$$C_\gamma \min\{T_\star, T^\star\}^{\frac{3}{4}} \|\gamma\|_{\dot{B}^{\frac{d}{2}-\frac{1}{4}}_{2,1}} \geq 1, \quad \text{if } d \geq 4,$$

$$C_{\gamma,\varepsilon} \min\{T_\star, T^\star\}^{\frac{3}{4}+\varepsilon} \|\gamma\|_{\dot{B}^{\frac{5}{4}+\varepsilon}_{2,1}} \geq 1, \quad \text{if } d = 3,$$

$$C_\gamma \min\{T_\star, T^\star\}^{\frac{7}{8}} \|\gamma\|_{\dot{B}^{\frac{7}{8}}_{2,1}} \geq 1, \quad \text{if } d = 2.$$

If $T_\star$ *or* $T^\star$ *is finite, then*

$$\limsup_{t \to T}\left(\|u(t)\|_{L^\infty} + \int_0^t \|\nabla u(t')\|_{L^\infty}\, dt'\right) = \infty$$

with $T = -T_\star$ *or* $T = T^\star$. *Moreover, if the initial data is such that* γ *belongs to* $\dot{B}^{s-1}_{2,r}$ *for some positive* s, *then* ∇u *is continuous with values in the space* $\dot{B}^{s-1}_{2,r}$.

As pointed out in the introduction, we shall instead solve a *truncated* quasilinear wave equation. More precisely, we fix some smooth function θ, compactly supported in $[-1, 1]$ and with value 1 near $[-1/2, 1/2]$. For any fixed positive time T, we then introduce the following equation:

$$(QW_T) \qquad \begin{cases} \partial_t^2 u - \Delta u - \partial \cdot (G_T(t, u) \cdot \partial u) = Q_T(t, u)(\nabla u, \nabla u) \\ \nabla u_{|t=0} = \gamma \end{cases}$$

with $G_T(t, u) \overset{\text{def}}{=} \theta\left(\dfrac{t}{T}\right)G(t, u)$ and $Q_T(t, u) \overset{\text{def}}{=} \theta\left(\dfrac{t}{T}\right)Q(t, u)$.

A quick glance at the definitions of A_0 and A_1 [see (9.2) and (9.13)] shows that the energy estimates of the preceding section may be made uniform with respect to the truncation parameter T, and that the value of the constant C_γ appearing in the forthcoming Theorem 9.12 may be made independent of T. As a consequence, the lifespan of (QW_T) may be bounded from below independently of T. Thus, if T is sufficiently small that the support of the function $\theta(\cdot T^{-1})$ is included in the interval of existence, then the solution is global. Indeed, (QW_T) reduces to the free linear wave equation with constant coefficients away from $[-T, T]$. Moreover, as the function θ has value 1 near $[-1/2, 1/2]$, the original problem (QW) is solved on $[-T/2, T/2]$.

Having global solutions greatly facilitates the implementation of the geometrical optics method which will be proposed in Section 9.5.2. In fact, this method requires the metric to be smooth with respect to both the space and time variables. Smoothness in the space variable can be achieved classically by spectral truncation. A similar method would work for the time variable; however, as it is nonlocal, this becomes quite unpleasant when solving an evolution equation. Now, if we deal only with functions with compact support in time (namely, G_T and Q_T), then using a cut-off function in the Fourier space for the time variable is quite harmless.

To simplify the presentation, we shall focus on the proof of the above theorem in dimension $d \geq 4$ and simply indicate what has to be changed for the case of dimension $d = 2, 3$. The proof of the theorem relies on an iterative method which is very much analogous to that of the first section. We define the sequence $(u_n)_{n \in \mathbb{N}}$ as follows. Start with the solution u_0 of

$$\partial_t u_0 - \Delta u_0 = 0 \quad \text{with} \quad \nabla u_0(0) = S_0 \gamma.$$

Once u_n has been constructed, we then define u_{n+1} as the solution of

$$\begin{cases} \partial_t u_{n+1} - \Delta u_{n+1} - \partial \cdot (G_T(t, u_n) \cdot \partial u_{n+1}) = Q_T(t, u_n)(\nabla u_n, \nabla u_n) \\ \qquad\qquad\qquad \nabla u_{n+1}(0) = S_{n+1} \gamma. \end{cases}$$

Let $L \overset{\text{def}}{=} [d/2 - 1/4, d/2 + 3/4]$. We first want to prove that if T is sufficiently small, then we have the property

$$(\mathcal{A}_{n,T}) \qquad \begin{cases} \|\nabla u_n\|_{T, s-1} \leq B_0 \|\gamma\|_{s-1} \quad \text{for all} \ \ s \in L \\ \|\nabla u_n\|_{L^2_T(L^\infty)} \leq C_\gamma \|\gamma\|_{\frac{d}{2} - \frac{1}{4}} T^{\frac{1}{4}}. \end{cases}$$

Once this has been proven, the rest of the proof will be more classical.

The property $(\mathcal{A}_{n,T})$ will be established by induction. As a first step, we show that our problem reduces to the proof of the $L^2_T(L^\infty)$ estimate for high frequencies, namely, for frequencies which are large with respect to T^{-1}. This reduction is the purpose of the next section.

9.3 Refined Paralinearization of the Wave Equation

In order to prove Strichartz estimates, we shall use geometrical optics. This method requires the smoothing out of the metric G_T. This may be achieved by means of a paralinearization procedure with respect to both the time and space variables. Here, we need a refinement of this procedure so as to get even better regularity of the coefficients involved in the paralinearization. Consequently, the remainders will be worse, as usual.

We now introduce the following definition.

Definition 9.6. *For δ in the interval $[0, 1]$, j in $\mathbb{Z}$, N_0 in $\mathbb{N}$, and $T > 0$, we set*

$$j_\delta \overset{def}{=} \left[j\delta - (1 - \delta) \log_2 T \right] - N_0 \quad and \quad S_j^\delta \overset{def}{=} \dot{S}_{j_\delta}^{(1+d)},$$

where $\dot{S}_k^{(1+d)}$ denotes the Littlewood–Paley low-frequency cut-off in $\mathbb{R}^{1+d}$ which was defined in Chapter 2.

The key result of this section is the following lemma.

Lemma 9.7. *Let L be a compact subinterval of $]0, \infty[$. Let u and v be two functions with space-time gradient in*

$$L_T^1(L^\infty) \cap L_T^\infty(\dot{B}_{2,1}^{s-1}) \quad for some \;\; s \in L.$$

If

$$\partial_t^2 u - \Delta u - \partial \cdot (G_T(t, v) \cdot \partial u) = f,$$

then

$$\partial_t^2 \dot{\Delta}_j u - \Delta \dot{\Delta}_j u - \partial \cdot (S_j^\delta(G_T(t, v) \cdot \partial \dot{\Delta}_j u) = \dot{\Delta}_j f + R_j^\delta(u, v)$$

with, if $2^j T$ is greater than 1,

$$2^{j(s-1)} \| R_j^\delta(u, v) \|_{L_T^1(L^2)} \le c_j C(\|v\|_{L^\infty([0,T] \times \mathbb{R}^d)})(2^j T)^{1-\delta} 2^{N_0} \tag{9.19}$$
$$\times \left(\|\partial u\|_{L_T^1(L^\infty)} \|\partial v\|_{T,s-1} + (1 + \|\nabla v\|_{L_T^1(L^\infty)}) \|\partial u\|_{T,s-1} \right)$$

and

$$2^{j(s-1)} \| R_j^\delta(u, v) \|_{L_T^1(L^2)} \le c_j C(\|v\|_{L^\infty([0,T] \times \mathbb{R}^d)})(2^j T)^{1-\delta} 2^{N_0} \tag{9.20}$$
$$\times \left(\|u\|_{L_T^1(L^\infty)} \|\partial v\|_{T,s} + (1 + \|\nabla v\|_{L_T^1(L^\infty)}) \|\partial u\|_{T,s-1} \right),$$

where C denotes a nondecreasing function from $\mathbb{R}^+$ to $\mathbb{R}^+$, dependent on L.

Proof. A straightforward modification of Corollary 9.4 implies that

$$\partial_t^2 \dot{\Delta}_j u - \Delta \dot{\Delta}_j u - \partial \cdot (G_T(t, v) \cdot \partial \dot{\Delta}_j u) = \dot{\Delta}_j f + R_j(u, v),$$

where the operator R_j is such that, for any t, there exists a sequence $(c_j)_{j \in \mathbb{Z}}$ in the unit sphere of $\ell^1(\mathbb{Z})$, where

$$2^{j(s-1)}\|R_j(u,v)\|_{L^1_T(L^2)} \leq c_j C(\|v\|_{L^\infty([0,T]\times\mathbb{R}^d)})$$
$$\times \left(\|u\|_{T,s}\|\partial v\|_{L^1_T(L^\infty)} + \|v\|_{T,s}\|\partial u\|_{L^1_T(L^\infty)}\right) \quad (9.21)$$

and

$$2^{j(s-1)}\|R_j(u,v)\|_{L^1_T(L^2)} \leq c_j C(\|v\|_{L^\infty([0,T]\times\mathbb{R}^d)})$$
$$\times \left(\|u\|_{T,s}\|\partial v\|_{L^1_T(L^\infty)} + \|v\|_{T,s+1}\|u\|_{L^1_T(L^\infty)}\right). \quad (9.22)$$

Noting that

$$R_j^\delta(u,v) = R_j(u,v) + \partial \cdot \left(((\mathrm{Id}-S_j^\delta)G_T(t,v)) \cdot \partial \dot{\Delta}_j u\right),$$

we see that we have to bound $\|(\mathrm{Id}-S_j^\delta)G_T(\cdot,v)\|_{L^1_T(L^\infty)}$. Now, the identity on page 52 (after an obvious rescaling) guarantees that there exist $d+1$ functions g_k in $L^1(\mathbb{R}^{1+d})$ such that for any $j' \in \mathbb{Z}$,

$$\dot{\Delta}_{j'}^{(1+d)}a = \sum_{k=0}^d 2^{-j'}2^{j'd}g_k(2^{j'}\cdot) \star \partial_k a.$$

From the anisotropic Young inequality, we thus infer that

$$\|\dot{\Delta}_{j'}^{(1+d)}a\|_{L^1(\mathbb{R};L^\infty)} \leq C2^{-j'}\|\nabla a\|_{L^1(\mathbb{R};L^\infty)}.$$

Thus, by summation over $j \geq j_\delta$, we get

$$\|(\mathrm{Id}-S_j^\delta)a\|_{L^1(\mathbb{R};L^\infty)} \leq C2^{-j}(2^jT)^{1-\delta}2^{N_0}\|\nabla a\|_{L^1(\mathbb{R};L^\infty)}. \quad (9.23)$$

We want to apply the above inequality with $a = G_T(\cdot,v)$. By definition of G_T, we have

$$\partial_t G_T(t,v) = \frac{1}{T}\theta'\left(\frac{t}{T}\right)G(t,v(t)) + \theta\left(\frac{t}{T}\right)\left(\partial_u G(t,v(t))\partial_t v(t) + \partial_t G(t,v(t))\right).$$

As the time cut-off commutes with the space derivative, we thus get

$$\|\nabla G_T(t,v)\|_{L^1(\mathbb{R},L^\infty)}$$
$$\leq C\left(\|\nabla v\|_{L^1_T(L^\infty)}\|\partial_u G\|_{L^\infty} + \|\theta'\|_{L^1}\|G\|_{L^\infty} + \|\partial_t G\|_{L^1_T(L^\infty)}\right).$$

Taking advantage of (9.23), we can then deduce that

$$\|(\mathrm{Id}-S_j^\delta)G_T(t,v)\|_{L^1(\mathbb{R};L^\infty)} \leq C2^{-j}(2^jT)^{1-\delta}2^{N_0}$$
$$\times \left(\|\nabla v\|_{L^1_T(L^\infty)}\|\partial_u G\|_{L^\infty} + \|\theta'\|_{L^1}\|G\|_{L^\infty} + \|\partial_t G\|_{L^1_T(L^\infty)}\right).$$

Therefore,

$$2^{j(s-1)}\|R_j^\delta(u,v)\|_{L^1_T(L^2)} \leq c_j C(\|v\|_{L^\infty([0,T]\times\mathbb{R}^d)})\Big(\big(\|u\|_{T,s}\|\partial v\|_{L^1_T(L^\infty)}$$
$$+\|v\|_{T,s}\|\partial u\|_{L^1_T(L^\infty)}\big) + 2^{j(s-1)}(2^jT)^{1-\delta}2^{N_0}\|\partial u\|_{T,s-1}\big(\|\nabla v\|_{L^1_T(L^\infty)}\|\partial_u G\|_{L^\infty}$$
$$+\|\theta'\|_{L^1}\|G\|_{L^\infty} + \|\partial_t G\|_{L^1_T(L^\infty)}\big)\Big).$$

The proof of the second inequality in Proposition 9.7 is similar. $\square$

Proposition 9.8. *If $(\mathcal{A}_{n,T})$ is satisfied and $C_\gamma \|\gamma\|_{\frac{d}{2}-\frac{1}{4}} T^{\frac{3}{4}}$ is sufficiently small, then for any $s \in L$, we have*

$$\|\nabla u_{n+1}\|_{T,s-1} \le \frac{B_0}{2} \|\gamma\|_{s-1} \left(1 + C_\gamma T^{\frac{1}{2}} \|\nabla u_{n+1}\|_{L_T^2(L^\infty)}\right).$$

Proof. We apply the inequality (9.11) of Corollary 9.4 and Lemma 9.7 under assumption $(\mathcal{A}_{n,T})$. By virtue of Lemma 9.1, this gives

$$\|\dot\Delta_j \nabla u_{n+1}\|_{L_T^\infty(L^2)} \le 2^{-j(s-1)} \mathcal{K}_j(T) e^{\mathcal{I}(T)} \quad \text{with} \qquad (9.24)$$

$$\mathcal{I}(T) \stackrel{\text{def}}{=} A_0 \int_0^T \|\partial_t(G_T(t, u_n(t, \cdot)))\|_{L^\infty} \, dt \quad \text{and}$$

$$\mathcal{K}_j(T) \stackrel{\text{def}}{=} A_0 2^{j(s-1)} \|\dot\Delta_j \gamma\|_{L^2} + C_\gamma \|\gamma\|_{s-1} \int_0^T c_j(t) \|\partial u_{n+1}(t)\|_{L^\infty} \, dt$$
$$+ c_j C_\gamma \|\gamma\|_{\frac{d}{2}-\frac{1}{4}} T^{\frac{3}{4}} \|\nabla u_{n+1}\|_{T,s-1}.$$

By definition of G_T, we have, thanks to $(\mathcal{A}_{n,T})$,

$$\mathcal{I}(T) \le A_0 \int_0^T \|\partial_t(\theta(tT^{-1})G(t, u_n(t, \cdot)))\|_{L^\infty} \, dt$$
$$\le A_0\left(\|\theta'\|_{L^1}\|G\|_{L^\infty} + \|\partial_u G\|_{L^\infty}\|\partial_t u_n\|_{L_T^1(L^\infty)} + \|\partial_t G\|_{L_T^1(L^\infty)}\right)$$
$$\le A_0\left(\|\theta'\|_{L^1}\|G\|_{L^\infty} + \|\partial_t G\|_{L_T^1(L^\infty)} + \|\partial_u G\|_{L^\infty} C_\gamma \|\gamma\|_{\frac{d}{2}-\frac{1}{4}} T^{\frac{3}{4}}\right).$$

Assume that

$$\|\theta'\|_{L^1}\|G\|_{L^\infty} + \|\partial_t G\|_{L_T^1(L^\infty)} + \|\partial_u G\|_{L^\infty} C_\gamma \|\gamma\|_{\frac{d}{2}-\frac{1}{4}} T^{\frac{3}{4}} \le A_1$$

and define

$$B_0 \stackrel{\text{def}}{=} 4A_0 \exp(A_0 A_1).$$

By summation over j in (9.24), we get

$$\|\nabla u_{n+1}\|_{T,s-1} \le C_\gamma \|\gamma\|_{\frac{d}{2}-\frac{1}{4}} T^{\frac{3}{4}} \|\nabla u_{n+1}\|_{T,s-1}$$
$$+ \frac{B_0}{4} \|\gamma\|_{s-1}\left(1 + C_\gamma \|\nabla u_{n+1}\|_{L_T^1(L^\infty)}\right).$$

Taking $C_\gamma \|\gamma\|_{\frac{d}{2}-\frac{1}{4}} T^{\frac{3}{4}}$ sufficiently small gives the result. □

Remark 9.9. Taking $s = d/2+3/4$ and using Bernstein's and Hölder's inequalities, we immediately deduce from Proposition 9.8 that for any integer j,

$$\|\dot S_j \nabla u_{n+1}\|_{L_T^2(L^\infty)} \le (2^j T)^{\frac{1}{4}} C_\gamma \|\gamma\|_{\frac{d}{2}-\frac{1}{4}} T^{\frac{1}{4}}\left(1 + T^{\frac{1}{2}} \|\nabla u_{n+1}\|_{L_T^2(L^\infty)}\right).$$

From Proposition 9.8 and Lemma 9.7, we easily deduce the following corollary, which will be needed in the next section.

Corollary 9.10. *Under the hypothesis $(\mathcal{A}_{n,T})$, if $C_\gamma \|\gamma\|_{\frac{d}{2}-\frac{1}{4}} T^{\frac{3}{4}}$ is sufficiently small, then we have, for any $s \in L$,*

$$\partial_t^2 \dot\Delta_j u_{n+1} - \Delta \dot\Delta_j u_{n+1} - \partial \cdot (S_j^\delta G_T(t, u_n) \cdot \partial \dot\Delta_j u_{n+1}) = R_j^\delta(n) \quad \text{with}$$
$$2^{j(s-1)} \|R_j^\delta(n)\|_{L_T^1(L^2)} \le c_j C_\gamma \|\gamma\|_{s-1} (2^j T)^{1-\delta} 2^{N_0}\left(1 + T^{\frac{1}{2}} \|\nabla u_{n+1}\|_{L_T^2(L^\infty)}\right).$$

9.4 Reduction to a Microlocal Strichartz Estimate

In this section, we complete the proof of Theorem 9.5. In view of what was proven in the preceding section, we first have to establish the second inequality of $(\mathcal{A}_{n+1,T})$. This will be based on the following *quasilinear Strichartz estimates* that we will temporarily assume to hold.

Theorem 9.11. *Let u_j be the solution of*

$$\partial_t^2 u_j - \Delta u_j - \partial \cdot \left(S_j^\delta G_T(\cdot, v) \cdot \partial u_j \right) = f_j$$

on the time interval $[0, T]$. We suppose that for any t, the support of the Fourier transform of $u_j(t, \cdot)$ is supported in an annulus $2^j \widetilde{\mathcal{C}}$. If $2^j T$ is sufficiently large, then it follows that:

– *If $d \geq 4$, then*

$$\|\nabla u_j\|_{L_T^2(L^\infty)} \leq C(\|G_T\|_{L^\infty}) 2^{j\left(\frac{d}{2} - \frac{1}{2}\right)}$$
$$\times \left((2^j T)^{\frac{\delta}{2}} \|\nabla u_j\|_{L_T^\infty(L^2)} + (2^j T)^{-\frac{\delta}{2}} \|f_j\|_{L_T^1(L^2)} \right).$$

– *If $d = 3$, then for all sufficiently small $\varepsilon > 0$,*

$$\|\nabla u_j\|_{L_T^2(L^\infty)} \leq C_\varepsilon(\|G_T\|_{L^\infty}) 2^j (2^j T)^{\frac{\varepsilon}{2}}$$
$$\times \left((2^j T)^{\frac{\delta}{2}} \|\nabla u_j\|_{L_T^\infty(L^2)} + (2^j T)^{-\frac{\delta}{2}} \|f_j\|_{L_T^1(L^2)} \right).$$

– *If $d = 2$, then*

$$\|\nabla u_j\|_{L_T^4(L^\infty)} \leq C(\|G_T\|_{L^\infty}) 2^{\frac{3}{4}j} \left((2^j T)^{\frac{\delta}{2}} \|\nabla u_j\|_{L_T^\infty(L^2)} + (2^j T)^{-\frac{\delta}{2}} \|f_j\|_{L_T^1(L^2)} \right).$$

Proof of the second inequality of $(\mathcal{A}_{n+1,T})$. For expository purposes, we restrict ourselves to the case $d \geq 4$. Applying Theorem 9.11 with $u_j = \dot{\Delta}_j u_{n+1}$ and $f_j = R_j^\delta(n)$ then gives

$$\|\nabla \dot{\Delta}_j u_{n+1}\|_{L_T^2(L^\infty)} \leq C_\gamma 2^{j\left(\frac{d}{2} - \frac{1}{2}\right)}$$
$$\times \left((2^j T)^{\frac{\delta}{2}} \|\nabla \dot{\Delta}_j u_{n+1}\|_{L_T^\infty(L^2)} + (2^j T)^{-\frac{\delta}{2}} \|R_j^\delta(n)\|_{L_T^1(L^2)} \right).$$

Combining the assumption $(\mathcal{A}_{n,T})$, Corollary 9.10, and Proposition 9.8 with $s = d/2 + 3/4$, we get, for $2^j T$ sufficiently large,

$$\|\nabla \dot{\Delta}_j u_{n+1}\|_{L_T^2(L^\infty)} \leq c_j C_\gamma 2^{j\left(\frac{d}{2} - \frac{1}{2}\right)} \left((2^j T)^{\frac{\delta}{2}} 2^{-j\left(\frac{d}{2} - \frac{1}{4}\right)} \|\gamma\|_{\frac{d}{2} - \frac{1}{4}} \right.$$
$$\left. + (2^j T)^{-\frac{\delta}{2}} 2^{-j\left(\frac{d}{2} - \frac{1}{4}\right)} \|\gamma\|_{\frac{d}{2} - \frac{1}{4}} (2^j T)^{1 - \delta} 2^{N_0} \left(1 + T^{\frac{1}{2}} \|\nabla u_{n+1}\|_{L_T^2(L^\infty)} \right) \right).$$

Therefore,

$$\|\nabla \dot{\Delta}_j u_{n+1}\|_{L^2_T(L^\infty)} \le T^{\frac{1}{4}} c_j C_\gamma (2^j T)^{-\frac{1}{4}} \|\gamma\|_{\frac{d}{2}-\frac{1}{4}} \Big((2^j T)^{\frac{\delta}{2}}$$
$$+ (2^j T)^{1-\frac{3}{2}\delta}\big(1 + T^{\frac{1}{2}}\|\nabla u_{n+1}\|_{L^2_T(L^\infty)}\big)\Big).$$

The "best" choice for δ here corresponds to $\delta/2 = 1 - 3\delta/2$, namely, $\delta = 1/2$. By summation over j, this gives, if $2^j T$ is sufficiently large,

$$\|(\mathrm{Id} - \dot{S}_j)\nabla u_{n+1}\|_{L^2_T(L^\infty)} \le C_\gamma \|\gamma\|_{\frac{d}{2}-\frac{1}{4}} T^{\frac{1}{4}} \big(1 + T^{\frac{1}{2}}\|\nabla u_{n+1}\|_{L^2_T(L^\infty)}\big).$$

Remark 9.9 now ensures that for all $j \in \mathbb{Z}$, we have

$$\|\dot{S}_j \nabla u_{n+1}\|_{L^2_T(L^\infty)} \le (2^j T)^{\frac{1}{4}} C_\gamma \|\gamma\|_{\frac{d}{2}-\frac{1}{4}} T^{\frac{1}{4}} \big(1 + T^{\frac{1}{2}}\|\nabla u_{n+1}\|_{L^2_T(L^\infty)}\big).$$

Combining these two inequalities and taking $2^j T$ sufficiently large, we end up with

$$\|\nabla u_{n+1}\|_{L^2_T(L^\infty)} \le C_\gamma \|\gamma\|_{\frac{d}{2}-\frac{1}{4}} T^{\frac{1}{4}} \big(1 + T^{\frac{1}{2}}\|\nabla u_{n+1}\|_{L^2_T(L^\infty)}\big).$$

So, finally, choosing T such that

$$C_\gamma \|\gamma\|_{\frac{d}{2}-\frac{1}{4}} T^{\frac{3}{4}} \quad \text{is sufficiently small} \tag{9.25}$$

ensures that the assertion $(\mathcal{A}_{n+1,T})$ is fulfilled.

We can now proceed to the proof of existence in Theorem 9.5. We assume from now on that the condition (9.25) is satisfied. From the above estimates, we deduce that if, in addition, the data are such that γ belongs to $\dot{B}_{2,1}^{\frac{d}{2}-1} \cap \dot{B}_{2,1}^{\frac{d}{2}}$, then the sequence $(\nabla u_n)_{n \in \mathbb{N}}$ is bounded in $L^2_T(L^\infty)$. Therefore, we may proceed as in the first section of this chapter to prove the following result.

Theorem 9.12. *Under the hypothesis of Theorem 9.2, the maximal time of existence $T^\star$ satisfies*

$$T^{\star \frac{3}{4}} C_\gamma \|\gamma\|_{\frac{d}{2}-\frac{1}{4}} \ge 1.$$

We will now prove that $(u_n)_{n \in \mathbb{N}}$ is a Cauchy sequence in a suitable space. As already encountered in Chapters 4 and 7, and in the preceding section (and also in Chapter 6 for a more subtle case), owing to hyperbolicity, we lose one space derivative in the stability estimates. Here, we shall prove that $(u_n)_{n \in \mathbb{N}}$ is a Cauchy sequence for the norm

$$\|v\|_T \overset{\text{def}}{=} \|\nabla v\|_{T, \frac{d}{2}-\frac{5}{4}} T^{\frac{1}{4}} + \|v\|_{L^2_T(L^\infty)}.$$

Proposition 9.13. *Let $\tilde{u}_n \overset{\text{def}}{=} u_{n+1} - u_n$. If $C_\gamma \|\gamma\|_{\frac{d}{2}-\frac{1}{4}} T^{\frac{3}{4}}$ is sufficiently small, then we have*

$$T^{\frac{1}{4}} \|\nabla \tilde{u}_n\|_{T, \frac{d}{2}-\frac{5}{4}} \le C 2^{-n} \|\gamma\|_{\frac{d}{2}-\frac{1}{4}} T^{\frac{1}{4}} + C_\gamma \|\gamma\|_{\frac{d}{2}-\frac{1}{4}} T^{\frac{3}{4}} \big(\|\tilde{u}_{n-1}\|_T + \|\tilde{u}_n\|_T\big).$$

Proof. As in Section 9.1, we write that $\widetilde{u}_n$ and $\widetilde{u}_{n-1}$ satisfy

$$(QW_n) \qquad \partial_t^2 \widetilde{u}_n - \Delta \widetilde{u}_n - \partial(G_T(\cdot, u_n) \cdot \partial \widetilde{u}_n) = f_n \overset{\text{def}}{=} \sum_{k=1}^{3} f_{k,n}$$

$$\text{with} \quad \begin{cases} f_{1,n} \overset{\text{def}}{=} (Q_T(\cdot, u_n) - Q_T(\cdot, u_{n-1}))(\nabla u_n, \nabla u_n) \\ f_{2,n} \overset{\text{def}}{=} Q_T(\cdot, u_{n-1})(\nabla \widetilde{u}_{n-1}, \nabla u_n + \nabla u_{n-1}) \\ f_{3,n} \overset{\text{def}}{=} \partial \cdot ((G_T(\cdot, u_n) - G_T(\cdot, u_{n-1})) \cdot \partial u_n). \end{cases}$$

Taking advantage of the law of products, we obtain that

$$\|f_{1,n}\|_{\frac{d}{2}-\frac{5}{4}} \leq \|Q_T(u_n) - Q_T(u_{n-1})\|_{\frac{d}{2}-\frac{1}{4}} \|(\nabla u_n, \nabla u_n)\|_{\frac{d}{2}-1}.$$

Corollary 2.66 page 97 thus implies that

$$\begin{aligned} T^{\frac{1}{4}} \|f_{1,n}\|_{\frac{d}{2}-\frac{5}{4}} &\leq C_\gamma T^{\frac{1}{4}} \|\widetilde{u}_{n-1}\|_{\frac{d}{2}-\frac{1}{4}} \|\nabla u_n\|_{L^\infty} \|\nabla u_n\|_{\frac{d}{2}-1} \\ &\leq C_\gamma \|\widetilde{u}_{n-1}\|_T \|\nabla u_n\|_{L^\infty} \|\nabla u_n\|_{\frac{d}{2}-1}. \end{aligned}$$

By virtue of $(\mathcal{A}_{n,T})$, the above inequality may be rewritten as

$$T^{\frac{1}{4}} 2^{j(\frac{d}{2}-\frac{5}{4})} \|\dot{\Delta}_j f_{1,n}\|_{L^1_T(L^2)} \leq c_j C_\gamma \|\gamma\|_{\frac{d}{2}-\frac{1}{4}} T^{\frac{3}{4}} \|\widetilde{u}_{n-1}\|_T. \qquad (9.26)$$

We shall now estimate $f_{2,n}$. From the usual product laws, we deduce that

$$\begin{aligned} \|\nabla \widetilde{u}_{n-1}(\nabla u_n + \nabla u_{n-1})\|_{\frac{d}{2}-\frac{5}{4}} &\leq C\Big((\|\nabla u_n\|_{L^\infty} + \|\nabla u_{n-1}\|_{L^\infty})\|\nabla \widetilde{u}_{n-1}\|_{\frac{d}{2}-\frac{5}{4}} \\ &\quad + \|\widetilde{u}_{n-1}\|_{L^\infty}(\|\nabla u_n\|_{\frac{d}{2}-\frac{1}{4}} + \|\nabla u_{n-1}\|_{\frac{d}{2}-\frac{1}{4}})\Big). \end{aligned}$$

As $Q_T(t, u_n(t)) - Q_T(t, 0)$ is bounded in $\dot{B}_{2,1}^{\frac{d}{2}}$, we thus have

$$\begin{aligned} T^{\frac{1}{4}} \|f_{2,n}\|_{\frac{d}{2}-\frac{5}{4}} &\leq C_\gamma (\|\nabla u_n\|_{L^\infty} + \|\nabla u_{n-1}\|_{L^\infty}) \|\widetilde{u}_{n-1}\|_T \\ &\quad + C_\gamma \|\gamma\|_{\frac{d}{2}-\frac{1}{4}} \|\widetilde{u}_{n-1}\|_{L^\infty} T^{\frac{1}{4}}. \end{aligned}$$

Hence, according to $(\mathcal{A}_{n,T})$,

$$T^{\frac{1}{4}} 2^{j(\frac{d}{2}-\frac{5}{4})} \|\dot{\Delta}_j f_{2,n}\|_{L^1_T(L^2)} \leq c_j C_\gamma \|\gamma\|_{\frac{d}{2}-\frac{1}{4}} T^{\frac{3}{4}} \|\widetilde{u}_{n-1}\|_T. \qquad (9.27)$$

Finally, the laws of product and composition lead to

$$\begin{aligned} \|f_{3,n}(t)\|_{\frac{d}{2}-\frac{5}{4}} &\leq \|(G_T(t, u_n(t)) - G_T(t, u_{n-1}(t))) \cdot \partial u_n(t)\|_{\frac{d}{2}-\frac{1}{4}} \\ &\leq C\|\widetilde{u}_{n-1}(t)\|_{L^\infty} \|\partial u_n(t)\|_{\frac{d}{2}-\frac{1}{4}} \\ &\quad + C_\gamma \|\partial u_n(t)\|_{L^\infty} \|\nabla \widetilde{u}_{n-1}(t)\|_{\frac{d}{2}-\frac{5}{4}}. \end{aligned}$$

Therefore,

$$T^{\frac{1}{4}}2^{j\left(\frac{d}{2}-\frac{5}{4}\right)}\|\dot{\Delta}_j f_{3,n}\|_{L^1_T(L^2)} \le c_j C_\gamma \|\gamma\|_{\frac{d}{2}-\frac{1}{4}} T^{\frac{3}{4}}\|\widetilde{u}_{n-1}\|_T.$$

Together with (9.26) and (9.27), this gives

$$T^{\frac{1}{4}}2^{j\left(\frac{d}{2}-\frac{5}{4}\right)}\|\dot{\Delta}_j f_n\|_{L^1_T(L^2)} \le c_j C_\gamma \|\gamma\|_{\frac{d}{2}-\frac{1}{4}} T^{\frac{3}{4}}\|\widetilde{u}_{n-1}\|_T. \tag{9.28}$$

Now, according to the second part of Corollary 9.4, we have

$$\partial_t^2 \dot{\Delta}_j \widetilde{u}_n - \Delta \dot{\Delta}_j \widetilde{u}_n - \partial\cdot(G_T(\cdot,u_n)\cdot\partial\dot{\Delta}_j\widetilde{u}_n) = \dot{\Delta}_j f_n + R_j(\widetilde{u}_n, u_n)$$

with

$$2^{j\left(\frac{d}{2}-\frac{5}{4}\right)}\|R_j(\widetilde{u}_n, u_n)(t)\|_{L^2} \le c_j(t) C(\|u_n(t)\|_{L^\infty})$$
$$\times \left(\|\partial\widetilde{u}_n(t)\|_{\frac{d}{2}-\frac{5}{4}}\|\partial u_n(t)\|_{L^\infty} + \|\partial u_n(t)\|_{\frac{d}{2}-\frac{1}{4}}\|\widetilde{u}_n(t)\|_{L^\infty}\right).$$

Thus, taking the L^1 norm in time of the above inequality and multiplying by $T^{\frac{1}{4}}$, we deduce that

$$T^{\frac{1}{4}}2^{j\left(\frac{d}{2}-\frac{5}{4}\right)}\|R_j(\widetilde{u}_n, u_n)\|_{L^1_T(L^2)} \le c_j C_\gamma \|\gamma\|_{\frac{d}{2}-\frac{1}{4}} T^{\frac{3}{4}}\|\widetilde{u}_n\|_T. \tag{9.29}$$

Taking advantage of the energy estimate stated in Lemma 9.1, it is now easy to complete the proof of the proposition.   □

Remark 9.14. From Bernstein's and Hölder's inequalities, we may deduce that for any integer j,

$$\|\dot{S}_j\widetilde{u}_n\|_{L^2_T(L^\infty)} \le C(2^j T)^{\frac{1}{4}}\sum_{j'<j} T^{\frac{1}{4}}\|\dot{\Delta}_{j'}\nabla\widetilde{u}_n\|_{L^\infty_T(L^2)}2^{j'\left(\frac{d}{2}-\frac{5}{4}\right)}.$$

Therefore, Proposition 9.13 yields

$$\|\dot{S}_j\widetilde{u}_n\|_{L^2_T(L^\infty)} \le C(2^j T)^{\frac{1}{4}}\left(2^{-n}\|\gamma\|_{\frac{d}{2}-\frac{1}{4}}T^{\frac{1}{4}} + C_\gamma\|\gamma\|_{\frac{d}{2}-\frac{1}{4}}T^{\frac{3}{4}}(\|\widetilde{u}_{n-1} + \|\widetilde{u}_n\|_T)\right).$$

This will be used to complete the proof of the theorem.

We now resume the proof of convergence of the sequence $(u_n)_{n\in\mathbb{N}}$. Applying the second inequality of Lemma 9.7 with $\delta = 1/2$ and $s = d/2 - 1/4$, we get

$$\partial_t^2 \dot{\Delta}_j\widetilde{u}_n - \Delta\dot{\Delta}_j\widetilde{u}_n - \partial\cdot(S_j^{\frac{1}{2}}(G_T(\cdot,u_n)\cdot\partial\dot{\Delta}_j\widetilde{u}_n) = \dot{\Delta}_j f_n + R_j^{\frac{1}{2}}(n)$$

with, if $2^j T$ is greater than or equal to 1,

$$2^{j\left(\frac{d}{2}-\frac{5}{4}\right)}\|R_j^{\frac{1}{2}}(n)\|_{L^1_T(L^\infty)} \le c_j C(\|u_n\|_{L^\infty([0,T]\times\mathbb{R}^d)})(2^j T)^{\frac{1}{2}}2^{N_0}$$
$$\times \left(\left(1+\|\nabla u_n\|_{L^1_T(L^\infty)}\right)\|\nabla\widetilde{u}_n\|_{T,\frac{d}{2}-\frac{5}{4}} + \|\widetilde{u}_n\|_{L^1_T(L^\infty)}\|\nabla u_n\|_{T,\frac{d}{2}-\frac{1}{4}}\right).$$

Thanks to Proposition 9.13, we obtain

$$T^{\frac{1}{4}}2^{j\left(\frac{d}{2}-\frac{5}{4}\right)}\|R_j^{\frac{1}{2}}(n)\|_{L_T^1(L^\infty)} \le c_j C_\gamma (2^j T)^{\frac{1}{2}} 2^{N_0} \|\gamma\|_{\frac{d}{2}-\frac{1}{4}}$$
$$\times \left(CT^{\frac{1}{4}}2^{-n} + T^{\frac{3}{4}}(\|\widetilde{u}_n\|_T + \|\widetilde{u}_{n-1}\|_T)\right). \qquad (9.30)$$

Bernstein's inequality and Theorem 9.11 give, for sufficiently large $2^j T$,

$$\|\dot{\Delta}_j \widetilde{u}_n\|_{L_T^2(L^\infty)} \le 2^{-j}\|\dot{\Delta}_j \nabla \widetilde{u}_n\|_{L_T^2(L^\infty)}$$
$$\le 2^{j\left(\frac{d}{2}-\frac{3}{2}\right)}\left((2^j T)^{\frac{1}{4}}\|\nabla \dot{\Delta}_j \widetilde{u}_n\|_{L_T^\infty(L^2)}\right.$$
$$\left.+ (2^j T)^{-\frac{1}{4}}\left(\|\dot{\Delta}_j f_n\|_{L_T^1(L^2)} + \|R_j^{\frac{1}{2}}(n)\|_{L_T^1(L^2)}\right)\right).$$

From (9.28) and (9.30), we infer that, for sufficiently large $2^j T$,

$$\|\dot{\Delta}_j \widetilde{u}_n\|_{L_T^2(L^\infty)} \le c_j \left(2^{-n}\|\gamma\|_{\frac{d}{2}-\frac{1}{4}}T^{\frac{1}{4}} + T^{\frac{1}{4}}\|\nabla \widetilde{u}_n\|_{T,\frac{d}{2}-\frac{5}{4}}\right.$$
$$\left.+ C_\gamma\|\gamma\|_{\frac{d}{2}-\frac{1}{4}}T^{\frac{3}{4}}(\|\widetilde{u}_n\|_T + \|\widetilde{u}_{n-1}\|_T)\right).$$

Note that the second term on the right-hand side may be bounded according to Proposition 9.13. Hence, if $2^j T$ is large enough, then

$$\|\dot{\Delta}_j \widetilde{u}_n\|_{L_T^2(L^\infty)} \le c_j \left(2^{-n}\|\gamma\|_{\frac{d}{2}-\frac{1}{4}}T^{\frac{1}{4}} + C_\gamma\|\gamma\|_{\frac{d}{2}-\frac{1}{4}}T^{\frac{3}{4}}(\|\widetilde{u}_n\|_T + \|\widetilde{u}_{n-1}\|_T)\right).$$

By summation, we thus infer that there exists some $M > 0$ such that if $2^j T \ge M$, and $C_\gamma\|\gamma\|_{\frac{d}{2}-\frac{1}{4}}T^{\frac{3}{4}}$ is sufficiently small, then

$$\|(\mathrm{Id}-\dot{S}_j)\widetilde{u}_n\|_{L_T^2(L^\infty)} \le 2^{-n}\|\gamma\|_{\frac{d}{2}-\frac{1}{4}}T^{\frac{1}{4}} + C_\gamma\|\gamma\|_{\frac{d}{2}-\frac{1}{4}}T^{\frac{3}{4}}(\|\widetilde{u}_n\|_T + \|\widetilde{u}_{n-1}\|_T).$$

Using Proposition 9.13 and Remark 9.14, we deduce that

$$\|\widetilde{u}_n\|_T \le T^{\frac{1}{4}}\|\nabla \widetilde{u}_n\|_{T,\frac{d}{2}-\frac{5}{4}} + \|\dot{S}_j\widetilde{u}_n\|_{L_T^2(L^\infty)} + \|(\mathrm{Id}-\dot{S}_j)\widetilde{u}_n\|_{L_T^2(L^\infty)}$$
$$\le C\left(1+(2^j T)^{\frac{1}{4}}\right)\left(2^{-n}\|\gamma\|_{\frac{d}{2}-\frac{1}{4}}T^{\frac{1}{4}}\right.$$
$$\left.+ C_\gamma\|\gamma\|_{\frac{d}{2}-\frac{1}{4}}T^{\frac{3}{4}}(\|\widetilde{u}_n\|_T + \|\widetilde{u}_{n-1}\|_T)\right).$$

We now choose T such that $(1 + M^{\frac{1}{4}})C_\gamma\|\gamma\|_{\frac{d}{2}-\frac{1}{4}}T^{\frac{3}{4}}$ is sufficiently small, then $j \in \mathbb{N}$ such that $M \le 2^j T < 2M$. The above inequality then ensures that $(u_n)_{n\in\mathbb{N}}$ is a Cauchy sequence in $L_T^\infty(\dot{B}_{2,1}^{\frac{d}{2}-\frac{1}{4}}) \cap L_T^2(L^\infty)$. This completes the proof of the existence part of Theorem 9.5.

To conclude, we shall say a few words about the proof of uniqueness. Unsurprisingly, we proceed as for the proof that $(u_n)_{n\in\mathbb{N}}$ is a Cauchy sequence. So, consider two solutions, u and v, of (QW) with the same initial data γ and defined on some interval $[0, T^\star]$. The difference $w = v - u$ between these two solutions satisfies

$$\partial_t^2 w - \Delta w - \partial \cdot (G(\cdot, v) \cdot \partial w) = Q(\cdot, v)(\nabla w, \nabla u + \nabla v)$$
$$- (Q(\cdot, u) - Q(\cdot, v))(\nabla u, \nabla u) - \partial \cdot (G(\cdot, u) - G(\cdot, v))\partial u).$$

We now introduce a cut-off function θ supported in $[0, 1]$ and with value 1 near $[0, 1/2[$. Let T be a positive time less than $T^\star$. If

$$G_T(t, v) \stackrel{\text{def}}{=} \widetilde{\theta}\Big(\frac{t}{T}\Big) G(t, v),$$

then w satisfies

$$\partial_t^2 w - \Delta w - \partial(G_T(\cdot, v) \cdot \partial w) = Q(\cdot, v)(\nabla w, \nabla u + \nabla v)$$
$$- (Q(\cdot, u) - Q(\cdot, v))(\nabla u, \nabla u) - \partial \cdot (G(\cdot, u) - G(\cdot, v)) \cdot \partial u)$$

on the interval $[0, T/2]$.

Mimicking the proof of the convergence of $(u_n)_{n \in \mathbb{N}}$ then shows that $w \equiv 0$ on $[0, T/2]$ if T is sufficiently small. The usual connectivity argument yields uniqueness on the whole interval $[0, T^\star]$. The continuation criterion is based on the inequality (9.17), as in the smooth case.

So, up to Theorem 9.11 (which we assumed), this completes the proof of Theorem 9.5. The proof of Theorem 9.11 rests on the following microlocal Strichartz estimates that will be established in the next section of this chapter. This theorem is "microlocal", inasmuch as it holds true on a time interval, the length of which depends on the size of the frequency we are working with.

Theorem 9.15. *Let u_j satisfy*

$$\partial_t^2 u_j - \Delta u_j - \partial \cdot \big(S_j^\delta G_T(\cdot, v) \cdot \partial u_j\big) = f_j \quad on \ \ [0, T] \times \mathbb{R}^d \,.$$

Assume that for any t in $[0, T]$, the support of the Fourier transform of $u_j(t, \cdot)$ is supported in the annulus $2^j \widetilde{\mathcal{C}}$. Let $I = [I^-, I^+]$ be a subinterval of $[0, T]$ such that for some sufficiently small ε_0,

$$|I| \leq \varepsilon_0 T(2^j T)^{-\delta}. \tag{9.31}$$

We then have, for sufficiently large $2^j T$ (and all sufficiently small positive ε, if $d = 3$),

$$\|\nabla u_j\|_{L^2(I;L^\infty)} \leq C 2^{j\left(\frac{d}{2} - \frac{1}{2}\right)}(\|\nabla u_j(I^-)\|_{L^2} + \|f_j\|_{L^1(I;L^2)}), \quad if \ \ d \geq 4,$$
$$\|\nabla u_j\|_{L^2(I;L^\infty)} \leq C_\varepsilon 2^j (2^j T)^\varepsilon (\|\nabla u_j(I^-)\|_{L^2} + \|f_j\|_{L^1(I;L^2)}), \quad if \ \ d = 3,$$
$$\|\nabla u_j\|_{L^4(I;L^\infty)} \leq C 2^{j\frac{3}{4}}(\|\nabla u_j(I^-)\|_{L^2} + \|f_j\|_{L^1(I;L^2)}), \quad if \ \ d = 2.$$

Proof of Theorem 9.11. This consists in splitting the original interval $[0, T]$ into subintervals $I_{j,k}$ on which the microlocal Strichartz estimates apply. Computing the total number of these subintervals is the key to the proof. In order to do this, we introduce a small parameter λ, the value of which will be chosen later. We want that, for each interval $I_{j,k}$,

$$|I_{j,k}| \le \varepsilon_0 T (2^j T)^{-\delta} \quad \text{and} \quad \|f_j\|_{L^1(I_{j,k};L^2)} \le \lambda \|f_j\|_{L_T^1(L^2)}. \tag{9.32}$$

This is satisfied whenever

$$(2^j T)^\delta \frac{1}{\varepsilon_0 T} \int_{I_{j,k}} dt + \frac{1}{\lambda \|f_j\|_{L_T^1(L^2)}} \int_{I_{j,k}} \|f_j(t)\|_{L^2}\, dt \le 1. \tag{9.33}$$

We shall prove by induction that such a finite decomposition exists and then control the number of intervals. Assume that we have constructed an increasing family $(t_\ell)_{0 \le \ell \le k}$ of times in $[0,T]$ such that $t_0 = 0$, $t_\ell < T$, and, for any $\ell \le k-1$,

$$(2^j T)^\delta \frac{1}{\varepsilon_0 T}(t_{\ell+1} - t_\ell) + \frac{1}{\lambda \|f_j\|_{L_T^1(L^2)}} \int_{t_\ell}^{t_{\ell+1}} \|f_j(t)\|_{L^2}\, dt = 1.$$

Define

$$F_k(t) \overset{\text{def}}{=} (2^j T)^\delta \frac{1}{\varepsilon_0 T}(t - t_k) + \frac{1}{\lambda \|f_j\|_{L_T^1(L^2)}} \int_{t_k}^{t} \|f_j(t')\|_{L^2}\, dt'.$$

This function is increasing and continuous. As $F_k(t_k) = 0$, either a unique t_{k+1} exists in $]t_k, T[$ such that $F_k(t_{k+1}) = 1$, or else the interval $[t_k, T]$ satisfies the condition (9.33), in which case we set $t_{k+1} = T$, and the procedure stops. This defines a sequence $(t_\ell)_{0 \le \ell \le k}$. As long as t_k is less than T, we have, by summation,

$$k = (2^j T)^\delta \frac{1}{\varepsilon_0 T} t_{k+1} + \frac{1}{\lambda \|f_j\|_{L_T^1(L^2)}} \int_0^{t_{k+1}} \|f_j(t)\|_{L^2}\, dt \le (2^j T)^\delta \frac{1}{\varepsilon_0} + \frac{1}{\lambda}.$$

Thus, the number N_j of intervals is finite and

$$N_j \le \frac{(2^j T)^\delta}{\varepsilon_0} + \frac{1}{\lambda}.$$

Taking $\lambda = (2^j T)^{-\delta}$ and applying Theorem 9.15 in the case $d \ge 4$ gives, for any interval $I_{j,\ell}$,

$$\|\nabla u_j\|_{L^2(I_{j,\ell};L^\infty)} \le C 2^{j\left(\frac{d}{2} - \frac{1}{2}\right)}\left(\|\nabla u_j\|_{L_T^\infty(L^2)} + \|f_j\|_{L^1(I_{j,\ell};L^2)}\right)$$

$$\le C 2^{j\left(\frac{d}{2} - \frac{1}{2}\right)}\left(\|\nabla u_j\|_{L_T^\infty(L^2)} + (2^j T)^{-\delta}\|f_j\|_{L_T^1(L^2)}\right).$$

We now write that

$$\|\nabla u_j\|_{L_T^2(L^\infty)}^2 = \sum_{\ell=0}^{N_j-1} \|\nabla u_j\|_{L^2(I_{j,\ell};L^\infty)}^2$$

$$\le C 2^{j(d-1)} N_j \left(\|\nabla u_j\|_{L_T^\infty(L^2)}^2 + (2^j T)^{-2\delta}\|f_j\|_{L_T^1(L^2)}^2\right).$$

As $N_j \le C(2^j T)^\delta$, we get the desired inequality for $d \ge 4$.

The case $d = 2, 3$ follows from similar arguments. It is only a matter of multiplying the above right-hand side by $(2^j T)^\varepsilon$ if $d = 3$ and of changing $2^{j(d-1)}$ to $2^{\frac{3}{4}j}$ if $d = 2$. In the latter case, we must also replace the L^2 time integration by an L^4 time integration. □

9.5 Microlocal Strichartz Estimates

This section is dedicated to the proof of the microlocal Strichartz estimates in Theorem 9.15. These will arise as a consequence of a much more general statement pertaining to a class of smooth variable coefficients linear wave equations (see Theorem 9.16 below) which are of independent interest.

9.5.1 A Rather General Statement

In order to define the class of linear wave equations that we shall consider, we first introduce a family $(G_\Lambda)_{\Lambda \geq \Lambda_0}$ of smooth functions from $\mathbb{R}^{1+d}$ to the space of symmetric matrices on $\mathbb{R}^d$ such that for some positive constant c_0, we have $\mathrm{Id} + G_\Lambda \geq c_0$ for all $\Lambda \geq \Lambda_0$, and

$$\forall k \in \mathbb{N}, \ \mathcal{G}_k \stackrel{\mathrm{def}}{=} \sup_{\Lambda \geq \Lambda_0} \Lambda^k \|\nabla^k G_\Lambda\|_{L^\infty(\mathbb{R}^{1+d})} < \infty. \tag{9.34}$$

Note that in the particular case where the support of the space-time Fourier transform of G_Λ is included in $\Lambda^{-1}\mathcal{B}$, where $\mathcal{B}$ stands for some fixed ball of $\mathbb{R}^{1+d}$, we have

$$\mathcal{G}_k \leq C^{k+1}\mathcal{G}_0. \tag{9.35}$$

Theorem 9.16. *Consider an external force f and initial data (u_0, u_1) such that $\widehat{f}(t, \cdot)$, $\widehat{u}_0$, and $\widehat{u}_1$ are supported in some annulus $\mathcal{C}$. Define $\gamma \stackrel{\mathrm{def}}{=} (u_1, \partial u_0)$, and let $(u_\Lambda)_{\Lambda \geq \Lambda_0}$ denote the family of solutions to*

$$(LW_\Lambda) \quad \begin{cases} \Box_\Lambda u = f \\ \nabla u_{|t=0} = \gamma \end{cases} \quad \text{with} \quad \Box_\Lambda u \stackrel{\mathrm{def}}{=} \partial_t^2 u - \Delta u - \sum_{1 \leq k, \ell \leq d} \partial_k (G_\Lambda^{k,\ell} \partial_\ell u).$$

Let $I_\Lambda \stackrel{\mathrm{def}}{=} [0, \varepsilon_0 \Lambda]$. If $d \geq 4$, then we have, for all $\Lambda \geq \Lambda_0$,

$$\|u_\Lambda\|_{L^2_{I_\Lambda}(L^\infty)} \leq C(\|\gamma\|_{L^2} + \|f\|_{L^1_{I_\Lambda}(L^2)}).$$

If $d = 3$, then we have, for all $\Lambda \geq \Lambda_0$,

$$\|u_\Lambda\|_{L^2_{I_\Lambda}(L^\infty)} \leq C(\log \Lambda)^{\frac{1}{2}}(\|\gamma\|_{L^2} + \|f\|_{L^1_{I_\Lambda}(L^2)}).$$

If $d = 2$, then we have, for all $\Lambda \geq \Lambda_0$,

$$\|u_\Lambda\|_{L^4_{I_\Lambda}(L^\infty)} \leq C(\|\gamma\|_{L^2} + \|f\|_{L^1_{I_\Lambda}(L^2)}).$$

In order to show that the above theorem implies Theorem 9.15, we have to perform a convenient rescaling in the family $(S_j^\delta G_T)$. We consider

$$(S_j^\delta G_T)_{res}(\tau, y) \stackrel{\text{def}}{=} S_j^\delta G_T(2^{-j}\tau, 2^{-j}y).$$

Obviously, we have

$$\|\nabla^k_{\tau, y}(S_j^\delta G_T)_{res}\|_{L^\infty(\mathbb{R}^{1+d})} = 2^{-jk}\|\nabla^k_{t,x}S_j^\delta G_T\|_{L^\infty(\mathbb{R}^{1+d})},$$

and, hence, according to the localization properties of the operator S_j^δ and Bernstein's inequality, there exists some positive constant C such that for all $k \in \mathbb{N}$,

$$\|\Lambda^k \nabla^k(S_j^\delta G_T)_{res}\|_{L^\infty(\mathbb{R}^{1+d})} \leq C^k \quad \text{with} \quad \Lambda \stackrel{\text{def}}{=} (2^j T)^{1-\delta}.$$

Hence, the inequality (9.34) holds true for this family.

Now, defining $u_{j,res} \stackrel{\text{def}}{=} u_j(2^{-j}\cdot, 2^{-j}\cdot)$ and $f_{j,res} \stackrel{\text{def}}{=} f_j(2^{-j}\cdot, 2^{-j}\cdot)$, we note that

$$\partial_\tau^2 u_{j,res} - \Delta_y u_{j,res} - \partial_y \cdot ((S_j^\delta G_T)_{res} \cdot \partial_y u_{j,res}) = 2^{-2j} f_{j,res}.$$

So, applying Theorem 9.16 to the family $(u_{j,res})$ (with j sufficiently large) and performing suitable changes of variable in the integrals involved in the inequalities, we readily get Theorem 9.15.   □

The rest of this chapter is devoted to proving Theorem 9.16. Compared to the case of the constant coefficients wave equation investigated in the previous chapter, the main difficulty is that here, we do not have any explicit representation of the solution. The naive idea consists in writing out an approximate representation by means of the geometrical optics method which is presented in the next subsection.

9.5.2 Geometrical Optics

In this section, we explain how geometrical optics may be used to approximate a solution of the variable coefficients linear wave equation

$$(W_g) \quad \begin{cases} \Box_g u = 0 \\ (u, \partial_t u)_{|t=0} = (u_0, u_1) \end{cases} \quad \text{with} \quad \Box_g u \stackrel{\text{def}}{=} \partial_t^2 - \sum_{1 \leq k, \ell \leq d} \partial_k(g^{k,\ell}\partial_\ell u),$$

in the case where g is a smooth function of the variables t and x with values in $\mathcal{S}_d^+(\mathbb{R})$.

For $g = \text{Id}$, we saw in Chapter 8 that the solution can be computed explicitly, namely,

$$u(t,x) = \frac{1}{(2\pi)^d} \sum_{\pm} \int e^{i(x|\xi) \pm it|\xi|} \widehat{\gamma}^{\pm}(\xi)\, d\xi \quad \text{with}$$

$$\widehat{\gamma}^{\pm} \stackrel{\text{def}}{=} \frac{1}{2}\left(\widehat{u}_0(\xi) \pm \frac{1}{i|\xi|}\widehat{u}_1(\xi)\right).$$

In the variable coefficients case, we look for an approximation of the solution of the form

$$u(t,x) = \sum_{\ell=0}^{1} \sum_{\pm} \mathcal{I}(\Phi^{\pm}, \sigma^{\pm,\ell}, u_\ell) \quad \text{with}$$

$$\mathcal{I}(\Phi, \sigma, a) \stackrel{\text{def}}{=} \int e^{i\Phi(t,x,\xi)} \sigma(t,x,\xi)\widehat{a}(\xi)\, d\xi. \tag{9.36}$$

Of course, initially, the *phase functions* $\Phi^{\pm}$ have to satisfy $\Phi^{\pm}(0,x,\xi) = (x|\xi)$, while the *modulus functions* $\sigma^{\pm,\ell}$ have to be chosen so that

$$\sigma^{\pm,0}(0,x,\xi) = \frac{1}{2}(2\pi)^{-d} \quad \text{and} \quad \sigma^{+,1}(0,x,\xi) + \sigma^{-,1}(0,x,\xi) = 0.$$

The action of $\Box_g$ on such quantities is described by the following lemma, which is an obvious consequence of the chain rule.

Lemma 9.17. *We have*

$$e^{-i\Phi}\Box_g(e^{i\Phi}\sigma) = \left(-(\partial_t\Phi)^2 + g(\partial_x\Phi, \partial_x\Phi)\right)\sigma + 2i\mathcal{L}_\Phi\sigma - \sigma\Box_g\Phi + R(\Phi,\sigma)$$

with

$$\mathcal{L}_\Phi \stackrel{\text{def}}{=} \partial_t\Phi\,\partial_t - \sum_{1 \le k,\ell \le d} g^{k,\ell}\partial_k\Phi\partial_\ell \quad \text{and} \quad R(\Phi,\sigma) \stackrel{\text{def}}{=} \Box_g\sigma. \tag{9.37}$$

Taking for granted that the remainder $R(\Phi,\sigma)$ is of lower order (in some sense that will be specified later), we are left with solving the *eikonal equation*

$$(EE) \qquad \begin{cases} (\partial_t\Phi)^2 = g(\partial_x\Phi, \partial_x\Phi) \\ \Phi(0,x,\xi) = (x|\xi) \end{cases}$$

and the cascade of transport equations

$$2i\mathcal{L}_\Phi\sigma_0 - \sigma_0\Box_g\Phi = 0 \quad \text{and} \quad 2i\mathcal{L}_\Phi\sigma_{n+1} - \sigma_{n+1}\Box_g\Phi + R(\Phi,\sigma_n) = 0. \tag{9.38}$$

9.5.3 The Solution of the Eikonal Equation

In all that follows, we fix two annuli, $\mathcal{C}$ and $\widetilde{\mathcal{C}}$, with $\mathcal{C} \subset \widetilde{\mathcal{C}}$ and $d(\partial\mathcal{C}, \partial\widetilde{\mathcal{C}}) > 0$, and consider the following family of eikonal equations:

$$(\widetilde{HJ}^{\pm}_\Lambda) \qquad \begin{cases} \partial_t\Phi^{\pm}_\Lambda(t,x,\xi) = F^{\pm}_\Lambda\left(t,x,\partial_x\Phi^{\pm}_\Lambda(t,x,\xi)\right) \\ \Phi^{\pm}_\Lambda(0,x,\xi) = (x|\xi), \end{cases}$$

where the family $(F^{\pm}_\Lambda)_{\Lambda \ge \Lambda_0}$ satisfies

$$\|\Lambda^k\nabla^k_{t,x}\partial^\ell_p F_\Lambda\|_{L^\infty(\mathbb{R}^{1+d}\times\widetilde{\mathcal{C}})} \le C_{k,\ell}. \tag{9.39}$$

Proposition 9.18. *There exists a constant ε_0 such that for any $\xi \in \mathcal{C}$ there exists a unique smooth solution $\Phi_\Lambda^\pm(\cdot, \cdot, \xi)$ of the equation $(\widetilde{HJ}_\Lambda^\pm)$ on the interval I_Λ.*

Moreover, $\partial_x \Phi_\Lambda^\pm$ is valued in $\widetilde{\mathcal{C}}$, and for any integer k there exists a nondecreasing function C_k from $\mathbb{R}^+$ into itself such that the family of solutions $(\Phi_\Lambda^\pm)_{\Lambda \geq \Lambda_0}$ satisfies

$$\sup_{\Lambda \geq \Lambda_0} \|\nabla_\Lambda^k \nabla_{t,x} \Phi_\Lambda^\pm\|_{L^\infty(I_\Lambda \times \mathbb{R}^d \times \mathcal{C})} \leq C_k \quad \text{with} \quad \nabla_\Lambda \stackrel{def}{=} (\Lambda \nabla_{t,x}, \partial_\xi). \tag{9.40}$$

Proof. From the classical theory of Hamilton–Jacobi equations (see, e.g., [15]), we infer that the equation $(\widetilde{HJ}_\Lambda^\pm)$ has a unique maximal smooth solution on some nontrivial time interval $[0, T_\Lambda^{\pm,\star}[$. In addition, if $T_\Lambda^{\pm,\star}$ is finite, then we have

$$\limsup_{t \to T_\Lambda^{\pm,\star}} \|\partial_x^2 \Phi_\Lambda^\pm(t, \cdot)\|_{L^\infty(\mathbb{R}^d \times \mathcal{C})} = +\infty. \tag{9.41}$$

To simplify notation, we omit the index $\pm$ in the rest of the proof. Let T_Λ denote the supremum of times $T < \min(T_\Lambda^\star, \varepsilon_0 \Lambda)$ such that

$$\|\partial_x^2 \Phi_\Lambda\|_{L^\infty([0,T] \times \mathbb{R}^d \times \mathcal{C})} \leq \Lambda^{-1} \quad \text{and} \quad \partial_x \Phi_\Lambda([0,T] \times \mathbb{R}^d \times \mathcal{C}) \subset \widetilde{\mathcal{C}}.$$

We note that differentiating the equation with respect to the variable x and setting $Z_\Lambda \stackrel{def}{=} -\partial_p F_\Lambda \cdot \partial_x$ gives

$$\begin{cases} \partial_t \partial_x \Phi_\Lambda + Z_\Lambda \cdot \partial_x \Phi_\Lambda = \partial_x F_\Lambda(t, x, \partial_x \Phi_\Lambda) \\ \partial_x \Phi_\Lambda(0, x, \xi) = \xi. \end{cases} \tag{9.42}$$

Hence, using (9.39) and integration, we get that for any $t < T_\Lambda$,

$$|\partial_x \Phi_\Lambda(t, x, \xi) - \xi| \leq C_1 \varepsilon_0. \tag{9.43}$$

As ξ is in $\mathcal{C}$, taking ε_0 sufficiently small obviously ensures that $\partial_x \Phi_\Lambda$ is valued in $\widetilde{\mathcal{C}}$.

Differentiating the equation once more with respect to the variable x and multiplying by Λ gives

$$\begin{cases} \partial_t \Lambda \partial_x^2 \Phi_\Lambda + Z_\Lambda \cdot \Lambda \partial_x^2 \Phi_\Lambda = R_\Lambda \\ \Lambda \partial_x^2 \Phi_\Lambda(0, x, \xi) = 0 \end{cases} \tag{9.44}$$

with $R_\Lambda \stackrel{def}{=} \Lambda \partial_x^2 F_\Lambda + 2\Lambda \partial_x \partial_p F_\Lambda \partial_x^2 \Phi_\Lambda + \Lambda \partial_p^2 F_\Lambda(\partial_x^2 \Phi_\Lambda, \partial_x^2 \Phi_\Lambda)$.

By the estimate (9.39) and the definition of T_Λ, for any $T < T_\Lambda$ we get

$$\|R_\Lambda\|_{L^\infty([0,T]\times\mathbb{R}^d\times\mathcal{C})} \leq C_2\Lambda^{-1}.$$

By integration, this gives

$$\Lambda\|\partial_x^2\Phi_\Lambda\|_{L^\infty([0,T]\times\mathbb{R}^d\times\mathcal{C})} \leq C_2 \quad \text{for all } T < T_\Lambda.$$

The blow-up criterion (9.41) implies that $T_\Lambda^\star \geq \varepsilon_0\Lambda$. Moreover, from the equation (9.42) we readily get

$$\|\partial_t\partial_x\Phi_\Lambda\|_{L^\infty(I_\Lambda\times\mathbb{R}^d\times\mathcal{C})} \leq C_2\Lambda^{-1}. \tag{9.45}$$

We now differentiate (9.42) with respect to the variable ξ. This gives

$$\begin{cases} \partial_t\partial_x\partial_\xi\Phi_\Lambda + Z_\Lambda\cdot\partial_x\partial_\xi\Phi_\Lambda = \widetilde{R}_\Lambda \\ \partial_x\partial_\xi\Phi_\Lambda(0,x,\xi) = \mathrm{Id} \end{cases}$$

with $\widetilde{R}_\Lambda \overset{\text{def}}{=} \partial_p^2 F_\Lambda(\partial_x^2\Phi_\Lambda,\partial_x\partial_\xi\Phi_\Lambda) + \partial_x\partial_p F_\Lambda\cdot\partial_x\partial_\xi\Phi_\Lambda$.

Now, by virtue of (9.39) we have

$$\|\widetilde{R}_\Lambda(t,\cdot)\|_{L^\infty(\mathbb{R}^d\times\mathcal{C})} \leq C_2\Lambda^{-1}\|\partial_x\partial_\xi\Phi_\Lambda(t,\cdot)\|_{L^\infty(\mathbb{R}^d\times\mathcal{C})}.$$

Therefore,

$$\|\partial_x\partial_\xi\Phi_\Lambda(t,\cdot)\|_{L^\infty(I_\Lambda\times\mathbb{R}^d\times\mathcal{C})} \leq C_2. \tag{9.46}$$

Combining (9.44), (9.45), and (9.46), we may thus conclude that (9.40) is satisfied for $k = 1$.

In order to prove the estimate (9.40) for all k, we proceed by induction. For the sake of simplicity we do not consider time derivatives as they may be recovered from the equation (9.42). We define

$$D_\Lambda \overset{\text{def}}{=} (\Lambda\partial_x, \partial_\xi).$$

Note that as the function F_Λ does not depend on ξ, the inequalities (9.39) can be written as

$$\|D_\Lambda^k\partial_p^\ell F_\Lambda\|_{L^\infty(I_\Lambda\times\mathbb{R}^d\times\mathcal{C})} \leq C_{k,\ell}. \tag{9.47}$$

We shall now prove by induction that for any $k \in \mathbb{N}$,

$$q_k \overset{\text{def}}{=} \sup_\Lambda \|D_\Lambda^k\partial_x\Phi_\Lambda\|_{L^\infty(I_\Lambda\times\mathbb{R}^d\times\mathcal{C})} \leq C_{k+1}. \tag{9.48}$$

We know that the inequality (9.48) holds true for $k = 1$. Assume that the property holds for $1 \leq j \leq k$. Now, applying the operator D_Λ^{k+1} to (9.42) gives

$$\partial_t D_\Lambda^{k+1} \partial_x \Phi_\Lambda + Z_\Lambda \cdot \partial_x D_\Lambda^{k+1} \Phi_\Lambda = \sum_{\ell=0}^{3} R_\Lambda^\ell \tag{9.49}$$

with, for some suitable nonnegative integers $A_{k_1,\ldots,k_r}^k$ and $B_{k_1,\ldots,k_r}^k$,

$$R_\Lambda^0 \overset{\text{def}}{=} \Lambda^{k+1} \partial_x^{k+2} F_\Lambda,$$

$$R_\Lambda^1 \overset{\text{def}}{=} \sum_{\substack{k_1+\cdots+k_r \leq k \\ k_j \geq 1}} A_{k_1,\ldots,k_r}^k D_\Lambda^{k+1-k_1-\cdots-k_r} \partial_x \partial_p^r F_\Lambda \big(D_\Lambda^{k_1} \partial_x \Phi_\Lambda, \ldots, D_\Lambda^{k_r} \partial_x \Phi_\Lambda \big),$$

$$R_\Lambda^2 \overset{\text{def}}{=} \sum_{\substack{k_0+k_1+\cdots+k_r \leq k \\ k_j \geq 1,\ k_0 < k}} B_{k_0,k_1,\ldots,k_r}^k$$
$$\times D_\Lambda^{k+1-k_1-\cdots-k_r} \partial_p^{r+1} F_\Lambda \big(D_\Lambda^{k_0} \partial_x^2 \Phi_\Lambda, D_\Lambda^{k_1} \partial_x \Phi_\Lambda, \ldots, D_\Lambda^{k_r} \partial_x \Phi_\Lambda \big),$$

$$R_\Lambda^3 \overset{\text{def}}{=} \partial_x \partial_p F_\Lambda D_\Lambda^{k+1} \partial_x \Phi_\Lambda + D_\Lambda \partial_p F_\Lambda \cdot D_\Lambda^k \partial_x^2 \Phi_\Lambda + \partial_p^2 F_\Lambda \big(\partial_x^2 \Phi_\Lambda, D_\Lambda^{k+1} \partial_x \Phi_\Lambda \big).$$

The inequality (9.39) readily implies that

$$\| R_\Lambda^0 \|_{L^\infty(I_\Lambda \times \mathbb{R}^d \times \mathcal{C})} \leq C_{k+1} \Lambda^{-1}.$$

Using the induction hypothesis and (9.39), we have

$$\big\| D_\Lambda^{k+1-k_1-\cdots-k_r} \partial_x \partial_p^r F_\Lambda \big(D_\Lambda^{k_1} \partial_x \Phi_\Lambda, \ldots, D_\Lambda^{k_r} \partial_x \Phi_\Lambda \big) \big\|_{L^\infty(I_\Lambda \times \mathbb{R}^d \times \mathcal{C})} \leq C_{k+1} \Lambda^{-1},$$

$$\big\| D_\Lambda^{k+1-k_1-\cdots-k_r} \partial_p^{r+1} F_\Lambda \big(D_\Lambda^{k_0} \partial_x^2 \Phi_\Lambda, D_\Lambda^{k_1} \partial_x \Phi_\Lambda, \ldots, D_\Lambda^{k_r} \partial_x \Phi_\Lambda \big) \big\|_{L^\infty(I_\Lambda \times \mathbb{R}^d \times \mathcal{C})}$$
$$\leq C_{k+1} \Lambda^{-1}.$$

Thus,

$$\| R_\Lambda^1 \|_{L^\infty(I_\Lambda \times \mathbb{R}^d \times \mathcal{C})} + \| R_\Lambda^2 \|_{L^\infty(I_\Lambda \times \mathbb{R}^d \times \mathcal{C})} \leq C_{k+1} \Lambda^{-1}.$$

From the property (9.40) with rank $k = 1$ and the inequality (9.39), we also infer that

$$\| R_\Lambda^3 \|_{L^\infty(I_\Lambda \times \mathbb{R}^d \times \mathcal{C})} \leq C_{k+1} \Lambda^{-1} \| D_\Lambda^{k+1} \partial_x \Phi_\Lambda \|_{L^\infty(I_\Lambda \times \mathbb{R}^d \times \mathcal{C})}.$$

Gronwall's lemma allows us to complete the proof of the inequality (9.48) with rank $k + 1$. This completes the proof of the proposition. □

In order to prove Theorem 9.16, we shall consider the following Hamilton–Jacobi equations:

$$(EE_\Lambda^\pm) \qquad \begin{cases} \partial_t \Phi_\Lambda^\pm = \pm \Big(\sum_{1 \leq j,k \leq d} (\delta^{k,\ell} + G_\Lambda^{j,k}) \partial_{x_j} \Phi_\Lambda^\pm \partial_{x_k} \Phi_\Lambda^\pm \Big)^{\frac{1}{2}} \\ \Phi_\Lambda^\pm(0, y, \eta) = (y|\eta). \end{cases}$$

Observe that if we consider some family $(G_\Lambda)_{\Lambda \geq \Lambda_0}$ such that (9.34) holds true, then these equations become part of the class of eikonal equations that have been considered in this subsection: It is only a matter of setting

$$F_\Lambda^\pm(t,x,p) \stackrel{\text{def}}{=} \pm\big(|p|^2 + G_\Lambda(t,x)(p,p)\big)^{\frac{1}{2}} \quad \text{for} \quad (t,x,p) \in \mathbb{R} \times \mathbb{R}^d \times \mathbb{R}^d. \quad (9.50)$$

Indeed, as we only have to consider those ξ's which belong to some annulus $\widetilde{\mathcal{C}}$, we can substitute for the square root in the above formula a convenient smooth function defined everywhere. Hence, the inequality (9.34) implies the inequality (9.39) and Proposition 9.18 applies.

9.5.4 The Transport Equation

Proving suitable a priori estimates for the transport equations considered in the geometrical optics method is the next step. More precisely, setting

$$\mathcal{L}_\Lambda^\pm \stackrel{\text{def}}{=} \partial_t \Phi_\Lambda^\pm \partial_t - \sum_{k,\ell} (\delta^{k,\ell} + G_\Lambda^{j,k}) \partial_{x_j} \Phi_\Lambda^\pm \partial_{x_k} \quad \text{and} \quad (9.51)$$

$$\mathcal{A}_\Lambda^\pm \stackrel{\text{def}}{=} \Box_\Lambda \Phi^\pm, \quad (9.52)$$

we wish to consider the following transport equations:

$$(T_\Lambda^\pm) \qquad \begin{cases} \mathcal{L}_\Lambda^\pm \cdot \nabla \sigma_\Lambda^\pm + \mathcal{A}_\Lambda^\pm \sigma_\Lambda^\pm = \rho_\Lambda \\ \sigma_{\Lambda|t=0}^\pm = \sigma_\Lambda^{(0)}. \end{cases}$$

Before going into further detail, we need to define a class of symbols.

Definition 9.19. *For any real number m, we denote by S^m the set of families $\sigma = (\sigma_\Lambda)_{\Lambda \geq \Lambda_0}$ of smooth functions from $I_\Lambda \times \mathbb{R}^d \times \mathcal{C}$ to $\mathbb{C}$ such that for any integer k,*

$$\|\sigma\|_{k,S^m} \stackrel{\text{def}}{=} \sup_{\Lambda \geq \Lambda_0} \Lambda^{-m} \|\nabla_\Lambda^k \sigma_\Lambda\|_{L^\infty(I_\Lambda \times \mathbb{R}^d \times \mathcal{C})} < \infty \quad \text{with} \quad \nabla_\Lambda \stackrel{\text{def}}{=} (\Lambda \nabla_{t,x}, \partial_\xi).$$

Remark 9.20. The inequality (9.40) implies that $(\nabla_{t,x} \Phi_\Lambda^\pm)$ belongs to S^0.

Remark 9.21. The above quantities define seminorms which endow S^m with the structure of a Fréchet space. Moreover, it is obvious that the operator ∇_Λ^k continuously maps S^m into S^m. This implies that $\nabla_{t,x}$ maps S^m into S^{m-1}. We also emphasize that the (numerical) product continuously maps $S^{m_1} \times S^{m_2}$ into $S^{m_1+m_2}$ and that if ϕ is a function of the Schwartz class $\mathcal{S}$, then $\phi(D)$ continuously maps S^0 into S^0. Finally, if f is a function of $\mathcal{D}$ and $\sigma \in S^0$, then $f \circ \sigma \stackrel{\text{def}}{=} (f(\sigma_\Lambda))_{\Lambda \geq \Lambda_0} \in S^0$. More precisely, for any integer k there exists a locally bounded function C such that $C(0) = 0$ and

$$\|(f \circ \sigma)\|_{k,S^0} \leq C\big(k, \sup_{j \leq k} \|\sigma\|_{j,S^0}\big).$$

Remark 9.22. The families $\mathcal{L}_\Lambda^{\pm,j}$ ($0 \leq j \leq d$) of coefficients of the vector field $\mathcal{L}_\Lambda^\pm$ defined in (9.51) are in S^0. We also emphasize that $(\mathcal{A}_\Lambda)$ belongs to S^{-1}.

From now on, we denote by C_k a generic increasing function depending on $\sup_{j \leq k} \mathcal{G}_j$. The following lemma pertaining to the class of transport equations considered in (9.38) will help us to construct approximate solutions of the variable coefficients wave equation.

Lemma 9.23. *Let m be a real number. Consider (ρ_Λ), a family in S^{m-1}. If the initial family $(\sigma_\Lambda^{(0)})_{\Lambda \geq \Lambda_0}$ satisfies*

$$\sup_{\Lambda \geq \Lambda_0} \Lambda^{-m} \|D_\Lambda^k \sigma_\Lambda^{(0)}\|_{L^\infty(\mathbb{R}^d \times \mathcal{C})} < \infty,$$

then the corresponding family $(\sigma_\Lambda^\pm)_{\Lambda \geq \Lambda_0}$ of solutions of $(T_\Lambda^\pm)$ belongs to S^m, and the map $(\sigma_\Lambda^{(0)}) \mapsto (\sigma_\Lambda^\pm)$ is continuous.

Proof. Recall that the family of symbols $(\mathcal{L}_\Lambda^{\pm,0})$ belongs to S^0. In addition, applying the inequality (9.40) with $k = 1$ to the Hamilton–Jacobi equation $(EE_\Lambda^\pm)$, we discover that there exists some positive real number c such that

$$\frac{1}{c} \geq |\mathcal{L}_\Lambda^{\pm,0}| \geq c \quad \text{for all} \quad \Lambda \geq \Lambda_0.$$

Hence, Remark 9.21 entails that the family $(\mathcal{L}_\Lambda^{\pm,0})^{-1}$ belongs to S^0 and the equation $(T_\Lambda^\pm)$ can be rewritten as

$$(\widetilde{T}_\Lambda^\pm) \qquad \begin{cases} \partial_t \sigma_\Lambda^\pm + \widetilde{\mathcal{L}}_\Lambda^\pm \cdot \partial \sigma_\Lambda^\pm + \widetilde{\mathcal{A}}_\Lambda^\pm \sigma_\Lambda^\pm = \widetilde{\rho}_\Lambda \\ \sigma_{\Lambda|t=0}^\pm = \sigma_\Lambda^{(0)} \end{cases}$$

with $\widetilde{\mathcal{L}}_\Lambda^{\pm,j} \overset{\text{def}}{=} \dfrac{\mathcal{L}_\Lambda^{\pm,j}}{\mathcal{L}_\Lambda^{\pm,0}}$, $\quad \widetilde{\mathcal{A}}_\Lambda^\pm \overset{\text{def}}{=} \dfrac{\mathcal{A}_\Lambda^\pm}{\mathcal{L}_\Lambda^{\pm,0}}$, and $\widetilde{\rho}_\Lambda^\pm \overset{\text{def}}{=} \dfrac{\rho_\Lambda}{\mathcal{L}_\Lambda^{\pm,0}}$.

According to Remarks 9.20–9.22, we have

$$(\widetilde{\mathcal{L}}_\Lambda^{\pm,j}) \in S^0, \quad (\widetilde{\mathcal{A}}_\Lambda^\pm) \in S^{-1}, \quad \text{and} \quad (\widetilde{\rho}_\Lambda^\pm) \in S^{m-1}.$$

Thus, Gronwall's lemma implies that

$$\|\sigma_\Lambda^\pm\|_{L^\infty(I_\Lambda \times \mathbb{R}^d \times \mathcal{C})} \leq C\Lambda^m.$$

Now, to estimate $\nabla_\Lambda^k \sigma_\Lambda^\pm$ we proceed as in the preceding subsection. We do not have to worry about time derivatives since they may be computed from the equation $(\widetilde{T}_\Lambda^\pm)$. Assume that for any $j \leq k$

$$q_j \overset{\text{def}}{=} \sup_\Lambda \Lambda^{-m} \|D_\Lambda^j \sigma_\Lambda^\pm\|_{L^\infty(I_\Lambda \times \mathbb{R}^d \times \mathcal{C})} \quad \text{is finite.}$$

Applying the operator D_Λ^{k+1} to the equation $(\widetilde{T}_\Lambda^\pm)$ then gives

$$\partial_t D_\Lambda^{k+1} \sigma_\Lambda^\pm + \widetilde{\mathcal{L}}_\Lambda^\pm \cdot \partial D_\Lambda^{k+1} \sigma_\Lambda^\pm + \widetilde{\mathcal{A}}_\Lambda^\pm D_\Lambda^{k+1} \sigma_\Lambda^\pm = D_\Lambda^{k+1} \widetilde{\rho}_\Lambda^\pm + \sum_{\ell=1}^{3} R_\Lambda^\ell$$

with, for some suitable integer values A_{k_1,k_2}^k and B_{k_1,k_2}^k,

$$R_\Lambda^1 \stackrel{\text{def}}{=} \Lambda^{-1} \sum_{\substack{k_1+k_2=k \\ k_2<k}} A_{k_1,k_2}^k D_\Lambda^{k_1} \widetilde{\mathcal{L}}_\Lambda^\pm \Lambda \partial_x D_\Lambda^{k_2} \sigma_\Lambda^\pm,$$

$$R_\Lambda^2 \stackrel{\text{def}}{=} \sum_{\substack{k_1+k_2=k+1 \\ k_2\leq k}} B_{k_1,k_2} D_\Lambda^{k_1} \widetilde{\mathcal{A}}_\Lambda^\pm D_\Lambda^{k_2} \sigma_\Lambda^\pm,$$

$$R_\Lambda^3 \stackrel{\text{def}}{=} \Lambda^{-1} D_\Lambda \mathcal{L}_\Lambda^\pm \Lambda \partial_x D_\Lambda^k \sigma_\Lambda^\pm.$$

The induction hypothesis implies that for $\ell \in \{1,2\}$ we have

$$\Lambda^m \|R_\Lambda^\ell\|_{L^\infty(I_\Lambda \times \mathbb{R}^d \times \mathcal{C})} \leq C\left(\sup_{j\leq k} \underline{q}_j\right) \Lambda^{-1}.$$

Moreover, we have

$$\|R_\Lambda^3(t,\cdot)\|_{L^\infty(\mathbb{R}^d \times \mathcal{C})} \leq C_k \Lambda^{-1} \|D_\Lambda^{k+1} \sigma_\Lambda^\pm(t,\cdot)\|_{L^\infty(\mathbb{R}^d \times \mathcal{C})}.$$

Gronwall's lemma then allows us to complete the proof. $\qquad\square$

9.5.5 The Approximation Theorem

We can now return to the initial problem of approximating the solutions of a family of variable coefficients wave equations. We consider the family of wave equations

$$(LW_\Lambda) \qquad \begin{cases} \Box_\Lambda u = 0 \\ (u, \partial_t u)_{|t=0} = (u_0, u_1), \end{cases}$$

where u_0 and u_1 are L^2 functions with Fourier transforms supported in $\mathcal{C}$.

The following statement ensures the existence of an arbitrarily accurate approximate solution. We shall see in the next subsection that keeping only the main order term suffices to prove the microlocal Strichartz estimates we are interested in.

Theorem 9.24. *There exist four families of sequences of symbols* $(\sigma_{n,\Lambda}^{\pm,\ell})_{n\in\mathbb{N}}$ *(with ℓ in $\{0,1\}$) such that $\sigma_{n,\Lambda}^{\pm,\ell}$ belongs to S^{-n} and that, for any $(k,N) \in \mathbb{N}^2$, a constant C exists such that*

$$\left\|\partial_x^k(u_\Lambda - u_{app,N,\Lambda})\right\|_{L_{I_\Lambda}^\infty(L^2)} \leq C\Lambda^{-N-1}\|(u_0,u_1)\|_{L^2} \quad \text{with}$$

$$u_{app,N,\Lambda} \stackrel{\text{def}}{=} \sum_{\ell,\pm} \sum_{n=0}^N \mathcal{I}(\Phi_\Lambda^\pm, \sigma_{n,\Lambda}^{\pm,\ell}, u_\ell),$$

where the function $\mathcal{I}$ is defined by the formula (9.36), and $\Phi_\Lambda^\pm$ is the solution of $(EE_\Lambda^\pm)$.

Proof. Note that the equation $(EE_\Lambda^\pm)$ implies that

$$\partial_t \Phi_{\Lambda|t=0}^\pm = \pm|\xi|_\Lambda \quad \text{with} \quad |\xi|_\Lambda \stackrel{\text{def}}{=} \left(|\xi|^2 + G_\Lambda(0, x)(\xi, \xi)\right)^{\frac{1}{2}}.$$

Bearing in mind that we want the true solution u_Λ of (LW_Λ) to satisfy the initial conditions $u_{\Lambda|t=0} = u_0$ and $\partial_t u_{\Lambda|t=0} = u_1$, we define the sequence $\sigma_{n,\Lambda}^{\pm,\ell}$ by means of the following induction:

– The function $\sigma_{0,\Lambda}^{\pm,\ell}$ is the solution of

$$2\mathcal{L}_\Lambda^\pm \cdot \nabla\sigma_{0,\Lambda}^{\pm,\ell} + i\sigma_{0,\Lambda}^{\pm,\ell}\square_\Lambda\Phi_\Lambda^\pm = 0 \quad \text{with}$$

$$\sigma_{0,\Lambda|t=0}^{\pm,0} = \frac{1}{2(2\pi)^d} \quad \text{and} \quad \sigma_{0,\Lambda|t=0}^{\pm,1} = \pm\frac{1}{2i|\xi|_\Lambda(2\pi)^d}.$$

– Once the function $\sigma_{n,\Lambda}^{\pm,\ell}$ has been defined, we set $\sigma_{n+1,\Lambda}^{\pm,\ell}$ to be the solution of

$$2\mathcal{L}_\Lambda^\pm \cdot \nabla\sigma_{n+1,\Lambda}^{\pm,\ell} + i\sigma_{n+1,\Lambda}^{\pm,\ell}\square_\Lambda\Phi_\Lambda^\pm = i\square_\Lambda\sigma_{n,\Lambda}^{\pm,\ell} \quad \text{with}$$

$$\sigma_{n+1,\Lambda|t=0}^{\pm,\ell} = \mp\frac{1}{2i|\xi|_\Lambda}\partial_t\sigma_{n,\Lambda|t=0}^{\pm,\ell}.$$

Using Lemma 9.23 and performing an omitted induction, we observe that the family $(\sigma_n^{\pm,\ell})$ belongs to S^{-n}. Further, as we have

$$\nabla\mathcal{I}(\Phi_\Lambda^\pm, \sigma_\Lambda, a) = \mathcal{I}(\Phi_\Lambda^\pm, i\sigma\nabla\Phi_\Lambda^\pm + \nabla\sigma_\Lambda, a) \tag{9.53}$$

for any family of symbols (σ_Λ), we discover that

$$u_{app,N,\Lambda|t=0} = u_0 \quad \text{and} \quad \partial_t u_{app,N,\Lambda|t=0} = u_1 + \left(\sum_{\ell,\pm}\mathcal{I}(\Phi_\Lambda^\pm, \sigma_{N+1,\Lambda}^{\pm,\ell}, u_\ell)\right)_{|t=0}.$$

Using Lemma 9.17, we then infer from the definition of the symbols $\sigma_n^{\pm,\ell}$ that

$$\square_\Lambda(u_\Lambda - u_{app,N,\Lambda}) = f_{N,\Lambda} \stackrel{\text{def}}{=} i\sum_{\ell,\pm}\mathcal{I}(\Phi_\Lambda^\pm, \square_\Lambda\sigma_{N,\lambda}^{\pm,\ell}, u_\ell) \quad \text{with}$$

$$\nabla(u_\Lambda - u_{app,N,\Lambda})_{|t=0} = \gamma_{N,\Lambda} \stackrel{\text{def}}{=} \left(\left(\sum_{\ell,\pm}\mathcal{I}(\Phi_\Lambda^\pm, \sigma_{N+1,\Lambda}^{\pm,\ell}, u_\ell)\right)_{|t=0}, 0\right).$$

Using Proposition 8.17 and the relation (9.53), we get, for any k in $\mathbb{N}$, that

$$\Lambda\|\partial_x^k f_{N,\Lambda}\|_{L_{T_\Lambda}^\infty(L^2)} + \|\partial_x^k\gamma_{N,\Lambda}\|_{L^2} \le C_N\Lambda^{-N-1}\|\gamma\|_{L^2}. \tag{9.54}$$

Performing an H^k energy estimate for the wave operator $\square_\Lambda$ (in the spirit of, e.g., the one used to prove Lemma 4.5 page 173) then allows us complete the proof of the theorem. The details are left to the reader. □

9.5.6 The Proof of Theorem 9.16

This final subsection is devoted to the proof of the general microlocal Strichartz estimates stated in Theorem 9.16. Recall that we consider the solution u_Λ of (LW_Λ) in the case where the external force f and the initial data (u_0, u_1) are such that $\widehat{f}(t, \cdot)$, $\widehat{u}_0$, and $\widehat{u}_1$ are supported in some annulus $\widetilde{\mathcal{C}}$.

For the time being, we will assume that $f \equiv 0$. Applying Theorem 9.24 with $N = 0$ ensures that four families of symbols $\sigma = (\sigma_\Lambda^{\pm,\ell})$ exist in S^0 such that for any integer k, there exists a constant C_k such that for any $\Lambda \geq \Lambda_0$, the solution u_Λ satisfies

$$\left\| \partial_x^k \left(u_\Lambda - \sum_{\pm,\ell} \mathcal{I}(\Phi_\Lambda^\pm, \sigma_\Lambda^{\pm,\ell}, u_\ell) \right) \right\|_{L^\infty_{I_\Lambda}(L^2)} \leq C\Lambda^{-1} \|\gamma\|_{L^2},$$

where $\Phi_\Lambda^\pm$ is the solution of $(EE_\Lambda^\pm)$. As

$$u_\Lambda = \left(u_\Lambda - \sum_{\pm,\ell} \mathcal{I}(\Phi_\Lambda^\pm, \sigma_\Lambda^{\pm,\ell}, u_\ell) \right) + \sum_{\pm,\ell} \mathcal{I}(\Phi_\Lambda^\pm, \sigma_\Lambda^{\pm,\ell}, u_\ell),$$

taking the $L^2_{I_\Lambda}(L^\infty)$ norm of both sides and using Sobolev embedding and the fact that the length of the interval I_Λ is less than $\varepsilon_0 \Lambda$ implies that

$$\|u_\Lambda\|_{L^2_{I_\Lambda}(L^\infty)} \leq C\|\gamma\|_{L^2} + \sum_{\pm,\ell} \|\mathcal{I}(\Phi_\Lambda^\pm, \sigma_\Lambda^{\pm,\ell}, u_\ell)\|_{L^2_{I_\Lambda}(L^\infty)}. \tag{9.55}$$

For notational simplicity, we omit the exponent $\pm$ in what follows. We first use the "$TT^\star$ duality argument" presented in Section 8.2. We write that

$$\|\mathcal{I}(\Phi_\Lambda, \sigma_\Lambda, u_\ell)\|_{L^2_{I_\Lambda}(L^\infty)} = \sup_{\psi \in \mathcal{B}_\Lambda} \int \mathcal{I}(\Phi_\Lambda, \sigma_\Lambda, u_\ell)(t, x) \psi(t, x)\, dt\, dx,$$

where $\mathcal{B}_\Lambda$ denotes the set of functions ψ such that $\|\psi\|_{L^2_{I_\Lambda}(L^1)} \leq 1$. By the definition of $\mathcal{I}(\Phi_\Lambda, \sigma_\Lambda, u_\ell)$, we have

$$\mathcal{J}_\Lambda(\psi) \overset{\text{def}}{=} \int \mathcal{I}(\Phi_\Lambda, \sigma_\Lambda, u_\ell)(t, x)\, \psi(t, x)\, dt\, dx$$

$$= \int \widehat{u}_\ell(\xi) \left(\int e^{i\Phi_\Lambda(t, x, \xi)} \sigma_\Lambda(t, x, \xi) \psi(t, x)\, dt\, dx \right) d\xi.$$

Using the Cauchy–Schwarz inequality, we get

$$\mathcal{J}_\Lambda(\psi) \leq \|\widehat{\gamma}\|_{L^2} \left\| \int e^{i\Phi_\Lambda(t, x, \cdot)} \sigma_\Lambda(t, x, \cdot) \psi(t, x)\, dt\, dx \right\|_{L^2(\mathbb{R}^d; d\xi)}.$$

By the definition of the L^2 norm, we have

$$\left\|\int e^{i\Phi_\Lambda(t,x,\cdot)}\sigma_\Lambda(t,x,\cdot)\psi(t,x)\,dt\,dx\right\|_{L^2(\mathbb{R}^d;d\xi)}^2$$

$$= \int K_\Lambda(t,t',x,y)\psi(t,x)\overline{\psi}(t',y)\,dt\,dt'\,dx\,dy$$

with

$$K_\Lambda(t,t',x,y) \overset{\text{def}}{=} \int_{\mathcal{C}} e^{i\Phi_\Lambda(t,x,\xi)-i\Phi_\Lambda(t',y,\xi)}\sigma_\Lambda(t,x,\xi)\overline{\sigma}_\Lambda(t',y,\xi)\,d\xi.$$

If we prove that

$$\forall (t,t',x,y) \in I_\Lambda^2 \times \mathbb{R}^{2d}\,,\ |K_\Lambda(t,t',x,y)| \leq \frac{C}{|t-t'|^{\frac{d-1}{2}}}\,, \qquad (9.56)$$

then Theorems 8.18 and 8.30 imply that for $\ell = 0,1$,

$$\|\mathcal{I}(\Phi_\Lambda,\sigma_\Lambda,u_\ell)\|_{L^2_{I_\Lambda}(L^\infty)} \leq C\|\gamma\|_{L^2}, \text{ if } d \geq 4, \qquad (9.57)$$

$$\|\mathcal{I}(\Phi_\Lambda,\sigma_\Lambda,u_\ell)\|_{L^2_{I_\Lambda}(L^\infty)} \leq C(\log\Lambda)^{\frac{1}{2}}\|\gamma\|_{L^2}, \text{ if } d = 3, \qquad (9.58)$$

$$\|\mathcal{I}(\Phi_\Lambda,\sigma_\Lambda,u_\ell)\|_{L^4_{I_\Lambda}(L^\infty)} \leq C\|\gamma\|_{L^2}, \text{ if } d = 2. \qquad (9.59)$$

Now, according to the mean value formula, we have

$$\Phi_\Lambda(t,x,\xi) - \Phi_\Lambda(t',y,\xi) = \big(x - y|\theta_\Lambda(t,t',x,y,\xi)\big) + (t - t')\Psi_\Lambda(t,t',x,y,\xi)$$

with

$$\theta_\Lambda(t,t',x,y,\xi) \overset{\text{def}}{=} \int_0^1 \partial_x\Phi_\Lambda(t' + s(t - t'),y + s(x - y),\xi)\,ds \quad \text{and}$$

$$\Psi_\Lambda(t,t',x,y,\xi) \overset{\text{def}}{=} \int_0^1 \partial_t\Phi_\Lambda(t' + s(t - t'),y + s(x - y),\xi)\,ds\,.$$

As $\partial_x\Phi_{\Lambda|t=0} = \xi$, we can write that $\theta_\Lambda(t,t',x,y,\xi) = \xi + \widetilde{\theta}_\Lambda(t,t',x,y,\xi)$ with

$$\widetilde{\theta}_\Lambda(t,t',x,y,\xi) \overset{\text{def}}{=} \int_0^1 \int_0^1 (t' + s(t - t'))$$

$$\times \partial_t\partial_x\Phi_\Lambda(vt' + sv(t - t'),y + s(x - y),\xi)\,ds\,dv.$$

Thanks to the inequality (9.40), we have, for all $\Lambda \geq \Lambda_0$,

$$\|\partial_\xi\widetilde{\theta}_\Lambda\|_{L^\infty(I_\Lambda^2 \times \mathbb{R}^{2d} \times \mathcal{C})} \leq C_2\frac{|I_\Lambda|}{\Lambda} \leq C_2\varepsilon_0, \qquad (9.60)$$

$$\|\partial_\xi^k\widetilde{\theta}_\Lambda\|_{L^\infty(I_\Lambda^2 \times \mathbb{R}^{2d} \times \mathcal{C})} \leq \leq C_{k+1}\varepsilon_0\Lambda_0^{1-k} \quad \text{if } k \geq 2. \qquad (9.61)$$

Assuming ε_0 to be sufficiently small, this implies that (up to an omitted finite decomposition of $\mathcal{C}$) the map

$$\xi \longmapsto \widetilde{\xi} \stackrel{\text{def}}{=} \theta_\Lambda(t, t', x, y, \xi)$$

is a smooth diffeomorphism from $\mathcal{C}$ onto its range, denoted by $\mathcal{C}_\Lambda(t, t', x, y)$. We denote by θ_Λ^{-1} the inverse diffeomorphism. Note that $\mathcal{C}_\Lambda(t, t', x, y)$ is included in some fixed annulus $\widetilde{\mathcal{C}}$. Performing the above change of variable, we eventually get

$$K_\Lambda(t, t', x, y) = \int_{\mathcal{C}_\Lambda(t, t', x, y)} e^{i(t-t')\left((z|\widetilde{\xi}) + \widetilde{\Psi}_\Lambda(t, t', x, y, \widetilde{\xi})\right)} \widetilde{\sigma}_\Lambda(t, t', x, y, \widetilde{\xi}) \, d\widetilde{\xi}$$

with $z \stackrel{\text{def}}{=} \dfrac{x - y}{t - t'}$,

$$\widetilde{\Psi}_\Lambda(t, t', x, y, \widetilde{\xi}) \stackrel{\text{def}}{=} \Psi_\Lambda\left(t, t', x, y, \theta_\Lambda^{-1}(t, t', x, y, \widetilde{\xi})\right),$$

$$\widetilde{\sigma}_\Lambda(t, t', x, y, \widetilde{\xi}) \stackrel{\text{def}}{=} \sigma_\Lambda\left(t, x, \theta_\Lambda^{-1}(t, t', x, y, \widetilde{\xi})\right) \overline{\sigma}_\Lambda\left(t', y, \theta_\Lambda^{-1}(t, t', x, y, \widetilde{\xi})\right)$$
$$\times J_\Lambda\left(t, t', x, y, \theta_\Lambda^{-1}(t, t', x, y, \widetilde{\xi})\right).$$

Above, J_Λ stands for the Jacobian of the diffeomorphism θ_Λ^{-1}.

Now, the inequalities (9.40), (9.60), and (9.61) imply that for all $(k, \ell) \in \mathbb{N}^2$,

$$\sup_{\Lambda \geq \Lambda_0} \sup_{\substack{\widetilde{\xi} \in \mathcal{C}_\Lambda(t, t', x, y) \\ (t, t', x, y) \in I_\Lambda^2 \times \mathbb{R}^{2d}}} |\partial_{\widetilde{\xi}}^\ell \widetilde{\sigma}_\Lambda(t, t', x, y, \widetilde{\xi})| < \infty \quad \text{and} \qquad (9.62)$$

$$\sup_{\Lambda \geq \Lambda_0} \Lambda^k \sup_{\substack{\widetilde{\xi} \in \mathcal{C}_\Lambda(t, t', x, y) \\ (t, t', x, y) \in I_\Lambda^2 \times \mathbb{R}^{2d}}} |\partial_{\widetilde{\xi}}^\ell \nabla_{t, t', x, y}^k \widetilde{\Psi}_\Lambda(t, t', x, y, \widetilde{\xi})| \leq C_{k, \ell}. \qquad (9.63)$$

Theorem 8.12 page 342 and the estimates (9.40) imply that a constant C exists such that for all (t, t', x, y) in $I_\Lambda^2 \times \mathbb{R}^{2d}$,

$$|K_\Lambda(t, t', x, y)| \leq \frac{C}{|t - t'|^{\frac{d-1}{2}}} + \int_{\widetilde{\mathcal{C}}_\Lambda(t, t', x, y)} \frac{d\widetilde{\xi}}{\left(1 + |t - t'| |z + \partial_{\widetilde{\xi}} \widetilde{\Psi}_\Lambda(t, t', x, y, \widetilde{\xi})|^2\right)^d},$$

where $\widetilde{\mathcal{C}}_\Lambda(t, t', x, y)$ denotes the set of $\widetilde{\xi} \in \mathcal{C}_\Lambda(t, t', x, y)$ such that

$$\left| \frac{x - y}{|t - t'|} + \partial_{\widetilde{\xi}} \Psi_\Lambda(t, t', x, y, \widetilde{\xi}) \right| \leq 1.$$

Hence, the inequality (9.56) reduces to proving that the Hessian of $\widetilde{\Psi}_\Lambda$ is at least of rank $d - 1$, uniformly in (t, t', x, y), and in $\widetilde{\xi} \in \mathcal{C}_\Lambda(t, t', x, y)$. The equation $(\widetilde{HJ}_\Lambda)$ and Proposition 9.18 now imply that

$$\sup_{\Lambda \geq \Lambda_0} \sup_{\substack{\widetilde{\xi} \in \mathcal{C}_\Lambda(t, t', x, y) \\ (t, t', x, y) \in I_\Lambda^2 \times \mathbb{R}^{2d}}} |\partial_{\widetilde{\xi}} \widetilde{\Psi}_\Lambda(t, t', x, y, \widetilde{\xi})| = C_0 < \infty.$$

We thus have, for any (t, t', x, y, ξ) such that $\widetilde{\xi} \in \widetilde{\mathcal{C}}_\Lambda(t, t', x, y)$,

$$\frac{|x - y|}{|t - t'|} \leq \left| \frac{x - y}{|t - t'|} + \partial_{\widetilde{\xi}} \widetilde{\Psi}_\Lambda(t, t', x, y, \widetilde{\xi}) \right| + C_0 \leq C_0 + 1.$$

In particular, we have $|x - y| \leq (C_0 + 1)|I_\Lambda|$. Therefore, the estimate (9.63) and Taylor's inequality give

$$\widetilde{\Psi}_\Lambda(t, t', x, y, \widetilde{\xi}) = \partial_t \Phi_\Lambda(0, y, \widetilde{\xi}) + R_\Lambda(t, t', x, y, \widetilde{\xi}) \quad \text{with}$$

$$\sup_{\substack{\Lambda \geq \Lambda_0 \\ (t, t', x, y) \in I_\Lambda^2 \times \mathbb{R}^{2d}}} \sup_{\widetilde{\xi} \in \widetilde{\mathcal{C}}_\Lambda(t, t', x, y)} |\partial_\xi^\ell R_\Lambda(t, t', x, y)| \leq C_\ell \varepsilon_0 \quad \text{for all } \ell \in \mathbb{N}. \quad (9.64)$$

Using $\widetilde{(HJ_\Lambda)}$, we have (dropping the tilde in what follows)

$$\partial_t \Phi_\Lambda(0, y, \xi) = (G_\Lambda(0, y)(\xi, \xi))^{\frac{1}{2}}.$$

For any positive quadratic form q, we have

$$D_\xi^2 (q(\xi, \xi))^{\frac{1}{2}} (h_1, h_2) = \frac{1}{(q(\xi, \xi))^{\frac{1}{2}}} \left((h_1 | h_2)_q - \frac{(h_1 | \xi)_q (h_2 | \xi)_q}{q(\xi, \xi)} \right),$$

where $(\cdot | \cdot)_q$ stands for the bilinear form associated with q.

This implies that $D_\xi^2 (q(\xi, \xi))^{\frac{1}{2}}$ restricted to the orthogonal set V of ξ (in the sense of q) is a positive quadratic form. More precisely,

$$D_\xi^2 (q(\xi, \xi))^{\frac{1}{2}}_{|V \times V} = \frac{1}{(q(\xi, \xi))^{\frac{1}{2}}} q_{|V \times V}.$$

As there exists a constant c_0 such that, on the orthogonal set V_y of ξ for $G_\Lambda(0, y)$,

$$\inf_{\Lambda \geq \Lambda_0} \inf_{y \in \mathbb{R}^d} (\text{Id} + G_\Lambda(0, y))(\xi, \xi) \geq c_0 |\xi|^2,$$

we have, for any $h \in V_y$,

$$D_\xi^2 \partial_t \Phi_\Lambda(0, y, \xi)(h, h) \geq c_0 |h|^2.$$

If we take ε_0 to be sufficiently small, the estimate (9.64) thus implies that

$$D^2 \widetilde{\Psi}_\Lambda(0, y, \xi)(h, h) \geq \frac{c_0}{2} |h|^2 \quad \text{for any } h \in V_y.$$

Thus, the inequality (9.56) is proved: It is only a matter of reproducing the end of the proof of Proposition 8.15.

In order to conclude, we denote by $A_\Lambda(t')$ the operator defined by

$$\Box_\Lambda (A_\Lambda(t') v_\Lambda) = 0,$$
$$(A_\Lambda(t') v_\Lambda, \partial_t A_\Lambda(t') v_\Lambda)_{|t = t'} = (0, v_\Lambda).$$

The solution of

$$\Box_\Lambda v_\Lambda = f_\Lambda \quad \text{and} \quad \nabla v_{\Lambda|t=0} = 0$$

is of the form

$$v_\Lambda(t,y) = \int_0^t (A_\Lambda(t')f_\Lambda(t'))(t,y)\,dt'.$$

Therefore, for all $t \in I_\Lambda$, we have

$$\|v_\Lambda(t,\cdot)\|_{L^\infty} \leq \int_0^t \|(A_\Lambda(t')f_\Lambda(t'))(t,\cdot)\|_{L^\infty}\,dt'$$

$$\leq \int_{I_\Lambda} \|(A_\Lambda(t')f_\Lambda(t'))(t,\cdot)\|_{L^\infty}\,dt'.$$

Taking the L^2 norm on I_Λ and using (9.55) and (9.57), we end up with

$$\|v_\Lambda\|_{L^2_{I_\Lambda}(L^\infty)} \leq \int_{I_\Lambda} \|A_\Lambda(t')f_\Lambda(t')\|_{L^2_{I_\Lambda}(L^\infty)}\,dt'$$

$$\leq \int_{I_\Lambda} \|f_\Lambda(t')\|_{L^2}\,dt'.$$

The cases $d = 2$ and $d = 3$ can be treated along the same lines. The details are left to the reader. This completes the proof of Theorem 9.16. ☐

9.6 References and Remarks

Motivated by the study of the Einstein equations in relativity theory, there are a number of works dedicated to the local well-posedness issue for the quasilinear wave equation. The first papers on this equation were mainly devoted to the study of the lifespan for solutions generated by smooth, compactly supported, small initial data (see, in particular, the pioneering work by S. Klainerman in [182], the book by L. Hörmander [168], and the more recent papers by S. Alinhac in [7–10] and by Klainerman and Rodnianski [185, 188]).

In this chapter we focused on the question of the lowest regularity for which local well-posedness holds true. One of the motivations for this study is that in the low-dimensional case, we may hope to achieve the level of regularity corresponding to a conserved quantity (such as, e.g., the energy) and thus get global existence.

The results of the first section belong to the mathematical folklore (at least in the framework of Sobolev spaces). The main novelty here is that we strive to find scaling invariant estimates. The other sections rely on ideas introduced by the first two authors in [18, 19], where Theorem 9.5 is proved. The lowest index for which local well-posedness holds true in dimension $d \geq 4$ was improved to $d/2 + 1/2 + 1/6$ by D. Tataru in [281] (compared with $d/2 + 1/2 + 1/4$ in this chapter). We emphasize that in the simpler case of the semilinear wave equation (i.e., $G \equiv 0$) with quadratic nonlinearity Q, the best index of regularity for which local well-posedness holds true is $d/2 + 1/2$ if $d \geq 3$ and $7/4$ if $d = 2$ (see the work by G. Ponce and T. Sideris in [253]). Actually, even in the semilinear case there is no hope of going below $d/2 +$

1/2 for general quadratic nonlinearities Q (see the counterexample by H. Lindblad in [209]).

We should also mention that combining the method presented in this chapter with the refined Strichartz estimate proved by S. Klainerman and D. Tataru in [194] is relevant to the study of the quasilinear wave equation in the case where the metric $G(u)$ satisfies the equation $\Delta G(u) = Q(\nabla u, \nabla u)$ for some quadratic form Q. In this framework, it was shown in [20] by the first two authors that the level of regularity of γ for which the equation may be solved locally falls to $d/2 - 1 + 1/6$.

Proving Strichartz estimates for the wave equation with variable coefficients is one of the main ingredients of Theorem 9.5. This question has been addressed by L. Kapitanski [175] in the smooth case and by H. Smith in [268] for coefficients in $C^{1,1}$. Alternatively, Strichartz estimates for the wave equation may be obtained by the method of commuting vector fields which was introduced by S. Klainerman in [182] for proving global existence results for small smooth initial data. This method was also used in [183, 184] by S. Klainerman for the smooth variable coefficients wave equation. This idea is the basis of the major work by S. Klainerman and I. Rodnianski, who proved in [189–192] that the Einstein equations are well posed for initial data in the Sobolev space $H^{2+\varepsilon}(\mathbb{R}^3)$ for some arbitrarily small ε. Other methods have proven to be efficient for solving (QW): For an approach based on the Fourier–Bros–Iagolnitzer transform, see the work [280] by D. Tataru; for an approach based on wave packets, see the work [269] by H. Smith and D. Tataru.

The idea of performing a refined paralinearization to study (QW) is borrowed from the work by G. Lebeau in [203]. Finally, we mention that cutting the time interval into small intervals, the lengths of which depend on the frequency, has been used recently by N. Burq, P. Gérard, and N. Tzvetkov to prove Strichartz estimates in the context of the Schrödinger equation on compact manifolds (see [51]).

The use of high-frequency approximation of solutions of hyperbolic partial differential equations has a long history, beginning with the construction of the so-called *Lax parametrix* (see [202]). The reader may refer to the book by M. Taylor [284] for an exposition on this method in the (more general) framework of pseudodifferential operators.

10

The Compressible Navier–Stokes System

In this chapter, we show the benefits that may be gained from Fourier analysis methods when investigating fluid mechanics models more complex than those which have been hitherto considered in this book. We will present the so-called isentropic compressible Navier–Stokes system, which contains more physics than the incompressible models we have seen thus far but is still not too cumbersome.

The content of this chapter is twofold. First, we present a few results concerning local or global solvability in the spirit of the theorem of Fujita and Kato which was presented in Chapter 5. It turns out that scaling invariance still allows the appropriate functional framework to be found. Next, we show that when the *Mach number* (i.e., the ratio of the sound speed to the characteristic speed of the velocity) is sufficiently small, the solution of the compressible model tends to that of the incompressible Navier–Stokes equations. In all the results that we obtain, the use of Besov spaces and Littlewood–Paley decomposition turns out to be fundamental.

The chapter unfolds as follows. The first section is devoted to a short presentation of the model of viscous compressible flows that we shall consider. In the next section we prove a local well-posedness statement for data with critical regularity in the case where the density is a small perturbation of a positive constant. In Section 10.3, we consider slightly more regular data in order to remove the small perturbation assumption. Section 10.4 is dedicated to the proof of global well-posedness for small perturbations of an initial stable state $(\underline{\rho}, 0)$ with constant density. In the final section, we study the extent to which the incompressible Navier–Stokes equations are a good approximation for slightly compressible fluids.

10.1 About the Model

In this introductory section we briefly explain how the system of equations for the flow of a compressible fluid may be derived from basic physics. More

H. Bahouri et al., *Fourier Analysis and Nonlinear Partial Differential Equations*, Grundlehren der mathematischen Wissenschaften 343, DOI 10.1007/978-3-642-16830-7_10, © Springer-Verlag Berlin Heidelberg 2011

details may be found in physics books such as, for example, [29], or in the
introduction of [213].

10.1.1 General Overview

We assume that the fluid fills the whole space (i.e., boundary effects are neglected), and that it may be described at every material point x in $\mathbb{R}^d$ and
time $t \in \mathbb{R}$ by:

- its *velocity field* $u \stackrel{\mathrm{def}}{=} u(t, x)$,
- its *density* $\rho \stackrel{\mathrm{def}}{=} \rho(t, x)$,
- its *internal energy* $e \stackrel{\mathrm{def}}{=} e(t, x)$,
- its *entropy by unit mass* $s \stackrel{\mathrm{def}}{=} s(t, x)$.

To any subdomain Ω of $\mathbb{R}^d$, we may associate:

- the *mass* $M(\Omega) \stackrel{\mathrm{def}}{=} \displaystyle\int_\Omega \rho \, dx$,
- the *momentum* $P(\Omega) \stackrel{\mathrm{def}}{=} \displaystyle\int_\Omega \rho u \, dx$,
- the *energy* $E(\Omega) \stackrel{\mathrm{def}}{=} \displaystyle\int_\Omega \left(\frac{1}{2}\rho|u|^2 + \rho e\right) dx$,
- the *entropy* $S(\Omega) \stackrel{\mathrm{def}}{=} \displaystyle\int_\Omega \rho s \, dx$.

Let ψ_t be the flow of u (see Chapter 3) and $\Omega_t \stackrel{\mathrm{def}}{=} \psi_t(\Omega)$. Assuming that there
is neither production nor loss of mass, the mass conservation translates as

$$\frac{d}{dt} M(\Omega_t) = \frac{d}{dt} \int_{\Omega_t} \rho \, dx = 0. \tag{10.1}$$

For the momentum, we have

$$\frac{d}{dt} P(\Omega_t) = \frac{d}{dt} \int_{\Omega_t} \rho u \, dx = \int_{\Omega_t} \rho f \, dx + \int_{\partial\Omega_t} (\sigma \cdot n) \, d\Sigma, \tag{10.2}$$

where the first term on the right-hand side represents external body forces
with density f (such as, e.g., gravity), and the second term represents surface
forces. In the absence of mass couples, the *angular momentum*

$$\int_{\Omega_t} x \wedge (\rho u)(t, x) \, dx$$

is also conserved. This can be shown to entail that σ is a symmetric tensor
(see, e.g., [29]).

Next, the energy conservation can be written as

$$\frac{d}{dt}E(\Omega_t) = \frac{d}{dt}\int_{\Omega_t} \rho\left(e + \frac{|u|^2}{2}\right) dx,$$

$$= \int_{\Omega_t} \rho f \cdot u \, dx + \int_{\partial\Omega_t} (\sigma \cdot n) \cdot n \, d\Sigma - \int_{\partial\Omega_t} q \cdot n \, d\Sigma, \quad (10.3)$$

where the last integral represents the amount of heat lost across the boundary, and q is the so-called *heat flux vector*.

Finally, introducing the temperature T, the entropy balance can be written

$$\frac{d}{dt}S(\Omega_t) = \frac{d}{dt}\int_{\Omega_t} \rho s \, dx \geq -\int_{\partial\Omega_t} \left(\frac{q \cdot n}{T}\right) d\Sigma: \quad (10.4)$$

We assume from now on that the fluid is Newtonian, that is:

– The tensor σ is a linear function of Du, invariant under rigid transforms.
– The fluid is isotropic [in other words, the physical quantities depend only on (t, x)].

As a consequence, it may be shown (see, e.g., [29]) that σ can be written as

$$\sigma = \tau - p \, \mathrm{Id} \quad \text{with} \quad \tau \overset{\text{def}}{=} \lambda \, \mathrm{div} \, u \, \mathrm{Id} + 2\mu D(u).$$

The scalar function $p = p(t, x)$ is called the *pressure* and $\tau = \tau(t, x)$ is called the *viscous stress tensor*. The real numbers λ and μ are the *viscosity coefficients* and $D(u) \overset{\text{def}}{=} \frac{1}{2}(Du + {}^tDu)$ is the *deformation tensor*.

From the global conservation laws (10.1)–(10.4), we may obtain a system of partial differential equations involving ρ, u, e, and s. This is a consequence of the following classical (formal) lemma.

Lemma 10.1. *Let Ω be an open subdomain of D, ψ the flow of u, and $\Omega_t \overset{\text{def}}{=} \psi_t(\Omega)$. Let b be a scalar function. We then have*

$$\frac{d}{dt}\int_{\Omega_t} b \, dx = \int_{\Omega_t} (\partial_t b + \mathrm{div}(bu)) \, dx = \int_{\Omega_t} \partial_t b \, dx + \int_{\partial\Omega_t} (b \, u \cdot n) \, d\Sigma.$$

If we assume for simplicity that the *Fourier law* $q = -k\nabla T$ is satisfied and that the coefficients k, λ, and μ are constant real numbers, then Lemma 10.1 implies that

$$\partial_t \rho + \mathrm{div}(\rho u) = 0,$$

$$\partial_t(\rho u) + \mathrm{div}(\rho u \otimes u) - \mu \Delta u - (\lambda + \mu)\nabla \, \mathrm{div} \, u + \nabla p = \rho f,$$

$$\partial_t\left(\rho\left(e + \frac{|u|^2}{2}\right)\right) + \mathrm{div}\left(\rho\left(e + \frac{|u|^2}{2}\right)u\right) + \mathrm{div} \, pu - k\Delta T$$
$$= \rho f \cdot u + \mathrm{div}(\tau \cdot u) - \mathrm{div} \, q,$$

$$\partial_t(\rho s) + \mathrm{div}(\rho s u) \geq k \, \mathrm{div}\left(\frac{\nabla T}{T}\right).$$

If we assume, in addition, that the so-called *Gibbs relation*

$$T ds = de + p\, d\Big(\frac{1}{\rho}\Big)$$

is satisfied, then combining the mass, momentum, and energy equations, we get

$$\partial_t(\rho s) + \mathrm{div}(\rho s u) = k \,\mathrm{div}\Big(\frac{\nabla T}{T}\Big) + \frac{\tau : D(u)}{T} + k\frac{|\nabla T|^2}{T^2}. \tag{10.5}$$

Hence, according to the entropy inequality, we must have

$$\tau : D(u) + k\frac{|\nabla T|^2}{T} \geq 0.$$

As, obviously,

$$\tau : D(u) = \lambda(\mathrm{div}\, u)^2 + 2\mu\mathrm{Tr}\,(D(u))^2$$

and, owing to the Cauchy–Schwarz inequality,

$$\big(\mathrm{Tr}\, D(u)\big)^2 \leq d\,\mathrm{Tr}\,(D(u))^2,$$

this yields the following constraints on λ, μ, and k:

$$k \geq 0, \quad \mu \geq 0, \quad \text{and} \quad 2\mu + d\lambda \geq 0.$$

We give a few examples:

- Monoatomic gases in dimension $d = 3$ satisfy $2\mu + 3\lambda = 0$.
- Inviscid fluids are such that $\mu = \lambda = 0$.
- Nonconducting fluids satisfy $k = 0$.

In order to solve the system, another two *state equations* involving p, ρ, e, s, and T are needed. We can assume that $p = P(\rho, T)$ and $e = \varepsilon(\rho, T)$ for some given functions P and ε depending on the nature of the fluid.

10.1.2 The Barotropic Navier–Stokes Equations

In what follows, we focus on a simplified model for compressible fluids, the so-called *barotropic Navier–Stokes equations*,

$$\begin{cases} \partial_t \rho + \mathrm{div}(\rho u) = 0 \\ \partial_t(\rho u) + \mathrm{div}(\rho u \otimes u) - \mu\Delta u - (\lambda + \mu)\nabla\,\mathrm{div}\, u + \nabla p = \rho f, \end{cases}$$

where it is assumed that $p \overset{\mathrm{def}}{=} P(\rho)$ for some given smooth function P.

The above system may be derived from the general model under the assumptions that s is a constant and $k = 0$. Note that in the viscous case (which we will consider in the next sections), the assumption of constant entropy is somewhat inconsistent with (10.5) for the term $\tau : D(u)$ may be positive. From a mathematical viewpoint, however, the barotropic (or isentropic) model retains many features of the full model.

In this chapter, we restrict our study to fluids with positive density tending to some positive constant at infinity (say 1, to simplify the notation). Letting $a = \rho - 1$, the barotropic system for sufficiently smooth solutions reduces to

$$(NSC) \qquad \begin{cases} \partial_t a + u \cdot \nabla a = -(1+a) \operatorname{div} u \\ \partial_t u - (1+a)^{-1} \mathcal{A} u + u \cdot \nabla u + \nabla g = f, \end{cases}$$

where $\mathcal{A} \overset{\text{def}}{=} \mu \Delta + (\lambda + \mu) \nabla \operatorname{div}$ is the viscosity operator, and $g \overset{\text{def}}{=} G(a)$ stands for the *chemical potential* expressed in terms of a. The function G is assumed to be conveniently smooth and, with no loss of generality, to vanish at 0.

The modified viscosity coefficients

$$\nu \overset{\text{def}}{=} \lambda + 2\mu, \quad \underline{\nu} \overset{\text{def}}{=} \min(\mu, \lambda + 2\mu), \quad \text{and} \quad \overline{\nu} \overset{\text{def}}{=} \mu + |\lambda + \mu|$$

will also play an important role.

Throughout this chapter, we consider only *viscous* fluids, those for which $\mu > 0$ and $\nu > 0$. This implies that the coefficients $\underline{\nu}$ and $\overline{\nu}$ are also positive.

10.2 Local Theory for Data with Critical Regularity

In Chapter 5 we proved global well-posedness for the incompressible Navier–Stokes equations with small initial data and local well-posedness for large initial data (see Theorem 5.6 page 209, Theorem 5.27 page 222, Theorem 5.35 page 229, and Theorem 5.40 page 234). In this section and the two which follow, we seek to establish similar results for compressible flows.

10.2.1 Scaling Invariance and Statement of the Main Result

As in Chapter 5, scaling invariance is the main thread for finding an appropriate functional framework. More precisely, we note that for all $\ell > 0$, (NSC) is invariant with respect to the rescaling $(a, u) \mapsto (a_\ell, u_\ell)$ defined by

$$a_\ell(t, x) = a(\ell^2 t, \ell x) \quad \text{and} \quad u_\ell(t, x) = \ell u(\ell^2 t, \ell x), \tag{10.6}$$

provided that the chemical potential g has been changed to $\ell^2 g$.

Hence, it may be appropriate to solve the system (NSC) in a function space whose norm is invariant for all ℓ (up to an irrelevant constant) with respect to the transform (10.6). Therefore, if we consider homogeneous Besov spaces, the data (a_0, u_0) have to be taken in $\dot{B}_{p_1, r_1}^{\frac{d}{p_1}} \times \left(\dot{B}_{p_2, r_2}^{\frac{d}{p_2} - 1} \right)^d$ for some $p_1, p_2, r_1, r_2 \geq 1$. In order to guarantee that the density is positive, however, an L^∞ control on a is needed. Hence, we have to assume that $r_1 = 1$ so that

$\dot{B}_{p_1,r_1}^{\frac{d}{p_1}} \hookrightarrow L^\infty$ (see Proposition 2.39 page 79). Next, owing to the smoothing properties of the heat flow, we expect ∇u to be in $\widetilde{L}_T^1(\dot{B}_{p_2,r_2}^{\frac{d}{p_2}})$ [see the inequality (3.39) page 157]. Now, as a satisfies a transport equation, preserving its Besov regularity requires that $\nabla u \in L_T^1(L^\infty)$ (see Theorem 3.14 page 133). Hence, we must also take $r_2 = 1$. Finally, owing to the coupling between the equations for a and for u, it is also natural to assume that $p_1 = p_2$.

For simplicity, we shall only consider the case $p_1 = p_2 = 2$. We thus wish to solve (NSC) in the function space

$$E_T \overset{\text{def}}{=} \left\{ (a, u) \in \widetilde{C}_T(\dot{B}_{2,1}^{\frac{d}{2}}) \times \left(\widetilde{C}_T(\dot{B}_{2,1}^{\frac{d}{2}-1}) \cap L_T^1(\dot{B}_{2,1}^{\frac{d}{2}+1}) \right)^d \right\},$$

where we agree that from now on, $\widetilde{C}_T(\dot{B}_{2,1}^s) \overset{\text{def}}{=} C([0,T]; \dot{B}_{2,1}^s) \cap \widetilde{L}_T^\infty(\dot{B}_{2,1}^s)$.

For the time being, we focus on small perturbations of a constant density state. For such data, our main local well-posedness result reads as follows.

Theorem 10.2. *If $d \geq 2$, then there exists a positive constant η such that for all u_0 in $\dot{B}_{2,1}^{\frac{d}{2}-1}$, f in $L_{loc}^1(\mathbb{R}^+; \dot{B}_{2,1}^{\frac{d}{2}-1})$, and $a_0 \in \dot{B}_{2,1}^{\frac{d}{2}}$ with*

$$\|a_0\|_{\dot{B}_{2,1}^{\frac{d}{2}}} \leq \eta\underline{\nu}/\bar{\nu}, \tag{10.7}$$

there exists a positive time T such that (NSC) has a solution (a, u) on $[0,T] \times \mathbb{R}^d$ which belongs to E_T.

Moreover, uniqueness holds true in E_T whenever

$$\|a\|_{L_T^\infty(\dot{B}_{2,1}^{\frac{d}{2}})} \leq \eta\underline{\nu}/\bar{\nu}, \quad \text{if } d \geq 3, \quad \text{and} \quad \|a\|_{\widetilde{L}_T^\infty(\dot{B}_{2,1}^1)} \leq \eta\underline{\nu}/\bar{\nu}, \quad \text{if } d = 2. \tag{10.8}$$

The rest of this section is devoted to proving Theorem 10.2. Before explaining how we shall proceed, we should point out that, in contrast with the incompressible Navier–Stokes equations, owing to the hyperbolic nature of the mass conservation equation, the system (NSC) *cannot* be solved by means of the Picard fixed point theorem. In fact, although a priori estimates for (NSC) may be proven directly in the space E_T, the term $u \cdot \nabla a$ in the mass equation induces a loss of one derivative in the stability estimates. For that reason, we shall instead use a Friedrichs method similar to that of Chapter 4 for hyperbolic quasilinear systems. Indeed, if T is taken to be sufficiently small, then it turns out to be possible to prove uniform estimates in E_T for the corresponding sequence $(a^n, u^n)_{n \in \mathbb{N}}$ of approximate solutions.

At this point, it would be natural to prove that $(a^n, u^n)_{n \in \mathbb{N}}$ is a Cauchy sequence for a weaker norm than that of E_T. This method would work in dimension $d \geq 3$, but is bound to fail in dimension 2, owing to the low regularity of the functions we work with. Therefore, we shall instead use compactness arguments (based on compact embeddings in Besov spaces and Ascoli's theorem) to show the convergence of $(a^n, u^n)_{n \in \mathbb{N}}$ up to extraction. This will enable us to prove the existence part of the above statement. Uniqueness will be obtained later by independent arguments (here, again, the case $d = 2$ turns out to be more tricky). In the last part of this section, we shall state a continuation criterion which will be useful for proving global existence in Section 10.4.

10.2.2 A Priori Estimates

For the time being, as we focus on local results, the gradient of the pressure may be considered as a lower order term. Therefore, the coupling between the mass and momentum equations is not so important, and the two equations may be treated (almost) separately. More precisely, in order to get a priori estimates for (NSC), it suffices to combine estimates for the transport equation (as stated in Theorem 3.14) and for the following heat system with convection terms:[1]

$$\partial_t u + v \cdot \nabla u + u \cdot \nabla w - \mathcal{A}u = f. \tag{10.9}$$

For this latter system, we have the following result.

Proposition 10.3. *Let* $s \in]-\frac{d}{2}, \frac{d}{2}]$. *There exists a universal constant* κ, *and a constant* C *depending only on* d *and* s, *such that*

$$\|u\|_{\widetilde{L}^\infty_t(\dot{B}^s_{2,1})} + \kappa\underline{\nu}\|u\|_{L^1_t(\dot{B}^{s+2}_{2,1})} \leq \left(\|u_0\|_{\dot{B}^s_{2,1}} + \|f\|_{L^1_t(\dot{B}^s_{2,1})}\right)$$

$$\times \exp\left(C\int_0^t \left(\|\nabla v\|_{\dot{B}^{\frac{d}{2}}_{2,1}} + \|\nabla w\|_{\dot{B}^{\frac{d}{2}}_{2,1}}\right) dt'\right).$$

If v *and* w *depend linearly on* u, *then the following inequality is true for all positive* s:

$$\|u\|_{\widetilde{L}^\infty_t(\dot{B}^s_{2,1})} + \kappa\underline{\nu}\|u\|_{L^1_t(\dot{B}^{s+2}_{2,1})} \leq \left(\|u_0\|_{\dot{B}^s_{2,1}} + \|f\|_{L^1_t(\dot{B}^s_{2,1})}\right) \exp\left(C\int_0^t \|\nabla u\|_{L^\infty} dt'\right).$$

Proof. As usual, the desired estimate will be obtained after localizing the equation (10.9) by means of the homogeneous Littlewood–Paley decomposition. More precisely, applying $\dot{\Delta}_j$ to (10.9) yields

$$\partial_t u_j + v \cdot \nabla u_j - \mathcal{A}u_j = f_j - \dot{\Delta}_j(u \cdot \nabla w) + R_j$$

with $u_j \stackrel{\text{def}}{=} \dot{\Delta}_j u$, $f_j \stackrel{\text{def}}{=} \dot{\Delta}_j f$, and $R_j \stackrel{\text{def}}{=} \sum_k [v^k, \dot{\Delta}_j]\partial_k u$.

---

[1] In fact, if we are only interested in proving well-posedness for (NSC), the convection terms may be included in the source term f. The main interest in keeping them on the left is that we get a more accurate estimate (note that the right-hand side in the Proposition 10.3 does not depend on the viscosity) which will be used to state a continuation criterion at the end of this section.

Taking the L^2 inner product of the above equation with u_j, we easily get

$$\frac{1}{2}\frac{d}{dt}\|u_j\|_{L^2}^2 - \frac{1}{2}\int |u_j|^2 \operatorname{div} v\, dx + \int \left(\mu|\nabla u_j|^2 + (\lambda+\mu)|\operatorname{div} u_j|^2\right) dx$$
$$\leq \|u_j\|_{L^2}\left(\|f_j\|_{L^2} + \|\dot{\Delta}_j(u\cdot\nabla w)\|_{L^2} + \|R_j\|_{L^2}\right).$$

Note that we have

$$\int \left(\mu|\nabla u_j|^2 + (\lambda+\mu)|\operatorname{div} u_j|^2\right) dx \geq \underline{\nu}\int |\nabla u_j|^2\, dx.$$

Indeed, the above inequality is obvious if $\lambda+\mu \geq 0$. Otherwise, it follows from the following chain of inequalities based on integration by parts:

$$\int (\operatorname{div} u_j)^2\, dx = \sum_{i,k}\int \partial_i u_j^i\, \partial_k u_j^k = \sum_{i,k}\int \partial_k u_j^i\, \partial_i u_j^k \leq \int |\nabla u_j|^2\, dx.$$

Hence, according to Bernstein's inequality, we get, for some universal constant κ,

$$\frac{1}{2}\frac{d}{dt}\|u_j\|_{L^2}^2 + 2\kappa\underline{\nu}2^{2j}\|u_j\|_{L^2}^2$$
$$\leq \|u_j\|_{L^2}\left(\|f_j\|_{L^2} + \|\dot{\Delta}_j(u\cdot\nabla w)\|_{L^2} + \|R_j\|_{L^2} + \frac{1}{2}\|\operatorname{div} v\|_{L^\infty}\|u_j\|_{L^2}\right).$$

According to Theorems 2.82 and 2.85 page 104, and to Lemma 2.100 page 112, we have the following estimates for $\dot{\Delta}_j(u\cdot\nabla w)$ and R_j:

$$\left\|\dot{\Delta}_j(u\cdot\nabla w)\right\|_{L^2} \leq Cc_j 2^{-js}\|\nabla w\|_{\dot{B}_{2,1}^{\frac{d}{2}}}\|u\|_{\dot{B}_{2,1}^s}, \quad \text{if } -d/2 < s \leq d/2, \quad (10.10)$$

$$\|R_j\|_{L^2} \leq Cc_j 2^{-js}\|\nabla v\|_{\dot{B}_{2,1}^{\frac{d}{2}}}\|u\|_{\dot{B}_{2,1}^s}, \quad \text{if } -d/2 < s \leq d/2+1, \quad (10.11)$$

where $(c_j)_{j\in\mathbb{Z}}$ denotes a positive sequence such that $\sum_j c_j = 1$.

Formally[2] dividing both sides of the inequality by $\|u_j\|_{L^2}$ and integrating over $[0,t]$ thus yields

$$\|u_j(t)\|_{L^2} + 2\kappa\underline{\nu}2^{2j}\int_0^t \|u_j\|_{L^2}\, dt' \leq \|u_j(0)\|_{L^2} + \int_0^t \|f_j\|_{L^2}\, dt'$$
$$+ C2^{-js}\int_0^t c_j\left(\|\nabla v\|_{\dot{B}_{2,1}^{\frac{d}{2}}} + \|\nabla w\|_{\dot{B}_{2,1}^{\frac{d}{2}}}\right)\|u\|_{\dot{B}_{2,1}^s}\, dt'.$$

Now, multiplying both sides by 2^{js} and summing over j, we end up with

$$\|u\|_{\widetilde{L}_t^\infty(\dot{B}_{2,1}^s)} + \kappa\underline{\nu}\|u\|_{L_t^1(\dot{B}_{2,1}^{s+2})} \leq \|u_0\|_{\dot{B}_{2,1}^s} + \|f\|_{L_t^1(\dot{B}_{2,1}^s)}$$
$$+ C\int_0^t \left(\|\nabla v\|_{\dot{B}_{2,1}^{\frac{d}{2}}} + \|\nabla w\|_{\dot{B}_{2,1}^{\frac{d}{2}}}\right)\|u\|_{\dot{B}_{2,1}^s}\, dt'$$

---

[2] Here, we may proceed exactly as in the proof of (4.31) page 194.

for some constant C depending only on d and s. Applying Gronwall's lemma then completes the proof.

If, in addition, we assume that v and w depend linearly on u, then we may take $w \equiv 0$ (with no loss of generality) and use the inequality (2.54) page 112 to bound R_j. We then easily get the last part of the statement.   □

Combining the above estimates with Theorem 3.14 page 133 will enable us to prove the following result for smooth solutions of (NSC).

Corollary 10.4. *Let (a, u) satisfy (NSC) on $[0, T] \times \mathbb{R}^d$. Suppose that $a \in \mathcal{C}^1([0, T]; \dot{B}_{2,1}^{\frac{d}{2}})$ and $u \in \mathcal{C}^1([0, T]; \dot{B}_{2,1}^{\frac{d}{2}-1} \cap \dot{B}_{2,1}^{\frac{d}{2}+1})^d$. Assume, in addition, that there exists a function $u_L \in \mathcal{C}^1([0, T]; \dot{B}_{2,1}^{\frac{d}{2}-1} \cap \dot{B}_{2,1}^{\frac{d}{2}+1})^d$ such that*

$$\partial_t u_L - \mathcal{A} u_L = f, \qquad u_{L|t=0} = u_0. \tag{10.12}$$

Let $\overline{U}(t) \overset{def}{=} \|\overline{u}\|_{\widetilde{L}_t^\infty(\dot{B}_{2,1}^{\frac{d}{2}-1})} + \underline{\nu}\|\overline{u}\|_{L_t^1(\dot{B}_{2,1}^{\frac{d}{2}+1})}$ with $\overline{u} \overset{def}{=} u - u_L$ and $U_0(t) \overset{def}{=} \|u_0\|_{\dot{B}_{2,1}^{\frac{d}{2}-1}} + \|f\|_{L_t^1(\dot{B}_{2,1}^{\frac{d}{2}-1})}$.

There exist two constants, η and G, depending only on d and G, respectively, such that if

$$\overline{\nu}\|a_0\|_{\dot{B}_{2,1}^{\frac{d}{2}}} \le \eta\underline{\nu} \quad and \quad C\left(\left(\overline{\nu} + \frac{\overline{\nu}U_0(T)}{\underline{\nu}}\right)\|u_L\|_{L_T^1(\dot{B}_{2,1}^{\frac{d}{2}+1})} + T\right) \le \underline{\nu}, \tag{10.13}$$

then we have, for all $t \in [0, T]$,

$$\|a\|_{\widetilde{L}_t^\infty(\dot{B}_{2,1}^{\frac{d}{2}})} \le 2\|a_0\|_{\dot{B}_{2,1}^{\frac{d}{2}}} + \eta\underline{\nu}/\overline{\nu}, \qquad \|a\|_{L^\infty([0,t]\times\mathbb{R}^d)} \le 3/4,$$
$$\overline{U}(t) \le C\left((U_0(t) + \underline{\nu}\eta)\|u_L\|_{L_t^1(\dot{B}_{2,1}^{\frac{d}{2}+1})} + \eta t\underline{\nu}/\overline{\nu}\right). \tag{10.14}$$

Proof. Defining $I(a) \overset{def}{=} a/(1 + a)$, we see that $(a, \overline{u})$ satisfies

$$\begin{cases} \partial_t a + u \cdot \nabla a + (1 + a)\operatorname{div} u = 0 \\ \partial_t \overline{u} + u \cdot \nabla \overline{u} + \overline{u} \cdot \nabla u_L - \mathcal{A}\overline{u} = -u_L \cdot \nabla u_L - I(a)\mathcal{A}u - \nabla(G(a)) \\ a_{|t=0} = a_0, \qquad \overline{u}_{|t=0} = 0. \end{cases}$$

Theorem 3.14 and Remark 3.16 page 134 enable us to bound a: We get

$$\|a\|_{\widetilde{L}_t^\infty(\dot{B}_{2,1}^{\frac{d}{2}})} \le \exp\left(C\int_0^t \|u\|_{\dot{B}_{2,1}^{\frac{d}{2}+1}} dt'\right)$$
$$\times \left(\|a_0\|_{\dot{B}_{2,1}^{\frac{d}{2}}} + \int_0^t \exp\left(-C\int_0^{t'} \|u\|_{\dot{B}_{2,1}^{\frac{d}{2}+1}} dt''\right)\|(1 + a)\operatorname{div} u\|_{\dot{B}_{2,1}^{\frac{d}{2}}} dt'\right).$$

Now, since $\dot{B}_{2,1}^{\frac{d}{2}}$ is an algebra (see Corollary 2.54 page 90), we may write

$$\|(1+a)\operatorname{div}u\|_{\dot{B}_{2,1}^{\frac{d}{2}}} \le C\big(1+\|a\|_{\dot{B}_{2,1}^{\frac{d}{2}}}\big)\|\operatorname{div}u\|_{\dot{B}_{2,1}^{\frac{d}{2}}}$$

so that, combining the previous inequality with Gronwall's lemma yields (for some larger constant C)

$$\|a\|_{\widetilde{L}_t^\infty(\dot{B}_{2,1}^{\frac{d}{2}})} \le \|a_0\|_{\dot{B}_{2,1}^{\frac{d}{2}}}\exp\Big(C\int_0^t\|u\|_{\dot{B}_{2,1}^{\frac{d}{2}+1}}\,dt'\Big) + \exp\Big(C\int_0^t\|u\|_{\dot{B}_{2,1}^{\frac{d}{2}+1}}\,dt'\Big) - 1.$$

Let C_0 be the norm of the embedding $\dot{B}_{2,1}^{\frac{d}{2}}\hookrightarrow L^\infty$. From the previous inequality, we see that if we assume that

$$\|a_0\|_{\dot{B}_{2,1}^{\frac{d}{2}}} \le \frac{1}{4C_0} \tag{10.15}$$

and if we have

$$\int_0^T\big(\|\overline{u}\|_{\dot{B}_{2,1}^{\frac{d}{2}+1}} + \|u_L\|_{\dot{B}_{2,1}^{\frac{d}{2}+1}}\big)\,dt \le \frac{\eta\nu}{2\overline{\nu}} \tag{10.16}$$

for some sufficiently small η, then $(10.14)_1$ is satisfied.

In order to bound $\overline{u}$, we may apply Proposition 10.3. We get

$$\overline{U}(t) \le C\exp\Big(C\int_0^t\big(\|u_L\|_{\dot{B}_{2,1}^{\frac{d}{2}+1}} + \|\overline{u}\|_{\dot{B}_{2,1}^{\frac{d}{2}+1}}\big)\,dt'\Big)$$
$$\times\int_0^t\Big(\|u_L\cdot\nabla u_L\|_{\dot{B}_{2,1}^{\frac{d}{2}-1}} + \|I(a)\mathcal{A}u\|_{\dot{B}_{2,1}^{\frac{d}{2}-1}} + \|\nabla(G(a))\|_{\dot{B}_{2,1}^{\frac{d}{2}-1}}\Big)\,dt'.$$

The right-hand side may be bounded by resorting to the product and composition estimates proved in Chapter 2. We get

$$\|u_L\cdot\nabla u_L\|_{\dot{B}_{2,1}^{\frac{d}{2}-1}} \le C\|u_L\|_{\dot{B}_{2,1}^{\frac{d}{2}-1}}\|\nabla u_L\|_{\dot{B}_{2,1}^{\frac{d}{2}}},$$
$$\|I(a)\mathcal{A}u\|_{\dot{B}_{2,1}^{\frac{d}{2}-1}} \le C\|a\|_{\dot{B}_{2,1}^{\frac{d}{2}}}\|\mathcal{A}u\|_{\dot{B}_{2,1}^{\frac{d}{2}-1}},$$
$$\|\nabla(G(a))\|_{\dot{B}_{2,1}^{\frac{d}{2}-1}} \le C\|a\|_{\dot{B}_{2,1}^{\frac{d}{2}}}.$$

Therefore, under the hypothesis (10.16) we have, by virtue of (10.14),

$$\overline{U}(t) \le C\Big(\|u_L\|_{L_t^\infty(\dot{B}_{2,1}^{\frac{d}{2}-1})}\|u_L\|_{L_t^1(\dot{B}_{2,1}^{\frac{d}{2}+1})}$$
$$+\overline{\nu}(\|a_0\|_{\dot{B}_{2,1}^{\frac{d}{2}}}+\eta\nu/\overline{\nu})\big(\nu^{-1}\overline{U}(t) + \|u_L\|_{L_t^1(\dot{B}_{2,1}^{\frac{d}{2}+1})}\big) + (\|a_0\|_{\dot{B}_{2,1}^{\frac{d}{2}}}+\eta\nu/\overline{\nu})t\Big).$$

Now, from Proposition 10.3 we have

$$\|u_L\|_{L_t^\infty(\dot{B}_{2,1}^{\frac{d}{2}-1})} \le U_0(t), \tag{10.17}$$

so if we assume

$$\overline{\nu}\|a_0\|_{\dot{B}^{\frac{d}{2}}_{2,1}} \leq \underline{\nu}\eta \qquad (10.18)$$

on a_0 with $\eta = \min(1/(2C), 1/(4C_0))$ [note that this implies the condition (10.15)], then we get

$$\overline{U}(t) \leq C\Big((U_0(t) + \underline{\nu}\eta)\|u_L\|_{L^1_t(\dot{B}^{\frac{d}{2}+1}_{2,1})} + \eta\underline{\nu}t/\overline{\nu}\Big).$$

Completing the proof of the corollary follows from a standard bootstrap argument: Let

$$I \overset{\text{def}}{=} \{t \in [0,T] \,/\, (10.16) \text{ is satisfied on } [0,t]\}.$$

By using the time continuity of the solution, we see that I is a nonempty closed subset of $[0,T]$. Now, if $T^* \in I$ and we assume that T has been chosen such that

$$4C\Big(\Big(1 + \frac{U_0(T)}{\underline{\nu}}\Big)\|u_L\|_{L^1_T(\dot{B}^{\frac{d}{2}+1}_{2,1})} + \frac{\eta T}{\overline{\nu}}\Big) \leq \eta\underline{\nu}/\overline{\nu},$$

then the inequality (10.16) is strict at time T^*. Again using the time continuity of the solution, we see that this entails that I is also an open subset of $[0,T]$. Hence, $T^* = T$. $\qquad\square$

Finally, we prove an a priori estimate for the nonstationary Stokes equation with convection terms,[3]

$$\partial_t u + \mathbb{P}(v \cdot \nabla u) + \mathbb{P}(u \cdot \nabla w) - \mu\Delta u = f. \qquad (10.19)$$

That estimate will be needed in the last section of this chapter, where we investigate the incompressible limit.

Proposition 10.5. *Let $s \in \,]-\frac{d}{2}, \frac{d}{2}]$ and let u be a solution of (10.19) with divergence-free data u_0 in $\dot{B}^s_{2,1}$ and $f \in L^1([0,T]; \dot{B}^s_{2,1})$. There exists a universal constant κ, and a constant C depending only on d and s, such that*

$$\|u\|_{\widetilde{L}^\infty_t(\dot{B}^s_{2,1})} + \kappa\mu\|u\|_{L^1_t(\dot{B}^{s+2}_{2,1})} \leq \big(\|u_0\|_{\dot{B}^s_{2,1}} + \|f\|_{L^1_t(\dot{B}^s_{2,1})}\big)$$

$$\times \exp\Big(C\int_0^t \big(\|\nabla v\|_{\dot{B}^{\frac{d}{2}}_{2,1}} + \|\nabla w\|_{\dot{B}^{\frac{d}{2}}_{2,1}}\big)\,dt'\Big).$$

If v and w are multiples of u, then for all positive s, the argument of the exponential term may be replaced with $C\int_0^t \|\nabla u\|_{L^\infty}\,dt'$.

---

[3] Recall that $\mathbb{P}$ stands for the Leray projector over divergence-free vector fields defined in (5.4) page 206.

Proof. The proof works in almost the same way as that of Proposition 10.3. The evolution equation for $u_j \stackrel{\text{def}}{=} \dot{\Delta}_j u$ now reads

$$\partial_t u_j + \mathbb{P}(v \cdot \nabla u_j) - \mu \Delta u_j = f_j - \dot{\Delta}_j \mathbb{P}(u \cdot \nabla w) + \mathbb{P}\, R_j.$$

Taking the L^2 inner product with u_j is the next step. Since $\operatorname{div} u_j = 0$ and $\mathbb{P}^2 = \mathbb{P}$, the operator $\mathbb{P}$ may be "omitted" in the computations so that by proceeding along the lines of the proof of Proposition 10.3, we get the desired inequality. $\square$

10.2.3 Existence of a Local Solution

In order to prove the existence part of Theorem 10.2, we proceed as follows:

- First, we approximate (NSC) by a sequence of ordinary differential equations, by means of the Friedrichs method.
- Second, we prove uniform a priori estimates in E_T (for suitably small T) for those solutions.
- Third, we establish further boundedness properties involving Hölder regularity with respect to time for the approximate solutions.
- Fourth, we use the previous steps to show compactness, hence convergence up to extraction.
- Finally, we show that the limit is indeed a solution of (NSC), and that it belongs to E_T.

First Step: Friedrichs Approximation

Let $\dot{L}_n^2$ be the set of L^2 functions spectrally supported in the annulus $\mathcal{C}_n \stackrel{\text{def}}{=} \left\{ \xi \in \mathbb{R}^d \,/\, n^{-1} \le |\xi| \le n \right\}$ and let Ω_n be the set of functions (a, u) of $(\dot{L}_n^2)^{d+1}$ such that $\inf_{x \in \mathbb{R}^d} a > -1$. The linear space $\dot{L}_n^2$ is endowed with the standard L^2 topology. Note that, owing to the Bernstein inequality, the L^∞ topology on $\dot{L}_n^2$ is weaker than the L^2 topology, so Ω_n is an open set of $(\dot{L}_n^2)^{d+1}$.

Let $\dot{\mathbb{E}}_n : L^2 \longrightarrow \dot{L}_n^2$ be the *Friedrichs projector*, defined by

$$\mathcal{F} \dot{\mathbb{E}}_n U(\xi) \stackrel{\text{def}}{=} \mathbf{1}_{\mathcal{C}_n}(\xi) \mathcal{F} U(\xi) \quad \text{for all } \xi \in \mathbb{R}^d.$$

We aim to solve the system of ordinary differential equations

$$(NSC_n) \qquad \frac{d}{dt}\begin{pmatrix} a \\ u \end{pmatrix} = \begin{pmatrix} F_n(a, \overline{u}) \\ G_n(a, \overline{u}) \end{pmatrix}, \qquad \begin{pmatrix} a \\ u \end{pmatrix}_{|t=0} = \begin{pmatrix} \dot{\mathbb{E}}_n \, a_0 \\ 0 \end{pmatrix}$$

in $(\dot{L}_n^2)^{d+1}$ with

$$F_n(a, \overline{u}) \stackrel{\text{def}}{=} -\dot{\mathbb{E}}_n \operatorname{div}\big((1 + a)u\big),$$
$$G_n(a, \overline{u}) \stackrel{\text{def}}{=} \dot{\mathbb{E}}_n \, \mathcal{A}\overline{u} - \dot{\mathbb{E}}_n\big(u \cdot \nabla u\big) - \dot{\mathbb{E}}_n\big(I(a)\,\mathcal{A}u\big) - \dot{\mathbb{E}}_n \nabla\big(G(a)\big).$$

Above, we agree that $u = \bar{u} + u_L$, where u_L is the solution of (10.12).[4]

Note that if $\|a_0\|_{\dot{B}^{\frac{d}{2}}_{2,1}}$ is small, then $1 + \dot{\mathbb{E}}_n\, a_0 > 0$ for large n. Hence, the initial data of (NSC_n) are in Ω_n. Therefore, to solve the system it suffices to check that the map

$$(a, \bar{u}) \mapsto \left(F_n(a, \bar{u}), G_n(a, \bar{u}) \right)$$

is in $\mathcal{C}(\mathbb{R}^+ \times \Omega_n; (\dot{L}^2_n)^{d+1})$ and is locally Lipschitz with respect to the variable $(a, \bar{u})$. The proof of that is left to the reader. The main two points are that $u^n_L \overset{\text{def}}{=} \dot{\mathbb{E}}_n\, u_L$ is in $\mathcal{C}^\infty(\mathbb{R}^+; H^\infty)$, so the time dependency is smooth, and that, owing to the low-frequency cut-off $\dot{\mathbb{E}}_n$, all the Sobolev norms are equivalent. Hence, if we restrict ourselves to nonnegative times, then the above system has a unique maximal solution $(a^n, \bar{u}^n)$ in the space $\mathcal{C}^1([0, T^*_n[; \Omega_n)$.

Second Step: Uniform Estimates

We claim that T^*_n may be bounded from below by the supremum T of all the times satisfying (10.13), and that $(a^n, u^n)_{n \geq 1}$ is bounded in E_T.

The key point is that since $\dot{\mathbb{E}}_n$ is an L^2 orthogonal projector, it has no effect on the energy estimates which were used in the proof of Corollary 10.4. Hence, the corollary applies to our approximate solution (a^n, u^n). Note, also, that the dependence on n in the condition (10.13) and in the inequalities (10.14) may be omitted. Now, as $(a^n, \bar{u}^n)$ is spectrally supported in $\mathcal{C}_n$, the inequalities (10.14) ensure that it is bounded in $L^\infty([0, T]; \dot{L}^2_n)$. So, finally, the standard continuation criterion for ordinary differential equations implies that T^*_n is greater than any time T satisfying (10.13) and that we have, for all $n \geq 1$,

$$\|a^n\|_{\widetilde{L}^\infty_T(\dot{B}^{\frac{d}{2}}_{2,1})} \leq 3\eta\nu/\bar{\nu}, \quad \|a^n\|_{L^\infty([0,T]\times\mathbb{R}^d)} \leq 3/4,$$

$$\|\bar{u}^n\|_{\widetilde{L}^\infty_T(\dot{B}^{\frac{d}{2}-1}_{2,1})} + \nu\|\bar{u}^n\|_{L^1_T(\dot{B}^{\frac{d}{2}+1}_{2,1})} \leq C\left(\eta\underline{\nu}T/\bar{\nu} \right. \tag{10.20}$$

$$\left. + \left(\|u_0\|_{\dot{B}^{\frac{d}{2}-1}_{2,1}} + \|f\|_{L^1_T(\dot{B}^{\frac{d}{2}-1}_{2,1})} + \nu\eta \right)\|u_L\|_{L^1_T(\dot{B}^{\frac{d}{2}+1}_{2,1})} \right).$$

Of course, because $u^n_L = \dot{\mathbb{E}}_n\, u_L$, the sequence $(u^n_L)_{n\in\mathbb{N}}$ is uniformly bounded in $\widetilde{C}_T(\dot{B}^{\frac{d}{2}-1}_{2,1}) \cap L^1_T(\dot{B}^{\frac{d}{2}+1}_{2,1})$. We further note that, using interpolation, the inequality (10.20) implies that $(\bar{u}^n)_{n\in\mathbb{N}}$ is bounded in $\widetilde{L}^r_T(\dot{B}^{\frac{d}{2}-1+\frac{2}{r}}_{2,1})$ for all $r \in [1, \infty]$, a property which will be used several times in the next steps. The same holds for u^n_L.

---

[4] Proving the existence of u_L involves the same arguments as for the ordinary heat equation.

Third Step: Time Derivatives

The following lemma will supply the compactness property needed to pass to the limit in (NSC_n).

Lemma 10.6. *Let* $\bar{a}^n \overset{def}{=} a^n - \dot{\mathbb{E}}_n a_0$. *Then, the sequence* $(\bar{a}^n)_{n\geq 1}$ *is bounded in*

$$\mathcal{C}([0,T]; \dot{B}_{2,1}^{\frac{d}{2}}) \cap \mathcal{C}^{\frac{1}{2}}([0,T]; \dot{B}_{2,1}^{\frac{d}{2}-1}),$$

and the sequence $(\bar{u}^n)_{n\geq 1}$ *is bounded in*

$$\mathcal{C}([0,T]; \dot{B}_{2,1}^{\frac{d}{2}-1}) \cap \mathcal{C}^{\frac{1}{4}}([0,T]; \dot{B}_{2,1}^{\frac{d}{2}-1} + \dot{B}_{2,1}^{\frac{d}{2}-\frac{3}{2}}).$$

Proof. The result for $(\bar{a}^n)_{n\geq 1}$ follows from the fact that $\bar{a}^n(0) = 0$ and

$$\partial_t \bar{a}^n = -\dot{\mathbb{E}}_n \operatorname{div}\left(u^n(1+a^n)\right).$$

Indeed, as $\dot{B}_{2,1}^{\frac{d}{2}}$ is an algebra, and as u^n and a^n are bounded in $L_T^2(\dot{B}_{2,1}^{\frac{d}{2}})$ and $L_T^\infty(\dot{B}_{2,1}^{\frac{d}{2}})$, respectively, the right-hand side is bounded in $L_T^2(\dot{B}_{2,1}^{\frac{d}{2}-1})$.

As regards $(\bar{u}^n)_{n\geq 1}$, it suffices to prove that $(\partial_t \bar{u}^n)_{n\geq 1}$ is bounded in $L^{\frac{4}{3}}([0,T]; \dot{B}_{2,1}^{\frac{d}{2}-1} + \dot{B}_{2,1}^{\frac{d}{2}-\frac{3}{2}})$. This follows from the fact that

$$\partial_t \bar{u}^n = -\dot{\mathbb{E}}_n\left(u^n \cdot \nabla u^n + I(a^n)\mathcal{A}u^n - \mathcal{A}\bar{u}^n + \nabla(G(a^n))\right).$$

Indeed, by using the fact that $(u^n)_{n\geq 1}$ and $(\bar{u}^n)_{n\geq 1}$ are bounded in $L_T^{\frac{4}{3}}(\dot{B}_{2,1}^{\frac{d}{2}+\frac{1}{2}}) \cap L_T^\infty(\dot{B}_{2,1}^{\frac{d}{2}-1})$, and that $(a^n)_{n\geq 1}$ is bounded in $L_T^\infty(\dot{B}_{2,1}^{\frac{d}{2}})$, we easily deduce that the first three terms on the right-hand side are in $L_T^{\frac{4}{3}}(\dot{B}_{2,1}^{\frac{d}{2}-\frac{3}{2}})$, and that the last one is in $L_T^\infty(\dot{B}_{2,1}^{\frac{d}{2}-1})$, uniformly. This is a simple consequence of the product and composition laws for homogeneous Besov spaces, as stated in Chapter 2. □

Remark 10.7. If $d \geq 3$, we can also prove that $(\partial_t \bar{u}^n)_{n\in\mathbb{N}}$ is bounded in $L^2([0,T]; \dot{B}_{2,1}^{\frac{d}{2}-1} + \dot{B}_{2,1}^{\frac{d}{2}-2})$.

Fourth Step: Compactness and Convergence

We introduce a sequence $(\phi_p)_{p\in\mathbb{N}}$ of smooth functions with values in $[0,1]$, supported in the ball $B(0,p+1)$ and equal to 1 on $B(0,p)$. Recall that, according to the previous lemma and step 2,

$$(\bar{a}^n)_{n\geq 1} \text{ is bounded in } \mathcal{C}^{\frac{1}{2}}([0,T]; \dot{B}_{2,1}^{\frac{d}{2}-1}) \cap \mathcal{C}([0,T]; \dot{B}_{2,1}^{\frac{d}{2}}).$$

Therefore, by virtue of Proposition 2.93 page 108,[5]

$(\phi_p \bar{a}^n)_{n\geq 1}$ is bounded in $\mathcal{C}^{\frac{1}{2}}([0,T]; B_{2,1}^{\frac{d}{2}-1}) \cap \mathcal{C}([0,T]; B_{2,1}^{\frac{d}{2}})$ for all $p \in \mathbb{N}$.

Now, according to Theorem 2.94, the map $z \longmapsto \phi_p z$ is compact from $B_{2,1}^{\frac{d}{2}}$ to $B_{2,1}^{\frac{d}{2}-1}$. Therefore, Ascoli's theorem ensures that there exists some function $\bar{a}_p$ such that, up to extraction, $(\phi_p \bar{a}^n)_{n\geq 1}$ converges to $\bar{a}_p$ in $\mathcal{C}([0,T]; B_{2,1}^{\frac{d}{2}-1})$. Using the Cantor diagonal process, we can then find a subsequence of $(\bar{a}^n)_{n\geq 1}$ [still denoted by $(\bar{a}^n)_{n\geq 1}$] such that for all $p \in \mathbb{N}$, $\phi_p \bar{a}^n$ converges to $\bar{a}_p$ in $\mathcal{C}([0,T]; B_{2,1}^{\frac{d}{2}-1})$. As $\phi_p \phi_{p+1} = \phi_p$, we have, in addition, $\bar{a}_p = \phi_p \bar{a}_{p+1}$. From that, we can easily deduce that there exists some function $\bar{a}$ such that for all $\phi \in \mathcal{D}$, $\phi \bar{a}^n$ tends to $\phi \bar{a}$ in $\mathcal{C}([0,T]; B_{2,1}^{\frac{d}{2}-1})$.

A similar argument, based on the bounds stated in step 2 for the velocity and on the second part of Lemma 10.6, allows us to show that there exists a vector field $\bar{u}$ such that, up to extraction, for any function $\phi \in \mathcal{D}$, we have $\phi \bar{u}^n \longrightarrow \phi \bar{u}$ in $\mathcal{C}([0,T]; B_{2,1}^{\frac{d}{2}-\frac{3}{2}})$.

Final Step: Completion of the Proof

Combining the uniform bounds that we proved in the second step and the Fatou property for Besov spaces (see Theorem 2.25), we readily get

$$(\bar{a}, \bar{u}) \in \widetilde{L}_T^\infty(\dot{B}_{2,1}^{\frac{d}{2}}) \times \left(\widetilde{L}_T^\infty(\dot{B}_{2,1}^{\frac{d}{2}-1})\right)^d.$$

Proving that $\bar{u}$ also belongs to $L_T^1(\dot{B}_{2,1}^{\frac{d}{2}+1})$ requires some attention. Indeed, having $(\bar{u}^n)_{n\in\mathbb{N}}$ bounded in $L_T^1(\dot{B}_{2,1}^{\frac{d}{2}+1})$ only ensures that $\bar{u}$ belongs to the set $\mathcal{M}_T(\dot{B}_{2,1}^{\frac{d}{2}+1})$ of bounded measures on $[0,T]$ with values in the space $\dot{B}_{2,1}^{\frac{d}{2}+1}$, and that

$$\int_0^T d\|\bar{u}(t)\|_{\dot{B}_{2,1}^{\frac{d}{2}+1}} \leq C_T,$$

where C_T stands for the right-hand side of (10.20).

It is now clear that the same inequality holds for $\dot{\mathbb{E}}_n \bar{u}$, for all $n \geq 1$. In addition, as $\bar{u} \in L_T^\infty(\dot{B}_{2,1}^{\frac{d}{2}-1})$, we obviously have $\dot{\mathbb{E}}_n \bar{u} \in L_T^1(\dot{B}_{2,1}^{\frac{d}{2}+1})$. Finally, then, we may write

$$\int_0^T \|\dot{\mathbb{E}}_n \bar{u}\|_{\dot{B}_{2,1}^{\frac{d}{2}+1}} dt \leq C_T \quad \text{for all } n \geq 1.$$

Using the definition of the norm in $\dot{B}_{2,1}^{\frac{d}{2}+1}$, the above inequality implies that

---

[5] In the case $d = 2$, as $d/2 - 1 = 0$, we also use the fact that $\dot{B}_{2,1}^0 \hookrightarrow B_{2,1}^0$.

$$\lim_{N\to\infty} \sum_{|j|\le N} 2^{j(\frac{d}{2}+1)} \int_0^T \|\dot{\Delta}_j \overline{u}\|_{L^2}\, dt \le C_T.$$

Therefore, $\overline{u} \in L_T^1(\dot{B}_{2,1}^{\frac{d}{2}+1})$.

Let $(a, u) \overset{\text{def}}{=} (a_0 + \overline{a}, u_L + \overline{u})$. By interpolating between the convergence results that we have obtained so far and the uniform bounds of step 2, we get better convergence results for $(\overline{a}^n, \overline{u}^n)$ so that we may pass to the limit in (NSC_n). As an example, we explain how the nonlinear term $\dot{\mathbb{E}}_n\big(I(a^n)\mathcal{A}u^n\big)$ may be handled. Fix some $\phi \in \mathcal{D}(\mathbb{R}^d)$ and some $p \ge 1$ sufficiently large so as to satisfy $\phi_p \equiv 1$ in a neighborhood of Supp ϕ. Using the symmetry of $\dot{\mathbb{E}}_n$ and the support properties of ϕ and ϕ_p, we may write

$$\big\langle \dot{\mathbb{E}}_n\big(I(a^n)\mathcal{A}u^n\big) - I(a)\mathcal{A}u, \phi \big\rangle = \big\langle I(a^n)\mathcal{A}u^n, (\dot{\mathbb{E}}_n - \mathrm{Id})\phi \big\rangle$$
$$+ \big\langle I(\phi_p a^n)\mathcal{A}(\phi_p u^n) - I(\phi_p a)\mathcal{A}(\phi_p u), \phi \big\rangle.$$

Combining the bounds of step 2 and product laws in Besov spaces, we see that $(I(a^n)\mathcal{A}u^n)_{n\ge 1}$ is bounded in $L_T^1(\dot{B}_{2,1}^{\frac{d}{2}-1})$. Hence, we can deduce from duality properties (see Proposition 2.29) and the smoothness of ϕ that the first term of the above equality tends to 0.

For the second term, it suffices to use the fact that for any $\varepsilon > 0$,

- $\phi_p a^n \longrightarrow \phi_p a$ in $L_T^\infty(B_{2,1}^{\frac{d}{2}-\varepsilon})$,
- $\mathcal{A}(\phi_p a^n) \longrightarrow \mathcal{A}(\phi_p u)$ in $L_T^1(B_{2,1}^{\frac{d}{2}+1-\varepsilon})$,

which, in view of the product properties in Besov spaces, suffices to show the convergence. Treating the other terms in (NSC_n) is left to the reader.

Finally, then, we have constructed a solution (a, u) of (NSC) with data (a_0, u_0), which satisfies

$$(a, u) \in \widetilde{L}_T^\infty(\dot{B}_{2,1}^{\frac{d}{2}}) \times \Big(\widetilde{L}_T^\infty(\dot{B}_{2,1}^{\frac{d}{2}-1}) \cap L_T^1(\dot{B}_{2,1}^{\frac{d}{2}+1})\Big)^d,$$

and the bounds of step 2 are satisfied.

In order to establish the properties of continuity with respect to time, it suffices to observe that

$$(\partial_t + u \cdot \nabla)a \in L_T^1(\dot{B}_{2,1}^{\frac{d}{2}}) \quad \text{and} \quad \partial_t u \in L_T^1(\dot{B}_{2,1}^{\frac{d}{2}-1}).$$

The second property obviously ensures that $u \in \mathcal{C}([0, T]; \dot{B}_{2,1}^{\frac{d}{2}-1})$, while according to Theorem 3.19, the first one guarantees that $a \in \mathcal{C}([0, T]; \dot{B}_{2,1}^{\frac{d}{2}})$.

Remark 10.8. Combining (10.13) with Lemma 2.4 page 54, we may deduce a lower bound for the lifespan T^* of the solution. Defining $u_j(T) \overset{\text{def}}{=}$

$2^{j(\frac{d}{2}-1)}\left(\|\dot{\Delta}_j u_0\|_{L^2}+\|\dot{\Delta}_j f\|_{L^1_T(L^2)}\right)$, we find that there exists some constant η, depending only on d and on G, such that

$$T^* \geq \sup\left\{T \in\,]0,\eta\underline{\nu}] \;/\; \sum_{j\in\mathbb{Z}}\left(1-e^{-\underline{\nu}T2^{2j}}\right)u_j(T) \leq \frac{\eta\underline{\nu}^2}{\overline{\nu}+\overline{\nu}U_0(T)/\underline{\nu}}\right\}.$$

10.2.4 Uniqueness

Assume that we are given (a^1,u^1) and (a^2,u^2), two solutions of (NSC) (with the same data) satisfying the regularity assumptions of Theorem 10.2. In order to show that these two solutions coincide, we shall give estimates for $(\delta a,\delta u) \stackrel{\text{def}}{=} (a^2-a^1, u^2-u^1)$. These estimates will be based on Proposition 10.3 and Theorem 3.14 applied to the following system satisfied by $(\delta a,\delta u)$:

$$\begin{cases} \partial_t\delta a + u^2\cdot\nabla\delta a + \sum_{i=1}^3 \delta F_i = 0 \\ \partial_t\delta u + u^2\cdot\nabla\delta u + \delta u\cdot\nabla u^1 - \mathcal{A}\delta u = \sum_{i=1}^3 \delta G_i \end{cases} \tag{10.21}$$

with $\quad \delta F_1 \stackrel{\text{def}}{=} \delta u\cdot\nabla a^1, \quad \delta F_2 \stackrel{\text{def}}{=} \delta a\;\mathrm{div}\,u^2, \quad \delta F_3 \stackrel{\text{def}}{=} (1+a^1)\,\mathrm{div}\,\delta u,$

$\delta G_1 \stackrel{\text{def}}{=} (I(a^1)-I(a^2))\mathcal{A}u^2, \quad \delta G_2 \stackrel{\text{def}}{=} -I(a^1)\,\mathcal{A}\delta u, \quad \delta G_3 \stackrel{\text{def}}{=} -\nabla(G(a^2)-G(a^1)).$

Note that, owing to the hyperbolic nature of the mass equation, we could not avoid a loss of one derivative in the stability estimates (because the term δF_1 in the first equation of (10.21) cannot be better than $L^\infty([0,T];\dot{B}_{2,1}^{\frac{d}{2}-1})$, for we only know that $\nabla a^1 \in L^\infty([0,T];\dot{B}_{2,1}^{\frac{d}{2}-1}))$. In addition, because of the coupling between the equations for δa and δu, this loss of one derivative also induces a loss of one derivative when bounding δu. Hence, we expect to prove uniqueness in the function space

$$F_T \stackrel{\text{def}}{=} \mathcal{C}([0,T];\dot{B}_{2,1}^{\frac{d}{2}-1}) \times \left(\mathcal{C}([0,T];\dot{B}_{2,1}^{\frac{d}{2}-2})\cap L^1_T(\dot{B}_{2,1}^{\frac{d}{2}})\right)^d.$$

We first consider the case $d \geq 3$, which is easier to deal with. We have to check that $(\delta a,\delta u)$ belongs to F_T, a fact which is not entirely obvious since *homogeneous* Besov spaces are involved. We note that both $\partial_t a^1$ and $\partial_t a^2$ are in $L^2_T(\dot{B}_{2,1}^{\frac{d}{2}-1})$ (just follow the proof of Lemma 10.6). As $a^1(0) = a^2(0)$, we thus have $\delta a \in \mathcal{C}([0,T];\dot{B}_{2,1}^{\frac{d}{2}-1})$, as desired.

In order to show that $\delta u \in \mathcal{C}([0,T];\dot{B}_{2,1}^{\frac{d}{2}-2})$, we introduce $\overline{u}^i \stackrel{\text{def}}{=} u^i - \overline{u}_L$, where $\overline{u}_L$ is the solution of

$$\partial_t\overline{u}_L - \mathcal{A}\overline{u}_L = f - \nabla(G(a_0)), \qquad \overline{u}_{L|t=0} = u_0. \tag{10.22}$$

We obviously have $\overline{u}^i(0) = 0$ and

$$\partial_t \overline{u}^i = \mathcal{A}\overline{u}^i - I(a^i)\mathcal{A}u^i - u^i \cdot \nabla u^i - \nabla\big(G(a^i) - G(a_0)\big).$$

Since $\overline{a}^i \in L_T^\infty(\dot{B}_{2,1}^{\frac{d}{2}-1})$ and $(a^i, u^i) \in E_T$, the right-hand side belongs to $L_T^2(\dot{B}_{2,1}^{\frac{d}{2}-2})$ (use Section 2.6). Hence, $\overline{u}^i$ belongs to $\mathcal{C}([0,T]; \dot{B}_{2,1}^{\frac{d}{2}-2})$, and we can now conclude that $(\delta a, \delta u) \in F_T$.

In order to get an estimate for δa, we apply Theorem 3.14 to the first equation of (10.21). For $\overline{T} \le T$ we get

$$\|\delta a\|_{L_{\overline{T}}^\infty(\dot{B}_{2,1}^{\frac{d}{2}-1})} \le e^{C\|u^2\|_{L_{\overline{T}}^1(\dot{B}_{2,1}^{\frac{d}{2}+1})}} \sum_{i=1}^{3} \|\delta F_i\|_{L_{\overline{T}}^1(\dot{B}_{2,1}^{\frac{d}{2}-1})}.$$

Easy computations based on Theorems 2.47 and 2.52 page 88 yield

$$\|\delta F_1\|_{\dot{B}_{2,1}^{\frac{d}{2}-1}} \le C\|\delta u\|_{\dot{B}_{2,1}^{\frac{d}{2}}} \|\nabla a^1\|_{\dot{B}_{2,1}^{\frac{d}{2}-1}},$$

$$\|\delta F_2\|_{\dot{B}_{2,1}^{\frac{d}{2}-1}} \le C\|\operatorname{div} u^2\|_{\dot{B}_{2,1}^{\frac{d}{2}}} \|\delta a\|_{\dot{B}_{2,1}^{\frac{d}{2}-1}},$$

$$\|\delta F_3\|_{\dot{B}_{2,1}^{\frac{d}{2}-1}} \le C\big(1 + \|a^1\|_{\dot{B}_{2,1}^{\frac{d}{2}}}\big)\|\delta u\|_{\dot{B}_{2,1}^{\frac{d}{2}}}.$$

Hence, using Gronwall's lemma and interpolation, we discover that there exists some constant C_T, independent of $\overline{T}$, such that

$$\|\delta a\|_{L_{\overline{T}}^\infty(\dot{B}_{2,1}^{\frac{d}{2}-1})} \le C_T\left(\|\delta u\|_{L_{\overline{T}}^1(\dot{B}_{2,1}^{\frac{d}{2}})} + \|\delta u\|_{L_{\overline{T}}^\infty(\dot{B}_{2,1}^{\frac{d}{2}-2})}\right). \tag{10.23}$$

Next, applying Proposition 10.3 to the second equation of (10.21) yields

$$\|\delta u\|_{L_{\overline{T}}^\infty(\dot{B}_{2,1}^{\frac{d}{2}-2})} + \|\delta u\|_{L_{\overline{T}}^1(\dot{B}_{2,1}^{\frac{d}{2}})}$$
$$\le Ce^{C\int_0^{\overline{T}}(\|u^1\|_{\dot{B}_{2,1}^{\frac{d}{2}+1}} + \|u^2\|_{\dot{B}_{2,1}^{\frac{d}{2}+1}})\,dt} \sum_{i=1}^{3} \|\delta G_i\|_{L_{\overline{T}}^1(\dot{B}_{2,1}^{\frac{d}{2}-2})}.$$

Because $\dot{B}_{2,1}^{\frac{d}{2}}(\mathbb{R}^d) \hookrightarrow \mathcal{C}(\mathbb{R}^d)$, we have $a^i \in \mathcal{C}([0,T] \times \mathbb{R}^d)$. Hence, for sufficiently small $\overline{T}$,

$$\|a^i\|_{L^\infty([0,\overline{T}]\times\mathbb{R}^d)} \le \frac{1}{2} \quad \text{for} \quad i = 1, 2. \tag{10.24}$$

Therefore, applying Theorems 2.47, 2.52, 2.61 and Corollary 2.66 yields

$$\|\delta G_1\|_{\dot{B}_{2,1}^{\frac{d}{2}-2}} \le C\overline{\nu}\big(1 + \|a^1\|_{\dot{B}_{2,1}^{\frac{d}{2}}} + \|a^2\|_{\dot{B}_{2,1}^{\frac{d}{2}}}\big)\|\delta a\|_{\dot{B}_{2,1}^{\frac{d}{2}-1}}\|u^1\|_{\dot{B}_{2,1}^{\frac{d}{2}+1}},$$

$$\|\delta G_2\|_{\dot{B}_{2,1}^{\frac{d}{2}-2}} \le C\overline{\nu}\|a^1\|_{\dot{B}_{2,1}^{\frac{d}{2}}}\|\delta u\|_{\dot{B}_{2,1}^{\frac{d}{2}}},$$

$$\|\delta G_3\|_{\dot{B}_{2,1}^{\frac{d}{2}-2}} \le C(1 + \|a^1\|_{\dot{B}_{2,1}^{\frac{d}{2}}} + \|a^2\|_{\dot{B}_{2,1}^{\frac{d}{2}}})\|\delta a\|_{\dot{B}_{2,1}^{\frac{d}{2}-1}}.$$

If η has been chosen to be sufficiently small in the condition (10.8), then δG_2 may be absorbed by the left-hand side of the inequality for δu. Therefore, we can conclude that there exists a constant C_T, *independent of* $\overline{T}$, such that

$$\|\delta u\|_{L_T^\infty(\dot{B}_{2,1}^{\frac{d}{2}-2})} + \|\delta u\|_{L_T^1(\dot{B}_{2,1}^{\frac{d}{2}})} \leq C_T \left(\bar{T} + \|u^1\|_{L_T^1(\dot{B}_{2,1}^{\frac{d}{2}})}\right) \|\delta a\|_{L_T^\infty(\dot{B}_{2,1}^{\frac{d}{2}-1})}.$$

Note that the factor $\bar{T} + \|u^1\|_{L_T^1(\dot{B}_{2,1}^{\frac{d}{2}})}$ decays to 0 when $\bar{T}$ goes to zero. Hence, plugging the inequality (10.23) into the above inequality, we conclude that $(\delta a, \delta u) \equiv 0$ on a nontrivial time interval $[0, \bar{T}]$. In order to show that we may take $\bar{T} = T$, we introduce the set

$$I \stackrel{\text{def}}{=} \left\{t \in [0, T] / (a^2, u^2) \equiv (a^1, u^1) \text{ on } [0, t]\right\}.$$

Obviously, I is a nonempty closed subset of $[0, T]$. In addition, the above arguments may be carried over to any $t \in I \cap [0, T[$, which ensures that I is an open subset of $[0, T]$. Therefore, $I \equiv [0, T]$, and the proof is complete in the case $d \geq 3$.

In the two-dimensional case, the above proof fails because, when estimating some terms on the right-hand side of the equation for δu (such as, e.g., δG_3), the sum of the indices of regularity is zero. Hence, we must use the endpoint inequalities of Proposition 2.52 [adapted to $\widetilde{L}_T^\rho(\dot{B}_{2,r}^s)$ spaces], but we then obtain a bound in the larger space $\widetilde{L}_T^1(\dot{B}_{2,\infty}^{-1})$, instead of $L_T^1(\dot{B}_{2,1}^{-1})$. At this point, we may be tempted to estimate $(\delta a, \delta u)$ in

$$F_T \stackrel{\text{def}}{=} L_T^\infty(\dot{B}_{2,\infty}^0) \times \left(L_T^\infty(\dot{B}_{2,\infty}^{-1}) \cap \widetilde{L}_T^1(\dot{B}_{2,\infty}^1)\right)^2,$$

but we then have to face the lack of control on δu in $L_T^1(L^\infty)$ (because, in contrast to $\dot{B}_{2,1}^1$, the space $\dot{B}_{2,\infty}^1$ is *not* embedded in L^∞) so that we run into trouble when estimating δF_1. In order to bypass this difficulty, we shall use the logarithmic interpolation inequality

$$\|w\|_{L_T^1(\dot{B}_{2,1}^1)} \leq C\|w\|_{\widetilde{L}_T^1(\dot{B}_{2,\infty}^1)} \log\left(e + \frac{\|w\|_{\widetilde{L}_T^1(\dot{B}_{2,\infty}^0)} + \|w\|_{\widetilde{L}_T^1(\dot{B}_{2,\infty}^2)}}{\|w\|_{\widetilde{L}_T^1(\dot{B}_{2,\infty}^1)}}\right), \quad (10.25)$$

the proof of which is similar to that of (2.104), except that we now have to split w into three parts,

$$w = \sum_{j < -M} \dot{\Delta}_j w + \sum_{j=-M}^N \dot{\Delta}_j w + \sum_{j > N} \dot{\Delta}_j w,$$

and choose the "best" nonnegative integers M and N. At this stage, it will be possible to conclude that we have uniqueness by taking advantage of Osgood's lemma.

We now give some more details. We omit the proof that $(\delta a, \delta u)$ is indeed in $F_{\bar{T}}$ as it is only a matter of repeating the arguments that were used for $d \geq 3$. Next, bounding δa may be achieved by combining Theorem 3.14 page 133 with Propositions 2.47 page 87 and 2.52 page 88, and by using the embedding $\dot{B}^1_{2,1} \hookrightarrow \dot{B}^1_{2,\infty} \cap L^\infty$. After a few computations, we get

$$\|\delta a\|_{L^\infty_t(\dot{B}^0_{2,\infty})} \le C \exp\left(C \int_0^t \|u^2\|_{\dot{B}^2_{2,1}}\, dt'\right)$$
$$\times \int_0^t \left(\|\delta a\|_{\dot{B}^0_{2,\infty}}\|\mathrm{div}\, u^2\|_{\dot{B}^1_{2,1}} + \|\delta u\|_{\dot{B}^1_{2,1}}\left(1 + \|a^1\|_{\dot{B}^1_{2,1}}\right)\right) dt',$$

from which it follows, according to Gronwall's inequality, that

$$\|\delta a\|_{L^\infty_t(\dot{B}^0_{2,\infty})} \le C \exp\left(C \int_0^t \|u^2\|_{\dot{B}^2_{2,1}}\, dt'\right)\left(1 + \|a^1\|_{L^\infty_t(\dot{B}^1_{2,1})}\right)\|\delta u\|_{L^1_t(\dot{B}^1_{2,1})}.$$

Making use of the inequality (10.25) with $w = \delta u$, we end up with

$$\|\delta a\|_{L^\infty_t(\dot{B}^0_{2,\infty})} \le C_T\|\delta u\|_{\widetilde{L}^1_t(\dot{B}^1_{2,\infty})} \log\left(e + \frac{\|\delta u\|_{\widetilde{L}^1_t(\dot{B}^0_{2,\infty})} + \|\delta u\|_{\widetilde{L}^1_t(\dot{B}^2_{2,\infty})}}{\|\delta u\|_{\widetilde{L}^1_t(\dot{B}^1_{2,\infty})}}\right)$$

for some constant C_T depending only on the bounds of the solutions in E_T.

Note that since $\widetilde{L}^\infty_t(\dot{B}^0_{2,1}) \hookrightarrow L^1_t(\dot{B}^0_{2,1})$ for finite t, we have

$$\forall t \in [0,T],\ \|\delta u\|_{\widetilde{L}^1_t(\dot{B}^0_{2,\infty})} + \|\delta u\|_{\widetilde{L}^1_t(\dot{B}^2_{2,\infty})} \le V(t) \overset{\mathrm{def}}{=} V_1(t) + V_2(t) < \infty$$

with

$$V_i'(t) \overset{\mathrm{def}}{=} \|u^i(t)\|_{\dot{B}^0_{2,1}} + \|u^i(t)\|_{\dot{B}^2_{2,1}} \in L^1([0,T]).$$

Therefore, V is in $L^\infty([0,T])$ and

$$\|\delta a\|_{L^\infty_t(\dot{B}^0_{2,\infty})} \le C_T\|\delta u\|_{\widetilde{L}^1_t(\dot{B}^1_{2,\infty})} \log\left(e + \frac{V(t)}{\|\delta u\|_{\widetilde{L}^1_t(\dot{B}^1_{2,\infty})}}\right). \tag{10.26}$$

We now bound δu. Making use of (an obvious generalization of) the inequality (3.39) page 157, we get

$$\|\delta u\|_{L^\infty_T(\dot{B}^{-1}_{2,\infty})} + \nu\|\delta u\|_{\widetilde{L}^1_T(\dot{B}^1_{2,\infty})}$$
$$\le C\left(\|u^2 \cdot \nabla \delta u\|_{\widetilde{L}^1_T(\dot{B}^{-1}_{2,\infty})} + \|\delta u \cdot \nabla u^1\|_{\widetilde{L}^1_T(\dot{B}^{-1}_{2,\infty})} + \sum_{i=1}^{3} \|\delta G_i\|_{\widetilde{L}^1_T(\dot{B}^{-1}_{2,\infty})}\right).$$

In order to bound the terms on the right-hand side, we may exploit Propositions 2.47, 2.52 and Corollary 2.66 page 97 [recall that (10.24) is satisfied], adapted to the spaces $\widetilde{L}^1_t(\dot{B}^s_{p,r})$. Since $L^1_t(\dot{B}^{-1}_{2,\infty}) \hookrightarrow \widetilde{L}^1_t(\dot{B}^{-1}_{2,\infty})$, we get, for all $t \le \bar{T}$,

$$\|u^2 \cdot \nabla \delta u\|_{\widetilde{L}^1_t(\dot{B}^{-1}_{2,\infty})} \le C\|u^2\|_{\widetilde{L}^2_t(\dot{B}^1_{2,1})}\|\nabla \delta u\|_{\widetilde{L}^2_t(\dot{B}^{-1}_{2,\infty})},$$

$$\|\delta u \cdot \nabla u^1\|_{\widetilde{L}^1_t(\dot{B}^{-1}_{2,\infty})} \le C \int_0^t \|\nabla u^1\|_{\dot{B}^1_{2,1}}\|\delta u\|_{\dot{B}^{-1}_{2,\infty}}\, dt',$$

$$\|\delta G_2\|_{\widetilde{L}^1_t(\dot{B}^{-1}_{2,\infty})} \le C\bar{\nu}\|a^1\|_{\widetilde{L}^\infty_t(\dot{B}^1_{2,1})}\|\nabla^2 \delta u\|_{\widetilde{L}^1_t(\dot{B}^{-1}_{2,\infty})}.$$

In order to bound the terms δG_1 and δG_2, we need to generalize Corollary 2.66 to the case of regularity index 0. For this, it suffices to note that for any sufficiently smooth function H, we have

$$H(a^2) - H(a^1) = \left(H'(0) + \int_0^1 \left(H'(a^1 + \tau \delta a) - H'(0) \right) d\tau \right) \delta a.$$

Hence, combining the product laws in Besov spaces and Theorem 2.61, we have

$$\| H(a^2) - H(a^1) \|_{\dot{B}_{2,\infty}^0} \leq C \| \delta a \|_{\dot{B}_{2,\infty}^0} \left(|H'(0)| + \| a^1 \|_{\dot{B}_{2,1}^1} + \| a^2 \|_{\dot{B}_{2,1}^1} \right).$$

So, finally, we get

$$\| \delta G_1 \|_{\widetilde{L}_t^1(\dot{B}_{2,\infty}^{-1})} \leq C \bar{\nu} \int_0^t \left(1 + \| a^1 \|_{\dot{B}_{2,1}^1} + \| a^2 \|_{\dot{B}_{2,1}^1} \right) \| \nabla^2 u^2 \|_{\dot{B}_{2,1}^0} \| \delta a \|_{\dot{B}_{2,\infty}^0} \, dt',$$

$$\| \delta G_3 \|_{\widetilde{L}_t^1(\dot{B}_{2,\infty}^{-1})} \leq C \int_0^t \left(1 + \| a^1 \|_{\dot{B}_{2,1}^1} + \| a^2 \|_{\dot{B}_{2,1}^1} \right) \| \delta a \|_{\dot{B}_{2,\infty}^0} \, dt'$$

and can conclude that

$$\| \delta u \|_{L_t^\infty(\dot{B}_{2,\infty}^{-1})} + \underline{\nu} \| \delta u \|_{\widetilde{L}_t^1(\dot{B}_{2,\infty}^1)} \leq C \| u^2 \|_{\widetilde{L}_t^2(\dot{B}_{2,1}^1)} \| \delta u \|_{\widetilde{L}_t^2(\dot{B}_{2,\infty}^0)}$$

$$+ \bar{\nu} \| \delta u \|_{\widetilde{L}_t^1(\dot{B}_{2,\infty}^1)} \| a^1 \|_{\widetilde{L}_t^\infty(\dot{B}_{2,1}^1)} + \int_0^t \Big[\| u^1 \|_{\dot{B}_{2,1}^2} \| \delta u \|_{\dot{B}_{2,\infty}^{-1}}$$

$$+ \left(1 + \| a^1 \|_{\dot{B}_{2,1}^1} + \| a^2 \|_{\dot{B}_{2,1}^1} \right) \left(1 + \bar{\nu} \| u^2 \|_{\dot{B}_{2,1}^2} \right) \| \delta a \|_{\dot{B}_{2,\infty}^0} \Big] \, dt'.$$

Now, if we take a sufficiently small constant η in the inequality (10.8), then the second term on the right-hand side may be absorbed by the left-hand side. Next, we note that by virtue of the Lebesgue dominated convergence theorem, $\| u^2 \|_{\widetilde{L}_t^2(\dot{B}_{2,1}^1)}$ tends to 0 when t goes to 0, and hence there exists a positive $\bar{T}$ such that the first term on the right-hand side may also be absorbed[6] for all $t \in [0, \bar{T}]$. We end up with the following inequality:

$$\| \delta u \|_{L_t^\infty(\dot{B}_{2,\infty}^{-1})} + \underline{\nu} \| \delta u \|_{\widetilde{L}_t^1(\dot{B}_{2,\infty}^1)}$$

$$\leq C \int_0^t \left(\| u^1 \|_{\dot{B}_{2,1}^2} \| \delta u \|_{\dot{B}_{2,\infty}^{-1}} + (1 + \underline{\nu} \| u^2 \|_{\dot{B}_{2,1}^2}) \| \delta a \|_{\dot{B}_{2,\infty}^0} \right) dt'.$$

We plug (10.26) into this inequality. Defining $X(t) \overset{\text{def}}{=} \| \delta u \|_{L_t^\infty(\dot{B}_{2,\infty}^{-1})} + \| \delta u \|_{\widetilde{L}_t^1(\dot{B}_{2,\infty}^1)}$, we get, for any t in $[0, T]$ and some constant C_T depending only on $\underline{\nu}$, $\bar{\nu}$, and the norms of the solutions (a^1, u^1) and (a^2, u^2) in E_T,

---

[6] By interpolation, we easily get $\| \delta u \|_{\widetilde{L}_t^2(\dot{B}_{2,\infty}^0)} \leq \| \delta u \|_{\widetilde{L}_t^1(\dot{B}_{2,\infty}^1)}^{\frac{1}{2}} \| \delta u \|_{\widetilde{L}_t^\infty(\dot{B}_{2,\infty}^{-1})}^{\frac{1}{2}}.$

$$X(t) \le C_T \int_0^t (1+V'(t'))X(t') \log\left(e + \frac{V(T)}{X(t')}\right) dt'.$$

As

$$V' \in L^1([0,T]) \quad \text{and} \quad \int_0^1 \frac{dr}{r \log\left(e + \frac{V(T)}{r}\right)} = \infty,$$

Osgood's lemma entails that $X \equiv 0$ on $[0,\bar{T}]$. This means that (a^1, u^1) and (a^2, u^2) coincide on $[0,\bar{T}]$. Appealing to the connectivity argument used in the case $d \ge 3$ then completes the proof. $\qquad \square$

Remark 10.9. Having a tilde in the condition (10.8) in the critical case $d = 2$ is necessary for conveniently bounding the term δG_4.

10.2.5 A Continuation Criterion

This section is devoted to the proof of the following continuation criterion.

Proposition 10.10. *Under the hypotheses of Theorem 10.2, assume that the system (NSC) has a solution (a,u) on $[0,T[\times \mathbb{R}^d$ which belongs to $E_{T'}$ for all $T' < T$ and satisfies*

$$\int_0^T \|\nabla u\|_{L^\infty}\, dt < \infty \quad \text{and} \quad \begin{cases} \|a\|_{L_T^\infty(\dot{B}_{2,1}^{\frac{d}{2}})} \le \eta\nu/\bar{\nu}, & \text{if } d \ge 3, \\ \|a\|_{\widetilde{L}_T^\infty(\dot{B}_{2,1}^1)} \le \eta\nu/\bar{\nu}, & \text{if } d = 2. \end{cases}$$

There exists some $T^* > T$ such that (a,u) may be continued on $[0,T^*] \times \mathbb{R}^d$ to a solution of (NSC) which belongs to E_{T^*}.

Proof. Note that u satisfies

$$\partial_t u + u \cdot \nabla u - \mathcal{A}u = f - \nabla(G(a)) - I(a)\mathcal{A}u, \qquad u_{|t=0} = u_0.$$

Hence, applying Proposition 10.3 and taking advantage of the smallness of a to absorb the term $I(a)\mathcal{A}u$, we get, for some constant C, depending only on d, and for all $t < T$,

$$\|u\|_{\widetilde{L}_t^\infty(\dot{B}_{2,1}^{\frac{d}{2}-1})} \le Ce^{C\int_0^t \|\nabla u\|_{L^\infty}\, dt'}$$

$$\times \left(\|u_0\|_{\dot{B}_{2,1}^{\frac{d}{2}-1}} + \|f\|_{L_t^1(\dot{B}_{2,1}^{\frac{d}{2}-1})} + t\|a\|_{L_t^\infty(\dot{B}_{2,1}^{\frac{d}{2}})} \right).$$

Hence, u belongs to $\widetilde{L}_T^\infty(\dot{B}_{2,1}^{\frac{d}{2}-1})$. Now, replacing $\|\dot{\Delta}_j u_0\|_{L^2}$ by $\|\dot{\Delta}_j u\|_{L_T^\infty(L^2)}$ and $\|\dot{\Delta}_j f\|_{L_T^1(L^2)}$ by $\|\dot{\Delta}_j f\|_{L_{1+T}(L^2)}$ in Remark 10.8, we get some $\varepsilon > 0$ such that for any $T' \in [0,T[$, the system (NSC) with data $a(T')$, $u(T')$, and $f(\cdot + T')$ has a solution on $[0,\varepsilon]$. Taking $T' = T - \varepsilon/2$ and using the fact that the solution (a,u) is unique on $[0,T[$, we thus get a continuation of (a,u) beyond T. $\qquad \square$

10.3 Local Theory for Data Bounded Away from the Vacuum

We next consider initial data (ρ_0, u_0) which do not satisfy the smallness condition (10.7), that is, the density need not be almost a constant function. Since having strict parabolicity in the momentum equation is fundamental, however, we shall always assume that ρ_0 is bounded away from 0, an assumption which will be shown to be conserved for sufficiently small times. In order to simplify the presentation, we assume that the data are more regular than needed according to our scaling considerations.

We now state the main result of this section.

Theorem 10.11. *Assume that the space dimension is $d \geq 2$ and that the data (a_0, u_0, f) satisfy, for some $\alpha \in \,]0, 1]$,*

$$a_0 \in \dot{B}_{2,1}^{\frac{d}{2}} \cap \dot{B}_{2,1}^{\frac{d}{2}+\alpha}, \quad u_0 \in \dot{B}_{2,1}^{\frac{d}{2}-1} \cap \dot{B}_{2,1}^{\frac{d}{2}-1+\alpha}, \quad and \quad f \in L_{loc}^1(\mathbb{R}^+; \dot{B}_{2,1}^{\frac{d}{2}-1} \cap \dot{B}_{2,1}^{\frac{d}{2}-1+\alpha}).$$

If, in addition, $\inf_x a_0(x) > -1$, then there exists a positive time T such that (NSC) has a unique solution (a, u) on $[0, T] \times \mathbb{R}^d$ which belongs to

$$E_T^\alpha \overset{def}{=} \widetilde{\mathcal{C}}_T(\dot{B}_{2,1}^{\frac{d}{2}} \cap \dot{B}_{2,1}^{\frac{d}{2}+\alpha}) \times \left(\widetilde{\mathcal{C}}_T(\dot{B}_{2,1}^{\frac{d}{2}-1} \cap \dot{B}_{2,1}^{\frac{d}{2}-1+\alpha}) \cap L^1([0, T]; \dot{B}_{2,1}^{\frac{d}{2}+1} \cap \dot{B}_{2,1}^{\frac{d}{2}+1+\alpha})\right)^d$$

and satisfies $\inf_{t,x} a(t, x) > -1$.

10.3.1 A Priori Estimates for the Linearized Momentum Equation

As in the previous section, since we are only interested in local results, at the linear level, the mass and momentum equations may be treated separately. For the mass equation, using Theorem 3.14 page 133 turns out to be still appropriate. As for the momentum equation, we now have to consider a linearization which allows for *nonconstant* coefficients, namely,

$$\partial_t u + v \cdot \nabla u + u \cdot \nabla w - b\mathcal{A}u = f, \tag{10.27}$$

where b is a given positive function depending on (t, x) and tending to (say) 1 when x goes to infinity.

In this subsection, we shall prove that the estimates of Proposition 10.3 may be adapted to this new framework, provided that $b - 1$ is in $L_T^\infty(\dot{B}_{2,1}^{\frac{d}{2}+\alpha})$ *for some positive α.*

Proposition 10.12. *Let $\alpha \in \,]0, 1]$ and $s \in \,]-\frac{d}{2}, \frac{d}{2}]$. Assume that $b = 1 + c$ with $c \in L_T^\infty(\dot{B}_{2,1}^{\frac{d}{2}+\alpha})$ and that*

$$b_* := \inf_{(t,x)\in[0,T]\times\mathbb{R}^d} b(t, x) > 0. \tag{10.28}$$

There exists a universal constant κ, and a constant C depending only on d, α, and s, such that for all $t \in [0, T]$,

$$\|u\|_{\widetilde{L}^\infty_t(\dot{B}^s_{2,1})} + \kappa b_* \underline{\nu}\|u\|_{L^1_t(\dot{B}^{s+2}_{2,1})} \le \Big(\|u_0\|_{\dot{B}^s_{2,1}} + \|f\|_{L^1_t(\dot{B}^s_{2,1})}\Big)$$

$$\times \exp\Big(C\int_0^t \Big(\|v\|_{\dot{B}^{\frac{d}{2}+1}_{2,1}} + \|w\|_{\dot{B}^{\frac{d}{2}+1}_{2,1}} + b_*\,\underline{\nu}\Big(\frac{\overline{\nu}}{b_*\underline{\nu}}\Big)^{\frac{2}{\alpha}}\|c\|_{\dot{B}^{\frac{d}{2}+\alpha}_{2,1}}^{\frac{2}{\alpha}}\Big)\,dt'\Big).$$

If v and w depend linearly on u, then the above inequality is true for all $s \in\ ]0, \frac{d}{2} + \alpha]$, and the argument of the exponential term may be replaced with

$$C\int_0^t \Big(\|\nabla u\|_{L^\infty} + b_*\,\underline{\nu}\Big(\frac{\overline{\nu}}{b_*\underline{\nu}}\Big)^{\frac{2}{\alpha}}\|c\|_{\dot{B}^{\frac{d}{2}+\alpha}_{2,1}}^{\frac{2}{\alpha}}\Big)\,dt'.$$

Proof. We first consider the case $\lambda + \mu \ge 0$. Applying the spectral cut-off operator $\dot{\Delta}_j$ to (10.27) then yields

$$\partial_t u_j + v\cdot\nabla u_j - \mu\,\mathrm{div}(b\nabla u_j) - (\lambda+\mu)\nabla(b\,\mathrm{div}\,u_j) = f_j + \dot{\Delta}_j(u\cdot\nabla w) + R_j + \widetilde{R}_j$$

with $u_j \overset{\text{def}}{=} \dot{\Delta}_j u$, $f_j \overset{\text{def}}{=} \dot{\Delta}_j f$, $R_j \overset{\text{def}}{=} [v^i, \dot{\Delta}_j]\partial_i u$, and

$$\widetilde{R}_j \overset{\text{def}}{=} \mu\big(\dot{\Delta}_j(c\Delta u) - \mathrm{div}(c\nabla\dot{\Delta}_j u)\big) + (\lambda+\mu)\big(\dot{\Delta}_j(c\nabla\,\mathrm{div}\,u) - \nabla(c\,\mathrm{div}\,\dot{\Delta}_j u)\big).$$

Taking the L^2 inner product of the above equation with u_j, we get, after a few calculations and integrations by parts,

$$\frac{1}{2}\frac{d}{dt}\|u_j\|_{L^2}^2 - \frac{1}{2}\int |u_j|^2\,\mathrm{div}\,v\,dx + \int b\big(\mu|\nabla u_j|^2 + (\lambda+\mu)|\,\mathrm{div}\,u_j|^2\big)\,dx$$

$$\le \|u_j\|_{L^2}\Big(\|f_j\|_{L^2} + \|\dot{\Delta}_j(u\cdot\nabla w)\|_{L^2} + \|R_j\|_{L^2} + \|\widetilde{R}_j\|_{L^2}\Big).$$

Under the assumption that $\lambda + \mu \ge 0$, the term $(\lambda+\mu)|\,\mathrm{div}\,u_j|^2$ may be omitted. Therefore, by virtue of Bernstein's inequality and (10.28), the above inequality entails that

$$\|u_j(t)\|_{L^2} + 2\kappa b_*\,\mu 2^{2j}\int_0^t \|u_j\|_{L^2}\,dt' \le \|u_j(0)\|_{L^2} + \int_0^t \|f_j\|_{L^2}\,dt'$$

$$+ \int_0^t \Big(\|\dot{\Delta}_j(u\cdot\nabla w)\|_{L^2} + \|R_j\|_{L^2} + \|\widetilde{R}_j\|_{L^2} + \frac{1}{2}\|\mathrm{div}\,v\|_{L^\infty}\|u_j\|_{L^2}\Big)\,dt \quad (10.29)$$

for some universal constant κ.

We will temporarily assume that $\widetilde{R}_j$ satisfies

$$\|\widetilde{R}_j\|_{L^2} \le Cc_j\overline{\nu}2^{-js}\|c\|_{\dot{B}^{\frac{d}{2}+\alpha}_{2,1}}\|\nabla u\|_{\dot{B}^{s+1-\alpha}_{2,1}} \quad \text{for } -d/2 < s \le d/2+\alpha, \quad (10.30)$$

where $(c_j)_{j\in\mathbb{Z}}$ denotes a positive sequence such that $\sum_j c_j = 1$. Then, using the inequalities (10.10) and (10.11), multiplying both sides of (10.29) by 2^{js}, and summing over j, we end up with

$$\|u\|_{\widetilde{L}_t^\infty(\dot{B}_{2,1}^s)} + 2\kappa b_* \,\mu \|u\|_{L_t^1(\dot{B}_{2,1}^{s+2})} \le \|u_0\|_{\dot{B}_{2,1}^s} + \|f\|_{L_t^1(\dot{B}_{2,1}^s)}$$

$$+ C \int_0^t \Big(\|v\|_{\dot{B}_{2,1}^{\frac{d}{2}+1}} + \|w\|_{\dot{B}_{2,1}^{\frac{d}{2}+1}} \Big) \|u\|_{\dot{B}_{2,1}^s} \, dt'$$

$$+ C\bar{\nu} \int_0^t \|c\|_{\dot{B}_{2,1}^{\frac{d}{2}+\alpha}} \|u\|_{\dot{B}_{2,1}^{s+2-\alpha}} \, dt' \tag{10.31}$$

for a constant C depending only on α, d, and s.

Next, by combining interpolation and Young's inequality, we easily get

$$C\bar{\nu}\|c\|_{\dot{B}_{2,1}^{\frac{d}{2}+\alpha}} \|u\|_{\dot{B}_{2,1}^{s+2-\alpha}} \le \kappa b_* \,\mu \|u\|_{\dot{B}_{2,1}^{s+2}} + C^{\frac{2}{\alpha}} \bar{\nu}^{\frac{2}{\alpha}} (\kappa b_* \,\mu)^{1-\frac{2}{\alpha}} \|c\|_{\dot{B}_{2,1}^{\frac{d}{2}+\alpha}}^{\frac{2}{\alpha}} \|u\|_{\dot{B}_{2,1}^s}.$$

Plugging this into (10.31) and making use of Gronwall's inequality completes the proof of the first inequality of Proposition 10.12 in the case where $\lambda + \mu \ge 0$. The case where v and w depend linearly on u follows from a slight modification of (10.11) [see the inequality (2.54) page 112].

The case $\lambda + \mu < 0$ works in almost the same way. We just have to write the equation for u_j in a slightly different way, namely, for all $i \in \{1, \dots, d\}$ (with the summation convention),

$$\partial_t u_j^i + v \cdot \nabla u_j^i - \mu \operatorname{div}(b\nabla u_j^i) - (\lambda + \mu)\partial_k(b\partial_i u_j^k) = f_j^i + \dot{\Delta}_j(u \cdot \nabla w^i) + R_j^i + \check{R}_j^i$$

with

$$\check{R}_j^i \overset{\text{def}}{=} \mu\big(\dot{\Delta}_j(c\Delta u^i) - \operatorname{div}(c\nabla\dot{\Delta}_j u^i)\big) + (\lambda + \mu)\big(\dot{\Delta}_j(c\partial_i \operatorname{div} u) - \partial_k(c\partial_i \dot{\Delta}_j u^k)\big).$$

Taking the L^2 inner product of the above equation with u_j, we get, after a few calculations and integrations by parts,

$$\frac{1}{2}\frac{d}{dt}\|u_j\|_{L^2}^2 - \frac{1}{2}\int |u_j|^2 \operatorname{div} v \, dx + \int b\big(\mu|\nabla u_j|^2 + (\lambda + \mu)\nabla u_j : \nabla u_j\big) \, dx$$

$$\le \|u_j\|_{L^2}\Big(\|f_j\|_{L^2} + \|\dot{\Delta}_j(u \cdot \nabla w)\|_{L^2} + \|R_j\|_{L^2} + \|\check{R}_j\|_{L^2}\Big).$$

Note that, according to the Cauchy–Schwarz inequality,

$$\nabla u_j : \nabla u_j \le |\nabla u_j|^2.$$

As we have $\lambda + \mu < 0$ and $\lambda + 2\mu > 0$, we thus get

$$\|u_j(t)\|_{L^2} + 2\kappa b_* \,\nu 2^{2j}\int_0^t \|u_j\|_{L^2} \, dt' \le \|u_j(0)\|_{L^2} + \int_0^t \|f_j\|_{L^2} \, dt'$$

$$+ \int_0^t \Big(\|\dot{\Delta}_j(u \cdot \nabla w)\|_{L^2} + \|R_j\|_{L^2} + \|\check{R}_j\|_{L^2} + \frac{1}{2}\|\operatorname{div} v\|_{L^\infty}\|u_j\|_{L^2}\Big) \, dt'.$$

Now, by virtue of Lemma 10.13 below, the new commutator $\check{R}_j$ also satisfies (10.30). Hence, the proof may be completed exactly as in the case $\lambda + \mu \ge 0$.

$\square$

For the sake of completeness, we now prove the inequality (10.30). This readily follows from the following lemma.

Lemma 10.13. *Let $\alpha \in \,]1-\frac{d}{2}, 1]$ and $\sigma \in \,]-\frac{d}{2}, \frac{d}{2}+\alpha[$. Define*

$$\mathcal{R}_j^k \stackrel{def}{=} \dot{\Delta}_j(c\partial_k w) - \partial_k(c\dot{\Delta}_j w) \quad for \quad k \in \{1,\ldots,d\}.$$

There exists some $C = C(\alpha, d, \sigma)$ such that

$$\sum_j 2^{j\sigma} \|\mathcal{R}_j^k\|_{L^2} \le C\|c\|_{\dot{B}_{2,1}^{\frac{d}{2}+\alpha}} \|\nabla w\|_{\dot{B}_{2,1}^{\sigma-\alpha}}. \tag{10.32}$$

Proof. The proof is based on Bony's decomposition: We split $\mathcal{R}_j^k$ as

$$\mathcal{R}_j^k = \underbrace{\partial_k[\dot{\Delta}_j, T_c]w}_{\mathcal{R}_j^{k,1}} - \underbrace{\dot{\Delta}_j T_{\partial_k c} w}_{\mathcal{R}_j^{k,2}} + \underbrace{\dot{\Delta}_j T_{\partial_k w} c}_{\mathcal{R}_j^{k,3}} + \underbrace{\dot{\Delta}_j R(\partial_k w, c)}_{\mathcal{R}_j^{k,4}} - \underbrace{\partial_k T'_{\dot{\Delta}_j w} c}_{\mathcal{R}_j^{k,5}}.$$

Using the fact that, owing to Proposition 2.10 page 59, we have

$$\mathcal{R}_j^{k,1} = \sum_{j'=j-4}^{j+4} \partial_k[\dot{\Delta}_j, \dot{S}_{j'-1}c]\dot{\Delta}_{j'}w,$$

Bernstein's inequality and Lemma 2.97 page 110 entail (under the hypothesis that $\alpha \le 1$) that

$$\sum_j 2^{j\sigma} \|\mathcal{R}_j^{k,1}\|_{L^2} \le C\|\nabla c\|_{\dot{B}_{\infty,1}^{\alpha-1}} \|\nabla w\|_{\dot{B}_{2,1}^{\sigma-\alpha}}. \tag{10.33}$$

Continuity results for the paraproduct (see Proposition 2.82) ensure that $\mathcal{R}_j^{k,2}$ satisfies (10.33) if $\alpha \le 1$, and that

$$\sum_j 2^{j\sigma} \|\mathcal{R}_j^{k,3}\|_{L^2} \le C\|\nabla w\|_{\dot{B}_{\infty,1}^{\sigma-\alpha-\frac{d}{2}}} \|c\|_{\dot{B}_{2,1}^{\frac{d}{2}+\alpha}} \quad if \quad \sigma - \alpha - \frac{d}{2} \le 0. \tag{10.34}$$

Next, Proposition 2.85 guarantees that under the hypothesis $\sigma > -\frac{d}{2}$, we have

$$\sum_j 2^{j\sigma} \|\mathcal{R}_j^{k,4}\|_{L^2} \le C\|\nabla w\|_{\dot{B}_{2,1}^{\sigma-\alpha}} \|c\|_{\dot{B}_{2,1}^{\frac{d}{2}+\alpha}}. \tag{10.35}$$

To bound $\mathcal{R}_j^{k,5}$, we use the decomposition

$$\mathcal{R}_j^{k,5} = \sum_{j'\ge j-3} \partial_k(\dot{S}_{j'+2}\dot{\Delta}_j w\,\dot{\Delta}_{j'}c),$$

which leads (after a suitable use of Bernstein and Hölder inequalities) to

$$2^{j\sigma}\left\|\mathcal{R}_j^{k,5}\right\|_{L^2} \leq C \sum_{j' \geq j-3} 2^{(j-j')(\alpha+\frac{d}{2}-1)} 2^{j(\sigma-\alpha)} \|\dot{\Delta}_j \nabla w\|_{L^2} 2^{j'(\frac{d}{2}+\alpha)} \|\dot{\Delta}_{j'} c\|_{L^2}.$$

Hence, since $\alpha + \frac{d}{2} - 1 > 0$, we have

$$\sum_j 2^{j\sigma} \|\mathcal{R}_j^{k,5}\|_{L^2} \leq C\|c\|_{\dot{B}_{2,1}^{\frac{d}{2}+\alpha}} \|\nabla w\|_{\dot{B}_{2,1}^{\sigma-\alpha}}.$$

Combining the last inequality with (10.33), (10.34), and (10.35) and using the embedding $\dot{B}_{2,1}^r \hookrightarrow \dot{B}_{\infty,1}^{r-\frac{d}{2}}$ for $r = \frac{d}{2}+\alpha-1$ and $\sigma - \alpha$ completes the proof of (10.32). □

Corollary 10.14. *Let (a, u) satisfy (NSC) on $[0,T] \times \mathbb{R}^d$. Assume that there exist two positive constants, b_* and b^*, such that*

$$b_* \leq 1 + a_0 \leq b^*, \tag{10.36}$$

and that $a \in \mathcal{C}^1([0,T]; \dot{B}_{2,1}^{\frac{d}{2}} \cap \dot{B}_{2,1}^{\frac{d}{2}+\alpha})$ and $u \in \mathcal{C}^1([0,T]; \dot{B}_{2,1}^{\frac{d}{2}-1} \cap \dot{B}_{2,1}^{\frac{d}{2}+1+\alpha})^d$. Also, suppose that (10.12) has a solution u_L in $\mathcal{C}^1([0,T]; \dot{B}_{2,1}^{\frac{d}{2}-1} \cap \dot{B}_{2,1}^{\frac{d}{2}+1+\alpha})^d$.

We introduce the following notation:

$$A_0^\alpha \stackrel{def}{=} \|a_0\|_{\dot{B}_{2,1}^{\frac{d}{2}} \cap \dot{B}_{2,1}^{\frac{d}{2}+\alpha}}, \quad A^\alpha(t) \stackrel{def}{=} \|a\|_{L_t^\infty(\dot{B}_{2,1}^{\frac{d}{2}} \cap \dot{B}_{2,1}^{\frac{d}{2}+\alpha})},$$

$$U_0^\alpha(t) \stackrel{def}{=} \|u_0\|_{\dot{B}_{2,1}^{\frac{d}{2}-1} \cap \dot{B}_{2,1}^{\frac{d}{2}-1+\alpha}} + \|f\|_{L_t^1(\dot{B}_{2,1}^{\frac{d}{2}-1} \cap \dot{B}_{2,1}^{\frac{d}{2}-1+\alpha})},$$

$$U_L^\alpha(t) \stackrel{def}{=} \|u_L\|_{L_t^1(\dot{B}_{2,1}^{\frac{d}{2}+1} \cap \dot{B}_{2,1}^{\frac{d}{2}+\alpha+1})},$$

$$\overline{U}^\alpha(t) \stackrel{def}{=} \|\overline{u}\|_{\widetilde{L}_t^\infty(\dot{B}_{2,1}^{\frac{d}{2}-1} \cap \dot{B}_{2,1}^{\frac{d}{2}-1+\alpha})} + b_*\nu\|\overline{u}\|_{L_t^1(\dot{B}_{2,1}^{\frac{d}{2}+1} \cap \dot{B}_{2,1}^{\frac{d}{2}+1+\alpha})} \quad with \quad \overline{u} \stackrel{def}{=} u - u_L.$$

There exist two constants, η and C, depending only on d, α, and G, such that if

$$\begin{cases} b_*\nu\left(\frac{\nu}{b_*\nu}\right)^{\frac{2}{\alpha}} T\left(A_0^\alpha + 1\right)^{\frac{2}{\alpha}} \leq \eta \\ U_0^\alpha(T)U_L^\alpha(T) + (A_0^\alpha + 1)(T + \overline{\nu}U_L^\alpha(T)) \leq \eta b_*\nu, \end{cases} \tag{10.37}$$

then we have

$$b_*/2 \leq 1 + a(t,x) \leq 2b^* \quad for~all \quad (t,x) \in [0,T] \times \mathbb{R}^d, \tag{10.38}$$

$$A^\alpha(T) \leq 2A_0^\alpha+1, \quad \overline{U}^\alpha(T) \leq C\Big((A_0^\alpha+1)(T+\overline{\nu}U_L^\alpha(T))+U_0^\alpha(T)U_L^\alpha(T)\Big). \tag{10.39}$$

Proof. We may write the system satisfied by $(a, \overline{u})$ as follows:

$$\begin{cases} \partial_t a + u \cdot \nabla a + (1+a)\,\mathrm{div}\,u = 0 \\ \partial_t \overline{u} + u \cdot \nabla \overline{u} + \overline{u} \cdot \nabla u_L - \frac{1}{1+a}\mathcal{A}\overline{u} = -u_L \cdot \nabla u_L - I(a)\mathcal{A}u_L - \nabla(G(a)) \\ (a, \overline{u})_{|t=0} = (a_0, 0). \end{cases}$$

We first bound a. Applying the product laws in Besov spaces (see Corollary 2.86 page 104), we get

$$\|(1+a)\operatorname{div} u\|_{\dot{B}_{2,1}^{\frac{d}{2}} \cap \dot{B}_{2,1}^{\frac{d}{2}+\alpha}} \leq C\big(1+\|a\|_{\dot{B}_{2,1}^{\frac{d}{2}} \cap \dot{B}_{2,1}^{\frac{d}{2}+\alpha}}\big)\|\operatorname{div} u\|_{\dot{B}_{2,1}^{\frac{d}{2}} \cap \dot{B}_{2,1}^{\frac{d}{2}+\alpha}}.$$

Hence, Theorem 3.14 (adapted to the homogeneous framework, see page 134) combined with Gronwall's lemma yields, for all $t \in [0, T]$,

$$A^\alpha(t) \leq A_0^\alpha \exp\Big(C\int_0^t \|\nabla u\|_{\dot{B}_{2,1}^{\frac{d}{2}} \cap \dot{B}_{2,1}^{\frac{d}{2}+\alpha}}\, dt'\Big)$$
$$+ \exp\Big(C\int_0^t \|\nabla u\|_{\dot{B}_{2,1}^{\frac{d}{2}} \cap \dot{B}_{2,1}^{\frac{d}{2}+\alpha}}\, dt'\Big) - 1. \qquad (10.40)$$

In order to ensure that the condition (10.38) is satisfied, we may use the fact that

$$(\partial_t + u \cdot \nabla)(1+a)^{\pm 1} \pm (1+a)^{\pm 1} \operatorname{div} u = 0.$$

Hence, taking advantage of Gronwall's lemma, we get

$$\big\|(1+a)^{\pm 1}(t)\big\|_{L^\infty} \leq \big\|(1+a_0)^{\pm 1}\big\|_{L^\infty} \exp\Big(\int_0^t \|\operatorname{div} u\|_{L^\infty}\, dt'\Big).$$

Therefore, the condition (10.38) is satisfied on $[0, t]$ whenever

$$\int_0^t \|\operatorname{div} u\|_{L^\infty}\, dt' \leq \log 2. \qquad (10.41)$$

In order to bound $\overline{u}$, we use Proposition 10.12 with $c = -I(a)$. As, according to Theorems 2.47, 2.52, and 2.61 page 94, we have, for all $\beta \in [0, \alpha]$,

$$\|u_L \cdot \nabla u_L\|_{\dot{B}_{2,1}^{\frac{d}{2}-1+\beta}} \leq C\|\nabla u_L\|_{\dot{B}_{2,1}^{\frac{d}{2}-1+\beta}}\|u_L\|_{\dot{B}_{2,1}^{\frac{d}{2}}},$$
$$\|\nabla G(a)\|_{\dot{B}_{2,1}^{\frac{d}{2}-1+\beta}} \leq C\|a\|_{\dot{B}_{2,1}^{\frac{d}{2}+\beta}},$$
$$\|I(a)\mathcal{A}u_L\|_{\dot{B}_{2,1}^{\frac{d}{2}-1+\beta}} \leq C\overline{\nu}\|I(a)\|_{\dot{B}_{2,1}^{\frac{d}{2}+\beta} \cap L^\infty}\|\nabla^2 u_L\|_{\dot{B}_{2,1}^{\frac{d}{2}-1+\beta}},$$
$$\|I(a)\|_{\dot{B}_{2,1}^{\frac{d}{2}+\beta} \cap L^\infty} \leq C\|a\|_{\dot{B}_{2,1}^{\frac{d}{2}+\beta} \cap L^\infty},$$

we get

$$\overline{U}^\alpha(T) \leq Ce^{C\int_0^T \big(\|\overline{u}\|_{\dot{B}_{2,1}^{\frac{d}{2}+1}} + \|u_L\|_{\dot{B}_{2,1}^{\frac{d}{2}+1}} + b_* \, \nu\big(\frac{\overline{\nu}}{b_*\nu}\big)^{\frac{2}{\alpha}} \|a\|_{\dot{B}_{2,1}^{\frac{d}{2}+\alpha}}^{\frac{2}{\alpha}}\big)\, dt'}$$
$$\times \Bigg(\|u_L\|_{L_T^1(\dot{B}_{2,1}^{\frac{d}{2}+1})}\|u_L\|_{L_T^\infty(\dot{B}_{2,1}^{\frac{d}{2}-1} \cap \dot{B}_{2,1}^{\frac{d}{2}+\alpha-1})}$$
$$+ \|a\|_{L_T^\infty(\dot{B}_{2,1}^{\frac{d}{2}} \cap \dot{B}_{2,1}^{\frac{d}{2}+\alpha})}\Big(T + \overline{\nu}\|u_L\|_{L_T^1(\dot{B}_{2,1}^{\frac{d}{2}+1} \cap \dot{B}_{2,1}^{\frac{d}{2}+\alpha+1})}\Big)\Bigg).$$

Note that Proposition 10.3 implies that

$$\|u_L\|_{L_T^\infty(\dot{B}_{2,1}^{\frac{d}{2}-1} \cap \dot{B}_{2,1}^{\frac{d}{2}+\alpha-1})} \leq C U_0^\alpha(T).$$

So, finally, defining

$$U_L^\alpha(T) \overset{\text{def}}{=} \|u_L\|_{L_T^1(\dot{B}_{2,1}^{\frac{d}{2}+1} \cap \dot{B}_{2,1}^{\frac{d}{2}+\alpha+1})} \quad \text{and} \quad A^\alpha(T) \overset{\text{def}}{=} \|a\|_{L_T^\infty(\dot{B}_{2,1}^{\frac{d}{2}} \cap \dot{B}_{2,1}^{\frac{d}{2}+\alpha})},$$

we conclude that

$$A^\alpha(T) \leq A_0^\alpha \exp\left(C\left(U_L^\alpha(T) + \frac{\overline{U}^\alpha(T)}{b_* \underline{\nu}}\right)\right) + \exp\left(C\left(U_L^\alpha(T) + \frac{\overline{U}^\alpha(T)}{b_* \underline{\nu}}\right)\right) - 1,$$

$$\overline{U}^\alpha(T) \leq C\left(U_0^\alpha(T)U_L^\alpha(T) + A^\alpha(T)(T + \overline{\nu}U_L^\alpha(T))\right)$$

$$\exp\left(C\left(U_L^\alpha(T) + \frac{\overline{U}^\alpha(T)}{b_* \underline{\nu}} + b_* \underline{\nu}\left(\frac{\overline{\nu}}{b_* \underline{\nu}}\right)^{\frac{2}{\alpha}} T(A^\alpha(T))^{\frac{2}{\alpha}}\right)\right).$$

Now, if T is sufficiently small so as to satisfy $\exp\left(C U_L^\alpha(T)\right) \leq \sqrt{2}$,

$$\exp\left(C\frac{\overline{U}^\alpha(T)}{b_* \underline{\nu}}\right) \leq \sqrt{2}, \quad \text{and} \quad \exp\left(Cb_* \underline{\nu}\left(\frac{\overline{\nu}}{b_* \underline{\nu}}\right)^{\frac{2}{\alpha}} T(A^\alpha(T))^{\frac{2}{\alpha}}\right) \leq 2,$$

then we have (10.39) and (10.41).

So, if we choose T such that (10.37) is satisfied for some sufficiently small constant η, then both (10.39) and the above conditions are satisfied *with a strict inequality.* It is now easy to complete the proof by means of a bootstrap argument similar to that of Corollary 10.4.

10.3.2 Existence of a Local Solution

The main steps of the proof of the existence are exactly the same as in the critical case.

First Step: Friedrichs Approximation

Using the notation introduced on page 440, we aim to solve the following system of ordinary differential equations:

$$(\widetilde{NSC_n}) \cdot \qquad \frac{d}{dt}\begin{pmatrix} a \\ u \end{pmatrix} = \begin{pmatrix} F_n(a, \overline{u}) \\ G_n(a, \overline{u}) \end{pmatrix}, \qquad \begin{pmatrix} a \\ u \end{pmatrix}_{|t=0} = \begin{pmatrix} \dot{\mathbb{E}}_n a_0 \\ 0 \end{pmatrix}$$

with $F_n(a, \overline{u}) \overset{\text{def}}{=} -\dot{\mathbb{E}}_n \operatorname{div}\left((1+a)u\right)$, $u \overset{\text{def}}{=} \overline{u} + u_L$, and

$$G_n(a, \overline{u}) \overset{\text{def}}{=} \dot{\mathbb{E}}_n\left((1+a)^{-1}\mathcal{A}\overline{u}\right) - \dot{\mathbb{E}}_n\left(u \cdot \nabla u\right) - \dot{\mathbb{E}}_n\left(I(a)\mathcal{A}u_L\right) - \dot{\mathbb{E}}_n \nabla\left(G(a)\right).$$

Note that if $1 + a_0$ is bounded away from zero, then so is $1 + \dot{\mathbb{E}}_n\, a_0$ for sufficiently large n, hence the data are in Ω_n, and it is not difficult to check that the map

$$(a, \overline{u}) \mapsto \big(F_n(a, \overline{u}), G_n(a, \overline{u})\big)$$

is in $\mathcal{C}(\mathbb{R}^+ \times \Omega_n; (\dot{L}_n^2)^{d+1})$ and is locally Lipschitz with respect to the variable $(a, \overline{u})$. Hence, the system $(\widetilde{NSC_n})$ has a unique maximal solution $(a^n, \overline{u}^n)$ in the space $\mathcal{C}^1([0, T_n^*[; \Omega_n)$.

Second Step: Uniform Estimates

We note that $(a^n, \overline{u}^n)$ satisfies

$$\begin{cases} \partial_t a^n + \dot{\mathbb{E}}_n\big(u^n \cdot \nabla a^n\big) + \dot{\mathbb{E}}_n\big((1 + a^n)\,\mathrm{div}\,u^n\big) = 0 \\ \partial_t \overline{u}^n - \dot{\mathbb{E}}_n\big((1 + a^n)^{-1}\mathcal{A}\overline{u}^n\big) + \dot{\mathbb{E}}_n\big(u^n \cdot \nabla u^n\big) + \nabla\dot{\mathbb{E}}_n\big(G(a^n)\big) = 0 \end{cases}$$

with initial data $a^n_{|t=0} = \dot{\mathbb{E}}_n\, a_0$ and $\overline{u}^n_{|t=0} = 0$, and where $u^n \overset{\text{def}}{=} u^n_L + \overline{u}^n$.

For the reasons already mentioned when treating the critical case on page 441, the results of Corollary 10.14 remain true for the approximate solution (a^n, u^n). Therefore, T_n^* may be bounded from below by any time T satisfying (10.37), and the inequalities (10.38), (10.39) are satisfied by (a^n, u^n). In particular, $(a^n, u^n)_{n \in \mathbb{N}}$ is bounded in E_T^α.

Third Step: Time Derivatives

The following lemma will supply the compactness needed to pass to the limit in (NSC_n).

Lemma 10.15. *Let $\overline{a}^n \overset{\text{def}}{=} a^n - \dot{\mathbb{E}}_n\, a_0$. The sequence $(\overline{a}^n)_{n \geq 1}$ is then bounded in*

$$\mathcal{C}([0, T]; \dot{B}_{2,1}^{\frac{d}{2}} \cap \dot{B}_{2,1}^{\frac{d}{2}+\alpha}) \cap \mathcal{C}^{\frac{1}{2}}([0, T]; \dot{B}_{2,1}^{\frac{d}{2}-1} \cap \dot{B}_{2,1}^{\frac{d}{2}-1+\alpha}),$$

and the sequence $(\overline{u}^n)_{n \geq 1}$ is bounded in

$$\mathcal{C}([0, T]; \dot{B}_{2,1}^{\frac{d}{2}-1} \cap \dot{B}_{2,1}^{\frac{d}{2}-1+\alpha}) \cap \mathcal{C}^{\frac{1}{2}}([0, T]; \dot{B}_{2,1}^{\frac{d}{2}-1} + \dot{B}_{2,1}^{\frac{d}{2}-2+\alpha}).$$

Proof. To get the result for $(\overline{a}^n)_{n \geq 1}$, it suffices to check whether $(\partial_t \overline{a}^n)_{n \geq 1}$ is bounded in $L^2([0, T]; \dot{B}_{2,1}^{\frac{d}{2}-1} \cap \dot{B}_{2,1}^{\frac{d}{2}-1+\alpha})$. As

$$\partial_t \overline{a}^n = -\dot{\mathbb{E}}_n\,\mathrm{div}\big(u^n(1 + a^n)\big),$$

this is a mere consequence of the bounds from step 2 and of the product laws in Besov spaces stated in Section 2.6.

As regards $(\overline{u}^n)_{n\geq 1}$, it suffices to prove that $(\partial_t \overline{u}^n)_{n\geq 1}$ is bounded in $L^2([0,T]; \dot{B}_{2,1}^{\frac{d}{2}-1+\beta} + \dot{B}_{2,1}^{\frac{d}{2}-2+\beta})$ for $\beta \in \{0, \alpha\}$. This follows from the fact that

$$\partial_t \overline{u}^n = -\dot{\mathbb{E}}_n (u^n \cdot \nabla u^n) + \dot{\mathbb{E}}_n ((1+a^n)^{-1} \mathcal{A}\overline{u}^n) - \nabla \dot{\mathbb{E}}_n (G(a^n)) - \dot{\mathbb{E}}_n (I(a^n)\mathcal{A}u_L^n).$$

Indeed, as $(u^n)_{n\geq 1}$ and $(\overline{u}^n)_{n\geq 1}$ are bounded in $L_T^2(\dot{B}_{2,1}^{\frac{d}{2}} \cap \dot{B}_{2,1}^{\frac{d}{2}+\alpha}) \cap L_T^\infty(\dot{B}_{2,1}^{\frac{d}{2}-1} \cap \dot{B}_{2,1}^{\frac{d}{2}-1+\alpha})$ and $(\overline{a}^n)_{n\geq 1}$ is bounded in $L_T^\infty(\dot{B}_{2,1}^{\frac{d}{2}} \cap \dot{B}_{2,1}^{\frac{d}{2}+\alpha})$, we deduce that the first three terms on the right-hand side belong to $L_T^2(\dot{B}_{2,1}^{\frac{d}{2}-2} \cap \dot{B}_{2,1}^{\frac{d}{2}-2+\alpha})$ and that the last one is in $L_T^\infty(\dot{B}_{2,1}^{\frac{d}{2}-1} \cap \dot{B}_{2,1}^{\frac{d}{2}-1+\alpha})$, uniformly. $\qquad\square$

Fourth Step: Compactness and Convergence

According to the previous lemma, the sequence $(\overline{a}^n)_{n\geq 1}$ is bounded in the space $\mathcal{C}^{\frac{1}{2}}([0,T]; \dot{B}_{2,1}^{\frac{d}{2}-1} \cap \dot{B}_{2,1}^{\frac{d}{2}-1+\alpha})$. By combining Proposition 2.93 and Theorem 2.94 page 108, it is easy to check that for all $\phi \in \mathcal{D}$, the map $u \mapsto \phi u$ is compact from $\dot{B}_{2,1}^{\frac{d}{2}+\alpha} \cap \dot{B}_{2,1}^{\frac{d}{2}-1+\alpha}$ to $B_{2,1}^{\frac{d}{2}-1+\alpha}$. Therefore, arguing as in the proof of existence for Theorem 10.2, we deduce that there exists some function $\overline{a}$ such that for all $\phi \in \mathcal{D}$, the sequence $(\phi \overline{a}^n)_{n\geq 1}$ converges (up to a subsequence independent of ϕ) to $\phi\overline{a}$ in $\mathcal{C}^{\frac{1}{2}}([0,T]; \dot{B}_{2,1}^{\frac{d}{2}-1+\alpha})$.

Likewise, since the map $u \mapsto \phi u$ is compact from $\dot{B}_{2,1}^{\frac{d}{2}-1} \cap \dot{B}_{2,1}^{\frac{d}{2}-1+\alpha}$ to $B_{2,1}^{\frac{d}{2}-2+\alpha}$, there exists a vector field $\overline{u}$ such that (up to extraction), for all $\phi \in \mathcal{D}$, the sequence $(\phi \overline{u}^n)_{n\geq 1}$ converges to $\phi\overline{u}$ in $\mathcal{C}^{\frac{1}{2}}([0,T]; \dot{B}_{2,1}^{\frac{d}{2}-2+\alpha})$.

Next, the uniform bounds supplied by the second step and the Fatou property together ensure that, in addition, $1 + a$ is positive and

$$(a, u) \in \widetilde{L}_T^\infty(\dot{B}_{2,1}^{\frac{d}{2}} \cap \dot{B}_{2,1}^{\frac{d}{2}+\alpha}) \times \left(\widetilde{L}_T^\infty(\dot{B}_{2,1}^{\frac{d}{2}-1} \cap \dot{B}_{2,1}^{\frac{d}{2}-1+\alpha}) \cap L_T^1(\dot{B}_{2,1}^{\frac{d}{2}+1} \cap \dot{B}_{2,1}^{\frac{d}{2}+1+\alpha})\right)^d.$$

The proof is similar to that of the critical case and is thus left to the reader. Interpolating with the above convergence results, we may get better convergence results for $(\overline{a}^n, \overline{u}^n)$ and pass to the limit in $(\widetilde{NSC_n})$. Defining $(a, u) \overset{\text{def}}{=} (a_0 + \overline{a}, u_L + \overline{u})$, we thus get a solution (a, u) of (NSC) with data (a_0, u_0). Using the equation and the product laws, we also have

$$(\partial_t + u \cdot \nabla)a \in L_T^1(\dot{B}_{2,1}^{\frac{d}{2}} \cap \dot{B}_{2,1}^{\frac{d}{2}+\alpha}) \quad \text{and} \quad \partial_t u \in L_T^1(\dot{B}_{2,1}^{\frac{d}{2}-1} \cap \dot{B}_{2,1}^{\frac{d}{2}-1+\alpha}).$$

Theorem 3.19 therefore guarantees that $a \in \widetilde{\mathcal{C}}_T(\dot{B}_{2,1}^{\frac{d}{2}} \cap \dot{B}_{2,1}^{\frac{d}{2}+\alpha})$ and, obviously, $u \in \widetilde{\mathcal{C}}_T(\dot{B}_{2,1}^{\frac{d}{2}-1} \cap \dot{B}_{2,1}^{\frac{d}{2}+\alpha-1})$.

Remark 10.16. Combining (10.13) with the properties of the heat semigroup described in Lemma 2.4 yields a rather explicit lower bound on the lifespan T^* of the solution. Indeed, using the fact that

$$1 - e^{-\underline{\nu}T2^{2j}} \le (\underline{\nu}T)^{\frac{\alpha}{2}}2^{j\alpha},$$

we may find some constant c, depending only on d, b_*, b^*, α, λ, and μ, and such that

$$T^* \ge \sup\left\{T \in \left]0, \frac{c}{(A_0^\alpha)^{\frac{2}{\alpha}}+1}\right[\Big/ (\underline{\nu}T)^{\frac{\alpha}{2}}U_0^\alpha(T) \le \frac{c}{1 + (A_0^\alpha)^{\frac{2}{\alpha}} + U_0^\alpha(T)}\right\}.$$

10.3.3 Uniqueness

Let (a^1, u^1) and (a^2, u^2) be two solutions in E_T^α of (NSC) with the same data. We can assume, without loss of generality, that (a^2, u^2) is the solution constructed in the previous subsection so that

$$1 + \inf_{(t,x)\in[0,T]\times\mathbb{R}^d} a^2(t,x) > 0.$$

We want to prove that $(a^2, u^2) \equiv (a^1, u^1)$ on $[0,T] \times \mathbb{R}^d$. For this, we shall estimate $(\delta a, \delta u) \overset{\text{def}}{=} (a^2 - a^1, u^2 - u^1)$ with respect to a suitable norm, having noted that

$$\begin{cases} \partial_t \delta a + u^2 \cdot \nabla \delta a + \delta F = 0 \\ \partial_t \delta u + u^2 \cdot \nabla \delta u + \delta u \cdot \nabla u^1 - (1+a^2)^{-1}\mathcal{A}\delta u = \delta G_1 + \delta G_2 \end{cases} \tag{10.42}$$

with $$\delta F \overset{\text{def}}{=} \delta u \cdot \nabla a^1 + \delta a \,\mathrm{div}\, u^2 + (1+a^1)\,\mathrm{div}\,\delta u,$$

$$\delta G_1 \overset{\text{def}}{=} \left(I(a^1) - I(a^2)\right)\mathcal{A}u^1, \qquad \delta G_2 \overset{\text{def}}{=} \nabla(G(a^1) - G(a^2)).$$

For the same reasons as in the critical case, the uniqueness is going to be proven in a larger function space, namely,

$$F_T^\alpha \overset{\text{def}}{=} \mathcal{C}([0,T]; \dot{B}_{2,1}^{\frac{d}{2}-1+\alpha}) \times \left(\mathcal{C}([0,T]; \dot{B}_{2,1}^{\frac{d}{2}-2+\alpha}) \cap L_T^1(\dot{B}_{2,1}^{\frac{d}{2}+\alpha})\right)^d.$$

Of course, we first have to establish that $(\delta a, \delta u)$ belongs to F_T^α. For δa, this is easy because, arguing as in Lemma 10.15, we get $a^i - a_0 \in \mathcal{C}^{\frac{1}{2}}([0,T]; \dot{B}_{2,1}^{\frac{d}{2}-1+\alpha})$.

To deal with δu, we again introduce $\bar{u}^i \overset{\text{def}}{=} u^i - \bar{u}_L$, where $\bar{u}_L$ is the solution of (10.22). We obviously have $\bar{u}^i(0) = 0$ and

$$\partial_t \bar{u}^i = \mathcal{A}\bar{u}^i - I(a^i)\mathcal{A}u^i - u^i \cdot \nabla u^i - \nabla(G(a^i) - G(a_0)).$$

Since $\bar{a}^i \in L_T^\infty(\dot{B}_{2,1}^{\frac{d}{2}-1+\alpha})$ and $(a^i, u^i) \in E_T^\alpha$, the right-hand side belongs to $L_T^2(\dot{B}_{2,1}^{\frac{d}{2}-2+\alpha})$. Hence, $\bar{u}^i$ belongs to $\mathcal{C}([0,T]; \dot{B}_{2,1}^{\frac{d}{2}-2+\alpha})$, and we can now conclude that $(\delta a, \delta u) \in F_T^\alpha$.

Next, applying Theorem 3.14 to the first equation of (10.42), we get

$$\|\delta a\|_{L_{\bar{T}}^{\infty}(\dot{B}_{2,1}^{\frac{d}{2}-1+\alpha})} \leq \exp\left(C\|u^2\|_{L_{\bar{T}}^1(\dot{B}_{2,1}^{\frac{d}{2}+1})}\right)\|\delta F\|_{L_{\bar{T}}^1(\dot{B}_{2,1}^{\frac{d}{2}-1+\alpha})} \quad \text{for all} \ \ \bar{T} \in [0,T].$$

Easy computations based on Theorems 2.47 and 2.52 yield

$$\|\delta F\|_{\dot{B}_{2,1}^{\frac{d}{2}-1+\alpha}} \leq C\|\delta u\|_{\dot{B}_{2,1}^{\frac{d}{2}}}\|\nabla a^1\|_{\dot{B}_{2,1}^{\frac{d}{2}-1+\alpha}}$$
$$+\|\operatorname{div} u^2\|_{\dot{B}_{2,1}^{\frac{d}{2}}}\|\delta a\|_{\dot{B}_{2,1}^{\frac{d}{2}-1+\alpha}} + \left(1+\|a^1\|_{\dot{B}_{2,1}^{\frac{d}{2}}}\right)\|\delta u\|_{\dot{B}_{2,1}^{\frac{d}{2}+\alpha}}.$$

Hence, using Gronwall's lemma and interpolation, we discover that there exists some constant C_T, independent of $\bar{T}$, such that

$$\|\delta a\|_{L_{\bar{T}}^{\infty}(\dot{B}_{2,1}^{\frac{d}{2}-1+\alpha})} \leq C_T\left(\|\delta u\|_{L_{\bar{T}}^1(\dot{B}_{2,1}^{\frac{d}{2}+\alpha})} + \|\delta u\|_{L_{\bar{T}}^{\infty}(\dot{B}_{2,1}^{\frac{d}{2}-2+\alpha})}\right). \tag{10.43}$$

Next, applying Proposition 10.12 to the second equation of (10.42) yields, for some constant C depending only on d, λ, μ, and α,

$$\|\delta u\|_{L_{\bar{T}}^{\infty}(\dot{B}_{2,1}^{\frac{d}{2}-2+\alpha})} + \|\delta u\|_{L_{\bar{T}}^1(\dot{B}_{2,1}^{\frac{d}{2}+\alpha})}$$
$$\leq Ce^{C\int_0^{\bar{T}}(\|u^1\|_{\dot{B}_{2,1}^{\frac{d}{2}+1}}+\|u^2\|_{\dot{B}_{2,1}^{\frac{d}{2}+1}}+\|a^2\|_{\dot{B}_{2,1}^{\frac{d}{2}+\alpha}}^{\frac{2}{\alpha}})\,dt}\sum_{i=1}^2\|\delta G_i\|_{L_{\bar{T}}^1(\dot{B}_{2,1}^{\frac{d}{2}-2+\alpha})}.$$

Because $\dot{B}_{2,1}^{\frac{d}{2}}(\mathbb{R}^d)\hookrightarrow\mathcal{C}(\mathbb{R}^d)$, we have $a^1 \in \mathcal{C}([0,T]\times\mathbb{R}^d)$. Hence, for sufficiently small $\bar{T}$, a^1 also satisfies (10.38). Therefore, applying Theorems 2.47, 2.52 and Corollary 2.66 page 97 yields

$$\|\delta G_1\|_{L_{\bar{T}}^1(\dot{B}_{2,1}^{\frac{d}{2}-2+\alpha})} \leq C\left(1+\|a^1\|_{L_{\bar{T}}^{\infty}(\dot{B}_{2,1}^{\frac{d}{2}})}+\|a^2\|_{L_{\bar{T}}^{\infty}(\dot{B}_{2,1}^{\frac{d}{2}})}\right)$$
$$\times\|\delta a\|_{L_{\bar{T}}^{\infty}(\dot{B}_{2,1}^{\frac{d}{2}-1+\alpha})}\|u^1\|_{L_{\bar{T}}^1(\dot{B}_{2,1}^{\frac{d}{2}+\alpha+1})},$$
$$\|\delta G_2\|_{L_{\bar{T}}^1(\dot{B}_{2,1}^{\frac{d}{2}-2+\alpha})} \leq C\bar{T}(1+\|a^1\|_{L_{\bar{T}}^{\infty}(\dot{B}_{2,1}^{\frac{d}{2}})}+\|a^2\|_{L_{\bar{T}}^{\infty}(\dot{B}_{2,1}^{\frac{d}{2}})})\|\delta a\|_{L_{\bar{T}}^{\infty}(\dot{B}_{2,1}^{\frac{d}{2}-1+\alpha})}.$$

Hence, for some constant C_T independent of $\bar{T}$, we have

$$\|\delta u\|_{L_{\bar{T}}^{\infty}(\dot{B}_{2,1}^{\frac{d}{2}-2+\alpha})} + \|\delta u\|_{L_{\bar{T}}^1(\dot{B}_{2,1}^{\frac{d}{2}+\alpha})} \leq C_T\left(\bar{T} + \|u^1\|_{L_{\bar{T}}^1(\dot{B}_{2,1}^{\frac{d}{2}})}\right)\|\delta a\|_{L_{\bar{T}}^{\infty}(\dot{B}_{2,1}^{\frac{d}{2}-1+\alpha})}.$$

Note that the factor $\bar{T}+\|u^1\|_{L_{\bar{T}}^1(\dot{B}_{2,1}^{\frac{d}{2}})}$ decays to 0 when $\bar{T}$ goes to zero. Hence, plugging the inequality (10.43) into the above inequality, we conclude that $(\delta a, \delta u) \equiv 0$ on a small time interval $[0,\bar{T}]$. The same connectivity arguments as those used in the proof of uniqueness in the critical case then yield uniqueness on the whole interval $[0,T]$. This completes the proof of Theorem 10.11. $\quad\square$

10.3.4 A Continuation Criterion

Proposition 10.17. *Under the hypotheses of Theorem 10.11, assume that the system* (NSC) *has a solution* (a, u) *on* $[0, T[\times \mathbb{R}^d$ *which belongs to* $E_{T'}^\alpha$ *for all* $T' < T$ *and satisfies*

$$a \in L_T^\infty(\dot{B}_{2,1}^{\frac{d}{2}+\alpha} \cap \dot{B}_{2,1}^{\frac{d}{2}}), \qquad 1 + \inf_{(t,x) \in [0,T[\times \mathbb{R}^d} a(t,x) > 0, \qquad \int_0^T \|\nabla u\|_{L^\infty}\, dt < \infty.$$

There then exists some $T^* > T$ *such that* (a, u) *may be continued on* $[0, T^*] \times \mathbb{R}^d$ *to a solution of* (NSC) *which belongs to* $E_{T^*}^\alpha$.

Proof. Note that u satisfies

$$\partial_t u + u \cdot \nabla u - (1+a)^{-1} \mathcal{A} u = f - \nabla(G(a)), \qquad u_{|t=0} = u_0.$$

Hence, applying Proposition 10.12, we get, for all $\beta \in [0, \alpha]$ and $T' < T$,

$$\|u\|_{\widetilde{L}_{T'}^\infty(\dot{B}_{2,1}^{\frac{d}{2}-1+\beta})} + \nu\|u\|_{L_{T'}^1(\dot{B}_{2,1}^{\frac{d}{2}+1+\beta})} \leq C e^{C \int_0^{T'} \left(\|\nabla u\|_{L^\infty} + \|a\|_{\dot{B}_{2,1}^{\frac{d}{2}+\alpha}}^{\frac{2}{\alpha}} \right) dt}$$
$$\times \left(\|u_0\|_{\dot{B}_{2,1}^{\frac{d}{2}-1+\beta}} + \|f\|_{L_{T'}^1(\dot{B}_{2,1}^{\frac{d}{2}-1+\beta})} + \int_0^{T'} \|a\|_{\dot{B}_{2,1}^{\frac{d}{2}+\beta}}\, dt \right)$$

for some constant C depending only on d, α, and the viscosity coefficients.

Hence, u is bounded in $\widetilde{L}_T^\infty(\dot{B}_{2,1}^{\frac{d}{2}-1} \cap \dot{B}_{2,1}^{\frac{d}{2}-1+\alpha})$. Replacing $\|\dot{\Delta}_j u_0\|_{L^2}$ by $\|\dot{\Delta}_j u\|_{L_T^\infty(L^2)}$, and $\|\dot{\Delta}_j f\|_{L_T^1(L^2)}$ by (say) $\|\dot{\Delta}_j f\|_{L_{1+T}^1(L^2)}$ in Remark 10.16, we get some $\varepsilon > 0$ such that for any $T' \in [0, T[$, the system (NSC) with data $a(T')$, $u(T')$, and $f(\cdot + T')$ has a solution on $[0, \varepsilon]$. Taking $T' = T - \varepsilon/2$ and using the fact that the solution (a, u) is unique on $[0, T[$, we thus get a continuation of (a, u) beyond T. $\qquad\square$

10.4 Global Existence for Small Data

Thus far, the gradient of the pressure has been considered as a lower order source term in the a priori estimates. This rough analysis entails a linear growth in time in the bounds for the solution, thus hindering the global closure of the estimates and any attempt to get a global existence result.

In this section, we shall see that including the main order part of the pressure term in the linearized equations (under the physically relevant assumption that the pressure law is an increasing function of the density) leads to a global existence statement in the same spirit as the theorem of Fujita and Kato (see Theorem 5.6 page 209).

10.4.1 Statement of the Results

For reasons which will become clear after our analysis of the linearized compressible Navier–Stokes equation (see Section 10.4.2), we introduce the following family of *hybrid Besov spaces* with different indices of regularity for low and high frequencies.

Definition 10.18. *For $\alpha > 0$, $r \in [1, \infty]$, and $s \in \mathbb{R}$, define*

$$\|u\|_{\widetilde{B}_\alpha^{s,r}} \overset{def}{=} \sum_{j \in \mathbb{Z}} 2^{js} \max\{\alpha, 2^{-j}\}^{1-2/r} \|\dot{\Delta}_j u\|_{L^2}.$$

Remark 10.19. We point out that here, the index r has nothing to do with the third index of classical Besov spaces. Hybrid Besov spaces carry different information for low and high frequencies, and the index r controls this difference: The low frequencies have the $\dot{B}_{2,1}^{s-1+\frac{2}{r}}$ regularity, while the high frequencies belong to $\dot{B}_{2,1}^s$. In particular,

$$\frac{1}{2}(\|u\|_{\dot{B}_{2,1}^{s-1}} + \alpha\|u\|_{\dot{B}_{2,1}^s}) \leq \|u\|_{\widetilde{B}_\alpha^{s,\infty}} \leq \|u\|_{\dot{B}_{2,1}^{s-1}} + \alpha\|u\|_{\dot{B}_{2,1}^s},$$

$$\frac{1}{2}\min(1,\alpha)\|u\|_{\dot{B}_{2,1}^s + \dot{B}_{2,1}^{s+1}} \leq \|u\|_{\widetilde{B}_\alpha^{s,1}} \leq \min\left(\frac{1}{\alpha}\|u\|_{\dot{B}_{2,1}^s}, \|u\|_{\dot{B}_{2,1}^{s+1}}\right).$$

Of course, we have $\widetilde{B}_\alpha^{s,2} = \dot{B}_{2,1}^s$.

We now define the space in which the solution is going to be constructed. From now on, we agree that if I is an interval of $\mathbb{R}$, and X is a Banach space, then the notation $\mathcal{C}_b(I; X)$ designates the set of bounded and continuous functions on I with values in X.

Definition 10.20. *The space E_α^s is the set of functions (b, v) in*

$$\left(L^1(\mathbb{R}^+; \widetilde{B}_\alpha^{s,1}) \cap \mathcal{C}_b(\mathbb{R}^+; \widetilde{B}_\alpha^{s,\infty})\right) \times \left(L^1(\mathbb{R}^+; \dot{B}_{2,1}^{s+1}) \cap \mathcal{C}_b(\mathbb{R}^+; \dot{B}_{2,1}^{s-1})\right)^d$$

endowed with the norm[7]

$$\|(b, v)\|_{E_\alpha^s} \overset{def}{=} \|b\|_{L^\infty(\widetilde{B}_\alpha^{s,\infty})} + \|v\|_{L^\infty(\dot{B}_{2,1}^{s-1})} + \nu\|b\|_{L^1(\widetilde{B}_\alpha^{s,1})} + \underline{\nu}\|v\|_{L^1(\dot{B}_{2,1}^{s+1})}.$$

We denote by $E_{T,\alpha}^s$ the space E_α^s restricted to functions over $[0,T] \times \mathbb{R}^d$.

We can now state our main global well-posedness result for small data with critical regularity.

---

[7] For notational simplicity, we omit the dependence on the viscosity coefficients.

Theorem 10.21. *Assume that $\underline{\nu} > 0$ and $P'(1) > 0$. There exist two positive constants, η and M, depending only on d and on the function G, such that if a_0 belongs to $\dot{B}_{2,1}^{\frac{d}{2}-1} \cap \dot{B}_{2,1}^{\frac{d}{2}}$, u_0 belongs to $\dot{B}_{2,1}^{\frac{d}{2}-1}$, f belongs to $L^1(\mathbb{R}^+; \dot{B}_{2,1}^{\frac{d}{2}-1})$, and*

$$\|a_0\|_{\dot{B}_{2,1}^{\frac{d}{2}-1}} + \nu\|a_0\|_{\dot{B}_{2,1}^{\frac{d}{2}}} + \|u_0\|_{\dot{B}_{2,1}^{\frac{d}{2}-1}} + \|f\|_{L^1(\dot{B}_{2,1}^{\frac{d}{2}-1})} \leq \eta\, \nu\underline{\nu}/\bar{\nu},$$

then the system (NSC) has a unique global solution (a, u) in $E_\nu^{\frac{d}{2}}$ which satisfies

$$\|(a,u)\|_{E_\nu^{\frac{d}{2}}} \leq M\left(\|a_0\|_{\dot{B}_{2,1}^{\frac{d}{2}-1}} + \nu\|a_0\|_{\dot{B}_{2,1}^{\frac{d}{2}}} + \|u_0\|_{\dot{B}_{2,1}^{\frac{d}{2}-1}} + \|f\|_{L^1(\dot{B}_{2,1}^{\frac{d}{2}-1})} \right).$$

10.4.2 A Spectral Analysis of the Linearized Equation

In what follows, we shall assume (with no loss of generality) that $P'(1) = 1$. Let $\mathbb{P}$ (resp., $\mathbb{P}^\perp$) be the L^2 projector over divergence-free (resp., potential) vector fields. Applying $\mathbb{P}$ and $\mathbb{P}^\perp$ to the momentum equation, the system (NSC) can be rewritten as

$$\begin{cases} \partial_t a + \mathrm{div}\, \mathbb{P}^\perp u = -\,\mathrm{div}(u\,a) \\ \partial_t \mathbb{P}^\perp u - \nu\Delta\mathbb{P}^\perp u + \nabla a = \mathbb{P}^\perp\left(f - u\cdot\nabla u - I(a)\mathcal{A}u + K(a)\nabla a \right) \\ \partial_t\mathbb{P}u - \mu\Delta\mathbb{P}u = \mathbb{P}\left(f - u\cdot\nabla u - I(a)\mathcal{A}u \right). \end{cases} \tag{10.44}$$

The function K is smooth, vanishes at 0, and may be explicitly computed in terms of the function G.

On the one hand, up to nonlinear terms, the last equation reduces to the standard heat equation, which is well understood. On the other hand, there is a linear hyperbolic/parabolic coupling between the first two equations that we have to investigate further.

Introducing the function $v \overset{\text{def}}{=} |D|^{-1}\,\mathrm{div}\, u$, the linear part of the first two equations reduces to the 2×2 system

$$\begin{cases} \partial_t a + |D|v = F \\ \partial_t v - \nu\Delta v - |D|a = G. \end{cases} \tag{10.45}$$

According to Duhamel's formula, the solution of (10.45) is of the form

$$\begin{pmatrix} a(t) \\ v(t) \end{pmatrix} = e^{A(D)t} \begin{pmatrix} a_0 \\ v_0 \end{pmatrix} + \int_0^t e^{A(D)(t-t')} \begin{pmatrix} F(t') \\ G(t') \end{pmatrix} dt', \tag{10.46}$$

where $A(D)$ stands for the matrix-valued pseudodifferential operator

$$\begin{pmatrix} 0 & -|D| \\ |D| & -\nu|D|^2 \end{pmatrix}.$$

In the low-frequency regime $\nu|\xi| < 2$, the eigenvalues of $A(\xi)$ are

$$\lambda^{\pm}(\xi) = -\frac{\nu}{2}|\xi|^2 \left(1 \pm i\sqrt{\frac{4}{\nu^2|\xi|^2} - 1}\right),$$

so a parabolic damping for low frequencies of a and v is expected.

For high frequencies (i.e., $\nu|\xi| > 2$), the eigenvalues are of the form

$$\lambda^{\pm}(\xi) = -\frac{\nu}{2}|\xi|^2 \left(1 \pm \sqrt{1 - \frac{4}{\nu^2|\xi|^2}}\right),$$

so a parabolic mode and a damped mode coexist. More precisely, performing the change of functions

$$\widehat{v}^{-}(\xi) = \frac{\nu|\xi|}{2}\left(1 + \sqrt{1 - \frac{4}{\nu^2|\xi|^2}}\right)\widehat{a}(\xi) - \widehat{v}(\xi),$$

$$\widehat{v}^{+}(\xi) = \frac{\nu|\xi|}{2}\left(1 + \sqrt{1 - \frac{4}{\nu^2|\xi|^2}}\right)\widehat{v}(\xi) - \widehat{a}(\xi),$$

we get

$$\widehat{v}^{\pm}(t,\xi) = e^{\lambda^{\pm}(\xi)t}\,\widehat{v}^{\pm}(0,\xi).$$

In the asymptotics $|\xi|$ going to infinity, v^{-} (resp., v^{+}) tends to be collinear with a (resp., u), and we have

$$\lambda^{-}(\xi) \sim -1/\nu \quad \text{and} \quad \lambda^{+}(\xi) \sim -\nu|\xi|^2.$$

Hence, the mode v^{-} is damped, whereas v^{+} has parabolic behavior.

In short, according to the above analysis, we may expect a parabolic smoothing for the velocity u to (NSC), whereas a should be damped for high frequencies and exhibit a parabolic behavior for low frequencies.

In fact, solving (10.45) explicitly yields the following estimates [which have to be compared to those for the heat equation stated in (3.39) page 157].

Proposition 10.22. *Let (a, v) be a solution of (10.45) on $[0, T] \times \mathbb{R}^d$. There exists a constant C, depending only on d, such that for all r in $[1, \infty]$ and s in $\mathbb{R}$, we have*

$$\|a\|_{L_T^r(\widetilde{B}_\nu^{s,r})} + \|v\|_{L_T^r(\dot{B}_{2,1}^{s-1+\frac{2}{r}})}$$

$$\leq C\left(\|a_0\|_{\widetilde{B}_\nu^{s,\infty}} + \|v_0\|_{\dot{B}_{2,1}^{s-1}} + \|F\|_{L_T^1(\widetilde{B}_\nu^{s,\infty})} + \|G\|_{L_T^1(\dot{B}_{2,1}^{s-1})}\right).$$

According to Remark 10.19, we have

$$\|\cdot\|_{\widetilde{B}_\nu^{s,\infty}} \approx \|\cdot\|_{\dot{B}_{2,1}^{s-1}} + \nu\|\cdot\|_{\dot{B}_{2,1}^{s}} \quad \text{and} \quad \|\cdot\|_{\widetilde{B}_\nu^{s,2}} = \|\cdot\|_{\dot{B}_\nu^{s}} \qquad (10.47)$$

so that if $a_0 \in \dot{B}_{2,1}^{\frac{d}{2}} \cap \dot{B}_{2,1}^{\frac{d}{2}-1}$, $v_0 \in \dot{B}_{2,1}^{\frac{d}{2}-1}$, $F \in L^1([0,T]; \dot{B}_{2,1}^{\frac{d}{2}} \cap \dot{B}_{2,1}^{\frac{d}{2}-1})$, and $G \in L^1([0,T]; \dot{B}_{2,1}^{\frac{d}{2}-1})$, then the above proposition provides us with estimates for a in $L^\infty([0,T]; \dot{B}_{2,1}^{\frac{d}{2}} \cap \dot{B}_{2,1}^{\frac{d}{2}-1})$ and also in $L^2([0,T]; \dot{B}_{2,1}^{\frac{d}{2}})$. The latter estimate is the key to getting a global control on the term $K(a)\nabla a$.

At the nonlinear level, however, the above estimates cannot be used for the compressible Navier–Stokes equation because no matter how smooth u is, the convection term $u \cdot \nabla a$ is one derivative less regular than a. Hence, we first have to adapt the statement of Proposition 10.22 to the following, more general, linear system:

$$\begin{cases} \partial_t a + v \cdot \nabla a + \operatorname{div} u = F \\ \partial_t u + v \cdot \nabla u - \mathcal{A}u + \nabla a = G. \end{cases} \tag{10.48}$$

10.4.3 A Priori Estimates for the Linearized Equation

For technical reasons, we shall instead consider a *paralinearized* version of the system (10.48) and introduce a (small) parameter ε, the so-called Mach number that will play an important role in the last section of this chapter.

Finally, then, the system we want to study in this section is of the form

$$\begin{cases} \partial_t a + \operatorname{div}(T_v a) + \dfrac{\operatorname{div} u}{\varepsilon} = F \\ \partial_t u + T_v \cdot \nabla u - \mathcal{A}u + \dfrac{\nabla a}{\varepsilon} = G \end{cases} \tag{LPH^ε}$$

with $\operatorname{div}(T_v a) \overset{\text{def}}{=} \partial_i(T_{v^i} a)$ and $T_v \cdot \nabla u \overset{\text{def}}{=} T_{v^i} \partial_i u$.

Proposition 10.23. *Let $\varepsilon > 0$, $s \in \mathbb{R}$, $1 \le p, r < \infty$, and (a,u) be a solution of (LPH^ε). There then exists a constant C, depending only on d, p, r, and s, and such that the following estimate holds:*

$$\|a(t)\|_{\widetilde{B}_{\varepsilon\nu}^{s,\infty}} + \|u(t)\|_{\dot{B}_{2,1}^{s-1}} + \int_0^t \left(\nu\|a\|_{\widetilde{B}_{\varepsilon\nu}^{s,1}} + \underline{\nu}\|u\|_{\dot{B}_{2,1}^{s+1}} \right) dt' \le C e^{CV_\varepsilon^{p,r}(t)}$$

$$\times \left(\|a_0\|_{\widetilde{B}_{\varepsilon\nu}^{s,\infty}} + \|u_0\|_{\dot{B}_{2,1}^{s-1}} + \int_0^t e^{-CV_\varepsilon^{p,r}(t')} \left(\|F\|_{\widetilde{B}_{\varepsilon\nu}^{s,\infty}} + \|G\|_{\dot{B}_{2,1}^{s-1}} \right) dt' \right),$$

where $V_\varepsilon^{p,r}(t) \overset{\text{def}}{=} \begin{cases} \displaystyle\int_0^t \left(\underline{\nu}^{1-p}\|\nabla v\|_{\dot{B}_{\infty,\infty}^{\frac{2}{p}-2}}^p + (\varepsilon^2\nu)^{r-1} \|\nabla v\|_{L^\infty}^r \right) dt', & \text{if } p > 1, \\ \displaystyle\int_0^t \left(\|\nabla v\|_{L^\infty} + (\varepsilon^2\nu)^{r-1} \|\nabla v\|_{L^\infty}^r \right) dt', & \text{if } p = 1. \end{cases}$

Remark 10.24. Only the case where $\varepsilon = 1$ and $p = r = 1$ is of interest for proving Theorem 10.21. The other cases will be needed in the last section of this chapter when investigating the low Mach number limit for large data.

Proof. Setting

$$\widetilde{a}(t,x) \stackrel{\text{def}}{=} \varepsilon a(\varepsilon^2 t, \varepsilon x), \qquad \widetilde{u}(t,x) \stackrel{\text{def}}{=} \varepsilon u(\varepsilon^2 t, \varepsilon x),$$

$$\widetilde{F}(t,x) \stackrel{\text{def}}{=} \varepsilon^3 F(\varepsilon^2 t, \varepsilon x), \qquad \widetilde{G}(t,x) \stackrel{\text{def}}{=} \varepsilon^3 G(\varepsilon^2 t, \varepsilon x)$$

reduces the study to the case $\varepsilon = 1$. It is left to the reader to demonstrate that the change of variables has the desired effect on the norms involved in the inequality we want to establish (this may be argued as in Proposition 2.18 page 64).

So, we assume that $\varepsilon = 1$. To avoid a tedious distinction between the cases $p > 1$ and $p = 1$, it will be intended throughout that $\|\nabla u\|_{\dot{B}^{\frac{2}{p}-2}_{\infty,\infty}}$ stands for $\|\nabla u\|_{L^\infty}$ if $p = 1$. We shall also denote by p' the conjugate exponent of p. We now find that $(a_j, u_j) \stackrel{\text{def}}{=} (\dot{\Delta}_j a, \dot{\Delta}_j u)$ satisfies

$$\begin{cases} \partial_t a_j + \operatorname{div}(v_j\, a_j) + \operatorname{div} \mathbb{P}^\perp u_j = f_j \\ \partial_t \mathbb{P}^\perp u_j + v_j \cdot \nabla \mathbb{P}^\perp u_j - \nu \Delta \mathbb{P}^\perp u_j + \nabla a_j = g_j^\perp \\ \partial_t \mathbb{P}\, u_j + v_j \cdot \nabla \mathbb{P}\, u_j - \mu \Delta \mathbb{P}\, u_j = g_j, \end{cases} \qquad (LPH_j)$$

where $v_j \stackrel{\text{def}}{=} \dot{S}_{j-1} v$,

$$f_j \stackrel{\text{def}}{=} F_j + \operatorname{div}\big(v_j\, a_j - \dot{\Delta}_j T_v a\big) \qquad \text{with} \quad F_j \stackrel{\text{def}}{=} \dot{\Delta}_j F,$$

$$g_j^\perp \stackrel{\text{def}}{=} G_j^\perp + v_j \cdot \nabla \dot{\Delta}_j \mathbb{P}^\perp u - \mathbb{P}^\perp \dot{\Delta}_j T_v \cdot \nabla u \quad \text{with} \quad G_j^\perp \stackrel{\text{def}}{=} \mathbb{P}^\perp \dot{\Delta}_j G,$$

$$g_j \stackrel{\text{def}}{=} G_j + v_j \cdot \nabla \dot{\Delta}_j \mathbb{P}\, u - \mathbb{P}\, \dot{\Delta}_j T_v \cdot \nabla u \qquad \text{with} \quad G_j \stackrel{\text{def}}{=} \mathbb{P}\, \dot{\Delta}_j G.$$

First Step: The Incompressible Part

We first consider the equation for $\mathbb{P}\, u_j$, which is easier to handle. Taking the L^2 inner product of the last equation of (LPH_j) with $\mathbb{P}\, u_j$ and integrating by parts, we get

$$\frac{1}{2}\frac{d}{dt} \|\mathbb{P}\, u_j\|_{L^2}^2 + \mu \|\nabla \mathbb{P}\, u_j\|_{L^2}^2 = \frac{1}{2}\int \operatorname{div} v_j\, |\mathbb{P}\, u_j|^2\, dx + \int g_j \cdot \mathbb{P}\, u_j\, dx.$$

The commutator in g_j may be bounded according to Lemma 10.25 below. We find that

$$\frac{1}{2}\frac{d}{dt} \|\mathbb{P}\, u_j\|_{L^2}^2 + \mu \|\nabla \mathbb{P}\, u_j\|_{L^2}^2$$

$$\leq \|\mathbb{P}\, u_j\|_{L^2} \left(\|G_j\|_{L^2} + C \sum_{|j'-j|\leq 4} 2^{\frac{2}{p'}j'} \|\nabla v\|_{\dot{B}^{-\frac{2}{p'}}_{\infty,\infty}} \|u_{j'}\|_{L^2} \right).$$

Now, using the Young inequality

$$2^{\frac{2}{p'}j'}\|\nabla v\|_{\dot{B}_{\infty,\infty}^{-\frac{2}{p'}}} \leq \frac{1}{p}\left(\frac{K_1}{\nu}\right)^{p-1}\|\nabla v\|_{\dot{B}_{\infty,\infty}^{-\frac{2}{p'}}}^{p} + \frac{\nu}{p'K_1}2^{2j'}$$

and integrating in time, we get, for some universal constant $\kappa > 0$ and all $K_1 > 0$,

$$\|\mathbb{P}\,u_j(t)\|_{L^2} + \kappa\mu 2^{2j}\int_0^t \|\mathbb{P}\,u_j\|_{L^2}\,dt' \leq \|\mathbb{P}\,u_j(0)\|_{L^2} + \int_0^t \|G_j\|_{L^2}\,dt'$$

$$+C\sum_{|j'-j|\leq 4}\int_0^t\left(\frac{\nu}{K_1 p'}2^{2j'} + \frac{1}{p}\left(\frac{K_1}{\nu}\right)^{p-1}\|\nabla v\|_{\dot{B}_{\infty,\infty}^{-\frac{2}{p'}}}^{p}\right)\|u_{j'}\|_{L^2}\,dt'. \quad (10.49)$$

Second Step: The Compressible Part

We now want to establish a similar inequality for the "compressible" mode $(a, \mathbb{P}^\perp u)$. According to the analysis in the previous section, we must bound the quantity $2^{j(s-1)}k_j$ with

$$k_j = \begin{cases} \|a_j\|_{L^2} + \|\mathbb{P}^\perp u_j\|_{L^2}, & \text{if } \nu 2^j \leq 1, \\ \nu\|\nabla a_j\|_{L^2} + \|\mathbb{P}^\perp u_j\|_{L^2}, & \text{if } \nu 2^j > 1. \end{cases}$$

Owing to the fact that the linear operator associated with (LPH) may not be diagonalized in a basis *independent* of ξ, however, bounding k_j cannot be achieved by means of basic energy arguments. We could introduce a convenient symmetrizer, whose definition depends on $|\xi|$. Here, though, we shall instead consider the following quantity (for some suitable $\alpha > 0$, to be chosen later):

$$Y_j \stackrel{\text{def}}{=} \begin{cases} \sqrt{\|a_j\|_{L^2}^2 + \|\mathbb{P}^\perp u_j\|_{L^2}^2 + \alpha\nu(\nabla a_j|u_j)}, & \text{for } j \leq j_0 - 1, \\ \sqrt{\|\nu\nabla a_j\|_{L^2}^2 + 2\|\mathbb{P}^\perp u_j\|_{L^2}^2 + 2(\nu\nabla a_j|u_j)}, & \text{for } j \geq j_0, \end{cases}$$

where j_0 is the unique integer such that

$$2 \leq c\nu 2^{j_0} < 4, \quad (10.50)$$

and c stands for a constant (e.g., $c = \frac{3}{4}$) such that

$$c2^j\|\dot{\Delta}_j z\|_{L^2} \leq \|\nabla\dot{\Delta}_j z\|_{L^2} \leq 2c^{-1}2^j\|\dot{\Delta}_j z\|_{L^2}. \quad (10.51)$$

We point out that if $\alpha < c$, then this choice ensures that $k_j \approx Y_j$.

We first derive an a priori estimate for the low frequencies of $(a, \mathbb{P}^\perp u)$. Take the L^2 inner product of the first equation of (LPH_j) with a_j and of the second one with u_j, integrate by parts, and add the two equalities together. Using the fact that the contributions of the skew-symmetric first order terms cancel out, we obtain

$$\frac{1}{2}\frac{d}{dt}\left(\|a_j\|_{L^2}^2 + \|\mathbb{P}^\perp u_j\|_{L^2}^2\right) + \nu\left\|\nabla\mathbb{P}^\perp u_j\right\|_{L^2}^2$$

$$= \int\left(f_j a_j + g_j^\perp \cdot \mathbb{P}^\perp u_j + \frac{1}{2}\left(|\mathbb{P}^\perp u_j|^2 - a_j^2\right)\text{ div }v_j\right)dx. \quad (10.52)$$

In order to determine the low-frequency parabolic behavior of a, we now write an equality involving the quantity $(\nabla a_j | u_j)$. We have

$$\frac{d}{dt}(\nabla a_j | u_j) - \left\|\nabla\mathbb{P}^\perp u_j\right\|_{L^2}^2 + \|\nabla a_j\|_{L^2}^2 - \nu(\Delta\mathbb{P}^\perp u_j | \nabla a_j)$$

$$= (\nabla f_j | u_j) + (g_j^\perp | \nabla a_j) + \int a_j \nabla v_j : \nabla u_j \, dx. \quad (10.53)$$

Now, if we assume that $j < j_0$, then the term $\nu(\Delta\mathbb{P}^\perp u_j | \nabla a_j)$ is of lower order because, due to (10.50), (10.51), and Young's inequality, we have

$$\nu(\Delta\mathbb{P}^\perp u_j | \nabla a_j) \leq \frac{1}{2}\|\nabla a_j\|_{L^2}^2 + \frac{\nu^2}{2}\|\Delta\mathbb{P}^\perp u_j\|_{L^2}^2 \leq \frac{1}{2}\|\nabla a_j\|_{L^2}^2 + \frac{8}{c^4}\|\nabla\mathbb{P}^\perp u_j\|_{L^2}^2.$$

Hence, adding the equality (10.53) times $\alpha/2$ to the equality (10.52) and using Lemma 10.25, we find that if $\alpha < c$, then

$$\frac{1}{2}\frac{d}{dt}Y_j^2 + \frac{\alpha\nu}{4}\|\nabla a_j\|_{L^2}^2 + \nu\left(1 - \frac{\alpha}{2} - \frac{4\alpha}{c^4}\right)\|\nabla\mathbb{P}^\perp u_j\|_{L^2}^2$$

$$\leq CY_j\left(\|F_j\|_{L^2} + \|G_j^\perp\|_{L^2} + \sum_{|j'-j|\leq 4} 2^{j'\frac{2}{p'}}\|\nabla v\|_{\dot{B}_{\infty,\infty}^{-\frac{2}{p'}}}X_{j'}\right)$$

with C depending only on d and p, and $X_j^2 \overset{\text{def}}{=} \|\mathbb{P}u_j\|_{L^2}^2 + Y_j^2$.

Take $\alpha = c^4/(c^4+8)$. From the previous inequality, Bernstein's lemma, and the fact that $k_j \approx Y_j$, we then deduce that there exists a universal constant κ such that

$$\frac{1}{2}\frac{d}{dt}Y_j^2 + \kappa\nu 2^{2j}Y_j^2 \leq CY_j\left(\|F_j\|_{L^2} + \|G_j^\perp\|_{L^2}\right.$$

$$\left. + \sum_{|j'-j|\leq 4} 2^{j'\frac{2}{p'}}\|\nabla v\|_{\dot{B}_{\infty,\infty}^{-\frac{2}{p'}}}X_{j'}\right). \quad (10.54)$$

We now aim to bound Y_j for $j \geq j_0$. For this, we may combine the three equalities

$$\frac{1}{2}\frac{d}{dt}\|\nabla a_j\|_{L^2}^2 + (\Delta u_j | \nabla a_j)$$

$$= (\nabla f_j | \nabla a_j) + \int a_j D^2 a_j : \nabla v_j \, dx - \frac{1}{2}\int |\nabla a_j|^2 \text{ div } v_j \, dx,$$

$$\frac{1}{2}\frac{d}{dt}\left\|\mathbb{P}^\perp u_j\right\|_{L^2}^2 + (\nabla a_j|u_j) + \nu\left\|\nabla\mathbb{P}^\perp u_j\right\|_{L^2}^2 = (g_j^\perp|\mathbb{P}^\perp u_j) + \frac{1}{2}\int|\mathbb{P}^\perp u_j|^2\,\mathrm{div}\,v_j\,dx,$$

$$\frac{d}{dt}(\nabla a_j|u_j) - \left\|\nabla\mathbb{P}^\perp u_j\right\|_{L^2}^2 + \left\|\nabla a_j\right\|_{L^2}^2 - \nu\left(\Delta\mathbb{P}^\perp u_j|\nabla a_j\right)$$
$$= (g_j^\perp|\nabla a_j) + (\nabla f_j|u_j) - \int a_j\nabla v_j : \nabla u_j\,dx$$

and get

$$\frac{1}{2}\frac{d}{dt}Y_j^2 + \nu\left\|\nabla a_j\right\|_{L^2}^2 + \nu\left\|\nabla\mathbb{P}^\perp u_j\right\|_{L^2}^2 + 2(\nabla a_j|u_j)$$
$$= \nu^2\,(\nabla f_j|\nabla a_j) + 2(\mathbb{P}^\perp u_j|g_j^\perp) + \nu(g_j^\perp|\nabla a_j) + \nu(\nabla f_j|u_j)$$
$$+ \int a_j D^2 a_j : \nabla v_j\,dx + \int\left(|\mathbb{P}^\perp u_j|^2 - \frac{\nu^2}{2}|\nabla a_j|^2\right)\mathrm{div}\,v_j\,dx - \int a_j\nabla v_j : \nabla u_j\,dx.$$

As $(\nabla a_j|\mathbb{P}\,u_j) = 0$, we get, according to the definition of j_0 and Bernstein's inequality, for all $j \geq j_0$,

$$2|(\nabla a_j|u_j)| \leq \frac{\nu}{2}\|\nabla a_j\|_{L^2}^2 + 2c^{-2}2^{-2j_0}\nu^{-2}\left(\nu\left\|\nabla\mathbb{P}^\perp u_j\right\|_{L^2}^2\right)$$
$$\leq \frac{\nu}{2}\|\nabla a_j\|_{L^2}^2 + \frac{\nu}{2}\left\|\nabla\mathbb{P}^\perp u_j\right\|_{L^2}^2.$$

Next, we note that for $j \geq j_0$, we also have

$$\nu\left\|\nabla\mathbb{P}^\perp u_j\right\|_{L^2}^2 \geq \kappa\nu^{-1}\left\|\mathbb{P}^\perp u_j\right\|_{L^2}^2.$$

Therefore, for all $j \geq j_0$, by virtue of Lemma 10.25, we get

$$\frac{1}{2}\frac{d}{dt}Y_j^2 + \frac{\kappa}{\nu}Y_j^2 \leq CY_j\left(\nu\|\nabla F_j\|_{L^2} + \left\|G_j^\perp\right\|_{L^2} + \|\nabla v\|_{L^\infty}\sum_{|j'-j|\leq 4} X_{j'}\right). \qquad (10.55)$$

Finally, inserting the following Young's inequalities (with $K_2 > 0$ and $K_3 > 0$)

$$2^{\frac{2}{p'}j'}\|\nabla v\|_{\dot{B}_{\infty,\infty}^{-\frac{2}{p'}}} \leq \frac{1}{p}\left(\frac{K_2}{\nu}\right)^{p-1}\|\nabla v\|_{\dot{B}_{\infty,\infty}^{-\frac{2}{p'}}}^p + \frac{\nu}{p'K_2}2^{2j},$$

$$\|\nabla v\|_{L^\infty} \leq \frac{1}{r}(K_3\nu)^{r-1}\|\nabla v\|_{L^\infty}^r + \frac{1}{r'\nu K_3},$$

into (10.54), (10.55) and performing a time integration, we get, for some universal positive constant κ and all positive K_2 and K_3,

$$Y_j(t) + \kappa\nu\min(2^{2j},\nu^{-2})\int_0^t Y_j\,dt' \leq Y_j(0) + C\left(\max(1,2^j\nu)\int_0^t\|F_j\|_{L^2}\,dt'\right.$$

$$+ \int_0^t\left\|G_j^\perp\right\|_{L^2}\,dt' + \nu\max\left(\frac{1}{K_2},\frac{1}{K_3}\right)\min(2^{2j},\nu^{-2})\sum_{|j'-j|\leq 4}\int_0^t X_{j'}\,dt'$$

$$\left.+ \sum_{|j'-j|\leq 4}\int_0^t\left[\left(\frac{K_2}{\nu}\right)^{p-1}\|\nabla v\|_{\dot{B}_{\infty,\infty}^{-\frac{2}{p'}}}^p + (\nu K_3)^{r-1}\|\nabla v\|_{L^\infty}^r\right]X_{j'}\,dt'\right).$$

Third Step: Global A Priori Estimates

Bounding $\|a(t)\|_{\widetilde{B}^{s,\infty}_\nu} + \|u(t)\|_{\dot{B}^{s-1}_{2,1}}$ and exhibiting a time decay for a and the low frequencies of u is our next task. To achieve this, we may add (10.49) to the above inequality. We get

$$X_j(t) + \kappa \int_0^t \left(\nu \min(2^{2j}, \nu^{-2})Y_j + \mu 2^{2j} \|\mathbb{P} u_j\|_{L^2}\right) dt' \le X_j(0)$$

$$+C \int_0^t \left(\max(1, 2^j \nu)\|\dot{\Delta}_j F\|_{L^2} + \|\dot{\Delta}_j G\|_{L^2}\right) dt'$$

$$+C\nu \max\left(\frac{1}{K_2}, \frac{1}{K_3}\right) \min(2^{2j}, \nu^{-2}) \sum_{|j'-j|\le 4} \int_0^t X_{j'}\, dt'$$

$$+\frac{C\nu}{K_1} \sum_{|j'-j|\le 4} \int_0^t 2^{2j'} \|\mathbb{P} u_{j'}\|_{L^2}\, dt'$$

$$+C \int_0^t \left[\left(\left(\frac{K_1}{\nu}\right)^{p-1} + \left(\frac{K_2}{\nu}\right)^{p-1}\right)\|\nabla v\|^p_{\dot{B}^{-\frac{2}{p'}}_{\infty,\infty}} + (\nu K_3)^{r-1} \|\nabla v\|^r_{L^\infty}\right] X_{j'}\, dt'.$$

Let $u^\ell \overset{\text{def}}{=} \sum_{j\le j_0} \dot{\Delta}_j u$. Multiply both sides of the above inequality by $2^{j(s-1)}$ and sum over $\mathbb{Z}$. Using the fact that $X_j \approx \|u_j\|_{L^2} + \max(1, \nu 2^j) \|a_j\|_{L^2}$ and choosing K_1, K_2, and K_3 to be sufficiently large, we infer that there exists some constant C, depending only on s, p, r, and d, such that

$$\|a(t)\|_{\widetilde{B}^{s,\infty}_\nu} + \|u(t)\|_{\dot{B}^{s-1}_{2,1}} + \nu \int_0^t \|a\|_{\widetilde{B}^{s,1}_\nu}\, dt' + \underline{\nu} \int_0^t \|u^\ell\|_{\dot{B}^{s+1}_{2,1}}\, dt'$$

$$\le C\left(\|a_0\|_{\widetilde{B}^{s,\infty}_\nu} + \|u_0\|_{\dot{B}^{s-1}_{2,1}} + \int_0^t \left(\|F\|_{\widetilde{B}^{s,\infty}_\nu} + \|G\|_{\dot{B}^{s-1}_{2,1}}\right) dt'\right.$$

$$\left.+ \int_0^t (V^{p,r}_1)'(t') \left(\|a\|_{\widetilde{B}^{s,\infty}_\nu} + \|u\|_{\dot{B}^{s-1}_{2,1}}\right) dt'\right).$$

Thanks to Gronwall's inequality, we conclude that

$$\|a(t)\|_{\widetilde{B}^{s,\infty}_\nu} + \|u(t)\|_{\dot{B}^{s-1}_{2,1}} + \nu \int_0^t \|a\|_{\widetilde{B}^{s,1}_\nu}\, dt' + \underline{\nu} \int_0^t \|u^\ell\|_{\dot{B}^{s+1}_{2,1}}\, dt' \le Ce^{CV^{p,r}_1(t)}$$

$$\times \left(\|a_0\|_{\widetilde{B}^{s,\infty}_\nu} + \|u_0\|_{\dot{B}^{s-1}_{2,1}} + \int_0^t e^{-CV^{p,r}_1(t')}\left(\|F\|_{\widetilde{B}^{s,\infty}_\nu} + \|G\|_{\dot{B}^{s-1}_{2,1}}\right) dt'\right). \quad (10.56)$$

Fourth Step: The Parabolic Behavior of u

To complete the proof of Proposition 10.23, we still have to determine the parabolic gain of regularity for the high-frequency part of u. This is the aim of the present step.

Applying $\dot{\Delta}_j$ to the second equation of (LPH^1) yields

$$\partial_t u_j + v_j \cdot \nabla u_j - \mathcal{A} u_j + \nabla a_j = G_j + \left(v_j \cdot \nabla u_j - \dot{\Delta}_j (T_v \cdot \nabla u) \right).$$

Taking the L^2 inner product with u_j, we easily get

$$\frac{1}{2} \frac{d}{dt} \| u_j \|_{L^2}^2 - \frac{1}{2} \int |u_j|^2 \operatorname{div} v_j \, dx + \mu \| \nabla u_j \|_{L^2}^2 + (\lambda + \mu) \| \operatorname{div} u_j \|_{L^2}^2$$

$$= \int (G_j - \nabla a_j) \cdot u_j \, dx + \int \left(v_j \cdot \nabla u_j - \dot{\Delta}_j (T_v \cdot \nabla u) \right) \cdot u_j \, dx.$$

The last integral may be bounded, thanks to Lemma 10.25. After a few calculations, we get, for all positive K,

$$\| u_j(t) \|_{L^2} + \kappa \underline{\nu} 2^{2j} \int_0^t \| u_j \|_{L^2} \, dt' \leq \| u_j(0) \|_{L^2}$$

$$+ \int_0^t \left(\| G_j \|_{L^2} + c^{-1} 2^j \| a_j \|_{L^2} \right) dt' + \frac{C\nu}{p'K} \sum_{|j'-j| \leq 4} 2^{2j'} \int_0^t \| u_{j'} \|_{L^2} \, dt'$$

$$+ \frac{1}{p} \left(\frac{K}{C\underline{\nu}} \right)^{p-1} \sum_{|j'-j| \leq 4} \int_0^t \| \nabla v \|_{\dot{B}_{\infty,\infty}^{-\frac{2}{p'}}}^p \| u_{j'} \|_{L^2} \, dt'.$$

We may now multiply both sides of the above inequality by $2^{j(s-1)}$ and sum over $j \geq j_0$. Choosing K to be suitably large, we eventually get

$$\| u^h(t) \|_{\dot{B}_{2,1}^{s-1}} + \underline{\nu} \int_0^t \| u^h \|_{\dot{B}_{2,1}^{s+1}} \, dt' \leq C \| u^h(0) \|_{\dot{B}_{2,1}^{s-1}}$$

$$+ \int_0^t \underline{\nu}^{1-p} \| \nabla u \|_{\dot{B}_{\infty,\infty}^{-\frac{2}{p'}}}^p \| u \|_{\dot{B}_{2,1}^{s-1}} \, dt' + \int_0^t \left(\| a^h \|_{\dot{B}_{2,1}^{s}} + \| G^h \|_{\dot{B}_{2,1}^{s-1}} \right) dt'$$

with $u^h = \sum_{j \geq j_0} \dot{\Delta}_j u$, $a^h = \sum_{j \geq j_0} \dot{\Delta}_j a$, and $G^h = \sum_{j \geq j_0} \dot{\Delta}_j G$.

Finally, using the fact that $\| a^h \|_{\dot{B}_{2,1}^s} \leq C\nu \| a \|_{\tilde{B}_\nu^{s,1}}$ and plugging (10.56) into the above inequality completes the proof of the proposition. $\qquad \square$

Lemma 10.25. *Let $m \in \mathbb{R}$ and $\alpha \geq 0$. Let $A(D)$ be a smooth homogeneous multiplier of degree m, in the sense of Proposition 2.30. There exists a constant C, depending only on α, m, A, and d, such that for all $p \in [1, \infty]$ the following inequality holds true:*

$$\left\| A(D) \dot{\Delta}_j T_a b - \dot{S}_{j-1} a \, \dot{\Delta}_j A(D) b \right\|_{L^p}$$

$$\leq C 2^{j(m+\alpha-1)} \left(\sum_{|j'-j| \leq 4} \| \dot{\Delta}_{j'} b \|_{L^p} \right) \times \begin{cases} \| \nabla a \|_{L^\infty}, & \text{if} \quad \alpha = 0, \\ \| \nabla a \|_{\dot{B}_{\infty,\infty}^{-\alpha}}, & \text{if} \quad \alpha > 0. \end{cases}$$

Proof. This is based on the following relation, the proof of which is similar to that of the equality (4.17):

$$A(D)\dot{\Delta}_j T_a b - \dot{S}_{j-1} a \dot{\Delta}_j A(D) b$$
$$= \sum_{|j'-j|\leq 4} \left(A(D)\dot{\Delta}_j \left((\dot{S}_{j'-1} - \dot{S}_{j-1}) a \, \dot{\Delta}_{j'} b \right) + [\dot{\Delta}_j A(D), \dot{S}_{j-1} a] \dot{\Delta}_{j'} b \right).$$

By taking advantage of Bernstein's inequality and Lemma 2.97 page 110 to bound the last commutator, we get, from the above formula, that

$$\left\| A(D)\dot{\Delta}_j T_a b - \dot{S}_{j-1} a \dot{\Delta}_j A(D) b \right\|_{L^p}$$
$$\leq C 2^{j(m-1)} \sum_{j-4 \leq j',j'' \leq j+4} \left(\|\nabla \dot{\Delta}_{j''} a\|_{L^\infty} + \|\nabla \dot{S}_{j-1} a\|_{L^\infty} \right) \|\dot{\Delta}_{j'} b\|_{L^p},$$

which obviously entails the desired inequality.   □

10.4.4 Proof of Global Existence

We first note that if η has been chosen to be sufficiently small in the statement of Theorem 10.21, then Theorem 10.2 supplies a *local* solution (a, u) with

$$a \in \widetilde{\mathcal{C}}_T(\dot{B}_{2,1}^{\frac{d}{2}}) \quad \text{and} \quad u \in \widetilde{\mathcal{C}}_T(\dot{B}_{2,1}^{\frac{d}{2}-1}) \cap L_T^1(\dot{B}_{2,1}^{\frac{d}{2}+1}).$$

As we have assumed, in addition, that $a_0 \in \dot{B}_{2,1}^{\frac{d}{2}-1}$, we easily deduce that a is also in $\mathcal{C}([0,T]; \dot{B}_{2,1}^{\frac{d}{2}-1})$ (use Theorem 3.19). So, finally, we have $(a, u) \in E_{T,\nu}^{\frac{d}{2}}$ for some positive T.

Let T^* be the lifespan of the solution (a, u). We want to show that $T^* = \infty$. For this, we shall use Proposition 10.23 and the fact that (a, u) verifies

$$\begin{cases} \partial_t a + \operatorname{div} T_u a + \operatorname{div} u = -\operatorname{div} T'_a u \\ \partial_t u + T_u \cdot \nabla u - \mathcal{A}u + \nabla a = K(a)\nabla a - \sum_j T'_{\partial_j u} u^j - I(a)\mathcal{A}u + f \end{cases}$$

for some smooth function K vanishing at 0.

We now introduce the notation

$$X(t) \stackrel{\text{def}}{=} \|a(t)\|_{\widetilde{B}_\nu^{\frac{d}{2},\infty}} + \|u(t)\|_{\dot{B}_{2,1}^{\frac{d}{2}-1}} + \int_0^t \left(\nu\|a\|_{\widetilde{B}_\nu^{\frac{d}{2},1}} + \underline{\nu}\|u\|_{\dot{B}_{2,1}^{\frac{d}{2}+1}} \right) dt'$$

$$X_0(t) \stackrel{\text{def}}{=} \|a_0\|_{\widetilde{B}_\nu^{\frac{d}{2},\infty}} + \|u_0\|_{\dot{B}_{2,1}^{\frac{d}{2}-1}} + \int_0^t \|f\|_{\dot{B}_{2,1}^{\frac{d}{2}-1}} dt'.$$

Applying Proposition 10.23 with $\varepsilon = 1$ and $p = r = 1$ yields

$$X(t) \leq C e^{C \int_0^t \|\nabla u\|_{L^\infty} dt'} \left(X_0(t) + \int_0^t \left(\|\operatorname{div} T'_a u\|_{\widetilde{B}_\nu^{\frac{d}{2},\infty}} \right. \right.$$
$$\left. \left. + \|K(a)\nabla a\|_{\dot{B}_{2,1}^{\frac{d}{2}-1}} + \sum_j \|T'_{\partial_j u} u^j\|_{\dot{B}_{2,1}^{\frac{d}{2}-1}} + \|I(a)\mathcal{A}u\|_{\dot{B}_{2,1}^{\frac{d}{2}-1}} \right) dt' \right). \quad (10.57)$$

Remark 10.19, combined with the standard product laws for the remainder and the paraproduct (see Chapter 2), now yields

$$\|\operatorname{div} T'_a u\|_{\tilde{B}_\nu^{\frac{d}{2},\infty}} \leq C \|a\|_{\tilde{B}_\nu^{\frac{d}{2},\infty}} \|u\|_{\dot{B}_{2,1}^{\frac{d}{2}+1}}. \tag{10.58}$$

Next, assuming that

$$\|a\|_{\tilde{L}_T^\infty(\dot{B}_{2,1}^{\frac{d}{2}})} \leq \eta \underline{\nu}/\overline{\nu}, \tag{10.59}$$

where η stands for the constant appearing in Proposition 10.10, we get the following inequalities:

$$\|K(a)\nabla a\|_{\dot{B}_{2,1}^{\frac{d}{2}-1}} \leq C \|a\|_{\dot{B}_{2,1}^{\frac{d}{2}}}^2,$$

$$\|T'_{\partial_j u} u^j\|_{\dot{B}_{2,1}^{\frac{d}{2}-1}} \leq C \|u\|_{\dot{B}_{2,1}^{\frac{d}{2}-1}} \|u\|_{\dot{B}_{2,1}^{\frac{d}{2}+1}},$$

$$\|I(a)\,\mathcal{A}u\|_{\dot{B}_{2,1}^{\frac{d}{2}-1}} \leq C\,\nu^{-1}\overline{\nu}\|a\|_{\tilde{B}_\nu^{\frac{d}{2},\infty}} \|u\|_{\dot{B}_{2,1}^{\frac{d}{2}+1}}.$$

By making use of interpolation, we see that

$$\int_0^t \|a\|_{\dot{B}_{2,1}^{\frac{d}{2}}}^2 \, dt' \leq \|a\|_{L_t^\infty(\tilde{B}_\nu^{\frac{d}{2},\infty})} \|a\|_{L_t^1(\tilde{B}_\nu^{\frac{d}{2},1})}.$$

Hence, inserting the above inequalities into (10.57), we get

$$X(t) \leq C\left(X_0(t) + \frac{\overline{\nu}}{\nu\underline{\nu}} X^2(t)\right) \exp\left(C \int_0^t \|\nabla u(t')\|_{L^\infty} \, dt'\right).$$

Now, if we assume, in addition, that

$$C \int_0^T \|\nabla u(t)\|_{L^\infty} \, dt \leq \log 2, \tag{10.60}$$

then we get

$$X(t) \leq 4C X_0(t) \quad \text{whenever} \quad 4C\overline{\nu}X(t) \leq \nu\underline{\nu}.$$

Because we have $\dot{B}_{2,1}^{\frac{d}{2}} \hookrightarrow L^\infty$, a standard bootstrap argument ensures that the conditions (10.59), (10.60), and $X(T^*) \leq 4C X_0(T^*)$ are satisfied, provided that

$$X_0(\infty) \leq c\nu\underline{\nu}/\overline{\nu}$$

for some sufficiently small constant c.

Applying the continuation criterion stated in Proposition 10.10 completes the proof of global existence.   □

10.5 The Incompressible Limit

It is common sense that *slightly compressible* flows should not differ much from *incompressible* flows. In fact, the incompressible Navier–Stokes equations

$$(NSI) \qquad \begin{cases} \partial_t v + v \cdot \nabla v - \mu \Delta v + \nabla \Pi = g \\ \operatorname{div} v = 0 \\ v_{|t=0} = v_0 \end{cases}$$

are often considered to be relevant for describing compressible barotropic fluids in the low Mach number regime.

This may be justified formally by rescaling the time variable by $t^\varepsilon = \varepsilon t$ (where ε denotes the *Mach number*) and performing the change of unknown $(\rho, u)(t, x) = (\rho^\varepsilon, \varepsilon u^\varepsilon)(t^\varepsilon, x)$. The system for $(\rho^\varepsilon, u^\varepsilon)$ is of the form

$$\begin{cases} \partial_t \rho^\varepsilon + \operatorname{div} \rho^\varepsilon u^\varepsilon = 0, \\ \partial_t(\rho^\varepsilon u^\varepsilon) + \operatorname{div}(\rho^\varepsilon u^\varepsilon \otimes u^\varepsilon) - \mu \Delta u^\varepsilon - (\lambda + \mu) \nabla \operatorname{div} u^\varepsilon + \dfrac{\nabla P^\varepsilon}{\varepsilon^2} = f^\varepsilon, \end{cases}$$

where $P^\varepsilon \overset{\text{def}}{=} P(\rho^\varepsilon)$ stands for the pressure.

At the formal level, it is clear that if $(\rho^\varepsilon, u^\varepsilon)$ tends to some function (ρ, v), then we must have $\nabla P^\varepsilon \to 0$ when ε goes to 0. Hence, if P' does not vanish, the limit density has to be a constant. Now, passing to the limit in the mass equation, we discover that v is divergence-free. Returning to the momentum equation, we can now conclude that v must satisfy (NSI) for some suitable data v_0 and g.

We aim to rigorously justify the above heuristic. If we assume that the data are *well prepared* (i.e., "almost incompressible"), then performing appropriate asymptotic expansions yields the result. In this section, we focus on the case of *ill prepared* data so that acoustic waves have to be considered. More precisely, we assume that the data $(\rho_0^\varepsilon \overset{\text{def}}{=} 1 + \varepsilon b^\varepsilon, u_0^\varepsilon, f^\varepsilon)$ are bounded and that $(u_0^\varepsilon, \mathbb{P} f^\varepsilon)$ tends to some (v_0, g) in a sense which will be made clear later.

10.5.1 Main Results

Writing $\rho^\varepsilon = 1 + \varepsilon b^\varepsilon$, it can be seen that $(b^\varepsilon, u^\varepsilon)$ satisfies

$$(NSC_\varepsilon) \qquad \begin{cases} \partial_t b^\varepsilon + \dfrac{\operatorname{div} u^\varepsilon}{\varepsilon} = -\operatorname{div}(b^\varepsilon u^\varepsilon) \\ \partial_t u^\varepsilon + u^\varepsilon \cdot \nabla u^\varepsilon - \dfrac{\mathcal{A} u^\varepsilon}{1 + \varepsilon b^\varepsilon} + \dfrac{P'(1 + \varepsilon b^\varepsilon)}{1 + \varepsilon b^\varepsilon} \dfrac{\nabla b^\varepsilon}{\varepsilon} = f^\varepsilon \\ (b^\varepsilon, u^\varepsilon)_{|t=0} = (b_0^\varepsilon, u_0^\varepsilon). \end{cases}$$

The main difficulty that is encountered when studying the asymptotics for ε going to 0 is that we have to face the propagation of acoustic waves with the speed ε^{-1}, a phenomenon that does not occur in the case of "well prepared" data. Nevertheless, in this section we prove that satisfactory results may be obtained for quite general data. More precisely, we shall get two types of results concerning the low Mach number limit:

– A global-in-time result for small data with critical regularity.
– A local-in-time result for large data with some extra regularity.

Before stating our first result, we introduce the following function space:

$$F^s \overset{\text{def}}{=} \mathcal{C}_b(\mathbb{R}^+; \dot{B}_{2,1}^{s-1}) \cap L^1(\mathbb{R}^+; \dot{B}_{2,1}^{s+1}). \tag{10.61}$$

We denote by F_T^s the set of functions of F^s restricted to $[0, T]$.

Theorem 10.26. *There exist two positive constants, η and M (depending only on the dimension d and the function G), such that if*

$$\|b_0\|_{\widetilde{B}_{\varepsilon\nu}^{\frac{d}{2},\infty}} + \|u_0\|_{\dot{B}_{2,1}^{\frac{d}{2}-1}} + \|f\|_{L^1(\dot{B}_{2,1}^{\frac{d}{2}-1})} \le \eta\, \nu\underline{\nu}/\overline{\nu}, \tag{10.62}$$

then the system (NSC_ε) with data (b_0, u_0, f) and the system (NSI) with data $(\mathbb{P}u_0, \mathbb{P}f)$ have respective unique global solutions $(b^\varepsilon, u^\varepsilon)$ and v in the spaces $E_{\varepsilon\nu}^{\frac{d}{2}}$ and $F^{\frac{d}{2}}$, respectively, and

$$\|(b^\varepsilon, u^\varepsilon)\|_{E_{\varepsilon\nu}^{\frac{d}{2}}} \le M\big(\|b_0\|_{\widetilde{B}_{\varepsilon\nu}^{\frac{d}{2},\infty}} + \|u_0\|_{\dot{B}_{2,1}^{\frac{d}{2}-1}} + \|f\|_{L^1(\dot{B}_{2,1}^{\frac{d}{2}-1})}\big),$$

$$\|v\|_{F^{\frac{d}{2}}} \le M\big(\|\mathbb{P}u_0\|_{\dot{B}_{2,1}^{\frac{d}{2}-1}} + \|\mathbb{P}f\|_{L^1(\dot{B}_{2,1}^{\frac{d}{2}-1})}\big).$$

Moreover, there exists a Banach space $E \subset \mathcal{S}'(\mathbb{R}^d)$ and an exponent $p \in [2, \infty[$ (both depending on the dimension d) such that $\mathbb{P}u^\varepsilon$ tends to v, and $(b^\varepsilon, \mathbb{P}^\perp u^\varepsilon)$ tends to 0 in $L^p(\mathbb{R}^+; E)$.

A more accurate statement for slightly more general data will be given in the forthcoming Theorem 10.29. There, a rate of convergence involving explicit norms will be obtained.

We now give a (simplified) statement concerning the case of large data with more regularity.

Theorem 10.27. *Assume that $b_0 \in \dot{B}_{2,1}^{\frac{d}{2}-1} \cap \dot{B}_{2,1}^{\frac{d}{2}+\alpha}$, $u_0 \in \dot{B}_{2,1}^{\frac{d}{2}-1} \cap \dot{B}_{2,1}^{\frac{d}{2}-1+\alpha}$, and $f \in L_{loc}^1(\mathbb{R}^+; \dot{B}_{2,1}^{\frac{d}{2}-1} \cap \dot{B}_{2,1}^{\frac{d}{2}-1+\alpha})$ for some sufficiently small positive α. Suppose that the incompressible system (NSI) with initial datum $\mathbb{P}u_0$ and external force $\mathbb{P}f$ has a solution $v \in F_{T_0}^{\frac{d}{2}} \cap F_{T_0}^{\frac{d}{2}+\alpha}$ on the time interval $[0, T_0[$ for some finite or infinite T_0.*

Then, for all sufficiently small $\varepsilon > 0$, the system (NSC_ε) has a unique solution $(b^\varepsilon, u^\varepsilon)$ in $E_{\varepsilon\nu,T_0}^{\frac{d}{2}} \cap E_{\varepsilon\nu,T_0}^{\frac{d}{2}+\alpha}$ (with bounds independent of ε). In addition, $\mathbb{P}u_\varepsilon$ tends to v in $F_{T_0}^{\frac{d}{2}} \cap F_{T_0}^{\frac{d}{2}+\alpha}$, and there exist an exponent $p \in [2, \infty[$ and a Banach space E such that $(b^\varepsilon, \mathbb{P}^\perp u_\varepsilon)$ tends to 0 in $L^p(\mathbb{R}^+; E)$.

Remark 10.28. Here, the exponents p and function space E may also be determined explicitly. In addition, an upper bound may be given for the rate of convergence. For more details, the reader is referred to Theorem 10.31.

We also point out that in the case $d = 2$, the above statement implies that the solution of (NSC_ε) is globally defined for all sufficiently small ε. This is a simple consequence of the fact that the corresponding solution of the two-dimensional Navier–Stokes equations is global.

10.5.2 The Case of Small Data with Critical Regularity

We now explain the basic ideas of the proof of Theorem 10.26. First, an appropriate change of variables enables us to apply Theorem 10.21. Under the smallness assumption (10.62), we get a global solution $(b^\varepsilon, u^\varepsilon)$ with uniform bounds in $E_{\varepsilon\nu}^{\frac{d}{2}}$. The existence of a global solution for the limit system (NSI) follows from classical arguments, similar to those in Chapter 5, and will thus be omitted.

While, up to this point, our method also works in the periodic setting, our proof of strong convergence is specific to $\mathbb{R}^d$. In effect, it relies on the dispersive properties of the acoustic wave operator in the whole space. To make our discussion more accurate, we resume the spectral analysis of Section 10.4.2. The linear system we now have to deal with is

$$\begin{cases} \partial_t b + \dfrac{\operatorname{div} u}{\varepsilon} = f \\ \partial_t u - \mathcal{A}u + \dfrac{\nabla b}{\varepsilon} = g. \end{cases} \tag{10.63}$$

As in the case $\varepsilon = 1$, the incompressible part of the velocity satisfies an ordinary heat equation with constant diffusion μ. As for $(b, v \overset{\text{def}}{=} |D|^{-1} \operatorname{div} \mathbb{P}^\perp u)$, we have

$$\frac{d}{dt}\begin{pmatrix} b \\ v \end{pmatrix} = \begin{pmatrix} F \\ G \end{pmatrix} + A_\varepsilon(D)\begin{pmatrix} b \\ v \end{pmatrix} \quad \text{with} \quad A_\varepsilon(D) \overset{\text{def}}{=} \begin{pmatrix} 0 & -\varepsilon^{-1}|D| \\ \varepsilon^{-1}|D| & -\nu|D|^2 \end{pmatrix}.$$

The low-frequency regime corresponds to those ξ's which satisfy $\nu\varepsilon|\xi| < 2$. The corresponding eigenvalues are

$$\lambda_\varepsilon^\pm(\xi) = -\frac{\nu|\xi|^2}{2}\left(1 \pm i\sqrt{\frac{4}{\varepsilon^2\nu^2|\xi|^2} - 1}\right)$$

so that in the limit where $\nu\varepsilon|\xi|$ goes to 0, we expect the system to behave like the linear operator

$$\frac{d}{dt} - \frac{\nu}{2}\Delta \pm \frac{i}{\varepsilon}|D|.$$

In other words, the low frequencies of (b, v) behave as if they were solutions to a heat equation plus a half-wave equation with propagation speed ε^{-1}.

In the high-frequency regime $\nu\varepsilon|\xi| > 2$, we have

$$\lambda_\varepsilon^\pm(\xi) = -\frac{\nu|\xi|^2}{2}\left(1 \pm \sqrt{1 - \frac{4}{\varepsilon^2\nu^2|\xi|^2}}\right),$$

which means that a parabolic mode and a damped mode coexist.

In the analysis which has been presented thus far for the case $\varepsilon = 1$, we have made extensive use of the parabolic properties of the system but have not taken advantage of its dispersive properties in low frequencies. In effect, using an L^2 approach precludes us from using the skew-symmetry of the first order terms. It turns out that in the "whole space" case we are interested in, the proof of convergence for ε going to 0 is intimately entangled with the dispersive properties of the system (10.63) (see Proposition 10.30 below).

We now give the full statement of our convergence result for small data.

Theorem 10.29. *There exist two positive constants, η and M, depending only on d and G, such that if*

$$
\begin{aligned}
C_0^{\varepsilon\nu} &\overset{def}{=} \|b_0^\varepsilon\|_{\dot{B}_{2,1}^{\frac{d}{2}-1}} + \varepsilon\nu\|b_0^\varepsilon\|_{\dot{B}_{2,1}^{\frac{d}{2}}} + \|u_0^\varepsilon\|_{\dot{B}_{2,1}^{\frac{d}{2}-1}} + \|f^\varepsilon\|_{L^1(\dot{B}_{2,1}^{\frac{d}{2}-1})} \le \eta\nu\underline{\nu}/\overline{\nu}, \\
C_0 &\overset{def}{=} \|v_0\|_{\dot{B}_{2,1}^{\frac{d}{2}-1}} + \|g\|_{L^1(\dot{B}_{2,1}^{\frac{d}{2}-1})} \le \eta\mu \quad with \quad \operatorname{div} v_0 = \operatorname{div} g = 0,
\end{aligned}
\tag{10.64}
$$

then the following results hold:

(i) *Existence:*

(a) *The system (NSC_ε) has a unique global solution $(b^\varepsilon, u^\varepsilon)$ in $E_{\varepsilon\nu}^{\frac{d}{2}}$ such that*

$$\|(b^\varepsilon, u^\varepsilon)\|_{E_{\varepsilon\nu}^{\frac{d}{2}}} \le MC_0^{\varepsilon\nu}.$$

(b) *The incompressible Navier–Stokes equations (NSI) with data v_0 and g have a unique solution v in $F^{\frac{d}{2}}$ such that*

$$\mu\|v\|_{L^1(\dot{B}_{2,1}^{\frac{d}{2}+1})} + \|v\|_{L^\infty(\dot{B}_{2,1}^{\frac{d}{2}-1})} \le MC_0.$$

(ii) *Convergence: For any $\alpha \in \,]0, 1/2]$ if $d \ge 4$, $\alpha \in \,]0, 1/2[$ if $d = 3$, and $\alpha \in \,]0, 1/6]$ if $d = 2$, $\mathbb{P}u^\varepsilon$ tends to v in $C(\mathbb{R}^+; \dot{B}_{\infty,1}^{-1-\alpha})$ when ε goes to 0. Moreover:*

(a) *Case $d \ge 4$: For all $p \in [p_d, \infty]$ with $p_d \overset{def}{=} 2(d-1)/(d-3)$, we have*

$$\|\mathbb{P}^\perp u^\varepsilon\|_{\widetilde{L}^2(\dot{B}_{p,1}^{\frac{d}{p}-\frac{1}{2}})} + \|b^\varepsilon\|_{\widetilde{L}^2(\dot{B}_{p,1}^{\frac{d}{p}-\frac{1}{2}})} \le M(1 + \overline{\nu}/\underline{\nu})C_0^{\varepsilon\nu}\varepsilon^{\frac{1}{2}},$$

$$\|\mathbb{P}u^\varepsilon - v\|_{L^1(\dot{B}_{p,1}^{\frac{d}{p}+\frac{1}{2}})} + \|\mathbb{P}u^\varepsilon - v\|_{L^\infty(\dot{B}_{p,1}^{\frac{d}{p}-\frac{3}{2}})}$$

$$\le M\left(C_0^{\varepsilon\nu}\varepsilon^{\frac{1}{2}} + \|\mathbb{P}u_0^\varepsilon - v_0\|_{\dot{B}_{p,1}^{\frac{d}{p}-\frac{3}{2}}} + \|\mathbb{P}f^\varepsilon - g\|_{L^1(\dot{B}_{p,1}^{\frac{d}{p}-\frac{3}{2}})}\right).$$

(b) Case $d = 3$: For all $p \in [2, \infty[$, we have

$$\|\mathbb{P}^{\perp} u^{\varepsilon}\|_{\tilde{L}^{\frac{2p}{p-2}}(\dot{B}_{p,1}^{\frac{2}{p}-\frac{1}{2}})} + \|b^{\varepsilon}\|_{\tilde{L}^{\frac{2p}{p-2}}(\dot{B}_{p,1}^{\frac{2}{p}-\frac{1}{2}})} \leq M(1 + \bar{\nu}/\underline{\nu})C_0^{\varepsilon\nu}\varepsilon^{\frac{1}{2}-\frac{1}{p}},$$

$$\|\mathbb{P} u^{\varepsilon} - v\|_{L^1(\dot{B}_{p,1}^{\frac{4}{p}+\frac{1}{2}})} + \|\mathbb{P} u^{\varepsilon} - v\|_{L^{\infty}(\dot{B}_{p,1}^{\frac{4}{p}-\frac{3}{2}})}$$

$$\leq M\left(C_0^{\varepsilon\nu}\varepsilon^{\frac{1}{2}-\frac{1}{p}} + \|\mathbb{P} u_0^{\varepsilon} - v_0\|_{\dot{B}_{p,1}^{\frac{4}{p}-\frac{3}{2}}} + \|\mathbb{P} f^{\varepsilon} - g\|_{L^1(\dot{B}_{p,1}^{\frac{4}{p}-\frac{3}{2}})}\right).$$

(c) Case $d = 2$: For all $p \in [2, 6]$, we have

$$\|\mathbb{P}^{\perp} u^{\varepsilon}\|_{\tilde{L}^{\frac{4p}{p-2}}(\dot{B}_{p,1}^{\frac{3}{2p}-\frac{3}{4}})} + \|b^{\varepsilon}\|_{\tilde{L}^{\frac{4p}{p-2}}(\dot{B}_{p,1}^{\frac{3}{2p}-\frac{3}{4}})} \leq M(1 + \bar{\nu}/\underline{\nu})C_0^{\varepsilon\nu}\varepsilon^{\frac{1}{4}-\frac{1}{2p}},$$

$$\|\mathbb{P} u^{\varepsilon} - v\|_{L^1(\dot{B}_{p,1}^{\frac{5}{2p}+\frac{3}{4}})} + \|\mathbb{P} u^{\varepsilon} - v\|_{L^{\infty}(\dot{B}_{p,1}^{\frac{5}{2p}-\frac{5}{4}})}$$

$$\leq M\left(C_0^{\varepsilon\nu}\varepsilon^{\frac{1}{4}-\frac{1}{2p}} + \|\mathbb{P} u_0^{\varepsilon} - v_0\|_{\dot{B}_{p,1}^{\frac{5}{2p}-\frac{5}{4}}} + \|\mathbb{P} f^{\varepsilon} - g\|_{L^1(\dot{B}_{p,1}^{\frac{5}{2p}-\frac{5}{4}})}\right).$$

Proof. We shall proceed as follows:

- First, we prove that the system (NSC_{ε}) has a unique global solution which satisfies uniform estimates.
- Second, we combine those estimates with the dispersive properties of (10.63) to prove that $(b^{\varepsilon}, \mathbb{P}^{\perp} u^{\varepsilon})$ converges to zero in some suitable function space.
- Third, we use the fact that $\mathbb{P} u^{\varepsilon} - v$ satisfies a heat equation with source terms which are small because they depend, at least linearly, on $(b^{\varepsilon}, \mathbb{P}^{\perp} u^{\varepsilon})$. This yields $\mathbb{P} u^{\varepsilon} \to v$ in some function space.

Step 1. Proof of Global Existence with Uniform Estimates

Making the changes of function

$$c^{\varepsilon}(t, x) \stackrel{\text{def}}{=} \varepsilon b^{\varepsilon}(\varepsilon^2 t, \varepsilon x), \quad v^{\varepsilon}(t, x) \stackrel{\text{def}}{=} \varepsilon u^{\varepsilon}(\varepsilon^2 t, \varepsilon x), \text{ and } h^{\varepsilon}(t, x) \stackrel{\text{def}}{=} \varepsilon^3 f^{\varepsilon}(\varepsilon^2 t, \varepsilon x),$$

we note that $(b^{\varepsilon}, u^{\varepsilon})$ solves (NSC_{ε}) if and only if $(c^{\varepsilon}, v^{\varepsilon})$ solves (NSC) with rescaled data $\varepsilon b_0^{\varepsilon}(\varepsilon\cdot)$, $\varepsilon u_0^{\varepsilon}(\varepsilon\cdot)$, and h^{ε}. Hence, Theorem 10.21 ensures global existence: There exist two positive constants, η and M, depending only on d and G, and such that (NSC) has a solution $(c^{\varepsilon}, v^{\varepsilon})$ in $E_{\nu}^{\frac{d}{2}}$ whenever

$$\|\varepsilon b_0^{\varepsilon}(\varepsilon\cdot)\|_{\tilde{B}_{\nu}^{\frac{d}{2}, \infty}} + \|\varepsilon u_0^{\varepsilon}(\varepsilon\cdot)\|_{\dot{B}_{2,1}^{\frac{d}{2}-1}} + \|h^{\varepsilon}\|_{L^1(\dot{B}_{2,1}^{\frac{d}{2}-1})} \leq \eta\nu\underline{\nu}/\bar{\nu}.$$

Furthermore,

$$\|(c^{\varepsilon}, v^{\varepsilon})\|_{E_{\nu}^{\frac{d}{2}}} \leq M\left(\|\varepsilon b_0^{\varepsilon}(\varepsilon\cdot)\|_{\tilde{B}_{\nu}^{\frac{d}{2}, \infty}} + \|\varepsilon u_0^{\varepsilon}(\varepsilon\cdot)\|_{\dot{B}_{2,1}^{\frac{d}{2}-1}} + \|h^{\varepsilon}\|_{L^1(\dot{B}_{2,1}^{\frac{d}{2}-1})}\right).$$

Now, arguing as in Proposition 2.18 page 64, we may check that

$$\|(c^\varepsilon, v^\varepsilon)\|_{E_\nu^{\frac{d}{2}}} \approx \|(b^\varepsilon, u^\varepsilon)\|_{E_{\varepsilon\nu}^{\frac{d}{2}}} \quad \text{and that}$$

$$\|\varepsilon b_0^\varepsilon(\varepsilon\cdot)\|_{\widetilde{B}_\nu^{\frac{d}{2},\infty}} + \|\varepsilon u_0^\varepsilon(\varepsilon\cdot)\|_{\dot{B}_{2,1}^{\frac{d}{2}-1}} + \|h^\varepsilon\|_{L^1(\dot{B}_{2,1}^{\frac{d}{2}-1})}$$

$$\approx \|b_0^\varepsilon\|_{\widetilde{B}_{\varepsilon\nu}^{\frac{d}{2},\infty}} + \|u_0^\varepsilon\|_{\dot{B}_{2,1}^{\frac{d}{2}-1}} + \|f^\varepsilon\|_{L^1(\dot{B}_{2,1}^{\frac{d}{2}-1})},$$

which yields the first part of Theorem 10.29.

Step 2. Convergence to Zero of the Compressible Modes

The proof relies on dispersive inequalities for the following (reduced) system of acoustics:

$$(W_\varepsilon) \qquad \begin{cases} \partial_t b + \varepsilon^{-1}|D|v = F \\ \partial_t v - \varepsilon^{-1}|D|b = G \\ (b, v)_{|t=0} = (b_0, v_0). \end{cases}$$

Proposition 10.30. *Let* (b, v) *be a solution of* (W_ε). *Then, for any* $s \in \mathbb{R}$ *and positive* T *(possibly infinite), the following estimate holds:*

$$\|(b, v)\|_{\widetilde{L}_T^r(\dot{B}_{p,1}^{s+d(\frac{1}{p}-\frac{1}{2})+\frac{1}{r}})} \leq C\varepsilon^{\frac{1}{r}}\|(b_0, v_0)\|_{\dot{B}_{2,1}^s} + \varepsilon^{1+\frac{1}{r}-\frac{1}{\bar{r}'}}\|(F, G)\|_{\widetilde{L}_T^{\bar{r}'}(B_{\bar{p}',1}^{s+d(\frac{1}{\bar{p}'}-\frac{1}{2})+\frac{1}{\bar{r}'}})}$$

with $p \geq 2$, $\dfrac{2}{r} \leq \min(1, \gamma(p))$, $(r, p, d) \neq (2, \infty, 3)$,

$$\bar{p} \geq 2, \quad \frac{2}{\bar{r}} \leq \min(1, \gamma(\bar{p})), \quad (\bar{r}, \bar{p}, d) \neq (2, \infty, 3),$$

where $\gamma(q) \overset{def}{=} (d-1)\left(\dfrac{1}{2} - \dfrac{1}{q}\right)$, $\dfrac{1}{p} + \dfrac{1}{p'} = 1$, *and* $\dfrac{1}{\bar{r}} + \dfrac{1}{\bar{r}'} = 1$.

Proof. Define $\Phi \overset{def}{=} {}^t(c, v)$ and $H \overset{def}{=} {}^t(F, G)$. Setting $\Psi(t, x) = \Phi(\varepsilon t, x)$ and $\mathcal{H}(t, x) = \varepsilon H(\varepsilon t, x)$, we easily check that Ψ solves (W_1) with the external force $\mathcal{H}$. Hence, the general case $\varepsilon > 0$ follows from the particular case $\varepsilon = 1$. Let $U(t)$ be the group associated with the system (W_1). We have, in Fourier variables,

$$\mathcal{F}(U(t)\Phi)(\xi) = \begin{pmatrix} \cos(|\xi|t) & -\sin(|\xi|t) \\ \sin(|\xi|t) & \cos(|\xi|t) \end{pmatrix} \mathcal{F}\Phi(\xi).$$

Exactly as for the wave equation (see Proposition 8.15), we deduce that for any function $\Phi \in L^1(\mathbb{R}^d; \mathbb{R}^2)$ with $\mathcal{F}\Phi$ supported in the annulus $\mathcal{C}$,

$$\begin{aligned} \|U(t)\Phi\|_{L^2} &\leq \|\Phi\|_{L^2}, \\ \|U(s)U^\star(t)\Phi\|_{L^\infty} &\leq C(1 + |t - s|)^{-\frac{d-1}{2}}\|\Phi\|_{L^1}. \end{aligned}$$

Applying Theorem 8.18 page 349 yields Proposition 10.30 in the special case where the spectrum of the data is supported in the annulus $\mathcal{C}$.

More generally, for all $j \in \mathbb{Z}$, we have

$$2^{j\left(d\left(\frac{1}{p}-\frac{1}{2}\right)+\frac{1}{r}\right)}\|\dot{\Delta}_j\Phi\|_{L^r_T(L^p)} \leq C\|\dot{\Delta}_j\Phi_0\|_{L^2} + 2^{j\left(d\left(\frac{1}{p'}-\frac{1}{2}\right)+\frac{1}{r'}\right)}\|\dot{\Delta}_j H\|_{L^{r'}_T(L^{p'})}.$$

This may be deduced from the case $j = 0$ (where the spectrum is supported in the annulus $\mathcal{C}$) after implementing a suitable change of variable. So, finally, multiplying both sides of the above inequality by 2^{js} and performing a summation over $\mathbb{Z}$ completes the proof. $\qquad\square$

In order to prove the convergence to 0 for $(b^\varepsilon, \mathbb{P}^\perp u^\varepsilon)$, we may use the fact that

$$\begin{cases} \partial_t b^\varepsilon + \varepsilon^{-1}\operatorname{div}\mathbb{P}^\perp u^\varepsilon = F^\varepsilon \\ \partial_t \mathbb{P}^\perp u^\varepsilon + \varepsilon^{-1}\nabla b^\varepsilon = G^\varepsilon \end{cases} \tag{10.65}$$

with $F^\varepsilon \stackrel{\text{def}}{=} -\operatorname{div}(b^\varepsilon u^\varepsilon)$ and

$$G^\varepsilon \stackrel{\text{def}}{=} -\mathbb{P}^\perp\operatorname{div}\left(u^\varepsilon\cdot\nabla u^\varepsilon + \frac{1}{1+\varepsilon b^\varepsilon}\mathcal{A}u^\varepsilon + \frac{K(\varepsilon b^\varepsilon)\nabla b^\varepsilon}{\varepsilon} + f^\varepsilon\right).$$

Obviously, the dispersive estimates stated in Proposition 10.30 are also true for the system (10.65) since b^ε and $v^\varepsilon \stackrel{\text{def}}{=} |D|^{-1}\operatorname{div}\mathbb{P}^\perp u^\varepsilon$ satisfy (W_ε) with source terms F^ε and $|D|^{-1}\operatorname{div}G^\varepsilon$, and $|D|^{-1}\operatorname{div}$ is a homogeneous multiplier of degree 0. Hence, taking $\bar{p} = 2$, $r = \infty$, and

- $s = d/2-1$ and $r = 2$, if $d \geq 4$,
- $2 \leq p < \infty$ and $r = 2p/(p-2)$, if $d = 3$,
- $2 \leq p \leq \infty$ and $r = 4p/(p-2)$, if $d = 2$,

we see that it is enough to prove that

$$\|(F^\varepsilon, G^\varepsilon)\|_{L^1(\dot{B}^{\frac{d}{2}-1}_{2,1})} \leq C(1+\overline{\nu}/\underline{\nu})C_0^{\varepsilon\nu}.$$

This inequality follows from the uniform estimates of step one. Indeed, combining Hölder's inequality with the usual product and composition laws in Besov spaces yields

$$\|F^\varepsilon\|_{L^1(\dot{B}^{\frac{d}{2}-1}_{2,1})} \leq C\|b^\varepsilon\|_{L^2(\dot{B}^{\frac{d}{2}}_{2,1})}\|u^\varepsilon\|_{L^2(\dot{B}^{\frac{d}{2}}_{2,1})} \leq CC_0^{\varepsilon\nu},$$

$$\|u^\varepsilon\cdot\nabla u^\varepsilon\|_{L^1(\dot{B}^{\frac{d}{2}-1}_{2,1})} \leq C\|u^\varepsilon\|_{L^2(\dot{B}^{\frac{d}{2}}_{2,1})}\|\nabla u^\varepsilon\|_{L^2(\dot{B}^{\frac{d}{2}-1}_{2,1})} \leq CC_0^{\varepsilon\nu},$$

$$\begin{aligned}\left\|\frac{1}{1+\varepsilon b^\varepsilon}\mathcal{A}u^\varepsilon\right\|_{L^1(\dot{B}^{\frac{d}{2}-1}_{2,1})} &\leq C\left(1+\varepsilon\|b^\varepsilon\|_{L^\infty(\dot{B}^{\frac{d}{2}}_{2,1})}\right)\|\mathcal{A}u^\varepsilon\|_{L^1(\dot{B}^{\frac{d}{2}-1}_{2,1})} \\ &\leq C\,\overline{\nu}\left(1+\nu^{-1}\|b^\varepsilon\|_{L^\infty(\tilde{B}^{\frac{d}{2},\infty}_{\varepsilon\nu})}\right)\|u^\varepsilon\|_{L^1(\dot{B}^{\frac{d}{2}+1}_{2,1})} \\ &\leq C\,\overline{\nu}\underline{\nu}^{-1}C_0^{\varepsilon\nu}\end{aligned}$$

$$\|K(\varepsilon b^\varepsilon)\nabla b^\varepsilon\|_{L^1(\dot{B}^{\frac{d}{2}-1}_{2,1})} \leq C\varepsilon\|b^\varepsilon\|_{L^2(\dot{B}^{\frac{d}{2}}_{2,1})}\|\nabla b^\varepsilon\|_{L^2(\dot{B}^{\frac{d}{2}-1}_{2,1})} \leq C\varepsilon C_0^{\varepsilon\nu}.$$

Step 3. Convergence of the Incompressible Part

Let $w^\varepsilon \overset{\text{def}}{=} \mathbb{P}u^\varepsilon - v$. Applying Leray's projector to the second equation of (NSC_ε) and subtracting (NSI) from it yields the following equation for w^ε:

$$\begin{cases} \partial_t w^\varepsilon - \mu \Delta w^\varepsilon = H^\varepsilon + h^\varepsilon \\ w^\varepsilon_{|t=0} = w^\varepsilon_0 \end{cases} \tag{10.66}$$

with $w^\varepsilon_0 \overset{\text{def}}{=} \mathbb{P}u^\varepsilon_0 - v_0$, $h^\varepsilon \overset{\text{def}}{=} \mathbb{P}f^\varepsilon - g$, and

$$H^\varepsilon \overset{\text{def}}{=} -\mathbb{P}(w^\varepsilon \cdot \nabla v) - \mathbb{P}(u^\varepsilon \cdot \nabla w^\varepsilon) - \mathbb{P}(\mathbb{P}^\perp u^\varepsilon \cdot \nabla v) - \mathbb{P}(u^\varepsilon \cdot \nabla \mathbb{P}^\perp u^\varepsilon) - \mathbb{P}\left(I(\varepsilon b^\varepsilon)\mathcal{A}u^\varepsilon\right).$$

We will just treat the case $d \geq 4$, which is easier to handle. Let

$$Y_p \overset{\text{def}}{=} \|w^\varepsilon\|_{L^1(\dot{B}^{\frac{d}{p}+\frac{1}{2}}_{p,1})} + \|w^\varepsilon\|_{L^\infty(\dot{B}^{\frac{d}{p}-\frac{3}{2}}_{p,1})}.$$

We claim that for all $p \in [p_d, \infty]$, we have

$$Y_p \leq C\left(C^{\varepsilon\nu}_0 \varepsilon^{\frac{1}{2}} + \|w^\varepsilon_0\|_{\dot{B}^{\frac{d}{p}-\frac{3}{2}}_{p,1}} + \|h^\varepsilon\|_{L^1(\dot{B}^{\frac{d}{p}-\frac{3}{2}}_{p,1})}\right). \tag{10.67}$$

Indeed, by virtue of the inequality (3.39) page 157, we have

$$Y_p \leq C\left(\|w^\varepsilon_0\|_{\dot{B}^{\frac{d}{p}-\frac{3}{2}}_{p,1}} + \|h^\varepsilon\|_{L^1(\dot{B}^{\frac{d}{p}-\frac{3}{2}}_{p,1})} + \|H^\varepsilon\|_{L^1(\dot{B}^{\frac{d}{p}-\frac{3}{2}}_{p,1})}\right). \tag{10.68}$$

From Proposition 2.54, the previous step, and (10.64), we deduce that

$$\begin{aligned}
\|\mathbb{P}(w^\varepsilon \cdot \nabla v)\|_{L^1(\dot{B}^{\frac{d}{p}-\frac{3}{2}}_{p,1})} &\leq C \|\nabla v\|_{L^2(\dot{B}^{\frac{d}{2}-1}_{2,1})} \|w^\varepsilon\|_{L^2(\dot{B}^{\frac{d}{p}-\frac{1}{2}}_{p,1})} \\
&\leq C\, \eta \|w^\varepsilon\|_{L^2(\dot{B}^{\frac{d}{p}-\frac{1}{2}}_{p,1})}, \\
\|\mathbb{P}(u^\varepsilon \cdot \nabla w^\varepsilon)\|_{L^1(\dot{B}^{\frac{d}{p}-\frac{3}{2}}_{p,1})} &\leq C \|u^\varepsilon\|_{L^2(\dot{B}^{\frac{d}{2}}_{2,1})} \|\nabla w^\varepsilon\|_{L^2(\dot{B}^{\frac{d}{p}-\frac{3}{2}}_{p,1})} \\
&\leq C\, \eta \|w^\varepsilon\|_{L^2(\dot{B}^{\frac{d}{p}-\frac{1}{2}}_{p,1})},
\end{aligned}$$

$$\begin{aligned}
\|\mathbb{P}(\mathbb{P}^\perp u^\varepsilon \cdot \nabla v)\|_{L^1(\dot{B}^{\frac{d}{p}-\frac{3}{2}}_{p,1})} &\leq C \|\nabla v\|_{L^2(\dot{B}^{\frac{d}{2}-1}_{2,1})} \|\mathbb{P}^\perp u^\varepsilon\|_{L^2(\dot{B}^{\frac{d}{p}-\frac{1}{2}}_{p,1})} \\
&\leq C\, \eta^2 \varepsilon^{\frac{1}{2}}, \\
\|\mathbb{P}(u^\varepsilon \cdot \nabla \mathbb{P}^\perp u^\varepsilon)\|_{L^1(\dot{B}^{\frac{d}{p}-\frac{3}{2}}_{p,1})} &\leq C \|u^\varepsilon\|_{L^2(\dot{B}^{\frac{d}{2}}_{2,1})} \|\nabla \mathbb{P}^\perp u^\varepsilon\|_{L^2(\dot{B}^{\frac{d}{p}-\frac{3}{2}}_{p,1})} \\
&\leq C\, \eta^2 \varepsilon^{\frac{1}{2}}.
\end{aligned}$$

Note that all the above product estimates are justified since $d/2 + d/p - 3/2$ is always positive for any $p \leq \infty$ when $d \geq 4$. Thanks to the embedding $\dot{B}^{\frac{d}{2}-\frac{3}{2}}_{2,1} \hookrightarrow \dot{B}^{\frac{d}{p}-\frac{3}{2}}_{p,1}$ and the definition of hybrid Besov norms, we also have

$$\|\mathbb{P}\left(I(\varepsilon b^{\varepsilon})\mathcal{A}u^{\varepsilon}\right)\|_{L^1(\dot{B}_{p,1}^{\frac{d}{p}-\frac{3}{2}})} \leq C \, \|I(\varepsilon b^{\varepsilon})\mathcal{A}u^{\varepsilon}\|_{L^1(\dot{B}_{2,1}^{\frac{d}{2}-\frac{3}{2}})}$$

$$\leq C \, \|\varepsilon b\|_{L^4(\dot{B}_{2,1}^{\frac{d}{2}})}\|\mathcal{A}u^{\varepsilon}\|_{L^{\frac{4}{3}}(\dot{B}_{2,1}^{\frac{d}{2}-\frac{3}{2}})}$$

$$\leq C \, \overline{\nu}\varepsilon^{\frac{1}{2}}\|b^{\varepsilon}\|_{L^4(\widetilde{B}_{\varepsilon\nu}^{\frac{d}{2},4})}\|u^{\varepsilon}\|_{L^{\frac{4}{3}}(\dot{B}_{2,1}^{\frac{d}{2}+\frac{1}{2}})}$$

$$\leq C \, \eta^2\varepsilon^{\frac{1}{2}}.$$

Plugging all the above estimates into (10.68), we end up with

$$Y_p \leq C\Big(\eta^2\varepsilon^{\frac{1}{2}} + \|w_0^{\varepsilon}\|_{\dot{B}_{p,1}^{\frac{d}{p}-\frac{3}{2}}} + \|h^{\varepsilon}\|_{L^1(\dot{B}_{p,1}^{\frac{d}{p}-\frac{3}{2}})} + \eta Y_p\Big),$$

so that we can conclude that (10.67) holds, provided that the constant η has been chosen to be sufficiently small.

In dimension $d = 2, 3$, the dispersive properties given in Proposition 10.30 are not as good as in dimension $d \geq 4$. Hence, we cannot get uniform estimates for $\varepsilon^{-\frac{1}{2}}\mathbb{P}^{\perp}u^{\varepsilon}$ in $L^2(\mathbb{R}^+; \dot{B}_{p,1}^{\frac{d}{p}-\frac{1}{2}})$. However, we can interpolate the uniform estimates for u^{ε} in $L^1(\mathbb{R}^+; \dot{B}_{2,1}^{\frac{d}{2}+1})$, given by step one, with the dispersive inequalities proved in the second step. This still gives some decay in ε. The reader is referred to [97] for more details.   □

10.5.3 The Case of Large Data with More Regularity

In the case of large data, the problem of global existence in dimension $d \geq 3$ for the incompressible Navier–Stokes system (NSI) [which is expected to be the limit system for (NSC_{ε})] is open. Therefore, there are few chances to get a global result for the system (NSC_{ε}), which is more complicated.

In this subsection, we want to establish that the system (NSC_{ε}) with suitably small ε has a strong solution on the time interval I (possibly infinite) whenever the limit system (NSI) has a strong solution on I. This result is of particular interest in dimension $d = 2$ since the limit system is globally well-posed for any divergence-free data in L^2.

Theorem 10.31. *Let $b_0 \in \dot{B}_{2,1}^{\frac{d}{2}-1} \cap \dot{B}_{2,1}^{\frac{d}{2}+\alpha}$, $u_0 \in \dot{B}_{2,1}^{\frac{d}{2}-1} \cap \dot{B}_{2,1}^{\frac{d}{2}-1+\alpha}$, and $f \in L^1_{loc}(\mathbb{R}^+; \dot{B}_{2,1}^{\frac{d}{2}-1} \cap \dot{B}_{2,1}^{\frac{d}{2}-1+\alpha})$ with $\alpha \in \,]0, 1/2]$ if $d \geq 4$, $\alpha \in \,]0, 1/2[$ if $d = 3$, and $\alpha \in \,]0, 1/6]$ if $d = 2$. Let $T_0 \in \,]0, \infty]$. Suppose that the incompressible system (NSI) with initial datum $\mathbb{P}u_0$ and external force $\mathbb{P}f$ has a solution $v \in F_{T_0}^{\frac{d}{2}} \cap F_{T_0}^{\frac{d}{2}+\alpha}$. Let $V \overset{def}{=} \|v\|_{F_{T_0}^{\frac{d}{2}} \cap F_{T_0}^{\frac{d}{2}+\alpha}}$ and*

$$X_0 \overset{def}{=} \|b_0\|_{\dot{B}_{2,1}^{\frac{d}{2}-1} \cap \dot{B}_{2,1}^{\frac{d}{2}+\alpha}} + \|\mathbb{P}^{\perp}u_0\|_{\dot{B}_{2,1}^{\frac{d}{2}-1} \cap \dot{B}_{2,1}^{\frac{d}{2}-1+\alpha}} + \|\mathbb{P}^{\perp}f\|_{L^1_{T_0}(\dot{B}_{2,1}^{\frac{d}{2}-1} \cap \dot{B}_{2,1}^{\frac{d}{2}-1+\alpha})}.$$

There exist two positive constants, ε_0 and C, depending only on d, α, λ, μ, P, V, and X_0, and such that the following results hold true:

(i) For all $0 < \varepsilon \leq \varepsilon_0$, the system (NSC_ε) has a unique solution $(b^\varepsilon, u^\varepsilon)$ in $E_{\varepsilon\nu,T_0}^{\frac{d}{2}} \cap E_{\varepsilon\nu,T_0}^{\frac{d}{2}+\alpha}$ such that

$$\|(b^\varepsilon, u^\varepsilon)\|_{E_{\varepsilon\nu,T_0}^{\frac{d}{2}} \cap E_{\varepsilon\nu,T_0}^{\frac{d}{2}+\alpha}} \leq C.$$

(ii) The vector field $\mathbb{P}\, u_\varepsilon$ tends to v in $F_{T_0}^{\frac{d}{2}} \cap F_{T_0}^{\frac{d}{2}+\alpha}$ and

$$\|\mathbb{P}\, u^\varepsilon - v\|_{F_{T_0}^{\frac{d}{2}} \cap F_{T_0}^{\frac{d}{2}+\alpha}} \leq C\varepsilon^{\frac{2\alpha}{2+d+2\alpha}}.$$

(iii) The couple $(b^\varepsilon, \mathbb{P}^\perp u_\varepsilon)$ tends to 0, in the following sense:

$$\|(b^\varepsilon, \mathbb{P}^\perp u^\varepsilon)\|_{L_T^2(\dot{B}_{\infty,1}^{\alpha-\frac{1}{2}})} \leq C\varepsilon^{\frac{1}{2}}, \quad \text{if} \quad d \geq 4,$$

$$\|(b^\varepsilon, \mathbb{P}^\perp u^\varepsilon)\|_{L_T^p(\dot{B}_{\infty,1}^{\alpha-1+\frac{1}{p}})} \leq C\varepsilon^{\frac{1}{p}}, \quad \text{if} \quad d = 3 \quad \text{and} \quad p > 2,$$

$$\|(b^\varepsilon, \mathbb{P}^\perp u^\varepsilon)\|_{L_T^4(\dot{B}_{\infty,1}^{\alpha-\frac{3}{4}})} \leq C\varepsilon^{\frac{1}{4}}, \quad \text{if} \quad d = 2.$$

Remark 10.32. For simplicity, we have assumed that the data do not depend on ε. It goes without saying that more general data may be considered.

Proof. Unsurprisingly, owing to the fact that less dispersive inequalities are available, the proof in dimension two or three is more technical. Here, to simplify the presentation, we shall prove the theorem only in the case $d \geq 4$. The reader is referred to [97] for the cases $d = 2, 3$.

The existence of a solution of (NSC^ε) on a small time interval (which may depend on ε) is ensured by Theorem 10.11, regardless of the size of the data: The only assumption that we need is that $1 + \varepsilon b_0$ is bounded away from zero. Since $b_0 \in \dot{B}_{2,1}^{\frac{d}{2}}$ and $\dot{B}_{2,1}^{\frac{d}{2}} \hookrightarrow L^\infty$, this is certainly true for sufficiently small ε. We therefore assume that we are given two times, T and T_0 (possibly infinite), such that $0 < T \leq T_0$ and a solution $(b^\varepsilon, u^\varepsilon)$ of (NSC_ε) belonging to $E_T \overset{\text{def}}{=} E_{\varepsilon\nu,T}^{\frac{d}{2}} \cap E_{\varepsilon\nu,T}^{\frac{d}{2}+\alpha}$ for some $\alpha \in\,]0, 1/2]$ and satisfying $\|\varepsilon b^\varepsilon\|_{L^\infty} \leq 3/4$. We shall likewise assume that the corresponding incompressible solution v is defined on $[0, T_0]$ if $T_0 < \infty$ (on $\mathbb{R}^+$ if $T = \infty$) and that it belongs to $F_{T_0}^{\frac{d}{2}} \cap F_{T_0}^{\frac{d}{2}+\alpha}$.

In the first step of the proof, we take advantage of Proposition 10.30 to bound a suitable norm of $(b^\varepsilon, \mathbb{P}^\perp u^\varepsilon)$ by the norm of $(b^\varepsilon, u^\varepsilon)$ in E_T times some positive power of ε. In the second step, we derive a priori bounds for $\varepsilon^{-\beta}(\mathbb{P}\, u^\varepsilon - v)$ (for a suitable $\beta > 0$) in terms of $\|(b^\varepsilon, u^\varepsilon)\|_{E_T}$ and $\|v\|_{F_T}$. These bounds may be obtained by combining estimates for the nonstationary Stokes equation with first order terms (see Proposition 10.5) and using paradifferential calculus. The third step is devoted to proving uniform bounds for $\|(b^\varepsilon, u^\varepsilon)\|_{E_T}$ in terms of v and the initial data. The key to this step is Proposition 10.23.

At this stage, we may use a bootstrap argument (fourth step) to close the estimates of the first three steps, and a continuity argument (last step) completes the proof.

Throughout, we shall use the following notation:

$$X_\beta(T) \overset{\text{def}}{=} \|b^\varepsilon\|_{L^1_T(\widetilde{B}^{\frac{d}{2}+\beta,1}_{\varepsilon\nu})} + \|\mathbb{P}^\perp u^\varepsilon\|_{L^1_T(\dot{B}^{\frac{d}{2}+1+\beta}_{2,1})}$$
$$+ \|b^\varepsilon\|_{L^\infty_T(\widetilde{B}^{\frac{d}{2}+\beta,\infty}_{\varepsilon\nu})} + \|\mathbb{P}^\perp u^\varepsilon\|_{L^\infty_T(\dot{B}^{\frac{d}{2}-1+\beta}_{2,1})},$$

$$V_\beta(T) \overset{\text{def}}{=} \|v\|_{L^1_T(\dot{B}^{\frac{d}{2}+1+\beta}_{2,1})} + \|v\|_{L^\infty_T(\dot{B}^{\frac{d}{2}-1+\beta}_{2,1})},$$

$$W_\beta(T) \overset{\text{def}}{=} \|w^\varepsilon\|_{L^1_T(\dot{B}^{\frac{d}{2}+1+\beta}_{2,1})} + \|w^\varepsilon\|_{L^\infty_T(\dot{B}^{\frac{d}{2}-1+\beta}_{2,1})} \quad \text{with} \quad w^\varepsilon \overset{\text{def}}{=} \mathbb{P}u^\varepsilon - v,$$

$$Y_\beta(T) \overset{\text{def}}{=} \|b^\varepsilon\|_{L^2_T(\dot{B}^{\beta-\frac{1}{2}}_{\infty,1})} + \|\mathbb{P}^\perp u^\varepsilon\|_{L^2_T(\dot{B}^{\beta-\frac{1}{2}}_{\infty,1})}.$$

We shall also use the notation $P_\beta(T) = V_\beta(T) + W_\beta(T)$ and

$$X^0_\beta \overset{\text{def}}{=} \|b_0\|_{\widetilde{B}^{\frac{d}{2}+\beta,\infty}_{\varepsilon\nu}} + \|\mathbb{P}^\perp u_0\|_{\dot{B}^{\frac{d}{2}-1+\beta}_{2,1}} + \|\mathbb{P}^\perp f\|_{L^1_{T_0+1}(\dot{B}^{\frac{d}{2}-1+\beta}_{2,1})}.$$

The argument T will sometimes be omitted, and β will always stand for 0 or α.

First Step: Dispersive Estimates for $(b^\varepsilon, \mathbb{P}^\perp u^\varepsilon)$

Applying Proposition 10.30 to the system (10.65), we get, for $d \geq 4$,

$$Y_\alpha \leq C\varepsilon^{\frac{1}{2}}\left(\|(b_0, \mathbb{P}^\perp u_0)\|_{\dot{B}^{\frac{d}{2}-1+\alpha}_{2,1}} + \|\mathbb{P}^\perp f\|_{L^1_T(\dot{B}^{\frac{d}{2}-1+\alpha}_{2,1})}\right.$$
$$\left. + \|\operatorname{div}(b^\varepsilon u^\varepsilon)\|_{L^1_T(\dot{B}^{\frac{d}{2}-1+\alpha}_{2,1})} + \|G\|_{L^1_T(\dot{B}^{\frac{d}{2}-1+\alpha}_{2,1})}\right). \qquad (10.69)$$

From Corollary 2.54, Theorem 2.61, and (10.19), we easily deduce that

$$\|\operatorname{div}(b^\varepsilon u^\varepsilon)\|_{L^1_T(\dot{B}^{\frac{d}{2}-1+\alpha}_{2,1})} \leq C\left(\|b^\varepsilon\|_{L^2_T(\dot{B}^{\frac{d}{2}}_{2,1})}\|u^\varepsilon\|_{L^2_T(\dot{B}^{\frac{d}{2}+\alpha}_{2,1})}\right.$$
$$\left. + \|u^\varepsilon\|_{L^2_T(\dot{B}^{\frac{d}{2}}_{2,1})}\|b^\varepsilon\|_{L^2_T(\dot{B}^{\frac{d}{2}+\alpha}_{2,1})}\right)$$
$$\leq C\left(X_0(X_\alpha + P_\alpha) + X_\alpha(X_0 + P_0)\right),$$

$$\|\mathbb{P}^\perp(u^\varepsilon \cdot \nabla u^\varepsilon)\|_{L^1_T(\dot{B}^{\frac{d}{2}-1+\alpha}_{2,1})} \leq \|u^\varepsilon\|_{L^2_T(\dot{B}^{\frac{d}{2}}_{2,1})}\|\nabla u^\varepsilon\|_{L^2_T(\dot{B}^{\frac{d}{2}-1+\alpha}_{2,1})}$$
$$\leq C(X_0 + P_0)(X_\alpha + P_\alpha),$$

$$\|\mathbb{P}^\perp(I(b^\varepsilon)\mathcal{A}u^\varepsilon)\|_{L^1_T(\dot{B}^{\frac{d}{2}-1+\alpha}_{2,1})} \leq C\varepsilon\|b^\varepsilon\|_{L^\infty_T(\dot{B}^{\frac{d}{2}}_{2,1})}\|\mathcal{A}u^\varepsilon\|_{L^1_T(\dot{B}^{\frac{d}{2}-1+\alpha}_{2,1})}$$
$$\leq CX_0(X_\alpha + P_\alpha),$$

$$\|K(\varepsilon b^\varepsilon)\nabla b^\varepsilon\|_{L^1_T(\dot{B}^{\frac{d}{2}-1+\alpha}_{2,1})} \leq C\,\|\varepsilon b^\varepsilon\|_{L^2_T(\dot{B}^{\frac{d}{2}}_{2,1})}\|\nabla b^\varepsilon\|_{L^2_T(\dot{B}^{\frac{d}{2}-1+\alpha}_{2,1})}$$
$$\leq C\,\varepsilon X_0 X_\alpha.$$

Plugging the above inequalities into (10.69), we conclude that

$$Y_\alpha \leq C\varepsilon^{\frac{1}{2}}\big(X^0_\alpha + X_\alpha + (X_0 + P_0)(X_\alpha + P_\alpha)\big). \qquad (10.70)$$

Second Step: Estimates for w^ε

From the momentum equation of (NSC_ε) and (NSI), we get

$$\partial_t w^\varepsilon + \mathbb{P}\,(A^\varepsilon \cdot \nabla w^\varepsilon) + \mathbb{P}\,(w^\varepsilon \cdot \nabla A^\varepsilon) - \mu\Delta w^\varepsilon = \mathbb{P}F^\varepsilon$$

with $A^\varepsilon \stackrel{\mathrm{def}}{=} \mathbb{P}^\perp u^\varepsilon + v$ and

$$F^\varepsilon \stackrel{\mathrm{def}}{=} -\big(\mathbb{P}^\perp u^\varepsilon \cdot \nabla v + v \cdot \nabla \mathbb{P}^\perp u^\varepsilon + w^\varepsilon \cdot \nabla w^\varepsilon + I(\varepsilon b^\varepsilon)\mathcal{A}u^\varepsilon\big).$$

Applying Proposition 10.5 with $s = d/2 - 1 + \beta$ yields

$$W_\beta \leq Ce^{C(V_0+X_0)}\|F^\varepsilon\|_{L^1_T(\dot{B}^{\frac{d}{2}-1+\beta}_{2,1})}. \qquad (10.71)$$

We now bound F^ε. We readily have

$$\|w^\varepsilon \cdot \nabla w^\varepsilon\|_{L^1_T(\dot{B}^{\frac{d}{2}-1+\beta}_{2,1})} \leq C\|w^\varepsilon\|_{L^2_T(\dot{B}^{\frac{d}{2}}_{2,1})}\|\nabla w^\varepsilon\|_{L^2_T(\dot{B}^{\frac{d}{2}-1+\beta}_{2,1})} \leq CW_0 W_\beta. \qquad (10.72)$$

Next, by interpolation and according to (10.19), we have

$$\|b^\varepsilon\|_{\dot{B}^{\frac{d}{2}}_{2,1}} \leq \|b^\varepsilon\|^\alpha_{\dot{B}^{\frac{d}{2}+\alpha-1}_{2,1}}\|b^\varepsilon\|^{1-\alpha}_{\dot{B}^{\frac{d}{2}+\alpha}_{2,1}} \leq (\varepsilon\nu)^{\alpha-1}\|b^\varepsilon\|_{\widetilde{B}^{\frac{d}{2}+\alpha,\infty}_{\varepsilon\nu}}. \qquad (10.73)$$

From this we deduce that

$$\|I(\varepsilon b^\varepsilon)\|_{L^1_T(\dot{B}^{\frac{d}{2}-1+\beta}_{2,1})} \leq C\,\varepsilon\|b^\varepsilon\|_{L^\infty_T(\dot{B}^{\frac{d}{2}}_{2,1})}\|\mathcal{A}u^\varepsilon\|_{L^1_T(\dot{B}^{\frac{d}{2}-1+\beta}_{2,1})}$$
$$\leq C\,\varepsilon^\alpha\|b^\varepsilon\|_{L^\infty_T(\widetilde{B}^{\frac{d}{2}+\alpha,\infty}_{\varepsilon\nu})}\|u^\varepsilon\|_{L^1_T(\dot{B}^{\frac{d}{2}+1+\beta}_{2,1})}$$
$$\leq C\,\varepsilon^\alpha X_\alpha(V_\beta + W_\beta + X_\beta). \qquad (10.74)$$

According to step one, $\mathbb{P}^\perp u^\varepsilon$ is small in $L^2([0,T];\dot{B}^0_{\infty,1})$. Indeed, using interpolation and embeddings, we have

$$\|\mathbb{P}^\perp u^\varepsilon\|_{L^2_T(\dot{B}^0_{\infty,1})} \leq \|\mathbb{P}^\perp u^\varepsilon\|^{2\alpha}_{L^2_T(\dot{B}^{\alpha-\frac{1}{2}}_{\infty,1})}\|\mathbb{P}^\perp u^\varepsilon\|^{1-2\alpha}_{L^2_T(\dot{B}^\alpha_{\infty,1})}$$
$$\leq C\varepsilon^\alpha\left(\varepsilon^{-\frac{1}{2}}\|\mathbb{P}^\perp u^\varepsilon\|_{L^2_T(\dot{B}^{\alpha-\frac{1}{2}}_{\infty,1})}\right)^{2\alpha}\|\mathbb{P}^\perp u^\varepsilon\|^{1-2\alpha}_{L^2_T(\dot{B}^{\frac{d}{2}+\alpha}_{2,1})}$$
$$\leq C\varepsilon^\alpha(X_\alpha + \varepsilon^{-\frac{1}{2}}Y_\alpha). \qquad (10.75)$$

A judicious use of paradifferential calculus will enable us to guarantee some smallness for $\mathbb{P}^\perp u^\varepsilon \cdot \nabla v$ and $v \cdot \nabla \mathbb{P}^\perp u^\varepsilon$. For $\mathbb{P}^\perp u^\varepsilon \cdot \nabla v$, we may use the following modified Bony decomposition for $\eta \in \,]0,1[$:

$$\mathbb{P}^\perp u^\varepsilon \cdot \nabla v = \underbrace{\sum_{j \in \mathbb{Z}} \dot{\Delta}_j \mathbb{P}^\perp u^\varepsilon \cdot \dot{S}_{j-1+[\log_2 \eta]} \nabla v}_{T_1} + \underbrace{\sum_{j \in \mathbb{Z}} \dot{S}_{j+2-[\log_2 \eta]} \mathbb{P}^\perp u^\varepsilon \cdot \dot{\Delta}_j \nabla v}_{T_2}.$$

Recall that for any $k \in \mathbb{Z}$, we have

$$\|\dot{S}_k \nabla v\|_{L^\infty} \leq C 2^{2k} \|\nabla v\|_{\dot{B}^{-2}_{\infty,1}}.$$

Therefore,

$$\left\| \dot{\Delta}_j \mathbb{P}^\perp u^\varepsilon \cdot \dot{S}_{j-1+[\log_2 \eta]} \nabla v \right\|_{L^2} \leq C \left\| \dot{S}_{j-1+[\log_2 \eta]} \nabla v \right\|_{L^\infty} \left\| \dot{\Delta}_j \mathbb{P}^\perp u^\varepsilon \right\|_{L^2}$$

$$\leq C \eta^2 2^{-j(\frac{d}{2}+\beta-1)} \|\nabla v\|_{\dot{B}^{-2}_{\infty,1}} \left(2^{j(\frac{d}{2}+\beta+1)} \left\| \dot{\Delta}_j \mathbb{P}^\perp u^\varepsilon \right\|_{L^2} \right).$$

As the functions $\dot{\Delta}_j \mathbb{P}^\perp u^\varepsilon \cdot \dot{S}_{j-1+[\log_2 \eta]} \nabla v$ are spectrally supported in dyadic annuli $2^j C(0, R_1, R_2)$ with R_1 and R_2 independent of η, Lemma 2.23 yields

$$\|T_1\|_{\dot{B}^{\frac{d}{2}-1+\beta}_{2,1}} \leq C \eta^2 \|\nabla v\|_{\dot{B}^{\frac{d}{2}-2}_{2,1}} \|\mathbb{P}^\perp u^\varepsilon\|_{B^{\frac{d}{2}+1+\beta}_{2,1}}. \tag{10.76}$$

Next, according to Proposition 2.10, we have, for all $k \in \mathbb{Z}$,

$$\dot{\Delta}_k T_2 = \sum_{j \geq k-2+[\log_2 \eta]} \dot{\Delta}_k \left(\dot{S}_{j+2-[\log_2 \eta]} \mathbb{P}^\perp u^\varepsilon \cdot \dot{\Delta}_j \nabla v \right).$$

Therefore,

$$2^{k(\frac{d}{2}+\beta-1)} \|\dot{\Delta}_k T_2\|_{L^2} \leq C \left\| \mathbb{P}^\perp u^\varepsilon \right\|_{L^\infty} \sum_{j \geq k-2+[\log_2 \eta]} 2^{(k-j)(\frac{d}{2}+\beta-1)} 2^{j(\frac{d}{2}+\beta-1)} \|\dot{\Delta}_j \nabla v\|_{L^2}$$

$$\leq C \eta^{1-\beta-\frac{d}{2}} \left\| \mathbb{P}^\perp u^\varepsilon \right\|_{L^\infty} \|\nabla v\|_{\dot{B}^{\frac{d}{2}+\beta-1}_{2,1}},$$

from which it follows that

$$\|T_2\|_{\dot{B}^{\frac{d}{2}+\beta-1}_{2,1}} \leq C \eta^{1-\beta-\frac{d}{2}} \|v\|_{\dot{B}^{\frac{d}{2}+\beta}_{2,1}} \left\| \mathbb{P}^\perp u^\varepsilon \right\|_{L^\infty}. \tag{10.77}$$

From (10.76), (10.77), and Hölder's inequality, we thus get

$$\|\mathbb{P}^\perp u^\varepsilon \cdot \nabla v\|_{L^1_T(\dot{B}^{\frac{d}{2}-1+\beta}_{2,1})} \leq C \left(\eta^2 \|v\|_{L^\infty_T(\dot{B}^{\frac{d}{2}-1}_{2,1})} \|\mathbb{P}^\perp u^\varepsilon\|_{L^1_T(\dot{B}^{\frac{d}{2}+1+\beta}_{2,1})} \right.$$

$$\left. + \eta^{1-\beta-\frac{d}{2}} \|v\|_{L^2_T(\dot{B}^{\frac{d}{2}+\beta}_{2,1})} \|\mathbb{P}^\perp u^\varepsilon\|_{L^2_T(L^\infty)} \right).$$

Since $\dot{B}^0_{\infty,1} \hookrightarrow L^\infty$, by choosing $\eta = \varepsilon^{\frac{2\alpha}{2+d+2\beta}}$ and using (10.75), we can now conclude that

$$\|\mathbb{P}^\perp u^\varepsilon \cdot \nabla v\|_{L^1_T(\dot{B}^{\frac{d}{2}-1+\beta}_{2,1})} \leq C\varepsilon^{\frac{4\alpha}{2+d+2\beta}}\left(V_0 X_\beta + V_\beta(X_\alpha + \varepsilon^{-\frac{1}{2}}Y_\alpha)\right). \tag{10.78}$$

The term $v \cdot \nabla\mathbb{P}^\perp u^\varepsilon$ may be treated similarly. Indeed, we have

$$v\cdot\nabla\mathbb{P}^\perp u^\varepsilon = \underbrace{\sum_{j\in\mathbb{Z}}\dot{S}_{j-1+[\log_2\eta]}v\cdot\dot{\Delta}_j\nabla\mathbb{P}^\perp u^\varepsilon}_{\widetilde{T}_1} + \underbrace{\sum_{j\in\mathbb{Z}}\dot{\Delta}_j v\cdot\dot{S}_{j+2-[\log_2\eta]}\nabla\mathbb{P}^\perp u^\varepsilon}_{\widetilde{T}_2}.$$

Now, following the proof of (10.76) and (10.77), we readily get that

$$\|\widetilde{T}_1\|_{L^1_T(\dot{B}^{\frac{d}{2}-1+\beta}_{2,1})} \leq C\,\eta\|v\|_{L^\infty_T(\dot{B}^{\frac{d}{2}-1}_{2,1})}\|\nabla\mathbb{P}^\perp u^\varepsilon\|_{L^1_T(\dot{B}^{\frac{d}{2}+\beta}_{2,1})},$$

$$\|\widetilde{T}_2\|_{L^1_T(\dot{B}^{\frac{d}{2}-1+\beta}_{2,1})} \leq C\,\eta^{-\frac{d}{2}-\beta}\|v\|_{L^2_T(\dot{B}^{\frac{d}{2}-1}_{2,1})}\|\nabla\mathbb{P}^\perp u^\varepsilon\|_{L^2_T(\dot{B}^{-1}_{\infty,1})}.$$

Choosing $\eta = \varepsilon^{\frac{2\alpha}{2+d+2\beta}}$, we conclude that

$$\|v\cdot\nabla\mathbb{P}^\perp u^\varepsilon\|_{L^1_T(\dot{B}^{\frac{d}{2}-1+\beta}_{2,1})} \leq C\varepsilon^{\frac{2\alpha}{2+d+2\beta}}\left(V_0 X_\beta + V_\beta(X_\alpha + \varepsilon^{-\frac{1}{2}}Y_\alpha)\right). \tag{10.79}$$

Plugging the inequalities (10.72), (10.74), (10.78), and (10.79) into (10.71), we eventually get

$$W_\beta \leq Ce^{C(V_0+X_0)}\big(W_0 W_\beta + \varepsilon^\alpha X_\alpha(V_\beta+X_\beta+W_\beta)$$
$$+ \varepsilon^{\frac{2\alpha}{2+d+2\beta}}\big(V_0 X_\beta + V_\beta(X_\alpha+\varepsilon^{-\frac{1}{2}}Y_\alpha)\big)\big). \tag{10.80}$$

Third Step: Estimates for $(b^\varepsilon, u^\varepsilon)$ in $E^{\frac{d}{2}+\beta}_{\varepsilon\nu,T}$

We use the fact that $(b^\varepsilon, u^\varepsilon)$ satisfies

$$\begin{cases} \partial_t b^\varepsilon + \operatorname{div} T_{u^\varepsilon} b^\varepsilon + \dfrac{\operatorname{div} u^\varepsilon}{\varepsilon} = F^\varepsilon \\ \partial_t u^\varepsilon + T_{u^\varepsilon}\cdot\nabla u^\varepsilon - \mathcal{A}u^\varepsilon + \dfrac{\nabla b^\varepsilon}{\varepsilon} = f + G^\varepsilon \end{cases}$$

with

$$F^\varepsilon \overset{\mathrm{def}}{=} -\operatorname{div}\big(T'_{b^\varepsilon}u^\varepsilon\big) \quad \text{and} \quad G^\varepsilon \overset{\mathrm{def}}{=} K(\varepsilon b^\varepsilon)\frac{\nabla b^\varepsilon}{\varepsilon} - I(\varepsilon b^\varepsilon)\mathcal{A}u^\varepsilon - T'_{\partial_j u^\varepsilon}(u^\varepsilon)^j.$$

According to Proposition 10.23, we have, for any $p, r > 1$,

$$X_\beta(T) \leq Ce^{U^{p,r}_\varepsilon(T)}\left(\|b_0\|_{\tilde{B}^{\frac{d}{2}+\beta,\infty}_{\varepsilon\nu}} + \|u_0\|_{\dot{B}^{\frac{d}{2}-1+\beta}_{2,1}}\right.$$

$$+\|f\|_{L^1_T(\dot{B}^{\frac{d}{2}-1+\beta}_{2,1})} + \|F^\varepsilon\|_{L^1_T(\tilde{B}^{\frac{d}{2}+\beta,\infty}_{\varepsilon\nu})} + \|G^\varepsilon\|_{L^1_T(\dot{B}^{\frac{d}{2}-1+\beta}_{2,1})}\Big) \quad (10.81)$$

with

$$U^{p,r}_\varepsilon(t) \stackrel{\text{def}}{=} \int_0^t \Big(\nu^{1-p}\|\nabla u^\varepsilon\|^p_{\dot{B}^{\frac{2}{p}-2}_{\infty,1}} + (\varepsilon^2\nu)^{r-1}\|\nabla u^\varepsilon\|^r_{L^\infty}\Big)\,dt'. \quad (10.82)$$

We first bound F^ε. According to (10.19), we have

$$\|F^\varepsilon\|_{L^1_T(\tilde{B}^{\frac{d}{2}+\beta,\infty}_{\varepsilon\nu})} \leq C\Big(\|F^\varepsilon\|_{L^1_T(\dot{B}^{\frac{d}{2}-1+\beta}_{2,1})} + \varepsilon\nu\|F^\varepsilon\|_{L^1_T(\dot{B}^{\frac{d}{2}+\beta}_{2,1})}\Big). \quad (10.83)$$

From Theorems 2.47 and 2.52, we have

$$\begin{aligned}
\|F^\varepsilon\|_{L^1_T(\dot{B}^{\frac{d}{2}-1+\beta}_{2,1})} &\leq C\,\|T'_{b^\varepsilon}u^\varepsilon\|_{L^1_T(\dot{B}^{\frac{d}{2}-1+\beta}_{2,1})}\\
&\leq C\,\|b^\varepsilon\|_{L^2_T(\dot{B}^0_{\infty,1})}\|u^\varepsilon\|_{L^2_T(\dot{B}^{\frac{d}{2}+\beta}_{2,1})}.
\end{aligned}$$

However, replacing $\mathbb{P}^\perp u^\varepsilon$ by b^ε in the proof of (10.75), we also get

$$\|b^\varepsilon\|_{L^2_T(\dot{B}^0_{\infty,1})} \leq C\varepsilon^\alpha(X_\alpha + \varepsilon^{-\frac{1}{2}}Y_\alpha) \quad (10.84)$$

so that

$$\|F^\varepsilon\|_{L^1_T(\tilde{B}^{\frac{d}{2}+\beta,\infty}_{\varepsilon\nu})} \leq C\varepsilon^\alpha(X_\beta + P_\beta)(X_\alpha + \varepsilon^{-\frac{1}{2}}Y_\alpha). \quad (10.85)$$

We now bound G^ε. First, by virtue of Theorems 2.47, 2.52, 2.61 and the inequality (10.84), we may write

$$\begin{aligned}
\|K(\varepsilon b^\varepsilon)\nabla b^\varepsilon\|_{L^1_T(\dot{B}^{\frac{d}{2}-1+\beta}_{2,1})} &\leq C\Big(\|\varepsilon b^\varepsilon\|_{L^2_T(L^\infty)}\|\nabla b^\varepsilon\|_{L^2_T(\dot{B}^{\frac{d}{2}-1+\beta}_{2,1})}\\
&\qquad\qquad + \|\nabla b^\varepsilon\|_{L^2_T(\dot{B}^{-1}_{\infty,1})}\|\varepsilon b^\varepsilon\|_{L^2_T(\dot{B}^{\frac{d}{2}+\beta}_{2,1})}\Big)\\
&\leq C\varepsilon^{1+\alpha}X_\beta(X_\alpha + \varepsilon^{-\frac{1}{2}}Y_\alpha).
\end{aligned}$$

Next, we decompose $T'_{\partial_j u^\varepsilon}(u^\varepsilon)^j$ as

$$T'_{\partial_j u^\varepsilon}(u^\varepsilon)^j = T'_{\partial_j \mathbb{P}^\perp u^\varepsilon}(u^\varepsilon)^j + T'_{\partial_j \mathbb{P} u^\varepsilon}(\mathbb{P}u^\varepsilon)^j + T'_{\partial_j \mathbb{P} u^\varepsilon}(\mathbb{P}^\perp u^\varepsilon)^j.$$

According to the inequality (10.84), we have

$$\begin{aligned}
\|T'_{\partial_j \mathbb{P}^\perp u^\varepsilon}(u^\varepsilon)^j\|_{L^1_T(\dot{B}^{\frac{d}{2}-1+\beta}_{2,1})} &\leq C\,\|\nabla \mathbb{P}^\perp u^\varepsilon\|_{L^2_T(\dot{B}^{-1}_{\infty,1})}\|u^\varepsilon\|_{L^2_T(\dot{B}^{\frac{d}{2}+\beta}_{2,1})}\\
&\leq C\,\varepsilon^\alpha\Big(X_\alpha + \varepsilon^{-\frac{1}{2}}Y_\alpha\Big)(X_\beta + P_\beta).
\end{aligned}$$

For the next term, we simply write

$$\|T'_{\partial_j \mathbb{P}u^\varepsilon}(\mathbb{P}u^\varepsilon)^j\|_{L^1_T(\dot{B}^{\frac{d}{2}-1+\beta}_{2,1})} \leq C\|\mathbb{P}u^\varepsilon\|_{L^2_T(\dot{B}^{\frac{d}{2}}_{2,1})}\|\mathbb{P}u^\varepsilon\|_{L^2_T(\dot{B}^{\frac{d}{2}+\beta}_{2,1})} \leq CP_0 P_\beta.$$

Arguing as for the bounding of T_1 in step 2 (with v replaced by $\mathbb{P}u^\varepsilon$), we get

$$\|T'_{\partial_j \mathbb{P}u^\varepsilon}(\mathbb{P}^\perp u^\varepsilon)^j\|_{L^1_T(\dot{B}^{\frac{d}{2}-1+\beta}_{2,1})} \leq C\varepsilon^{\frac{4\alpha}{2+d+2\beta}}\left(P_0 X_\beta + P_\beta(X_\alpha + \varepsilon^{-\frac{1}{2}}Y_\alpha)\right).$$

So, finally, we have

$$\|G^\varepsilon\|_{L^1_T(\dot{B}^{\frac{d}{2}-1+\beta}_{2,1})} \leq C\Big(P_0 P_\beta + \varepsilon^\alpha X_\beta(X_\alpha + \varepsilon^{-\frac{1}{2}}Y_\alpha)$$
$$+ \varepsilon^{\frac{4\alpha}{2+d+2\beta}}\left(P_0 X_\beta + P_\beta(X_\alpha + \varepsilon^{-\frac{1}{2}}Y_\alpha)\right)\Big). \quad (10.86)$$

Now, we take $p = 1/\alpha$ and $r = 2/(2-\alpha)$ in (10.81). Using interpolation and embeddings, we have

$$\|\nabla u^\varepsilon\|_{L^{\frac{1}{\alpha}}_T(\dot{B}^{2\alpha-2}_{\infty,1})} \leq C\|\nabla\mathbb{P}^\perp u^\varepsilon\|^{2\alpha}_{L^2_T(\dot{B}^{\alpha-\frac{3}{2}}_{\infty,1})}\|\nabla\mathbb{P}^\perp u^\varepsilon\|^{1-2\alpha}_{L^\infty_T(\dot{B}^{\alpha-2}_{\infty,1})} + \|\nabla\mathbb{P}u\|_{L^{\frac{1}{\alpha}}_T(\dot{B}^{\frac{d}{2}+2\alpha-2}_{2,1})}$$
$$\leq C\left(\varepsilon^\alpha(X_\alpha + \varepsilon^{-\frac{1}{2}}Y_\alpha) + P_0\right)$$

and

$$\|\nabla u^\varepsilon\|_{L^{\frac{2}{2-\alpha}}_T(L^\infty)} \leq C\|u^\varepsilon\|_{L^{\frac{2}{2-\alpha}}_T(\dot{B}^{\frac{d}{2}+1}_{2,1})}$$
$$\leq C\left(P_\alpha + X_\alpha\right).$$

According to (10.82), we thus have

$$U^{\frac{1}{\alpha},\frac{2}{2-\alpha}}_\varepsilon \leq C\left(P_0^{\frac{1}{\alpha}} + \varepsilon(X_\alpha + \varepsilon^{-\frac{1}{2}}Y_\alpha)^{\frac{1}{\alpha}} + (\varepsilon^\alpha(P_\alpha + X_\alpha))^{\frac{2}{2-\alpha}}\right). \quad (10.87)$$

Plugging this inequality, (10.85), and (10.86) into (10.81), we eventually get

$$X_\beta \leq Ce^{C\left(P_0^{\frac{1}{\alpha}} + \varepsilon(X_\alpha + \varepsilon^{-\frac{1}{2}}Y_\alpha)^{\frac{1}{\alpha}} + (\varepsilon^\alpha(P_\alpha + X_\alpha))^{\frac{2}{2-\alpha}}\right)}\Big(X^0_\beta + P_0 P_\beta$$
$$+ \varepsilon^{2\alpha_d}\left(P_0 X_\beta + (X_\beta + P_\beta)(X_\alpha + \varepsilon^{-\frac{1}{2}}Y_\alpha)\right)\Big) \quad (10.88)$$

with $\alpha_d \overset{\text{def}}{=} 2\alpha/(2+d+2\alpha)$.

Fourth Step: Bootstrap

Let $X \overset{\text{def}}{=} X_0 + X_\alpha$, $V \overset{\text{def}}{=} V_0 + V_\alpha$, $W \overset{\text{def}}{=} W_0 + W_\alpha$, and $X^0 \overset{\text{def}}{=} X^0_0 + X^0_\alpha$. Combining estimates (10.70), (10.80), and (10.88) yields

$$W \leq Ce^{C(V+X)}\left(\varepsilon^{\alpha_d}\left(X^2 + V(X_0 + X + X^2 + V^2)\right) + W^2(1+\varepsilon^{\alpha_d}V)\right), \quad (10.89)$$

$$X \leq C e^{C\left(\varepsilon(X+X^2)^{\frac{1}{\alpha}}+(\varepsilon^\alpha X)^{\frac{2}{2-\alpha}}\right)} e^{C\left((V+W)^{\frac{1}{\alpha}}+\varepsilon(X^0+(V+W)2)^{\frac{1}{\alpha}}+\varepsilon^{\frac{2\alpha}{2-\alpha}}(V+W)^{\frac{2}{2-\alpha}}\right)}$$
$$\times\left(X^0+(V+W)(V+W+\varepsilon^{\alpha_d}(X^0+V^2+W^2))\right.$$
$$\left.+\varepsilon^{\alpha_d}X(X^0+V+W+X+X^2)\right). \quad (10.90)$$

A bootstrap argument will enable us to get a bound for $(b^\varepsilon, u^\varepsilon)$ from the two estimates above. More precisely, we have the following lemma.

Lemma 10.33. *Suppose that $v \in F_{T_0}^{\frac{d}{2}} \cap F_{T_0}^{\frac{d}{2}+\alpha}$ for some finite or infinite T_0. There then exists an $\varepsilon_0 > 0$, depending only on $\alpha, d, V(T_0)$, and the norm of $(b_0, \mathbb{P}^\perp u_0, \mathbb{P}^\perp f)$ in*

$$\dot{B}_{2,1}^{\frac{d}{2}-1} \cap \dot{B}_{2,1}^{\frac{d}{2}+\alpha} \times (\dot{B}_{2,1}^{\frac{d}{2}-1} \cap \dot{B}_{2,1}^{\frac{d}{2}-1+\alpha})^d \times L^1(\mathbb{R}^+; (\dot{B}_{2,1}^{\frac{d}{2}-1} \cap \dot{B}_{2,1}^{\frac{d}{2}-1+\alpha})^d)$$

and such that if $\varepsilon \leq \varepsilon_0$, $(b^\varepsilon, u^\varepsilon) \in E_{\varepsilon\nu,T}^{\frac{d}{2}} \cap E_{\varepsilon\nu,T}^{\frac{d}{2}+\alpha}$, and $\varepsilon|b^\varepsilon| \leq 3/4$ for some $T \leq T_0$, then the following estimates hold with the constant $C = C(d, \mu, \lambda, P, \alpha)$ appearing in (10.89) and (10.90):

$$X(T) \leq X_M \stackrel{\text{def}}{=} 16Ce^{CV^{\frac{1}{\alpha}}(T_0)}\left(X^0 + V^2(T_0)\right),$$

$$\varepsilon^{-\alpha_d}W(T) \leq W_M \stackrel{\text{def}}{=} 4Ce^{C(V(T_0)+X_M)}$$
$$\times\left(X_M^2 + V(T_0)(X^0 + X_M + X_M^2 + V^2(T_0))\right).$$

Proof. Let $I \stackrel{\text{def}}{=} \left\{t \leq T \mid X(t) \leq X_M \text{ and } W(t) \leq \varepsilon^{\alpha_d}W_M\right\}$. Obviously, X and W are continuous nondecreasing functions so that if, say, $C \geq 1$, then I is a closed interval of $\mathbb{R}^+$ with lower bound 0.

Let $T^\star \stackrel{\text{def}}{=} \sup I$. Choose ε sufficiently small so that the following conditions are fulfilled:

$$Ce^{C(V(T_0)+X_M)}\varepsilon^{\alpha_d}W_M(1+\varepsilon^{\alpha_d}V(T_0)) \leq 1/2,$$

$$e^{C\left(\varepsilon(X_M+X_M^2)^{\frac{1}{\alpha}}+(\varepsilon^\alpha X_M)^{\frac{2}{2-\alpha}}\right)} \leq 2,$$

$$e^{C\left((V(T_0)+\varepsilon^{\alpha_d}W_M)^{\frac{1}{\alpha}}+\varepsilon(X^0+(V(T_0)+\varepsilon^{\alpha_d}W_M)2)^{\frac{1}{\alpha}}+\varepsilon^{\frac{2\alpha}{2-\alpha}}(V(T_0)+\varepsilon^{\alpha_d}W_M)^{\frac{2}{2-\alpha}}\right)}$$
$$\leq 2e^{CV^{\frac{1}{\alpha}}(T_0)},$$

$$X^0+(V(T_0)+\varepsilon^{\alpha_d}W_M)(V(T_0)+\varepsilon^{\alpha_d}W_M+\varepsilon^{\alpha_d}(X^0+V^2(T_0)+\varepsilon^{2\alpha_d}W_M^2))$$
$$\leq 2(X^0+V^2(T_0)),$$

$$Ce^{CV^{\frac{1}{\alpha}}(T_0)}\varepsilon^{\alpha_d}(X^0 + V(T_0) + \varepsilon^{\alpha_d}W_M + X_M + X_M^2) \leq 1/12.$$

From the inequalities (10.89) and (10.90), we get

$$X(T^\star) \leq 12Ce^{CV^{\frac{1}{\alpha}}(T_0)}\left(X^0 + V^2(T)\right),$$
$$W(T^\star) \leq 2\varepsilon^{\alpha_d}Ce^{C(V(T_0)+X_M)}\left(X_M^2 + V(T_0)(X^0 + X_M + X_M^2 + V^2(T_0))\right).$$

In other words, at time $T^\star$ the desired inequalities are strict. Hence, we must have $T^\star = T$. $\qquad\square$

Last Step: Continuation Argument

First, we have to establish the existence of a local solution in $E_{\varepsilon\nu,T}^{\frac{d}{2}} \cap E_{\varepsilon\nu,T}^{\frac{d}{2}+\alpha}$. Making the change of function $a^\varepsilon = \varepsilon b^\varepsilon$, we see that it suffices to apply Theorem 10.11. We readily get a local solution $(b^\varepsilon, u^\varepsilon)$ on $[0,T] \times \mathbb{R}^d$ which belongs to

$$\mathcal{C}([0,T]; \dot{B}_{2,1}^{\frac{d}{2}} \cap \dot{B}_{2,1}^{\frac{d}{2}+\alpha}) \times \left(\mathcal{C}([0,T]; \dot{B}_{2,1}^{\frac{d}{2}-1} \cap \dot{B}_{2,1}^{\frac{d}{2}-1+\alpha}) \cap L^1([0,T]; \dot{B}_{2,1}^{\frac{d}{2}+1} \cap \dot{B}_{2,1}^{\frac{d}{2}+\alpha})\right)^d$$

and satisfies $1 + \varepsilon \inf_{t,x} b^\varepsilon(t,x) > 0$.

Since, in addition, b_0 belongs to $\dot{B}_{2,1}^{\frac{d}{2}-1}$ and $\partial_t b^\varepsilon + u^\varepsilon \cdot \nabla b^\varepsilon$ is in $L^1([0,T]; \dot{B}_{2,1}^{\frac{d}{2}-1})$, we deduce that $b^\varepsilon \in \mathcal{C}([0,T]; \dot{B}_{2,1}^{\frac{d}{2}-1})$. Therefore, $(b^\varepsilon, u^\varepsilon) \in E_{\varepsilon\nu,T}^{\frac{d}{2}} \cap E_{\varepsilon\nu,T}^{\frac{d}{2}+\alpha}$.

Now, suppose that we have $v \in F_{T_0}^{\frac{d}{2}} \cap F_{T_0}^{\frac{d}{2}+\alpha}$ for some $T_0 \in]0, +\infty]$. We will show that the lifespan T_ε, defined as the supremum of the set

$$\left\{T \in \mathbb{R}^+ / (b^\varepsilon, u^\varepsilon) \in E_{\varepsilon\nu,T}^{\frac{d}{2}} \cap E_{\varepsilon\nu,T}^{\frac{d}{2}+\alpha} \text{ and } \forall(t,x) \in [0,T] \times \mathbb{R}^d, \, |\varepsilon b^\varepsilon(t,x)| \le 3/4\right\},$$

satisfies $T_\varepsilon \ge T_0$ if ε is sufficiently small.

We assume (with the aim of arriving at a contradiction) that T_ε is finite and satisfies $T_\varepsilon \le T_0$. According to Lemma 10.33, we have, for any $T < T_\varepsilon$ and $\varepsilon \le \varepsilon_0$,

$$X(T) \le X_M \quad \text{and} \quad W(T) \le \varepsilon^{\alpha_d} W_M.$$

From the first inequality and (10.73), we deduce that

$$\varepsilon \|b^\varepsilon\|_{L_T^\infty(\dot{B}_{2,1}^{\frac{d}{2}})} \le \varepsilon^\alpha \nu^{\alpha-1} X_M.$$

Obviously, changing ε_0 once more if necessary, this entails that

$$1 + \varepsilon \inf_{(t,x) \in ([0,T]_\varepsilon \times \mathbb{R}^d)} |b^\varepsilon(t,x)| > 0.$$

As $b^\varepsilon \in L^\infty([0,T_\varepsilon[; \dot{B}_{2,1}^{\frac{d}{2}} \cap \dot{B}_{2,1}^{\frac{d}{2}+\alpha})$ and $\nabla u^\varepsilon \in L^1([0,T_\varepsilon[; L^\infty)$, the continuation criterion stated in Proposition 10.17 ensures that $(b^\varepsilon, u^\varepsilon)$ may be continued beyond T_ε. This stands in contradiction to the definition of T_ε. Therefore, $T_\varepsilon \ge T_0$ for $\varepsilon \le \varepsilon_0$. □

10.6 References and Remarks

There is a huge literature devoted to the one-dimensional compressible Navier–Stokes equations. Since the usual methods are quite far from our own, we will not elaborate on what is known in this case. However, we should mention the pioneering work by A. Kazhikhov and V. Shelukhin in [179] and the recent paper by A. Mellet and A. Vasseur [227] wherein the existence of global strong solutions is established for a large class of initial data bounded away from the vacuum.

In the multidimensional case, to the best of our knowledge, the first mathematical work devoted to the Cauchy problem for the full compressible Navier–Stokes system is the paper by J. Nash [238] in 1962. There, the existence of local-in-time classical solutions is proved. By using an L^p approach based on parabolic maximal regularity, in [272], V. Solonnikov has stated local well-posedness results in the case of a smooth bounded domain. An extensive study of the compressible Navier–Stokes equations in two-dimensional domains with corners has been undertaken by J.R. Kweon (see, e.g., [200]).

Global existence for small smooth perturbations of a stable equilibrium was stated in 1980 by A. Matsumura and T. Nishida in [225] in the $\mathbb{R}^3$ framework and extended to the half-space, exterior, or bounded smooth domains with Dirichlet conditions in [226] (see also [125, 293] and the more recent work by P. Mucha and W. Zajączkowski [237] for another approach). More general boundary conditions have been considered in, for example, [294].

The work by P.-L. Lions concerning weak solutions of the isentropic compressible Navier–Stokes system had a great impact on the subject and may be seen as the natural continuation of the seminal work by J. Leray in [207] (see Chapter 5) for incompressible viscous fluids. The construction of weak solutions relies on the following formal energy identity (here, we take $f \equiv 0$ in order to simplify matters):

$$\|(\sqrt{\rho}u)(t)\|_{L^2}^2 + \int \pi(t,x)\,dx + 2\int_0^t \left(\mu\,\|\nabla u\|_{L^2}^2 + (\lambda+\mu)\,\|\mathrm{div}\,u\|_{L^2}^2\right) dt'$$
$$= \left\|\sqrt{\rho_0}u_0\right\|_{L^2}^2 + \int \pi_0(x)\,dx, \quad (10.91)$$

where π stands for the free energy per unit volume.[8] This suggests proving the existence of global solutions for data (ρ_0, u_0) such that the right-hand side of (10.91) is finite. However, both constructing approximate solutions satisfying the energy inequality and passing to the limit is much more involved than in the incompressible case (see the original work by P.-L. Lions in [215] for more details). More regular weak solutions have been constructed by B. Desjardins in [114]. For the presentation of a few recent improvements, the reader is referred to the review paper by E. Feireisl in [123] and the book by A. Novotný and I. Straškraba [242]. We should also mention that, following some ideas from Lions' book, E. Feireisl has developed a complete theory of so-called *variational weak solutions* for the full Navier–Stokes equations (see [124]).

The results presented in this chapter concerning local and global well-posedness are borrowed from recent works by the third author (see [93, 95], and [101]). We point out that for critical data, the smallness condition (10.7) is not needed (i.e. the statement of Theorem 10.11 is true for $\alpha = 0$). This was proven recently in [104] by the third author (see also [79] for another approach). We should also stress that local existence results in the spirit of Theorems 10.2 and 10.11 may be established for polytropic heat-conductive fluids and extended to the L^p framework (see [95, 79, 155]). In this case, the scaling invariant space for (a_0, u_0, θ_0) (where θ_0 stands for the discrepancy from a reference temperature) is

---

[8] Recall that if $P(\rho) = a\rho^\gamma$, then we have $\pi = a\left(\dfrac{\rho^\gamma - 1 - \gamma(\rho - 1)}{\gamma(\gamma - 1)}\right)$.

$$\dot{B}_{p,1}^{\frac{d}{p}} \times \left(\dot{B}_{p,1}^{\frac{d}{p}-1}\right)^d \times \dot{B}_{p,1}^{\frac{d}{p}-2}.$$

In addition, still in the critical framework, for polytropic fluids, a global well-posedness result in the spirit of Theorem 10.21 for small perturbations of a stable equilibrium has been proven in [94]. Theorem 10.21 has recently been extended to the L^p framework by F. Charve and the third author in [61]. In particular, as for the incompressible Navier–Stokes equations, the smallness condition for global existence involves Besov norms with a *negative* index of regularity. Hence, highly oscillating initial velocities with possibly large moduli give rise to global solutions. Finally, we point out that a similar approach works for fluids endowed with internal capillarity (see [107]).

Until now, even in the barotropic case, the question of weak-strong uniqueness for the compressible Navier–Stokes equations has remained open. More precisely, for sufficiently smooth data (ρ_0, u_0) bounded away from the vacuum, we can construct both a (unique) local smooth solution and, according to Lions' results, a global weak solution with finite energy. However, in contrast to what is known in the incompressible case, there is no evidence that the weak and strong solutions coincide, even for small time. One of the main difficulties that has to be faced is that Lions' theorem does not give any information on the possible appearance of vacuum, and such a control seems crucial to get uniqueness. We should mention here that in [80], Y. Cho, H.J. Choe, and H. Kim have obtained a result involving the existence *and* uniqueness of a special class of initial data where vacuum is not excluded (see also a promising recent result by P. Germain in [142] and the work [220] by T.-P. Liu and T. Yang concerning the inviscid case). For a particular class of barotropic fluids, D. Bresch and B. Desjardins have constructed "stronger" weak solutions with an additional H^1 control on the density (see [49]). Knowing that in dimension two, Theorem 10.2 provides strong solutions for data having almost the same regularity [i.e., $u_0 \in \dot{B}_{2,1}^0$ and $(\rho_0 - 1) \in \dot{B}_{2,1}^1$], it may be tempting to study whether we can bridge the gap between weak and strong solutions. Other types of weak solutions with possibly discontinuous data (including jumps across a hypersurface for the density) have been constructed by D. Hoff in [159, 162, 163].

It turns out that for smooth perturbations of a stable constant state, very accurate information may be obtained concerning the asymptotics of the global solution. Roughly, the time decay properties of the solution are the same as for the linearized system about the constant reference state. There is an important literature devoted to this subject (see, in particular, [164] and [195]).

A number of papers have been devoted to the study of the incompressible limit for the compressible Navier–Stokes equations. The earliest mathematical works are concerned with the case where $\rho^\varepsilon - 1 = \mathcal{O}(\varepsilon^2)$ and $\operatorname{div} u^\varepsilon = \mathcal{O}(\varepsilon)$. In that case, $\partial_t \rho^\varepsilon$ and $\partial_t u^\varepsilon$ are uniformly bounded so that time oscillations cannot occur. Starting from this simple consideration, different authors have studied $(\widetilde{NSC_\varepsilon})$ with data of the type $\rho_0^\varepsilon = 1 + \varepsilon^2 \rho_{0,1}^\varepsilon$ and $u_0^\varepsilon = u_0 + \varepsilon u_{0,1}^\varepsilon$ with $\operatorname{div} u_0 = 0$ and $(\rho_{0,1}^\varepsilon, u_{0,1}^\varepsilon)$ uniformly bounded in a suitable function space (here, we take $f^\varepsilon \equiv 0$ in order to simplify matters). In the usual partial differential equations terminology, such data are referred to as *well prepared*. Indeed, they are well prepared in the sense that they belong (up to lower order terms) to the kernel of the singular operator appearing in $(\widetilde{NSC_\varepsilon})$. Hence, they are unlikely to produce highly oscillating terms.

For such data, it is possible to calculate an asymptotic expansion for $(\rho^\varepsilon, u^\varepsilon)$ in terms of powers of ε. This approach has been adopted by a number of authors: S. Klainerman and A. Majda in [186, 187], H.-O. Kreiss, J. Lorenz, and M. Naughton in [199], and D. Hoff in [160, 161], among others.

As explained at the beginning of Section 10.5, in this chapter we considered *ill prepared* data (i.e., $\rho_0^\varepsilon = 1 + \varepsilon b_0^\varepsilon$ and with no particular assumption on u_0^ε) so that strong time oscillations have to be considered. Solving the problem of the low Mach number limit in this framework is rather recent. The first result is due to P.-L. Lions in his book [215] and deals with global weak solutions. The basic idea is that the energy equality associated with (NSC_ε) is of the form

$$\left\|(\sqrt{\rho^\varepsilon} u^\varepsilon)(t)\right\|_{L^2}^2 + \int \pi^\varepsilon(t,x)\, dx + 2\int_0^t \left(\mu \|\nabla u^\varepsilon\|_{L^2}^2 + (\lambda+\mu)\|\operatorname{div} u^\varepsilon\|_{L^2}^2\right) d\tau$$

$$= \left\|\sqrt{\rho_0^\varepsilon} u_0^\varepsilon\right\|_{L^2}^2 + \int \pi_0^\varepsilon(x)\, dx + 2\int_0^t \int \rho^\varepsilon f^\varepsilon \cdot u^\varepsilon(\tau,x)\, dx\, d\tau,$$

where $\pi^\varepsilon = a\dfrac{(\rho^\varepsilon)^\gamma - 1 - \gamma(\rho^\varepsilon - 1)}{\varepsilon^2 \gamma(\gamma - 1)}$ if $P(\rho) = a\rho^\gamma$.

Taking advantage of the uniform estimates provided by the above equality, it is possible to pass to the limit when ε goes to 0. However, the mathematical justification strongly depends on the boundary conditions. The reader may refer to [216] for the case of periodic boundary conditions, to [115] for the whole space, and to [116] for the case of bounded domain with homogeneous Dirichlet conditions. In a more general context, P.-L. Lions and N. Masmoudi have also proven *local* weak convergence results in [217]. We emphasize that, to the best of our knowledge, [115] is the first paper devoted to the incompressible limit where Strichartz estimates have been used.

As regards the study of the incompressible limit in the framework of strong solutions, we mention the works by S. Ukai [291], S. Schochet [257], G. Métivier and S. Schochet [228, 229] in the inviscid case, and the papers by I. Gallagher [133] and T. Hagström and J. Lorenz [152] in the viscous case. The two results on convergence presented here (namely, Theorems 10.29 and 10.31) are borrowed from [97].

We conclude this section with a few words on the case of periodic boundary conditions, which turns out to be quite different (the reader is referred to [98] for more details). Indeed, there is no dispersion whatsoever, so acoustic waves may interact. It turns out, however, that resonances are not so frequent, so it is possible to pass to the limit anyway. The mathematical study of the incompressible limit may be undertaken by means of the filtering method introduced by S. Schochet in [257]. More precisely, if $P'(1) = 1$, then the system (NSC_ε) can be rewritten as

$$\frac{d}{dt}\begin{pmatrix} b^\varepsilon \\ u^\varepsilon \end{pmatrix} + \frac{L}{\varepsilon}\begin{pmatrix} b^\varepsilon \\ u^\varepsilon \end{pmatrix} = \begin{pmatrix} -\operatorname{div}(b^\varepsilon u^\varepsilon) \\ f - u^\varepsilon \cdot \nabla u^\varepsilon - b^\varepsilon \nabla b^\varepsilon - \widetilde{K}(\varepsilon b^\varepsilon) b^\varepsilon \nabla b^\varepsilon - I(\varepsilon b^\varepsilon)\mathcal{A}u^\varepsilon \end{pmatrix},$$

where the function $\widetilde{K}$ vanishes at 0 and the skew-symmetric operator L is defined by

$$L\begin{pmatrix} b \\ u \end{pmatrix} \overset{\mathrm{def}}{=} \begin{pmatrix} \operatorname{div} u \\ \nabla b \end{pmatrix}.$$

The operator L generates a unitary group $e^{\tau L}$ such that $e^{\tau L}\begin{pmatrix} 0 \\ \mathbb{P}u \end{pmatrix} = \begin{pmatrix} 0 \\ \mathbb{P}u \end{pmatrix}$.

Defining $V^\varepsilon \overset{\text{def}}{=} e^{\frac{t}{\varepsilon} L} \begin{pmatrix} b^\varepsilon \\ \mathbb{P}^\perp u^\varepsilon \end{pmatrix}$, we deduce from the above system that

$$\partial_t V^\varepsilon + \mathcal{Q}_1^\varepsilon(\mathbb{P}u^\varepsilon, V^\varepsilon) + \mathcal{Q}_2^\varepsilon(V^\varepsilon, V^\varepsilon) - \nu \mathcal{A}_2^\varepsilon(D)V^\varepsilon \qquad (10.92)$$

$$= e^{\frac{t}{\varepsilon} L} \begin{pmatrix} 0 \\ \mathbb{P}^\perp(f - \mathbb{P}u^\varepsilon \cdot \nabla \mathbb{P}u^\varepsilon) \end{pmatrix} + o(1),$$

where $\mathcal{A}_2^\varepsilon(D)B$ is a linear operator, and $\mathcal{Q}_1^\varepsilon$ and $\mathcal{Q}_2^\varepsilon$ are bilinear operators which may be computed explicitly in terms of Fourier series. It may be shown that $\mathcal{A}_2^\varepsilon(D)B$ tends formally to $-\Delta B/2$ and that the operators $\mathcal{Q}_1^\varepsilon$ and $\mathcal{Q}_2^\varepsilon$ tend to some first order bilinear operators $\mathcal{Q}_1$ and $\mathcal{Q}_2$, respectively.

If $\mathbb{P}u^\varepsilon$ tends to some limit v, then the stationary phase theorem ensures that the right-hand side of (10.92) tends to 0 in the sense of distributions. We can thus expect $(\mathbb{P}u^\varepsilon, V^\varepsilon)$ to tend to some limit (v, V), where v is a solution of the incompressible Navier–Stokes equation

$$(NSI) \qquad \begin{cases} \partial_t v + \mathbb{P}(v \cdot \nabla v) - \mu \Delta v = \mathbb{P}f \\ v_{|t=0} = \mathbb{P}u_0, \end{cases}$$

and V satisfies

$$(LS) \qquad \begin{cases} \partial_t V + \mathcal{Q}_1(v, V) + \mathcal{Q}_2(V, V) - \frac{\nu}{2}\Delta V = 0 \\ V_{|t=0} = (b_0, \mathbb{P}^\perp u_0). \end{cases}$$

Up to the term $\mathcal{Q}_1(v, V)$ (which is linear with respect to v), the system (LS) has the same structure as (NSI). However, it was observed by N. Masmoudi in [223] that the term $\mathcal{Q}_2(V, V)$ is so sparse that the diffusion $-\frac{\nu}{2}\Delta V$ dominates *in any dimension*. Hence, V exists as long as v is defined.

In [98], it was shown that for any data $b_0 \in H^{\frac{d}{2}+\alpha}$ with zero average, $u_0 \in H^{\frac{d}{2}+\alpha-1}$, and $f \in L^1(\mathbb{R}^+; H^{\frac{d}{2}-1+\alpha})$ with div $f = 0$, the solution of (LS) is defined as long as the solution v of (NSI) is defined. Moreover, if v is defined on $[0, T]$ or on $\mathbb{R}^+$, then the same holds for $(b^\varepsilon, u^\varepsilon)$ for sufficiently small ε, and

$$(b^\varepsilon, u^\varepsilon) = (0, v) + e^{-\frac{t}{\varepsilon} L} V + o(1) \quad \text{in} \quad \widetilde{L}_T^\infty(H^{\frac{d}{2}+\alpha'}) \cap L_T^2(H^{\frac{d}{2}+\alpha'+1}) \quad \text{for all} \quad \alpha' < \alpha.$$

Owing to the appearance of small divisors when proving the convergence of V^ε, the exact meaning of $o(1)$ strongly depends on the quotients of the lengths of the periodic box $\mathbb{T}_a^N$ in which (NSC_ε) is solved. For special values of $(a_1, \ldots, a_n)$, the convergence may be slower than any power of ε.

For polytropic fluids, the study of the low Mach number limit turns out to be even richer. For more details, the reader may refer to the recent work by T. Alazard in [4].

References

1. H. Abidi and R. Danchin: Optimal bounds for the inviscid limit of Navier–Stokes equations, *Asymptotic Analysis*, **38**, 2004, pages 35–46.
2. S. Adachi and K. Tanaka: Trudinger type inequalities in $\mathbb{R}^N$ and their best exponents, *Proceedings of the American Mathematical Society*, **128**(7), 2000, pages 2051–2057.
3. R. Adams: *Sobolev Spaces*, Pure and Applied Mathematics, **65**. Academic Press, New York-London, 1975.
4. T. Alazard: Low Mach number limit of the full Navier–Stokes equations, *Archive for Rational Mechanics and Analysis*, **180**(1), 2006, pages 1–73.
5. S. Alinhac: Paracomposition et opérateurs paradifférentiels, *Communications in Partial Differential Equations*, **11**, 1986, pages 87–121.
6. S. Alinhac: Remarques sur l'instabilité du problème des poches de tourbillon, *Journal of Functional Analysis*, **98**(2), 1991, pages 361–379.
7. S. Alinhac: Temps de vie précisé et explosion géométrique pour des systèmes hyperboliques quasilinéaires en dimension un d'espace II, *Duke Mathematical Journal*, **73**, 1994, pages 543–560.
8. S. Alinhac: Temps de vie et comportement explosif des solutions d'équations d'ondes quasi-linéaires en dimension deux I, *Annales Scientifiques de l'École Normale Supérieure*, **28**, 1995, pages 225–251.
9. S. Alinhac: Blow up of small data solutions for a class of quasilinear wave equations in two space dimensions I, *Annals of Mathematics*, **149**, 1999, pages 97–127.
10. S. Alinhac: Blow up of small data solutions for a class of quasilinear wave equations in two space dimensions II, *Acta Mathematica*, **182**, 1999, pages 1–23.
11. S. Alinhac and P. Gérard: *Introduction à la théorie des opérateurs pseudo-différentiels et théorème de Nash-Moser*, Savoirs actuels, Interéditions, 1991.
12. L. Ambrosio: Transport equation and Cauchy problem for non-smooth vector fields. *Calculus of Variations and Nonlinear Partial Differential Equations*, 1–41, Lecture Notes in Math., **1927**, Springer, Berlin, 2008.

H. Bahouri et al., *Fourier Analysis and Nonlinear Partial Differential Equations*, Grundlehren der mathematischen Wissenschaften 343, DOI 10.1007/978-3-642-16830-7, © Springer-Verlag Berlin Heidelberg 2011

498   References

13. L. Ambrosio and P. Bernard: Uniqueness of signed measures solving the continuity equation for Osgood vector fields, *Atti Accad. Naz. Lincei Cl. Sci. Fis. Mat. Natur. Rend. Lincei (9) Mat. Appl.* **19**, 2008, pages 237–245.

14. S. Antontsev, A. Kazhikhov and V. Monakhov: *Boundary Value Problems in Mechanics of Nonhomogeneous Fluids*, Studies in Mathematics and its Applications, **22**, North-Holland Publishing Co., Amsterdam, 1990.

15. V. Arnold: *Mathematical Methods of Classical Mechanics*, Graduate Texts in Mathematics, **60**, Springer-Verlag, Berlin, 1989.

16. V. Arnold: Sur la géométrie différentiable des groupes de Lie de dimension infinie et ses applications à l'hydrodynamique des fluides parfaits, *Annales de l'Institut Fourier*, **16**, 1966, pages 319–361.

17. H. Bahouri and J.-Y. Chemin: Équations de transport relatives à des champs de vecteurs non-lipschitziens et mécanique des fluides, *Archive for Rational Mechanics and Analysis*, **127**, 1994, pages 159–182.

18. H. Bahouri and J.-Y. Chemin: Équations d'ondes quasilinéaires et inégalités de Strichartz, *American Journal of Mathematics*, **121**, 1999, pages 1337–1377.

19. H. Bahouri and J.-Y. Chemin: Équations d'ondes quasilinéaires et effet dispersif, *International Mathematical Research News*, **21**, 1999, pages 1141–1178.

20. H. Bahouri and J.-Y. Chemin: Microlocal analysis, bilinear estimates and cubic quasilinear wave equation, *Astérique*, 2003, pages 93–142.

21. H. Bahouri and J.-Y. Chemin: On global well-posedness for defocusing cubic wave equation, *International Mathematics Research Notices*, **20**, 2006, pages 1–12.

22. H. Bahouri, J.-Y. Chemin and I. Gallagher: Inégalités de Hardy precisées, *Notes aux Comptes-Rendus de l'Académie des Sciences de Paris*, **341**(1), 2005, pages 89–92.

23. H. Bahouri, J.-Y. Chemin and I. Gallagher: Refined Hardy inequalities, *Annali della Scuola Normale di Pisa*, **5**, 2006, pages 375–391.

24. H. Bahouri and A. Cohen: Refined Sobolev inequalities in Lorentz spaces, preprint.

25. H. Bahouri and B. Dehman: Remarques sur l'apparition de singularités dans les écoulements Eulériens incompressibles à données initiales Hölderiennes, *Journal de Mathématiques Pures et Appliquées*, **73**, 1994, pages 335–346.

26. H. Bahouri and J. Shatah: Global estimate for the critical semilinear wave equation, *Annales de l'Institut Henri Poincaré, Analyse non linéaire*, **15**(6), 1998, pages 783–789.

27. A. Baraket: Local existence and estimations for a semilinear wave equation in two dimension space, *Bollettino della Unione Matematica Italiana*, **8**(1) 2004, pages 1–21.

28. C. Bardos and S. Benachour: Domaine d'analyticité des solutions de l'équation d'Euler dans un ouvert de $\mathbb{R}^n$, *Annali della Scuola Normale Superiore di Pisa*, **4**(4), 1977, pages 647–687.

29. G.K. Batchelor: *An Introduction to Fluids Dynamics*, Cambridge University Press, Cambridge, 1967.

30. S. Benachour: Analyticité des solutions des équations d'Euler, *Archive for Rational Mechanics and Analysis*, **71**(3), 1979, pages 271–299.

31. J. Beale, T. Kato and A. Majda: Remarks on the breakdown of smoothness for the 3-D Euler equations, *Communication in Mathematical Physics*, **94**, 1984, pages 61–66.

32. J. Ben Ameur and R. Danchin: Limite non visqueuse pour les fluides incompressibles axisymétriques, *Nonlinear Partial Differential Equations and Their Applications*, Collège de France Seminar, Vol. XIV (Paris, 1997/1998), 29–55, North-Holland, Amsterdam, 2002.

33. S. Benzoni and D. Serre: *Multi-dimensional Hyperbolic Partial Differential Equations: First-order Systems and Applications*, Oxford University Press, Oxford, 2007.

34. J. Bergh and J. Löfström: *Interpolation Spaces, an Introduction*, Springer Verlag, Berlin, 1976.

35. A. Bertozzi and P. Constantin: Global regularity for vortex patches, *Communication in Mathematical Physics*, **152**(1), 1993, pages 19–26.

36. A. Bertozzi and A. Majda: *Vorticity and Incompressible Flow*, Cambridge Texts in Applied Mathematics, **27**, Cambridge University Press, Cambridge, 2002.

37. O. Besov: On a family of function spaces. Embedding theorems and applications. *Dokladi Akademiĭ Naỳk SSSR*, **126**, 1959, pages 1163–1165.

38. J. Bona and R. Smith: The initial-value problem for the Korteweg–de Vries equation, *Philosophical Transactions of the Royal Society of London. Series A*, **278**, 1975, pages 555–601.

39. J.-M. Bony: Calcul symbolique et propagation des singularités pour les équations aux dérivées partielles non linéaires, *Annales de l'École Normale Supérieure*, **14**, 1981, pages 209–246.

40. J.-M. Bony: *Cours d'analyse: théorie des distributions et analyse de Fourier*, Ecole Polytechnique, Palaiseau.

41. G. Bourdaud: Réalisations des espaces de Besov homogènes, *Arkiv för Matematik*, **26**, 1988, pages 41–54.

42. G. Bourdaud: Fonctions qui opèrent sur les espaces de Besov et de Triebel, *Annales de l'Institut Henri Poincaré, Analyse Non Linéaire*, **10**, 1993, pages 413–422.

43. G. Bourdaud and D. Kateb: Fonctions qui opèrent sur les espaces de Besov, *Mathematische Annalen*, **303**, 1995, pages 653–675.

44. J. Bourgain: Refinements of Strichartz' inequality and applications to 2D-NLS with critical nonlinearity, *International Mathematical Research Notices*, **5**, 1998, pages 253–283.

45. Y. Brenier: The least action principle and the related concept of generalized flow for incompressible perfect fluids, *Journal of the American Mathematical Society*, **2**(2), 1989, pages 225–255.

46. P. Brenner: On $L^p - L^{p'}$ estimates for the wave-equation, *Mathematische Zeitschrift*, **145**(3), 1975, pages 251–254.

47. P. Brenner: On L^p-decay and scattering for nonlinear Klein–Gordon equations, *Mathematica Scandinavica*, **51**(2), 1982, pages 333–360.

48. P. Brenner: On space-time means and everywhere defined scattering operators for nonlinear Klein–Gordon equations, *Mathematische Zeitschrift*, **186**, 1984, pages 383–391.

49. D. Bresch and B. Desjardins: On the existence of global weak solutions to the Navier–Stokes equations for viscous compressible and heat conducting fluids, *Journal de Mathématiques Pures et Appliquées*, **87**(1), 2007, pages 57–90.

50. H. Brezis and T. Gallouët: Nonlinear Schrödinger evolution equations, *Nonlinear Analysis. Theory, Methods & Applications*, **4**(4), 1980, pages 677–681.

51. N. Burq, P. Gérard and N. Tvzetkov: An instability property of the nonlinear Schrödinger equation on $\mathbb{S}^d$. *Mathematical Research Letters*, **9**(2–3), 2002, pages 323–335.

52. T. Buttke: The observation of singularities in the boundary of patches of vorticity, *Physical Fluids A*, **1**(7), 1989, pages 1283–1285.

53. T. Buttke: A fast adaptative vortex method for patches of constant vorticity, *Journal of Computational Physics*, **89**, 1990, pages 161–186.

54. L. Caffarelli, R. Kohn and L. Nirenberg: Partial regularity of suitable weak solutions of the Navier–Stokes equations, *Communications on Pure and Applied Mathematics*, **35**, 1982, pages 771–831.

55. C. Calderón: Existence of weak solutions for the Navier–Stokes equations with initial data in L^p, *Transactions of the American Mathematical Society*, **318**(1), 1990, pages 179–200.

56. R. Camassa and D. Holm: An integrable shallow water equation with peaked solitons, *Physical Review Letters*, **71**, 1993, pages 1661–1664.

57. M. Cannone: *Ondelettes, paraproduits et Navier–Stokes*, Diderot Éditeur, Arts et Sciences, Lyon, 1995.

58. M. Cannone, Y. Meyer and F. Planchon: Solutions autosimilaires des équations de Navier–Stokes, *Séminaire Équations aux Dérivées Partielles de l'École Polytechnique*, 1993–1994.

59. E. Carlen and M. Loss: Optimal smoothing and decay estimates for viscously damped conservation laws, with applications to the 2-D Navier–Stokes equation, *Duke Mathematical Journal*, **81**(1), 1995, pages 135–157.

60. T. Cazenave: *Semilinear Schrödinger equations*, Courant Lecture Notes in Mathematics, **10**, New York University, American Mathematical Society, Providence, 2003.

61. F. Charve and R. Danchin: A global existence result for the compressible Navier–Stokes equations in the critical L^p framework, *Archive for Rational Mechanics and Analysis*, **198**(1), 2010, pages 233–271.

62. J.-Y. Chemin: Calcul paradifférentiel précisé et application à des équations aux dérivées partielles non linéaires, *Duke Mathematical Journal*, **56**(1), 1988, pages 431–469.

63. J.-Y. Chemin: Dynamique des gaz à masse totale finie, *Asymptotic Analysis*, **3**, 1990, pages 215–220.

64. J.-Y. Chemin: Sur le mouvement des particules d'un fluide parfait incompressible bidimensionnel, *Inventiones Mathematicae*, **103**, 1991, pages 599–629.

65. J.-Y. Chemin: Persistance des structures géométriques liées aux poches de tourbillon, *Séminaire Equations aux Dérivées Partielles de l'École Polytechnique*, 1990–1991.

66. J.-Y. Chemin: Régularité des trajectoires des particules d'un fluide incompressible remplissant l'espace, *Journal de Mathématiques Pures et Appliquées*, **71**(5), 1992, pages 407–417.

67. J.-Y. Chemin: Remarques sur l'existence pour le système de Navier–Stokes incompressible, *SIAM Journal of Mathematical Analysis*, **23**, 1992, pages 20–28.

68. J.-Y. Chemin: Persistance de structures géométriques dans les fluides incompressibles bidimensionnels, *Annales de l'École Normale Supérieure*, **26**, 1993, pages 517–542.

69. J.-Y. Chemin: *Fluides parfaits incompressibles*, Astérisque, **230**, 1995.

70. J.-Y. Chemin: A remark on the inviscid limit for two-dimensional incompressible fluid, *Communication in Partial Differential Equations*, **21**, 1996, pages 1771–1779.

71. J.-Y. Chemin: About Navier–Stokes system. Prépublication du Laboratoire d'Analyse Numérique de l'Université Paris 6.

72. J.-Y. Chemin: Théorèmes d'unicité pour le système de Navier–Stokes tridimensionnel, *Journal d'Analyse Mathématique*, **77**, 1999, pages 27–50.

73. J.-Y. Chemin: Le système de Navier–Stokes incompressible soixante-dix ans après Jean Leray, *Séminaire et Congrès*, **9**, 2004, pages 99–123.

74. J.-Y. Chemin, B. Desjardins, I. Gallagher and E. Grenier: Fluids with anisotropic viscosity, *Modélisation Mathématique et Analyse Numérique*, **34**, 2000, pages 315–335.

75. J.-Y. Chemin, B. Desjardins, I. Gallagher and E. Grenier: *Mathematical Geophysics. An Introduction to Rotating Fluids and the Navier–Stokes Equations*, Oxford Lecture Series in Mathematics and its Applications, **32**, 2006.

76. J.-Y. Chemin and N. Lerner: Flot de champs de vecteurs non-lipschitziens et équations de Navier–Stokes, *Journal of Differential Equations*, **121**, 1995, pages 314–328.

77. J.-Y. Chemin and C.-J. Xu: Inclusions de Sobolev en calcul de Weyl–
Hörmander et systèmes sous-elliptiques, *Annales de l'École Normale
Supérieure*, **30**, 1997, pages 719–751.

78. J.-Y. Chemin and P. Zhang: On the global well-posedness to the 3-D
incompressible anisotropic Navier–Stokes equations, *Communications in
Mathematical Physics*, **272**, 2007, pages 529–566.

79. Q. Chen, C. Miao and Z. Zhang: Well-posedness in critical spaces for the
compressible Navier–Stokes equations with density dependent viscosities,
preprint.

80. Y. Cho, H. J. Choe and H. Kim: Unique solvability of the initial boundary
value problems for compressible viscous fluids, *Journal de Mathématiques
Pures et Appliquées*, **83**(2), 2004, pages 243–275.

81. A. Cohen and R. Danchin: Multiscale approximation of vortex patches,
SIAM Journal on Applied Mathematics, **60**, 2000, pages 477–502.

82. A. Cohen, Y. Meyer and F. Oru, Improved Sobolev embedding theorem,
Séminaire Équations aux Dérivées Partielles, 1997–1998.

83. F. Colombini and N. Lerner: Uniqueness of continuous solutions for BV
vector fields, *Duke Mathematical Journal*, **111**(2), 2002, pages 357–384.

84. A. Constantin and J. Escher: Global existence and blow-up for a shallow
water equation, *Annali della Scuola Normale Superiore di Pisa, Classe
di Scienze*, **26**, 1998, pages 303–328.

85. P. Constantin: Note on loss of regularity for solutions of the 3-D incompressible Euler and related equations, *Communications in Mathematical
Physics*, **104**, 1986, pages 311–326.

86. P. Constantin and C. Foias: *Navier–Stokes Equations*, Chicago University Press, Chicago, 1988.

87. P. Constantin and E. Titi: On the evolution of nearly circular vortex
patches, *Communications in Mathematical Physics*, **119**, 1988, pages
177–198.

88. P. Constantin and J. Wu: Inviscid limit of in incompressible fluids, *Nonlinearity*, **8**, 1995, pages 735–742.

89. R. Danchin: Évolution temporelle d'une poche de tourbillon singulière,
Communications in Partial Differential Equations, **22**, 1997, pages 685–
721.

90. R. Danchin: Poches de tourbillon visqueuses, *Journal de Mathématiques
Pures et Appliquées*, **76**, 1997, pages 609–647.

91. R. Danchin: Persistance de structures géométriques et limite non
visqueuse pour les fluides incompressibles en dimension quelconque, *Bulletin de la Société Mathématique de France*, **127**, 1999, pages 179–227.

92. R. Danchin: Évolution d'une singularité de type cusp dans une poche
de tourbillon, *Revista Matemática Iberoamericana*, **16**, 2000, pages 281–
329.

93. R. Danchin: Global existence in critical spaces for compressible Navier–
Stokes equations, *Inventiones Mathematicae*, **141**(3), 2000, pages 579–
614.

94. R. Danchin: Global existence in critical spaces for flows of compressible viscous and heat-conductive gases, *Archive for Rational Mechanics and Analysis*, **160**(1), 2001, pages 1–39.

95. R. Danchin: Local theory in critical spaces for compressible viscous and heat-conductive gases, *Communications in Partial Differential Equations*, **26**, 2001, pages 1183–1233, and Erratum, **27**, 2002, pages 2531–2532.

96. R. Danchin: A few remarks on the Camassa–Holm equation, *Differential and Integral Equations*, **14**, 2001, pages 953–988.

97. R. Danchin: Zero Mach number limit in critical spaces for compressible Navier–Stokes equations, *Annales Scientifiques de l'École Normale Supérieure*, **35**(1), 2002, pages 27–75.

98. R. Danchin: Zero Mach number limit for compressible flows with periodic boundary conditions, *American Journal of Mathematics*, **124**(6), 2002, pages 1153–1219.

99. R. Danchin: A note on well-posedness for Camassa–Holm equation, *Journal of Differential Equations*, **192**(2), 2003, pages 429–444.

100. R. Danchin: Local and global well-posedness results for flows of inhomogeneous viscous fluids, *Advances in Differential Equations*, **9**(3–4), 2004, pages 353–386.

101. R. Danchin: On the uniqueness in critical spaces for compressible Navier–Stokes equations, *Nonlinear Differential Equations and Applications*, **12**(1), 2005, pages 111–128.

102. R. Danchin: Estimates in Besov spaces for transport and transport-diffusion equations with almost Lipschitz coefficients, *Revista Matemática Iberoamericana*, **21**, 2005, pages 863–888.

103. R. Danchin: Uniform estimates for transport-diffusion equations, *Journal of Hyperbolic Differential Equations*, **4**, 2007, pages 1–17.

104. R. Danchin: Well-posedness in critical spaces for barotropic viscous fluids with truly nonconstant density, *Communications in Partial Differential Equations*, **32**, 2007, pages 1373–1397.

105. R. Danchin: Axisymmetric incompressible flows with bounded vorticity, *Russian Mathematical Surveys*, **62**(3), 2007, pages 73–94.

106. R. Danchin: On the well-posedness of the incompressible density-dependent Euler equations in the L^p framework, *Journal of Differential Equations*, **248**(8), 2010, pages 2130–2170.

107. R. Danchin and B. Desjardins: Existence of solutions for compressible fluid models of Korteweg type, *Annales de l'IHP, Analyse Non Linéaire*, **18**, 2001, pages 97–133.

108. C. De Lellis, T. Kappeler and P. Topalov: Low-regularity solutions of the periodic Camassa–Holm equation, *Communications in Partial Differential Equations*, **32**, 2007, pages 87–126.

109. J.-M. Delort: Existence de nappes de tourbillon en dimension deux, *Journal of the American Mathematical Society*, **4**, 1991, pages 553–586.

110. N. Depauw: Poche de tourbillon pour Euler 2D dans un ouvert à bord, *Journal de Mathématiques Pures et Appliquées*, **78**, 1999, pages 313–351.
111. N. Depauw: Non-unicité du transport par un champ de vecteurs presque BV, *Séminaire EDP*, 2002–2003, Exp. No. XIX, 9 pp., Ecole Polytechnique, Palaiseau, 2003.
112. B. Desjardins: A few remarks on ordinary differential equations, *Communications in Partial Differential Equations*, **21**, 1996, pages 1667–1703.
113. B. Desjardins: Linear transport equations with values in Sobolev spaces and application to the Navier–Stokes equations, *Differential and Integral Equations*, **10**, 1997, pages 577–586.
114. B. Desjardins: Regularity of weak solutions of the compressible isentropic Navier–Stokes equations, *Communications in Partial Differential Equations*, **22**, 1997, pages 977–1008.
115. B. Desjardins and E. Grenier: Low Mach number limit of compressible flows in the whole space, *Proceedings of the Royal Society of London, Series A*, **455**, 1999, pages 2271–2279.
116. B. Desjardins, E. Grenier, P.-L. Lions and N. Masmoudi: Incompressible limit for solutions of the isentropic Navier–Stokes equations with Dirichlet boundary conditions, *Journal de Mathématiques Pures et Appliquées*, **78** (5), 1999, pages 461–471.
117. R. Di Perna and P.-L. Lions: Ordinary differential equations, transport theory and Sobolev spaces, *Inventiones Mathematicae*, **98**(3), 1989, pages 511–549.
118. A. Dutrifoy: Existence globale en temps de solutions hélicoïdales des équations d'Euler, *Compte-rendu de l'Académie des Sciences, Paris, Série I*, **329**(7), 1999, pages 653–656.
119. A. Dutrifoy: Precise regularity results for the Euler equations, *Journal of Mathematical Analysis and Applications*, **282**(1), 2003, pages 177–200.
120. A. Dutrifoy: On 3-D vortex patches in bounded domains, *Communications in Partial Differential Equations*, **28**(7–8), 2003, pages 1237–1263.
121. D. Ebin and J. Marsden: Group of diffeomorphism and the motion of an incompressible fluid, *Annals of Mathematics*, **92**(2), 1970, pages 102–163.
122. L. C. Evans: *Partial Differential Equations*, Graduate Studies in Mathematics, AMS, Providence, 1998.
123. E. Feireisl: Viscous and/or heat conducting compressible fluids, *Handbook of Mathematical Fluid Dynamics*, Vol. I, 307–371, North-Holland, Amsterdam, 2002.
124. E. Feireisl: *Dynamics of Viscous Compressible Fluids*, Oxford Lecture Series in Mathematics and its Applications, **26**, Oxford University Press, Oxford, 2004.
125. W. Fiszdon and W. Zajączkowski: The initial-boundary value problem for the flow of a barotropic viscous fluid, global in time, *Applicable Analysis*, **15**, 1983, pages 91–114.

126. A. Fokas and B. Fuchssteiner: Symplectic structures, their Bäcklund transformation and hereditary symmetries, *Physica D*, **4**, 1988, pages 47–66.

127. C. Foias, C. Guillopé and R. Temam: Lagrangian representation of a flow, *Journal of Differential Equations*, **57**, 1985, pages 440–449.

128. D. Foschi: *On the regularity of multilinear forms associated to the wave equation*, Ph.D thesis, Princeton University, January 2000.

129. H. Fujita and T. Kato: On the Navier–Stokes initial value problem I, *Archive for Rational Mechanic Analysis*, **16**, 1964, pages 269–315.

130. G. Furioli, P.-G. Lemarié-Rieusset and E. Terraneo: Unicité des solutions mild des équations de Navier–Stokes dans $L^3(\mathbb{R}^3)$ et d'autres espaces limites, *Revista Matemática Iberoamericana*, **16**(3), 2000, pages 605–667.

131. E. Gagliardo: Proprietà di alcune classi di funzioni in più variabili, *Ricerche di Matematica, Napoli*, **7**, 1958, pages 102–137.

132. I. Gallagher: The tridimensional Navier–Stokes equations with almost bidimensional data: stability, uniqueness and life span, *International Mathematical Research Notices*, **18**, 1997, pages 919–935.

133. I. Gallagher: A remark on smooth solutions of the weakly compressible periodic Navier–Stokes equations, *Journal of Mathematics of Kyoto University*, **40**(3), 2000, pages 525–540.

134. I. Gallagher, D. Iftimie and F. Planchon: Asymptotics and stability for global solutions to the Navier–Stokes equations, *Annales de l'Institut Fourier*, **53**, 2003, pages 1387–1424.

135. I. Gallagher and F. Planchon: On global solutions to a defocusing semilinear wave equation, *Revista Matemática Iberoamericana*, **19**, 2003, pages 161–177.

136. P. Gamblin: Système d'Euler incompressible et régularité microlocale analytique, *Annales de l'Institut Fourier*, **44**, 1994, pages 1449–1475.

137. P. Gamblin and X. Saint-Raymond: On three-dimensional vortex patches, *Bulletin de la Société Mathématique de France*, **123**, 1995, pages 375–424.

138. P. Gérard: Résultats récents sur les fluides parfaits incompressibles bidimensionnels (d'après J.-Y. Chemin et J.-M. Delort), *Séminaire Bourbaki*, **757**, 1992.

139. P. Gérard: Description du défaut de compacité de l'injection de Sobolev, *ESAIM. Control, Optimisation and Calculus of Variations*, **3**, 1998, pages 213–233.

140. P. Gérard, Y. Meyer and F. Oru: Inégalités de Sobolev précisées, *Séminaire Équations aux Dérivées Partielles*, 1996–1997, École Polytechnique, Palaiseau.

141. P. Gérard and J. Rauch: Propagation de la régularité locale de solutions d'équations aux dérivées partielles non linéaires, *Annales de l'Institut Fourier*, **37**, 1987, pages 65–84.

506   References

142. P. Germain: Weak-strong uniqueness for the isentropic compressible Navier–Stokes system, *Journal of Mathematical Fluid Mechanics*, in press.

143. Z. Grujic and I. Kukavica: Space analycity for the Navier–Stokes and related equations with initial data in L^p, *Journal of Functional Analysis*, **152**, 1998, pages 247–466.

144. Y. Giga: Solutions for semilinear parabolic equations in L^p and regularity of weak solutions of the Navier–Stokes system, *Journal of Differential Equations*, **61**, 1986, pages 186–212.

145. Y. Giga and T. Miyakawa: Solutions in L^r of the Navier–Stokes initial value problem, *Archive for Rational Mechanics and Analysis*, **89**, 1985, pages 267–281.

146. D. Gilbarg and N. Trudinger: *Elliptic Partial Differential Equations of Second Order*, Classics in Mathematics, Reprint of the 1998 Edition, Springer.

147. J. Ginibre and G. Velo: Time decay of finite energy solutions of the nonlinear Klein–Gordon and Schrodinger equations, *Annales de l'Institut Henri Poincaré, Physique Théorique*, **43**, 1985, pages 399–442.

148. J. Ginibre and G. Velo: The global Cauchy problem for nonlinear Klein–Gordon equation, *Mathematische Zeitschrift*, **189**, 1985, pages 487–505.

149. J. Ginibre and G. Velo: Generalized Strichartz inequalities for the wave equations, *Journal of Functional Analysis*, **133**, 1995, pages 50–68.

150. L. Grafakos: *Classical and Modern Fourier Analysis*, Prentice Hall, New York, 2006.

151. R. Grauer and T. Sideris: Finite time singularities in ideal fluids with swirl, *Physica D*, **88**, 1995, pages 116–132.

152. T. Hagstrom and J. Lorenz: All-time existence of classical solutions for slightly compressible flows, *SIAM Journal on Mathematical Analysis*, **29**(3), 1998, pages 652–667.

153. G. H. Hardy: Note on a theorem of Hilbert, *Mathematische Zeitschrift*, **6**, 1920, pages 314–317.

154. G. H. Hardy: An inequality between integrals, *Messenger of Mathematics*, **54**, 1925, pages 150–156.

155. B. Haspot: Well-posedness in critical spaces for barotropic viscous fluids, preprint.

156. T. Hmidi: Régularité höldérienne des poches de tourbillon visqueuses, *Journal de Mathématiques Pures et Appliquées*, **84**(11), 2005, pages 1455–1495.

157. T. Hmidi and S. Keraani: Existence globale pour le système d'Euler incompressible 2-D dans $B^1_{\infty,1}$, *Compte-rendu de l'Académie des Sciences, Paris, Série I*, **341**(11), 2005, pages 655–658.

158. T. Hmidi and S. Keraani: Incompressible viscous flows in borderline Besov spaces, *Archive for Rational Mechanics and Analysis*, **189**, 2009, pages 283–300.

159. D. Hoff: Global solutions of the Navier–Stokes equations for multidimensional compressible flow with discontinuous initial data, *Journal of Differential Equations*, **120**(1), 1995, pages 215–254.

160. D. Hoff: Strong convergence to global solutions for multidimensional flows of compressible, viscous fluids with polytropic equations of state and discontinuous initial data, *Archive for Rational Mechanics and Analysis*, **132**(1), 1995, pages 1–14.

161. D. Hoff: The zero-Mach limit of compressible flows, *Communications in Mathematical Physics*, **192**(3), 1998, pages 543–554.

162. D. Hoff: Dynamics of singularity surfaces for compressible, viscous flows in two space dimensions, *Communications on Pure and Applied Mathematics*, **55**(11), 2002, pages 1365–1407.

163. D. Hoff: Uniqueness of weak solutions of the Navier–Stokes equations of multidimensional, compressible flow, *SIAM Journal on Mathematical Analysis*, **37**(6), 2006, pages 1742–1760.

164. D. Hoff and K. Zumbrun: Multi-dimensional diffusion waves for the Navier–Stokes equations of compressible flow, *Indiana University Mathematical Journal*, **44**(2), 1995, pages 603–676.

165. E. Hopf: Über die Anfangswertaufgabe für die hydrodynamischen Grundgleichungen, *Mathematische Nachrichten*, **4**, 1951, pages 213–231.

166. L. Hörmander: *Linear Partial Differential Operators*, Die Grundlehren der mathematischen Wissenschaften, Springer-Verlag, Berlin, 1963.

167. L. Hörmander: *The Analysis of Linear Partial Differential Operators. I. Distribution Theory and Fourier Analysis.* Reprint of the second (1990) edition. Springer-Verlag, Berlin, 2003.

168. L. Hörmander: *Lectures on Nonlinear Hyperbolic Differential Equations*, Mathematics and Applications, **26**, Springer, Berlin, 1996.

169. C. Huang: Singular integral system approach to regularity of 3D vortex patches, *Indiana University Mathematical Journal*, **50**(1), 2001, pages 509–552.

170. M. Hughes, K. Roberts and N. Zabusky: Contour dynamics for the Euler equations in two dimensions, *Journal of Computational Physics*, **30**, 1979, pages 96–106.

171. R. Hunt: On $L(p, q)$ spaces, *Enseignement mathématique*, **12**, 1966, pages 249–276.

172. D. Iftimie: The resolution of the Navier–Stokes equations in anisotropic spaces, *Revista Matemática Iberoamericana*, **15**, 1999, pages 1–36.

173. D. Iftimie: A uniqueness result for the Navier–Stokes equations with vanishing vertical viscosity, *SIAM Journal on Mathematical Analysis*, **33**(6), 2002, pages 1483–1493.

174. F. John and L. Nirenberg: On functions of bounded mean oscillations, *Communications on Pure and Applied Mathematics*, **14**, 1961, pages 415–426.

175. L. Kapitanski: Some generalization of the Strichartz–Brenner inequality, *Leningrad Mathematical Journal*, **1**, 1990, pages 693–721.

508   References

176. T. Kato: Nonstationary flows of viscous and ideal fluids in $\mathbb{R}^3$, *Journal of Functional Analysis*, **9**, 1972, pages 296–305.

177. T. Kato: Quasi-linear equations of evolution, with applications to partial differential equations, *Lecture Notes in Mathematics*, **448**, Springer-Verlag, Berlin, 1975, pages 25–70.

178. T. Kato and G. Ponce: Well-posedness of the Euler and Navier–Stokes equations in the Lebesgue spaces $L_s^p(\mathbb{R}^2)$, *Revista Matemática Iberoamericana*, **2**, 1986, pages 73–88.

179. A. Kazhikhov and V. Shelukhin: Unique global solution with respect to time of initial-boundary value problems for one-dimensional equations of a viscous gas, *Journal of Applied Mathematics and Mechanics*, **41**(2), 1977, pages 273–282.

180. M. Keel and T. Tao: Endpoint Strichartz estimates, *American Journal of Mathematics*, **120**, 1998, pages 995–980.

181. C. Kenig, G. Ponce and L. Vega: Global well-posedness for semilinear wave equations, *Communications in Partial Differential Equations*, **25**, 2000, pages 1741–1752.

182. S. Klainerman: Global existence for nonlinear wave equations, *Communications on Pure and Applied Mathematics*, **33**, 1980, pages 43–101.

183. S. Klainerman: Uniform decay estimates and the Lorentz invariance of the classical wave equation, *Communications on Pure and Applied Mathematics*, **38**, 1985, pages 321–332.

184. S. Klainerman: The null condition and global existence to non linear wave equations, *Nonlinear Systems of Partial Differential Equations in Applied Mathematics, Part 1*, pages 293–326, Lectures in Applied Mathematics, **23**, 1986.

185. S. Klainerman: The mathematics of general relativity and nonlinear wave equations. *Phase Space Analysis of Partial Differential Equations*. Pubbl. Cent. Ric. Mat. Ennio Giorgi, Scuola Normale Superiore di Pisa, 2004, Vol. II, page 271–343.

186. S.Klainerman and A. Majda: Singular limits of quasilinear hyperbolic systems and the incompressible limit of compressible fluids, *Communications on Pure and Applied Mathematics*, **34**(4), 1981, pages 481–524.

187. S.Klainerman and A. Majda: Compressible and incompressible fluids, *Communications on Pure and Applied Mathematics*, **35**(5), 1982, pages 629–651.

188. S. Klainerman and I. Rodnianski: Improved local well-posedness for quasilinear wave equations in dimension three, *Duke Mathematical Journal*, **117**, 2003, pages 1–124.

189. S. Klainerman and I. Rodnianski: Ricci defects of microlocalized Einstein metrics, *Journal of Hyperbolic Differential Equations*, **1**, 2004, pages 85–113.

190. S. Klainerman and I. Rodnianski: Causal geometry of Einstein-vacuum spacetimes with finite curvature flux, *Inventiones Mathematicae*, **159**, 2005, pages 437–529.

191. S. Klainerman and I. Rodnianski: The causal structure of microlocalized rough Einstein metrics, *Annals of Mathematics*, **161**, 2005, pages 1195–1243.

192. S. Klainerman and I. Rodnianski: Rough solutions of the Einstein-vacuum equations, *Annals of Mathematics*, **161**, 2005, pages 1143–1193.

193. S. Klainerman and S. Selberg: Remark on the optimal regularity for equations of wave maps type, *Communications in Partial Differential Equations*, **22**, 1997, pages 901–918.

194. S. Klainerman and D. Tataru: On the optimal local regularity for the Yang–Mills equations in $\mathbf{R}^{4+1}$, *Journal of the American Mathematical Society*, **12**, 1999, pages 93–116.

195. T. Kobayashi and Y. Shibata: Remark on the rate of decay of solutions to linearized compressible Navier–Stokes equations, *Pacific Journal of Mathematics*, **207**(1), 2002, pages 199–234.

196. H. Koch and D. Tataru: Well-posedness for the Navier–Stokes equations, *Advances in Mathematics*, **157**, 2001, pages 22–35.

197. H. Kozono and Y. Taniuchi: Limiting case of the Sobolev inequality in BMO, with application to the Euler equations, *Communications in Mathematical Physics*, **214**(1), 2000, pages 191–200.

198. H. Kozono and M. Yamazaki: Semilinear heat equations and the Navier–Stokes equations with distributions in new function spaces as initial data, *Communications in Partial Differential Equations*, **19**, 1994, pages 959–1014.

199. H.-O. Kreiss, J. Lorenz and M.J. Naughton: Convergence of the solutions of the compressible to the solutions of the incompressible Navier–Stokes equations, *Advances in Applied Mathematics*, **12**(2), 1991, pages 187–214.

200. J.R. Kweon: The evolution compressible Navier–Stokes system on polygonal domains, *Journal of Differential Equations*, **232**(2), 2007, pages 487–520.

201. O. Ladyženskaja: Solution "in the large" of the nonstationary boundary value problem for the Navier–Stokes system with two space variables, *Communications on Pure and Applied Mathematics*, **12**, 1959, pages 427–433.

202. P. Lax and R. Phillips: *Scattering Theory*, Pure and Applied Mathematics, **26**, Academic Press, New York-London, 1967.

203. G. Lebeau: Singularités de solutions d'équations d'ondes semi-linéaires, *Annales Scientifiques de l'École Normale Supérieure*, **25**, 1992, pages 201–231.

204. P.-G. Lemarié-Rieusset: Base d'ondelettes sur les groupes de Lie stratifiés, *Bulletin de la Société Mathématique de France*, **117**, 1989, pages 211–232.

205. P.-G. Lemarié-Rieusset: *Recent Developments in the Navier–Stokes Problem*, Chapman & Hall/CRC Research Notes in Mathematics, **431**, 2002.

206. J. Leray: Étude de diverses équations intégrales non linéaires et quelques problèmes que pose l'hydrodynamique, *Journal de Mathématiques Pures et Appliquées*, **9**(12), 1933, pages 1–82.

207. J. Leray: Essai sur le mouvement d'un liquide visqueux emplissant l'espace, *Acta Mathematica*, **63**, 1933, pages 193–248.

208. L. Lichtenstein: Über einige Existenzprobleme der Hydrodynamik homogenerunzusammendrückbarer, reibungsloser Flüßigkeiten und die Helmoltzochen Wirbelsatze, *Mathematische Zeitschrift*, **23**, 1925, pages 89–154; **26**, 1927, pages 196–323; **28**, 1928, pages 387–415 and **32**, 1930, pages 608–725.

209. H. Lindblad: A sharp counterexample to local existence of low regularity solutions to non linear wave equations, *Duke Mathematical Journal*, **72**, 1993, pages 503–539.

210. H. Lindblad and C. Sogge: On existence and scattering with minimal regularity for semilinear wave equations, *Journal of Functional Analysis*, **130**(2), 1995, pages 357–426.

211. J.-L. Lions and G. Prodi: Un théorème d'existence et unicité dans les équations de Navier–Stokes en dimension 2, *Comptes-Rendus de l'Académie des Sciences de Paris*, **248**, 1959, pages 3519–3521.

212. P.-L. Lions: The concentration-compactness principle in the calculus of variations. The limit case, *Revista Matemática Iberoamericana*, **1**, 1985, pages 12–45.

213. P.-L. Lions: Équations différentielles ordinaires et équations de transport avec des coefficients irréguliers, *Séminaire Équations aux Dérivées Partielles*, 1988–1989, École Polytechnique, Palaiseau.

214. P.-L. Lions: *Mathematical Topics in Fluid Mechanics. Incompressible Models*, Oxford Lecture Series in Mathematics and its Applications, **3**, 1996.

215. P.-L. Lions: *Mathematical Topics in Fluid Mechanics. Compressible Models*, Oxford Lecture Series in Mathematics and its Applications, **10**, 1998.

216. P.-L. Lions and N. Masmoudi: Incompressible limit for a viscous compressible fluid, *Journal de Mathématiques Pures et Appliquées*, **77**(6), 1998, pages 585–627.

217. P.-L. Lions and N. Masmoudi: Une approche locale de la limite incompressible, *Comptes-Rendus de l'Académie des Sciences, Paris*, **329**(5), 1999, pages 387–392.

218. J. Littlewood and R. Paley: Theorems on Fourier series and power series I, *Journal of the London Mathematical Society*, **6**, 1931, pages 230–233.

219. J. Littlewood and R. Paley: Theorems on Fourier series and power series II, *Proceedings of the London Mathematical Society*, **42**, 1936, pages 52–89.

220. T.-P. Liu and T. Yang: Compressible Euler equations with vacuum, *Journal of Differential Equations*, **140**, 1997, pages 223–237.

221. A. Majda: Vorticity and the mathematical theory of an incompressible fluid flow, *Communications on Pure and Applied Mathematics*, **38**, 1986, pages 187–220.

222. C. Marchioro and M. Pulvirenti: *Mathematical Theory of Incompressible Non Viscous Fluids*, Applied Mathematical Sciences, **96**, Springer Verlag, Berlin, 1994.

223. N. Masmoudi: Incompressible, inviscid limit of the compressible Navier–Stokes system, *Annales de l'IHP, Analyse Non Linéaire*, **18**(2), 2001, pages 199–224.

224. N. Masmoudi: Remarks about the inviscid limit of the Navier–Stokes system, *Communications in Mathematical Physics*, **270**(3), 2007, pages 777–788.

225. A. Matsumura and T. Nishida: The initial value problem for the equations of motion of viscous and heat-conductive gases, *Journal of Mathematics of Kyoto University*, **20**(1), 1980, pages 67–104.

226. A. Matsumura and T. Nishida: Initial-boundary value problems for the equations of motion of compressible viscous and heat-conductive fluids, *Communications in Mathematical Physics*, **89**(4), 1983, pages 445–464.

227. A. Mellet and A. Vasseur: Existence and uniqueness of global strong solutions for one-dimensional compressible Navier–Stokes equations, *SIAM Journal of Mathematical Analysis*, **39**(4), 2008, pages 1344–1365.

228. G. Métivier and S. Schochet: The incompressible limit of the non-isentropic Euler equations, *Archive for Rational Mechanics and Analysis*, **158**(1), 2001, pages 61–90.

229. G. Métivier and S. Schochet: Averaging theorems for conservative systems and the weakly compressible Euler equations, *Journal of Differential Equations*, **187**(1), 2003, pages 106–183.

230. Y. Meyer: *Ondelettes et opérateurs. I*, Hermann, Paris, 1990.

231. Y. Meyer: *Ondelettes et opérateurs. II*, Hermann, Paris, 1990.

232. Y. Meyer and R. Coifman: *Ondelettes et opérateurs. III*, Hermann, Paris, 1991.

233. S. Montgomery-Smith: Time decay for the bounded mean oscillation of solutions of the Schrödinger and wave equations, *Duke Mathematical Journal*, **91**(2), 1998, pages 393–408.

234. S. Montgomery-Smith: Finite time blow up for a Navier–Stokes like equation, *Proceedings of the American Mathematical Society*, **129**(10), 2001, pages 3025–3029.

235. C. Morrey: On the solutions of quasi-linear elliptic partial differential equations, *Transactions of the American Mathematical Society*, **43**, 1938, pages 126–166.

236. J. Moser: A sharp form of an inequality of N. Trudinger, *Indiana University Mathematical Journal*, **20**, 1971, pages 1077–1092.

237. P. Mucha and W. Zajączkowski: Global existence of solutions of the Dirichlet problem for the compressible Navier–Stokes equations,

Zeitschrift für Angewandte Mathematik und Mechanik, **84**(6), 2004, pages 417–424.

238. J. Nash: Le problème de Cauchy pour les équations différentielles d'un fluide général, *Bulletin de la Société Mathématique de France*, **90**, 1962, pages 487–497.

239. J. Neças, M. Ružička and V. Šverák: On Leray's self-similar solutions of the Navier–Stokes equations, *Acta Mathematica*, **176**, 1996, pages 283–294.

240. S. Nikol'skiǐ: *Approximation of Functions of Several Variables and Imbedding Theorems*, **205**. Springer-Verlag, New York-Heidelberg, 1975.

241. L. Nirenberg: On elliptic partial differential equations, *Annali di Pisa*, **13**, 1959, pages 116–162.

242. A. Novotný and I. Straškraba: *Introduction to the Mathematical Theory of Compressible Flow*. Oxford Lecture Series in Mathematics and Its Applications, **27**, Oxford University Press, Oxford, 2004.

243. F. Oru: PhD dissertation.

244. M. Paicu: Équation anisotrope de Navier–Stokes dans des espaces critiques, *Revista Matemática Iberoamericana*, **21**(1), 2005, pages 179–235.

245. H. C. Pak and Y. J. Park: Existence of solution for the Euler equations in a critical Besov space $B^1_{\infty,1}(\mathbb{R}^n)$, *Communications in Partial Differential Equations*, **29**, 2004, pages 1149–1166.

246. H. Pecher: L^p – Abschätzungen und klassische Lösungen für nichtlineare Wellengleichungen, *Mathematische Zeitschrift*, **150**, 1976, pages 159–183.

247. H. Pecher: Small data scattering for the wave and Klein–Gordon equation, *Mathematische Zeitschrift*, **185**, 1985, pages 445–457.

248. J. Pedlosky: *Geophysical Fluid Dynamics*, Second Edition, Springer-Verlag, Berlin, 1986.

249. J. Peetre: Sur les espaces de Besov, *Comptes-Rendus de l'Académie des Sciences*, **264**, 1967, pages 281–283.

250. J. Peetre: *New Thoughts on Besov Spaces*, Duke University Mathematics Series, No. 1, Mathematics Department, Duke University, Durham, N.C., 1976.

251. F. Planchon: Sur une inégalité de type Poincaré, *Notes aux Comptes-rendus de l'Académie des Sciences de Paris*, **330**, 2000, pages 21–23.

252. G. Ponce, R. Racke, T.C. Sideris, and E.S. Titi: Global stability of large solutions to the 3D Navier–Stokes equations, *Communications in Mathematical Physics*, **159**, 1994, pages 329–341.

253. G. Ponce and T. Sideris: Local regularity of non linear wave equations in three space dimensions, *Communications in Partial Differential Equations*, **18**, 1993, pages 169–177.

254. T. Runst and W. Sickel: *Sobolev Spaces of Fractional Order, Nemytskij Operators, and Nonlinear Partial Differential Equations*, Nonlinear Analysis and Applications, **3**. Walter de Gruyter & Co., Berlin, 1996.

255. M. Sablé-Tougeron: Régularité microlocale pour des problèmes aux limites non linéaires, *Annales de l'Institut Fourier*, **36**, 1986, pages 39–82.

256. X. Saint-Raymond: Remarks on axisymmetric solutions of the incompressible Euler system, *Communications in Partial Differential Equations*, **19**(1–2), 1994, pages 321–334.

257. S. Schochet: Fast singular limits of hyperbolic PDEs, *Journal of Differential Equations*, **114**, 1994, pages 476–512.

258. P. Serfati: Une preuve directe d'existence globale des vortex patches 2D, *Notes aux Comptes-Rendus de l'Académie des Sciences de Paris*, **318**(1), 1993, pages 515–518.

259. P. Serfati: Régularité stratifiée et équation d'Euler à temps grand, *Notes aux Comptes-Rendus de l'Académie des Sciences de Paris*, **318**(1), 1993, pages 925–928.

260. D. Serre: La croissance de la vorticité dans les écoulements parfaits incompressibles, *Notes aux Comptes-Rendus de l'Académie des Sciences de Paris*, **328**(6), 1999, pages 549–552.

261. D. Serre: Solutions classiques globales des équations d'Euler pour un fluide parfait compressible, *Annales de l'Institut Fourier*, **47**, 1997, pages 139–153.

262. D. Serre: *Systèmes de lois de conservation*, Diderot Editeur, Arts et Sciences, Paris–New York–Amsterdam.

263. J. Shatah and M. Struwe: Regularity results for nonlinear wave equations, *Annals of Mathematics*, **2**(138), 1993, pages 503–518.

264. J. Shatah and M. Struwe: Well-posedness in the energy space for semilinear wave equation with critical growth, *International Mathematics Research Notices*, **7**, 1994, pages 303–309.

265. T. Shirota and T. Yanagisawa: Note on global existence for axially symmetric solutions of the Euler system, *Proceedings of the Japan Academy, Series A*, **70**(10), 1994, pages 299–304.

266. A. Shnirelman: On the geometry of the group of diffeomorphisms and the dynamics of an ideal incompressible fluid, *Mat. USSR Sbornik*, **56**, 1987, pages 79–105.

267. T. Sideris: Global behavior of solutions to nonlinear wave equations in three dimensions, *Communications in Partial Differential Equations*, **12**, 1983, pages 1291–1323.

268. H. Smith: A parametrix construction for wave equation with $C^{1,1}$ coefficients, *Annales de l'Institut Fourier*, **48**, 1998, pages 797–835.

269. H. Smith and D. Tataru: Sharp local well-posedness results for the nonlinear wave equation, *Annals of Mathematics*, **162**, 2005, pages 291–366.

270. S. Sobolev: On a theorem of functional analysis, *Mat. Sbornik*, **4**, 1938, pages 471–497.

271. S. Sobolev: *Applications of Functional Analysis in Mathematical Physics*, Translations of Mathematical Monographs, **7**, American Mathematical Society, Providence, 1963.

272. V. Solonnikov: The solvability of the initial-boundary value problem for the equations of motion of a viscous compressible fluid. Investigations on linear operators and theory of functions, VI. *Zap. Naucn. Sem. Leningrad. Otdel. Mat. Inst. Steklov. (LOMI)*, **56**, 1976, pages 128–142, 197.

273. E. M. Stein: *Singular Integrals and Differentiability Properties of Functions*, Princeton University Press, Princeton, 1970.

274. E. M. Stein: Oscillatory integrals in Fourier analysis, *Beijing Lectures in Harmonic Analysis*, Princeton University Press, Princeton, 1986, pages 307–355.

275. E. M. Stein: *Harmonic Analysis*, Princeton University Press, Princeton, 1993.

276. R. Strichartz: Restriction Fourier transform of quadratic surfaces and decay of solutions of the wave equations, *Duke Mathematical Journal*, **44**, 1977, pages 705–714.

277. F. Sueur: Vortex internal transition layers for the Navier–Stokes equation, submitted.

278. T. Tao: Spherically averaged endpoint Strichartz estimates for the two-dimensional Schrödinger equation, *Communications in Partial Differential Equations*, **25**(7–8), 2000, pages 1471–1485.

279. D. Tataru: Local and global results for wave maps. I, *Communications in Partial Differential Equations*, **23**, 1998, pages 1781–1793.

280. D. Tataru: Strichartz estimates for operators with nonsmooth coefficients and the nonlinear wave equation, *American Journal of Mathematics*, **122**, 2000, pages 349–376.

281. D. Tataru: Strichartz estimates for second order hyperbolic operators with nonsmooth coefficients. III, *Journal of the American Mathamtical Society*, **15**, 2002, pages 419–442.

282. M.E. Taylor: *Partial Differential Equations I. Basic Theory*, **115**, Applied Mathematical Sciences, Springer.

283. M. Taylor: *Partial Differential Equations. III. Nonlinear Equations*, **117**, Applied Mathematical Sciences. Springer-Verlag, 1197.

284. M. Taylor: *Pseudodifferential Operators*, Princeton Mathematical Series, **34**, Princeton University Press, Princeton, 1981.

285. M. Taylor: *Tools for PDE: Pseudodifferential Operators, Paradifferential Operators and Layer Potentials*, Mathematical Surveys and Monographs, **81**, American Mathematical Society, Providence, 2000.

286. R. Temam: *Navier–Stokes Equations, Theory and Numerical Analysis*, North-Holland, Amsterdam, 1984.

287. A. Torchinsky: *Real Variable Methods in Harmonic Analysis*, Pure and Applied Mathematics, **123**, Academic Press, New York.

288. H. Triebel: *Interpolation Theory, Function Spaces, Differential Operators*, Second edition, Johann Ambrosius Barth, Heidelberg, 1995.

289. H. Triebel: *Theory of Function Spaces*, Birkhäuser, Basel, 1983.

290. N.S. Trudinger: On embedding into Orlicz spaces and some applications, *Journal of Mathematical Mechanics*, **17**, 1967, pages 473–484.

291. S. Ukai: The incompressible limit and the initial layer of the compressible Euler equation, *Journal of Mathematics of Kyoto University*, **26**(2), 1986, pages 323–331.

292. M. Ukhovskii and V. Yudovich: Axially symmetric flows of ideal and viscous fluids filling the whole space, *Journal of Applied Mathematics and Mechanics*, **32**, 1968, pages 52–61.

293. A. Valli: An existence theorem for compressible viscous fluids, *Annali di Matematica Pura ed Applicata*, **130**(4), 1982, pages 197–213.

294. A. Valli and W. Zajączkowski: Navier–Stokes equations for compressible fluids: global existence and qualitative properties of the solutions in the general case, *Communications in Mathematical Physics*, **103**(2), 1986, pages 259–296.

295. F. Vigneron: Spatial decay of the velocity field of an incompressible viscous field in $\mathbb{R}^d$, *Nonlinear Analysis*, **63**, 2005, pages 525–549.

296. M. Vishik: Hydrodynamics in Besov spaces, *Archive for Rational Mechanics and Analysis*, **145**, 1998, pages 197–214.

297. M. Vishik: Incompressible flows of an ideal fluid with vorticity in borderline spaces of Besov type, *Annales Scientifiques de l'École Normale Supérieure*, **32**, 1999, pages 769–812.

298. A. Vol'pert and S. Hudjaev: On the Cauchy problem for composite systems of nonlinear differential equations, *Mathematics of the USSR-Sbornik*, **16**, 1972, pages 517–544.

299. W. von Wahl: *The Equations of Navier–Stokes and Abstract Parabolic Equations*, Aspect der Mathematik, Vieweg & Sohn, Wiesbaden, 1985.

300. W. Wolibner: Un théorème d'existence du mouvement plan d'un fluide parfait, homogène, incompressible, pendant un temps infiniment long, *Mathematische Zeitschrift*, **37**, 1933, pages 698–726.

301. Z. Xin and P. Zhang: On the weak solutions to a shallow water equations, *Communications on Pure and Applied Mathematics*, **53**, 2000, pages 1411–1433.

302. V. Yudovich: Non stationary flows of an ideal incompressible fluid, *Zhurnal Vycislitelnoi Matematiki i Matematiceskoi Fiziki*, **3**, 1963, pages 1032–1066.

303. V. Yudovich: Uniqueness theorem for the basic non stationary problem in the dynamics of an ideal incompressible fluid, *Mathematical Research Letters*, **2**, 1995, pages 27–38.

304. A. Zygmund: Smooth functions, *Duke Mathematical Journal*, **12**, 1945, pages 47–76.

List of Notations

1_A, 3

$\mathcal{A}$, 433
$\mathcal{A}_k$, 169
div $\mathcal{A}$, 170
$|\alpha|$, 17
$\approx$, 136

$B^1_{\infty,\infty}$, 116
$B^s_{p,r}$, 99
$\mathcal{B}^{0,\frac{1}{2}}, \mathcal{B}^{-\frac{1}{2},\frac{1}{2}}_4$, 256
$\mathcal{B}^{0,\frac{1}{2}}(T), \mathcal{B}^{-\frac{1}{2},\frac{1}{2}}_4(T)$, 257
$\mathcal{B}^d$, 392
$\dot{B}^{-\sigma}$, 30
$\overline{B}^\sigma_{p,r}$, 300
$\dot{B}^s_{p,r}$, 63
$\|\cdot\|_{\dot{B}^s_{p,r}}$, 63
$\widetilde{B}^{s,r}_\alpha$, 463
$B_\Gamma(\mathbb{R}^d)$, 117
$\mathcal{B}_h, \mathcal{B}_v$, 255
$B^\infty_{p,r}$, 137
$B^s_{p,r}(K)$, 108
$BMO(\mathbb{R}^d)$, 36
$\|\cdot\|_{BMO}$, 36
$|\cdot|_{B^s_{2,r}}$, 186
$\|\cdot\|_{B^s_{p,r}}$, 99
$\square$, 344
$\square_g$, 414
$\square_\Lambda$, 413
$\dot{\mathcal{B}}^s_{p,r}$, 101
$\mathcal{C}_b(I;X)$, 463

c_d, 292
C_γ, 392
$\mathcal{C}_h, \mathcal{C}_v$, 255
$\mathcal{C}^\infty_b$, 327
$C^{k,\rho}(\mathbb{R}^d)$, 37, 99
$\|\cdot\|_{C^{k,\rho}}$, 37
$C_\mu(X)$, 117
$u \cdot \nabla$, 203
$\widetilde{C}_T$, 434
$\mathcal{C}_w([0,T];X)$, 137

$|D|^{-2}$, 295
D, 158
∇, 158
$\partial^\alpha f$, 17
$\dot{\Delta}_j$, 61
$D^2\Phi(\theta,\theta)$, 342
Δ_h, 245
δ_{jk}, 206
$(-\Delta)^s$, 71
$\partial \cdot (g \cdot \partial)$, 391
div, 203
div_h, 246
Δ, 203
Δ_j, 61
$\widetilde{\Delta}_j$, 79
D^{vi}_j, S^{vi}_j, 277
$\mathcal{D}_K$, 340
D_λ, 417
$\mathcal{D}(\mathbb{R}^d)$, 17
D_t, 325
$\partial_0, \partial, \nabla$, 359, 389

H. Bahouri et al., *Fourier Analysis and Nonlinear Partial Differential Equations*, Grundlehren der mathematischen Wissenschaften 343, DOI 10.1007/978-3-642-16830-7, © Springer-Verlag Berlin Heidelberg 2011

517

Index

H. Bahouri et al., *Fourier Analysis and Nonlinear Partial Differential Equations*, Grundlehren der mathematischen Wissenschaften 343, DOI 10.1007/978-3-642-16830-7, © Springer-Verlag Berlin Heidelberg 2011